Lecture Notes in Computer Science 10392

Commenced Publication in 1973
Founding and Former Series Editors:
Gerhard Goos, Juris Hartmanis, and Jan van Leeuwen

More information about this series at http://www.springer.com/series/7407

Yixin Cao · Jianer Chen (Eds.)

Computing and Combinatorics

23rd International Conference, COCOON 2017
Hong Kong, China, August 3–5, 2017
Proceedings

 Springer

Editors
Yixin Cao (iD)
Department of Computing
Hong Kong Polytechnic University
Hong Kong
China

Jianer Chen
Texas A&M University
College Station, TX
USA

ISSN 0302-9743 ISSN 1611-3349 (electronic)
Lecture Notes in Computer Science
ISBN 978-3-319-62388-7 ISBN 978-3-319-62389-4 (eBook)
DOI 10.1007/978-3-319-62389-4

Library of Congress Control Number: 2017945277

LNCS Sublibrary: SL1 – Theoretical Computer Science and General Issues

Printed on acid-free paper

This Springer imprint is published by Springer Nature
The registered company is Springer International Publishing AG
The registered company address is: Gewerbestrasse 11, 6330 Cham, Switzerland

Preface

This volume contains the papers presented at the 23rd International Computing and Combinatorics Conference (COCOON 2017), held during August 3–5, 2017, in Hong Kong, China. COCOON 2017 provided a forum for researchers working in the areas of algorithms, theory of computation, computational complexity, and combinatorics related to computing.

The technical program of the conference included 48 contributed papers selected by the Program Committee from 111 full submissions received in response to the call for papers. All the papers were peer reviewed by at least three Program Committee members or external reviewers. The papers cover various topics, including algorithms and data structures, complexity theory and computability, algorithmic game theory, computational learning theory, cryptography, computational biology, computational geometry and number theory, graph theory, and parallel and distributed computing. Some of the papers will be selected for publication in special issues of *Algorithmica*, *Theoretical Computer Science* (TCS), and *Journal of Combinatorial Optimization* (JOCO). It is expected that the journal version of the papers will be in a more complete form.

The conference also included three invited presentations, delivered by Dániel Marx (Hungarian Academy of Sciences), Shang-Hua Teng (University of Southern California), and Virginia Vassilevska Williams (Massachusetts Institute of Technology). Abstracts of their talks are included in this volume. Also included are eight contributed talks from a workshop on computational social networks (CSoNet 2017) co-located with COCOON 2017. The papers of CSoNet 2017 were selected by an independent Program Committee, chaired by Donghyun Kim, Miloš Kudělka, and R.N. Uma. We appreciate the work by the CSoNet Program Committee that helped enrich the conference topics.

We wish to thank all the authors who submitted extended abstracts for consideration, the members of the Program Committee for their scholarly efforts, and all external reviewers who assisted the Program Committee in the evaluation process. We thank the Hong Kong Polytechnic University for hosting COCOON 2017. We are also grateful to all members of the Organizing Committee and to their supporting staff.

The conference-management system EasyChair was used to handle the submissions, to conduct the electronic Program Committee meetings, and to assist with the assembly of the proceedings.

August 2017

Yixin Cao
Jianer Chen

Organization

Conference Chair

Jiannong Cao — Hong Kong Polytechnic University, China

Program Chairs

Yixin Cao — Hong Kong Polytechnic University, China
Jianer Chen — Texas A&M University, USA

Program Committee

Jarek Byrka	University of Wroclaw, Poland
Liming Cai	University of Georgia, USA
Hubert Chan	University of Hong Kong, China
Rajesh Chitnis	Weizmann Institute, Israel
Vida Dujmovic	University of Ottawa, Canada
Funda Ergun	Indiana University Bloomington, USA
Siyao Guo	Chinese University of Hong Kong, China
Magnús Már Halldórsson	Reykjavik University, Iceland
Pinar Heggernes	University of Bergen, Norway
Juraj Hromkovic	ETH Zurich, Switzerland
Kazuo Iwama	Kyoto University, Japan
Iyad Kanj	DePaul University, USA
Bingkai Lin	University of Tokyo, Japan
Daniel Lokshtanov	University of Bergen, Norway
Monaldo Mastrolilli	IDSIA, Switzerland
Ross McConnell	Colorado State University, USA
Vangelis Th. Paschos	University of Paris-Dauphine, France
Youming Qiao	National University of Singapore, Singapore
Piotr Sankowski	University of Warsaw, Poland
Ryuhei Uehara	Japan Advanced Institute of Science and Technology, Japan
Magnus Wahlström	Royal Holloway, University of London, UK
Gerhard J. Woeginger	RWTH Aachen, Germany
Mary Wootters	Stanford University, USA
Chee Yap	New York University, USA
Grigory Yaroslavtsev	University of Pennsylvania, USA
Neal Young	University of California, Riverside, USA
Huacheng Yu	Stanford University, USA
Shengyu Zhang	Chinese University of Hong Kong, China
Daming Zhu	Shandong University, China

Additional Reviewers

Abboud, Amir
Abdelkader, Ahmed
Agrawal, Akanksha
Akutsu, Tatsuya
Alman, Josh
Applegate, David
Ásgeirsson, Eyjólfur Ingi
Banerjee, Sandip
Barba, Luis
Barcucci, Elena
Bei, Xiaohui
Belmonte, Rémy
Bienkowski, Marcin
Biswas, Arindam
Boeckenhauer, Hans-Joachim
Bose, Prosenjit
Burjons Pujol, Elisabet
Casel, Katrin
Chen, Shiteng
Cheng, Hao-Chung
Cousins, Benjamin
Curticapean, Radu
Dean, Brian
Della Croce, Federico
Eiben, Eduard
Emura, Keita
Epstein, Leah
Erlebach, Thomas
Fan, Chenglin
Fagerberg, Rolf
Fefferman, Bill
Flatland, Robin
Frei, Fabian
Fujiwara, Hiroshi
Ghosal, Pratik
Giannakos, Aristotelis
Giannopoulos, Panos
Golovach, Petr
Golovnev, Alexander
Gordinowicz, Przemysław
Gourves, Laurent
Gurjar, Rohit
Haraguchi, Kazuya

Hatami, Pooya
Hoffmann, Michael
Huang, Xin
Hubáček, Pavel
Im, Sungjin
Izumi, Taisuke
Jahanjou, Hamidreza
Jeż, Łukasz
Jones, Mark
Jowhari, Hossein
Kalaitzis, Christos
Kamiyama, Naoyuki
Khan, Arindam
Kim, Eun Jung
Klein, Kim-Manuel
Kolay, Sudeshna
Komargodski, Ilan
Komm, Dennis
Kozma, Laszlo
Kralovic, Rastislav
Krithika, R.
Kumar, Mrinal
Kurpisz, Adam
Lampis, Michael
van Leeuwen, Erik Jan
Leppänen, Samuli
Levin, Asaf
Li, Wenjun
Li, Yinan
Liedloff, Mathieu
Lin, Chengyu
Marcinkowski, Jan
Markakis, Evangelos
Matsuda, Takahiro
Meulemans, Wouter
Michail, Dimitrios
Misra, Pranabendu
Miyazaki, Shuichi
Monnot, Jerome
Mori, Ryuhei
Morin, Pat
Murat, Cécile
Myrvold, Wendy

Nagami Coregliano, Leonardo
Narayanan, Hariharan
Nasre, Meghana
Nilsson, Bengt J.
Nishimura, Harumichi
Nisse, Nicolas
Oum, Sang-Il
Pankratov, Denis
Papadopoulos, Charis
Pergola, Elisa
Pilipczuk, Michał
Rai, Ashutosh
Reyzin, Lev
Rohwedder, Lars
Rote, Günter
Sahu, Abhishek
Saurabh, Saket
Schmidt, Paweł
Seddighin, Saeed
Sheng, Bin
Sikora, Florian
Skopalik, Alexander
Spoerhase, Joachim
Srivastav, Abhinav
Staals, Frank
Stamoulis, Georgios

van Stee, Rob
Suzuki, Akira
Tale, Praffulkumar
Talmon, Nimrod
Tonoyan, Tigran
Tsang, Hing Yin
Ueno, Shuichi
Unger, Walter
Vaze, Vikrant
Vempala, Santosh
Verdugo, Victor
Wang, Guoming
Wang, Joshua
Wehner, David
Wrochna, Marcin
Xia, Ge
Xu, Zhisheng
Yamanaka, Katsuhisa
Yan, Li
Yao, Penghui
Ye, Junjie
Yeo, Anders
Yu, Nengkun
Zhang, Chihao
Zhang, Yihan
Zheng, Chaodong

Invited Talks

The Optimality Program
for Parameterized Algorithms

Dániel Marx

Institute for Computer Science and Control,
Hungarian Academy of Sciences, Budapest, Hungary

Parameterized complexity analyzes the computational complexity of NP-hard combinatorial problems in finer detail than classical complexity: instead of expressing the running time as a univariate function of the size n of the input, one or more relevant parameters are defined and the running time is analyzed as a function depending on both the input size and these parameters. The goal is to obtain algorithms whose running time depends polynomially on the input size, but may have arbitrary (possibly exponential) dependence on the parameters. Moreover, we would like the dependence on the parameters to be as slowly growing as possible, to make it more likely that the algorithm is efficient in practice for small values of the parameters. In recent years, advances in parameterized algorithms and complexity have given us a tight understanding of how the parameter has to influence the running time for various problems. The talk will survey results of this form, showing that seemingly similar NP-hard problems can behave in very different ways if they are analyzed in the parameterized setting.

Scalable Algorithms for Data and Network Analysis

Shang-Hua Teng

Computer Science and Mathematics,
University of Southern California, Los Angeles, USA

In the age of Big Data, efficient algorithms are in higher demand now more than ever before. While Big Data takes us into the asymptotic world envisioned by our pioneers, the explosive growth of problem size has also significantly challenged the classical notion of efficient algorithms: Algorithms that used to be considered efficient, according to polynomial-time characterization, may no longer be adequate for solving today's problems. It is not just desirable, but essential, that efficient algorithms should be scalable. In other words, their complexity should be nearly linear or sub-linear with respect to the problem size. Thus, scalability, not just polynomial-time computability, should be elevated as the central complexity notion for characterizing efficient computation.

In this talk, I will discuss a family of algorithmic techniques for the design of provably-good scalable algorithms, focusing on the emerging Laplacian Paradigm, which has led to breakthroughs in scalable algorithms for several fundamental problems in network analysis, machine learning, and scientific computing. These techniques include local network exploration, advanced sampling, sparsification, and graph partitioning. Network analysis subject include four recent applications: (1) PageRank Approximation (and identification of network nodes with significant PageRanks). (2) Social-Influence Shapley value. (3) Random-Walk Sparsification. (4) Scalable Newton's Method for Gaussian Sampling.

Solutions to these problems exemplify the fusion of combinatorial, numerical, and statistical thinking in network analysis.

Fine-Grained Complexity of Problems in P

Virginia Vassilevska Williams

Massachusetts Institute of Technology, Cambridge, USA

A central goal of algorithmic research is to determine how fast computational problems can be solved in the worst case. Theorems from complexity theory state that there are problems that, on inputs of size n, can be solved in $t(n)$ time but not in $t(n)^{1-\varepsilon}$ time for $\varepsilon > 0$. The main challenge is to determine where in this hierarchy various natural and important problems lie. Throughout the years, many ingenious algorithmic techniques have been developed and applied to obtain blazingly fast algorithms for many problems. Nevertheless, for many other central problems, the best known running times are essentially those of their classical algorithms from the 1950s and 1960s.

Unconditional lower bounds seem very difficult to obtain, and so practically all known time lower bounds are conditional. For years, the main tool for proving hardness of computational problems have been NP-hardness reductions, basing hardness on $P \neq NP$. However, when we care about the exact running time (as opposed to merely polynomial vs non-polynomial), NP-hardness is not applicable, especially if the problem is already solvable in polynomial time. In recent years, a new theory has been developed, based on "fine-grained reductions" that focus on exact running times. Mimicking NP-hardness, the approach is to (1) select a key problem X that is conjectured to require essentially $t(n)$ time for some t, and (2) reduce X in a fine-grained way to many important problems. This approach has led to the discovery of many meaningful relationships between problems, and even sometimes to equivalence classes.

The main key problems used to base hardness on have been: the 3SUM problem, the CNF-SAT problem (based on the Strong Exponential TIme Hypothesis (SETH)) and the All Pairs Shortest Paths Problem. Research on SETH-based lower bounds has flourished in particular in recent years showing that the classical algorithms are optimal for problems such as Approximate Diameter, Edit Distance, Frechet Distance, Longest Common Subsequence, many dynamic graph problems, etc.

In this talk I will give an overview of the current progress in this area of study, and will highlight some exciting new developments.

Contents

CSoNet Papers

COCOON 2017

A Time-Space Trade-Off for Triangulations of Points in the Plane

Hee-Kap Ahn[1], Nicola Baraldo[2], Eunjin Oh[1(✉)], and Francesco Silvestri[2]

[1] Pohang University of Science and Technology, Pohang, Korea
{heekap,jin9082}@postech.ac.kr
[2] University of Padova, Padova, Italy
nicola.baraldo@gmail.com, silvestri@dei.unipd.it

Abstract. In this paper, we consider time-space trade-offs for reporting a triangulation of points in the plane. The goal is to minimize the amount of working space while keeping the total running time small. We present the first multi-pass algorithm on the problem that returns the edges of a triangulation with their adjacency information. This even improves the previously best known random-access algorithm.

1 Introduction

There are two optimization goals in the design of algorithms: the time complexity and the space complexity. However, one cannot achieve both goals at the same time in general. This can be seen in a time-space tradeoff of algorithmic efficiency that an algorithm has to use more space to improve its running time and it has to spend more time with less amount of space. With this reason, time-space trade-offs for a number of problems were considered even as early as in 1980s. For example, Frederickson presented optimal time-space trade-offs for sorting and selection problems in 1987 [11]. After this work, a significant amount of research has been done for time-space trade-offs in the design of algorithms.

In this paper, we consider time-space trade-offs for one of fundamental geometric problems, reporting a triangulation of points in the plane. We assume that the points are given in a read-only memory. This assumption has been considered in applications where the input is required to be retained in its original state. Many time-space tradeoffs for fundamental problems have been studied under this read-only assumption. For instance, a few read-only sorting algorithms have been presented under the assumption [6,14].

There are two typical access models to the read-only input, a *random-access model* and a *multi-pass model*. In the *multi-pass model*, the only way to access elements in the input array is to scan the array from the beginning, and algorithms are allowed to make multiple passes over the input. A single pass is a special case of the multi-pass model. Multi-pass algorithms are more restrictive than algorithms under the *random-access* for any element in the input array.

This work was supported by the NRF grant 2011-0030044 (SRC-GAIA) funded by the government of Korea.

Y. Cao and J. Chen (Eds.): COCOON 2017, LNCS 10392, pp. 3–12, 2017.
DOI: 10.1007/978-3-319-62389-4_1

The multi-pass model has applications where large data sets are stored some-where such as an external memory and it is more efficient in I/O to read them sequentially in a few passes. Multi-pass algorithms have been studied recently in areas including geometry [1,15] and graphs [9,10].

The goal of our problem is to minimize the amount of working space while keeping the total running time small. More precisely, we are allowed to use $O(s)$ words as working space in addition to the memory for input and output for a positive integer parameter s which is determined by users. We assume that a word is large enough to store a number and a pointer. While processing input, we send the answer to a write-only output stream without repetition. An algorithm designed in this setting is called an *s-workspace algorithm*.

1.1   Related Works

A triangulation of a set S of n points in the plane is defined to be a maximal subdivision of the plane whose vertices are in S and faces are triangles, except for the unbounded face. The unbounded face of a triangulation of S is the region outside of the convex hull of S. Thus the sorting problem which asks for sorting n numbers reduces to this problem. Similarly, the problem of computing the convex hull of n points in the plane reduces to this problem. In the following, we simply call these problems the *sorting problem* and the *convex hull problem*, respectively.

The optimal trade-offs for the sorting problem and the convex hull problem are known for both models (the random-access model and the multi-pass stream-ing model.) Under the random-access model, both problems can be solved in $O(n^2/(s\log n)+n\log(s\log n))$ time[1] using $O(s)$ words of workspace [8,14]. Under the multi-pass streaming model, both problems can be solved in $O(n^2/\log n + n\log s)$ time [7,13] using $O(s)$ words of workspace and $O(n/s)$ passes of the input array.

With linear-size working space, a triangulation of S can be computed in $O(n\log n)$ time. For the case that a space is given as a positive integer parame-ter s at most n, several results are known for the random-access model while no result is known for the multi-pass streaming model. Korman et al. presented an s-workspace algorithm for computing a triangulation of S in $O(n^2/s+n\log n\log s)$ time [12]. In the same paper, they presented an s-workspace algorithm for computing the Delaunay triangulation of S in $O((n^2/s)\log s + n\log s\log^* s)$ expected time. Recently, it is improved to $O(n^2\log n/s)$ deterministic time [4]. Combining [4,12], a triangulation of S can be computed in $O(\min\{n^2/s + n\log n\log s, n^2\log n/s\})$ time.

The problem of computing a triangulation of a simple polygon has also been studied under the random-access model. Aronov et al. [2] presented an s-workspace algorithm for computing a triangulation of a simple n-gon. Their algorithm returns the edges of a triangulation without repetition in

---

[1] They state that their running time is $O(n^2/s + n\log s)$ for s bits of workspace, but we measure workspace in words.

$O(n^2/s + n \log s \log^5 n/s)$ expected time. Moreover, their algorithm can be modified to report the resulting triangles of a triangulation together with their adjacency information. For a monotone n-gon, Barba et al. [5] presented an $(s \log_s n)$-workspace algorithm for triangulating the polygon in $O(n \log_s n)$ time for a parameter $s \in \{1, \ldots, n\}$. Later, Asano and Kirkpatrick [3] showed how to reduce the working space to $O(s)$ words without increasing the running time.

1.2 Our Results

We present an s-workspace $O(n^2/s + n \log s)$-time algorithm for computing a triangulation of a set of n points in the plane. Our algorithm uses $O(n/s)$ passes over the input array. To our best knowledge, this is the first result on the problem under the multi-pass model. These bounds are asymptotically optimal, which can be shown by a reduction from the sorting problem [13].

Our multi-pass algorithm also improves the previously best known algorithm under the random-access model by Korman et al. which takes $O(\min\{n^2/s + n \log n \log s, n^2 \log n/s\})$ time [4,12] although the multi-pass model is more restrictive than the random-access model. It seems unclear whether the algorithm by Korman et al. [12] can be extended to a multi-pass streaming algorithm.

Our algorithm has an additional advantage compared to the previously best one. Our algorithm can be extended to report the triangles together with adjacency information as well as the edges of a triangulation without increasing the running time and space. The edge adjacency is essential information in representing and reconstructing the triangulation. In contrast, the algorithms in [4,12] report the edges of a triangulation in an arbitrary order with no adjacency information of them. Furthermore, the algorithm by Korman et al. [12] uses the algorithm by Asano and Kirkpatrick [3] as a subprocedure, but it is unclear how to modify the subprocedure to report a triangulation together with edge adjacency information [2].

2 Reporting the Edges of a Triangulation

In this section, we present an s-workspace $O(n^2/s + n \log s)$-time algorithm to compute a triangulation of a set S of n points in the plane using $O(n/s)$ passes. Our algorithm is based on the multi-pass streaming algorithm by Chan and Chen [7] for computing the convex hull of a set of points in the plane. For a subset S' of S, we use $\text{CH}(S')$ to denote the convex hull of S'.

Chan and Chen presented an algorithm to compute the convex hull of a set S of points in the plane by scanning the points $O(n/s)$ times. They consider $\lceil n/s \rceil$ disjoint vertical slabs each of which contains exactly s points of S, except for the last vertical slab. They use two passes to compute the boundary of $\text{CH}(S)$ contained in each vertical slab. For one pass, they find the points of S contained in the vertical slab using Lemma 1. Then they compute the convex hull of them in $O(s \log s)$ time. For the other pass, they find the part of the boundary of the convex hull which appears on the boundary of $\text{CH}(S)$. In total, their algorithm takes $O(n^2/s + n \log s)$ time and uses $O(s)$ words of working space.

Lemma 1 ([7]). *Given a point $p \in S$, we can compute the leftmost s points lying to the right of p in $O(n)$ time using $O(s)$ words of working space in a single pass.*

2.1 Our Algorithm

Imagine $\lceil n/s \rceil$ disjoint vertical slabs each of which contains exactly s points of S, except for the last vertical slabs. Let S_i be the set of points of S contained in the ith slab for $i \in \{1, 2, \ldots, \lceil n/s \rceil\}$. By Lemma 1, we can compute all points in S_i by scanning the points in S once using $O(s)$ words of working space if we have the leftmost point of S_i. The pseudocode of the overall algorithm can be found in Algorithm 1.

Algorithm 1. Computing a triangulation of S

1: **procedure** TRIANGULATION(S)
2:     $s \leftarrow$ the leftmost point of S
3:     **for** $i \leftarrow 1$ to $\lceil n/s \rceil$ **do**
4:         Compute S_i by scanning all points in S once.
5:         Report all edges of a triangulation of S_i.
6:         Let T_i be the set of points lying to the left of any point in S_i.
7:         $a \leftarrow$ the rightmost point of T_i and $b \leftarrow$ the leftmost point of S_i
8:         LOWERTRIANGULATION(a, b, S_i)
9:         UPPERTRIANGULATION(a, b, S_i)

The overall algorithm (Algorithm 1). We consider all vertical slabs from left to right one by one. After we process a vertical slab, we guarantee that we report all edges of a triangulation of the points of S contained in the union of all previous vertical slabs. First, we compute S_1 explicitly in $O(n)$ time by applying Lemma 1, and compute a triangulation of S_1 in $O(s \log s)$ time. Assume that we just considered S_{i-1} for some $i \in \{2, \ldots, \lceil n/s \rceil\}$ and we computed a triangulation of $S_1 \cup \ldots \cup S_{i-1}$. Let T_i be the set of points lying to the left of any point in S_i, that is, $T_i = S_1 \cup \ldots \cup S_{i-1}$.

Now we handle S_i. We compute S_i explicitly in $O(n)$ time by applying Lemma 1, and compute a triangulation of S_i. Let a be the rightmost point of T_i and b be the leftmost point of S_i (See Fig. 1(a).).

For two convex polygons C_1 and C_2, we say a line segment c_1c_2 for $c_1 \in C_1$ and $c_2 \in C_2$ a *bridge* of C_1 and C_2 if it appears on the boundary of the convex hull of C_1 and C_2. If a bridge connects the lower chains of C_1 and C_2, we call the bridge the *lower bridge*. Otherwise, we call it the *upper bridge*.

We traverse the boundary of CH(T_i) from a in clockwise order and traverse the boundary of CH(S_i) from b in counterclockwise order until we find the lower bridge of CH(T_i) and CH(S_i). Let L_i be the polygon whose boundary consists of the chains we visited, ab and the lower bridge. During the traversal, we compute a triangulation of L_i. See Fig. 1(a). The pseudocode of this procedure can be found in Algorithm 2. We call this procedure LOWERTRIANGULATION.

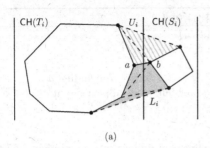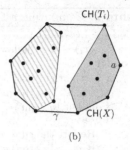

Fig. 1. (a) Starting from ab, we report the edges of a triangulation of L_i (the gray region) and the edges of a triangulation of U_i (the dashed region). (b) The edge γ is the lower bridge of $\mathrm{CH}(X)$ and the convex hull of the points lying to the left of any point of X.

Similarly, we find the upper bridge of $\mathrm{CH}(T_i)$ and $\mathrm{CH}(S_i)$ by traversing the boundaries of $\mathrm{CH}(T_i)$ and $\mathrm{CH}(S_i)$. Let U_i be the polygon whose boundary consists of the chains we visited, ab and the upper bridge. We call this procedure UPPERTRIANGULATION. This can be done a way similar to LOWERTRIANGULATION. Note that $L_i \cup U_i = \mathrm{CH}(T_{i+1}) \setminus (\mathrm{CH}(T_i) \cup \mathrm{CH}(S_i))$. We show how to compute a triangulation of L_i only because a triangulation of U_i can be computed analogously.

Computing a triangulation of L_i (Algorithm 2). We can construct and traverse the boundary of $\mathrm{CH}(S_i)$ in $O(s \log s)$ time since we can store $\mathrm{CH}(S_i)$ explicitly. However, this does not hold for $\mathrm{CH}(T_i)$ since the size of $\mathrm{CH}(T_i)$ might exceed $O(s)$. To traverse the boundary of $\mathrm{CH}(T_i)$ using $O(s)$ words of working space, we first find the rightmost s points of T_i using one pass by applying Lemma 1. Let X denote the set of such s points. We store X explicitly.

Then we compute $\mathrm{CH}(X)$ and compute the lower bridge of $\mathrm{CH}(X)$ and the convex hull of points of S lying to the left of any point of X. See Fig. 1(b). We can compute the lower bridge in $O(n)$ time by considering all points of S lying to the left of any point of X one by one. Due to this bridge, we can decide which part of the boundary of $\mathrm{CH}(X)$ appears on the boundary of $\mathrm{CH}(T_i)$. We traverse the part of $\mathrm{CH}(X)$ appearing on the boundary of $\mathrm{CH}(T_i)$ until we find the lower bridge of $\mathrm{CH}(T_i)$ and $\mathrm{CH}(S_i)$.

Once we reach the most clockwise vertex of $\mathrm{CH}(X)$ appearing on the boundary of $\mathrm{CH}(T_i)$, we again find the rightmost s points lying to the left of the endpoint of γ not in X, where γ is the lower bridge of $\mathrm{CH}(X)$ and the convex hull of the points lying to the left of any point of X. Then we update X to the set of these points. Note that X may not be S_j for any $1 \le j \le i$ in this case. By construction, the rightmost point of X appears on the lower chain of $\mathrm{CH}(T_i)$. We do this until we find the lower bridge of $\mathrm{CH}(T_i)$ and $\mathrm{CH}(S_i)$.

Algorithm 2. Computing a triangulation of L_i

1: **procedure** LOWERTRIANGULATION(a,b,S_i)
2:     $\langle x_1, \ldots, x_t \rangle \leftarrow$ be the lower hull of S_i (from right to left).
3:     **repeat**
4:         $X \leftarrow$ the rightmost s points lying to the left of a including a
5:         $\langle y_1, \ldots, y_{t'} \rangle \leftarrow$ the part of the lower hull of X appearing on CH(T_i)
            (from left to right)
6:         **repeat**
7:             Report the edge $x_t y_{t'}$.
8:             **if** $y_{t'-s}$ lies to the left of the line containing $x_t y_{t'}$ in direction $\overrightarrow{x_t y_{t'}}$ **then**
9:                 $t' \leftarrow t' - 1$
10:            **else**
11:                $t \leftarrow t - 1$
12:        **until** all points in $X \cup S_i$ lie above the line containing $x_t y_{t'}$
13:        $\gamma \leftarrow$ the bridge of CH(X) and the convex hull of points lying to the left of
           any point of X
14:        Report γ.
15:        $a \leftarrow$ the endpoint of γ not in X
16:    **until** all points in T_i lie above the line containing γ

2.2   Analyses

In Algorithm 1, Line 2 can be done in $O(n)$ time using a single pass. Line 4 and 6 can be done in $O(n)$ time using $O(1)$ passes due to Lemma 1. Line 5 can be done in $O(s \log s)$ time since we compute S_i explicitly. Thus, the total running time is $O(n^2/s + n \log s + \sum_i \tau_i)$, where τ_i is the running time of LOWERTRIANGULATION($\cdot, \cdot, S_i$).

Now consider Algorithm 2. Let t_i be the number of updates of X for S_i. Line 4 takes $O(n)$ time using a single pass due to Lemma 1. Line 5 takes $O(s \log s + n)$ time using a single pass. For Lines 6–12, we compute an edge of a triangulation in each iteration. And each iteration takes $O(1)$ time. Note that Line 12 can be also done in $O(1)$ time since it suffices to consider the boundaries of CH(S_i) and CH(X) locally. Thus Lines 6–12 can be done in $O(n)$ time in total for all S_i's. Lines 13–15 take $O(n)$ time using $O(1)$ passes. Therefore, for a fixed i, the running time of Algorithm 2, except Lines 6–12, is $O(s \log s + t_i n)$. Since Lines 6–12 can be done in $O(n)$ time for all indices i, the total running time of Algorithm 1 is $O(n^2/s + n \log s + \sum_i \tau_i) = O(n^2/s + n \log s + n \sum_i t_i)$.

We claim that the sum of t_i over all i's is $O(n/s)$, which implies that Algorithm 1 takes $O(n^2/s + n \log s)$ time using $O(n/s)$ passes. Assume that X is set to $A_1, A_2, \ldots, A_k$ in order when we handle S_i. No point in A_ℓ appears on the lower chain of CH(T_i) for $\ell = 1, 2, \ldots, k - 1$. Thus, no point in A_ℓ appears on the lower chain of CH(T_j) for any $j \geq i$. Recall that X is set to a point set whose the rightmost point appears on the lower chain of CH(T_j) when we handle S_j. Therefore, no point in A_t is contained in X at any time after we handle S_i for $t = 2, 3, \ldots, k - 1$. Therefore, the sum of t_i is $O(n/s)$, and the total running time is $O(n^2/s + n \log s)$.

Theorem 1. *Given a set S of n points in the plane, we can report the edges of a triangulation of S in $O(n^2/s + n \log s)$ time using $O(s)$ words of working space and $O(n/s)$ passes.*

3   Reporting the Triangles with Adjacency Information

Let T be the triangulation of S computed by the algorithm in Sect. 2.1. In this section, we show how to modify the algorithm to report the triangles of T together with their adjacency information in addition to the edges of T. That is, we report every pair (τ, τ') of the triangles of T such that τ and τ' are adjacent to each other in T.

We say a triangle τ of T is an *inner-slab triangle* if all three corners of τ are in the same vertical slab S_i for some $i = 1, \ldots, \lceil n/s \rceil$. Otherwise, we say τ is a *cross-slab triangle*. Note that if two inner-slab triangle are adjacent to each other in T, they are contained in the same vertical slab. Moreover, for an inner-slab triangle τ and a cross-slab triangle τ', we compute τ' after computing τ. In this case, we report the adjacency between τ and τ' when we compute and report τ'.

Reporting an inner-slab triangle. Consider an inner-slab triangle τ with corners in S_i for some i. Recall that we compute all points in S_i explicitly, and compute a triangulation of them. When we compute a triangulation of them, we also report the triangles of it with their adjacency information. For a cross-slab triangle τ' of T adjacent to τ, we will report their adjacency information when we compute and report τ'.

Reporting a cross-slab triangle. Consider a cross-slab triangle τ' whose rightmost corner lies on S_i for some i. This triangle comes from a triangulation of $L_i \cup U_i$. We compute τ' while we traverse the boundaries of $\text{CH}(T_i)$ and $\text{CH}(S_i)$. A cross-slab triangle adjacent to τ' is also computed during this traversal, thus the adjacency information between them can be computed during the traversal.

Each corner of τ' lies on the boundary of $\text{CH}(S_i)$ or the boundary of $\text{CH}(T_i)$. If two corners lie on the boundary of $\text{CH}(S_i)$, there is an inner-slab triangle adjacent to τ' contained in S_i. The adjacency information between them can be computed without increasing the running time because we compute the convex hull of S_i explicitly.

Now assume that two corners a and b lie on the boundary of $\text{CH}(T_i)$. Let τ'' be the triangle of T incident to ab other than τ'. To report the adjacency information between τ' and τ'', we compute τ'' together with ab when we traverse the boundary of $\text{CH}(T_i)$. To do this, we specify a way to triangulate $L_i \cup U_i$ as described in Algorithm 2.

For L_i, we initially set a' to the rightmost point of X and b' to the leftmost point of S_i for each set X. Then we move a' along the part of the boundary of $\text{CH}(X)$ appearing on $\text{CH}(T_i)$ in clockwise direction as much as possible until $a'b'$ intersects the boundaries of $\text{CH}(X)$ and $\text{CH}(S_i)$. Then we move b' one step along the boundary of $\mathrm{CH}(S_i)$ in counterclockwise direction, and move a' again. We do this until we all points in $X \cup S_i$ lie above the line containing $a'b'$. For U_i, we can compute a triangulation similarly.

Then we have the following lemma.

Lemma 2. *Given the convex hull of the set of points in S_j for some $j = 1, \ldots, \lceil n/s \rceil$, we can find the lowest triangle of $\mathcal{T}$ contained in L_i in $O(n)$ time using $O(s)$ words of workspace and $O(1)$ passes.*

Proof. The lowest triangle of $\mathcal{T}$ contained in L_i is incident to the lower bridge of $\mathrm{CH}(T_j)$ and $\mathrm{CH}(S_j)$. By scanning the points of S once, we compute the lower bridge of $\mathrm{CH}(T_j)$ and $\mathrm{CH}(S_j)$. Let a and b be the endpoints of the lower bridge such that $a \in \mathrm{CH}(T_j)$ and $b \in \mathrm{CH}(S_j)$. Then, by scanning the points in S once again, we compute the counterclockwise neighbor a' of a along the boundary of $\mathrm{CH}(T_i)$. Since we maintain the set S_i explicitly, we can compute the clockwise neighbor b' of b along the boundary of $\mathrm{CH}(S_i)$ in constant time without scanning the points of S.

By construction, the triangle with corners a, b, b' is the lowest triangle of $\mathcal{T}$ contained in L_i if b and b' lie below the line passing through a and a'. Otherwise, the triangle with corners a, a', b is the lowest triangle of $\mathcal{T}$ contained in L_i. In any case, we can report the lowest triangle of $\mathcal{T}$ in L_i using $O(s)$ words of workspace and $O(1)$ passes.   □

We modify Algorithm 2 as follows. For Line 4, we set X to the set of points in S_j lying to the left of a for $a \in S_j$ using Lemma 3. Then we can obtain the triangles of $\mathcal{T}$ incident to the part of the lower hull of X appearing on $\mathrm{CH}(T_i)$ by applying the algorithm for triangulating S_j we use in Line 5 of Algorithm 1. To compute the triangle of $\mathcal{T}$ incident to γ and contained in $\mathrm{CH}(T_i)$, we apply Lemma 2.

Lemma 3. *Given any point $p \in S_j$ for some $j = 1, \ldots, \lceil n/s \rceil$, we can compute S_i in $O(n)$ time using $O(s)$ words of workspace and $O(1)$ passes.*

Proof. For the first pass, we compute the number of points in S lying to the left of p. This determines the value of j with $p \in S_j$. Then we find the rightmost s points lying to the left of p in $O(n)$ time using a single pass by applying Lemma 1. One of them is the leftmost point of S_j, and we can find it in $O(s)$ time. We can compute S_i using a single pass by applying Lemma 1 again.   □

The following theorem summarizes this section.

Theorem 2. *Given a set S of n points in the plane, we can report the triangles of a triangulation of S with their adjacency information in $O(n^2/s + n \log s)$ time using $O(s)$ words of working space and $O(n/s)$ passes.*

4   Conclusion

In this paper, we present an s-workspace $O(n^2/s + s \log n)$-time algorithm for computing a triangulation of a set of n points in the plane under the multi-pass model. Our algorithm uses $O(n/s)$ passes over the input array. It is not only the first algorithm for this problem under the multi-pass model, but it also improves the previously best known random-access algorithm [12]. Moreover, its running time is optimal under the multi-pass model.

One interesting open problem remaining from our work is whether our algorithm can be improved under the random-access model. Under this model, the best known lower bound is $\Omega(n^2/(s \log n) + n \log(s \log n))$ for $s \le n/\log n$.

References

1. Agarwal, P.K., Krishnan, S., Mustafa, N.H., Venkatasubramanian, S.: Streaming geometric optimization using graphics hardware. In: Battista, G., Zwick, U. (eds.) ESA 2003. LNCS, vol. 2832, pp. 544–555. Springer, Heidelberg (2003). doi:10.1007/978-3-540-39658-1_50
2. Aronov, B., Korman, M., Pratt, S., van Ressen, A., Roeloffzen, M.: Time-space trade-offs for triangulating a simple polygon. In: Proceedings of the 15th Scandinavian Symposium and Workshops on Algorithm Theory (SWAT 2016), vol. 53, pp. 30:1–30:12 (2016)
3. Asano, T., Kirkpatrick, D.: Time-space tradeoffs for all-nearest-larger-neighbors problems. In: Dehne, F., Solis-Oba, R., Sack, J.-R. (eds.) WADS 2013. LNCS, vol. 8037, pp. 61–72. Springer, Heidelberg (2013). doi:10.1007/978-3-642-40104-6_6
4. Banyassady, B., Korman, M., Mulzer, W., van Renssen, A., Roeloffzen, M., Seiferth, P., Stein, Y.: Improved time-space trade-offs for computing Voronoi diagrams. In: Proceedings of the 34th Symposium on Theoretical Aspects of Computer Science (STACS 2017), vol. 66, pp. 9:1–9:14 (2017)
5. Barba, L., Korman, M., Langerman, S., Sadakane, K., Silveira, R.I.: Space-time trade-offs for stack-based algorithms. Algorithmica 72(4), 1097–1129 (2015)
6. Borodin, A., Cook, S.: A time-space tradeoff for sorting on a general sequential model of computation. SIAM J. Comput. 11(2), 287–297 (1982)
7. Chan, T.M., Chen, E.Y.: Multi-pass geometric algorithms. Discret. Comput. Geom. 37(1), 79–102 (2007)
8. Darwish, O., Elmasry, A.: Optimal time-space tradeoff for the 2D convex-hull problem. In: Schulz, A.S., Wagner, D. (eds.) ESA 2014. LNCS, vol. 8737, pp. 284–295. Springer, Heidelberg (2014). doi:10.1007/978-3-662-44777-2_24
9. Feigenbaum, J., Kannan, S., McGregor, A., Suri, S., Zhang, J.: On graph problems in a semi-streaming model. In: Díaz, J., Karhumäki, J., Lepistö, A., Sannella, D. (eds.) ICALP 2004. LNCS, vol. 3142, pp. 531–543. Springer, Heidelberg (2004). doi:10.1007/978-3-540-27836-8_46
10. Feigenbaum, J., Kannan, S., McGregor, A., Suri, S., Zhang, J.: Graph distances in the data-stream model. SIAM J. Comput. 38(5), 1709–1727 (2009)
11. Frederickson, G.N.: Upper bounds for time-space trade-offs in sorting and selection. J. Comput. Syst. 34(1), 19–26 (1987)
12. Korman, M., Mulzer, W., Renssen, A., Roeloffzen, M., Seiferth, P., Stein, Y.: Time-space trade-offs for triangulations and Voronoi diagrams. In: Dehne, F., Sack, J.-R., Stege, U. (eds.) WADS 2015. LNCS, vol. 9214, pp. 482–494. Springer, Cham (2015). doi:10.1007/978-3-319-21840-3_40

13. Munro, J., Paterson, M.: Selection and sorting with limited storage. Theoret. Comput. Sci. **12**(3), 315–323 (1980)
14. Pagter, J., Rauhe, T.: Optimal time-space trade-offs for sorting. In: Proceedings of the 39th Annual Symposium on Foundations of Computer Science (FOCS 1998), pp. 264–268 (1998)
15. Suri, S., Toth, C.D., Zhou, Y.: Range counting over multidimensional data streams. Discret. Comput. Geom. **36**(4), 633–655 (2006)

An FPTAS for the Volume of Some $\mathcal{V}$-polytopes—It is Hard to Compute the Volume of the Intersection of Two Cross-Polytopes

Ei Ando[1]([✉]) and Shuji Kijima[2,3]

[1] Sojo University, 4-22-1, Ikeda, Nishi-Ku, Kumamoto 860-0082, Japan
ando-ei@cis.sojo-u.ac.jp
[2] Kyushu University, 744 Motooka, Nishi-ku, Fukuoka 819-0395, Japan
kijima@inf.kyushu-u.ac.jp
[3] JST PRESTO, 744 Motooka, Nishi-ku, Fukuoka 819-0395, Japan

Abstract. Given an n-dimensional convex body by a membership oracle in general, it is known that any polynomial-time *deterministic* algorithm cannot approximate its volume within ratio $(n/\log n)^n$. There is a substantial progress on *randomized* approximation such as Markov chain Monte Carlo for a high-dimensional volume, and for many #P-hard problems, while only a few #P-hard problems are known to yield deterministic approximation. Motivated by the problem of deterministically approximating the volume of a $\mathcal{V}$-polytope, that is a polytope with a small number of vertices and (possibly) exponentially many facets, this paper investigates the problem of computing the volume of a "knapsack dual polytope," which is known to be #P-hard due to Khachiyan (1989). We reduce an approximate volume of a knapsack dual polytope to that of the *intersection of two cross-polytopes*, and give FPTASs for those volume computations. Interestingly, computing the volume of the intersection of two cross-polytopes (i.e., L_1-balls) is #P-hard, unlike the cases of L_∞-balls or L_2-balls.

Keywords: #P-hard · Deterministic approximation · FPTAS · $\mathcal{V}$-polytope · Intersection of L_1-balls

1 Introduction

1.1 Approximation of a High Dimensional Volume: Randomized vs. Deterministic

A high dimensional volume is hard to compute, even for approximation. When an n-dimensional convex body is given by a *membership oracle*, no polynomial-time *deterministic* algorithm can approximate its volume within ratio $(n/\log n)^n$ [3,6,10,21]. The impossibility comes from the fact that the volume of an n-dimensional L_∞-ball (i.e., hypercube) is exponentially large to

© Springer International Publishing AG 2017
Y. Cao and J. Chen (Eds.): COCOON 2017, LNCS 10392, pp. 13–24, 2017.
DOI: 10.1007/978-3-319-62389-4_2

the volume of its inscribed L_2-ball or L_1-ball, despite that the L_2-ball (L_1-ball as well) is convex and touches each facet of the L_∞-ball (see e.g., [23]). Lovász said in [21] for a convex body K that "*If K is a polytope, then there may be much better ways to compute* Vol(K)." Unfortunately, computing an exact volume is often #P-hard, even for a relatively simple polytope. For instance, computing the *volume* of a knapsack polytope $K(b) = \{\boldsymbol{x} \in [0,1]^n \mid \sum_{i=1}^{n} a_i x_i \leq b\}$, where $a_1, \ldots, a_n \in \mathbb{Z}_{\geq 0}$ is the "item sizes" and $b \in \mathbb{Z}_{\geq 0}$ is the "knapsack capacity", is a well-known #P-hard problem [8].

The difficulty caused by the exponential gap between L_∞-ball and L_1-ball also does harm a simple Monte Carlo algorithm. Then, the Markov chain Monte Carlo (MCMC) method achieves a great success for approximating the high dimensional volume. Dyer, Frieze and Kannan [9] gave the first fully polynomial-time randomized approximation scheme (FPRAS) for the volume computation of a general convex body[1]. They employed a *grid-walk*, which is efficiently implemented with a membership oracle, and showed it is rapidly mixing, then they gave an FPRAS runs in $O^*(n^{23})$ time where O^* ignores poly$(\log n)$ and $1/\epsilon$ factors. After several improvements, Lovász and Vempala [22] improved the time complexity to $O^*(n^4)$ in which they employ hit-and-run walk, and recently Cousins and Vempala [5] gave an $O^*(n^3)$-time algorithm. Many randomized techniques, including MCMC, also have been developed for designing FPRAS for #P-hard problems.

In contrast, the development of *deterministic* approximations for #P-hard problems is a current challenge, and not many results seem to be known. A remarkable progress is the *correlation decay* argument due to Weitz [25]; he designed a *fully polynomial time approximation scheme* (*FPTAS*) for counting independent sets in graphs whose maximum degree is at most 5. A similar technique is independently presented by Bandyopadhyay and Gamarnik [2], and there are several recent developments on the technique, e.g., [4,11,17,18,20]. For counting knapsack solutions[2], Gopalan, Klivans and Meka [12], and Štefankovič, Vempala and Vigoda [24] gave deterministic approximation algorithms based on dynamic programming (see also [13]), in a similar way to the simple random sampling algorithm by Dyer [7]. (He showed a deterministic dynamic programming and a random sampling algorithm in [7].) Modifying dynamic programming in [24], Li and Shi [19] gave an FPTAS that can approximate the volume of a knapsack polytope. Their algorithm runs in $O((n^3/\epsilon^2)$poly $\log b)$ time where b is the knapsack capacity. Motivated by a different approach, Ando and Kijima [1] gave another FPTAS for the volume of a knapsack polytope.

---

[1] Precisely, they are concerned with a "well-rounded" convex body, after an affine transformation of a general finite convex body.

[2] Given $\boldsymbol{a} \in \mathbb{Z}_{>0}^n$ and $b \in \mathbb{Z}_{>0}$, the problem is to compute $|\{\boldsymbol{x} \in \{0,1\}^n \mid \sum_{i=1}^{n} a_i x_i \leq b\}|$. Remark that it is computed in polynomial time when all the inputs a_i ($i = 1, \ldots, n$) and b are bounded by poly(n), using a version of the standard dynamic programming for knapsack problem (see e.g., [7,13]). It should be worth noting that [12,24] needed special techniques, different from ones for optimization problems, to design FPTASs for the counting problem.

Their scheme is based on a classical approximate convolution and runs in $O(n^3/\epsilon)$ time. The running time is independent of the size of items and the knapsack capacity if we assume that the basic arithmetic operations can be performed in constant time.

1.2 $\mathcal{H}$-polytope and $\mathcal{V}$-polytope

An $\mathcal{H}$-polyhedron is an intersection of finitely many closed half-spaces in $\mathbb{R}^n$. An $\mathcal{H}$-polytope is a bounded $\mathcal{H}$-polyhedron. A $\mathcal{V}$-polytope is a convex hull of a finite point set in $\mathbb{R}^n$ [23]. From the view point of computational complexity, a major difference between an $\mathcal{H}$-polytope and a $\mathcal{V}$-polytope is the measure of their 'input size.' An $\mathcal{H}$-polytope given by linear inequalities defining half-spaces may have vertices exponentially many to the number of the inequalities, e.g., an n-dimensional hypercube is given by $2n$ linear inequalities as an $\mathcal{H}$-polytope, and has 2^n vertices. In contrast, a $\mathcal{V}$-polytope given by a point set may have facets exponentially many to the number of vertices, e.g., an n-dimensional cross-polytope (that is an L_1-ball, in fact) is given by a set of $2n$ points as a $\mathcal{V}$-polytope, and it has 2^n facets.

There are many interesting properties between $\mathcal{H}$-polytope and $\mathcal{V}$-polytope [23]. A membership query is polynomial time for both $\mathcal{H}$-polytope and $\mathcal{V}$-polytope. It is still unknown about the complexity of a query if a given pair of $\mathcal{V}$-polytope and $\mathcal{H}$-polytope are identical. Linear programming (LP) on a $\mathcal{V}$-polytope is trivially polynomial time since it is sufficient to check the objective value of all vertices and hence LP is usually concerned with an $\mathcal{H}$-polytope.

1.3 Volume of $\mathcal{V}$-polytope

Motivated by a hardness of the volume computation of a $\mathcal{V}$-polytope, Khachiyan [15] is concerned with the following $\mathcal{V}$-polytope: Suppose a vector $a = (a_1, \ldots, a_n) \in \mathbb{Z}_{\geq 0}^n$ is given. Then let

$$P_a \stackrel{\text{def}}{=} \text{conv}\{\pm e_1, \ldots, \pm e_n, a\} \tag{1}$$

where $e_1, \ldots, e_n$ are the standard basis vectors in $\mathbb{R}^n$. This paper calls P_a *knapsack dual polytope*[3]. Khachiyan [15] showed that computing $\text{Vol}(P_a)$ is #P-hard[4]. The hardness is given by a Cook reduction from counting set partitions, of which the decision version is a celebrated *weakly* NP-hard problem. It is not known if

---

[3] See [23] for the duality of polytopes. In fact, P_a itself is not the dual of a knapsack polytope in a canonical form, but it is obtained by an affine transformation from a dual of knapsack polytope under some assumptions. Khachiyan [16] says that computing $\text{Vol}(P_a)$ 'is "polar" to determining the volume of the intersection of a cube and a halfspace.' .

[4] If all a_i $(i = 1, \ldots, n)$ are bounded by $\text{poly}(n)$, it is computed in polynomial time, so did the counting knapsack solutions. See also footnote 1 for counting knapsack solutions.

we can have an efficient approximation algorithm for computing $\mathrm{Vol}(P_a)$ immediately from the approximation algorithm in e.g., [1] by exploiting that P_a is a dual of a knapsack polytope.

1.4   Contribution

Motivated by a development of techniques for *deterministic* approximation of the volumes of $\mathcal{V}$-polytopes, this paper investigates the knapsack dual polytope P_a given by (1). The main goal of the paper is to establish the following theorem.

Theorem 1. *For any ϵ ($0 < \epsilon < 1$), there exists a* deterministic *algorithm that outputs a value $\widehat{V}$ satisfying $(1 - \epsilon)\mathrm{Vol}(P_a) \leq \widehat{V} \leq (1 + \epsilon)\mathrm{Vol}(P_a)$ in $\mathrm{O}(n^{10}\epsilon^{-6})$ time.*

As far as we know, this is the first result on designing an FPTAS for computing the volume of a $\mathcal{V}$-polytope which is known to be #P-hard. We also discuss some topics related to the volume of $\mathcal{V}$-polytopes appearing in the proof process. Let us briefly explain the outline of the paper.

Technique/Organization. The first step for Theorem 1 is a transformation of the *approximation problem* to another one: An approximate volume of P_a is reduced to the volume of a union of geometric sequence of cross-polytopes (Sect. 3.1), and then it is reduced to the volume of the intersection of two cross-polytopes (Sect. 3.2). We remark that the former reduction is just for approximation, and is useless for proving #P-hardness. A technical point of this step is that the latter reduction is based on a subtraction—if you are familiar with an approximation, you may worry that a subtraction may destroy an approximation ratio[5]. It requires careful tuning of a parameter (β in Sect. 3) which plays conflicting functions in Sects. 3.1 and 3.2: the larger β, the better approximation in Sect. 3.1, while the smaller β, the better in Sect. 3.2. Then, Sect. 3.3 claims by giving an appropriate β that if we have an FPTAS for the volume of an *intersection of two cross-polytopes* then we have an FPTAS of $\mathrm{Vol}(P_a)$.

Section 4 shows an FPTAS for the volume of the intersection of two cross-polytopes (i.e., L_1-balls). The scheme is based on a modified version of the technique developed in [1], which is based on a classical approximate convolution. At a glance, the volume of the intersection of two-balls may seem easy. It is true for two L_∞-balls (i.e., axis-aligned hypercubes), or L_2-balls (i.e., Euclidean balls). However, we show in Sect. 5 that computing the volume of the intersection of cross-polytopes is #P-hard. Intuitively, this interesting fact may come from the fact that the $\mathcal{V}$-polytope, meaning that an n-dimensional cross-polytope, has 2^n facets. In Sect. 6, we extend the technique in Sect. 4 to the intersection of any constant number of cross-polytopes.

---

[5] Suppose you know that x is approximately 49 within 1% error. Then, you know that $x + 50$ is approximately 99 within 1% error. However, it is difficult to say $50 - x$ is approximately 1. Even when additionally you know that x does not exceed 50, $50 - x$ may be 2, 1, 0.1 or smaller than 0.001, meaning that the approximation ratio is unbounded.

2 Preliminaries

This section presents some notation. Let $\mathrm{conv}(S)$ denote the convex hull of $S \subseteq \mathbb{R}^n$, where S is not restricted to a finite point set. A *cross-polytope* $C(\boldsymbol{c}, r)$ of radius $r \in \mathbb{R}_{>0}$ centered at $\boldsymbol{c} \in \mathbb{R}^n$ is given by

$$C(\boldsymbol{c}, r) \overset{\text{def}}{=} \mathrm{conv}\{\boldsymbol{c} \pm r\boldsymbol{e}_i \; i = 1, \ldots, n\}$$

where $\boldsymbol{e}_1, \ldots, \boldsymbol{e}_n$ are the standard basis vectors in $\mathbb{R}^n$. Clearly, $C(\boldsymbol{c}, r)$ has $2n$ vertices. In fact, $C(\boldsymbol{c}, r)$ is an L_1-ball in $\mathbb{R}^n$ described by

$$C(\boldsymbol{c}, r) = \{\boldsymbol{x} \in \mathbb{R}^n \mid \|\boldsymbol{x} - \boldsymbol{c}\|_1 \leq r\} = \{\boldsymbol{x} \in \mathbb{R}^n \mid \langle \boldsymbol{x} - \boldsymbol{c}, \boldsymbol{\sigma} \rangle \leq r \; (\forall \boldsymbol{\sigma} \in \{-1, 1\}^n)\}$$

where $\|\boldsymbol{u}\|_1 = \sum_{i=1}^n |u_i|$ for $\boldsymbol{u} = (u_1, \ldots, u_n) \in \mathbb{R}^n$ and $\langle \boldsymbol{u}, \boldsymbol{v} \rangle = \sum_{i=1}^n u_i v_i$ for $\boldsymbol{u}, \boldsymbol{v} \in \mathbb{R}^n$. Note that $C(\boldsymbol{c}, r)$ has 2^n facets. It is not difficult to see that the volume of a cross-polytope in n-dimension is $\mathrm{Vol}(C(\boldsymbol{c}, r)) = \frac{2^n}{n!} r^n$ for any $r \geq 0$ and $\boldsymbol{c} \in \mathbb{R}^n$, where $\mathrm{Vol}(S)$ for $S \subseteq \mathbb{R}^n$ denotes the (n-dimensional) volume of S.

3 FPTAS for Knapsack Dual Polytope

This section reduces an approximation of $\mathrm{Vol}(P_{\boldsymbol{a}})$ to that of the intersection of two cross-polytopes. In Sect. 4, we will give an FPTAS for the volume of the intersection of two cross-polytopes, accordingly we obtain Theorem 1.

3.1 Reduction to a Geometric Series of Cross-Polytopes

Let β be a parameter[6] satisfying $0 < \beta < 1$, and let $Q_0, Q_1, Q_2, \ldots$ be a sequence of cross-polytopes defined by

$$Q_k \overset{\text{def}}{=} C((1 - \beta^k)\boldsymbol{a}, \beta^k) \tag{2}$$

for $k = 0, 1, 2, \ldots$. Remark that $Q_0 = C(\boldsymbol{0}, 1), Q_1 = C((1 - \beta)\boldsymbol{a}, \beta), Q_\infty = C(\boldsymbol{a}, 0) = \{\boldsymbol{a}\}$. The goal of Sect. 3.1 is to establish the following. Here $1 \pm \epsilon$ is the final relative approximation ratio that we aim to achieve.

Lemma 1. *Let ϵ satisfy $0 < \epsilon < 1$. If $1 - \beta \leq \dfrac{c_1 \epsilon}{n\|\boldsymbol{a}\|_1}$ where $0 < c_1 < 1$, then*

$$(1 - c_1\epsilon)\mathrm{Vol}(P_{\boldsymbol{a}}) \leq \mathrm{Vol}\left(\bigcup_{k=0}^{\infty} Q_k\right) \leq \mathrm{Vol}(P_{\boldsymbol{a}}).$$

Figure 1 illustrates the approximation of $P_{\boldsymbol{a}}$ by this infinite sequence of cross-polytopes. The second inequality in Lemma 1 is relatively easy by the following lemma. [7]

Lemma 2. $\bigcup_{k=0}^{\infty} Q_k \subseteq P_{\boldsymbol{a}}$

To prove the first inequality in Lemma 1, we need the following lemmas.

Lemma 3. $\bigcup_{k=0}^{\infty} \mathrm{conv}(Q_k \cup Q_{k+1}) \cup \{\boldsymbol{a}\} \supseteq P_{\boldsymbol{a}}$

Lemma 4. *If $1 - \beta \leq \dfrac{c_1 \epsilon}{n\|\boldsymbol{a}\|_1}$, then $\mathrm{Vol}\left(\bigcup_{k=0}^{\infty} Q_k\right) \geq (1 - c_1\epsilon)\mathrm{Vol}(P_{\boldsymbol{a}})$.*

---

[6] We will set $\beta = 1 - \dfrac{\epsilon}{2n\|\boldsymbol{a}\|_1}$, later.

[7] Most of the proofs cannot be included due to the space limit.

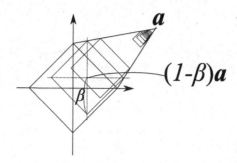

Fig. 1. Approximating P_a by an infinite sequence of cross-polytopes.

3.2   Reduction to the Intersection of Two Cross-Polytopes

We here claim the following.

Lemma 5. $\mathrm{Vol}\left(\bigcup_{k=0}^{\infty} Q_k\right) = \frac{1}{1-\beta^n}\left(\frac{2^n}{n!} - \mathrm{Vol}(Q_1 \cap Q_0)\right)$

The first step of the proof is the following recursive formula.

Lemma 6. $\bigcup_{k=0}^{m} Q_k = \left(\dot{\bigcup}_{k=0}^{m-1} Q_k \setminus Q_{k+1}\right) \dot{\cup} Q_m$ *where* $A \dot{\cup} B$ *denotes the disjoint union of* A *and* B, *meaning that* $A \dot{\cup} B = A \cup B$ *and* $A \cap B = \emptyset$.

The second step is the following lemma.

Lemma 7. $\mathrm{Vol}(Q_k \setminus Q_{k+1}) = \beta^{nk}\mathrm{Vol}(Q_0 \setminus Q_1)$.

By using Lemmas 6 and 7, we can prove Lemma 5 as follows.

Proof (Proof of Lemma 5).

$$
\mathrm{Vol}\left(\dot{\bigcup}_{k=0}^{\infty} Q_k\right) = \mathrm{Vol}\left(\left(\dot{\bigcup}_{k=0}^{\infty} Q_k \setminus Q_{k+1}\right) \dot{\cup} Q_\infty\right)
$$

$$
= \sum_{k=0}^{\infty} \mathrm{Vol}(Q_k \setminus Q_{k+1}) + \mathrm{Vol}(Q_\infty) = \sum_{k=0}^{\infty} \beta^{nk}\mathrm{Vol}(Q_0 \setminus Q_1)
$$

$$
= \frac{1}{1-\beta^n}\mathrm{Vol}(Q_0 \setminus Q_1) = \frac{1}{1-\beta^n}\left(\frac{2^n}{n!} - \mathrm{Vol}(Q_1 \cap Q_0)\right)
$$

$\square$

A reader who are familiar with approximation may worry about the subtraction $\frac{2^n}{n!} - \mathrm{Vol}(Q_0 \cap Q_1)$ in Lemma 5. We claim the following.

Lemma 8. *If* $1 - \beta \geq \dfrac{c_2\epsilon}{n\|a\|_1}$ *and* $0 < c_2\epsilon < 1$, *then* $\mathrm{Vol}(Q_0 \cap Q_1) \leq \dfrac{1}{1 + \frac{c_2\epsilon}{2n}} \dfrac{2^n}{n!}$.

Intuitively, Lemma 8 implies that $\frac{2^n}{n!} - \mathrm{Vol}(Q_0 \cap Q_1)$ is large enough, and an approximation of $\mathrm{Vol}(Q_0 \cap Q_1)$ provides a good approximation of $\mathrm{Vol}(\bigcup_{k=0}^{\infty} Q_k)$, and hence $\mathrm{Vol}(P_a)$.

3.3   Approximation Algorithm and Analysis

Based on Lemma 1 in Sect. 3.1 and Lemma 5 in Sect. 3.2, we give an FPTAS for $\mathrm{Vol}(P_a)$ where we assume an algorithm to approximate $\mathrm{Vol}(Q_0 \cap Q_1)$.

Algorithm 1 ($(1 \pm \epsilon)$-approximation $(0 < \epsilon \le 1/2)$)
Input: $a \in \mathbb{Z}_+^n$;

1. Set parameter $\beta := 1 - \dfrac{\epsilon}{2n\|a\|_1}$;

2. Approximate $I \overset{\text{def}}{=} \mathrm{Vol}(C(\mathbf{0}, 1) \cap C((1-\beta)a, \beta))$ by Z such that $I \le Z \le \left(1 + \frac{\epsilon^2}{4n}\right) I$;

3. Output $\widehat{V} = \frac{1+\epsilon}{1-\beta^n}\left(\frac{2^n}{n!} - Z\right)$.

Lemma 9. *The output $\widehat{V}$ of Algorithm 1 satisfies*

$$(1 - \epsilon)\,\mathrm{Vol}(P_a) \le \widehat{V} \le (1 + \epsilon)\mathrm{Vol}(P_a).$$

4   The Volume of the Intersection of Two Cross-Polytopes

This section gives an FPTAS for the volume of the intersection of two cross-polytopes in the n-dimensional space. Without loss of generality[8], we are concerned with $\mathrm{Vol}(C(\mathbf{0}, 1) \cap C(c, r))$ for $c \ge \mathbf{0}$ and r $(0 < r \le 1)$. This section establishes the following.

Theorem 2. *For any δ $(0 < \delta < 1)$, there exists a deterministic algorithm which outputs a value Z satisfying $\mathrm{Vol}(C(\mathbf{0}, 1) \cap C(c, r)) \le Z \le (1 + \delta)\mathrm{Vol}(C(\mathbf{0}, 1) \cap C(c, r))$ for any input $c \ge \mathbf{0}$ and r $(0 < r \le 1)$ satisfying $\|c\|_1 \le r$, and runs in $O(n^7\delta^{-3})$ time.*

The assumption that $\|c\|_1 \le r$ implies both centers $\mathbf{0}$ and c are contained in the intersection $C(\mathbf{0}, 1) \cap C(c, r)$. Note that the assumption does not harm to our main goal Theorem 1 (recall Algorithm 1 in Sect. 3.3). We show in Sect. 5 that Computing $\mathrm{Vol}(C(\mathbf{0}, 1) \cap C(c, r))$ remains #P-hard even on the assumption.

4.1   Preliminaries: Convolution for the Volume

As a preliminary step, Sect. 4.1 gives a convolution which provides $\mathrm{Vol}(C(\mathbf{0}, 1) \cap C(c, r))$. Let $\Psi_0 \colon \mathbb{R}^2 \to \mathbb{R}$ be given by $\Psi_0(u, v) = 1$ if $u \ge 0$ and $v \ge 0$, otherwise $\Psi_0(u, v) = 0$. Inductively, we define $\Psi_i \colon \mathbb{R}^2 \to \mathbb{R}$ for $i = 1, 2, \ldots, n$ by

$$\Psi_i(u, v) \overset{\text{def}}{=} \int_{-1}^{1} \Psi_{i-1}(u - |s|, v - |s - c_i|)\mathrm{d}s \tag{3}$$

for $u, v \in \mathbb{R}$. We remark that $\Psi_i(u, v) = 0$ holds if $u \le 0$ or $v \le 0$, for any $i = 1, 2, \ldots, n$ by the definition.

---

[8]   Remark that $\mathrm{Vol}(C(c, r) \cap C(c', r')) = r^n\mathrm{Vol}\left(C(\mathbf{0}, 1) \cap C\left(\frac{(c-c')^+}{r}, \frac{r'}{r}\right)\right)$ holds for any $c, c' \in \mathbb{R}^n$ and $r, r' \in \mathbb{R}_{>0}$, where $(c - c')^+ = (|c_1 - c_1'|, |c_2 - c_2'|, \ldots, |c_n - c_n'|)$.

Lemma 10. $\Psi_n(1, r) = \mathrm{Vol}(C(\mathbf{0}, 1) \cap C(\mathbf{c}, r))$

To prove Lemma 10, it might be helpful to introduce a probability space. Let $\mathbf{X} = (X_1, \ldots, X_n)$ be a uniform random variable over $[-1, 1]^n$, i.e., X_i $(i = 1, \ldots, n)$ are mutually independent. Then, $\Pr[\mathbf{X} \in C(\mathbf{0}, 1) \cap C(\mathbf{c}, r)] = \frac{\mathrm{Vol}(C(\mathbf{0},1) \cap C(\mathbf{c}, r))}{\mathrm{Vol}([-1,1]^n)} = \frac{1}{2^n} \mathrm{Vol}(C(\mathbf{0}, 1) \cap C(\mathbf{c}, r))$ holds.

Lemma 11. *For any $u, v \in \mathbb{R}$ and any $i = 1, 2, \ldots, n$,*

$$\frac{1}{2^i} \Psi_i(u, v) = \Pr\left[\left(\sum_{j=1}^{i} |X_j| \le u \right) \wedge \left(\sum_{j=1}^{i} |X_j - c_j| \le v \right) \right].$$

Now, Lemma 10 is easy from Lemma 11.

4.2 Idea for Approximation

Our FPTAS is based on an approximation of $\Psi_i(u, v)$. Let $G_0(u, v) = \Psi_0(u, v)$ for any $u, v \in \mathbb{R}$, i.e., $G_0(u, v) = 1$ if $u \ge 0$ and $v \ge 0$, otherwise $G_0(u, v) = 0$. Inductively assuming $G_{i-1}(u, v)$, we define

$$\overline{G}_i(u, v) \stackrel{\text{def}}{=} \int_{-1}^{1} G_{i-1}(u - |s|, v - |s - c_i|) \mathrm{d}s \tag{4}$$

for $u, v \in \mathbb{R}$, for convenience. Then, let $G_i(u, v)$ be a staircase approximation of $\overline{G}_i(u, v)$, given by

$$G_i(u, v) \stackrel{\text{def}}{=} \begin{cases} \overline{G}_i\left(\frac{1}{M}k, \frac{r}{M}\ell\right) & \left(\begin{matrix} \text{if } \frac{1}{M}(k-1) < u \le \frac{1}{M}k \ (k = 1, 2, \ldots), \text{ and} \\ \frac{r}{M}(\ell - 1) < v \le \frac{r}{M}\ell \ (\ell = 1, 2, \ldots). \end{matrix} \right) \\ 0 & \text{(otherwise)} \end{cases} \tag{5}$$

for any $u, v \in \mathbb{R}$. Thus, we remark that

$$G_i(u, v) = G_i\left(\frac{1}{M}\lceil Mu \rceil, \frac{r}{M}\lceil \frac{M}{r}v \rceil\right) \tag{6}$$

holds for any $u, v \in \mathbb{R}$, by the definition. Section 4.3 will show that $G_i(u, v)$ approximates $\Psi_i(u, v)$ well.

In the rest of Sect. 4.2, we briefly comment on the computation of G_i. First, remark that (4) implies that $\overline{G}_i(u, v)$ is computed only from $G_{i-1}(u', v')$ for $u' \le u$ and $v' \le v$, i.e., we do not need to know $G_{i-1}(u', v')$ for $u' > u$ or $v' > v$. Second, remark that (6) implies that $G_i(u, v)$ for $u \le 1$ and $v \le r$ takes (at most) $(M+1)^2$ different values. Precisely, let

$$\Gamma \stackrel{\text{def}}{=} \left\{ \frac{1}{M}(k, r\ell) \mid k = 0, 1, 2, \ldots, M, \ \ell = 0, 1, 2, \ldots, M \right\}$$

then $G_i(u, v)$ for $(u, v) \in \Gamma$ provides all possible values of $G_i(u, v)$ for $u \leq 1$ and $v \leq r$, since (6).

Then, we explain how to compute $G_i(u, v)$ for $(u, v) \in \Gamma$ from G_{i-1}. For an arbitrary $(u, v) \in \Gamma$, let

$$S(u) \stackrel{\text{def}}{=} \left\{ s \in [-1, 1] \mid u - |s| = \tfrac{1}{M}k \ (k = 0, 1, 2, \ldots, M) \right\}$$
$$= \left\{ s \in [-1, 1] \mid s = \pm(u - \tfrac{1}{M}k) \ (k = 0, 1, 2, \ldots, M) \right\},$$

let

$$S_i(v) \stackrel{\text{def}}{=} \left\{ s \in [-1, 1] \mid v - |s - c_i| = \tfrac{r}{M}\ell \ (\ell = 0, 1, 2, \ldots, M) \right\}$$
$$= \left\{ s \in [-1, 1] \mid s = c_i \pm (v - \tfrac{r}{M}\ell) \ (\ell = 0, 1, 2, \ldots, M) \right\},$$

and let $T_i(u, v) \stackrel{\text{def}}{=} S(u) \cup S_i(v) \cup \{-1, 0, c_i, 1\}$. Suppose $t_0, t_1, \ldots, t_m$ be an ordering of all elements of $T_i(u, v)$ such that $t_i \leq t_{i+1}$ for any $i = 0, 1, \ldots, m$, where $m = |T_i(u, v)|$. Then, we can compute $G_i(u, v)$ for any $(u, v) \in \Gamma$ by $G_i(u, v) = \overline{G}_i(u, v)$, which can be transformed into

$$\overline{G}_i(u, v) = \int_{-1}^{1} G_{i-1}(u - |s|, v - |s - c_i|) \mathrm{d}s$$

$$= \sum_{j=0}^{m-1} (t_{j+1} - t_j) G_{i-1} \left(\tfrac{1}{M} \lceil M(u - |t_{j+1}|) \rceil, \tfrac{r}{M} \left\lceil \tfrac{M}{r}(v - |t_{j+1} - c_i|) \right\rceil \right) \quad (7)$$

where we remark again that the terms of (7) consist of $G_{i-1}(u, v)$ for $(u, v) \in \Gamma$.

4.3   Algorithm and Analysis

Based on the arguments in Sect. 4.2, our algorithm is described as follows.

Algorithm 2 (for $(1 + \delta)$-approximation $(0 < \delta \leq 1)$)
Input: $c \in \mathbb{Q}_{\geq 0}^n$, $r \in \mathbb{Q}$ $(0 \leq r \leq 1)$;
1. Set $M := \lceil 4n^2\delta^{-1} \rceil$;
2. Set $G_0(u, v) := 1$ for $(u, v) \in \Gamma$, otherwise $G_0(u, v) := 0$;
3. For $i := 1, \ldots, n$,
4.   For $(u, v) \in \Gamma$,
5.   Compute $G_i(u, v)$ from G_{i-1} by (7);
6. Output $G_n(1, r)$.

Lemma 12. *The running time of Algorithm 2 is $\mathrm{O}(n^7\delta^{-3})$.*

Theorem 2 is immediate from Lemma 12 and the following Lemma 13.

Lemma 13. *$\Psi_n(1, r) \leq G_n(1, r) \leq (1 + \delta)\Psi_n(1, r)$.*

The proof sketch of Lemma 13 is the following.

Proof (Proof Sketch of Lemma 13). The first inequality is immediate. Then, we show the latter inequality. We can prove that

$$\frac{\Psi_n(1,r)}{\Psi_n(1+\frac{n}{M}, r(1+\frac{n}{M}))} \geq \left(\frac{M}{M+n}\right)^{2n} = \left(\frac{1}{1+\frac{n}{M}}\right)^{2n} \geq \left(1-\frac{n}{M}\right)^{2n}$$

$$\geq \left(1-\frac{\delta}{4n}\right)^{2n} \geq 1 - 2n\frac{\delta}{4n} = 1 - \frac{\delta}{2}.$$

Then, $\frac{\Psi_n(1+\frac{n}{M}, r(1+\frac{n}{M}))}{\Psi_n(1,r)} \leq \frac{1}{1-\frac{\delta}{2}} \leq 1 + \delta$ for any $\delta \leq 1$, and we obtain the claim. $\square$

5 Hardness of the Volume of the Intersection of Two Cross-Polytopes

This section establishes the following.

Theorem 3. *Given a vector $c \in \mathbb{Z}^n_{\geq 0}$ and integers $r_1, r_2 \in \mathbb{Z}_{>0}$, computing the volume of $C(\mathbf{0}, r_1) \cap C(c, r_2)$ is #P-hard, even when each cross-polytopes contains the center of the other one, i.e., $\mathbf{0} \in C(c, r_2)$ and $c \in C(\mathbf{0}, r_1)$.*

The proof of Theorem 3 is a reduction of counting set partitions, which is a well-known #P-hard problem. To be precise, we reduce the following problem, which is a version of counting set partition (for the #P-hardness of counting set partition, see e.g., [14]).

Problem 1 (#LARGE SET). Given an integer vector $a \in \mathbb{Z}^n_{\geq 0}$ such that $\|a\|_1$ is even, meaning that $\|a\|_1/2$ is an integer, the problem is to compute

$$|\{\sigma \in \{-1,1\}^n \mid \langle \sigma, a \rangle > 0\}|. \tag{8}$$

Note that $|\{\sigma \in \{-1,1\}^n \mid \langle \sigma, a \rangle = 0\}| = \left|\left\{S \subseteq \{1,\ldots,n\} \mid \sum_{i \in S} a_i = \frac{\|a\|_1}{2}\right\}\right|$ holds: if $\sigma \in \{-1,1\}^n$ satisfies $\langle \sigma, a \rangle = 0$, then let $S \subseteq \{1,\ldots,n\}$ be the set of indices of $\sigma_i = 1$ then $\sum_{i \in S} a_i = \|a\|_1/2$ holds. Using the following simple observation, we see that Problem 1 is equivalent to counting set partitions.

Observation 1. *For any $\sigma \in \{-1,1\}^n$, $\langle \sigma, a \rangle > 0$ if and only if $\langle -\sigma, a \rangle < 0$.*

By Observation 1, we see that $|\{\sigma \in \{-1,1\}^n \mid \langle \sigma, a \rangle = 0\}|$ is equal to $2^n - 2|\{\sigma \in \{-1,1\}^n \mid \langle \sigma, a \rangle > 0\}|$.

In the following, let $a \in \mathbb{Z}^n_{\geq 0}$ be an instance of Problem 1. Roughly speaking, our proof of Theorem 3 claims that the volume of $(C(\delta a, 1) \cap C(\mathbf{0}, 1+\epsilon)) \setminus C(\mathbf{0}, 1)$ is proportional to the answer of #LARGE SET when $0 < \epsilon < \delta \ll 1/\|a\|_1$. If we could compute the volume of the intersection of two cross-polytopes exactly, then we obtain $\text{Vol}((C(\delta a, 1) \cap C(\mathbf{0}, 1+\epsilon)) \setminus C(\mathbf{0}, 1)) = \text{Vol}(C(\delta a, 1) \cap C(\mathbf{0}, 1+\epsilon)) - \text{Vol}(C(\delta a, 1) \cap C(\mathbf{0}, 1))$, which would solve #LARGE SET.

6 Intersection of a Constant Number of Cross-Polytopes

Let $p_i \in \mathbb{R}^n$, $r_i \in \mathbb{R}_{\geq 0}$ and $C(p_i, r_i)$ for $i = 1, \ldots, k$, where $C(p, r)$ is a cross-polytope (L_1-ball) with center $p \in \mathbb{R}^n$ and radius $r \in \mathbb{R}_{\geq 0}$. Then, we are to compute the following polytope given by $S(\Pi, r) = \bigcap_{i=1}^{k} C(p_i, r_i)$, where Π is an $n \times k$ matrix $\Pi = (p_1, \ldots, p_k)$ and $r = (r_1, \ldots, r_k)$. For the analysis, we assume that $p_1, \ldots, p_k$ are internal points of $S(\Pi, r)$.

Theorem 4. *There is an algorithm that outputs an approximation Z of* $\mathrm{Vol}(S(\Pi, r))$ *in* $O(k^{k+2} n^{2k+3} / \delta^{k+1})$ *time satisfying* $\mathrm{Vol}(S(\Pi, r)) \leq Z \leq (1 + \delta)\mathrm{Vol}(S(\Pi, r))$.

7 Conclusion

Motivated by the problem of deterministically approximating the volume of a $\mathcal{V}$-polytope, this paper gave an FPTAS for the volume of the knapsack dual polytope $\mathrm{Vol}(P_a)$. In the process, we showed that computing the volume of the intersection of L_1-balls is #P-hard, and gave an FPTAS. As we remarked, the volume of the intersection of two L_q-balls are easy for $q = 2, \infty$. The complexity of the volume of the intersection of two L_q-balls for other $q > 0$ is interesting. The problem seems difficult even for approximation in the case of $q \in (0, 1)$, since L_q-ball is no longer convex. Our FPTAS for the intersection of two cross-polytopes assumes that each cross-polytope contains the center of the other one. It is open if an FPTAS exists without the assumption.

Acknowledgments. This work is partly supported by Grant-in-Aid for Scientific Research on Innovative Areas MEXT Japan "Exploring the Limits of Computation (ELC)" (No. 24106008, 24106005) and by JST PRESTO Grant Number JPMJPR16E4, Japan.

References

1. Ando, E., Kijima, S.: An FPTAS for the volume computation of 0–1 knapsack polytopes based on approximate convolution. Algorithmica **76**(4), 1245–1263 (2016)
2. Bandyopadhyay, A., Gamarnik, D.: Counting without sampling: asymptotics of the log-partition function for certain statistical physics models. Random Struct. Algorithms **33**, 452–479 (2008)
3. Bárány, I., Füredi, Z.: Computing the volume is difficult. Discrete Comput. Geom. **2**, 319–326 (1987)
4. Bayati, M., Gamarnik, D., Katz, D., Nair, C., Tetali, P.: Simple deterministic approximation algorithms for counting matchings. In: Proceedings of STOC 2007, pp. 122–127 (2007)
5. Cousins, B., Vempala, S., Bypassing, K.L.S.: Gaussian cooling and an $O^*(n^3)$ volume algorithm. In: Proceedings of STOC 2015, pp. 539–548 (2015)
6. Dadush, D., Vempala, S.: Near-optimal deterministic algorithms for volume computation via M-ellipsoids. Proc. Natl. Acad. Sci. USA **110**(48), 19237–19245 (2013)

7. Dyer, M.: Approximate counting by dynamic programming. In: Proceedings of STOC 2003, pp. 693–699 (2003)
8. Dyer, M., Frieze, A.: On the complexity of computing the volume of a polyhedron. SIAM J. Comput. **17**(5), 967–974 (1988)
9. Dyer, M., Frieze, A., Kannan, R.: A random polynomial-time algorithm for approximating the volume of convex bodies. J. Assoc. Comput. Mach. **38**(1), 1–17 (1991)
10. Elekes, G.: A geometric inequality and the complexity of computing volume. Discrete Comput. Geom. **1**, 289–292 (1986)
11. Gamarnik, D., Katz, D.: Correlation decay and deterministic FPTAS for counting list-colorings of a graph. In: Proceedings of SODA 2007, pp. 1245–1254 (2007)
12. Gopalan, P., Klivans, A., Meka, R.: Polynomial-time approximation schemes for knapsack and related counting problems using branching programs. arXiv:1008.3187v1 (2010)
13. Gopalan, P., Klivans, A., Meka, R., Štefankovič, D., Vempala, S., Vigoda, E.: An FPTAS for #knapsack and related counting problems. In: Proceedings of FOCS 2011, pp. 817–826 (2011)
14. Karp, R.: Reducibility among combinatorial problems. In: Miller, R.E., Thatcher, J.W. (eds.) Complexity of Computer Computations, pp. 85–103. Plenum Press, New York (1972)
15. Khachiyan, L.: The problem of computing the volume of polytopes is #P-hard. Uspekhi Mat. Nauk. **44**, 199–200 (1989)
16. Khachiyan, L.: Complexity of polytope volume computation. In: Pach, J. (ed.) New Trends in Discrete and Computational Geometry, pp. 91–101. Springer, Berlin (1993)
17. Li, L., Lu, P., Yin, Y.: Approximate counting via correlation decay in spin systems. In: Proceedings of SODA 2012, pp. 922–940 (2012)
18. Li, L., Lu, P., Yin, Y.: Correlation decay up to uniqueness in spin systems. In: Proceedings of SODA 2013, pp. 67–84 (2013)
19. Li, J., Shi, T.: A fully polynomial-time approximation scheme for approximating a sum of random variables. Oper. Res. Lett. **42**, 197–202 (2014)
20. Lin, C., Liu, J., Lu, P.: A simple FPTAS for counting edge covers. In: Proceedings of SODA 2014, pp. 341–348 (2014)
21. Lovász, L.: An Algorithmic Theory of Numbers, Graphs and Convexity. Applied Mathematics. SIAM Society for Industrial, Philadelphia (1986)
22. Lovász, L., Vempala, S.: Simulated annealing in convex bodies and an $O^*(n^4)$ volume algorithm. J. Comput. Syst. Sci. **72**, 392–417 (2006)
23. Matoušek, J.: Lectures on Discrete Geometry. Springer, New York (2002)
24. Štefankovič, D., Vempala, S., Vigoda, E.: A deterministic polynomial-time approximation scheme for counting knapsack solutions. SIAM J. Comput. **41**(2), 356–366 (2012)
25. Weitz, D.: Counting independent sets up to the tree threshold. In: Proceedings of STOC 2006, pp. 140–149 (2006)

Local Search Strikes Again: PTAS for Variants of Geometric Covering and Packing

Pradeesha Ashok[1]([⊠]), Aniket Basu Roy[2]([⊠]), and Sathish Govindarajan[2]([⊠])

[1] International Institute of Information Technology, Bangalore, India
pradeesha@iiitb.ac.in
[2] Computer Science and Automation, Indian Institute of Science, Bangalore, India
{aniket.basu,gsat}@csa.iisc.ernet.in

Abstract. Geometric Covering and Packing problems have been extensively studied in the last few decades and have applications in diverse areas. Several variants and generalizations of these problems have been studied recently. In this paper, we look at the following covering variants where we require that each point is "uniquely" covered, i.e., it is covered by exactly one object: *Unique Coverage problem*, where we want to maximize the number of uniquely covered points and *Exact Cover problem*, where we want to uniquely cover every point and minimize the number of objects used for covering. We also look at the following generalizations: *Multi-Cover problem*, a generalization of Set Cover, where we want to select the minimum subset of objects with the constraint that each input point is covered by at least k objects in the solution. And *Shallow Packing problem*, a generalization of Packing problem, where we want to select the maximum subset of objects with the constraint that any point in the plane is contained in at most k objects in the solution. The above problems are NP-hard even for unit squares in the plane. Thus, the focus has been on obtaining good approximation algorithms.

Local Search have been quite successful in the recent past in obtaining good approximation algorithms for a wide variety of problems. We consider the Unique Coverage and Multi-Cover problems on non-piercing objects, which is a broad class that includes squares, disks, pseudo-disks, etc. and show that the local search algorithm yields a PTAS approximation under the assumption that the depth of the input points is at most a constant. For the Shallow Packing problem, we show that the local search algorithm yields a PTAS approximation for objects with sub-quadratic union complexity, which is a very broad class of objects that even includes non-piercing objects. For the Exact Cover problem, we show that finding a feasible solution is NP-hard even for unit squares in the plane, thus negating the existence of polynomial time approximation algorithms.

Keywords: Packing · Covering · PTAS · Local search · Non-piercing regions

© Springer International Publishing AG 2017
Y. Cao and J. Chen (Eds.): COCOON 2017, LNCS 10392, pp. 25–37, 2017.
DOI: 10.1007/978-3-319-62389-4_3

1   Introduction

Set Covering and Packing problems form a class of well studied problems in theoretical computer science. Given a universe U of n elements and a family $\mathcal{F}$ of subsets of U, set cover problem asks for a subset of sets in $\mathcal{F}$ such that every element in U is contained in at least one of the chosen sets and the number of chosen sets is minimized. An element is said to be covered by a set if it is contained in that set. Similarly, Set Packing problem asks for subset of sets in $\mathcal{F}$ such that every element is contained in *at most* one of the chosen subset and the number of chosen sets is maximized. An important variant of these problems is when the universe is defined by set of points in $\mathbb{R}^d$ and the sets are subsets of points contained in geometric objects. Apart from the fact that geometric variants have a lot of real world applications, geometric setting usually leads to improved algorithms also.

We study variants of geometric covering and packing problems in the context of approximation algorithms. We look at variants of covering where points are *uniquely* covered i.e., a point is contained in *exactly one* of the chosen sets. Specifically, we look at the following variants:

Let U be a set of n points in the plane and $\mathcal{F}$ be a family of subsets of U defined by a class of geometric objects.

1. UNIQUE COVERAGE: Select a subset $\mathcal{F}' \subseteq \mathcal{F}$ such that the number of uniquely covered points in U is maximized.
2. EXACT COVER: Select a subset $\mathcal{F}' \subseteq \mathcal{F}$ such that all points in U are uniquely covered and the cardinality of $\mathcal{F}'$ is minimized.

We also look into a generalization of the geometric covering problem called Multi-Cover where each point has an integer demand associated with it.
MULTI COVER: Select a subset $\mathcal{F}' \subseteq \mathcal{F}$ such that each point in U is contained in at least as many sets in $\mathcal{F}'$ as its demand and the cardinality of $\mathcal{F}'$ is minimized.

Further we also study a generalization of geometric packing called Shallow Packing:
SHALLOW PACKING: Select a subset $\mathcal{F}' \subseteq \mathcal{F}$ such that every point in U is contained in at most k sets in $\mathcal{F}'$ and the cardinality of $\mathcal{F}'$ is maximized.

We study the UNIQUE COVERAGE, EXACT COVER, and MULTI COVER problems for a class of geometric objects called non-piercing regions[1] and SHALLOW PACKING problem for geometric objects with sub-quadratic union complexity[2].

UNIQUE COVERAGE problem was introduced by Demaine et al. [11] motivated by applications in wireless networks. They proved that for any $\epsilon > 0$, it is hard to approximate UNIQUE COVERAGE with ratio $O(\log^{\sigma(\epsilon)} n)$, assuming $NP \nsubseteq BPTIME(2^{n^\epsilon})$, where $\sigma(\epsilon)$ is some constant dependent on ϵ. Geometric variant of UNIQUE COVERAGE problem was studied in [13]. They prove that

---

[1] A set of regions is said to be non-piercing if for any pair of regions A and B, $A \setminus B$ and $B \setminus A$ are connected.

[2] Union complexity of a set of objects is the description complexity of the boundary of the union of the objects.

UNIQUE COVERAGE is NP-hard even when the sets are defined by unit-disks and give an 18-approximation algorithm for UNIQUE COVERAGE with unit disks, which was later improved to 4.31-factor by Ito et al. [20]. However, when the depth[3] of the input points is bounded with respect to input disks (bounded ply), [13] give an asymptotic FPTAS. PTAS for UNIQUE COVERAGE on unit squares was given in [21]. The parameterized complexity of UNIQUE COVERAGE problem was studied in [3,24].

EXACT COVER was one of the twenty one NP-complete problems in [17]. A related problem is the EXACT SAT or 1-IN-K SAT problem which is a well-studied problem [10,28]. The parameterized complexity of the geometric EXACT COVER problem is studied in [3]. Contrary to the FPT tractability results of UNIQUE COVERAGE, EXACT COVER with unit squares is proved to be W[1]-hard when parameterized by the number of sets in the solution.

MULTI COVER problem is a very natural generalization of the Set Cover problem. Chekuri et al. [7] gave an $O(\log OPT)$-approximation algorithm when the objects have bounded VC-dimension with unit weights. Bansal and Pruhs [5] gave an $O(\log \phi(n))$-approximation algorithm when the objects have *shallow cell complexity* which means the cell complexity is at most $n\phi(n)k^{O(1)}$. This means that for objects with linear union complexity viz., disks, pseudo-disks, halfspaces, etc., this algorithm returns solutions that are constant factor away from the optimum.

SHALLOW PACKING problem is again a natural generalization of the Packing problem. Aschner et al. [2] gave a PTAS when the objects are fat. Wheras, Har-Peled [19] gave QPTAS for pseudo-disks which was improved to PTAS by Govindarajan et al., [18]. Moreover, in [18] the proof also works for the discrete setting, i.e., there is a given point set and the depth of every given point with respect to the solution set should be some constant. Ene et al. [12], had given a $O(u(n)/n)$-factor approximation algorithm based on LP for the capacitated packing problem where $u(n)$ is the union complexity of the objects concerned and the capacities of the points need not be constants. On a related note, Pach and Walczak [26] proved that k-fold packing can be decomposed into p_k many (1-fold) packings for class of objects with sub-quadratic union complexity, where p_k is a function of k.

In this paper, we study *non-piercing regions* and objects with *sub-quadratic union complexity* in the plane. See Sect. 3 for precise definitions. Non-piercing regions are very broad class of objects viz., disks, squares, pseudo-disks, half-planes, homothets of a fixed convex object, etc. On the other hand, objects with sub-quadratic union complexity are even more general class of objects. They include non-piercing regions (having constant description complexity) as well as piercing objects like fat triangles. Note that axis-parallel rectangles have quadratic union complexity thus it belongs to neither class of the objects.

We study the approximability of the problems described above. To design approximation algorithms, more specifically PTASs, we use a well known design

---

[3] The depth of a point with respect to a set of objects is the number of objects in the set containing that point.

paradigm called *local search*. The use of local search algorithms to obtain approximation results has been proved to be quite fruitful in the recent times. Mustafa and Ray [25] and Chan and Har-Paled [6] used local search algorithm to obtain PTASs for the geometric hitting set and independent set problems. Local Search has been successful in obtaining PTAS for various problems like art gallery problems [4, 22], different facility location problems [9], k-means [8, 16], set cover and dominating set [18].

2   Our Results

In this paper we study the following problems and prove the corresponding theorems.

Unique Coverage. Given a finite set of non-piercing regions $\mathcal{R}$ and, a set of points P in the plane, compute a subset $\mathcal{R}' \subseteq \mathcal{R}$ such that the number of points in P that are contained by exactly one region in $\mathcal{R}'$ is maximized.

Theorem 1. *The Local Search algorithm yields a PTAS for the* UNIQUE COVERAGE *problem for non-piercing regions with the properties: Every object in $\mathcal{R}$ contains at most some constant c number of points from P and every point in P has depth at most some constant d.*

To the best of our knowledge, this is the first result based on local search technique for the UNIQUE COVERAGE problems where most of the previous results were based on the grid shifting technique [13, 21].

PTAS for the UNIQUE COVERAGE problem was known only for unit squares and disks with bounded depth [13, 21]. Our result gives a PTAS for a broader class of objects namely, non-piercing regions with the additional assumptions on the input as stated in the above theorem. We note that additional assumptions are needed to obtain a PTAS as the UNIQUE COVERAGE problem on disks is as hard as the general UNIQUE COVERAGE problem, and thus hard to approximate beyond poly-logarithmic factors [11].

Multi Cover. Given a finite set of non-piercing regions $\mathcal{R}$, a set of points P in the plane and a demand d_p of every point $p \in P$, compute the maximum cardinality subset $\mathcal{R}' \subseteq \mathcal{R}$ such that every point $p \in P$ is contained in at least d_p regions from $\mathcal{R}'$. We assume the demands are bounded by some constant.

Theorem 2. *The Local Search algorithm yields a PTAS for the* MULTI COVER *problem for non-piercing regions where depth of every input point is at most some constant k.*

The current state of the art result for MULTI COVER problem for non-piercing regions is a $O(1)$-approximation algorithm which also works in the weighted setting [5]. To our knowledge we do not know of any PTAS for the MULTI COVER problem for geometric objects in the plane even when the demands are all 2.

Shallow Packing. Given a finite set of regions $\mathcal{R}$ in the plane and a constant k, compute a subset $\mathcal{R}' \subseteq \mathcal{R}$ such that every point in the plane is contained in at most k regions from $\mathcal{R}'$. The objective is to maximize the cardinality of $\mathcal{R}'$.

Theorem 3. *The Local Search algorithm yields a PTAS for the* SHALLOW PACKING *problem for class of regions with sub-quadratic union complexity.*

Aschner et al. [2] gave a PTAS for the SHALLOW PACKING problem for fat objects using local search. Whereas, Govindarajan et al. [18] gave a PTAS again using local search but for the discrete case, which is a further generalization. In light of these two results we too give a PTAS using local search (in the continuous setting) but for a very general class of objects, i.e., objects with sub-quadratic union complexity that includes both fat objects and non-piercing regions.

Exact Cover. Given a finite set of regions $\mathcal{R}$ and, a set of points P in the plane, compute a subset $\mathcal{R}' \subseteq \mathcal{R}$ such that every point is contained by exactly one region in $\mathcal{R}'$. The objective is to minimize the cardinality of $\mathcal{R}'$.

Theorem 4. *Finding any feasible* EXACT COVER *for a set of unit squares is NP-hard.*

This NP-hardness result rules out the possibility of any polynomial time approximation algorithm, even for objects as special as unit squares. The proof follows the NP-hardness proof of the Set Cover problem for unit squares by Fowler et al. [14]. The details of the reduction will be presented in the full version of this paper.

Organization. Proofs of Theorems 1, 2, and 3 can be found in Sects. 5, 6, and 7, respectively.

3 Preliminaries

The UNIQUE COVERAGE, MULTI COVER problems are studied for the non-piercing regions and we formally define it below.

Definition 1 (Non-piercing Regions). *For a given set $\mathcal{R}$ of simply connected regions in the plane, if for every pair of regions $A, B \in \mathcal{R}$, $A \setminus B$ is a connected region and so is $B \setminus A$ then we call $\mathcal{R}$ to be a set of non-piercing regions.*

For technical reasons, we assume that if two regions intersect then their boundaries intersect at most a constant number of times.

Definition 2 (Union Complexity). *Given a set of objects $\mathcal{R}$ in the plane, the combinatorial complexity of the boundary of the union of objects in $\mathcal{R}$ is called its union complexity.*

For objects with constant description complexity the union complexity can be at most $O(n^2)$. For instance, the union complexity of arbitrary family of axis-parallel rectangles is $\Theta(n^2)$. Whereas, non-piercing regions (with constant complexity) have linear union complexity [29]. Examples of such class of objects are pseudo-disks, disks, homothets of a fixed convex object. Somewhere in between lies the union complexity of objects like α-fat triangles, which is $O((n/\alpha)\log\log n\log(1/\alpha))$. For a detailed exposition on the above results on union complexity we suggest the survey article by Agarwal et al. [1].

Definition 3 (Discrete Intersection Graph). *Given a set of regions $\mathcal{R}$ and a point set P, the discrete intersection graph of $\mathcal{R}$ with respect to P is the graph whose vertex set is $\mathcal{R}$ and there exists an edge (A, B) iff $A \cap B \cap P \neq \emptyset$ for $A, B \in \mathcal{R}$.*

Lemma 1 ([18]). *The discrete intersection graph of a set of non-piercing regions $\mathcal{R}$ with respect to a point set P in the plane, such that the depth of every point in P is at most some constant k, has a balanced separator of size sublinear in $|\mathcal{R}|$.*

4   Local Search Framework

We use the following local search algorithm as described in the following works [2,6,18,25]. Fix a parameter k that is polynomial in $1/\epsilon$. Start with some feasible solution. Add and remove at most k elements to and from our current solution retaining its feasibility. If the objective function improves due to this change we update the current solution to the changed one. We keep making the updates as long as possible. Once it can no longer be improved we return this solution and call it the local search solution. The running time of the algorithm is $n^{O(k)}$.

Let $\mathcal{A}$ be the local search solution and $\mathcal{O}$ be the optimum solution. If there exists a suitable bipartite graph on the elements of $\mathcal{A}$ and $\mathcal{O}$ then one can show that the sizes are within a multiplicative factor of $1 + \epsilon$ (PTAS). The below theorem explicitly states the properties required by the bipartite graph. For a detailed exposition see Aschner et al. [2].

Theorem 5 ([2,6,18,25]). *Consider a problem Π.*

1. *Suppose Π is a minimization problem. If there exists a bipartite graph $H = (\mathcal{O} \cup \mathcal{A}, E)$, that belongs to a family of graphs having a balanced vertex separator of sub-linear size, and it satisfies the* local-exchange property: *For any subset $\mathcal{A}' \subseteq \mathcal{A}$, $(\mathcal{A} \setminus \mathcal{A}') \cup \Gamma(\mathcal{A}')$ is a feasible solution. Then, the Local Search algorithm is a PTAS for Π. Here, $\Gamma(\mathcal{A}')$ denotes the set of neighbors of $\mathcal{A}'$ in H.*
2. *Suppose Π is a maximization problem. If there exists a bipartite graph $H = (\mathcal{O} \cup \mathcal{A}, E)$ that belongs to a family of graphs having a balanced vertex separator of sub-linear size, and it satisfies the* local-exchange property: *For any $\mathcal{O}' \subseteq \mathcal{O}$, $(\mathcal{A} \cup \mathcal{O}') \setminus \Gamma(\mathcal{O}')$ is a feasible solution. Then, the Local Search algorithm is a PTAS for Π. Here, as above $\Gamma(\mathcal{O}')$ denotes the set of neighbors of $\mathcal{O}'$ in H.*

The proof of the above theorem relies on the multi-separator theorem by Frederickson [15] whose generalized version we state as given in [2].

Theorem 6 ([2, 15]). *There are constants c_1, c_2 such that for any $G = (V, E) \in \mathcal{G}$ that has a balanced separator of size $cn^{1-\delta}$ and r be a parameter, $1 \leq r \leq n$, we can find a collection of $t = \Theta(n/r)$ pairwise disjoint subsets $V_1, \ldots, V_t$ such that the following properties hold.*

1. $|V_i| \leq c_1 r$.
2. $|\Gamma(V_i)| \leq c_2 r^{1-\delta}$, *where $\Gamma(V_i)$ is the neighbourhood of V_i.*
3. $\Gamma(V_i) \cap V_j = \emptyset$ *and* $\bigcup (V_i \cup \Gamma(V_i)) = V$.

Also, let $\mathcal{X} = \cup_{i=[t]} \Gamma(V_i)$ be called the set of boundary (or separator) vertices whose size is at most $t \cdot c_2 r^{1-\delta} = \Theta(n/r^\delta)$.

5 Unique Coverage of Non-piercing Regions with Bounded Depth and Degree

We are given a point set P and a set of non-piercing regions $\mathcal{R}$ in the plane. A point in P is said to be uniquely covered with respect to $\mathcal{R}$ if there exists a unique region in $\mathcal{R}$ containing that point. We need to output a subset $\mathcal{R}' \subseteq \mathcal{R}$ such that the number of uniquely covered points in P with respect to $\mathcal{R}'$ is maximized.

We make the following assumptions on the input.

1. Depth of every point in P with respect to $\mathcal{R}$ is at most some constant d.
2. Every region in $\mathcal{R}$ has at most some constant c number of points from P. We call this the degree of the region.

We denote $\mathsf{UC}_\mathcal{R}(\mathcal{S})$ to be the set of uniquely covered points in P covered by the regions in $\mathcal{S}$ with respect to $\mathcal{R}$, assuming $\mathcal{S} \subseteq \mathcal{R}$. We shall drop the subscript when $\mathcal{S} = \mathcal{R}$, i.e., $\mathsf{UC}(\mathcal{R})$.

We run the local search algorithm with parameter k. Let the solution returned by the algorithm be $\mathcal{A}$ and an optimal solution be $\mathcal{O}$. For comparing the quality of the two solutions we assume without loss of generality, that they are disjoint. If there is a region common to both the solutions, we make two copies, one for each solution. Note that the depth of any point may become at most double of what it was. Thus, the depth of every point still remains a constant as assumed before. The points that are uniquely covered by both the solutions are removed as they do not affect the analysis. Also, without loss of generality both $\mathcal{A}$ and $\mathcal{O}$ are minimal in size, i.e., every region in them uniquely cover at least one point in P.

Proof of Theorem 1: Consider the bipartite discrete intersection graph $G = (V, E)$ over $\mathcal{A} \uplus \mathcal{O}$, i.e., we put an edge between a region in $\mathcal{A}$ and a region in $\mathcal{O}$ if they both contain a common point from P. From Lemma 1 this graph[4] has a balanced and sub-linear separator since the depth of each point is bounded by a constant d. Thus we can apply Frederickson's separator theorem on this graph [15].

---

[4] Actually, Lemma 1 proves for the discrete intersection which is a super graph of the bipartite version.

Recall that Theorem 6 partitions V into $V_1, \ldots, V_t, \mathcal{X}$. For every $1 \leq i \leq t$, let $\mathcal{A}_i = \mathcal{A} \cap V_i$ and $\mathcal{O}_i = \mathcal{O} \cap V_i$. Let $\tilde{\mathcal{A}}_i$ be $\mathcal{A}_i \cup \Gamma(\mathcal{O}_i)$ and $\mathcal{A}^i$ be $(\mathcal{A} \cup \mathcal{O}_i) \setminus \tilde{\mathcal{A}}_i$. As $\mathcal{A}$ is locally optimal, $|\mathsf{UC}(\mathcal{A})| \geq |\mathsf{UC}(\mathcal{A}^i)|$. Observe that for every uniquely covered point there is a unique region that is covering that point. This region is in some part of the partition. Now we write $\mathsf{UC}(\mathcal{A})$ in terms of its constituent parts.

$$\mathsf{UC}(\mathcal{A}) = \bigcup_{j=1}^{t} \mathsf{UC}(\mathcal{A}_j) \bigcup \mathsf{UC}((\mathcal{A} \setminus \tilde{\mathcal{A}}_i) \cap \mathcal{X}) \bigcup \mathsf{UC}(\tilde{\mathcal{A}}_i \cap \mathcal{X})$$

Similarly,

$$\mathsf{UC}(\mathcal{A}^i) = \bigcup_{j \neq i} \mathsf{UC}(\mathcal{A}_j) \bigcup \mathsf{UC}(\mathcal{O}_i) \bigcup \mathsf{UC}((\mathcal{A} \setminus \tilde{\mathcal{A}}_i) \cap \mathcal{X})$$

Note that there are some uniquely covered points common to both $\mathcal{A}$ and $\mathcal{A}^i$. As $\mathcal{A}$ is locally optimal, the following inequality holds.

$$\sum_{j=1}^{t} |\mathsf{UC}_{\mathcal{A}}(\mathcal{A}_j)| + |\mathsf{UC}_{\mathcal{A}}((\mathcal{A} \setminus \tilde{\mathcal{A}}_i) \cap \mathcal{X})| + |\mathsf{UC}_{\mathcal{A}}(\tilde{\mathcal{A}}_i \cap \mathcal{X})| \geq$$

$$\sum_{j \neq i} |\mathsf{UC}_{\mathcal{A}^i}(\mathcal{A}_j)| + |\mathsf{UC}_{\mathcal{A}^i}(\mathcal{O}_i)| + |\mathsf{UC}_{\mathcal{A}^i}((\mathcal{A} \setminus \tilde{\mathcal{A}}_i) \cap \mathcal{X})| \quad (1)$$

$$\Rightarrow |\mathsf{UC}_{\mathcal{A}}(\mathcal{A}_i)| + |\mathsf{UC}_{\mathcal{A}}(\tilde{\mathcal{A}}_i \cap \mathcal{X})| \geq |\mathsf{UC}_{\mathcal{A}^i}(\mathcal{O}_i)|$$

Lemma 2. $\mathsf{UC}_{\mathcal{O}}(\mathcal{O}_i) \subseteq \mathsf{UC}_{\mathcal{A}^i}(\mathcal{O}_i)$

Proof. We need to show that every point in $\mathsf{UC}_{\mathcal{O}}(\mathcal{O}_i)$ is also contained in $\mathsf{UC}_{\mathcal{A}^i}(\mathcal{O}_i)$, i.e., a point that is uniquely covered by some region in $\mathcal{O}_i$ with respect to $\mathcal{O}$ remains uniquely covered by some region in $\mathcal{O}$ with respect to $\mathcal{A}^i$. As we are removing all the neighbours of $\mathcal{O}_i$ in $\mathcal{A}$, this is true. $\square$

Lemma 2 implies that $|\mathsf{UC}_{\mathcal{O}}(\mathcal{O}_i)| \leq |\mathsf{UC}_{\mathcal{A}^i}(\mathcal{O}_i)|$. Hence,

$$|\mathsf{UC}_{\mathcal{A}}(\mathcal{A}_i)| + |\mathsf{UC}_{\mathcal{A}}(\tilde{\mathcal{A}}_i \cap \mathcal{X})| \geq |\mathsf{UC}_{\mathcal{O}}(\mathcal{O}_i)|$$

Summing over all i, where $1 \leq i \leq t$,

$$\sum_i |\mathsf{UC}_{\mathcal{A}}(\mathcal{A}_i)| + \sum_i |\mathsf{UC}_{\mathcal{A}}(\tilde{\mathcal{A}}_i \cap \mathcal{X})| \geq \sum_i |\mathsf{UC}_{\mathcal{O}}(\mathcal{O}_i)|$$

$$\Rightarrow |\mathsf{UC}(\mathcal{A})| + \sum_i |\mathsf{UC}_{\mathcal{A}}(\tilde{\mathcal{A}}_i \cap \mathcal{X})| \geq |\mathsf{UC}(\mathcal{O})| - |\mathsf{UC}_{\mathcal{O}}(\mathcal{O} \cap \mathcal{X})|$$

Theorem 6 implies that every element in $V_i \cap \mathcal{X}$ is either in $\mathcal{O} \cap \mathcal{X}$ or is a neighbour of some element in $\mathcal{O}_i$. Therefore, $\sum_i |\mathsf{UC}_{\mathcal{A}}(\tilde{\mathcal{A}}_i \cap \mathcal{X})| + |\mathsf{UC}_{\mathcal{O}}(\mathcal{O} \cap \mathcal{X})| = \sum_i |(V_i \cap \mathcal{X})|$.

$$|\mathsf{UC}(\mathcal{A})| + c \sum_i |(V_i \cap \mathcal{X})| \geq |\mathsf{UC}(\mathcal{O})|$$

where c is the maximum number of points in P that a region can contain (Assumption 2). Also, from Theorem 6 we know $\sum_i |(V_i \cap \mathcal{X})| \leq \Theta(n/k^\delta)$, hence the following.

$$|UC(\mathcal{A})| + c'(|\mathcal{A}| + |\mathcal{O}|)/k^\delta \geq |UC(\mathcal{O})|$$

where c' is an appropriate constant. As, the number of uniquely covered points by a region is at least one,

$$|UC(\mathcal{A})| + c'(|UC(\mathcal{A})| + |UC(\mathcal{O})|)/k^\delta \geq |UC(\mathcal{O})|$$

$$\Rightarrow |UC(\mathcal{A})|(1 + \frac{c'}{k^\delta}) \geq |UC(\mathcal{O})|(1 - \frac{c'}{k^\delta})$$

For $\epsilon = 2c'/(k^\delta - c')$ we get the desired ratio.

$$|UC(\mathcal{A})|(1 + \epsilon) \geq |UC(\mathcal{O})|$$

$\square$

Corollary 1. *Given a family of unit disks and a point set in the plane with constant depth for every input point, the local search algorithm yields a PTAS.*

6 Multi-covering of Bounded Depth Points with Non-piercing Regions

Proof of Theorem 2: Recall we are given a set of regions $\mathcal{R}$ and a point set P in the plane where every $p \in P$ has a demand d_p. Consider the bipartite discrete intersection graph G over $\mathcal{A} \cup \mathcal{O}$. From Lemma 1 we know that G has a small and balanced separator when every point in P has a constant depth. Next we claim that G satisfies the local-exchange property as mentioned in Theorem 5.

Lemma 3. *Given the bipartite discrete intersection graph G, for every subset $\mathcal{A}' \subseteq \mathcal{A}$, $(\mathcal{A} \setminus \mathcal{A}') \cup \Gamma(\mathcal{A}')$ satisfies demands of every point in P.*

Proof. For every point $p \in P$, let $\mathcal{A}_p$ be the set of regions in $\mathcal{A}$ containing p. Similarly, let $\mathcal{O}_p$ be the set of regions in $\mathcal{O}$ containing p. Fix a point $p \in P$. Now there are two possibilities — either $\mathcal{A}' \cap \mathcal{A}_p = \emptyset$ or not. For the first case, p continues to be contained by at least d_p regions in $(\mathcal{A} \setminus \mathcal{A}') \cup \Gamma(\mathcal{A}')$. In case, if $\mathcal{A}' \cap \mathcal{A}_p \neq \emptyset$ then let $A \in \mathcal{A}' \cap \mathcal{A}_p$. As, G contains edges between every region in $\mathcal{A}_p$ to every region in $\mathcal{O}_p$, $\mathcal{O}_p \subseteq (\mathcal{A} \setminus \mathcal{A}') \cup \Gamma(\mathcal{A}')$. Thus, p is contained by at least d_p regions in this case as well. The same argument holds for every point in P. Thus, $(\mathcal{A} \setminus \mathcal{A}') \cup \Gamma(\mathcal{A}')$ satisfies demands of every point in P. $\square$

Thus, the bipartite discrete intersection G satisfies the properties required as mentioned in Theorem 5. Therefore, the Local Search algorithm yields a PTAS. $\square$

7   Shallow Packing of Regions with Sub-quadratic Union Complexity

In this section we consider the problem of shallow packing of regions whose union complexity is sub-quadratic. We are given a set of regions $\mathcal{D}$ with the hereditary property that any of its subset $\mathcal{D}'$ has its union complexity sub-quadratic in $|\mathcal{D}'|$. The problem is to find a subset $\mathcal{R} \subseteq \mathcal{D}$ such that every point in the plane is contained by at most k regions from $\mathcal{R}$, where k is a constant independent of the size of $\mathcal{D}$.

We use the local search algorithm and prove that the solution it returns is within a factor of $1 + \epsilon$. Our proof is on similar lines to that of the proof of PTAS for the capacitated region packing problem [18], where they show that the discrete intersection graph of $\mathcal{A} \cup \mathcal{O}$ has a balanced separator of sublinear size using a planar graph on the input points [27]. Note that the planar graph constructed in [27] works for non-piercing regions only. Since, we deal with a broader class of objects namely, objects with sub-quadratic union complexity, we use the arrangement graph on $\mathcal{R}$ to show that our intersection graph, which is not discrete, has a small and balanced separator.

Given a set of regions in the plane, an arrangement graph is a plane multi-graph on the intersection points of the boundaries of the regions. We assume that regions are closed, simply connected and no three regions have their boundaries intersecting at a point. The part of the boundary of a region joining two such intersection points forms an edge between them.

Proof of Theorem 3: As $\mathcal{R}$ is embedded in the plane its arrangement graph H is planar. We use the separator of H to get a separator of G. Here onwards for every $R \in \mathcal{R}$ we shall interchangeably use it with the subgraph $H[R]$ induced over the vertices of H in R. Therefore, G is also the intersection graph over the subgraphs $H[R]$ for every $R \in \mathcal{R}$. Also, as R is a simple and connected region so $H[R]$ is connected too.

We put weights on the vertices of H. For every vertex p of H and $R \in \mathcal{R}$ we add $1/|R \cap V(H)|$ whenever $p \in R$. Thus, $wt(p) = \sum_{R|p \in R} 1/|R \cap V(H)|$. We apply the Lipton-Tarjan's separator theorem on H to get a separator S of size $O(\sqrt{n})$ where n is the number of vertices in H. Also the total weight of the separated parts A and B is at most $2W/3$ each, where $W = \sum_{p \in V(H)} wt(p)$.

We start with an empty set S. For every R that contains a vertex p from separator S, we add that to the set S. Note that the union of the vertices of $H[R]$ in S is a superset of S. We claim that S is a small balanced separator of G. The fact that S separates G into parts A and B follows because a region R containing a vertex from A and a vertex from B must contain a vertex from S.

Next we prove that the size of S is sub-linear in the size of $\mathcal{R}$. We refer n to be the number of vertices in H and m to be the number of regions in $\mathcal{R}$. We slightly generalize the exposition of the Clarkson-Shor technique as given by Matoušek [23, p. 141].

Lemma 4. *If the union complexity of $\mathcal{R}$ is sub-quadratic in its size, i.e., $O(m^{2-\delta})$ where m is the size of $\mathcal{R}$, and every subset of $\mathcal{R}$ also has sub-quadratic union complexity in its size, then the number of $\leq k$-level vertex points $N_{\leq k}$ is $O(k^{\delta}m^{2-\delta})$.*

Proof. Let the union complexity of $\mathcal{R}$ be $O(m^{2-\delta})$, where m is the number of regions in $\mathcal{R}$. We do a random sampling of regions from $\mathcal{R}$. For every region $R \in \mathcal{R}$ we pick it with probability p and refer the sample set as $\mathcal{R}'$. Consider the expected union complexity of $\mathcal{R}'$. We will upper and lower bound this quantity. The expectation can be upper bounded by $O((pm)^{2-\delta})$.

For every vertex point p whose depth is at most k we compute the probability of p being a boundary point in $\mathcal{R}'$. Let the depth of p with respect to $\mathcal{R}$ be $d(p)$, where $d(p) \leq k$. The probability of p being a boundary point is $p^2(1-p)^{d(p)-2}$ that is at least $p^2(1-p)^k$ as $p < 1$. Therefore, the expected number of vertex points whose depth is at most k in $\mathcal{R}$ that lies in the boundary of the union of regions in $\mathcal{R}'$ is at least $N_{\leq k}p^2(1-p)^k$. Thus, $N_{\leq k}p^2(1-p)^k \leq O((pm)^{2-\delta})$. Setting $p \leftarrow 1/k$ leads $N_{\leq k} \leq O(k^{\delta}m^{2-\delta})$.   □

This means in our context, $n \leq O(k^{\delta}m^{2-\delta})$. Also, we know that $|\mathcal{S}| \leq k|S|$ as depth of every vertex point of H is at most k. Together with the fact $|S| \leq O(\sqrt{n})$, it implies that $|\mathcal{S}| \leq O(k\sqrt{k^{\delta}m^{2-\delta}})$. As k is a constant independent of m, size of $\mathcal{S}$ is sublinear in m.

Now it only remains to show that the two parts $\mathcal{A}$ and $\mathcal{B}$ which are separated by $\mathcal{S}$ are balanced, i.e., $|\mathcal{A}| \leq 2m/3$ and $|\mathcal{B}| \leq 2m/3$. Since the weight of all regions in $\mathcal{A}$ are distributed among vertex points in A, $wt(A) \geq |\mathcal{A}|$. Again from planar separator theorem, $wt(A) \leq 2|\mathcal{R}|/3$. Hence, $|\mathcal{A}| \leq 2m/3$. The same holds for $\mathcal{B}$. Thus $\mathcal{S}$ is a balanced separator of sublinear size for the intersection graph of $\mathcal{R}$.

Lastly, it is easy to see that the local-exchange property is satisfied by the graph G [2,18]. As the properties of the graphs as mentioned in Theorem 5 are satisfied, PTAS follows.   □

References

1. Agarwal, P.K., Pach, J., Sharir, M.: State of the union (of geometric objects): a review (2007)
2. Aschner, R., Katz, M.J., Morgenstern, G., Yuditsky, Y.: Approximation schemes for covering and packing. In: Ghosh, S.K., Tokuyama, T. (eds.) WALCOM 2013. LNCS, vol. 7748, pp. 89–100. Springer, Heidelberg (2013). doi:10.1007/978-3-642-36065-7_10
3. Ashok, P., Kolay, S., Misra, N., Saurabh, S.: Unique covering problems with geometric sets. In: Xu, D., Du, D., Du, D. (eds.) COCOON 2015. LNCS, vol. 9198, pp. 548–558. Springer, Cham (2015). doi:10.1007/978-3-319-21398-9_43
4. Bandyapadhyay, S., Basu Roy, A.: Effectiveness of local search for art gallery problems. In: WADS (2017)
5. Bansal, N., Pruhs, K.: Weighted geometric set multi-cover via quasi-uniform sampling. JoCG 7(1), 221–236 (2016)

6. Chan, T.M., Har-Peled, S.: Approximation algorithms for maximum independent set of pseudo-disks. Discrete Comput. Geom. **48**(2), 373–392 (2012)
7. Chekuri, C., Clarkson, K.L., Har-Peled, S.: On the set multicover problem in geometric settings. ACM Trans. Algorithms (TALG) **9**(1), 9 (2012)
8. Cohen-Addad, V., Klein, P.N., Mathieu, C.: Local search yields approximation schemes for k-means and k-median in euclidean and minor-free metrics. In: FOCS, pp. 353–364 (2016)
9. Cohen-Addad, V., Mathieu, C.: Effectiveness of local search for geometric optimization. In: SoCG, pp. 329–343 (2015)
10. Dahllöf, V., Jonsson, P., Beigel, R.: Algorithms for four variants of the exact satisfiability problem. Theor. Comput. Sci. **320**(2–3), 373–394 (2004)
11. Demaine, E.D., Feige, U., Hajiaghayi, M., Salavatipour, M.R.: Combination can be hard: approximability of the unique coverage problem. SIAM J. Comput. **38**(4), 1464–1483 (2008)
12. Ene, A., Har-Peled, S., Raichel, B.: Geometric packing under non-uniform constraints. In: SoCG, pp. 11–20 (2012)
13. Erlebach, T., Van Leeuwen, E.J.: Approximating geometric coverage problems. In: Proceedings of the Nineteenth Annual ACM-SIAM Symposium on Discrete Algorithms, pp. 1267–1276. Society for Industrial and Applied Mathematics (2008)
14. Fowler, R.J., Paterson, M.S., Tanimoto, S.L.: Optimal packing and covering in the plane are np-complete. Inf. Process. Lett. **12**(3), 133–137 (1981)
15. Frederickson, G.N.: Fast algorithms for shortest paths in planar graphs, with applications. SIAM J. Comput. **16**(6), 1004–1022 (1987)
16. Friggstad, Z., Rezapour, M., Salavatipour, M.R.: Local search yields a PTAS for k-means in doubling metrics. In: FOCS, pp. 365–374 (2016)
17. Garey, M.R., Johnson, D.S.: Computers and intractability (1979)
18. Govindarajan, S., Raman, R., Ray, S., Basu Roy, A.: Packing and covering with non-piercing regions. In: ESA, pp. 47:1–47:17 (2016)
19. Har-Peled, S.: Quasi-polynomial time approximation scheme for sparse subsets of polygons. In: SoCG, pp. 120:120–120:129 (2014)
20. Ito, T., Nakano, S., Okamoto, Y., Otachi, Y., Uehara, R., Uno, T., Uno, Y.: A 4.31-approximation for the geometric unique coverage problem on unit disks. Theor. Comput. Sci. **544**, 14–31 (2014)
21. Ito, T., Nakano, S.-I., Okamoto, Y., Otachi, Y., Uehara, R., Uno, T., Uno, Y.: A polynomial-time approximation scheme for the geometric unique coverage problem on unit squares. In: Fomin, F.V., Kaski, P. (eds.) SWAT 2012. LNCS, vol. 7357, pp. 24–35. Springer, Heidelberg (2012). doi:10.1007/978-3-642-31155-0_3
22. Krohn, E., Gibson, M., Kanade, G., Varadarajan, K.: Guarding terrains via local search. J. Comput. Geom. **5**(1), 168–178 (2014)
23. Matoušek, J.: Lectures on Discrete Geometry. Springer, New York (2002)
24. Misra, N., Moser, H., Raman, V., Saurabh, S., Sikdar, S.: The parameterized complexity of unique coverage and its variants. Algorithmica **65**(3), 517–544 (2013)
25. Mustafa, N.H., Ray, S.: Improved results on geometric hitting set problems. Discrete Comput. Geom. **44**(4), 883–895 (2010)
26. Pach, J., Walczak, B.: Decomposition of multiple packings with subquadratic union complexity. Comb. Probab. Comput. **25**(1), 145–153 (2016)

27. Pyrga, E., Ray, S.: New existence proofs for ε-nets. In: SoCG, pp. 199–207 (2008)
28. Schaefer, T.J.: The complexity of satisfiability problems. In: Proceedings of the Tenth Annual ACM Symposium on Theory of Computing, pp. 216–226. ACM (1978)
29. Whitesides, S., Zhao, R.: K-admissible collections of jordan curves and offsets of circular arc figures. Technical report, McGill University, School of Computer Science (1990)

Depth Distribution in High Dimensions

Jérémy Barbay[1], Pablo Pérez-Lantero[2], and Javiel Rojas-Ledesma[1(✉)]

[1] Departamento de Ciencias de la Computación,
Universidad de Chile, Santiago, Chile
`jeremy@barbay.cl, jrojas@dcc.uchile.cl`
[2] Departamento de Matemática y Ciencia de la Computación,
Universidad de Santiago, Santiago, Chile
`pablo.perez.l@usach.cl`

Abstract. Motivated by the analysis of range queries in databases, we introduce the computation of the DEPTH DISTRIBUTION of a set $\mathcal{B}$ of axis aligned boxes, whose computation generalizes that of the KLEE'S MEASURE and of the MAXIMUM DEPTH. In the worst case over instances of fixed input size n, we describe an algorithm of complexity within $\mathcal{O}(n^{\frac{d+1}{2}} \log n)$, using space within $\mathcal{O}(n \log n)$, mixing two techniques previously used to compute the KLEE'S MEASURE. We refine this result and previous results on the KLEE'S MEASURE and the MAXIMUM DEPTH for various measures of difficulty of the input, such as the profile of the input and the degeneracy of the intersection graph formed by the boxes.

1 Introduction

Problems studied in Computational Geometry have found important applications in the processing and querying of massive databases [1], such as the computation of the MAXIMA of a set of points [2,4], or compressed data structures for POINT LOCATION and RECTANGLE STABBING [3]. In particular, we consider cases where the input or queries are composed of axis-aligned boxes in d dimensions: in the context of databases it corresponds for instance to a search for cars within the intersection of ranges in price, availability and security ratings range.

Consider a set $\mathcal{B}$ of n axis-parallel boxes in $\mathbb{R}^d$, for fixed d. The KLEE'S MEASURE of $\mathcal{B}$ is the volume of the union of the boxes in $\mathcal{B}$. Originally suggested on the line by Klee [18], its computation is well studied in higher dimensions [7–10,22], and can be done in time within $\mathcal{O}(n^{d/2})$, using an algorithm introduced by Chan [10] based on a new paradigm called "Simplify, Divide and Conquer". The MAXIMUM DEPTH of $\mathcal{B}$ is the maximum number of boxes overlapping at any point, and its computational complexity is similar to that of KLEE'S MEASURE's, converging to the same complexity within $\mathcal{O}(n^{d/2})$ [10].

Hypothesis. The known algorithms to compute these two measures are all strikingly similar. That would suggest a reduction from one to another, except that those two measures are completely distinct: KLEE'S MEASURE is a volume whose value can be a real number, while MAXIMUM DEPTH is a cardinality whose value

© Springer International Publishing AG 2017
Y. Cao and J. Chen (Eds.): COCOON 2017, LNCS 10392, pp. 38–49, 2017.
DOI: 10.1007/978-3-319-62389-4_4

is an integer in the range $[1..n]$. **Is there any way to formalize the relationship between the computation of these two measures?**

Our Results. We describe a first step towards such a formalization, in the form of **a new problem**, which we show to be intermediary in terms of the techniques being used, between the KLEE'S MEASURE and the MAXIMUM DEPTH, slightly more costly in time and space, and with interesting applications and results of its own.

We introduce the notion of the DEPTH DISTRIBUTION of a set $\mathcal{B}$ of n axis-parallel boxes in $\mathbb{R}^d$, formed by the vector of n values $(V_1, \ldots, V_n)$, where V_i corresponds to the volume covered by exactly i boxes from $\mathcal{B}$. The DEPTH DISTRIBUTION of a set $\mathcal{B}$ can be interpreted as a probability distribution function (hence the name): if a point p is selected from the union of the boxes in $\mathcal{B}$ uniformly at random, the probability that p intersects exactly k boxes is $(V_k / \sum_{i=1}^{n} V_i)$, for all $k \in [1..n]$.

The DEPTH DISTRIBUTION **refines both** KLEE'S MEASURE and MAXIMUM DEPTH. It is a measure finer than KLEE'S MEASURE in the sense that the KLEE'S MEASURE of a set $\mathcal{B}$ can be obtained in time linear in the size n of $\mathcal{B}$ by summing the components of the DEPTH DISTRIBUTION of $\mathcal{B}$. Similarly, the DEPTH DISTRIBUTION is a measure finer than the MAXIMUM DEPTH in the sense that the MAXIMUM DEPTH of a set $\mathcal{B}$ can be obtained in linear time by finding the largest $i \in [1..n]$ such that $V_i \neq 0$. In the context of a database, when receiving multidimensional range queries (e.g. about cars), the DEPTH DISTRIBUTION of the queries yields valuable information to the database owner (e.g. a car dealer) about the repartition of the queries in the space of the data, to allow for informed decisions on it (e.g. to orient the future purchase of cars to resell based on the clients' desires, as expressed by their queries).

In the classical computational complexity model where one studies the **worst case over instances of fixed size** n, we combine techniques previously used to compute the KLEE'S MEASURE [10,22] to compute the DEPTH DISTRIBUTION in time within $\mathcal{O}(n^{\frac{d+1}{2}} \log n)$, using space within $\mathcal{O}(n \log n)$ (in Sect. 3.1). This solution is slower by a factor within $\mathcal{O}(\sqrt{n} \log n)$ than the best known algorithms for computing the KLEE'S MEASURE and the MAXIMUM DEPTH: we show in Sect. 3.2 that such a gap might be ineluctable, via a reduction from the computation of MATRIX MULTIPLICATION.

In the refined computational complexity model where one studies the worst case complexity taking advantage of **additional parameters describing the difficulty** of the instance [4,17,21], we introduce (in Sect. 4) new measures of difficulty such as the *profile* and the *degeneracy* of the intersection graph of the boxes, and describe algorithms in these new models to compute the DEPTH DISTRIBUTION, KLEE'S MEASURE and MAXIMUM DEPTH of a set $\mathcal{B}$.

After a short overview of the known results on the computation of the KLEE'S MEASURE and MAXIMUM DEPTH (in Sect. 2), we describe in Sect. 3 the results in the worst case over instances of fixed size. In Sect. 4, we describe results on refined partitions of the instance universe, both for DEPTH DISTRIBUTION and for the previously known problems, KLEE'S MEASURE and MAXIMUM

DEPTH. We conclude in Sect. 5 with a discussion on discrete variants and further refinements of the analysis.

2   Background

The techniques used to compute the KLEE'S MEASURE have evolved over time, and can all be used to compute the MAXIMUM DEPTH. We retrace some of the main results, which will be useful for the definition of an algorithm computing the DEPTH DISTRIBUTION (in Sect. 3), and for the refinements of the analysis for DEPTH DISTRIBUTION, KLEE'S MEASURE and MAXIMUM DEPTH (in Sect. 4).

The computation of the KLEE'S MEASURE of a set $\mathcal{B}$ of n axis-aligned d-dimensional boxes was first posed by Klee [18] in 1977. After some initial progresses [6,15,18], Overmars and Yap [22] described a solution running in time within $\mathcal{O}(n^{d/2} \log n)$. This remained the best solution for more than 20 years until 2013, when Chan [10] presented a simpler and faster algorithm running in time within $\mathcal{O}(n^{d/2})$.

The algorithms described by Overmars and Yap [22] and by Chan [10], respectively, take both advantage of solutions to the special case of the problem where all the boxes are slabs. A box b is said to be a *slab* within another box Γ if $b \cap \Gamma = \{(x_1, \ldots, x_d) \in \Gamma \mid \alpha \le x_i \le \beta\}$, for some integer $i \in [1..d]$ and some real values α, β (see Fig. 1 for an illustration). Overmars and Yap [22] showed that, if all the boxes in $\mathcal{B}$ are slabs inside the domain box Γ, then the KLEE'S MEASURE of $\mathcal{B}$ within Γ can be computed in linear time (provided that the boxes have been pre-sorted in each dimension).

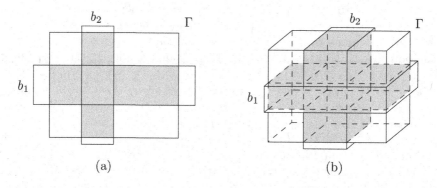

(a)   (b)

Fig. 1. An illustration in dimensions 2 (a) and 3 (b) of two boxes b_1, b_2 equivalent to slabs when restricted to the box Γ. The KLEE'S MEASURE of $\{b_1, b_2\}$ within Γ is the area (resp. volume) of the shadowed region in (a) (resp. (b)).

Overmars and Yap's algorithm [22] is based on a technique originally described by Bentley [6]: solve the static problem in d dimensions by combining a data structure for the dynamic version of the problem in $d-1$ dimensions with a plane sweep over the d-th dimension. The algorithm starts by partitioning the space into $\mathcal{O}(n^{d/2})$ rectangular cells such that the boxes in $\mathcal{B}$ are equivalent to slabs when restricted to each of those cells. Then, the algorithm builds a *tree-like* data structure whose leaves are the cells of the partition, supporting insertion and deletion of boxes while keeping track of the KLEE'S MEASURE of the boxes.

Chan's algorithm [10] is a simpler *divide-and-conquer* algorithm, where the slabs are *simplified* and removed from the input before the recursive calls (Chan [10] named this technique *Simplify, Divide and Conquer, SDC* for short). To obtain the recursive subproblems, the algorithm assigns a constant weight of $2^{\frac{i+j}{2}}$ to each $(d\text{-}2)$-face intersecting the domain and orthogonal to the i-th and j-th dimensions, $i, j \in [1..d]$. Then, the domain is partitioned into two sub-domains by the hyperplane $x_1 = m$, where m is the weighted median of the $(d\text{-}2)$-faces orthogonal to the first dimension. This yields a decrease by a factor of $2^{2/d}$ in the total weight of the $(d\text{-}2)$-faces intersecting each sub-domain. Chan [10] uses this, and the fact that slabs have no $(d\text{-}2)$-face intersecting the domain, to prove that the SDC algorithm runs in time within $\mathcal{O}(n^{d/2})$.

Unfortunately, there are sets of boxes which require partitions of the space into a number of cells within $\Omega(n^{d/2})$ to ensure that, when restricted to each cell, all the boxes are equivalent to slabs. Hence, without a radically new technique, any algorithm based on this approach will require running time within $\Omega(n^{d/2})$. Chan [10] conjectured that any *combinatorial* algorithm computing the KLEE'S MEASURE requires within $\Omega(n^{d/2})$ operations, via a reduction from the parameterized K-CLIQUE problem, in the worst case over instances of fixed size n. As a consequence, recent work have focused on the study of special cases which can be solved faster than $\Omega(n^{d/2})$, like for instance when all the boxes are *orthants* [10], α-*fat boxes* [7], or cubes [8]. In turn, we show in Sect. 4 that there are measures which gradually separate *easy* instances for these problems from the *hard* ones.

In the next section, we present an algorithm for the computation of the DEPTH DISTRIBUTION inspired by a combination of the approaches described above, outperforming naive applications of those techniques.

3   Computing the Depth Distribution

We describe in Sect. 3.1 an algorithm to compute the DEPTH DISTRIBUTION of a set of n boxes. The running time of this algorithm in the worst case over d-dimensional instances of fixed size n is within $\mathcal{O}(n^{\frac{d+1}{2}} \log n)$, using space within $\mathcal{O}(n \log n)$. This running time is worse than that of computing only the KLEE'S MEASURE (or the MAXIMUM DEPTH) by a factor within $\mathcal{O}(\sqrt{n} \log n)$: we argue in Sect. 3.2 that computing the DEPTH DISTRIBUTION is computationally harder than the special cases of computing the KLEE'S MEASURE and the MAXIMUM DEPTH, unless computing MATRIX MULTIPLICATION is much easier than usually assumed.

3.1   Upper Bound

We introduce an algorithm to compute the DEPTH DISTRIBUTION inspired by a combination of the techniques introduced by Chan [10], and by Overmars and

Yap [22], for the computation of the KLEE'S MEASURE (described in Sect. 2). As in the approaches mentioned previously, the algorithm partitions the domain Γ into $\mathcal{O}(n^{d/2})$ cells where the boxes of $\mathcal{B}$ are equivalent to slabs, and then combines the solution within each cell to obtain the final answer. Two main issues must be addressed: how to compute the DEPTH DISTRIBUTION when the boxes are slabs, and how to partition the domain efficiently.

We address first the special case of slabs. We show in Lemma 1 that computing the DEPTH DISTRIBUTION of a set of n d-dimensional slabs within a domain Γ can be done via a multiplication of d polynomials of degree at most n.

Lemma 1. *Let $\mathcal{B}$ be a set of n axis-parallel d-dimensional axis aligned boxes, with $d \geq 2$, that, when restricted to a domain box Γ, are equivalent to slabs. The computation of the DEPTH DISTRIBUTION $(V_1, \ldots, V_n)$ of $\mathcal{B}$ within Γ can be performed via a multiplication of d polynomials of degree at most n.*

Proof. For all $i \in [1..d]$, let $\mathcal{B}_i$ be the subset of slabs that are orthogonal to the i-th dimension, and let $\left(V_1^i, \ldots, V_n^i\right)$ be the DEPTH DISTRIBUTION of the intervals resulting from projecting $\mathcal{B}_i$ to the i-th dimension within Γ. We associate a polynomial $P_i(x)$ of degree n with each $\mathcal{B}_i$ as follows:

- let Γ_i be the projection of the domain Γ into the i-th dimension, and
- let V_0^i be the length of the region of Γ_i not covered by a box in B_i (i.e., $V_0^i = (|\Gamma_i| - \sum_{j=1}^n V_j^i)$); then
- $P_i(x) = \sum_{j=0}^n V_j^i \cdot x^j$.

Since any slab entirely covers the domain in all the dimensions but the one to which it is orthogonal, any point p has depth k in $\mathcal{B}$ if and only if it has depth j_1 in $\mathcal{B}_1$, j_2 in $\mathcal{B}_2$, ..., and j_d in $\mathcal{B}_d$, such that $j_1 + j_2 + \ldots + j_d = k$. Thus, for all $k \in [0..n]$:

$$V_k = \sum_{\substack{0 \leq j_1, \ldots, j_d \leq n \\ j_1 + \ldots + j_d = k}} \left(\prod_{i=1}^d V_{j_i}^i \right),$$

which is precisely the $(k+1)$-th coefficient of $P_1(x) \cdot P_2(x) \cdot \ldots \cdot P_d(x)$. Thus, this product yields the DEPTH DISTRIBUTION $(V_1, \ldots, V_n)$ of $\mathcal{B}$ in Γ. $\square$

Using standard *Fast Fourier Transform* techniques, two polynomials can be multiplied in time within $O(n \log n)$ [12]. Moreover, the DEPTH DISTRIBUTION of a set of intervals (i.e., when $d = 1$) can be computed in linear time after sorting, by a simple scan-line algorithm, as for KLEE'S MEASURE [10]. Thus, as a consequence of Lemma 1, when the boxes in $\mathcal{B}$ are slabs when restricted to a domain box Γ, the DEPTH DISTRIBUTION of $\mathcal{B}$ within Γ can be computed in time within $\mathcal{O}(n \log n)$.

Corollary 2. *Let $\mathcal{B}$ be a set of n d-dimensional axis aligned boxes, whose intersections with Γ are slabs. The DEPTH DISTRIBUTION of $\mathcal{B}$ inside Γ can be computed in time within $\mathcal{O}(n \log n)$.*

A naive application of previous techniques [10, 22] to the computation of DEPTH DISTRIBUTION yields poor results. Combining the result in Corollary 2 with the partition of the space and the data structure described by Overmars and Yap [22] yields an algorithm to compute the DEPTH DISTRIBUTION running in time within $\mathcal{O}(n^{\frac{d+1}{2}} \log n)$, and using space within $\mathcal{O}(n^{d/2} \log n)$. Similarly, if the result in Corollary 2 is combined with Chan's partition of the space [10], one obtains an algorithm using space linear in the number of boxes, but running in time within $\mathcal{O}(n^{\frac{d}{2}+1} \log n)$ (i.e., paying an extra $\mathcal{O}(n^{\frac{1}{2}})$-factor for the reduction in space usage of Overmars and Yap [22]).

We combine these two approaches into an algorithm which achieves the best features of both: it runs in time within $\mathcal{O}(n^{\frac{d+1}{2}} \log n)$, and uses $\mathcal{O}(n \log n)$-space. As in Chan's approach [10] we use a divide and conquer algorithm, but we show in Theorem 3 that the running time is asymptotically the same as if using the partition and data structures described by Overmars and Yap [22] (see Algorithm 1 for a detailed description).

Algorithm 1. SDC-DDistribution$(\mathcal{B}, \Gamma, c, (V_1, \ldots, V_n))$

Input: A set $\mathcal{B}$ of n boxes in $\mathbb{R}^d$; a d-dimensional domain box Γ; the number c of boxes not in $\mathcal{B}$ but in the original set that completely contain Γ; and a vector $(V_1, \ldots, V_n)$ representing the DEPTH DISTRIBUTION computed so far.

1: **if** no box in $\mathcal{B}$ has a $(d\text{-}2)$-face intersecting Γ (i.e., all the boxes are slabs) **then**
2:    Compute the DEPTH DISTRIBUTION $(V_1', \ldots, V_{|\mathcal{B}|}')$ of $\mathcal{B}$ within Γ using Lemma 1
3:    **for** $i \in [1..|\mathcal{B}|]$ **do**
4:        $V_{i+c} \leftarrow V_{i+c} + V_i'$
5: **else**
6:    Let $\mathcal{B}^0 \subseteq \mathcal{B}$ be the subset of boxes completely containing Γ
7:    $c \leftarrow c + |\mathcal{B}^0|$
8:    Let $\mathcal{B}' = \mathcal{B} \setminus \mathcal{B}^0$
9:    Let m be the weighted median of the $(d\text{-}2)$-faces orthogonal to x_1
10:   Split Γ into Γ_L, Γ_R by the hyperplane $x_1 = m$
11:   Rename the dimensions so that $x_1, \ldots, x_d$ becomes $x_2, \ldots, x_d, x_1$
12:   Let $\mathcal{B}_L$ and $\mathcal{B}_R$ be the subsets of $\mathcal{B}'$ intersecting Γ_L and Γ_R respectively
13:   Call SDC-DDistribution$(\mathcal{B}_L, \Gamma_L, c, (V_1, \ldots, V_n))$
14:   Call SDC-DDistribution$(\mathcal{B}_R, \Gamma_R, c, (V_1, \ldots, V_n))$

Theorem 3. *Let $\mathcal{B}$ be a set of n axis-parallel boxes in $\mathbb{R}^d$. The* DEPTH DISTRIBUTION *of $\mathcal{B}$ can be computed in time within $\mathcal{O}(n^{\frac{d+1}{2}} \log n)$, using space within $\mathcal{O}(n \log n)$.*

Due to lack of space we defer the complete proof to the extended version [5].

The bound for the running time in Theorem 3 is worse than that for the computation of the KLEE'S MEASURE (and MAXIMUM DEPTH) by a factor within $\mathcal{O}(\sqrt{n} \log n)$, which raises the question of the optimality of the bound: we consider this matter in the next section.

3.2   Conditional Lower Bound

As for many problems handling high dimensional inputs, the best lower bound known for this problem is $\Omega(n \log n)$, which derives from the fact that the DEPTH DISTRIBUTION is a generalization of the KLEE'S MEASURE problem. This bound, however, is tight only when the input is a set of intervals (i.e., $d = 1$). For higher dimensions, the conjectured lower bound of $\Omega(n^{d/2})$ described by Chan in 2008 [9] for the computational complexity of computing the KLEE'S MEASURE can be extended analogously to the computation of the DEPTH DISTRIBUTION.

One intriguing question is whether in dimension $d = 2$, as for KLEE'S MEASURE, the DEPTH DISTRIBUTION can be computed in time within $\mathcal{O}(n \log n)$. We argue that doing so would imply breakthrough results in a long standing problem, MATRIX MULTIPLICATION. We show that any instance of MATRIX MULTIPLICATION can be solved using an algorithm which computes the DEPTH DISTRIBUTION of a set of rectangles in the plane. For this, we make use of the following simple observation:

Observation 1. Let A, B be two $n \times n$ matrices of real numbers, and let C_i denote the $n \times n$ matrix that results from multiplying the $n \times 1$ vector corresponding to the i-th column of A with the $1 \times n$ vector corresponding to the i-th row of B. Then, $AB = \sum_{i=1}^{n} C_i$.

We show in Theorem 4 that multiplying two $n \times n$ matrices can be done by transforming the input into a set of $\mathcal{O}(n^2)$ 2-dimensional boxes, and computing the DEPTH DISTRIBUTION of the resulting box set. Moreover, this transformation can be done in linear time, thus, the theorem yields a conditional lower bound for the computation of the DEPTH DISTRIBUTION.

Theorem 4. Let A, B be two $n \times n$ matrices of non-negative real numbers. There is a set $\mathcal{B}$ of rectangles of size within $O(n^2)$, and a domain rectangle Γ, such that the DEPTH DISTRIBUTION of $\mathcal{B}$ within Γ can be projected to obtain the value of the product AB.

Intuitively, we create a *gadget* to represent each matrix C_i. Within the i-th gadget, there will be a rectangular region for each component of C_i with the value of that component as volume. We arrange the boxes so that two distinct regions have the same depth if and only if they represent the same respective coefficients of two distinct matrices C_i and $C_{i'}$ (formally, they represent coefficients $(C_i)_{j,k}$ and $(C_{i'})_{j',k'}$, respectively, such that $i \neq i', j = j'$, and $k = k'$). Due to lack of space, we defer the complete proof to the extended version [5].

The optimal time to compute the product of two $n \times n$ matrices is still open. It can naturally be computed in time within $O(n^3)$. However, Strassen showed in 1969 that within $O(n^{2.81})$ arithmetic operations are enough [23]. This gave rise to a new area of research, where the central question is to determine the value of the exponent of the computational complexity of square matrix multiplication, denoted ω, and defined as the minimum value such that two $n \times n$ matrices can be multiplied using within $O(n^{\omega + \varepsilon})$ arithmetic operations for any $\varepsilon > 0$.

The result of Theorem 4 directly yields a conditional lower bound on the complexity of DEPTH DISTRIBUTION: in particular, DEPTH DISTRIBUTION in dimension as low as two, can be solved in time within $\mathcal{O}(n \log n)$, then MATRIX MULTIPLICATION can be computed in time within $\mathcal{O}(n^2)$, i.e. $\omega = 2$. However, this would be a great breakthrough in the area, the best known upper bound to date is approximately $\omega \leq 2.37$, when improvements in the last 30 years [11, 16] have been in the range $[2.3728, 2.3754]$.

Corollary 5 (Conditional lower bound). *Computing the* DEPTH DISTRIBU-TION *of a set* B *of* n *d-dimensional boxes requires time within* $\Omega(n^{1+c})$, *for some constant* $c > 0$, *unless two* $n \times n$ *matrices can be multiplied in time* $\mathcal{O}(n^{2+\varepsilon})$, *for any constant* $\varepsilon > 0$.

The running time of the algorithm that we described in Theorem 3 can be improved for large classes of instances (i.e. asymptotically infinite) by considering measures of the difficulty of the input other than its size. We describe two of these improved solutions in the next section.

4   Multivariate Analysis

Even though the asymptotic complexity of $\mathcal{O}(n^{\frac{d+1}{2}} \log n)$ is the best we know so far for computing the DEPTH DISTRIBUTION of a set of n d-dimensional boxes, there are many cases which can be solved faster. Some of those "easy" instances can be mere particular cases, but others can be hints of some hidden measures of difficulty of the DEPTH DISTRIBUTION problem. We show that, indeed, there are at least two such difficulty measures, gradually separating instances of the same size n into various classes of difficulty. Informally, the first one (the *profile* of the input set, Sect. 4.1) measures how separable the boxes are by axis-aligned hyperplanes, whereas the second one (the *degeneracy* of the intersection graph, Sect. 4.2) measures how "complex" the interactions of the boxes are in the set between them. Those measures inspire similar results for the computation of the KLEE'S MEASURE and of the MAXIMUM DEPTH.

4.1   Profile

The *i-th profile p_i* of a set of boxes B is the maximum number of boxes intersected by any hyperplane orthogonal to the i-th dimension; and the *profile p* of B is the minimum $p = \min_{i \in [1..d]} \{p_i\}$ of those over all dimensions. D'Amore [13] showed how to compute it in linear time (after sorting the coordinate of the boxes in each dimension). The following lemma shows that the DEPTH DISTRIBUTION can be computed in time sensitive to the profile of the input set.

Lemma 6. *Let B be a set of boxes with profile p, and Γ be a d-dimensional axis-aligned domain box. The* DEPTH DISTRIBUTION *of B within Γ can be computed in time within* $\mathcal{O}(n \log n + np^{\frac{d-1}{2}} \log p) \subseteq \mathcal{O}(n^{\frac{d+1}{2}} \log n)$.

Due to lack of space we defer the complete proof to the extended version [5].

The lemma above automatically yields refined results for the computation of the KLEE'S MEASURE and the MAXIMUM DEPTH of a set of boxes $\mathcal{B}$. However, applying the technique in an *ad-hoc* way to these problems yields better bounds:

Corollary 7. *Let $\mathcal{B}$ be a set of boxes with profile p, and Γ be a domain box. The KLEE'S MEASURE and MAXIMUM DEPTH of $\mathcal{B}$ within Γ can be computed in time within $\mathcal{O}(n \log n + np^{\frac{d-2}{2}} \log p) \subseteq \mathcal{O}(n^{d/2} \log n)$.*

The algorithms from Lemma 6 and Corollary 7 asymptotically outperform previous ones in the sense that their running time is never worse than previous algorithms by more than a constant factor, but can perform faster by more than a constant factor on specific families of instances.

An orthogonal approach is to consider how complex the interactions between the boxes are in the input set $\mathcal{B}$, analyzing, for instance, the intersection graph of $\mathcal{B}$. We study such a technique in the next section.

4.2   Intersections Graph Degeneracy

A *k-degenerate* graph is an undirected graph in which every subgraph has a vertex of degree at most k [19]. Every k-degenerate graph accepts an ordering of the vertices in which every vertex is connected with at most k of the vertices that precede it (we refer below to such an ordering as a *degenerate ordering*).

In the following lemma we show that this ordering can be used to compute the DEPTH DISTRIBUTION of a set $\mathcal{B}$ of n boxes in running time sensitive to the degeneracy of the intersection graph of $\mathcal{B}$.

Lemma 8. *Let $\mathcal{B}$ be a set of boxes and Γ be a domain box, and let k be the degeneracy of the intersection graph G of the boxes in $\mathcal{B}$. The DEPTH DISTRIBUTION of $\mathcal{B}$ within Γ can be computed in time within $\mathcal{O}(n \log^d n + e + nk^{\frac{d+1}{2}})$, where $e \in \mathcal{O}(n^2)$ is the number of edges of G.*

Proof. We describe an algorithm that runs in time within the bound in the lemma. The algorithm first computes the intersection graph G of $\mathcal{B}$ in time within $\mathcal{O}(n \log^d n + e)$ [14], as well as the k-degeneracy of this graph and a degenerate ordering O of the vertices in time within $\mathcal{O}(n + e)$ [20]. The algorithm then iterates over O maintaining the invariant that, after the i-th step, the DEPTH DISTRIBUTION of the boxes corresponding to the vertices $v_1, v_2, \ldots, v_i$ of the ordering has been correctly computed.

For any subset V of vertices of G, let $DD_{\mathcal{B}}^{\Gamma}(V)$ denote the DEPTH DISTRIBUTION within Γ of the boxes in $\mathcal{B}$ corresponding to the vertices in V. Also, for $i \in [1..n]$ let $O[1..i]$ denote the first i vertices of O, and $O[i]$ denote the i-th vertex of O. From $DD_{\mathcal{B}}^{\Gamma}(O[1..i\text{-}1])$ (which the algorithm "knows" after the $(i-1)$-th iteration), $DD_{\mathcal{B}}^{\Gamma}(O[1..i])$ can be obtained as follows: (*i.*) let P be the subset of $O[1..i\text{-}1]$ connected with $O[i]$; (*ii.*) compute $DD_{\mathcal{B}}^{O[i]}(P \cup \{O[i]\})$ in time within $\mathcal{O}(k^{\frac{d+1}{2}} \log k)$ using SDC-DDistribution (note that the domain this time is $O[i]$

itself, instead of Γ); (*iii.*) add to $(DD_{\mathcal{B}}^{\Gamma}(O[1..i]))_1$ the value of $(DD_{\mathcal{B}}^{O[i]}(P \cup O[i]))_1$; and (*iv.*) for all $j = [2..k{+}1]$, substract from $(DD_{\mathcal{B}}^{\Gamma}(O[1..i]))_{j-1}$ the value of $(DD_{\mathcal{B}}^{O[i]}(P \cup O[i]))_j$ and add it to $(DD_{\mathcal{B}}^{\Gamma}(O[1..i\text{-}1]))_j$.

Since the updates to the DEPTH DISTRIBUTION in each step take time within $\mathcal{O}(k^{\frac{d+1}{2}} \log k)$, and there are n such steps, the result of the lemma follows.   □

Unlike the algorithm sensitive to the profile, this one can run in time within $\mathcal{O}(n^{1+\frac{d+1}{2}})$ (e.g. when $k = n$), which is only better than the $\mathcal{O}(n^{\frac{d+1}{2}})$ complexity of SDC-DDistribution for values of the degeneracy k within $\mathcal{O}(n^{1-\frac{2}{d}})$.

Applying the same technique to the computation of KLEE'S MEASURE and MAXIMUM DEPTH yields improved solutions as well:

Corollary 9. *Let $\mathcal{B}$ be a set of boxes and Γ be a domain box, and let k be the degeneracy of the intersection graph G of the boxes in $\mathcal{B}$. The KLEE'S MEASURE and MAXIMUM DEPTH of $\mathcal{B}$ within Γ can be computed in time within $\mathcal{O}(n \log^d n + e + nk^{\frac{d}{2}})$, where $e \in \mathcal{O}(n^2)$ is the number of edges of G.*

Such refinements of the worst-case complexity analysis are only examples and can be applied to many other problems handling high dimensional data inputs. We discuss a selection in the next section.

5   Discussion

The DEPTH DISTRIBUTION captures many of the features in common between KLEE'S MEASURE and MAXIMUM DEPTH, so that new results on the computation of the DEPTH DISTRIBUTION will yield corresponding results for those two measures, and has its own applications of interest. Nevertheless, there is no direct reduction from KLEE'S MEASURE or MAXIMUM DEPTH to DEPTH DISTRIBUTION, as the latter is computationally more costly, and clarifying further the relationship between these problems will require finer models of computation. We discuss below some further issues to ponder about those measures.

Discrete variants. In practice, multidimensional range queries are applied to a database of multidimensional points. This yields discrete variants of each of the problems previously discussed [1,24]. In the DISCRETE KLEE'S MEASURE, the input is composed of not only a set $\mathcal{B}$ of n boxes, but also of a set S of m points. The problem is now to compute not the volume of the union of the boxes, but the number (and/or the list) of points which are covered by those boxes. Similarly, one can define a discrete version of the MAXIMUM DEPTH (which points are covered by the maximum number of boxes) and of the DEPTH DISTRIBUTION (how many and which points are covered by exactly i boxes, for $i \in [1..n]$). Interestingly enough, the computational complexity of these discrete variants is much less than that of their continuous versions when there are reasonably few points [24]: the discrete variant becomes hard only when there are many more points than boxes [1]. Nevertheless, "easy" configurations of the boxes also yield "easy" instances in the discrete case: it will be interesting to analyze the discrete variants of those problems according to the measures of *profile* and *k-degeneracy* introduced on the continuous versions.

Tighter Bounds. Chan [10] conjectured that a complexity of $\Omega(n^{d/2})$ is required to compute the KLEE'S MEASURE, and hence to compute the DEPTH DISTRIBUTION. However, the output of DEPTH DISTRIBUTION gives much more information than the KLEE'S MEASURE, of which a large part can be ignored during the computation of the KLEE'S MEASURE (while it is required for the computation of the DEPTH DISTRIBUTION). It is not clear whether even a lower bound of $\Omega(n^{d/2+\epsilon})$ can be proven on the computational complexity of the DEPTH DISTRIBUTION given this fact.

Funding. All authors were supported by the Millennium Nucleus "Information and Coordination in Networks" ICM/FIC RC130003. Jérémy Barbay and Pablo Pérez-Lantero were supported by the projects CONICYT Fondecyt/Regular nos 1170366 and 1160543 (Chile) respectively, while Javiel Rojas-Ledesma was supported by CONICYT-PCHA/Doctorado Nacional/2013-63130209 (Chile).

References

1. Khamis, M.A., Ngo, H.Q., Ré, C., Rudra, A.: Joins via geometric resolutions: worst-case and beyond. In: Proceedings of the 34th ACM Symposium on Principles of Database Systems (PODS), Melbourne, Victoria, Australia, 31 May–4 June, 2015, pp. 213–228 (2015)
2. Afshani, P.: Fast computation of output-sensitive maxima in a word RAM. In: Proceedings of the Twenty-Fifth Annual ACM-SIAM Symposium on Discrete Algorithms (SODA), Portland, Oregon, USA, January 5–7, 2014, pp. 1414–1423. SIAM (2014)
3. Afshani, P., Arge, L., Larsen, K.G.: Higher-dimensional orthogonal range reporting and rectangle stabbing in the pointer machine model. In: Symposuim on Computational Geometry (SoCG), Chapel Hill, NC, USA, June 17–20, 2012, pp. 323–332 (2012)
4. Afshani, P., Barbay, J., Chan, T.M.: Instance-optimal geometric algorithms. J. ACM (JACM) **64**(1), 3:1–3:38 (2017)
5. Barbay, J., Pérez-Lantero, P., Rojas-Ledesma, J.: Depth distribution in high dimension. ArXiv e-prints (2017)
6. Bentley, J.L.: Algorithms for Klee's rectangle problems. Unpublished notes (1977)
7. Bringmann, K.: An improved algorithm for Klee's measure problem on fat boxes. Comput. Geom. Theor. Appl. **45**(5–6), 225–233 (2012)
8. Bringmann, K.: Bringing order to special cases of Klee's measure problem. In: Chatterjee, K., Sgall, J. (eds.) MFCS 2013. LNCS, vol. 8087, pp. 207–218. Springer, Heidelberg (2013). doi:10.1007/978-3-642-40313-2_20
9. Chan, T.M.: A (slightly) faster algorithm for Klee's Measure Problem. In: Proceedings of the 24th ACM Symposium on Computational Geometry (SoCG), College Park, MD, USA, June 9–11, 2008, pp. 94–100 (2008)
10. Chan, T.M.: Klee's measure problem made easy. In: 54th Annual IEEE Symposium on Foundations of Computer Science (FOCS), Berkeley, CA, USA, 26–29 October, 2013, pp. 410–419 (2013)

11. Coppersmith, D., Winograd, S.: Matrix multiplication via arithmetic progressions. In: Proceedings of the 19th Annual ACM Symposium on Theory of Computing (STOC), New York, USA, pp. 1–6. ACM (1987)

12. Cormen, T.H., Leiserson, C.E., Rivest, R.L., Stein, C.: Introduction to Algorithms, 3rd edn. MIT Press, Cambridge (2009)

13. d'Amore, F., Nguyen, V.H., Roos, T., Widmayer, P.: On optimal cuts of hyperrectangles. Computing **55**(3), 191–206 (1995)

14. Edelsbrunner, H.: A new approach to rectangle intersections part I. J. Comput. Math. (JCM) **13**(3–4), 209–219 (1983)

15. Fredman, M.L., Weide, B.W.: On the complexity of computing the measure of $\cup_1^n [a_i; b_i]$. Commun. ACM (CACM) **21**(7), 540–544 (1978)

16. Gall, F.L.: Powers of tensors and fast matrix multiplication. In: International Symposium on Symbolic and Algebraic Computation (ISSAC), Kobe, Japan, July 23–25, 2014, pp. 296–303. ACM (2014)

17. Kirkpatrick, D.G., Seidel, R.: Output-size sensitive algorithms for finding maximal vectors. In: Proceedings of the First Annual Symposium on Computational Geometry (SoCG), Baltimore, Maryland, USA, June 5–7, 1985, pp. 89–96 (1985)

18. Klee, V.: Can the measure of $\cup_1^n [a_i; b_i]$ be computed in less than O(n log n) steps? Am. Math. Mon. (AMM) **84**(4), 284–285 (1977)

19. Lick, D.R., White, A.T.: k-Degenerate graphs. Can. J. Math. (CJM) **22**, 1082–1096 (1970)

20. Matula, D.W., Beck, L.L.: Smallest-last ordering and clustering and graph coloring algorithms. J. ACM (JACM) **30**(3), 417–427 (1983)

21. Moffat, A., Petersson, O.: An overview of adaptive sorting. Aust. Comput. J. (ACJ) **24**(2), 70–77 (1992)

22. Overmars, M.H., Yap, C.: New upper bounds in Klee's measure problem. SIAM J. Comput. (SICOMP) **20**(6), 1034–1045 (1991)

23. Strassen, V.: Gaussian elimination is not optimal. Numer. Math. **13**(4), 354–356 (1969)

24. Yildiz, H., Hershberger, J., Suri, S.: A discrete and dynamic version of Klee's measure problem. In: Proceedings of the 23rd Annual Canadian Conference on Computational Geometry (CCCG), Toronto, Ontario, Canada, August 10–12, 2011 (2011)

An Improved Lower Bound on the Growth Constant of Polyiamonds

Gill Barequet$^{(\boxtimes)}$, Mira Shalah, and Yufei Zheng

Department of Computer Science, Technion—Israel Institute of Technology,
32000 Haifa, Israel
{barequet,mshalah,yufei}@cs.technion.ac.il

Abstract. A polyiamond is an edge-connected set of cells on the triangular lattice. In this paper we provide an improved lower bound on the asymptotic growth constant of polyiamonds, proving that it is at least 2.8424. The proof of the new bound is based on a concatenation argument and on elementary calculus. We also suggest a nontrivial extension of this method for improving the bound further. However, the proposed extension is based on an unproven (yet very reasonable) assumption.

Keywords: Polyiamonds · Lattice animals · Growth constant

1 Introduction

A *polyomino* of size n is an edge-connected set of n cells on the square lattice $\mathbb{Z}^2$. Similarly, a *polyiamond* of size n is an edge-connected set of n cells on the two-dimensional triangular lattice. *Fixed* polyiamonds are considered distinct if they have different *shapes* or *orientations*. In this paper we consider only fixed polyiamonds, and so we refer to them in the sequel simply as "polyiamonds." Fig. 1 shows polyiamonds of size up to 5.

In general, a connected set of cells on a lattice is called a *lattice animal*. The fundamental combinatorial problem concerning lattice animals is "How many animals with n cells are there?" The study of lattice animals began in parallel more than half a century ago in two different communities. In statistical physics, Temperley [20] investigated the mechanics of macro-molecules, and Broadbent and Hammersley [6] studied percolation processes. In mathematics, Harary [11] composed a list of unsolved problems in the enumeration of graphs, and Eden [7] analyzed cell growth problems. Since then, counting animals has attracted much attention in the literature. However, despite serious efforts over the last 50 years, counting polyominoes is still far from being solved, and is considered [2] one of the long-standing open problems in combinatorial geometry.

The symbol $A(n)$ usually denotes the number of polyominoes of size n; See sequence A001168 in the On-line Encyclopedia of Integer Sequences (OEIS) [1].

Work on this paper by all authors has been supported in part by ISF Grant 575/15.

Y. Cao and J. Chen (Eds.): COCOON 2017, LNCS 10392, pp. 50–61, 2017.
DOI: 10.1007/978-3-319-62389-4_5

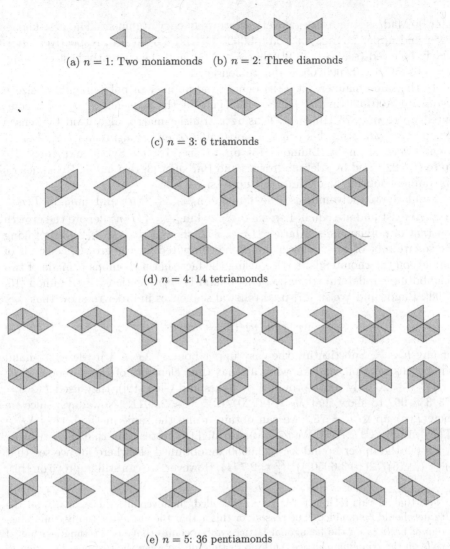

(a) $n = 1$: Two moniamonds   (b) $n = 2$: Three diamonds

(c) $n = 3$: 6 triamonds

(d) $n = 4$: 14 tetriamonds

(e) $n = 5$: 36 pentiamonds

Fig. 1. Polyiamonds of sizes $1 \leq n \leq 5$

Since no analytic formula for the number of animals is yet known for any nontrivial lattice, a great portion of the research has so far focused on efficient algorithms for *counting* animals on lattices, primarily on the square lattice. Elements of the sequence $A(n)$ are currently known up to $n = 56$ [12]. The growth constant of polyominoes was also treated extensively in the literature, and a few asymptotic results are known. Klarner [13] showed that the limit $\lambda := \lim_{n \to \infty} \sqrt[n]{A(n)}$ exists, and the main problem so far has been to evaluate this constant. The convergence of $A(n+1)/A(n)$ to λ (as $n \to \infty$) was proven only three decades later by Madras [16], using a novel pattern-frequency argument. The best-known lower and upper bounds on λ are 4.0025 [4] and 4.6496 [14], respectively. It is widely believed (see, e.g., [8,9]) that $\lambda \approx 4.06$, and the currently best estimate, $\lambda = 4.0625696 \pm 0.0000005$, is due to Jensen [12].

In the same manner, let $T(n)$ denote the number of polyiamonds of size n (sequence A001420 in the OEIS). Elements of the sequence $T(n)$ were computed up to $n = 75$ [10, p. 479] using a transfer-matrix algorithm by Jensen [12, p. 173], adapting his original polyomino-counting algorithm [12]. Earlier counts were given by Lunnon [15] up to size 16, by Sykes and Glen [19] up to size 22,[1] and by Aleksandrowicz and Barequet [3] (extending Redelmeier's polyomino-counting algorithm [18]) up to size 31.

Similarly to polyominoes, the limits $\lim_{n\to\infty} \sqrt[n]{T(n)}$ and $\lim_{n\to\infty} T(n+1)/T(n)$ exist and are equal. Let, then, $\lambda_T = \lim_{n\to\infty} \sqrt[n]{T(n)}$ denote the growth constant of polyiamonds. Klarner [13, p. 857] showed that $\lambda_T \geq 2.13$ by taking the square root of 4.54, a lower bound he computed for the growth constant of animals on the rhomboidal lattice, using the fact that a rhombus is made of two neighboring equilateral triangles. This bound is also mentioned by Lunnon [15, p. 98]. Rands and Welsh [17] used renewal sequences in order to show that

$$\lambda_T \geq (T(n)/(2(1+\lambda_T)))^{1/n} \tag{1}$$

for any $n \in \mathbb{N}$. Substituting the easy upper bound[2] $\lambda_T \leq 4$ in the right-hand side of this relation, and knowing at that time elements of the sequence $T(n)$ for $1 \leq n \leq 20$ only (data provided by Sykes and Glen [19]), they used $T(20) = 173,338,962$ to show that $\lambda_T \geq (T(20)/10)^{1/20} \approx 2.3011$.[3] Nowadays, since we know $T(n)$ up to $n = 75$,[4] we can obtain, using the same method, that $\lambda_T \geq (T(75)/10)^{1/75} \approx 2.7714$. We can even do slightly better than that. Substituting in Eq. (1) the upper bound $\lambda_T \leq 3.6050$ we obtained elsewhere [5], we see that $\lambda_T \geq (T(75)/(2(1+3.6050)))^{1/75} \approx 2.7744$. However, we can still improve on this.

---

[1] Note that in this reference the lattice is called "honeycomb" (hexagonal) and the terms should be doubled. The reason for this is that the authors actually count clusters of *vertices* on the hexagonal lattice, whose connectivity is the same as that of *cells* on the triangular lattice, with no distinction between the two possible orientations of the latter cells. This is why polyiamonds are often regarded in the literature as site animals on the hexagonal lattice, and polyhexes (cell animals on the hexagonal lattice) are regarded as site animals on the triangular lattice, which sometimes causes confusion.

[2] This easy upper bound, based on an idea of Eden [7] was described by Lunnon [15, p. 98]. Every polyiamond P can be built according to a set of $n-1$ "instructions" taken from a superset of size $2(n-1)$. Each instruction tells us how to choose a lattice cell c, neighboring a cell already in P, and add c to P. (Some of these instruction sets are illegal, and some other sets produce the same polyiamonds, but this only helps.) Hence, $\lambda_T \leq \lim_{n\to\infty} \binom{2(n-1)}{n-1}^{1/n} = 4$.

[3] We wonder why Rands and Welsh did not use $T(22) = 1,456,891,547$ (which was also available in their reference [19]) to show that $\lambda_T \geq (T(22)/10)^{1/22} \approx 2.3500$.

[4] $T(75) = 15,936,363,137,225,733,301,433,441,827,683,823$.

Based on existing data, it is believed [19] (but has never been proven) that $\lambda_T = 3.04 \pm 0.02$. In this paper we improve the lower bound on λ_T, showing that $\lambda_T \geq 2.8424$. The new lower bound is obtained by combining a concatenation argument with elementary calculus.

2   Preliminaries

We orient the triangular lattice as is shown in Fig. 2(a), and define a lexico-graphic order on the cells of the lattice as follows: A triangle t_1 is *smaller* than triangle $t_2 \neq t_1$ if the lattice column of t_1 is to the left of the column of t_2, or if t_1, t_2 are in the same column and t_1 lies below t_2. Triangles that look like a "left arrow" (Fig. 2(b)) are of Type 1, and triangles that look like a "right arrow" (Fig. 2(c)) are of Type 2. In addition, let $T_1(n)$ be the number of polyiamonds of size n whose *largest* (top-right) triangle is of Type 1, and let $T_2(n)$ be the number of polyiamonds of size n whose *largest* triangle is of Type 2.

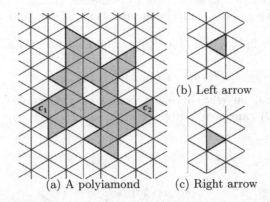

(b) Left arrow

(c) Right arrow

(a) A polyiamond

Fig. 2. Polyiamonds on the triangular lattice

Let $x(n)$ $(0 < x(n) < 1)$ denote the fraction of polyiamonds of Type 1 out of all polyiamonds of size n, that is, $T_1(n) = x(n)T(n)$ and $T_2(n) = (1 - x(n))T(n)$. In addition, let $y(n) = T_2(n)/T_1(n) = (1 - x(n))/x(n)$ denote the ratio between polyiamonds of Type 2 and polyiamonds of Type 1 (all of size n).

A concatenation of two polyiamonds P_1, P_2 is the union of P_1 and a translated copy of P_2, so that the largest triangle of P_1 is attached to the smallest triangle of P_2, and all cells of P_1 are smaller than the translates of cells of P_2. We use a concatenation argument in order to improve the lower bound on λ_T.

3   The Bound

Our proof of a lower bound on λ_T uses the division of polyiamonds into Type 1 and Type 2, but does not employ the asymptotic proportion between the two types.

Theorem 1. $\lambda_T \geq 2.8424$.

Proof. First note that by rotational symmetry, the number of polyiamonds of size n, whose *smallest* (bottom-left) triangle is of Type 2, is $T_1(n)$. Similarly, the number of polyiamonds, whose *smallest* triangle is of Type 1, is $T_2(n)$.

We proceed with a concatenation argument tailored to the specific case of the triangular lattice. Interestingly, not all pairs of polyiamonds of size n can be concatenated. In addition, there exist many polyiamonds of size $2n$ which cannot be represented as the concatenation of two polyiamonds of size n. Let us count carefully the amount of pairs of polyiamonds that can be concatenated.

- Polyiamonds, whose largest triangle is of Type 1, can be concatenated only to polyiamonds whose smallest triangle is of Type 2, and this can be done in two different ways (see Figs. 3(a–c)). There are $2(T_1(n))^2$ concatenations of this kind.
- Polyiamonds, whose largest triangle is of Type 2, can be concatenated, in a single way, only to polyiamonds whose smallest triangle is of Type 1 (see Figs. 3(d, e)). There are $(T_2(n))^2$ concatenations of this kind.

Altogether, we have $2(T_1(n))^2 + (T_2(n))^2$ possible concatenations, and, as argued above,

$$2(T_1(n))^2 + (T_2(n))^2 \leq T(2n). \tag{2}$$

Let us now find an efficient lower bound on the number of concatenations. Equation (2) can be rewritten as $T(2n) \geq 2(x(n)T(n))^2 + ((1 - x(n))$

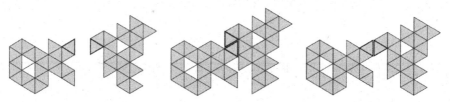

(a) Two polyiamonds   (b) Vertical concatenation (c) Horizontal concatenation
Two attachments of Type-1 and Type-2 triangles

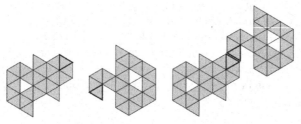

(d) Two polyiamonds   (e) Vertical concatenation
A single attachment of Type-2 and Type-1 triangles

Fig. 3. Possible concatenations of polyiamonds

$T(n))^2 = (3x^2(n) - 2x(n) + 1)T^2(n)$. Elementary calculus shows that the function $f(x) = 3x^2 - 2x + 1$ assumes its minimum at $x = 1/3$ and that $f(1/3) = 2/3$. Hence,

$$\frac{2}{3}T^2(n) \leq T(2n).$$

By simple manipulations of this relation, we obtain that

$$\left(\frac{2}{3}T(n)\right)^{1/n} \leq \left(\frac{2}{3}T(2n)\right)^{1/(2n)}.$$

This implies that the sequence $\left(\frac{2}{3}T(k)\right)^{1/k}, \left(\frac{2}{3}T(2k)\right)^{1/(2k)}, \left(\frac{2}{3}T(4k)\right)^{1/(4k)}, \ldots$ is monotone increasing for any value of k, and, as a subsequence of $\left(\left(\frac{2}{3}T(n)\right)^{1/n}\right)$, it converges to λ_T too. Therefore, any term of the form $\left(\frac{2}{3}T(n)\right)^{1/n}$ is a lower bound on λ_T. In particular, $\lambda_T \geq (\frac{2}{3}T(75))^{1/75} \approx 2.8424$. $\qquad\square$

4   Convergence, Interval Containment, and Monotonicity

First, we show a simple relation between the number of polyiamonds of Type 2 of size n and the number of polyiamonds of Type 1 of size $n+1$.

Observation 2. $T_2(n) = T_1(n + 1)$ *(for all $n \in \mathbb{N}$).*

This simple observation follows from the fact that if the largest triangle t of a polyiamond P is of Type 1, then the only possible neighboring triangle of t within P is the triangle immediately below it, and so, removing t from P will not break P into two parts. This implies a bijection between Type-1 polyiamonds of size $n+1$ and Type-2 polyiamonds of size n.

An immediate consequence is that $y(n) = T_1(n + 1)/T_1(n)$, a fact which we will use later frequently (see, e.g., the proof of Theorem 4(ii) below).

Next, we prove the convergence of the sequences $(T_1(n + 1)/T_1(n))$, $x(n)$, and $y(n)$, and find their limits.

Observation 3. *The sequence $(T_1(n + 1)/T_1(n))_{n=1}^{\infty}$ converges.*

Indeed, polyiamonds of Type 1 fulfill all the premises of Madras's Ratio Limit Theorem [16, Theorem 2.2]. (A more detailed explanation is given in the full version of the paper.) Hence, the limit $\lim_{n\to\infty} T_1(n+1)/T_1(n)$ exists and is equal to some constant μ, whose value is specified (indirectly) in the following theorem.

Theorem 4. *(i) $\mu = \lambda_T$; (ii) $\displaystyle\lim_{n\to\infty} y(n) = \lambda_T$; (iii) $\displaystyle\lim_{n\to\infty} x(n) = 1/(\lambda_T + 1)$.*

Proof.

(i) On the one hand, by Madras [16] we know that the limit $\lambda_T :=$ $\lim_{n\to\infty} T(n+1)/T(n)$ exists. On the other hand, we have that

$$\frac{T(n+1)}{T(n)} = \frac{T_1(n+1) + T_2(n+1)}{T_1(n) + T_2(n)} = \frac{T_1(n+1) + T_1(n+2)}{T_1(n) + T_1(n+1)}$$

$$= \frac{1 + \frac{T_1(n+2)}{T_1(n+1)}}{\frac{T_1(n)}{T_1(n+1)} + 1} \xrightarrow[n\to\infty]{} \frac{1+\mu}{\frac{1}{\mu}+1} = \mu.$$

(Note that the convergence in the last step relies on the fact that the sequence $(T_1(n+1)/T_1(n))$ has a limit μ.) Hence, $\mu = \lambda_T$.

(ii)

$$y(n) = \frac{T_2(n)}{T_1(n)} = \frac{T_1(n+1)}{T_1(n)} \xrightarrow[n\to\infty]{} \mu = \lambda_T.$$

(iii)

$$x(n) = \frac{T_1(n)}{T(n)} = \frac{T_1(n)}{T_1(n) + y(n)T_1(n)} = \frac{1}{1 + y(n)} \xrightarrow[n\to\infty]{} \frac{1}{\lambda_T + 1}. \qquad \square$$

Now, we show relations between bounds on the entire sequence $(x(n))$ and bounds on the entire sequence $(y(n)) \equiv (T(n+1)/T(n))$.

Theorem 5. *Let d, e be two constants.*

(i) $x(n) \geq 1/d$ for all $n \in \mathbb{N} \longleftrightarrow T(n+1)/T(n) \leq d-1$ for all $n \in \mathbb{N}$;
(ii) $x(n) \leq 1/e$ for all $n \in \mathbb{N} \longleftrightarrow T(n+1)/T(n) \geq e-1$ for all $n \in \mathbb{N}$.

Proof.
(i) First,

$$x(n) \geq \frac{1}{d} \longleftrightarrow \frac{T_1(n)}{T(n)} \geq \frac{1}{d} \longleftrightarrow \frac{T_1(n)}{T_1(n) + T_1(n+1)} \geq \frac{1}{d} \longleftrightarrow$$

$$\frac{T_1(n) + T_1(n+1)}{T_1(n)} \leq d \longleftrightarrow 1 + \frac{T_1(n+1)}{T_1(n)} \leq d \longleftrightarrow \frac{T_1(n+1)}{T_1(n)} \leq d-1.$$

Second,

$$\frac{T(n+1)}{T(n)} = \frac{T_1(n+1) + T_1(n+2)}{T_1(n) + T_1(n+1)} = \frac{1 + T_1(n+2)/T_1(n+1)}{T_1(n)/T_1(n+1) + 1}$$

$$\leq \frac{1 + (d-1)}{1/(d-1) + 1} = d-1.$$

(ii) The proof is completely analogous to that of item (i). $\qquad \square$

Finally, we observe the relation between the directions of monotonicity (if exist) of the sequences $(x(n))$ and $(y(n))$.

Observation 6. *If any of the two sequences $(x(n))$ and $(y(n))$ is monotone, then the other sequence is also monotone but in the opposite direction.*

Indeed, this follows immediately from the equality $y(n) = (1 - x(n))/x(n) = 1/x(n) - 1$. The available data suggest that $(x(n))$ be monotone decreasing and that $(y(n))$ be monotone increasing. Note that the monotonicity of $x(n)$ (or $y(n)$) neither implies, nor is implied by, the monotonicity of $(T(n + 1)/T(n))$.

5   Conditional Lower Bounds

Let $L(n)$ denote the number of animals on a lattice $\mathcal{L}$. It is widely believed (but has never been proven) that asymptotically

$$L(n) \sim C_{\mathcal{L}} n^{-\theta_{\mathcal{L}}} \lambda_{\mathcal{L}}, \tag{3}$$

where $C_{\mathcal{L}}$, $\theta_{\mathcal{L}}$, and $\lambda_{\mathcal{L}}$ are constants which depend on $\mathcal{L}$.[5] If this were true, then it would guarantee the existence of the limit $\lambda_L := \lim_{n \to \infty} L(n + 1)/L(n)$, the growth constant of animals on the lattice $\mathcal{L}$. As was mentioned in the introduction, the existence of this limit was proven by Madras [16] without assuming the relation in Eq. (3). Furthermore, it is widely believed that the "ratio sequence" $(L(n + 1)/L(n))$ is *monotone increasing.*[6] Available data (numbers of animals on various lattices) support this belief, but it was unfortunately never proven. Such monotonicity of the ratio sequence would imply that the entire sequence lies below its limit. (In fact, this may be the case even without monotonicity.) Consequently, every element of the ratio sequence would be a lower bound on the growth constant of animals on $\mathcal{L}$. In particular, it would imply the lower bound $\lambda_T \geq T(75)/(74) \approx 2.9959$, only 0.05 short of the estimated value.

Another plausible direction for setting a lower bound on λ_T would be to prove that the entire sequence $(T(n+1)/T(n))$ lies below some constant $c > \lambda_T$, that is, $T(n + 1)/T(n) \leq c$ for all $n \in \mathbb{N}$. As we demonstrate at the end of this section, proving this relation for *any* arbitrary constant c, even a very large one, would improve the lower bound on λ_T, provided in Sect. 3.

In the proof of Theorem 1, we considered the three possible kinds of concatenations of the largest and the smallest triangles in two polyiamonds of size n. In fact, more compositions which preserve the lexicographical order of the polyiamonds can be obtained by using horizontal attachments of Type-1 and Type-2

---

[5] In fact, it is widely believed (but not proven) that the constant θ is common to *all* lattices in the same dimension. In particular, there is evidence that $\theta = 1$ for all lattices in two dimensions.

[6] Madras [16, Proposition 4.2] proved "almost monotonicity" for all lattices, namely, that $L(n + 2)/L(n) \geq (L(n + 1)/L(n))^2 - \Gamma_{\mathcal{L}}/n$ for all sufficiently large values of n, where $\Gamma_{\mathcal{L}}$ is a constant which depends on $\mathcal{L}$. Note that if $\Gamma_{\mathcal{L}} = 0$, then we have $L(n + 2)/L(n + 1) \geq L(n + 1)/L(n)$, i.e., that the ratio sequence of $\mathcal{L}$ is monotone increasing.

triangles which are not the largest or smallest triangles in the respective polyiamonds. The following cases, shown in Fig. 4, are categorized by the type of
extreme triangles (t_1, t_2) in a pair of composed polyiamonds, where t_1 is the
largest triangle of the smaller polyiamond and t_2 is the smallest triangle of the
larger polyiamond. Therefore, these cases are distinct. In the figure, attached
triangles are colored red, the direction of the attachment is marked by a small
arrow, and an "×" sign marks an area free of triangles of the polyiamonds. The
triangles in gray are the largest (or smallest) in polyiamonds of size less than n,
which are then extended to full size-n polyiamonds. (In Case (c), the red and
gray triangles identify.) For the purpose of further computation, let x_i denote
the ratio $T_1(n-i)/T(n-i)$, for $i = 0, 1, 2, 3$, and let c be an upper bound on the
sequence $(T(n+1)/T(n))$. All compositions below do not appear in the proof of
Theorem 1.

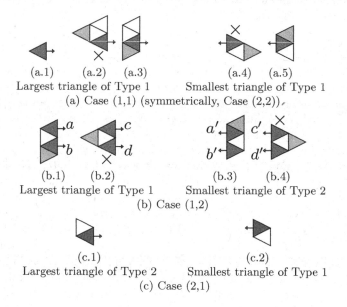

(a.1)   (a.2)   (a.3)   (a.4)   (a.5)
Largest triangle of Type 1   Smallest triangle of Type 1
(a) Case (1,1) (symmetrically, Case (2,2)),

(b.1)   (b.2)   (b.3)   (b.4)
Largest triangle of Type 1   Smallest triangle of Type 2
(b) Case (1,2)

(c.1)   (c.2)
Largest triangle of Type 2   Smallest triangle of Type 1
(c) Case (2,1)

Fig. 4. Additional compositions of polyiamonds (Color figure online)

(1,1) In this type of composition, each one of the polyiamond types shown
in Figs. 4(a.1–a.3) is composed with each one of the types shown in
Figs. 4(a.4, a.5). Specifically, the attached triangle in types (a.1) and (a.2,
a.3) is the largest and third largest, respectively, and the attached triangle in types (a.4, a.5) is the second smallest. The number of polyiamonds
whose largest triangle is of Type 1 (Fig. 4(a.1)) is $T_1(n)$, and among these
polyiamonds, the number of polyiamonds shown in Fig. 4(a.2) is $T_1(n-3)$.
One way to see it is that, for every polyiamond of of size $n-3$ whose largest
triangle (shown in gray) is of Type 1, a column-like polyiamond composed
of three triangles whose largest is of Type 1 is attached to the right of it.
With a similar reasoning, there are $T_2(n-3)$, $T_1(n-2)$, and $T_2(n-2)$

polyiamonds of the types shown in Figs. 4(a.3–a.5), respectively. Hence, the number of compositions in this category is

$$\big(T_1(n) + T_1(n-3) + T_2(n-3)\big)\big(T_1(n-2) + T_2(n-2)\big).$$

By Observation 2, we can substitute $T_2(n-i)$ by $T_1(n-i+1)$, and by the definition of x_i, replace $T_1(n-i)$ by $x_i T(n-i)$. Then, by the definition of c, we have $T(n-i) \geq \frac{T(n)}{c^i}$, and we conclude that in this category we have

$$\big(x_0 T(n) + x_3 T(n-3) + x_2 T(n-2)\big)\big(x_2 T(n-2) + x_1 T(n-1)\big)$$
$$\geq \Big(x_0 + \frac{x_2}{c^2} + \frac{x_3}{c^3}\Big)\Big(\frac{x_1}{c} + \frac{x_2}{c^2}\Big) T^2(n). \quad (4)$$

Notice that, in principal, the attachment of smaller and larger triangles (of the two polyiamonds, respectively) can be counted as well. However, in such cases, higher orders of $1/c$ would appear in Eq. 4. The contribution of such compositions is relatively insignificant, thus, they are not considered.

(1,2) In this case, we count attachments of the largest and the third largest triangles in the smaller polyiamonds, and the smallest and third smallest triangles in the larger polyiamonds; see Figs. 4(b.1–b.4). Table 1 lists all possible attachments and their corresponding numbers of compositions. To sum up, the total number of compositions of this category is

$$3T_1^2(n-3) + 6T_1(n-3)T_2(n-3) + 3T_2^2(n-3) \geq 3\Big(\frac{x_2}{c^2} + \frac{x_3}{c^3}\Big)^2 T^2(n).$$

(2,1) The numbers of polyiamonds shown in Figs. 4(c.1, c.2) are both $T_1(n-1)$. Therefore, there are

$$T_1^2(n-1) \geq \frac{x_1^2}{c^2} T^2(n)$$

compositions in this category.

(2,2) By rotational symmetry, the number of compositions in this category is the same as that of Case (1,1).

Table 1. Additional compositions in Case (1,2)

Attached triangles	Number of compositions
(a ↔ a',c')	$T_2^2(n-3) + T_1(n-3)T_2(n-3)$
(b ↔ a',b',c',d')	$2T_2^2(n-3) + 2T_1(n-3)T_2(n-3)$
(c ↔ a',c')	$T_1^2(n-3) + T_1(n-3)T_2(n-3)$
(d ↔ a',b',c',d')	$2T_1^2(n-3) + 2T_1(n-3)T_2(n-3)$

Summing up the four cases and the three cases in the proof of Theorem 1, we obtain the relation

$$\left(1 - 2x_0 + 3x_0{}^2 + 2\left(x_0 + \frac{x_2}{c^2} + \frac{x_3}{c^3}\right)\left(\frac{x_1}{c} + \frac{x_2}{c^2}\right) + \frac{x_1^2}{c^2} + 3\left(\frac{x_2}{c^2} + \frac{x_3}{c^3}\right)^2\right) T^2(n)$$

(5)

$$\leq T(2n)$$

In order to set a good (high) lower bound on λ_T, we need good lower bounds on x_0, x_1, x_2, x_3 and a good upper bound on c. Then we obtain a relation of the form $\mu \cdot T^2(n) \leq T(2n)$, where μ is a constant, and proceed in the same way as in the proof of Theorem 1. If we denote the 4-variable polynomial multiplying $T^2(n)$ by $f(x_0, x_1, x_2, x_3)$, we find that $f(\cdot, \cdot, \cdot, \cdot)$ is monotone decreasing when x_1, x_2, x_3 decrease, so its infimum is obtained when $x_1 = x_2 = x_3 = 0$, which is useless.

However, the actual data show that the sequence $(T(n+1)/T(n))$ is monotone increasing with the limit $\lambda_T \approx 3.04$. Assume, then, *without proof* that $(T(n+1)/T(n)) \leq 4$ for all $n \in \mathbb{N}$. By Theorem 5, this implies that $x(n) > 0.2$ for all values of n. Using these bounds, we see that the left-hand side of Eq. (5) becomes a quadratic function $g(x_0) = 3x_0^2 - 1.875x_0 + 1.00519$. (All computations were done with a much higher precision.) Elementary calculus shows that $g(x_0)$ assumes its minimum at $x_0 = 0.3125$ and that $g(0.3125) = 0.7122$. Hence,

$$0.7122 \cdot T^2(n) \leq T(2n).$$

With manipulations similar to those in the proof of Theorem 1, it can be shown that the sequence $\left(\left(0.7122 \cdot T(2^i k)\right)^{1/(2^i k)}\right)_{i=0}^{\infty}$ is monotone increasing for any value of k, and, as a subsequence of $\left(\left(0.7122 \cdot T(n)\right)^{1/n}\right)$, it converges to λ_T as well. Therefore, any term of the form $\left(0.7122 \cdot T(n)\right)^{1/n}$ is a lower bound on λ_T. In particular, $\lambda_T \geq (0.7122 \cdot T(75))^{1/75} \approx 2.8449$, which is a slight improvement over the lower bound on λ_T, shown in Sect. 3, and is pending only the correctness of the assumption that $T(n+1)/T(n) \leq 4$ for all $n \in \mathbb{N}$.

References

1. The On-Line Encyclopedia of Integer Sequences. http://oeis.org
2. The Open Problems Project. http://cs.smith.edu/~orourke/TOPP/
3. Aleksandrowicz, G., Barequet, G.: Counting d-dimensional polycubes and nonrectangular planar polyominoes. Int. J. Comput. Geometry Appl. **19**, 215–229 (2009)
4. Barequet, G., Rote, G., Shalah, M.: $\lambda > 4$: an improved lower bound on the growth constant of polyominoes. Comm. ACM **59**, 88–95 (2016)
5. Barequet, G., Shalah, M.: Improved bounds on the growth constant of polyiamonds. In: Proceedings of the 32nd European Workshop on Computational Geometry, Lugano, Switzerland, pp. 67–70, March-April 2016
6. Broadbent, S.R., Hammersley, J.M.: Percolation processes: I. Crystals and mazes. Proc. Camb. Philosophical Soc. **53**, 629–641 (1957)

7. Eden, M.: A two-dimensional growth process. In: Proceedings of the 4th Berkeley Symposium on Mathematical Statistics and Probability, IV, Berkeley, CA, pp. 223–239 (1961)

8. Gaunt, D.S.: The critical dimension for lattice animals. J. Phys. A Math. General **13**, L97–L101 (1980)

9. Gaunt, D.S., Sykes, M.F., Ruskin, H.: Percolation processes in d-dimensions. J. Phys. A Math. General **9**, 1899–1911 (1976)

10. Guttmann, A.J.: Polygons, Polyominoes, and Polycubes. Lecture Notes in Physics, vol. 775. Springer and Canopus Academic Publishing Ltd., Dordrecht (2009)

11. Harary, F.: Unsolved problems in the enumeration of graphs. Pub. Math. Inst. Hungarian Academy Sci. **5**, 1–20 (1960)

12. Jensen, I.: Counting polyominoes: a parallel implementation for cluster computing. In: Sloot, P.M.A., Abramson, D., Bogdanov, A.V., Gorbachev, Y.E., Dongarra, J.J., Zomaya, A.Y. (eds.) ICCS 2003. LNCS, vol. 2659, pp. 203–212. Springer, Heidelberg (2003). doi:10.1007/3-540-44863-2_21

13. Klarner, D.A.: Cell growth problems. Canadian J. Math. **19**, 851–863 (1967)

14. Klarner, D.A., Rivest, R.L.: A procedure for improving the upper bound for the number of n-ominoes. Canadian J. Math. **25**, 585–602 (1973)

15. Lunnon, W.F.: Counting hexagonal and triangular polyominoes. In: Read, R.C. (ed.) Graph Theory and Computing, pp. 87–100. Academic Press, New York (1972)

16. Madras, N.: A pattern theorem for lattice clusters. Ann. Comb. **3**, 357–384 (1999)

17. Rands, B.M.I., Welsh, D.J.A.: Animals, trees and renewal sequences. IMA J. Appl. Math. **27**, 1–17 (1981). Corrigendum **28**, 107 (1982)

18. Redelmeier, D.H.: Counting polyominoes: yet another attack. Discrete Math. **36**, 191–203 (1981)

19. Sykes, M.F., Glen, M.: Percolation processes in two dimensions: I. Low-density series expansions. J. Phys. A Math. General **9**, 87–95 (1976)

20. Temperley, H.N.V.: Combinatorial problems suggested by the statistical mechanics of domains and of rubber-like molecules. Phys. Rev. **2**(103), 1–16 (1956)

Constrained Routing Between
Non-Visible Vertices

Prosenjit Bose[1], Matias Korman[2], André van Renssen[3,4(✉)],
and Sander Verdonschot[1]

[1] School of Computer Science, Carleton University, Ottawa, Canada
jit@scs.carleton.ca, sander@cg.scs.carleton.ca
[2] Tohoku University, Sendai, Japan
mati@dais.is.tohoku.ac.jp
[3] National Institute of Informatics, Tokyo, Japan
andre@nii.ac.jp
[4] JST, ERATO, Kawarabayashi Large Graph Project, Tokyo, Japan

Abstract. Routing is an important problem in networks. We look at routing in the presence of line segment constraints (i.e., obstacles that our edges are not allowed to cross). Let P be a set of n vertices in the plane and let S be a set of line segments between the vertices in P, with no two line segments intersecting properly. We present the first 1-local $O(1)$-memory routing algorithm on the *visibility graph* of P with respect to a set of constraints S (i.e., it never looks beyond the direct neighbours of the current location and does not need to store more than $O(1)$-information to reach the target). We also show that when routing on any triangulation T of P such that $S \subseteq T$, no $o(n)$-competitive routing algorithm exists when only considering the triangles intersected by the line segment from the source to the target (a technique commonly used in the unconstrained setting). Finally, we provide an $O(n)$-competitive 1-local $O(1)$-memory routing algorithm on any such T, which is optimal in the worst case, given the lower bound.

1 Introduction

Routing is a fundamental problem in graph theory and networking. What makes this problem challenging is that often in a network the routing strategy must be *local*, i.e. the routing algorithm must decide which vertex to forward a message to based solely on knowledge of the current vertex, its neighbors and a constant amount of additional information (such as the source and destination vertex). Routing algorithms are considered *geometric* when the graph that is routed on is embedded in the plane, with edges being straight line segments connecting pairs

P. Bose is supported in part by NSERC. M. Korman was partially supported by MEXT KAKENHI Nos. 12H00855, 15H02665, and 17K12635. A. van Renssen was supported by JST ERATO Grant Number JPMJER1305, Japan. S. Verdonschot is supported in part by NSERC, the Ontario Ministry of Research and Innovation, and the Carleton-Fields Postdoctoral Award.

© Springer International Publishing AG 2017
Y. Cao and J. Chen (Eds.): COCOON 2017, LNCS 10392, pp. 62–74, 2017.
DOI: 10.1007/978-3-319-62389-4_6

of vertices and weighted by the Euclidean distance between their endpoints. Geometric routing algorithms are important in wireless sensor networks (see [11,12] for surveys of the area) since they offer routing strategies that use the coordinates of the vertices to guide the search, instead of the more traditional routing tables.

Most of the research on this problem has focused on the situation where the network is constructed by taking a subgraph of the complete Euclidean graph, i.e. the graph that contains an edge between every pair of vertices and the length of this edge is the Euclidean distance between the two vertices. We study this problem in a more general setting with the introduction of *line segment constraints* S. Specifically, let P be a set of n vertices in the plane and let S be a set of line segments between the vertices in P, with no two line segments properly intersecting (i.e., anywhere except at the endpoints). Two vertices u and v can *see each other* if and only if either the line segment uv does not properly intersect any constraint or uv is itself a constraint. If two vertices u and v can see each other, the line segment uv is a *visibility edge*. The *visibility graph* of P with respect to a set of constraints S, denoted $Vis(P,S)$, has P as vertex set and all visibility edges as edge set. In other words, it is the complete graph on P minus all non-constraint edges that properly intersect one or more constraints in S.

This setting has been studied extensively in the context of motion planning amid obstacles. Clarkson [8] was one of the first to study this problem. He showed how to construct a $(1 + \epsilon)$-spanner of $Vis(P,S)$ with a linear number of edges. A subgraph H of G is called a t-spanner of G (for $t \geq 1$) if for each pair of vertices u and v, the shortest path in H between u and v has length at most t times the shortest path between u and v in G. The smallest value t for which H is a t-spanner is the *spanning ratio* or *stretch factor* of H. Following Clarkson's result, Das [9] showed how to construct a spanner of $Vis(P,S)$ with constant spanning ratio and constant degree. Bose and Keil [4] showed that the Constrained Delaunay Triangulation is a 2.42-spanner of $Vis(P,S)$. Recently, the constrained half-Θ_6-graph (which is identical to the constrained Delaunay graph whose empty visible region is an equilateral triangle) was shown to be a plane 2-spanner of $Vis(P,S)$ [2] and all constrained Θ-graphs with at least 6 cones were shown to be spanners as well [7].

Spanners of $Vis(P,S)$ are desirable because they are sparse and the bounded stretch factor certifies that paths do not make large detours. However, it is not known how to route locally on them. To address this issue, we look at local routing algorithms in the constrained setting, i.e. routing algorithms that must decide which vertex to forward a message to based solely on knowledge of the source and destination vertex, the current vertex, all vertices that can be seen from the current vertex and a constant amount of memory. We define this model formally in the next section. Furthermore, we study *competitiveness* of our routing algorithms, i.e. the ratio of the length of the path followed by the routing algorithm and the length of the shortest path in the graph.

In the constrained setting, routing has not been studied much. Bose et al. [3] showed that it is possible to route locally and 2-competitively between any two visible vertices in the constrained Θ_6-graph. Additionally, an 18-competitive routing algorithm between any two visible vertices in the constrained half-Θ_6-graph was provided. While it seems like a serious shortcoming that these routing algorithms only route between pairs of visible vertices, in the same paper the authors also showed that no deterministic local routing algorithm is $o(\sqrt{n})$-competitive between all pairs of vertices of the constrained Θ_6-graph, regardless of the amount of memory it is allowed to use.

In this paper, we develop routing algorithms that work between any pair of vertices in the constrained setting. We provide a non-competitive 1-local routing algorithm on the visibility graph of P with respect to a set of constraints S. We also show that when routing on any triangulation T of P such that $S \subseteq T$, no $o(n)$-competitive routing algorithm exists when only considering the triangles intersected by the line segment from the source to the target (a technique commonly used in the unconstrained setting). Finally, we provide an $O(n)$-competitive 1-local routing algorithm on T, which is optimal in the worst case, given the lower bound. Prior to this work, no local routing algorithms were known to work between any pair of vertices in the constrained setting.

2 Preliminaries

The Θ_m-graph plays an important role in our routing strategy. We begin with their definitions. Define a *cone* C to be the region in the plane between two rays originating from a vertex referred to as the apex of the cone. When constructing a (constrained) Θ_m-graph, for each vertex u consider the rays originating from u with the angle between consecutive rays being $2\pi/m$. Each pair of consecutive rays defines a cone. The cones are oriented such that the bisector of some cone coincides with the vertical halfline through u that lies above u. Let this cone be C_0 of u and number the cones in clockwise order around u (see Fig. 1). The cones around the other vertices have the same orientation as the ones around u. We write C_i^u to indicate the i-th cone of a vertex u, or C_i if u is clear from the context. For ease of exposition, we only consider point sets in general position: no two points lie on a line parallel to one of the rays that define the cones, no two points lie on a line perpendicular to the bisector of a cone, and no three points are collinear.

Let vertex u be an endpoint of a constraint c and let the other endpoint v that lies in cone C_i^u (if any). The lines through all such constraints c split C_i^u into several *subcones* (see Fig. 2). We use $C_{i,j}^u$ to denote the j-th subcone of C_i^u (again, numbered in clockwise order). When a constraint $c = (u, v)$ splits a cone of u into two subcones, we define v to lie in both of these subcones. We consider a cone that is not split to be a single subcone.

We now introduce the constrained Θ_m-graph: for each subcone $C_{i,j}$ of each vertex u, add an edge from u to the closest vertex in that subcone that can see u, where distance is measured along the bisector of the original cone (*not the*

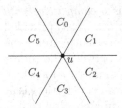 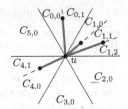 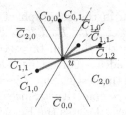

Fig. 1. The cones with apex u in the Θ_6-graph. All points of S have exactly six cones.

Fig. 2. The subcones with apex u in the constrained Θ_6-graph (constraints denoted as red thick segments). (Color figure online)

Fig. 3. If we consider the half-Θ_6-graph instead, we have the same amount of cones, but different notation.

subcone). More formally, we add an edge between two vertices u and v if v can see u, $v \in C_{i,j}^u$, and for all points $w \in C_{i,j}^u$ that can see u, $|uv'| \leq |uw'|$, where v' and w' denote the projection of v and w on the bisector of C_i^u and $|xy|$ denotes the length of the line segment between two points x and y. Note that our general position assumption implies that each vertex adds at most one edge per subcone.

Next, we define the constrained half-Θ_6-graph. This is a generalized version of the half-Θ_6-graph as described by Bonichon *et al.* [1]. The constrained half-Θ_6-graph is similar to the constrained Θ_6-graph with one major difference: edges are only added in every second cone. More formally, its cones are categorized as positive and negative. Let $(C_0, \overline{C_2}, C_1, \overline{C_0}, C_2, \overline{C_1})$ be the sequence of cones in counterclockwise order starting from the positive y-axis. The cones C_0, C_1, and C_2 are called *positive* cones and $\overline{C_0}$, $\overline{C_1}$, and $\overline{C_2}$ are called *negative* cones. Note that the positive cones coincide with the even cones of the constrained Θ_6-graph and the negative cones coincide with the odd ones. We add edges only in the positive cones (and their subcones). We use C_i^u and $\overline{C_i^u}$ to denote cones C_i and $\overline{C_i}$ with apex u. Note that, by the way in which cones are labeled, for any two vertices u and v, it holds that $v \in C_i^u$ if and only if $u \in \overline{C_i^v}$. Analogous to the subcones defined for the Θ_6-graph, constraints split cones into subcones. We call a subcone of a positive cone a positive subcone and a subcone of a negative cone a negative subcone (see Fig. 3). We look at the undirected version of these graphs, i.e. when an edge is added, both vertices are allowed to use it. This is consistent with previous work on Θ-graphs.

Finally, we define the constrained Delaunay triangulation. Given any two visible vertices p and q, the constrained Delaunay triangulation contains an edge between p and q if and only if pq is a constraint or there exists a circle O with p and q on its boundary such that there are no vertices of P in the interior of O is visible to both p and q. For simplicity, we assume that no four vertices lie on the boundary of any circle.

There are two notions of *competitiveness* of a routing algorithm. One is to look at the Euclidean shortest path between the two vertices, i.e. the shortest path in $Vis(P, S)$, and the other is to compare the routing path to the shortest path in the subgraph of $Vis(P, S)$. A routing algorithm is *c-competitive with respect to the Euclidean shortest path (resp. shortest path in the graph)* provided that the total distance traveled by the message is not more than c times the Euclidean shortest path length (resp. shortest path length) between the source and the destination. The *routing ratio* of an algorithm is the smallest c for which it is c-competitive. Since the shortest path in the graph between two vertices is at least as long as the Euclidean shortest path between them, an algorithm that is c-competitive with respect to the Euclidean shortest path is also c-competitive with respect to the shortest path in the graph. We use competitiveness with respect to the Euclidean shortest path when proving upper bounds and with respect to the shortest path in the graph when proving lower bounds.

We now define our routing model. Formally, a routing algorithm A is a deterministic k-local, m-memory routing algorithm, if the vertex to which a message is forwarded from the current vertex s is a function of s, t, $N_k(s)$, and M, where t is the destination vertex, $N_k(s)$ is the k-neighborhood of s and M is a memory of size m, stored with the message. The k-neighborhood of a vertex s is the set of vertices in the graph that can be reached from s by following at most k edges. For our purposes, we consider a unit of memory to consist of a $\log_2 n$ bit integer or a point in $\mathbb{R}^2$. Our model also assumes that the only information stored at each vertex of the graph is $N_k(s)$. Unless stated otherwise, when we say "local", we will assume that $k = 1$ and that $|M| \in O(1)$, i.e. our algorithms are 1-local and use a constant amount of memory. Since our graphs are geometric, we identify each vertex by its coordinates in the plane.

We say that a region R *contains* a vertex v if v lies in the interior or on the boundary of R. We call a region *empty* if it does not contain any vertex of P.

Lemma 2.1 [2]. *Let u, v, and w be three arbitrary points in the plane such that uw and vw are visibility edges and w is not the endpoint of a constraint intersecting the interior of triangle uvw. Then there exists a convex chain of visibility edges from u to v in triangle uvw, such that the polygon defined by uw, wv and the convex chain is empty and does not contain any constraints.*

3   Local Routing on the Visibility Graph

In the unconstrained setting, local routing algorithms have focused on subgraphs of the complete Euclidean graph such as Θ-graphs. There exists a very simple local routing algorithm that works on all Θ-graphs with at least 4 cones and is competitive when the graph uses at least 7 cones. This routing algorithm (often called the Θ-routing) always follows the edge to the closest vertex in the cone that contains the destination. In the constrained setting, however, a problem arises if we try to apply this strategy: even though a cone contains the destination it need not contain any visible vertices, since a constraint may block its visibility. Hence, the Θ-routing algorithm will often get stuck since it cannot follow any edge in that cone. In fact, given a set P of points in the plane and a set S of disjoint segments, no local routing algorithm is known for routing on $Vis(P, S)$.

When the destination t is visible to the source s, it is possible to route locally by essentially "following the segment st", since no constraint can intersect st. This approach was used to give a 2-competitive 1-local routing algorithm on the constrained half-Θ_6-graph, provided that t is in a positive cone of s [3]. In the case where t is in a negative cone of s, the algorithm is much more involved and the competitive ratio jumps to 18.

The stumbling block of all known approaches is the presence of constraints. In a nutshell, the problem is to determine how to "go around" a constraint in such a way as to reach the destination and not cycle. This poses the following natural question: does there exist a deterministic 1-local routing algorithm that always reaches the destination when routing on the visibility graph? In this section, we answer this question in the affirmative and provide such a 1-local algorithm. The main idea is to route on a planar subgraph of $Vis(P, S)$ that can be computed locally.

In [10] it was shown how to route locally on a plane geometric graph. Subsequently, in [6], a modified algorithm was presented that seemed to work better in practice. Both algorithms are described in detail in [6], where the latter algorithm is called FACE-2 and the former is called FACE-1. Neither of the algorithms is competitive. FACE-1 reaches the destination after traversing at most $\Theta(n)$ edges in the worst case and FACE-2 traverses $\Theta(n^2)$ edges in the worst case. Although FACE-1 performs better in the worst case, FACE-2 performs better on average in random graphs generated by points uniformly distributed in the unit square.

Coming back to our problem of routing locally from a source s to a destination t in $Vis(P, S)$, the main difficulty for using the above strategies is that the visibility graph is not plane. Its seems counter-intuitive that having more edges renders the problem of finding a path more difficult. Indeed, almost all local routing algorithms in the literature that guarantee delivery do so by routing on a plane subgraph that is computed locally. For example, in [6], a local routing algorithm is presented for routing on a unit disk graph and the algorithm actually routes on a planar subgraph known as the Gabriel graph. However, none of these algorithms guarantee delivery in the presence of constraints. In this section, we adapt the approach from [6] by showing how to locally identify the edges of a planar spanning subgraph of $Vis(P, S)$, which then allows us to use FACE-1 or FACE-2 to route locally on $Vis(P, S)$.

The graph in question is the constrained half-Θ_6-graph, which was shown to be a plane 2-spanner of $Vis(P, S)$ [2]. Therefore, if we can show how to locally identify the edges of this graph, we can apply FACE-1 or FACE-2. If we are at a vertex v and we know all visibility edges incident to v, then identifying the edges in v's positive cones is easy: they connect v to the endpoints in this cone whose projection on the bisector is closest. Thus, the hard part is deciding which of the visibility edges in v's negative cone are added by their endpoints. Next, we show that v has enough information to find these edges locally.

Lemma 3.1. *Let u and v be vertices such that $u \in \overline{C_0^v}$. Then uv is an edge of the constrained half-Θ_6-graph if and only if v is the vertex whose projection on* the bisector of C_0^u is closest to u, among all vertices in C_0^u visible to v and not blocked from u by constraints incident on v.

Proof. First, suppose that v is not closest to u among the vertices in C_0^u visible to v and not blocked by constraints incident on v (see Fig. 4a). Then there are such vertices whose projection on the bisector is closer to u. Among those vertices, let x be the one that minimizes the angle between vx and vu. Now v cannot be the endpoint of a constraint intersecting the interior of triangle uvx, since the endpoint of that constraint would lie inside the triangle, contradicting our choice of x. Since both uv and vx are visibility edges, Lemma 2.1 tells us that there is a convex chain of visibility edges connecting u and x inside triangle uvx. In particular, the first vertex y from u on this chain is visible from both u and v and is closer to u than v is (in fact, $y = x$ by our choice of x). Moreover, v must be in the same subcone of u as y, since the region between v and the chain is completely empty of both vertices and constraints. Thus, uv cannot be an edge of the half-Θ_6-graph.

Second, suppose that v is closest to u among the vertices visible to v and not blocked by constraints incident on v, but uv is not an edge of the half-Θ_6-graph. Then there is a vertex $x \in C_0^u$ and in the same subcone as v, who is visible to u, but not to v, and whose projection on the bisector is closer to u (see Fig. 4b). Since x and v are in the same subcone, u is not incident to any constraints that intersect the interior of triangle uvx, so we again apply Lemma 2.1 to the triangle formed by visibility edges uv and ux. This gives us that there is a convex chain of visibility edges connecting v and x, inside triangle uvx. In particular, the first point y from v on this chain must be visible to both u and v. And since y lies in triangle uvx, it lies in C_0^u and its projection is closer to u. But this contradicts our assumption that v was the closest vertex. Thus, if v is the closest vertex, uv must be an edge of the half-Θ_6-graph.

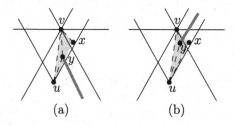

Fig. 4. (a) If v is not closest to u among the vertices visible to v, then uv is not in the half-Θ_6-graph. (b) If v is closest to u among the vertices visible to v, then uv must be in the half-Θ_6-graph.

With Lemma 3.1, we can compute 1-locally which of the edges of $Vis(P, S)$ incident on v are also edges of the half-Θ_6-graph. Therefore, we can apply FACE-1 or FACE-2 in order to route on $Vis(P, S)$ by routing on the half-Θ_6-graph.

Corollary 3.2. *We can 1-locally route on* $Vis(P, S)$ *by routing on the constrained half-Θ_6-graph.*

Although it was shown in [3] that no deterministic local routing algorithm is $o(\sqrt{n})$-competitive on all pairs of vertices of the constrained Θ_6-graph, regardless of the amount of memory it is allowed to use, the caveat to the above local routing algorithm is that the competitive ratio is not bounded by any function of n. In fact, by applying FACE-1, it is possible to visit almost every edge of the graph four times before reaching the destination. It is worse with FACE-2, where almost every edge may be visited a linear number of times before reaching the destination. In the next section, we present a 1-local routing algorithm that is $O(n)$-competitive and provide a matching worst-case lower bound.

4   Routing on Constrained Triangulations

Next, we look at routing on a given constrained triangulation: a graph in which all constraints are edges and all faces are triangles. Hence, we do not have to check that the graph is a triangulation and we can focus on the routing process.

4.1   Lower Bound

Given a triangulation G and a source vertex s and a destination vertex t, let H be the subgraph of G that contains all edges of G that are part of a triangle that is intersected by st. We first show that if G is a constrained Delaunay triangulation or a constrained half-Θ_6-graph, the shortest path in H can be a factor of $n/4$ times longer than that in G. This implies that any local routing algorithm that considers only the triangles intersected by st cannot be $o(n)$-competitive with respect to the shortest path in G on every constrained Delaunay triangulation or constrained half-Θ_6-graph on every pair of points. In the remainder of this paper, we use $\pi_G(u, v)$ to denote the shortest path from u to v in a graph G.

Lemma 4.1. *There exists a constrained Delaunay triangulation G with vertices s and t such that* $|\pi_H(s, t)| \geq \frac{n}{4} \cdot |\pi_G(s, t)|$.

Proof. We construct a constrained Delaunay graph with this property. For ease of description and calculation, we assume that the size of the point set is a multiple of 4. Note that we can remove this restriction by adding 1, 2, or 3 vertices "far enough away" from the construction so it does not influence the shortest path.

We start with two columns of $n/2 - 1$ points each, aligned on a grid. We add a constraint between every horizontal pair of points. Next, we shift every other row by slightly less than half a unit to the right (let $\varepsilon > 0$ be the small amount that we did not shift). We also add a vertex s below the lowest row and a vertex t above the highest row, centred between the two vertices on said row. Note that this placement implies that st intersects every constraint. Finally, we stretch the point set by an arbitrary factor $2x$ in the horizontal direction, for

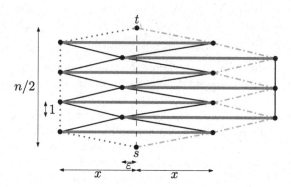

Fig. 5. Lower bound construction: the constraints are shown in thick red, the shortest path in G is shown in blue (dotted), and the shortest path in H is shown in orange (dash dotted). The remaining edges of G are shown in black (solid). (Color figure online)

some arbitrarily large constant x. When we construct the constrained Delaunay triangulation on this point set, we get the graph G shown in Fig. 5.

In order to construct the graph H, we note that all edges that are part of H lie on a face that has a constraint as an edge. In particular, H does not contain any of the vertical edges on the left and right boundary of G. Hence, all that remains is to compare the length of the shortest path in H to that in G.

Ignoring the terms that depend on ε, the shortest path in H uses $n/2$ edges of length x, hence it has length $x \cdot n/2$. Graph G on the other hand contains a path of length $2x + n/2 - 1$ (again, ignoring small terms that depend on ε), by following the path to the leftmost column and following the vertical path up. Hence, the ratio $|\pi_H(s,t)|/|\pi_G(s,t)|$ approaches $n/4$, since $\lim_{x\to\infty} \frac{x \cdot \frac{n}{2}}{2x + \frac{n}{2} - 1} = \frac{n}{4}$.

Note that the above construction is also the constrained half-Θ_6-graph of the given points and constraints.

Corollary 4.2. *There exist triangulations G such that no local routing algorithm that considers only the triangles intersected by st is $o(n)$-competitive when routing from s to t.*

4.2 Upper Bound

Next, we provide a simple local routing algorithm that is $O(n)$-competitive. To make it easier to bound the length of the routing path, we use an auxiliary graph H' defined as follows: let H' be the graph H, augmented with the edges of the convex hull of H and all visibility edges between vertices on the same internal face (after the addition of the convex hull edges). For these visibility edges, we only consider constraints with both endpoints in H. The different graphs G, H, and H' are shown in Fig. 6. Note that the gray region in Fig. 6c is one of the regions where visibility edges are added and note that edge uv is not added, since visibility is blocked by a constraint that has both endpoints in H. We first show that the length of the shortest path in H' is not longer than that in G.

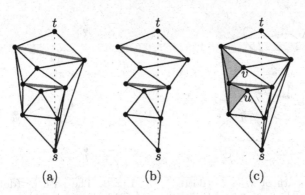

Fig. 6. The three different graphs: (a) The original triangulation G, (b) the subgraph H, (c) graph H' constructed by adding edges to H.

Lemma 4.3. *For any triangulation G, we have $|\pi_{H'}(s,t)| \leq |\pi_G(s,t)|$.*

Proof. If every vertex along $\pi_G(s,t)$ is part of H', we claim that every edge of $\pi_G(s,t)$ is also part of H'. Consider an edge uv of $\pi_G(s,t)$. If uv is part of a triangle intersected by st, it is already part of H and therefore of H'. If uv is not part of a triangle intersected by st, then u and v must lie on the same face of H' before we add the visibility edges, since otherwise the edge uv would violate the planarity of G. Furthermore, since uv is an edge of G, u and v can see each other. Hence, the edge uv is added to H' when the visibility edges are added. Therefore, every edge of $\pi_G(s,t)$ is part of H' and thus $|\pi_{H'}(s,t)| \leq |\pi_G(s,t)|$.

If not every vertex along $\pi_G(s,t)$ is part of H', we create subsequences of the edges of $\pi_G(s,t)$ such that each subpath satisfies either (*i*) all vertices are in H', or (*ii*) only the first and last vertex of the subpath are in H'. Using an argument analogous to the previous case, it can be shown that subpaths of $\pi_G(s,t)$ that satisfy (*i*) only use edges that are in H'.

To complete the proof, it remains to show that given a subpath π' that satisfies (*ii*), there exists a different path in H' that connects the two endpoints of H' and has length at most $|\pi'|$. Let u and v be the first and last vertex of this π' (see Fig. 7). If u and v lie on the same face of H' before the visibility edges are added, H' contains the geodesic path $\pi_{H'}$ between u and v with respect to the constraints using only vertices in H'. Note that this path can cross constraints that have at most 1 endpoint in H'. Path π' uses only edges of G which by definition do not cross any constraints. Hence, π' cannot be shorter than $\pi_{H'}$.

Finally, we consider the case in which u and v do not lie on the same face before the visibility edges are added. We construct $\pi_{H'}$ by following the geodesic paths in the faces of u and v, and using the convex hull edges for the remainder of the path. Since u and v do not lie on the same face, a convex hull vertex x lies on this path. Next, we look at π'. Recall that no vertex along π' (other than u and v) is in H'. This implies that π' cannot cross st and it must cross any ray originating from x that does not intersect the convex hull. Hence, π' goes around $\pi_{H'}$ and therefore H' contains a path from u to v of length at most $|\pi'|$.

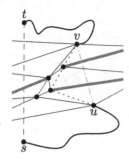

Fig. 7. A subpath of $\pi_G(s,t)$ (dotted blue) that satisfies condition (ii): no vertex other than its endpoints are in H'. (Color figure online)

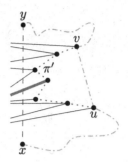

Fig. 8. The path π' (dotted blue) simulates uv on the shortest path in H' (dot dashed orange) and lies on the boundary of a pocket of H. (Color figure online)

Lemma 4.4. *For any triangulation G, we have $|\pi_H(s,t)| \le (n-1) \cdot |\pi_{H'}(s,t)|$.*

Proof. To prove the lemma, we show that for every edge uv on the shortest path in H', there is a path in H from u to v whose length is at most $|\pi_{H'}(s,t)|$. Consider an edge uv on the shortest path in H'. If uv is also an edge of H we use it in our path. Since uv is part of the shortest path in H' we have $|uv| \le |\pi_{H'}(s,t)|$.

It remains to consider the case in which uv is not part of H. Note that this implies that uv is either an edge of the convex hull of H or a visibility edge between two vertices of the same internal face. Instead of following uv, we *simulate* uv by following the path π' along the pocket of H from u to v (the part of the boundary that does not visit both sides of st; see Fig. 8).

The path $\pi_{H'}(s,t)$ must cross the segment st several times (at least once at s and once at t). Let x be the last intersection before u in $\pi_{H'}(s,t)$. Similarly, let y be the first intersection after v. Since π' lies on the boundary of a pocket, it cannot cross st and therefore it is contained in the polygon defined by the segment xy, and the portion of $\pi_{H'}(s,t)$ that lies between x and y. All edges of π' lie inside this polygon. In particular, each such edge has length at most the length of $\pi_{H'}(s,t)$ from x to y, which is at most $|\pi_{H'}(s,t)|$ (since a segment inside a polygon has length at most half the perimeter of the polygon).

Concatenate all our simulating paths and shortcut the resulting path from s to t such that every vertex is visited at most once. The result is a path which consists of at most $n-1$ edges, each of length at most $|\pi_{H'}(s,t)|$. $\square$

Theorem 4.5. *For any triangulation G, we have $|\pi_H(s,t)| \le (n-1) \cdot |\pi_G(s,t)|$.*

In order to route on the graph H, we apply the Find-Short-Path routing algorithm by Bose and Morin [5]. This routing algorithm is designed precisely to route on the graph created by the union of the triangles intersected by the line segment between the source and destination. The algorithm reaches t after having travelled at most 9 times the length of the shortest path from s to t in this union of triangles. Hence, applying Find-Short-Path to graph H yields a routing path of length at most $9(n-1) \cdot |\pi_G(s,t)|$.

Theorem 4.6. *For any triangulation, there exists a 1-local $O(n)$-competitive routing algorithm that visits only triangles intersected by the line segment between the source to destination.*

5 Conclusions

We presented the first local routing algorithm to route on the visibility graph. We then showed a local $O(n)$-competitive routing algorithm for any triangulation and showed that this is optimal when restricted to routing on the set of triangles intersected by the segment from the source to the destination. The competitiveness of our routing algorithm on $Vis(S, P)$ is not bounded by any function of n. On the other hand, our local $O(n)$-competitive routing algorithms require a triangulated subgraph of $Vis(S, P)$. Unfortunately, it is not known how to compute such a triangulation locally, which naturally leads to the following open problem: Can one locally compute a triangulation of $Vis(S, P)$? It is known that the constrained Delaunay triangulation cannot be computed locally and the constrained half-Θ_6-graph is not necessarily a triangulation.

Acknowledgements. We thank Luis Barba, Sangsub Kim, and Maria Saumell for fruitful discussions.

References

1. Bonichon, N., Gavoille, C., Hanusse, N., Ilcinkas, D.: Connections between theta-graphs, delaunay triangulations, and orthogonal surfaces. In: Thilikos, D.M. (ed.) WG 2010. LNCS, vol. 6410, pp. 266–278. Springer, Heidelberg (2010). doi:10.1007/978-3-642-16926-7_25
2. Bose, P., Fagerberg, R., Renssen, A., Verdonschot, S.: On plane constrained bounded-degree spanners. In: Fernández-Baca, D. (ed.) LATIN 2012. LNCS, vol. 7256, pp. 85–96. Springer, Heidelberg (2012). doi:10.1007/978-3-642-29344-3_8
3. Bose, P., Fagerberg, R., van Renssen, A., Verdonschot, S.: Competitive local routing with constraints. In: Elbassioni, K., Makino, K. (eds.) ISAAC 2015. LNCS, vol. 9472, pp. 23–34. Springer, Heidelberg (2015). doi:10.1007/978-3-662-48971-0_3
4. Bose, P., Keil, J.M.: On the stretch factor of the constrained Delaunay triangulation. In: ISVD, pp. 25–31 (2006)
5. Bose, P., Morin, P.: Competitive online routing in geometric graphs. Theor. Comput. Sci. **324**(2), 273–288 (2004)
6. Bose, P., Morin, P., Stojmenovic, I., Urrutia, J.: Routing with guaranteed delivery in ad hoc wireless networks. Wirel. Netw. **7**(6), 609–616 (2001)
7. Bose, P., Renssen, A.: Upper bounds on the spanning ratio of constrained theta-graphs. In: Pardo, A., Viola, A. (eds.) LATIN 2014. LNCS, vol. 8392, pp. 108–119. Springer, Heidelberg (2014). doi:10.1007/978-3-642-54423-1_10
8. Clarkson, K.: Approximation algorithms for shortest path motion planning. In: STOC, pp. 56–65 (1987)

9. Das, G.: The visibility graph contains a bounded-degree spanner. In: CCCG, pp. 70–75 (1997)
10. Kranakis, E., Singh, H., Urrutia, J.: Compass routing on geometric networks. In: CCCG, pp. 51–54 (1999)
11. Misra, S., Misra, S.C., Woungang, I.: Guide to Wireless Sensor Networks. Springer, Heidelberg (2009)
12. Räcke, H.: Survey on oblivious routing strategies. In: Ambos-Spies, K., Löwe, B., Merkle, W. (eds.) CiE 2009. LNCS, vol. 5635, pp. 419–429. Springer, Heidelberg (2009). doi:10.1007/978-3-642-03073-4_43

Deletion Graph Problems
Based on Deadlock Resolution

Alan Diêgo Aurélio Carneiro, Fábio Protti, and Uéverton S. Souza[⊠]

Fluminense Federal University, Niterói, RJ, Brazil
{aurelio,fabio,ueverton}@ic.uff.br

Abstract. A deadlock occurs in a distributed computation if a group
of processes wait indefinitely for resources from each other. In this paper
we study actions to be taken after deadlock detection, especially the
action of searching a small deadlock-resolution set. More precisely, given
a "snapshot" graph G representing a deadlocked state of a distributed
computation governed by a certain deadlock model $\mathbb{M}$, we investigate
the complexity of vertex/arc deletion problems that aim at finding minimum vertex/arc subsets whose removal turns G into a deadlock-free
graph (according to model $\mathbb{M}$). Our contributions include polynomial
algorithms and hardness proofs, for general graphs and for special graph
classes. Among other results, we show that the arc deletion problem in
the OR model can be solved in polynomial time, and the vertex deletion
problem in the OR model remains NP-Complete even for graphs with
maximum degree four, but it is solvable in $O(m\sqrt{n})$ time for graphs with
$\Delta \leq 3$.

Keywords: Deadlock resolution · Knot · Wait-for graphs ·
Computational complexity

1 Introduction

A set of processes is in *deadlock* if each process of the set is blocked, waiting for a
response from another process of the same set. In other words, the processes cannot proceed with their execution because of necessary events that only processes
in the same set can provide. Deadlock is a *stable* property, in the sense that
once it occurs in a global state Ψ of a distributed computation, it still holds for
all the states subsequent to Ψ. Deadlock avoidance and deadlock resolution are
fundamental problems in the study of distributed systems [1].

Distributed computations are usually represented by the so-called *wait-for
graphs*, where the behavior of processes is determined by a set of prescribed rules
(the *deadlock model* or *dependency model*). In a wait-for graph $G = (V, E)$, the
vertex set V represents processes in a distributed computation, and the set E
of directed arcs represents wait conditions [4]. An arc exists in E directed away
from $v_i \in V$ towards $v_j \in V$ if v_i is blocked, waiting a signal from v_j. The graph
G changes dynamically according to the deadlock model, as the computation

© Springer International Publishing AG 2017
Y. Cao and J. Chen (Eds.): COCOON 2017, LNCS 10392, pp. 75–86, 2017.
DOI: 10.1007/978-3-319-62389-4_7

progresses. In essence, the deadlock model specifies rules for vertices that are not *sinks* in G to become sinks [3]. (A sink is a vertex with out-degree zero.)

The main deadlock models investigated so far in the literature are presented below.

AND MODEL – In the AND model, a process v_i can only become a sink when it receives a signal from *all* the processes in O_i, where O_i stands for the set of outneighbors of v_i.

OR MODEL – In this model, it suffices for a process v_i to become a sink to receive a signal from *at least one* of the processes in O_i.

X-OUT-OF-Y MODEL – There are two integers, x_i and y_i, associated with a process v_i. Also, $y_i = |O_i|$, meaning that process v_i is in principle waiting for a signal from every process in O_i. However, in order to be relieved from its wait state, it suffices for v_i to receive a signal from any x_i of those y_i processes.

AND-OR MODEL – There are $t_i \geq 1$ subsets of O_i associated with process v_i. These subsets are denoted by $O_i^1, \ldots, O_i^{t_i}$ and must be such that $O_i = O_i^1 \cup \cdots \cup O_i^{t_i}$. It suffices for a process v_i to become a sink to receive a signal from all processes in at least one of $O_i^1, \ldots, O_i^{t_i}$. For this reason, these t_i subsets of O_i are assumed to be such that no one is contained in another.

The study of deadlocks is fundamental in computer science and can be divided in four fields: Deadlock Prevention, Deadlock Avoidance, Deadlock Detection, and Deadlock Resolution (or Deadlock Recovery). Although prevention, avoidance, and detection of deadlocks have been widely studied in the literature, only few studies have been dedicated to deadlock recovery [7,10,17,19], most of them considering only the AND model. One of the reasons for this is that prevention and avoidance of deadlocks provide rules that are designed to ensure that a deadlock will never occur in a distributed computation. As pointed out in [19], deadlock prevention and avoidance strategies are conservative solutions, whereas deadlock detection is optimistic. Whenever the prevention and avoidance techniques are not applied, and deadlocks are detected, they must be broken through some intervention such as aborting one or more processes to break the circular wait condition causing the deadlock, or preempting resources from one or more processes which are in deadlock. In this paper we consider such' a scenario where deadlock was detected in a system and some minimum cost deadlock-breaking set must be found and removed from the system.

Although distributed computations are dynamic, deadlock is a stable property; thus, whenever we refer to G, we mean the wait-for graph that corresponds to a *snapshot* of the distributed computation in the usual sense of a consistent global state [2,8].

The contribution of this work is to provide an analysis of the computational complexity of optimization problems (deletion graph problems) related to deadlock resolution. Given a deadlocked distributed computation represented by G graph, that operates according to a deadlock model $\mathbb{M} \in \{AND, OR, X-OUT-OF-Y, AND-OR\}$, we investigate vertex deletion and arc deletion problems whose goal is to obtain the minimum number of removals in order to turn G free of graph structures that characterize deadlocks. The complexity of such problems depends on the deadlock model that governs the computation as well as the type of structure to be removed. To the best of our knowledge, such a computational complexity mapping considering the particular combination of deletion operations and deadlock models is novel. We remark that deadlock detection can be done in polynomial time for any model $M \in \{AND, OR, X\text{-}OUT\text{-}OF\text{-}Y, AND\text{-}OR\}$ [16].

The results of this paper are NP-completeness results and polynomial algorithms for general and specific graph classes. In particular, we focus on the OR Model. We show that the arc deletion problem on the OR Model can be solved in polynomial time and the vertex deletion problem on the OR Model remains NP-Complete even for graphs with $\Delta(G) = 4$, but is solvable in polynomial time for graphs with $\Delta(G) \leq 3$.

Due to space constraints, some proofs are omitted.

Additional concepts and notations. Let $G = (V, E)$ be a directed graph. For $v_i \in V$, let D_i denote the set of descendants of v_i in G (nodes that are reachable from v_i, including itself). Let $O_i \subseteq D_i$ be the set of immediate descendants of $v_i \in G$ (descendants that are one arc away from v_i). The out-degree (resp., in-degree) of a vertex v is denoted by $deg^+(v)$ (resp., $deg^-(v)$). In addition, $deg^+(G)$ (resp., $deg^-(G)$) denotes the minimum out-degree (resp., in-degree) of a vertex in G. Additional graph concepts and notation can be found in [6].

1.1 Deletion Problems

We denote by $\lambda\text{-}DELETION(M)$ a generic optimization problem for deadlock resolution, where λ indicates the type of deletion operation to be used in order to break all the deadlocks, and $M \in \{AND, OR, X\text{-}OUT\text{-}OF\text{-}Y, AND\text{-}OR\}$ is the deadlock model of the input wait-for graph G.

The types of deletion operation considered in this work are given below:

1. **Arc:** The intervention is given by arc removal. For a given graph G, ARC–DELETION(M) consists of finding the minimum number of arcs to be removed from G in order to make it deadlock-free. The removal of an arc can be viewed as the preemption of a resource.
2. **Vertex:** The intervention is given by vertex removal. For a given graph G, VERTEX–DELETION(M) consists of finding the minimum number of vertices to be removed from G in order to make it deadlock-free. The removal of a vertex can be viewed as the abortion of one process.
3. **Output:** The intervention is given by removing all the out-arcs of a vertex. For a given graph G, OUTPUT–DELETION(M) consists of finding the minimum number of vertices to be transformed into sinks in order to make G deadlock-free. The removal of all the out-arcs of a vertex can be viewed as the immediate transformation of a blocked process into an executable process by preempting all its required resources.

The twelve possible combinatorial problems of the form $\lambda\text{-}DELETION(M)$ are shown in Table 1.

Table 1. Computational complexity questions for λ–Deletion(M).

λ–DELETION(M)				
λ\M	AND	OR	AND-OR	X-Out-Of-Y
Arc	?	?	?	?
Vertex	?	?	?	?
Output	?	?	?	?

2 Computational Complexities

In this section we present complexity results for all the problems listed in Table 1, except VERTEX–DELETION(OR), which receives a special analysis in the next section.

2.1 And Model and Generalizations

To determine if there is a deadlock in a graph G in the AND model, it is necessary and sufficient to check the existence of cycles. Therefore, it is easy to see that VERTEX–DELETION(AND) coincides with *Directed Feedback Vertex Set* and ARC–DELETION(AND) coincides with *Directed Feedback Arc Set*, well-known problems proved to be NP-Hard in [14]. In addition, the following lemma holds:

Lemma 1. OUTPUT–DELETION(AND) *is NP-Hard.*

Proof. Let G be a directed graph and S be a subset of $V(G)$. The transformation into sinks of all vertices in S can make G cycle-free if and only if $G[V \setminus S]$ is acyclic.   □

The AND-OR model is a generalization of the AND and OR models; therefore, every instance of a deadlock resolution problem for either the AND model or the OR model is also an instance for the AND-OR model. Also, the X-Out-Of-Y model also generalizes the AND and OR models. From this observation, it follows that:

Corollary 2. *For* M $\in$ {AND-OR, X-OUT-OF-Y}, *it holds that:*

- VERTEX–DELETION(M) *is NP-hard;*
- ARC–DELETION(M) *is NP-hard;*
- OUTPUT–DELETION(M) *is NP-hard.*

2.2 Or Model

To determine if there is a deadlock in a wait-for graph G on the OR model, it is sufficient and necessary to check the existence of a structure called *knot* [3]. A knot K is a strongly connected component (SCC) of order at least two where no vertex of K has an out-arc pointing to a vertex that is not in K. All the knots of a digraph can be identified in linear time as follows: first, find all the SCCs in linear time by running a depth-first search [9]; next, contract each SCC into a single vertex, obtaining an acyclic digraph H whose sinks represent the knots of G.

Lemma 3

(a) Let K be a knot in a wait-for graph G on the OR model. The minimum number of arcs to be removed in K to make it knot-free is $\delta^+(K)$.

(b) Let G be a wait-for graph on the OR model, and $\mathcal{K} = \{K_1, K_2, ..., K_p\}$ the non-empty set of all the existing knots in G. The minimum number of arcs to be removed from G to make it knot-free is $\sum_{i=1}^{p} \delta^+(K_i)$.

By Lemma 3 we can obtain in linear time a minimum set of arcs whose removal turn a given digraph G knot-free.

Corollary 4. ARC–DELETION(OR) *can be solved in linear time.*

In the preceding subsection we observed that OUTPUT–DELETION(AND) is closely related to VERTEX–DELETION(AND). The next result shows that OUTPUT–DELETION(OR) is closely related to ARC–DELETION(OR).

Lemma 5. *The minimum number of vertices to be transformed into sinks in order to make a graph G knot-free is equal to the number of knots of G.*

Corollary 6. OUTPUT–DELETION(OR) *can be solved in linear time.*

Table 2 presents the computational complexities of the problems presented so far. The complexity analysis of VERTEX–DELETION(OR) is presented in next section.

Table 2. Partial scenario of λ–DELETION(M) complexity.

λ–DELETION(M)				
$\lambda \setminus$ M	AND	OR	AND-OR	X-Out-Of-Y
Arc	NP-H	P	NP-H	NP-H
Vertex	NP-H	?	NP-H	NP-H
Output	NP-H	P	NP-H	NP-H

3 Vertex–Deletion(OR)

In this section we show that VERTEX–DELETION(OR) is NP-hard. In addition, we analyze the problem for some particular graph classes and present an interesting case solvable in polynomial time.

Lemma 7. VERTEX–DELETION(OR) *is NP-hard.*

Proof. Let F be an instance of 3-SAT [12] with n variables and having at most 3 literals per clause. From F we build a graph $G_F = (V, E)$ which will contain a set $S \subseteq V(G)$ such that $|S| = n$ and $G_F[V \setminus S]$ is knot-free if and only if F is satisfiable. The construction of G_F is described below:

1. For each variable x_i in F, create a directed cycle with two vertices ("variable cycle"), Tx_i and Fx_i, in G_F.
2. For each clause C_j in F create a directed cycle with three vertices ("clause cycle"), where each literal of C_j has a corresponding vertex in the cycle.
3. for each vertex v that corresponds to a literal of a clause C_j, create an arc from v to Tx_i if v represents the positive literal x_i, and create an arc from v to Fx_i if v represents the negative literal $\bar{x}_i$.

Figure 1 shows the graph G_F built from an instance F of 3-SAT.

Suppose that F admits a truth assignment A. We can determine a set of vertices S with cardinality n such that $G_F[V \setminus S]$ is knot-free as follows. For each variable of F, select a vertex of G_F according to the assignment A such that the selected vertex represents the opposite value in A, i.e., if the variable x_i is true in A, Fx_i is included in S, otherwise Tx_i is included in S. Since each knot corresponds to a variable cycle, it is easy to see that $G_F[V \setminus S]$ has exactly n sinks. Therefore, since A satisfies F, at least one vertex corresponding to a literal in each clause cycle will have an arc towards a sink (vertex that matches the assignment). Thus $G_F[V \setminus S]$ will be knot-free.

Conversely, suppose that G_F contains a set S with cardinality n such that $G_F[V \setminus S]$ is knot-free. By construction G_F contains n knots, each one associated with a variable of F. Hence, S has exactly one vertex per knot (one of Tx_i, Fx_i). As each cycle of $G_F[V \setminus S]$ corresponds to a clause of F, and $G_F[V \setminus S]$ is knot-free, each cycle of $G_F[V \setminus S]$ has at least one out-arc pointing to a sink. Thus, we can define a truth assignment A for F by setting $x_i = true$ if and only if $Tx_i \in \{V \setminus S\}$. Since at least one vertex corresponding to a literal in each clause cycle will have an arc towards a sink, we conclude that F is satisfiable. □

In general, a wait-for graph on the OR model can be viewed as a conglomerate of several strongly connected components. As observed in Subsect. 2.2, the problems that can be solved in polynomial time have a characteristic in common: it suffices to singly solve every knot in G because no other SCC will become a knot after these removals.

Corollary 8. VERTEX–DELETION(OR) *remains NP-Hard even if G is strongly connected, i.e., G is a single knot.*

Now we consider properties of the underlying undirected graph of G.

Since one of the most used architectures in distributed computation follows the user/server paradigm, a intuitively interesting graph class for distributed computation purposes are bipartite graphs. Planar graphs can also be interesting if physical settings must be considered; finally, bounded-degree graphs are very common in practice.

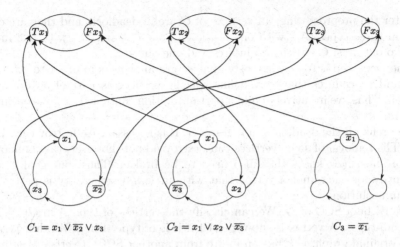

$$C_1 = x_1 \vee \overline{x_2} \vee x_3 \qquad C_2 = x_1 \vee x_2 \vee \overline{x_3} \qquad C_3 = \overline{x_1}$$

Fig. 1. Graph G_F built from $F = (x_1 \vee \overline{x_2} \vee x_3) \wedge (x_1 \vee x_2 \vee \overline{x_3}) \wedge (\overline{x_1})$.

Theorem 9. VERTEX–DELETION(OR) *remains NP-Hard even when the underlying undirected graph of G is bipartite, planar, and with maximum degree 4.*

Proof. Let F be an instance of Planar 3-SAT where each variable has at most three occurrences (at least one positive and at least one negative). This problem is known to be NP-complete [18]. Clearly, G_F as built in Lemma 7 will have maximum degree 4. The bipartition of G_F can be obtained by replacing each clause cycle (with size three) by another directed cycle of size six, such that no pair of vertices representing positive (resp. negative) literals are adjacent; moreover, in such cycles, if a vertex represents a positive literal and another vertex a negative literal then they are at an odd distance. Finally, the planarity property can be obtained by using the planar embedding of the variable-clause bipartite graph associated to F to coordinate the ideal position of the vertices inside the clause cycles as well as the proper position of variable cycles and clause cycles in order to obtain a planar embedding of G_F. □

3.1 Subcubic Graphs

Since VERTEX–DELETION(OR) remains NP-hard for graphs with $\Delta(G) = 4$, and is trivial for graphs with $\Delta(G) \leq 2$, an interesting question is to study the complexity of VERTEX–DELETION(OR) when the underlying undirected graph of G has maximum degree three, i.e., is subcubic.

To answer this question the first step is to phase out unnecessary vertices, i.e., vertices that will never belong to any solution, such as sources and sinks.

Preprocessing. Successively remove all source and sink vertices from G, until removals can no longer be applied. Using depth-first search the preprocessing can be done in $O(n + m)$ time.

After the preprocessing, all vertices of G are in deadlock, and they are classified in three types: **A** - with one in-arc and one out-arc; **B** - with one in-arc and two out-arcs; **C** - with two in-arcs and one out-arc.

The next step is to continuously analyse graph aspects in order to establish rules and procedures that may define specific vertices as part of an optimum solution. Thus, we iteratively build a partial solution contained in some optimum solution.

To break all the deadlocks in a graph G, it is necessary to destroy each knot in G. The removal of some vertices can destroy a knot; however, these removals may produce new knots that also need to be broken. Thus, our goal now is to identify for each knot a vertex that without loss of generality is part of an optimum solution.

Let W be a SCC of G. We can classify the vertices of type A in $G[W]$ into three sub-types. A vertex is sub-type $A.1$ if it is of type A in $G[W]$, but type C in the original graph, i.e., has an in-arc from another SCC; a vertex is subtype $A.2$ if it is the same type in both $G[W]$ and G; finally, a vertex is subtype $A.3$ when it is type A in $G[C]$ but type B in G. Note that in a knot there will never be vertices of type $A.3$. It is worth noting that in a subcubic graph every knot vertex has at most one external neighbor (in-neighbor).

The following lemma presents an interesting relation between vertices of type B and C.

Lemma 10. *Let Q be a strongly connected subcubic graph. The number of vertices type B in Q is equal to the number of vertices type C.*

At this point, we can identify some vertices of an optimal solution.

Theorem 11. *Let G be a subcubic graph and Q be a knot in G. If Q contains a vertex type B or C or $A.2$ then G has an optimal solution S for* VERTEX–DELETION(OR) *which contains exactly one vertex $v_i \in V(Q)$, and such a vertex can be found in linear time.*

Proof

(a) Since Q is strongly connected, any sink arising from a vertex removal will break Q. Since a vertex type B has no neighbor outside Q (otherwise Q would not be a knot) removing it will never create a knot in $G \backslash Q$. Thus, given a vertex v_i type B in Q, the removal of v_i will not turn its in-neighbor w_i into a sink only if w_i is also type B. In this case we repeat the same process for w_i. From Lemma 10, eventually we will find a vertex v_j type B whose in-neighbor w_j is not type B, otherwise, Q would be composed only by vertices type B.

(b) The proof for the case where Q contains a vertex type C follows direct from Lemma 10 and (a).

(c) Suppose that v_i is a vertex type $A.2$ in Q. Since such a vertex has no neighbours outside Q, removing it will never create a knot in $G \backslash Q$. In addition, the removal of v_i will not break Q only if its in-neighbor w_i is type B. In this case we can apply (a). $\square$

Corollary 12. *Let G be a subcubic graph. If G has no knot Q having a vertex v_i such that (i) $Q-v_i$ is knot-free, (ii) all the knots of $G-v_i$ are knots of G, and (iii) v_i has no in-neighbor outside Q, then any knot of G is a cycle composed only by vertices type A.1.*

Proof. By Theorem 11, a knot Q of G contains no vertex of type B, C, or $A.2$. Clearly, it cannot have any vertices type $A.3$. Thus, Q is a cycle of vertices type $A.1$.   □

Now, we can determine lower and upper bounds for the problem.

Lemma 13. *Let S be an optimal solution of a subcubic instance G of* VERTEX–DELETION(OR). *Then it holds that $k \leq |S| \leq 2k$, where k is the number of knots of G.*

Corollary 14. VERTEX–DELETION(OR) *on subcubic graphs can be 2-approximated in linear time.*

Proof. Follows from Theorem 11 and Lemma 13.   □

A Polynomial Time Algorithm. In order to obtain an optimum solution in polynomial time, there are some significant considerations to be made regarding the remaining graph.

Lemma 15. *Let G be a subcubic instance of* VERTEX–DELETION(OR). *Let $\mathcal{C} = \{C^1, C^2, \ldots, C^j\}$ be a set of non-knot SCCs of G, where for each $C_i \in \mathcal{C}$ there is a directed path from C^i to C^{i+1} and from C^j to Q. Then:*

(a) *No vertex in $C^1, C^2, \ldots, C^{j-1}$ is part of an optimal solution.*
(b) *If C^j has two or more out-arcs pointing to a knot Q or there is some $i < j$ for which C^i is directly connected to Q then there is an optimal solution S such that $V(Q) \cap S = \{v_i\}$, and such a vertex can be found in linear time.*

At this point, we are able to apply the first steps of our algorithm.

First steps of the algorithm. (i) Remove any SCC that is pointing to another non-knot SCC; (ii) If a non-knot SCC has at least two arcs pointing to a same knot then remove all such arcs but one; (iii) For each knot Q having vertices of type B, C, or $A.2$, find a vertex v_i that, without loss of generality, is in an optimal solution and remove it; (iv) Remove all vertices that are no longer in deadlock.

The correctness of the above steps follows from Theorem 11 and Lemma 15. Observe also that such routines can be performed in $O(n + m)$ time.

Now, consider the graph G obtained after applying the above steps. The knots of G are directed cycles composed by vertices of type $A.1$ (see Corollary 12), and each non-knot SCC of G is at distance one of a knot. At this point, any vertex v removed from a knot Q will break it, but potentially creates a new knot from the SCC W that has an out-arc to v. However, such new knot W will have a vertex of type $A.2$; therefore, W can be solved by the removal of another vertex $w \in V(W)$, called *solver vertex*.

Our final step is to minimize the number of solver vertices that actually need to be removed in order to make the graph knot-free. To achieve this purpose, we will consider the bipartite graph $B = (K \cup C, E)$, where K is the set of vertices representing a contracted knot, and C is the set of vertices representing contracted non-knot SCCs. Note that each arc from a vertex c_j in C to another vertex k_i in K represents a connection from a vertex w of the SCC represented by c_j to a vertex v of the knot represented by k_i, and indicates that w becomes a vertex type $A.2$ after the removal of v, which guarantees the existence of a solver vertex, that may or may not be used. Therefore, from B we seek a set M' of arcs such that:

1. Each vertex in K is adjacent to at most one arc of M'. (This arc indicates the vertex of the knot to be removed.)
2. Each vertex in C has at least one arc that does not belong to M'. (This arc indicates a path to a sink, a broken knot in K; thus, the SCC is deadlocked without removing internal vertices.)
3. M' is maximum.

From the set M' we can obtain an optimal solution S such that $G[V \setminus S]$ is knot-free (where G is prior to the contraction). In fact, M' indicates the maximum number of knots that can be broken without generating new knots, as well as the number of solver vertices needed. Solution S can be built as follows: for each vertex k_i in K, adjacent to M', include in S the associated knot vertex in G; for every vertex $k_i \in K$ such that k_i is not adjacent to M', choose an arc e and include in S the knot vertex of G associated with e; then, for the SCC C of G indicated by the arc e, include in S a solver vertex.

The set M' can be obtained by using the concept of (f, g)-*semi-matching*. The (f, g)-semi-matching is a generalization of semi-matching presented in [5]. Let $f : K \to \mathbb{N}$ and $g : C \to \mathbb{N}$ be functions. An (f, g)-semi-matching in a bipartite graph $G = (K \cup C, E)$ is a set of arcs $M \subseteq E$ such that each vertex $k_i \in K$ is incident with at most $f(k_i)$ arcs of M, and each vertex $c_j \in C$ is incident with at most $g(c_j)$ arcs of M. In fact, M' can be found by an $(1, g)$-semi-matching where $g(v) = d(v) - 1$.

Lemma 16 [5,11,15]. *Given a bipartite graph $G = (K \cup C, E)$ and two functions $f : K \to \mathbb{N}$ and $g : C \to \mathbb{N}$, finding a maximum (f, g)-semi-matching of G can be done in polynomial time, and a maximum $(1, g)$-semi-matching of G can be found in $O(m\sqrt{n})$ time.*

These algorithms use similar ideas to those used in the well known Hopcroft-Karp algorithm [13].

At this point, we have all elements to answer the raised question about the complexity of VERTEX-DELETION(OR).

Theorem 17. VERTEX-DELETION(OR) *restricted to subcubic graphs is solvable in $O(m\sqrt{n})$ time.*

4   Conclusions

We define λ–DELETION($\mathbb{M}$) as a generic optimization problem for deadlock resolution, where λ indicates the type of deletion operation to be used in order to break all the deadlocks, and $\mathbb{M} \in$ {AND, OR, X-OUT-OF-Y, AND-OR} is the deadlock model of the input wait-for graph G. VERTEX–DELETION(AND) and ARC–DELETION(AND) are equivalent to Direct Feedback Vertex Set and Direct Feedback Arc Set, respectively. We proved that ARC–DELETION(OR) and OUTPUT–DELETION(OR) are solvable in polynomial time. In addition, VERTEX–DELETION(OR) was shown to be NP-complete. Such results are summarized in the following Table 3.

Table 3. Computational complexity of λ–DELETION($\mathbb{M}$).

λ–DELETION($\mathbb{M}$)				
$\lambda \setminus \mathbb{M}$	AND	OR	AND-OR	X-Out-Of-Y
Arc	NP-H	P	NP-H	NP-H
Vertex	NP-H	NP-H	NP-H	NP-H
Output	NP-H	P	NP-H	NP-H

A study of the complexity of VERTEX–DELETION(OR) in different graph classes was also done. We proved that the problem remains NP-hard even for strongly connected graphs and planar bipartite graphs with maximum degree four. Furthermore, we proved that for graphs with maximum degree three the problem can be solved in polynomial time. Thus we have the following Table 4:

Table 4. Complexity of VERTEX–DELETION(OR) for some graph classes.

VERTEX–DELETION(OR)	
Instance	Complexity
Weakly connected	NP-Hard
Strongly connected	NP-Hard
Planar, bipartite, $\Delta(G) \geq 4$ and $\Delta(G)^+ = 2$	NP-Hard
$\Delta(G) = 3$	Polynomial

Acknowledgments. This research was partially supported by the Brazilian National Council for Scientific and Technological Development (CNPq), the Brazilian National Council for the Improvement of Higher Education (CAPES) and FAPERJ.

References

1. Atreya, R., Mittal, N., Kshemkalyani, A.D., Garg, V.K., Singhal, M.: Efficient detection of a locally stable predicate in a distributed system. J. Parallel Distrib. Comput. **67**(4), 369–385 (2007)
2. Barbosa, V.C.: An Introduction to Distributed Algorithms. The MIT Press, Cambridge (1996)
3. Barbosa, V.C.: The combinatorics of resource sharing. In: Corrêa, R., Dutra, I., Fiallos, M., Gomes, F. (eds.) Models for Parallel and Distributed Computation, pp. 27–52. Kluwer, Dordrecht (2002)
4. Barbosa, V.C., Benevides, M.R.: A graph-theoretic characterization of AND-OR deadlocks. Technical report COPPE-ES-472/98, Federal University of Rio de Janeiro, Rio de Janeiro, Brazil (1998)
5. Bokal, D., Brešar, B., Jerebic, J.: A generalization of hungarian method and hall's theorem with applications in wireless sensor networks. Discret. Appl. Math. **160**(4), 460–470 (2012)
6. Bondy, J.A., Murty, U.S.R.: Graph Theory with Applications, vol. 290. Macmilan, London (1976)
7. Chahar, P., Dalal, S.: Deadlock resolution techniques: an overview. Int. J. Sci. Res. Publ. **3**(7), 1–5 (2013)
8. Chandy, K.M., Lamport, L.: Distributed snapshots: determining global states of distributed systems. ACM Trans. Comput. Syst. **3**, 63–75 (1985)
9. Cormen, T.H., Leiserson, C.E., Rivest, R.L., Stein, C.: Introduction to Algorithms. MIT press, Cambridge (2009)
10. de Mendívil, J.G., Fariña, F., Garitagotia, J.R., Alastruey, C.F., Bernabeu-Auban, J.M.: A distributed deadlock resolution algorithm for the and model. IEEE Trans. Parallel Distrib. Syst. **10**(5), 433–447 (1999)
11. Galčík, F., Katrenič, J., Semanišin, G.: On computing an optimal semi-matching. In: Kolman, P., Kratochvíl, J. (eds.) WG 2011. LNCS, vol. 6986, pp. 250–261. Springer, Heidelberg (2011). doi:10.1007/978-3-642-25870-1_23
12. Gary, M.R., Johnson, D.S.: Computers and intractability: a guide to the theory of NP-completeness (1979)
13. Hopcroft, J.E., Karp, R.M.: An $n^{5/2}$ algorithm for maximum matchings in bipartite graphs. SIAM J. Comput. **2**(4), 225–231 (1973)
14. Karp, R.: Reducibility among combinatorial problems. In: Miller, R., Thatcher, J., Bohlinger, J. (eds.) Complexity of Computer Computations. The IBM Research Symposia Series, pp. 85–103. Springer, New York (1972)
15. Katrenic, J., Semanisin, G.: A generalization of hopcroft-karp algorithm for semi-matchings and covers in bipartite graphs. arXiv:1103.1091 (2011)
16. Kshemkalyani, A.D., Singhal, M.: Distributed Computing: Principles, Algorithms, and Systems. Cambridge University Press, Cambridge (2011)
17. Leung, J.Y.-T., Lai, E.K.: On minimum cost recovery from system deadlock. IEEE Trans. Comput. **9**(C–28), 671–677 (1979)
18. Penso, L.D., Protti, F., Rautenbach, D., dos Santos Souza, U.: Complexity analysis of P3-convexity problems on bounded-degree and planar graphs. Theoret. Comput. Sci. **607**, 83–95 (2015)
19. Terekhov, I., Camp, T.: Time efficient deadlock resolution algorithms. Inf. Process. Lett. **69**(3), 149–154 (1999)

Space-Efficient Algorithms for Maximum Cardinality Search, Stack BFS, Queue BFS and Applications

Sankardeep Chakraborty[1](✉) and Srinivasa Rao Satti[2]

[1] The Institute of Mathematical Sciences,
CIT Campus, Taramani, Chennai 600113, India
sankardeep@imsc.res.in

[2] Seoul National University, 1 Gwanak-ro, Gwanak-gu, Seoul, South Korea
ssrao@cse.snu.ac.kr

Abstract. Following the recent trends of designing space efficient algorithms for fundamental algorithmic graph problems, we present several time-space tradeoffs for performing Maximum Cardinality Search (MCS), Stack Breadth First Search (Stack BFS), and Queue Breadth First Search (Queue BFS) on a given input graph. As applications of these results, we also provide space-efficient implementations for testing if a given undirected graph is chordal, reporting an independent set, and a proper coloring of a given chordal graph among others. Finally, we also show how two other seemingly different graph problems and their algorithms have surprising connection with MCS with respect to designing space efficient algorithms.

1 Introduction

Space efficient algorithms are becoming increasingly important owing to their applications in the presence of rapid growth of "big data". Another reason for the importance of space efficient algorithms is the proliferation of specialized handheld devices and embedded systems that have a limited supply of memory. As a consequence, algorithms that are oblivious to space constraint are not desired in such scenario. Even if mobile devices and embedded systems are designed with large supply of memory, it might be useful to restrict the number of write operations. For example, on flash memory, writing is a costly operation in terms of speed, and it also reduces the reliability and longevity of the memory. Keeping all these constraints in mind, it makes sense to consider algorithms that do not modify the input and use only a limited amount of work space. Such computational model has been proposed in algorithmic literature to study space efficient algorithms, and is known as the read-only memory (ROM) model. Following the recent trend, in this article, we focus on the space requirement for implementing some fundamental graph algorithms in such settings.

There is already a rich history of designing space efficient algorithms in the read-only memory model. In computational complexity theory, L is the complexity class [1] containing decision problems that can be solved by a deterministic

© Springer International Publishing AG 2017
Y. Cao and J. Chen (Eds.): COCOON 2017, LNCS 10392, pp. 87–98, 2017.
DOI: 10.1007/978-3-319-62389-4_8

Turing machine using only logarithmic amount of work space for computation. There are several important algorithmic results for this class, the most celebrated being Reingold's method [20] for checking reachability between two vertices in an undirected graph. Barnes et al. [6] gave a slightly sublinear space (using $n/2^{\Theta(\sqrt{\lg n})}$ bits[1]) algorithm for checking s-t connectivity in a directed graph with polynomial running time. Other than these fundamental graph theoretical problems, researchers have also focused on designing space-efficient algorithms for the more classical selection and sorting problems [18], and problems in computational geometry [2,11]. Most of these graph algorithms using small space i.e., sublinear bits, often take time that is some polynomial of very high degree. For example, to the best of our knowledge, the exact running time of Reingold's algorithm [20] for undirected s-t connectivity is not analysed, yet we know it admits a large polynomial running time. In fact this phenomenon is not unusual, as Edmonds et al. [12] have shown in the so-called NNJAG model that only a slightly sublinear working-space bound is possible for an algorithm that solves the reachability problem when required to run in polynomial time. Tompa [24] showed a surprising result that for directed s-t connectivity, if the number of bits available is $o(n)$ then some natural algorithmic approaches to the problem require superpolynomial time. Motivated by these impossibility results from complexity theory and inspired by the practical applications of these fundamental graph algorithms, recently there has been a surge of interest in improving the space complexity of the fundamental graph algorithms without paying too much penalty in the running time i.e., reducing the working space of the classical graph algorithms to $O(n)$ bits with little or no penalty in running time. Generally these classical linear time graph algorithms take $O(n \lg n)$ bits.

Starting with the paper of Asano et al. [3] who showed how one can implement DFS using $O(n)$ bits, improving on the naïve $O(n \lg n)$-bit implementation, the recent series of papers [4,5,9,10] present space-efficient algorithms for few other basic graph problems: namely BFS, topological sort, (strongly) connected components, sparse spanning biconnected subgraph, biconnectivity, st-numbering, dynamic programming in bounded treewidth graphs among others. We add to this small yet growing body of space-efficient algorithm design literature by providing such algorithms for a few more fundamental and classical graph problems, namely performing maximum cardinality search, using this to recognize chordal graphs and solve some combinatorial problems like coloring, vertex cover, etc. in chordal graphs, and computing queue BFS and stack BFS among others.

Model of Computation. We use the same model of computation that is used in the recent research that focused on space-efficient graph algorithms [3–5,9,10]. More specifically, we assume that the input graph G with n vertices and m edges is given in a read-only memory. The algorithm produces the output on a separate write-only memory which cannot be read or rewritten again. In addition to the input and the output media, a limited amount of random-access workspace is available. The data on this workspace is manipulated at word level as in the standard word RAM model with word size $w = \Theta(\lg n)$. We count space in terms

---

[1] We use lg to denote logarithm to the base 2.

of the number of bits in the workspace used by the algorithms. We assume that the input undirected (directed) graph $G = (V, E)$ is represented using (in and out) *adjacency array*, i.e., given a vertex v and an integer k, we can access the k-th (in/out) neighbor of vertex v in constant time. This representation was used in [4,9,10] recently to design various other space efficient graph algorithms.

Organization of the paper. In Sect. 2 we introduce MCS and provide its implementation as described in [23]. This is followed by various space efficient implementations of MCS and the proofs are presented in Sect. 3. Next we show, in Sect. 4, some applications of our MCS algorithm by providing space efficient procedures to recognize chordal graphs, find Independent set and proper coloring, etc. in chordal graphs. Section 5 describes space efficient algorithms for two other popular graph search methods which are known as Stack BFS and Queue BFS in the literature.

2 Maximum Cardinality Search (MCS)

A widely used graph search method which is a restriction of breadth-first search (BFS) is lexicographic BFS (Lex-BFS), introduced by Rose et al. [21] under the name Lex-P. They used Lex-BFS to find a perfect elimination ordering (PEO) of the vertices of a graph G if G is chordal. A perfect elimination ordering in G is an ordering of the vertices such that, for each vertex v, v and the neighbors of v that occur after v in the order form a clique. Fulkerson et al. [13] showed that a graph is chordal if and only if it has a perfect elimination ordering. Thus to recognize a chordal graph, we can run the Lex-BFS algorithm, and test whether the resulting order is a perfect elimination order. Rose et al. [21] showed both the tasks of performing Lex-BFS of G and testing if the resulting order is PEO can be done in $O(m + n)$ time. Even though the Lex-BFS runs in linear time, its implementation is a bit involved, and it takes $O(m + n)$ words of space [21]. Later Tarjan, in an unpublished note [22], derived another simpler and alternate graph search method for finding a PEO of chordal graphs, known as Maximum Cardinality Search (MCS). Tarjan and Yannakakis [23] presented MCS and its applications to recognize chordal graphs and test acyclicity of hypergraphs, etc. We refer the interested readers to the excellent text of Golumbic [14] for thorough coverage of chordal graph recognition, MCS, Lex-BFS and many other related topics. In what follows, we provide space efficient algorithms for MCS and its many applications in chordal graphs.

2.1 The MCS Algorithm and Its Implementation

We start by briefly describing the MCS algorithm and its implementation as provided in [23]. The output of the MCS algorithm is a numbering of the vertices from 1 to n. During the execution of the algorithm, vertices are chosen by choosing an unnumbered vertex that is adjacent to the most numbered vertices.

Tarjan et al. [23] gave a $O(m+n)$ time implementation of MCS. Even though they did not explicitly analyse the space requirement of their algorithm, we show that it takes $O(n)$ words of space. This exact space bound is particularly interesting and worth mentioning since many of the subsequent papers actually cite this version of MCS result saying, the implementation of Tarjan et al. [23] takes $O(m + n)$ time and words of space. For example see Theorem 7.1 of [16] and Theorem 5.2 of [8]. In the rest of this section, we briefly describe the original algorithm of Tarjan et al. [23] and its time and space complexities.

The MCS algorithm of Tarjan et al. maintains an array of sets $set[i]$ for $0 \le i \le n - 1$ where $set[i]$ stores all unnumbered vertices adjacent to exactly i numbered vertices. So, at the beginning all the vertices belong to $set[0]$. They also maintain the largest index j such that $set[j]$ is non-empty. To implement an iteration of the MCS algorithm, they remove a vertex v from $set[j]$ and number it. For each unnumbered vertex w adjacent to v, w is moved from the set containing it, say $set[i]$, to $set[i + 1]$. If there is a new entry in $(j + 1)$, we move to $set[j + 1]$ and repeat the same. Otherwise when $set[j]$ becomes empty, they repeatedly decrement j till a non-empty set is found and in this set we repeat the same procedure. In order to delete easily, they implement each set as a doubly-linked list. In addition, for every vertex, they store the index of the set containing it. This completes the description of the implementation level details of MCS as provided by Tarjan et al. [23].

Now, observe that as it is described in above implementation, when the vertex w needs to be moved from $set[i]$ to $set[i + 1]$, we just know the index of $set[i]$ that w belongs to, but not w's location inside $set[i]$. To get overall linear time, we cannot afford to perform a linear search for w in $set[i]$ as this might be a costly operation. A simple way to fix this is to, instead of storing for every vertex v the set index where it belongs to, store a pair (i, j) if a vertex v belongs to the list $set[i]$ and j is the pointer to v's location inside $set[i]$. Then we can directly access v and move it to $set[i+1]$ from $set[i]$ in constant time. This concludes the description of our modified implementation. It is easy to see that every vertex appears only once in any of these sets, so array set takes at most $2n$ words in worst case. Clearly the running time is $O(m + n)$ and space required is $O(n)$ words. In the next section we provide space efficient implementations for MCS.

3   Space Efficient Implementations of MCS

Algorithm 1: Using $n + O(\lg n)$ bits. Here we show using just $n + O(\lg n)$ bits, albeit with increased time, we can perform MCS. Towards that we maintain a bit vector B (initialized with all 0 entries) of size n bits where the $B[i]$-th entry is 1 if and only if the vertex v_i has already been numbered. The algorithm works as follows: at each step it scans the whole B array to find all the zero entries and for each one of them, it goes over the adjacency array to find out how many of its neighbours are already marked '1'. Then the vertex v which has the maximum number of numbered neighbors (ties are broken arbitrarily) is marked '1' in B. We repeat this step until all the vertices are marked. This procedure uses $n + O(\lg n)$ bits. At each step, the algorithm spends $O(m)$ time to find a vertex to number, and this happens exactly n times. Hence, the running time is $O(mn)$.

Algorithm 2: Using $O(n)$ bits. By increasing the space bound by a constant factor, we can design a faster algorithm, by using similar ideas as in Tarjan et al.'s [23] algorithm with a few changes. We define the *label* of an unnumbered vertex (at any instance of the algorithm) as its number of numbered neighbors. The main idea is to maintain a doubly linked list, call it L, of size at most $n/\lg n$ at any point during the execution of the algorithm. Each element in L stores a label k and a pointer to a *sublist*. The sublist labeled k stores a set of vertices with label k, and is itself maintained as a doubly linked list. The sublists are in L are stored in the increasing order of their labels. Moreover, the sum of the sizes of all the sublists, at any time, is at most $n/\lg n$. Also, we maintain all the vertices that are currently stored in L in a balanced binary search tree, T, and store pointers from the nodes of T to the corresponding vertices in L.

At the beginning of the algorithm, we add $n/\lg n$ vertices into the tree T and the same vertices also into the list L in a sublist labeled 0 (with pointers to the corresponding vertices from T to L). As before, we maintain a bit vector B of length n to keep track of the numbered vertices. The i-th step of the algorithm, for $1 \le i \le n$, proceeds as follows. We select the first (arbitrary) element of the sublist with the largest label in L. Let v be the vertex stored in this element. Then, we first number the vertex v with i and delete v from both L and T. We then go through each unnumbered neighbor w of v and compute the label k of w. If k is greater than the label of the sublist with the smallest label, then we add w to L in the sublist labeled k; delete the first (arbitrary) element from the sublist with the smallest label, if necessary (to maintain the invariant that L has at most $n/\lg n$ elements); and also add (and delete) the corresponding vertices to the tree T. (Note that we can also add w to L if k is equal to the label of the sublist with the smallest label as long as the number of elements in L is at most $n/\lg n$; but this does not change the worst-case running time of the algorithm.) Also, if w is already stored in L (in the sublist labeled $k - 1$), then we move w from the sublist labeled $k - 1$ to the sublist labeled k.

Note that, after a while, due to deletion of vertices, the list L may become empty. At that time, we refill the list L with $n/\lg n$ unnumbered vertices with the highest labels (or refill with all the remaining vertices if there are fewer than $n/\lg n$ unnumbered vertices). This is done by scanning the bit vector B from left to right, and for each unnumbered vertex v, computing its label. The first $n/\lg n$ unnumbered vertices are simply inserted into the appropriate sublists in L. For the remaining unnumbered vertices, if the label is greater than the label of the smallest labeled sublist currently in L, then we insert the new vertex into the appropriate sublist, and delete the first vertex from the sublist with the smallest label. The cost of this refilling is dominated by the cost of computing the labels of the vertices which is $O(m)$. After constructing the list L, we insert each of the vertices in L into an initially empty binary search tree T and add pointers from the nodes in T to the corresponding elements in L, which takes $O(n)$ time. The refilling of the list L happens at most $O(\lg n)$ times since at least $n/\lg n$ vertices will be numbered after each refilling, and hence over the full execution of the algorithm it takes $O(m \lg n)$ time. Computing the label of a vertex v takes

$O(d_v)$ time, where d_v is the degree of v. During the execution of the algorithm, we need to compute the label of a vertex v at most $O(d_v)$ times (every time one of its neighbors is numbered). Thus, running time of computing the labels of vertices is bounded by, $O(\sum_{v \in V} d_v^2) = O(m^2/n)$. Finally, moving an element from one sublist to another takes $O(\lg n)$ time since we need to search for it in the binary search tree first, before moving it. Since a vertex v is moved at most d_v times, this step contributes $O(\sum_{v \in V} d_v \lg n) = O(m \lg n)$ time to the total running time. Thus overall the algorithm takes $O(m^2/n + m \lg n)$ time.

Algorithm 3: Using $O(n \lg \frac{m}{n})$ bits. We show that using $O(n \lg \frac{m}{n})$ bits we can design a significantly faster algorithm. Note that for sparse graphs ($m = O(n)$), this space is only $O(n)$ bits. We first scan the adjacency list of each vertex and construct a bitvector D as follows: starting with an empty bitvector D, for $1 \le i \le n$, if d_i is the degree of the vertex v_i, then we append the string $0^{\lceil \lg d_i \rceil - 1} 1$ to D. The length of D is $\sum_{i=1}^{n} \lceil \lg d_i \rceil$, which is bounded by $O(n \lg(m/n))$. We also construct auxiliary structures to support *select* queries on D in constant time [17]. Like the previous algorithm, we also maintain the current top $O(n/\lg n)$ values in the list of doubly linked list L along with all other auxiliary information. Finally, we maintain in a bitmap B marking all the already numbered and output vertices. Overall space usage is $O(n \lg \frac{m}{n})$ bits.

The algorithm is essentially same as earlier. The only difference is that, using the structure D, we can compute and update the labels of vertices in $O(1)$ time (instead of $O(d_v)$ time as in the earlier algorithm). Thus the running time of computing the labels of the vertices is now bounded by $O(m)$, and the rest of the computations is same as earlier. Hence the overall running time is $O(m \lg n)$.

The three algorithms described above can be summarized as follows.

Theorem 1. *Given a graph G, we can obtain an MCS ordering of G in (a) $O(mn)$ time using $n + O(\lg n)$ bits, or (b) $O(m^2/n + m \lg n)$ time using $O(n)$ bits, or (c) $O(m \lg n)$ time using $O(n \lg(m/n))$ bits.*

4   Applications of MCS

In this section, we provide several applications of our MCS algorithms presented in the previous section. We start with some surprising connection of our MCS algorithm with a few other totally unrelated graph problems and their algorithms with respect to designing space efficient algorithms. Next we show how to use MCS to provide space efficient solutions for solving some combinatorial problems in chordal graphs, and also recognition of chordal graphs. We start by describing the connection with MCS first.

4.1   Connection with Other Problems

In this section, we discuss two other seemingly different problems and their algorithms which have surprising connection with MCS with respect to designing space efficient algorithms. First is the problem of topologically sorting the vertices of a directed acyclic graph, and the second is that of finding the degeneracy of an undirected graph. The similarity of these two problems with MCS comes from the fact that the linear time algorithms for all three of these problems have a very natural greedy strategy. We start by briefly explaining the linear time greedy algorithms for these two problems.

One of the algorithms for topological sort works by maintaining the in-degree count of every vertex, and at each step, it deletes a vertex of in-degree zero and all of its outgoing edges. The order in which the vertices are deleted gives the topological sorted order. To efficiently implement the algorithm, all the vertices currently having in-degree zero are stored in a queue. This algorithm can also test if the given graph is acyclic or not at the same time. If it is acyclic, it produces a topological sort. Note the similairy of this algorithm with that of MCS. It is not hard to see that each of the solutions for MCS explained before could be made to work for topological sort with similar resource bounds.

The degeneracy of a graph G is the smallest value d such that every non-empty subgraph of G contains a vertex of degree at most d. Such graphs have a degeneracy ordering, i.e., an ordering in which each vertex has d or fewer neighbors that come later in the ordering. Degeneracy and degeneracy ordering, can be computed by a simple greedy strategy of repeatedly removing a vertex with smallest degree (and its incident edges) from the graph until it is empty. The degeneracy is the maximum of the degrees of the vertices at the time they are removed from the graph, and the degeneracy ordering is the order in which vertices are removed from the graph. The linear time implementation of this works almost in the same way as the MCS algorithm [7]. One can implement this algorithm space efficiently using similar ideas as that of MCS. We omit the relatively easy details. It would be intersting to find other problems with similar flavour. We want to conclude this section by remarking that, for any greedy algorithm of this flavour we can use similar technique to design space efficient implementation. We summarize our results in the theorem below.

Theorem 2. *Given a directed graph G, we can report whether G is acyclic or not, and if so, we can produce a topological sort ordering in (a) $O(mn)$ time using $n + O(\lg n)$ bits, or (b) $O(m^2/n + m\lg n)$ time using $O(n)$ bits, or (c) $O(m\lg n)$ time using $O(n\lg(m/n))$ bits. Using the same running time and space bounds, we can also test if an undirected graph G is d-degenerate, and if so, we can output a d-degenerate ordering of G.*

4.2   Finding Independent Set, Vertex Cover and Proper Coloring

In this section, we show that using Theorem 1 how one can solve some combinatorial problems on chordal graphs space efficiently. Tarjan et al. [23] showed that, if G is a chordal graph and σ is an MCS ordering of G, then the reverse of σ is a PEO of G. More specifically, given the graph G, and a vertex ordering σ of G, we define the following:

- The edge directions implied by σ is obtained by directing (v_i, v_j) as $v_i \to v_j$ if $i < j$ and $v_j \to v_i$ if $i > j$.
- If $v_i \to v_j$ is an edge implied by the order, then v_i is the predecessor of v_j and v_j is a successor of v_i.

Thus the theorem of Tarjan et al. [23] says that, the predecessor set of every vertex (i.e., set containing all the predecessors) in σ forms a clique, or equivalently in the reverse of σ, for each vertex v, v and its successor set form a clique if and only if G is chordal. To solve some combinatorial problems in chordal graphs, we need the reverse PEO whereas for some applications we need the PEO. For all applications where we need a PEO, we spend extra time for reversing the list produced by our MCS algorithm as, due to space restriction, we cannot store the PEO and reverse it in linear time. This seems to be a fundamental bottleneck while designing space efficient algorithms. We start with the problem of finding a maximum independent set (MIS), and for this, a simple greedy strategy works [14]. Given a reverse PEO of the input chordal graph G, the algorithm scans the vertices in order, and for every vertex v_i, it adds v_i to the solution set I if none of its predecessors has been added to I already. Note that using a bitmap S of size n bits, where $S[i]$ is set if the vertex v_i belongs to I, on top of the structures of Theorem 1, we can easily implement this to find an MIS with no extra time. Also the complement of the set S gives us a minimum vertex cover for G. Thus,

Theorem 3. *Given a chordal graph G, we can output a maximum independent set and/or a minimum vertex cover of G in (a) $O(mn)$ time using $2n + O(\lg n)$ bits, or (b) $O(m^2/n + m \lg n)$ time using $O(n)$ bits, or (c) $O(m + n \lg n)$ time using $O(n \lg(m/n))$ bits.*

In what follows, we discuss a space efficient implementation of finding a proper coloring of a chordal graph G. It is known that the natural greedy algorithm [14] for coloring yields the optimal number of colors iff the vertex order is a PEO. The algorithm works as follows: given a vertex order, it scans the vertices in order, and colors each vertex with the smallest color not used among its predecessors. Note that, we need a PEO here, but MCS produces a reverse one, thus we first need to reverse the list produced by the MCS algorithm. Also, it is easy to see that this coloring scheme assigns, for any vertex v, the color at most $max_i\{indeg(v_i) + 1\}$ where $indeg(v_i)$ refers to the neighbors of v_i which appeared before in PEO and have already been colored. Suppose we store the explicit colors for each vertex in an array B, then the length of B is $\sum_{i=1}^{n}\lceil \lg d_i + 1\rceil = O(n \lg(m/n))$. We construct auxiliary structures to support *select* queries on B in constant time [17]. Suppose we have $O(n \lg(m/n))$ bits at our disposal, then we run our MCS algorithm and store the last chunk of $O(n/\lg_{m/n} n)$ vertices in a queue Q. Now dequeue vertices from Q one by one and run the greedy coloring algorithm while storing the colors in B array explicitly, and continue. Once Q becomes empty, run the MCS algorithm to generate the previous chunk and store them in Q, and repeat the greedy coloring algorithm afterwards. This process is repeated at most $O(\lg_{m/n} n)$ times, and each time we run the MCS algorithm followed by the greedy coloring scheme, hence total running time is $O((m + n \lg n) \lg_{m/n} n)$. We summarize our result below.

Theorem 4. *Given a chordal graph G, we can output a proper coloring of G in $O((m + n \lg n) \lg_{m/n} n)$ time using $O(n \lg(m/n))$ bits.*

Note that with $O(n)$ bits, we cannot store the colors of all the vertices simultaneosuly, and this posses a challenge for the greedy algorithm. We leave open the problem to find a proper coloring of chordal graphs using $O(n)$ bits.

4.3   Recognition of Chordal Graphs

In what follows, we present a space efficient implementation for the recognition of chordal graphs. The idea is to apply MCS on G first to generate a vertex ordering, and then check whether the resulting vertex ordering is indeed a PEO. Let $v_1, v_2, \ldots, v_n$ be the ordering of the vertices reported by the MCS algorithm.

Chordal graph recongnition with $O(n \lg(m/n))$ bits. We first observe that one can compute the predecessor/successor of any vertex v_i during the MCS algorithm. Since we have $O(n \lg(m/n))$ bits, we can store one pointer with each vertex. We use this space to store a pointer to the last predecessor for each vertex. To test whether G is chordal, we need to test, for each vertex v_i, whether the neighbors of v_i numbered less than i form a clique. To perform this test, it is enough to test whether the predecessor set of v_i is a subset of the predecessor set of v_j where v_j is the last predecessor of v_i [14].

Now we run the MCS algorithm once again, and whenever we number a vertex v_j, we also generate its predecessor set P_j. For this purpose, we maintain a bit vector B to mark all the vertices that are numbered during the current MCS algorithm. Using this (dynamic) bit vector B, one can easily generate the set P_j by simply scanning the adjacency list of v_j and checking whether they are marked in B. This set P_j is stored as a bit vector Q of length n such that $Q[\ell] = 1$ iff $v_\ell \in P_j$. This helps us in checking the membership of a vertex in the set P_j in $O(1)$ time. The bit vector Q is initialized in $O(n)$ time at the beginning of the algorithm, and is used to store the predecessor set of the currently numbered vertex v_j, for $1 \leq j \leq n$. After generating the predecessor set P_j of v_j, we scan the adjacency list of v_j to check for any vertex v_i, where $i > j$, if v_j is the last predecessor of v_i (given v_i and v_j, we can check whether v_j is the last predecessor of v_i in constant time using the predecessor pointer of v_i). If v_j is the last predecessor of v_j, then we test whether the predecessor set of v_i, excluding v_j, is a subset of P_j. Note that since v_j is the last predecessor of v_i, all the other predecessors of v_i must have already been numbered when v_j is numbered, and hence are stored in the set represented by the bit vector B. Thus we can test whether the predecessor set of v_i is a subset of P_j in time proportional to the degree of v_i (with the aid of the bit vector Q). This completes the description of our algorithm for chordal graph recongnition.

The overall runtime of the algorithm is dominated by the runtime of the MCS algorithm. Thus if we can perform MCS on a graph G in $t(n, m)$ time using $s(m, n)$ bits, then we can also check whether G is chardal in $O(t(n, m))$ time using $s(m, n) + O(n \lg(m/n))$ bits.

Chordal graph recongnition with $O(n)$ bits. Since we cannot store all the n vertices of G (in the MCS order), we generate these vertices in $\lg n$ phases, where in the ℓ-th phase, we generate the vertices $V_\ell = \{v_{\ell k+1}, v_{\ell k+2}, \ldots v_{\ell(k+1)}\}$, for $1 \leq \ell \leq \lg$ and $k = n/\lg n$. In each phase, for each vertex v_i generated, we test the condition whether predecessor set of v_i is a subset of the predecessor set of v_j where v_j is the last predecessor v_i. To do this, we first perform another MCS, to compute the last predecessor of each vertex in V_ℓ (in fact, this step can be combined with the step where we generate the set V_ℓ). We maintain a bit vector to mark the set of predecessors of elements in V_ℓ to check their membership in $O(1)$ time, and another bit vector B to mark the numbered vertices. Now we start another MCS, and whenever a node v_j is numbered, we check to see if it is the last predecessor of some node v_i in V_ℓ. If so, we generate the predecessor sets of v_j and v_i with the aid of B, and check the required condition.

Thus if we can perform MCS on a graph G in $t(n, m)$ time using $s(m, n)$ bits, then we can also check whether G is chardal in $O(t(n, m) \cdot \lg n)$ time using $s(m, n) + O(n)$ bits. Thus we obtain the following,

Theorem 5. *Given an undirected graph G, if MCS on G can be performed in $t(n, m)$ time using $s(m, n)$ bits, then chordality of G can be tested in (a) $O(t(n, m))$ time using $s(m, n) + O(n \lg(m/n))$ bits, or (b) $O(t(n, m) \cdot \lg n)$ time using $s(m, n) + O(n)$ bits.*

5   Queue BFS and Stack BFS

The classical Breadth-first search (BFS) algorithm (following the nomenclature of [15], we call it Queue BFS) is one of the most popular graph search methods along with Depth-first search (DFS). The standard implementation of Queue BFS uses a queue to store the vertices which are to be explored next. The algorithm also uses three colors (white for unvisited, grey for partially explored and black for completely explored) for each vertex v to store the different state v could possible be at anytime during the execution of the algorithm. As there could be $O(n)$ vertices at anytime in the queue at worst case, the classical implementation takes $O(n \lg n)$ bits and $O(m + n)$ time. Recently Banerjee et al. [4] showed that using just $2n + o(n)$ bits, we can still perform a (modified) BFS traversal of a graph G in $O(m+n)$ time. To reduce the space from $O(n \lg n)$ bits used in the standard implementation, they crucially observe the following two properties of BFS: (i) elements in the queue are only from two consecutive levels of the BFS tree, and (ii) elements belonging to the same level can be processed in "any" order, but elements of the lower level must be processed before processing elements of the higher level. Thus they replace the queue by a space efficient "findany" data structure, for the two consecutive levels, which quickly (in $O(1)$ time) returns "any" element from these levels, and the algorithm proceeds with exploring this returned vertex and so on until all the vertices are exhausted. For details, the readers are referred to [4]. Although this algorithm actually suffices if we are only interested in the shortest path distance in any arbitrary unweighted undirected or directed graph by means of BFS, one shortcoming of this algorithm is that it looses the ordered structure of the standard BFS because of the second property. Greenlaw [15] defined another graph search method which he calls Stack BFS which is essentially a BFS traversal of the input graph but uses a stack instead of a queue. In what follows we briefly describe space efficient implementation for both of these graph search methods. Our main theorem is the following.

Theorem 6. *Given a directed or an undirected graph G, we can perform a Queue BFS and Stack BFS traversal of G in (a) $O(m \lg^2 n)$ time using $O(n)$ bits, or (b) $O(m + n \lg n)$ time using $O(n \lg(m/n))$ bits.*

Proof sketch: If we have only $O(n)$ bits, we store constant number of chunks of $O(n/\lg n)$ vertices along with the color array and some other auxilliary information using $O(n)$ bits so that we can reconstruct the required level of the partially explored BFS tree whenever needed. We pay for this repeated reconstruction by incurring extra time. For $O(n \lg(m/n))$ bit algorithm, we can maintain explicitly the parent pointers for every vertex, and this along with other stored information improves the time bound compared to what we can achieve with $O(n)$ bits. We omit the detailed proof of this theorem due to lack of space. The proof will appear in the full version of this paper.

6 Conclusions

We showed several time-space tradeoffs for performing three different graph search methods, namely MCS, Stack BFS, and Queue BFS. As applications of these results, we proposed space-efficient implementations for testing if a given undirected graph is chordal, reporting an independent set, and a proper coloring of a given chordal graph. Exploring other graph search methods such as Maximum Cardinality BFS, Maximum Cardinality DFS, Local MCS [19] and Lex DFS [8,16] is an interesting open problem. Also, obtaining a sublinear space implementation for any of these methods (including MCS) is a challenging open problem.

References

1. Arora, S., Barak, B.: Computational Complexity - A Modern Approach. Cambridge University Press, New York (2009)
2. Asano, T., Buchin, K., Buchin, M., Korman, M., Mulzer, W., Rote, G., Schulz, A.: Reprint of: Memory-constrained algorithms for simple polygons. Comput. Geom. **47**(3), 469–479 (2014)
3. Asano, T., et al.: Depth-first search using $O(n)$ bits. In: Ahn, H.-K., Shin, C.-S. (eds.) ISAAC 2014. LNCS, vol. 8889, pp. 553–564. Springer, Cham (2014). doi:10.1007/978-3-319-13075-0_44

4. Banerjee, N., Chakraborty, S., Raman, V.: Improved space efficient algorithms for BFS, DFS and applications. In: Dinh, T.N., Thai, M.T. (eds.) COCOON 2016. LNCS, vol. 9797, pp. 119–130. Springer, Cham (2016). doi:10.1007/978-3-319-42634-1_10

5. Banerjee, N., Chakraborty, S., Raman, V., Roy, S., Saurabh, S.: Time-space trade-offs for dynamic programming algorithms in trees and bounded treewidth graphs. In: Xu, D., Du, D., Du, D. (eds.) COCOON 2015. LNCS, vol. 9198, pp. 349–360. Springer, Cham (2015). doi:10.1007/978-3-319-21398-9_28

6. Barnes, G., Buss, J., Ruzzo, W., Schieber, B.: A sublinear space, polynomial time algorithm for directed s-t connectivity. SICOMP 27(5), 1273–1282 (1998)

7. Batagelj, V., Zaversnik, M.: An O(m) algorithm for cores decomposition of networks. CoRR cs.DS/0310049 (2003)

8. Berry, A., Krueger, R., Simonet, G.: Maximal label search algorithms to compute perfect and minimal elimination orderings. SIAM J. Discrete Math. 23(1), 428–446 (2009)

9. Chakraborty, S., Jo, S., Satti, S.R.: Improved space-efficient linear time algorithms for some classical graph problems. In: 15th CTW (2017)

10. Chakraborty, S., Raman, V., Satti, S.R.: Biconnectivity, chain decomposition and st-numbering using O(n) bits. In: 27th ISAAC, vol. 64. LIPIcs, pp. 22:1–22:13. Schloss Dagstuhl - Leibniz-Zentrum fuer Informatik (2016)

11. Darwish, O., Elmasry, A.: Optimal time-space tradeoff for the 2D convex-hull problem. In: Schulz, A.S., Wagner, D. (eds.) ESA 2014. LNCS, vol. 8737, pp. 284–295. Springer, Heidelberg (2014). doi:10.1007/978-3-662-44777-2_24

12. Edmonds, J., Poon, C.K., Achlioptas, D.: Tight lower bounds for st-connectivity on the NNJAG model. SIAM J. Comput. 28(6), 2257–2284 (1999)

13. Fulkerson, D.R., Gross, O.A.: Incidence matrices and interval graphs. Pacific J. Math 15, 835–855 (1965)

14. Golumbic, M.C.: Algorithmic Graph Theory and Perfect Graphs (2004)

15. Greenlaw, R.: A model classifying algorithms as inherently sequential with applications to graph searching. Inf. Comput. 97(2), 133–149 (1992)

16. Krueger, R., Simonet, G., Berry, A.: A general label search to investigate classical graph search algorithms. Discret. Appl. Math. 159(2–3), 128–142 (2011)

17. Munro, J.I.: Tables. In: Chandru, V., Vinay, V. (eds.) FSTTCS 1996. LNCS, vol. 1180, pp. 37–42. Springer, Heidelberg (1996). doi:10.1007/3-540-62034-6_35

18. Munro, J.I., Paterson, M.: Selection and sorting with limited storage. Theor. Comput. Sci. 12, 315–323 (1980)

19. Panda, B.S.: New linear time algorithms for generating perfect elimination orderings of chordal graphs. Inf. Process. Lett. 58(3), 111–115 (1996)

20. Reingold, O.: Undirected connectivity in log-space. J. ACM 55(4), 1–24 (2008)

21. Rose, D.J., Tarjan, R.E., Lueker, G.S.: Algorithmic aspects of vertex elimination on graphs. SIAM J. Comput. 5(2), 266–283 (1976)

22. Tarjan, R.E.: Maximum cardinality search and chordal graphs. Unpublished Lecture Notes CS 259

23. Tarjan, R.E., Yannakakis, M.: Simple linear-time algorithms to test chordality of graphs, test acyclicity of hypergraphs, and selectively reduce acyclic hypergraphs. SIAM J. Comput. 13(3), 566–579 (1984)

24. Tompa, M.: Two familiar transitive closure algorithms which admit no polynomial time, sublinear space implementations. SIAM J. Comput. 11(1), 130–137 (1982)

Efficient Enumeration of Non-Equivalent Squares in Partial Words with Few Holes

Panagiotis Charalampopoulos[1($\boxtimes$)], Maxime Crochemore[1,2],
Costas S. Iliopoulos[1], Tomasz Kociumaka[3], Solon P. Pissis[1],
Jakub Radoszewski[1,3], Wojciech Rytter[3], and Tomasz Waleń[3]

[1] Department of Informatics, King's College London, London, UK
{panagiotis.charalampopoulos,maxime.crochemore,
costas.iliopoulos,solon.pissis}@kcl.ac.uk
[2] Université Paris-Est, Marne-la-Vallée , France
[3] Faculty of Mathematics, Informatics and Mechanics,
University of Warsaw, Warsaw, Poland
{kociumaka,jrad,rytter,walen}@mimuw.edu.pl

Abstract. A word of the form WW for some word $W \in \Sigma^*$ is called a square, where Σ is an alphabet. A partial word is a word possibly containing holes (also called don't cares). The hole is a special symbol $\diamond \notin \Sigma$ which *matches* (agrees with) any symbol from $\Sigma \cup \{\diamond\}$. A *p-square* is a partial word matching at least one square WW without holes. Two p-squares are called *equivalent* if they match the same set of squares. We denote by $psquares(T)$ the number of non-equivalent p-squares which are factors of a partial word T. Let $\mathrm{PSQUARES}_k(n)$ be the maximum value of $psquares(T)$ over all partial words of length n with at most k holes. We show asymptotically tight bounds:

$$c_1 \cdot \min(nk^2, n^2) \leq \mathrm{PSQUARES}_k(n) \leq c_2 \cdot \min(nk^2, n^2)$$

for some constants $c_1, c_2 > 0$. We also present an algorithm that computes $psquares(T)$ in $\mathcal{O}(nk^3)$ time for a partial word T of length n with k holes. In particular, our algorithm runs in linear time for $k = \mathcal{O}(1)$ and its time complexity near-matches the maximum number of non-equivalent p-square factors in a partial word.

1 Introduction

A *word* is a sequence of letters from a given alphabet Σ. By Σ^* we denote the set of all words over Σ. A word of the form $U^2 = UU$, for some word U, is called a *square*. For a word W, a *square factor* is a factor of W which is a square. Enumeration of

P. Charalampopoulos—Supported by the Graduate Teaching Scholarship scheme of the Department of Informatics at King's College London.

T. Kociumaka—Supported by Polish budget funds for science in 2013–2017 as a research project under the 'Diamond Grant' program.

J. Radoszewski, W. Rytter and T. Waleń—Supported by the Polish National Science Center, grant no. 2014/13/B/ST6/00770.

© Springer International Publishing AG 2017
Y. Cao and J. Chen (Eds.): COCOON 2017, LNCS 10392, pp. 99–111, 2017.
DOI: 10.1007/978-3-319-62389-4_9

square factors in words is a well-studied topic, both from a combinatorial and from an algorithmic perspective. Obviously, a word W of length n may contain $\Theta(n^2)$ square factors (e.g. $W = a^n$), however, it is known that such a word contains only $\mathcal{O}(n)$ distinct square factors [14,17]; currently the best known upper bound is $\frac{11}{6}n$ [12]. Moreover, all distinct square factors of a word can be listed in $\mathcal{O}(n)$ time using the suffix tree [15] or the suffix array and the structure of runs (maximal repetitions) in the word [10].

A *partial word* is a sequence of letters from $\Sigma \cup \{\Diamond\}$, where $\Diamond$ denotes a *hole*, that is, a don't care symbol. We assume that Σ is non-unary. Two symbols $a, b \in \Sigma \cup \{\Diamond\}$ are said to *match* (denoted as $a \approx b$) if they are equal or one of them is a hole; note that this relation is not transitive. The relation of matching is extended in a natural way to partial words of the same length.

A partial word UV is called a *p-square* if $U \approx V$. Like in the context of words, a *p-square factor* of a partial word T is a factor being a p-square; see [2,7]. Alongside [2,6,7], we define a *solid square* (also called a *full square*) as a square of a word, and a *square subword* of a partial word T as a solid square that matches a factor of T.

We introduce the notion of *equivalence* of p-square factors in partial words. Let $sq\text{-}val(UV)$ denote the set of solid squares that match the partial word UV:

$$sq\text{-}val(UV) = \{WW : W \in \Sigma^*, WW \approx UV\}.$$

Example 1. $sq\text{-}val(a\Diamond b\ a\Diamond\Diamond) = \{(aab)^2, (abb)^2\}$, with $\Sigma = \{a, b\}$.

Then p-squares UV and $U'V'$ are called *equivalent* if $sq\text{-}val(UV) = sq\text{-}val(U'V')$ (denoted as $UV \equiv U'V'$). For example, $a\Diamond b\ a\Diamond\Diamond \equiv a\Diamond\Diamond\ \Diamond\Diamond b$, but $a\Diamond b\ a\Diamond\Diamond \not\equiv a\Diamond\Diamond\ \Diamond ab$.

Note that two p-square factors of a partial word T are equivalent in this sense if and only if they correspond to exactly the same set of square subwords. The number of non-equivalent p-square factors in a partial word T is denoted by $psquares(T)$. Our work is devoted to the enumeration of non-equivalent p-square factors in a partial word with a given number k of holes.

We say that $X^2 = XX$ is the *representative* (also called *general form*; see [6]) of a p-square UV, denoted as $repr(UV)$, if $XX \approx UV$ and $sq\text{-}val(XX) = sq\text{-}val(UV)$. (In other words, X is the "most general" partial word that matches both U and V.) It can be noted that the representative of a p-square is unique. Then $UV \equiv U'V'$ if and only if $repr(UV) = repr(U'V')$.

Example 2. $repr(a\Diamond b\ a\Diamond\Diamond) = (a\Diamond b)^2$, $repr(a\Diamond\Diamond\ \Diamond ab) = (aab)^2$.

Previous studies on squares in partial words were mostly focused on combinatorics. They started with the case of $k = 1$ [6], in which case distinct square subwords correspond to non-equivalent p-square factors. It was shown that a partial word with one hole contains at most $\frac{7}{2}n$ distinct square subwords [4] ($3n$ for binary partial words [16]). Also a generalization of the three squares lemma (see [11]) was proposed for partial words [5]. As for a larger number of holes, the existing literature is devoted mainly to counting the number of distinct square subwords of a partial word [2,6] or all occurrences of p-square factors [2,3]. On the algorithmic side, [21] proved that the problem of counting distinct square subwords of a partial word is #P-complete and [13,20] and [7] showed quadratic- and nearly-quadratic-time algorithms for finding all occurrences of p-square factors and primitively-rooted p-square factors of a partial word, respectively.

Our combinatorial results. We prove that a partial word of length n with k holes contains $\mathcal{O}(nk^2)$ non-equivalent p-square factors. We also construct a family of partial words that contain $\Omega(nk^2)$ non-equivalent p-square factors, for $k = \mathcal{O}(\sqrt{n})$. This proves the aforementioned asymptotic bounds for $\mathrm{PSQUARES}_k(n)$. Our work can be viewed as a generalization of the results on partial words with one hole [4,6,16] to $k \geq 1$ holes.

Our algorithmic results. We present an algorithm that reports all non-equivalent p-square factors in a partial word of length n with k holes in $\mathcal{O}(nk^3)$ time. In particular, our algorithm runs in linear time for $k = \mathcal{O}(1)$ and its time complexity near-matches the maximum number of non-equivalent p-square factors. We assume integer alphabet $\Sigma \subseteq \{1, \ldots, n^{\mathcal{O}(1)}\}$. The main tool in the algorithm are two new types of non-standard runs in partial words and relations between them. We also use recently introduced advanced data structures from [18].

2 Preliminary Notation for Words and Partial Words

For a word $W \in \Sigma^*$, by $|W| = n$ we denote the length of W, and by $W[i]$, for $i = 1, \ldots, n$, the ith letter of W. For $1 \leq i \leq j \leq n$, $W[i..j]$ denotes the *factor* of W equal to $W[i] \cdots W[j]$. A factor of the form $W[1..j]$ is called a *prefix*, a factor of the form $W[i..n]$ is called a *suffix*, and a factor that is both a prefix and a suffix of W is called a *border* of W. A positive integer q is called a *period* of W if $W[i] = W[i+q]$ for all $i = 1, \ldots, n - q$. In this case, $W[1..q]$ is called a *string period* of W. W has a period q if and only if it has a border of length $n - q$; see [8]. Two equal-length words V and W are called *cyclic shifts* if there are words X, Y such that $V = XY$ and $W = YX$. A word W is called *primitive* if there is no word U and integer $k > 1$ such that $U^k = W$. Note that the shortest string period of W is always primitive. Every primitive word W has the following *synchronization property*: W is not equal to any of its non-trivial cyclic shifts [8].

For a partial word T we use the same notation as for words: $|T| = n$ for its length, $T[i]$ for the ith letter, $T[i..j]$ for a factor. If T does not contain holes, then it is called *solid*. The relation of matching on $\Sigma \cup \{\diamond\}$ is defined as: $a \approx a$, $\diamond \approx a$, and $a \approx \diamond$ for all $a \in \Sigma \cup \{\diamond\}$. We define an operation $\wedge$ such that $a \wedge a = a \wedge \diamond = \diamond \wedge a = a$ for all $a \in \Sigma \cup \{\diamond\}$, and otherwise $a \wedge b$ is undefined. Two equal-length partial words T and S are said to *match* (denoted as $T \approx S$) if $T[i] \approx S[i]$ for all $i = 1, \ldots, n$. In this case, by $S \wedge T$ we denote the partial word $S[1] \wedge T[1], \ldots, S[n] \wedge T[n]$. If $U \approx T[i..i + |U| - 1]$ for a partial word U, then we say that U occurs in T at position i. Also note that if UV is a p-square, then $repr(UV) = (U \wedge V)^2$. A *quantum period* of T is a positive integer q such that

$T[i] \approx T[i + q]$ for all $i = 1, \ldots, n - q$. A *deterministic period* of T is an integer q such that there exists a word W such that $W \approx T$ and W has a period q. T is called *quantum (deterministically) periodic* if it has a quantum (deterministic) period q such that $2q \leq n$.

An integer j is an *ambiguous length* in the partial word T if there are two holes in T at distance $j/2$. A p-square is called *ambiguous* if its representative is non-solid. Note that if a p-square factor in T is ambiguous, then the p-square has an ambiguous length (the converse is not always true). The p-square factors of T of non-ambiguous length have solid representatives.

Example 3. Let $T = ab\diamond\diamond ba\diamond aaba\diamond b$. For T, 4 is a non-ambiguous length. T contains four non-equivalent classes of p-squares of length 4: $a\diamond aa$ with representative $(aa)^2$, $ab\diamond\diamond \equiv \diamond ba\diamond \equiv aba\diamond$ with representative $(ab)^2$, $\diamond\diamond ba \equiv ba\diamond a$ with representative $(ba)^2$, and $b\diamond\diamond b$ with representative $(bb)^2$. On the other hand, 6 is an ambiguous length in T. T contains four non-equivalent classes of p-squares of length 6: $aaba\diamond b$ with representative $(aab)^2$, $ab\diamond\diamond ba \equiv a\diamond aaba$ with representative $(aba)^2$, $\diamond aaba\diamond$ with representative $(baa)^2$, and $b\diamond\diamond ba\diamond$ with representative $(ba\diamond)^2$. Note that only the last one is an ambiguous p-square. Overall, T contains 14 non-equivalent p-squares.

3   Combinatorial Bounds

3.1   Lower Bound

We say that a set A of positive integers is an (m, t)-*cover* if the following conditions hold:

(1) For each $d \geq m$, A contains at most one pair of elements with difference d;
(2) $|\{ |j - i| \geq m : i, j \in A \}| \geq t$.

For a set $A \subseteq \{1, \ldots, n\}$ we denote by $w_{A,n}$ the partial word of length n over the alphabet Σ such that $w_{A,n}[i] = \diamond \Leftrightarrow i \in A$, and $w_{A,n}[i] = a$ otherwise.

Lemma 4. *Assume that* $A \subseteq \{1, \ldots, n\}$ *is an* (m, t)-*cover such that* $m = \Theta(n)$, $|A| = k$, *and* $t = \Omega(k^2)$. *Let* $\Sigma = \{a, b\}$ *be the alphabet. Then*

$$psquares(a^{n-2} \cdot w_{A,n} \cdot a^{n-2}) = \Omega(n \cdot k^2).$$

Proof. Each even-length factor of $a^{n-2} \cdot w_{A,n} \cdot a^{n-2}$ is a p-square. Let $\mathcal{Z}$ be the set of these factors X which contain two positions i, j containing holes with $|j - i| \geq m$ and $|X| = 2|j - i|$. As A is an (m, t)-cover, i and j are determined uniquely by $d = |j - i|$. Then all elements of $\mathcal{Z}$ are pairwise non-equivalent p-squares. The size of $\mathcal{Z}$ is $\Omega(nt)$ which is $\Omega(n \cdot k^2)$. This completes the proof. $\square$

Example 5. Let $n = 5$, $m = 4$, and $t = 1$. $aaa\diamond aaa\diamond aaa$ has 4 non-equivalent p-square factors of length 8 if $\Sigma = \{a, b\}$. If $\Sigma = \{a\}$, all of them are equivalent.

Theorem 6. *For every positive integer n and $k \leq \sqrt{2n}$, there is a partial word of length n with k holes that contains $\Omega(nk^2)$ non-equivalent p-square factors.*

Proof. Due to Lemma 4, it is enough to construct a suitable set A. By monotonicity, we may assume that k and n are even. We take:

$$A = \{1, \ldots, \tfrac{k}{2}\} \cup \{j \cdot \tfrac{k}{2} + \tfrac{n}{2} : 1 \leq j \leq \tfrac{k}{2}\}.$$

We claim that A is an $(\tfrac{n}{2}, \tfrac{k^2}{4})$-cover for $t = \Omega(k^2)$. Indeed, take any $i \in \{1, \ldots, \tfrac{k}{2}\}$ and j satisfying the above condition. Then $j \cdot \tfrac{k}{2} + \tfrac{n}{2} - i \geq \tfrac{n}{2}$ and all such values are distinct; hence, $t = \tfrac{k^2}{4}$. The thesis follows from the claim. $\square$

3.2 Upper Bound

Let T be a partial word of length n with k holes. The proof of the upper bound for ambiguous lengths is easy.

Lemma 7. *There are at most nk^2 p-square factors of ambiguous length in T.*

Proof. The number of ambiguous lengths is at most $\binom{k}{2}$, since we have $\binom{k}{2}$ possible distances between k holes. Consequently, the number of p-squares with such lengths is at most nk^2. $\square$

Each of the remaining p-square factors of T has a solid representative. We say that a solid square W^2 has a *solid occurrence* in T if T contains a factor *equal to* W^2. By the following fact, there are at most $2n$ non-equivalent p-square factors of T with solid occurrences.

Fact 8 ([12,14,17]). *Every position of a (solid) word contains at most two rightmost occurrences of squares.*

We say that a solid square is a *u-square* in T if it occurs in T, does not have a solid occurrence in T, and has a non-ambiguous length. We denote by $\mathcal{U}$ the set of u-squares for T.

Observation 9. Each u-square in T corresponds in a one-to-one way to an equivalence class of p-square factors of T which have non-ambiguous length and do not have a solid occurrence in T.

Thus it suffices to bound $|\mathcal{U}|$. This is the essential part of the proof. Let $\alpha = \frac{1}{2k+2}$ and

$$\mathcal{U}(\ell) = \{W^2 \in \mathcal{U} : 2\ell \leq |W|^2 \leq 2(\ell + \lfloor \ell\alpha \rfloor)\}.$$

Also denote by $\mathcal{U}_i(\ell)$ (and $\mathcal{U}last_i(\ell)$) the set of words of $\mathcal{U}(\ell)$ which have an occurrence (the last occurrence, respectively) at position i in T. The next lemma follows from the pigeonhole principle and periodicity of (solid) words.

Lemma 10. *Suppose that $\ell \geq \frac{1}{\alpha}$ and $|\mathcal{U}_i(\ell)| \geq 2$. Let $\Delta = \lfloor \ell\alpha \rfloor$. There exist positions s, s' such that:*

- *$s \in [i, i + \ell - 2\Delta]$,*
- *$s' \in [s + \ell, s + \ell + \Delta]$,*
- *$T[s..s + 2\Delta - 1] = T[s'..s' + 2\Delta - 1]$ is solid and periodic.*

Proof. Let $T[i..i + 2d - 1]$ be a u-square from $\mathcal{U}_i(\ell)$. Consider positions $x_j = i + 2j\Delta$ and $y_j = x_j + d$ for $0 \leq j \leq k$. Note that factors $X_j = T[x_j..x_j + 2\Delta - 1]$ and $Y_j = T[y_j..y_j + 2\Delta - 1]$ match; see Fig. 1. Moreover, factors $X_0, \ldots, X_k$ and $Y_0, \ldots, Y_k$ are disjoint because $2(k+1)\Delta \leq 2(k+1)\frac{\ell}{2k+2} = \ell$. By the pigeonhole principle, we can choose j so that X_j and Y_j are solid, i.e., $X_j = Y_j$. We set $s = x_j$ and $s' = y_j$.

It remains to prove that $X_j = Y_j$ is periodic. Let $T[i..i + 2d' - 1]$ (with $d' \neq d$) be another u-square in $\mathcal{U}_i(\ell)$, and let $Y'_j = T[x_j + d'..x_j + d' + 2\Delta - 1]$. Note that $Y'_j \approx X_j = Y_j$ and factors Y_j and Y'_j have an overlap of $2\Delta - |d - d'|$ positions being a border of Y_j. Consequently, $|d - d'| \leq \lfloor \ell\alpha \rfloor = \Delta$ is a period of Y_j.   □

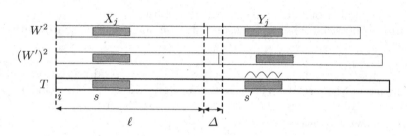

Fig. 1. Situation from the proof of Lemma 10; $W^2, (W')^2 \in \mathcal{U}_i(\ell)$. Occurrences of $X_j = Y_j$ are denoted by dark rectangles. $\mathcal{U}_i(\ell)$ is the set of all u-squares having an occurrence in i with center in the window of size Δ.

We denote by $\mathcal{I}_{i,\ell}$ the interval $[i, i + 2(\ell + \lfloor \ell\alpha \rfloor) - 1]$. Let $\#_\Diamond([a, b])$ denote the number of holes in $T[a..b]$. Our upper bound for partial words is based on the following key lemma; it is a property of partial words similar to Fact 8.

Lemma 11. $|\mathcal{U}last_i(\ell)| = \mathcal{O}(\#_\Diamond(\mathcal{I}_{i,\ell}))$.

Proof. Denote $k' = \#_\Diamond(\mathcal{I}_{i,\ell})$. If $k' = 0$, then $|\mathcal{U}last_i(\ell)| = 0$. From now we assume that $k' \geq 1$. Assume that $|\mathcal{U}last_i(\ell)| \geq 2$. Let p be the shortest period of the equal periodic factors $X = T[s..s + 2\Delta - 1]$ and $Y = T[s'..s' + 2\Delta - 1]$ from the previous lemma. We consider three types of u-squares $W^2 \in \mathcal{U}last_i(\ell)$:

Type (a): W^2 has period p;
Type (b): W has period p but W^2 does not have period p;
Type (c): W does not have period p.

At most 1 u-square of type (a). Observe that the length of W is a multiple of its shortest period p (this is due to the synchronization property for the string period of W). Consequently, if we have two u-squares of type (a) occurring at position i and with the same shortest period p, then the shorter u-square also occurs at position $i + p$. This contradicts the definition of $\mathcal{U}last_i(\ell)$.

At most $k' + 1$ u-squares of type (b). Suppose to the contrary that there are at least $k' + 2$ u-squares of type (b), of lengths $d_1 < \ldots < d_{k'+2}$. Note that $Y_j' := T[s + d_j..s + d_j + 2\Delta - 1]$ matches $X = Y$ due to a u-square of length $2d_j$. Moreover, the factors Y and Y_j' have an overlap of at least $\Delta \geq p$ positions, so the string periods of Y_j' and Y must be synchronized. Consequently, the values $d_j \bmod p$ are all the same (and non-zero, as these are not squares of type (a)).

Consider the shortest W^2 and the longest $(W')^2$ of these u-squares and the factor $Z = T[i + d_1..i + d_{k'+2} - 1]$. It matches a prefix P of length $d_{k'+2} - d_1$ of W and a suffix S of the same length of W'. Both P and S have period p; however, their string periods of length p are not equal (again, due to synchronization property), as p does not divide d_1. Consequently, in every factor of length p in Z there must be a hole. This yields $\lfloor |Z|/p \rfloor = (d_{k'+2} - d_1)/p \geq k' + 1$ holes in total, a contradiction.

At most $4k' + 2$ u-squares of type (c). Let $d = |W|$. Let us extend the occurrence of X in W at position $s - i + 1$ to a maximal factor $W[j'..j]$ with period p. Note that $j' > 1$ or $j < d$ as W^2 is not of type (b). Below, we assume $j < d$; the other subcase is handled in an analogous way. Consider the positions $j_1 = i + j$ and $j_2 = i + d + j$ of T. We will show that there are at most $2k' + 1$ possible pairs (j_1, j_2) across the u-squares $W \in \mathcal{U}last_i(\ell)$, i.e., at most $2k' + 1$ corresponding u-squares, as $d = j_2 - j_1$.

Positions $T[j_1]$ and $T[j_2]$ cannot both contain holes, as $2d$ is a non-ambiguous length. If $T[j_1]$ is not a hole, then it is determined uniquely as the first position where the deterministic period p breaks, starting from the position s, i.e., j_1 is the smallest index such that $T[s..j_1]$ does not have deterministic period p. The same holds for j_2 and s'; this is also due to the fact that Y and the occurrence of X at position $s + d$ have an overlap of at least $\Delta \geq p$ positions, so they are synchronized. Hence, if neither $T[j_1]$ nor $T[j_2]$ is a hole, then (j_1, j_2) is determined uniquely. Otherwise, if $T[j_1]$ or $T[j_2]$ is a hole, then the other position is determined uniquely, so there are at most $2k'$ choices. This concludes the proof. $\qquad\square$

Theorem 12. *The number of non-equivalent p-square factors in a partial word T of length n with k holes is $\mathcal{O}(\min(nk^2, n^2))$.*

Proof. The $\mathcal{O}(n^2)$ bound is obvious. Due to Lemma 7 there are at most nk^2 p-squares of ambiguous length in T. Let us consider p-squares of non-ambiguous lengths. By Fact 8, among them there are $\mathcal{O}(n)$ non-equivalent p-squares with a solid occurrence. From now on we count only non-equivalent non-ambiguous p-squares without a solid occurrence, i.e., different u-squares.

Clearly, there are $\mathcal{O}(nk)$ different u-squares of length smaller than $\frac{2}{\alpha}$. Let $\ell \geq \frac{1}{\alpha}$ and $r = 2(\ell + \lfloor \ell\alpha \rfloor)$. By Lemma 11:

$$|\mathcal{U}(\ell)| = \sum_{i=1}^{n} |\mathcal{U}last_i(\ell)| = \mathcal{O}\left(\sum_{i=1}^{n} \#_\diamond(\mathcal{I}_{i,\ell}) \right) = \mathcal{O}(k\ell). \tag{1}$$

The last equality is based on the fact that each of the k holes in T is counted in at most $2r$ terms $\#_\diamond(\mathcal{I}_{i,\ell})$.

Let us consider a family of endpoints $r_j = \left\lceil \frac{n}{(1+\alpha)^j} \right\rceil$ for $j \geq 0$ and let $t = \max\{j : r_j > 1\}$. One can check that $\mathcal{U} = \bigcup_{j=0}^{t} \mathcal{U}(r_{j+1})$.

By (1), the total number of u-squares of length at least $\frac{2}{\alpha}$ in T is at most:

$$\sum_{j=1}^{t+1} |\mathcal{U}(r_j)| = \mathcal{O}\left(\sum_{j=1}^{t+1} kr_j \right) = \mathcal{O}\left(k \sum_{j=1}^{t+1} \left(1 + \frac{n}{(1+\alpha)^j} \right) \right)$$

$$= \mathcal{O}\left(k\log_{1+\alpha} n + \sum_{j=0}^{\infty} \frac{nk}{(1+\alpha)^j} \right) = \mathcal{O}\left(\frac{k\log n}{\alpha} + \frac{nk}{1 - \frac{1}{1+\alpha}} \right) = \mathcal{O}(nk^2). \qquad \square$$

4   Runs Toolbox for Partial Words

A *run* (also called a maximal repetition) in a word W is a triple (a, b, q) such that $W[a..b]$ is periodic with period q $(2q \leq b - a + 1)$ and the interval $[a, b]$ cannot be extended to the left nor to the right without violating the above property, that is, $W[a-1] \neq W[a+q-1]$ and $W[b-q+1] \neq W[b+1]$, provided that the respective positions exist. The *exponent* of a run is defined as $\frac{b-a+1}{q}$. A word of length n has $\mathcal{O}(n)$ runs and they can all be computed in $\mathcal{O}(n)$ time [1,19].

From a run (a, b, q) we can produce all triples (a, b, kq) for integer $k \geq 1$ such that $2kq \leq b - a + 1$; we call such triples *generalized runs*. That is, the period of a generalized run need not be the shortest period. The number of generalized runs is also $\mathcal{O}(n)$ as the sum of exponents of runs is $\mathcal{O}(n)$ [1,19].

For a partial word T, we call a triple (a, b, q) a *quantum generalized run* (Q-run, for short) in T if $T[a..b]$ is quantum periodic with period q and none of the partial words $T[a-1..b]$ and $T[a..b+1]$ (if it exists) has the quantum period q; for an example see Fig. 2.

Fig. 2. A partial word together with all its Q-runs.

Generalized runs in words are strongly related to squares: (1) a square of length $2q$ belongs to a generalized run of period q and, moreover, (2) all factors of length $2q$ of a generalized run with period q are squares being each other's cyclic shifts. Unfortunately, Q-runs in partial words have only property (1). However, we introduce a type of run in partial words that has a property analogous to (2). A *pseudorun* is a triple (a, b, q) such that:

(a) $T[a..b]$ is quantum periodic with period q,
(b) $T[i - q] \wedge T[i] = T[i] \wedge T[i + q]$ for all i such that $i - q, i + q \in [a, b]$,
(c) none of the partial words $T[a - 1..b]$ and $T[a..b + 1]$ (if exists) satisfies the conditions (a) and (b).

We say that a p-square factor $T[c..d]$ is *induced* by the pseudorun (a, b, q) if $d - c + 1 = 2q$ and $[c, d] \subseteq [a, b]$.

Example 13. The partial word from Fig. 2 contains two Q-runs with period 2: $(1, 9, 2)$ that corresponds to factor $ab\diamond\diamond ba\diamond aa$ and $(9, 12, 2)$ that corresponds to factor $aba\diamond$. The partial word contains five pseudoruns with this period: $(1, 4, 2)$: $ab\diamond\diamond$, $(2, 5, 2)$: $b\diamond\diamond b$, $(3, 8, 2)$: $\diamond\diamond ba\diamond a$, $(6, 9, 2)$: $a\diamond aa$, and $(9, 12, 2)$: $aba\diamond$. All but one of these pseudoruns induce exactly one p-square; the pseudorun $(3, 8, 2)$ induces two non-equivalent p-squares: $\diamond\diamond ba$ and $\diamond ba\diamond$.

Observation 14. (1) Every p-square factor in T is induced by a pseudorun. (2) All factors of length $2q$ of a pseudorun with period q are p-squares and their representatives are each other's cyclic shifts.

5   The Algorithm

We design an $\mathcal{O}(nk^3)$-time algorithm for enumerating non-equivalent p-squares in a partial word T of length n with k holes. We assume that Σ is an ordered integer alphabet and that $\diamond$ is smaller than all the letters from Σ. Then any two factors of T can be lexicographically compared using the suffix array of T in $\mathcal{O}(1)$ time after $\mathcal{O}(n)$-time preprocessing [8]. The first two steps of the algorithm are computing all Q-runs in T and decomposing Q-runs into pseudoruns. The final phase consists in grouping pseudoruns in T by the representatives of induced p-squares, which lets us enumerate non-equivalent p-squares.

5.1   Computing Q-Runs

We classify Q-runs into *solid Q-runs* that do not contain a hole and the remaining *non-solid Q-runs*. A solid Q-run is a generalized run in a maximal solid factor of T that is not adjacent to a hole in T. Thus all solid Q-runs can be computed in $\mathcal{O}(n)$ time using any linear-time algorithm for computing runs in words [1,19].

The length of the *longest common compatible prefix* of two positions i, j, denoted $lccp(i, j)$, is the largest ℓ such that $T[i..i + \ell - 1] \approx T[j..j + \ell - 1]$. Symmetrically, we can define $lccs(i, j)$ as the length of the longest common compatible suffix of $T[1..i]$ and $T[1..j]$. After $\mathcal{O}(nk)$-time preprocessing, queries for $lccp$ (hence, queries for $lccs$) can be answered on-line in $\mathcal{O}(1)$ time [9].

For every position i containing a hole and integer $q \in \{1, \ldots, n\}$, we can use the $lccp$- and $lccs$-queries to check if there is a Q-run with period q containing the position i. If the Q-run is to contain i anywhere except for its last q positions, we can compute $a = i - lccs(i, i+q) + 1$, $b = i + q + lccp(i, i+q) - 1$ and check if $b - a + 1 \geq 2q$; if so, the sought Q-run is (a, b, q). A symmetric test with $i - q$ and i can be used to check for a Q-run containing i among its last q positions.

Clearly, this procedure works in $\mathcal{O}(nk)$ time. Therefore, the number of Q-runs is at most $\mathcal{O}(nk)$. The same Q-run may be reported several times; therefore, in the end we remove repeating triples (a, b, q) via radix sort. Together with the $\mathcal{O}(n)$-time computation of solid Q-runs we arrive at the following lemma.

Lemma 15. *A partial word of length n with k holes contains $\mathcal{O}(nk)$ Q-runs and they can all be computed in $\mathcal{O}(nk)$ time.*

5.2   Computing Pseudoruns

Q-runs correspond to maximal factors of T that satisfy only the condition (a) of a pseudorun. Hence, every pseudorun is a factor of a Q-run.

A position i inside a Q-run $\beta = (a, b, q)$ is called a *break point* if $a \leq i - q < i + q \leq b$ and $T[i - q] \wedge T[i] \neq T[i] \wedge T[i + q]$.

Observation 16. i is a break point for (a, b, q) if and only if $a \leq i - q < i + q \leq b$, $T[i] = \diamond$, and $T[i - q] \neq T[i + q]$.

By $\Gamma(\beta)$ we denote the set of all break points of a Q-run β. The Q-run can be decomposed into $|\Gamma(\beta)| + 1$ pseudoruns: if i is the first break point in β, then we have a pseudorun $(a, i + q - 1, q)$ and continue the decomposition for $(i - q + 1, b, q)$. Consecutive pseudoruns in the decomposition overlap by $2p - 1$ positions. See Fig. 3 for an abstract illustration.

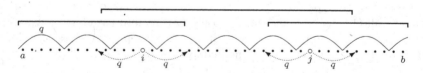

Fig. 3. A Q-run (a, b, q) with break points at positions i and j is decomposed into three pseudoruns: $(a, i + q - 1, q)$, $(i - q + 1, j + q - 1, q)$, and $(j - q + 1, b, q)$.

Lemma 17. $\sum_{\beta \in Q\text{-}runs(T)} |\Gamma(\beta)| \leq nk$.

Proof. Consider all Q-runs β of period q. Every two overlap by at most $q - 1$ positions, so the $\Gamma(\beta)$ sets are pairwise disjoint and their sizes sum up to at most k. Summing up over all $q = 1, \ldots, n/2$, we arrive at the conclusion. $\square$

Lemma 17 shows that there are $\mathcal{O}(nk)$ pseudoruns (we use the fact that, by Lemma 15, there are $\mathcal{O}(nk)$ Q-runs). They can be computed in $\mathcal{O}(nk^2)$ time by inspecting all the holes inside each Q-run β and checking which of them are break points in β.

Lemma 18. *A partial word of length n with k holes contains $\mathcal{O}(nk)$ pseudoruns and they can all be computed in $\mathcal{O}(nk^2)$ time.*

5.3 Grouping Pseudoruns and Reporting Squares

We define the *representative* of a pseudorun $\beta = (a, b, q)$ as

$$repr(\beta) = \text{lex-min}\{repr(T[i..i + 2q - 1]) : a \le i \le b - 2q + 1\}.$$

First, let us show how to group pseudoruns by equal representatives. This part of our algorithm builds upon the methods for grouping runs in words from [10].

We use a separate approach for solid and for non-solid pseudoruns. Each solid pseudorun corresponds to a solid Q-run. Hence, there are $\mathcal{O}(n)$ of them and they can all be grouped using the approach of [10] in $\mathcal{O}(n)$ time.

We say that a partial word U is a *d-fragment* of T if U is a factor of T with symbols at d positions substituted with other symbols. Obviously, a d-fragment can be represented in $\mathcal{O}(d)$ space. The following lemma is a consequence of Observation 18 from [18] and Theorem 23 from [18].

Lemma 19 ([18]). *For a word of length n, after $\mathcal{O}(n)$-time preprocessing:*

(a) Any two d-fragments can be compared lexicographically in $\mathcal{O}(d)$ time;
(b) The minimal cyclic shift of a d-fragment can be computed in $\mathcal{O}(d^2)$ time.

Lemma 20. *After $\mathcal{O}(n)$-time preprocessing, for any pseudorun β, $repr(\beta)$ represented as a k-fragment can be computed in $\mathcal{O}(k^2)$ time.*

Proof. Let $\beta = (a, b, q)$. Knowing the positions of holes in T, we can represent $repr(T[a..a + 2q - 1]) = U^2$ as a k-fragment (the positions with holes of the p-square are filled with single symbols). By Lemma 19(b), we can find the minimal cyclic shift of the k-fragment in $\mathcal{O}(k^2)$ time. The cyclic shift can be represented as a k-fragment as well. We apply this to find $(U')^2$, the minimal cyclic shift of U^2. Then $repr(\beta) = (U')^2$.   □

We group non-solid pseudoruns by their periods first; let $\mathcal{R}_q$ be the set of non-solid pseudoruns with period q. From what we have already observed, we see that every pseudorun from $\mathcal{R}_q$ can overlap with at most six other pseudoruns from $\mathcal{R}_q$: two that come from the same Q-run and two that come from each of the neighbouring Q-runs with period q. Hence, each hole position is contained in at most seven pseudoruns from $\mathcal{R}_q$, and $|\mathcal{R}_q| \le 7k$. The representatives of pseudoruns from $\mathcal{R}_q$ can be sorted using $\mathcal{O}(k)$-time comparison (Lemma 19(a)). Thus the time complexity for sorting and grouping all pseudoruns from $\mathcal{R}_q$ is $\mathcal{O}(k^2 \log k)$, which gives $\mathcal{O}(nk^2 \log k)$ in total.

By Observation 14, the representatives of all p-squares induced by a pseudorun β are cyclic shifts of $repr(\beta)$. Thus only pseudoruns from the same group may induce equivalent p-squares. For each pseudorun β we can specify an interval $I(\beta)$ of cyclic shift values of induced p-squares. Then all non-equivalent p-squares induced by pseudoruns in the same group can be reported by carefully processing the intervals $I(\beta)$ as in [10]. This processing takes time linear in the number of all intervals from all groups and n, i.e., $\mathcal{O}(nk)$ time. This concludes the algorithm.

Theorem 21. *All non-equivalent p-squares in a partial word of length n with k holes can be reported (as factors of the partial word) in $\mathcal{O}(nk^3)$ time.*

Proof. Lemma 18 shows that there are $\mathcal{O}(nk)$ pseudoruns in a partial word and they can all be computed in $\mathcal{O}(nk^2)$ time. Solid pseudoruns can be handled separately in $\mathcal{O}(n)$ time. Lemma 20 lets us find the representatives of non-solid pseudoruns in $\mathcal{O}(nk^3)$ time. In the end, we group those pseudoruns by the representatives in $\mathcal{O}(nk^2 \log k)$ time and use the approach from [10] to report all non-equivalent p-squares induced by each group in $\mathcal{O}(nk)$ time.   □

References

1. Bannai, H., I, T., Inenaga, S., Nakashima, Y., Takeda, M., Tsuruta, K.: A new characterization of maximal repetitions by Lyndon trees. In: Indyk, P. (ed.) 26th Annual ACM-SIAM Symposium on Discrete Algorithms, SODA 2015, pp. 562–571. SIAM (2015)
2. Blanchet-Sadri, F., Bodnar, M., Nikkel, J., Quigley, J.D., Zhang, X.: Squares and primitivity in partial words. Discrete Appl. Math. **185**, 26–37 (2015)
3. Blanchet-Sadri, F., Jiao, Y., Machacek, J.M., Quigley, J., Zhang, X.: Squares in partial words. Theor. Comput. Sci. **530**, 42–57 (2014)
4. Blanchet-Sadri, F., Mercaş, R.: A note on the number of squares in a partial word with one hole. Inform. Theor. Appl. **43**(4), 767–774 (2009)
5. Blanchet-Sadri, F., Mercaş, R.: The three-squares lemma for partial words with one hole. Theor. Comput. Sci. **428**, 1–9 (2012)
6. Blanchet-Sadri, F., Mercaş, R., Scott, G.: Counting distinct squares in partial words. Acta Cybern. **19**(2), 465–477 (2009)
7. Blanchet-Sadri, F., Nikkel, J., Quigley, J.D., Zhang, X.: Computing primitively-rooted squares and runs in partial words. In: Kratochvíl, J., Miller, M., Froncek, D. (eds.) IWOCA 2014. LNCS, vol. 8986, pp. 86–97. Springer, Cham (2015). doi:10.1007/978-3-319-19315-1_8
8. Crochemore, M., Hancart, C., Lecroq, T.: Algorithms on Strings. Cambridge University Press, Cambridge (2007)
9. Crochemore, M., Iliopoulos, C.S., Kociumaka, T., Kubica, M., Langiu, A., Radoszewski, J., Rytter, W., Szreder, B., Waleń, T.: A note on the longest common compatible prefix problem for partial words. J. Discrete Algorithms **34**, 49–53 (2015)
10. Crochemore, M., Iliopoulos, C.S., Kubica, M., Radoszewski, J., Rytter, W., Waleń, T.: Extracting powers and periods in a word from its runs structure. Theor. Comput. Sci. **521**, 29–41 (2014)

11. Crochemore, M., Rytter, W.: Squares, cubes, and time-space efficient string searching. Algorithmica **13**(5), 405–425 (1995)
12. Deza, A., Franek, F., Thierry, A.: How many double squares can a string contain? Discrete Appl. Math. **180**, 52–69 (2015)
13. Diaconu, A., Manea, F., Tiseanu, C.: Combinatorial queries and updates on partial words. In: Kutyłowski, M., Charatonik, W., Gębala, M. (eds.) FCT 2009. LNCS, vol. 5699, pp. 96–108. Springer, Heidelberg (2009). doi:10.1007/978-3-642-03409-1_10
14. Fraenkel, A.S., Simpson, J.: How many squares can a string contain? J. Comb. Theory, Ser. A **82**(1), 112–120 (1998)
15. Gusfield, D., Stoye, J.: Linear time algorithms for finding and representing all the tandem repeats in a string. J. Comput. Syst. Sci. **69**(4), 525–546 (2004)
16. Halava, V., Harju, T., Kärki, T.: On the number of squares in partial words. RAIRO Theor. Inform. Appl. **44**(1), 125–138 (2010)
17. Ilie, L.: A simple proof that a word of length n has at most $2n$ distinct squares. J. Comb. Theory, Ser. A **112**(1), 163–164 (2005)
18. Kociumaka, T.: Minimal suffix and rotation of a substring in optimal time. In: Grossi, R., Lewenstein, M. (eds.) Combinatorial Pattern Matching, CPM 2016. LIPIcs, vol. 54, pp. 28:1–28:12. Schloss Dagstuhl (2016)
19. Kolpakov, R.M., Kucherov, G.: Finding maximal repetitions in a word in linear time. In: 40th Annual Symposium on Foundations of Computer Science, FOCS 1999, pp. 596–604. IEEE Computer Society (1999)
20. Manea, F., Mercaş, R., Tiseanu, C.: An algorithmic toolbox for periodic partial words. Discrete Appl. Math. **179**, 174–192 (2014)
21. Manea, F., Tiseanu, C.: Hard counting problems for partial words. In: Dediu, A.-H., Fernau, H., Martín-Vide, C. (eds.) LATA 2010. LNCS, vol. 6031, pp. 426–438. Springer, Heidelberg (2010). doi:10.1007/978-3-642-13089-2_36

The Approximability of the p-hub Center Problem with Parameterized Triangle Inequality

Li-Hsuan Chen[1]([✉]), Sun-Yuan Hsieh[1], Ling-Ju Hung[1], and Ralf Klasing[2]

[1] Department of Computer Science and Information Engineering,
National Cheng Kung University, Tainan 701, Taiwan
{clh100p,hunglc}@cs.ccu.edu.tw, hsiehsy@mail.ncku.edu.tw
[2] CNRS, LaBRI, Université de Bordeaux, 351 Cours de la Libération,
33405 Talence Cedex, France
ralf.klasing@labri.fr

Abstract. A complete weighted graph $G = (V, E, w)$ is called Δ_β-metric, for some $\beta \geq 1/2$, if G satisfies the β-triangle inequality, *i.e.,* $w(u, v) \leq \beta \cdot (w(u, x) + w(x, v))$ for all vertices $u, v, x \in V$. Given a Δ_β-metric graph $G = (V, E, w)$ and an integer p, the Δ_β-pHUB CENTER PROBLEM (Δ_β-pHCP) is to find a spanning subgraph H^* of G such that (i) vertices (hubs) in $C^* \subset V$ form a clique of size p in H^*; (ii) vertices (non-hubs) in $V \backslash C^*$ form an independent set in H^*; (iii) each non-hub $v \in V \backslash C^*$ is adjacent to exactly one hub in C^*; and (iv) the diameter $D(H^*)$ is minimized. For $\beta = 1$, Δ_β-pHCP is NP-hard. (Chen *et al., CMCT 2016*) proved that for any $\varepsilon > 0$, it is NP-hard to approximate the Δ_β-pHCP to within a ratio $\frac{4}{3} - \varepsilon$ for $\beta = 1$. In the same paper, a $\frac{5}{3}$-approximation algorithm was given for Δ_β-pHCP for $\beta = 1$. In this paper, we study Δ_β-pHCP for all $\beta \geq \frac{1}{2}$. We show that for any $\varepsilon > 0$, to approximate the Δ_β-pHCP to a ratio $g(\beta) - \varepsilon$ is NP-hard and we give $r(\beta)$-approximation algorithms for the same problem where $g(\beta)$ and $r(\beta)$ are functions of β. If $\beta \leq \frac{3-\sqrt{3}}{2}$, we have $r(\beta) = g(\beta) = 1$, *i.e.,* Δ_β-pHCP is polynomial time solvable. If $\frac{3-\sqrt{3}}{2} < \beta \leq \frac{2}{3}$, we have $r(\beta) = g(\beta) = \frac{3\beta - 2\beta^2}{3(1-\beta)}$. For $\frac{2}{3} \leq \beta \leq \frac{5+\sqrt{5}}{10}$, $r(\beta) = g(\beta) = \beta + \beta^2$. Moreover, for $\beta \geq 1$, we have $g(\beta) = \beta \cdot \frac{4\beta - 1}{3\beta - 1}$ and $r(\beta) = 2\beta$, the approximability of the problem (*i.e.,* upper and lower bound) is linear in β.

Parts of this research were supported by the Ministry of Science and Technology of Taiwan under grants MOST 105–2221–E–006–164–MY3, and MOST 103–2221–E–006–135–MY3.

Li-Hsuan Chen is supported by the Ministry of Science and Technology of Taiwan under grant MOST 106–2811–E–006–008.

Ling-Ju Hung is supported by the Ministry of Science and Technology of Taiwan under grants MOST 105–2811–E–006–046.

Part of this work was done while Ralf Klasing was visiting the Department of Computer Science and Information Engineering at National Cheng Kung University. This study has been carried out in the frame of the "Investments for the future" Programme IdEx Bordeaux - CPU (ANR-10-IDEX-03-02). Research supported by the LaBRI under the "Projets émergents" program.

Y. Cao and J. Chen (Eds.): COCOON 2017, LNCS 10392, pp. 112–123, 2017.
DOI: 10.1007/978-3-319-62389-4_10

1   Introduction

The *hub location problems* have various applications in transportation and telecommunication systems. Variants of hub location problems have been defined and well-studied in the literatures (see the two survey papers [1,15]). Suppose that we have a set of demand nodes that want to communicate with each other through some hubs in a network. A *single allocation hub location problem* requests each demand node can only be served by exactly one hub. Conversely, if a demand node can be served by several hubs, then this kind of hub location problem is called *multi-allocation.* Classical hub location problems ask to minimize the total cost of all origin-destination pairs (see *e.g.,* [28]). However, minimizing the total routing cost would lead to the result that the poorest service quality is extremely bad. In this paper, we consider a single allocation hub location problem with min-max criterion, called Δ_β-p HUB CENTER PROBLEM which is different from the classic hub location problems. The min-max criterion is able to avoid the drawback of minimizing the total cost.

A complete weighted graph $G = (V, E, w)$ is called Δ_β-metric, for some $\beta \geq 1/2$, if the distance function $w(\cdot, \cdot)$ satisfies $w(v, v) = 0$, $w(u, v) = w(v, u)$, and the β-triangle inequality, *i.e.,* $w(u, v) \leq \beta \cdot (w(u, x) + w(x, v))$ for all vertices $u, v, x \in V$. (If $\beta > 1$ then we speak about *relaxed triangle inequality,* and if $\beta < 1$ we speak about *sharpened triangle inequality.*)

Lemma 1 ([8]). *Let $G = (V, E)$ be a Δ_β-metric graph for $\frac{1}{2} \leq \beta < 1$. For any two edges $(u, x), (v, x)$ with a common endvertex x in G, $w(u, x) \leq \frac{\beta}{1-\beta} \cdot w(v, x)$.*

Let u, v be two vertices in a graph H. Use $d_H(u, v)$ to denote the distance between u, v in H. Define $D(H) = \max_{u,v \in H} d_H(u, v)$ called the diameter of H. We give the definition of the Δ_β-pHUB CENTER PROBLEM as follows.

Δ_β-p HUB CENTER PROBLEM (Δ_β-pHCP)
Input: A Δ_β-metric graph $G = (V, E, w)$ and a positive integer p.
Output: A spanning subgraph H^* of G such that (i) vertices (hubs) in $C^* \subset V$
            form a clique of size p in H^*; (ii) vertices (non-hubs) in $V \backslash C^*$ form
            an independent set in H^*; (iii) each non-hub $v \in V \backslash C^*$ is adjacent
            to exactly one hub in C^*; and (iv) the diameter $D(H^*)$ is minimized.

The Δ_β-pHCP problem is a general version of the original p-HUB CENTER PROBLEM (pHCP) since the original problem assumes the input graph to be a metric graph, *i.e.,* $\beta = 1$. We use pHCP to denote the Δ_β-pHCP for $\beta = 1$.

The pHCP is NP-hard in metric graphs [23]. Several approaches for PHCP with linear and quadratic integer programming were proposed in the literature [14,19,23,25]. Many research efforts for solving the pHCP are focused on the development of heuristic algorithms, *e.g.,* [13,26,27,30–32]. Chen *et al.* [16] proved that for any $\varepsilon > 0$, it is NP-hard to approximate the pHCP to within a ratio $4/3 - \varepsilon$. In the same paper, a $\frac{5}{3}$-approximation algorithm was given for pHCP.

The Star p-Hub Center Problem (SpHCP) introduced in [29] is closely related to pHCP and well-studied in [17, 18, 24]. The difference between the two problems is that in SpHCP, the hubs are connected to a center rather than fully connected.

If $\beta = 1$, Δ_β-pHCP is NP-hard and even NP-hard to have a $(\frac{4}{3} - \varepsilon)$-approximation algorithm for any $\varepsilon > 0$ [16]. In this paper, we investigate the complexity of Δ_β-pHCP parameterized by β-triangle inequality. The motivation of this research for $\beta < 1$ is to investigate whether there exists a large subclass of input instances of Δ_β-pHCP that can be solved in polynomial time or admits polynomial-time approximation algorithms with a reasonable approximation ratio. For $\beta \geq 1$, it is an interesting issue to see whether there exists a polynomial-time approximation algorithm with an approximation ratio linear in β.

Our study uses the well-known concept of *stability of approximation* for hard optimization problems [9, 11, 21, 22]. The idea of this concept is similar to that of the stability of numerical algorithms. But instead of observing the size of the change in the output value according to a small change of the input value, one is interested in the size of the change of the approximation ratio according to a small change in the specification (some parameters, characteristics) of the set of problem instances considered. If the change of the approximation ratio is small for every small change in the set of problem instances, then the algorithm is called *stable*. The concept of *stability of approximation* has been successfully applied to several fundamental hard optimization problems. E.g. in [2–4, 8–10, 12] it was shown that one can partition the set of all input instances of the Traveling Salesman Problem into infinitely many subclasses according to the degree of violation of the triangle inequality, and for each subclass one can guarantee upper and lower bounds on the approximation ratio. Similar studies demonstrated that the β-triangle inequality can serve as a measure of hardness of the input instances for other problems as well, in particular for the problem of constructing 2-connected spanning subgraphs of a given complete edge-weighted graph [5], and for the problem of finding, for a given positive integer $k \geq 2$ and an edge-weighted graph G, a minimum k-edge- or k-vertex-connected spanning subgraph [6, 7].

Table 1. The main results where Δ_β-pHCP cannot be approximated within $g(\beta) - \varepsilon$ and has an $r(\beta)$-approximation algorithm.

β	Lower bound $g(\beta)$	Upper bound $r(\beta)$
$[\frac{1}{2}, \frac{3-\sqrt{3}}{2}]$	1	1
$(\frac{3-\sqrt{3}}{2}, \frac{2}{3}]$	$\frac{3\beta - 2\beta^2}{3(1-\beta)}$	$\frac{3\beta - 2\beta^2}{3(1-\beta)}$
$[\frac{2}{3}, \frac{5+\sqrt{5}}{10}]$	$\beta + \beta^2$	$\beta + \beta^2$
$[\frac{5+\sqrt{5}}{10}, \frac{3+\sqrt{29}}{10}]$	$\frac{4\beta^2 + 3\beta - 1}{5\beta - 1}$	$\beta + \beta^2$
$[\frac{3+\sqrt{29}}{10}, 1]$	$\frac{4\beta^2 + 3\beta - 1}{5\beta - 1}$	$\frac{4\beta^2 + 5\beta + 1}{5\beta + 1}$
$[1, \infty)$	$\beta \cdot \frac{4\beta - 1}{3\beta - 1}$	2β

In Table 1, we list the main results of this paper. We prove that for any $\varepsilon > 0$, to approximate Δ_β-pHCP to a ratio $g(\beta) - \varepsilon$ is NP-hard where $\beta > \frac{3-\sqrt{3}}{2}$ and $g(\beta)$ is a function of β. We give $r(\beta)$-approximation algorithms for Δ_β-pHCP. If $\beta \leq \frac{3-\sqrt{3}}{2}$, we have $r(\beta) = g(\beta) = 1$, i.e., Δ_β-pHCP is polynomial time solvable. If $\frac{3-\sqrt{3}}{2} < \beta \leq \frac{5+\sqrt{5}}{10}$, we have $r(\beta) = g(\beta)$. For $\frac{5+\sqrt{5}}{10} \leq \beta \leq 1$, $r(\beta) = \min\{\beta + \beta^2, \frac{4\beta^2+5\beta+1}{5\beta+1}\}$ and $g(\beta) = \frac{4\beta^2+3\beta-1}{5\beta-1}$. Moreover, for $\beta \geq 1$, we have $r(\beta) = 2\beta$ and $g(\beta) = \beta \cdot \frac{4\beta-1}{3\beta-1}$. For $\beta \geq 1$, the approximability of the problem (i.e., upper and lower bound) is linear in β.

We use C_H to denote the set of hub vertices in solution H. Let H^* be an optimal solution of Δ_β-pHCP in a given β-metric graph $G = (V, E, w)$. For a non-hub x in H^*, we use $f^*(x)$ to denote the hub adjacent to x in H^*. We use $\tilde{H}$ to denote the best solution among all solutions in $\mathcal{H}$ where $\mathcal{H}$ is the collection of all solutions satisfying that all non-hubs are adjacent to the same hub for Δ_β-pHCP in a given β-metric graph $G = (V, E, w)$.

2   Inapproximability Results

We show that for $\beta > \frac{3-\sqrt{3}}{2}$, to approximate Δ_β-pHCP to a factor $g(\beta) - \varepsilon$ is NP-hard where $g(\beta) = \frac{3\beta-2\beta^2}{3(1-\beta)}$ for $\frac{3-\sqrt{3}}{2} < \beta \leq \frac{2}{3}$, $g(\beta) = \beta + \beta^2$ for $\frac{2}{3} \leq \beta \leq \frac{5+\sqrt{5}}{10}$, $g(\beta) = \frac{4\beta^2+3\beta-1}{5\beta-1}$ for $\frac{5+\sqrt{5}}{10} \leq \beta \leq 1$, and $g(\beta) = \beta \cdot \frac{4\beta-1}{3\beta-1}$ for $\beta \geq 1$.

Theorem 1. *Let $\beta > \frac{3-\sqrt{3}}{2}$. For any $\varepsilon > 0$, to approximate Δ_β-pHCP to a factor $g(\beta) - \varepsilon$ is NP-hard where*

(i) $g(\beta) = \frac{3\beta-2\beta^2}{3(1-\beta)}$ if $\frac{3-\sqrt{3}}{2} < \beta \leq \frac{2}{3}$;

(ii) $g(\beta) = \beta + \beta^2$ if $\frac{2}{3} \leq \beta \leq \frac{5+\sqrt{5}}{10}$;

(iii) $g(\beta) = \frac{4\beta^2+3\beta-1}{5\beta-1}$ if $\frac{5+\sqrt{5}}{10} \leq \beta \leq 1$;

(vi) $g(\beta) = \beta \cdot \frac{4\beta-1}{3\beta-1}$ if $\beta \geq 1$.

Proof. Due to the limitation of space, we omit the proof of (ii)–(vi). In the following, we will prove that, if Δ_β-pHCP can be approximated to within a factor $\frac{3\beta-2\beta^2}{3(1-\beta)} - \varepsilon$ in polynomial time, for some $\varepsilon > 0$, then SET COVER can be solved in polynomial time. This will complete the proof, since SET COVER is well-known to be NP-hard [20].

Let $(\mathcal{S}, U)$ be an instance of SET COVER where U is the universal set, $|U| = n$, and $\mathcal{S} = \{S_1, S_2, \ldots, S_m\}$ is a collection of subsets of U, $|\mathcal{S}| = m$. The goal is to decide whether $\mathcal{S}$ has a subset $\mathcal{S}'$ of size k such that $\bigcup_{S_i \in \mathcal{S}'} S_i = U$. In the following, we construct a β-metric graph $G = (V \cup S \cup \{y\}, E, w)$ according to $(\mathcal{S}, U)$. For each element $v \in U$, construct a vertex $v \in V$, i.e., $V = U$. For each set $S_i \in \mathcal{S}$, construct a vertex $s_i \in S$, $|S| = |\mathcal{S}|$. We add a vertex y in G. The edge cost of G is defined in Table 2.

Table 2. The costs of edges (a, b) in G

$w(a,b)$	$b \in S$	$b \in V$	$b = y$
$a \in S$	1	1 if $b \in a$	$\frac{\beta}{1-\beta}$
		2β otherwise	
$a \in V$	1 if $a \in b$	2β	$\frac{\beta}{1-\beta}$
	2β otherwise		

Clearly, G can be constructed in polynomial time. It is easy to verify that G is a β-metric graph. Let G be the input of Δ_β-pHCP constructed according to (S, U) where $p = k + 1$.

Let $S' \subset S$ be a set cover of (S, U) of size $k > 1$. We then construct a solution H of Δ_β-pHCP according to S'. For each set $S_i \in S'$, collect the vertex $s_i \in S'$ in G. Let all vertices in $S' \cup \{y\}$ be hubs where $|S'| = |S'|$. For each $v \in V$, connect v to exactly one vertex $s_i \in S'$ satisfying that $v \in S_i$. Since each $v \in V$ is connected to a vertex $s_i \in S'$ satisfying that $v \in S_i$, we see that $w(v, s_i) = 1$. Hence $D(H) = 3$. Let H^* denote an optimal solution of Δ_β-pHCP in G. We have $D(H^*) \leq 3$.

Assume that there exists a polynomial time algorithm that finds a solution H of Δ_β-pHCP in G with $D(H) < \frac{3\beta - 2\beta^2}{1-\beta}$. W.l.o.g., assume that $C_H = S' \cup V' \cup Y'$ where $S' \subseteq S$, $V' \subseteq V$, and $Y' \subseteq \{y\}$. For any non-hub v in H, use $f(v)$ to denote the hub in H adjacent to v. Note that if v is a hub in H, let $f(v) = v$. For u, v in H, let $d_H(u, v) = w(u, f(u)) + w(f(u), f(v)) + w(v, f(v))$ be the distance between u and v in H.

Claim 1. The vertex y must be a hub.

Proof of Claim. Suppose that y is not a hub in H. There are two cases.

- If $f(y) \in S'$, then all vertices $v \in V$ must adjacent to $f(y)$ and satisfying $w(v, f(y)) = 1$; otherwise there exists $x \in V$ with

$$d_H(x, y) = d_H(x, f(y)) + w(f(y), y)$$
$$\geq 2\beta + \frac{\beta}{1-\beta} \quad \text{(since } \frac{3-\sqrt{3}}{2} < \beta \leq \frac{2}{3})$$
$$= \frac{3\beta - 2\beta^2}{1-\beta},$$

  a contradiction to the assumption that $D(H) < \frac{3\beta - 2\beta^2}{1-\beta}$. Thus the set in S with respect to the vertex $f(y)$ forms a set cover of (S, U). This contradicts to the assumption that the optimal solution of SET COVER is of size $k > 1$.
- If $f(y) \in V'$, then there exists $x \in V \backslash C_H$ with

$$d_H(x, y) = d_H(x, f(y)) + w(f(y), y)$$
$$\geq 2\beta + \frac{\beta}{1-\beta} \quad \text{(since } \frac{3-\sqrt{3}}{2} < \beta \leq \frac{2}{3})$$
$$\geq \frac{3\beta - 2\beta^2}{1-\beta},$$

  a contradiction to the assumption that $D(H) < \frac{3\beta - 2\beta^2}{1-\beta}$.

Thus, y must be a hub, *i.e.*, $Y' = \{y\}$.   ∎

Claim 2. The hub y is not adjacent to any non-hub in H.

Proof of Claim. Suppose that the hub y is adjacent to a non-hub $z \in (S \cup V) \backslash C_H$, then there exists $x \in C_H$ with

$$d_H(x, z) = w(x, y) + w(y, z) \geq \frac{\beta}{1-\beta} + \frac{\beta}{1-\beta} \geq \frac{3\beta - 2\beta^2}{1-\beta},$$

a contradiction to the assumption that $D(H) < \frac{3\beta - 2\beta^2}{1-\beta}$.

Thus, y is not adjacent to any non-hub in H.   ∎

Claim 3. No $v \in V \backslash V'$ is adjacent to any $u \in V'$.

Proof of Claim. Suppose that there exists $v \in V \backslash V'$ is adjacent to $u \in V'$ in H. We see that

$$d_H(v, y) = w(v, u) + w(u, y) = 2\beta + \frac{\beta}{1-\beta} \geq \frac{3\beta - 2\beta^2}{1-\beta},$$

a contradiction to the assumption that $D(H) < \frac{3\beta - 2\beta^2}{1-\beta}$. Thus, no $v \in V \backslash V'$ is adjacent to any $u \in V'$.   ∎

According to Claims 1, 2, and 3, in H all vertices $V \backslash V'$ must be adjacent to vertices in S'. If there exists $v \in V \backslash V'$ satisfying that $w(v, f(v)) = 2\beta$, then

$$d_H(v, y) = w(v, f(v)) + w(f(v), y) = 2\beta + \frac{\beta}{1-\beta} = \frac{3\beta - 2\beta^2}{1-\beta},$$

a contradiction to the assumption that $D(H) < \frac{3\beta - 2\beta^2}{1-\beta}$. Thus, each $v \in V \backslash V'$ satisfies that $w(v, f(v)) = 1$. We see that $\mathcal{S}' = S'$ forms a set cover of $V \backslash V'$. For each $u \in V'$, pick a set $S_i \in \mathcal{S}$ satisfying $u \in S_i$, call the collection of sets $\mathcal{S}''$. It is easy to see that $|\mathcal{S}''| = |V'|$ and $\mathcal{S}' \cup \mathcal{S}''$ forms a set cover of $V = U$ of size at most k. This shows that if Δ_β-pHCP has a solution H with $D(H) < \frac{3\beta - 2\beta^2}{1-\beta}$ that can be found in polynomial time, then SET COVER can be solved in polynomial time. However, SET COVER is a well-known NP-hard problem [20]. By the fact that SET COVER is NP-hard and $D(H^*) \leq 3$, this implies that for any $\varepsilon > 0$, to approximate Δ_β-pHCP to a factor $\frac{3\beta - 2\beta^2}{3(1-\beta)} - \varepsilon$ is NP-hard. This completes the proof of (i).   □

3 Polynomial-Time Algorithms

In this section, we show that for $\frac{1}{2} \leq \beta \leq \frac{3-\sqrt{3}}{2}$, Δ_β-pHCP can be solved in polynomial time. Besides, we give polynomial-time approximation algorithms for Δ_β-pHCP for $\beta > \frac{3-\sqrt{3}}{2}$. For $\frac{3-\sqrt{3}}{2} < \beta \leq \frac{5+\sqrt{5}}{10}$, our approximation algorithm achieves the factor that closes the gap between the upper and lower bounds of approximability for Δ_β-pHCP.

Due to the limitation of space, we omit some proofs in this section.

Lemma 2. *Let $\frac{1}{2} \le \beta < 1$. Then the following statements hold.*

(i) *There exists a solution $\tilde{H}$ satisfying that all non-hubs are adjacent to the same hub and $D(\tilde{H}) \le \max\{1, \min\{\frac{3\beta-2\beta^2}{3(1-\beta)}, \beta + \beta^2\}\} \cdot D(H^*)$.*

(ii) *There exists a polynomial-time algorithm to compute a solution H such that $D(H) = D(\tilde{H})$.*

According to Lemma 2, we obtain the following results.

Lemma 3. *Let $\frac{1}{2} \le \beta \le \frac{3+\sqrt{29}}{10}$. Then the following statements hold.*

1. *If $\beta \le \frac{3-\sqrt{3}}{2}$, then Δ_β-pHCP can be solved in polynomial time.*
2. *If $\frac{3-\sqrt{3}}{2} < \beta \le \frac{3+\sqrt{29}}{10}$, there is a $\min\{\frac{3\beta-2\beta^2}{3(1-\beta)}, \beta + \beta^2\}$-approximation algorithm for Δ_β-pHCP.*

Proof. Let H^* denote an optimal solution of the Δ_β-pHCP problem. According to Lemma 2, there is a polynomial-time algorithm for Δ_β-pHCP to compute a solution H such that $D(H) \le \max\{1, \min\{\frac{3\beta-2\beta^2}{3(1-\beta)}, \beta + \beta^2\}\} \cdot D(H^*)$.

If $\beta \le \frac{3-\sqrt{3}}{2}$, then $D(H) \le \max\{1, \min\{\frac{3\beta-2\beta^2}{3(1-\beta)}, \beta + \beta^2\}\} \cdot D(H^*) = D(H^*)$.

If $\frac{3-\sqrt{3}}{2} < \beta \le \frac{3+\sqrt{29}}{10}$, we see that

$$D(H) \le \max\{1, \min\{\frac{3\beta - 2\beta^2}{3(1 - \beta)}, \beta + \beta^2\}\} \cdot D(H^*) = \min\{\frac{3\beta - 2\beta^2}{3(1 - \beta)}, \beta + \beta^2\} \cdot D(H^*).$$

This completes the proof.   □

Algorithm 1. Approximation algorithm for Δ_β-pHCP (G, c).

(i) Run Algorithm APX1.
(ii) Run Algorithm APX2.
(iii) Return the best solution found by Algorithms APX1 and APX2.

Algorithm APX1

Guess the correct edge (y, z) where $w(y, z) = \ell$ is the largest edge cost in an optimal solution H^* with y as a hub and z as a non-hub. Let $U := V$ and $c_1 = y$. Let H_1 be the graph found by the following steps and C be the hub set in H_1. Initialize $C = \emptyset$.

(i) Let $C := C \cup \{c_1\}$, and let $U := U \setminus \{c_1\}$.
(ii) For $x \in U$, if $w(c_1, x) \le \ell$, add an edge (x, c_1) in H_1 and let $U := U \setminus \{x\}$.
(iii) While $i = |C| + 1 \le p$ and $U \ne \emptyset$,
   – choose $v \in U$, let $c_i = v$, connect c_i to all other vertices in C, let $U := U \setminus \{v\}$, and let $C := C \cup \{c_i\}$;
   – for $x \in U$, if $w(x, c_i) \le 2\beta\ell$, then add edge (x, c_i) in H_1 and $U := U \setminus \{x\}$.
(vi) If $|C| < p$ and $U = \emptyset$, we arbitrarily select $p - |C|$ non-hubs to be hubs and connect all edges between hubs.

Algorithm APX2

Guess all possible edges (y, z) to be a longest edge in H^* with one end vertex as a hub and the other end vertex as a non-hub $i.e.$, $f^*(z) = y$ and $w(z,y) \geq w(v, f^*(v))$ for all non-hubs v. Let H_2 be the graph found by the following steps and C' be the hub set of H_2.

(i) Connect y to all vertices in U.
(ii) Pick $(p - 1)$ vertices $\{v_1, v_2, \ldots, v_{p-1}\}$ with largest distance to y from $U\backslash\{y, z\}$. Let $C' = \{y, v_1, v_2, \ldots, v_{p-1}\}$.
(iii) Connect all pairs of vertices in C'.

It is not hard to see that Algorithm 1 runs in polynomial time. Let ℓ be the largest edge cost in H^* with one end vertex as a hub and the other end vertex as a non-hub. Note that both Algorithm APX1 and Algorithm APX2 guess all possible edges (y, z) to be the longest edge in H^* with y as a hub and z as a non-hub.

Lemma 4. *Let H_1 be the best solution returned by Algorithm APX1 and H^* be the optimal solution. Then $D(H_1) \leq D(H^*) + 4\beta\ell$ for $\beta \leq 1$.*

Proof. Let H^* be an optimal solution of Δ_β-pHCP and let $f(u)$ be the hub adjacent to vertex u in H_1 and $f(u) = u$ if u is a hub.

Removing edges with both end vertices in $C^* = \{s_1, s_2, \ldots, s_p\}$ from H^* obtains p components and each component is a star. Let $S_1, S_2, \ldots, S_p$ be the p stars and s_i be the center of star S_i for $i = 1, 2, \ldots, p$. W.l.o.g., assume that $c_1 = s_1$. Because for each $v \in V\backslash C^*$, $w(v, f^*(v)) \leq \ell$, we see that for $u, v \in S_i$, $d_{H^*}(u, v) \leq 2\beta\ell$. Since the algorithm adds edges (v, c_1) in H_1 if $w(v, c_1) \leq \ell$, we see that $S_1 \subset N_H[c_1]\backslash C$. Notice that for each S_j, $j \geq 2$, if there exists $v \in S_j$ specified as $c_i \in C$, then all the other vertices in S_j are connected to one of $c_1, c_2, \ldots, c_i$ in H_1. Moreover, for each c_i, $1 < i \leq |C|$, there exists S_j, $1 < j \leq p$, such that $c_i \in S_j$ and $S_j \cap C = \{c_i\}$. Notice that if there exists S_j, $1 < j \leq p$, $S_j \cap C = \emptyset$, then all vertices of S_j must be connected to one of vertices in C in H_1 and $|C| \leq p$. H_1 is a feasible solution if $|C| = p$. Suppose that $|C| < p$ and the algorithm selects $p - |C|$ vertices non-hubs to be hubs in Step (iv). Thus, H_1 returned by Algorithm APX1 is a feasible solution.

We then show that $D(H_1) \leq D(H^*) + 4\beta\ell$. For any two hubs u, v in H_1, $d_{H_1}(u, v) = w(u, v) \leq D(H^*)$ if $\beta \leq 1$. Notice that each non-hub v in H_1 is adjacent to a hub $f(v)$ in H_1 if $w(v, f(v)) \leq 2\beta\ell$. Thus, for u, v in H_1, $d_{H_1}(u, v) \leq D(H^*) + 4\beta\ell$ and $D(H_1) \leq D(H^*) + 4\beta\ell$. Since ℓ is the largest edge cost in H^* with y as a hub and z as a non-hub, the pairwise distances between non-hubs which are connected to the same hub in H^* are at most $4\beta\ell$. This completes the proof. $\qquad\square$

Lemma 5. *Let H_2 be the best solution returned by Algorithm APX2 and H^* be the optimal solution. Then $D(H_2) \leq \max\{D(H^*), (1 + \beta) \cdot (D(H^*) - \ell), 2\beta(D(H^*) - \ell)\}$.*

Proof. Let H^* be an optimal solution. For a non-hub v, use $f^*(v)$ to denote the hub adjacent to v in H^*. For a hub v in H^*, let $f^*(v) = v$. Notice that Algorithm APX2 guesses all possible edges (y, z) to be a longest edge in H^* with one end vertex as a hub and the other end vertex as a non-hub. In the following we assume that $w(y, z) = \ell$ is the largest edge cost in H^* with y as a hub and z as a non-hub. We see that for any hub v in H_2, $d_{H_2}(v, y) = w(v, y) \leq D(H^*) - \ell$.

For two non-hubs u, v in H_2, we have the following three cases.

- $f^*(u) = f^*(v) = y$, we see that $d_{H_2}(u, v) = d_{H^*}(u, v) \leq D(H^*)$.
- $f^*(u) = y$ and $f^*(v) \neq y$, we see that

$$d_{H_2}(u, v) = w(u, y) + w(v, y) \leq \ell + \beta \cdot d_{H^*}(v, y) \leq \ell + \beta \cdot (D(H^*) - \ell) \leq D(H^*).$$

- $f^*(u) \neq y$ and $f^*(v) \neq y$, we see that

$$d_{H_2}(u, v) = w(u, y) + w(v, y) \leq \beta \cdot d_{H^*}(u, y) + \beta \cdot d_{H^*}(v, y) \leq 2\beta(D(H^*) - \ell).$$

For a non-hub u and a hub v in H_2, there are two cases.

- If $f^*(u) = y$, we see that

$$d_{H_2}(u, v) = w(u, y) + d_{H_2}(v, y) \leq \ell + D(H^*) - \ell = D(H^*).$$

- If $f^*(u) \neq y$, we see that

$$d_{H_2}(u, v) = w(u, y) + d_{H_2}(v, y) \leq \beta \cdot (D(H^*) - \ell) + (D(H^*) - \ell).$$

For two hubs u, v in H_2, $u \neq y$ and $v \neq y$, we see that $d_{H_2}(u, v) = w(u, v) \leq D(H^*)$.

Thus, $D(H_2) \leq \max\{D(H^*), (1 + \beta) \cdot (D(H^*) - \ell), 2\beta(D(H^*) - \ell)\}$. This completes the proof. $\qquad\square$

Lemma 6. *Let* $\frac{3+\sqrt{29}}{10} \leq \beta \leq 1$. *Then, there is a* $(\frac{4\beta^2+5\beta+1}{5\beta+1})$-*approximation algorithm for* Δ_β-*pHCP.*

Proof. Let H^* be an optimal solution of Δ_β-pHCP. In this lemma, we show that for $\frac{3+\sqrt{29}}{10} \leq \beta \leq 1$, Algorithm 1 returns a solution H such that $D(H) \leq (\frac{4\beta^2+5\beta+1}{5\beta+1}) \cdot D(H^*)$.

By Lemmas 4 and 5, we see that the approximation ratio of Algorithm 1 is $r(\beta) = \min\{\frac{D(H_1)}{D(H^*)}, \frac{D(H_2)}{D(H^*)}\}$.

Note that if $\frac{\ell}{D(H^*)} \geq \frac{\beta}{1+\beta}$, then $D(H_2) = D(H^*)$. Assume that $\frac{\ell}{D(H^*)} < \frac{\beta}{1+\beta}$, we see that $D(H_2) \leq D(H^*) - \ell + \beta(D(H^*) - \ell)$.

The worst case approximation ratio of Algorithm 1 happens when $D(H_1) = D(H_2)$, i.e.,

$$D(H^*) + 4\beta\ell = \max\{(1 + \beta) \cdot (D(H^*) - \ell), 2\beta \cdot (D(H^*) - \ell)\}.$$

Since $\beta \leq 1$, we obtain that $\frac{\ell}{D(H^*)} = \frac{\beta}{5\beta+1}$. Thus,

$$r(\beta) = \min\{\frac{D(H_1)}{D(H^*)}, \frac{D(H_2)}{D(H^*)}\} \leq 1 + \frac{4\beta^2}{5\beta+1}.$$

This completes the proof. $\qquad\square$

In Lemma 7, we prove that if $\beta \geq 1$, Algorithm 1 is a 2β-approximation algorithm for Δ_β-$pHCP$.

Lemma 7. *Let $\beta \geq 1$. Then, there is a 2β-approximation algorithm for Δ_β-$pHCP$.*

Proof. Since $\ell \geq 0$, by Lemma 5, we have

$$D(H_2) \leq \max\{D(H^*), (1+\beta) \cdot (D(H^*) - \ell), 2\beta(D(H^*) - \ell)\} \leq 2\beta \cdot D(H^*).$$

This completes the proof. □

We close this section with the following theorem.

Theorem 2. *Let $\beta \geq \frac{1}{2}$. There exists a polynomial-time $r(\beta)$-approximation algorithm for Δ_β-$pHCP$ where*

(i) $r(\beta) = 1$ if $\beta \leq \frac{3-\sqrt{3}}{2}$;

(ii) $r(\beta) = \frac{3\beta-2\beta^2}{3(1-\beta)}$ if $\frac{3-\sqrt{3}}{2} < \beta \leq \frac{5+\sqrt{5}}{10}$;

(iii) $r(\beta) = \beta + \beta^2$ if $\frac{5+\sqrt{5}}{10} \leq \beta \leq \frac{3+\sqrt{29}}{10}$;

(vi) $r(\beta) = \frac{4\beta^2+5\beta+1}{5\beta+1}$ if $\frac{3+\sqrt{29}}{10} \leq \beta \leq 1$;

(v) $r(\beta) = 2\beta$ if $\beta \geq 1$.

4   Conclusion

In this paper, we have studied Δ_β-$pHCP$ for all $\beta \geq \frac{1}{2}$. We showed that for any $\varepsilon > 0$, to approximate Δ_β-$pHCP$ to a ratio $g(\beta) - \varepsilon$ is NP-hard where $g(\beta) = \frac{3\beta-2\beta^2}{3(1-\beta)}$ if $\frac{3-\sqrt{3}}{2} < \beta \leq \frac{2}{3}$; $g(\beta) = \beta + \beta^2$ if $\frac{2}{3} < \beta \leq \frac{5+\sqrt{5}}{10}$; $g(\beta) = \frac{4\beta^2+3\beta-1}{5\beta-1}$ if $\frac{5+\sqrt{5}}{10} < \beta \leq 1$; $g(\beta) = \beta \cdot \frac{4\beta-1}{3\beta-1}$ if $\beta \geq 1$. Moreover, we gave $r(\beta)$-approximation algorithms for the same problem. If $\beta \leq \frac{3-\sqrt{3}}{2}$, we have $r(\beta) = g(\beta) = 1$, *i.e.*, Δ_β-$pHCP$ is polynomial-time solvable for $\beta \leq \frac{3-\sqrt{3}}{2}$. If $\frac{3-\sqrt{3}}{2} < \beta \leq \frac{5+\sqrt{5}}{10}$, we have $r(\beta) = g(\beta)$. For $\frac{2}{3} \leq \beta \leq 1$, $r(\beta) = \min\{\beta + \beta^2, \frac{4\beta^2+5\beta+1}{5\beta+1}\}$. For $\beta \geq 1$, we have $r(\beta) = 2\beta$. In the future work, it is of interest to extend the range of β for Δ_β-$pHCP$ such that the gap between the upper and lower bounds of approximability can be reduced.

References

1. Alumur, S.A., Kara, B.Y.: Network hub location problems: the state of the art. Eur. J. Oper. Res. **190**, 1–21 (2008)
2. Andreae, T.: On the traveling salesman problem restricted to inputs satisfying a relaxed triangle inequality. Networks **38**, 59–67 (2001)
3. Andreae, T., Bandelt, H.-J.: Performance guarantees for approximation algorithms depending on parameterized triangle inequalities. SIAM J. Discr. Math. **8**, 1–16 (1995)

4. Bender, M.A., Chekuri, C.: Performance guarantees for the TSP with a parameterized triangle inequality. Inf. Process. Lett. **73**, 17–21 (2000)
5. Böckenhauer, H.-J., Bongartz, D., Hromkovič, J., Klasing, R., Proietti, G., Seibert, S., Unger, W.: On the Hardness of constructing minimal 2-connected spanning subgraphs in complete graphs with sharpened triangle inequality. In: Agrawal, M., Seth, A. (eds.) FSTTCS 2002. LNCS, vol. 2556, pp. 59–70. Springer, Heidelberg (2002). doi:10.1007/3-540-36206-1_7
6. Böckenhauer, H.-J., Bongartz, D., Hromkovič, J., Klasing, R., Proietti, G., Seibert, S., Unger, W.: On k-edge-connectivity problems with sharpened triangle inequality. In: Petreschi, R., Persiano, G., Silvestri, R. (eds.) CIAC 2003. LNCS, vol. 2653, pp. 189–200. Springer, Heidelberg (2003). doi:10.1007/3-540-44849-7_24
7. Böckenhauer, H.-J., Bongartz, D., Hromkovič, J., Klasing, R., Proietti, G., Seibert, S., Unger, W.: On k-connectivity problems with sharpened triangle inequality. J. Discr. Algorithms **6**(4), 605–617 (2008)
8. Böckenhauer, H.-J., Hromkovič, J., Klasing, R., Seibert, S., Unger, W.: Approximation algorithms for the TSP with sharpened triangle inequality. Inf. Process. Lett. **75**, 133–138 (2000)
9. Böckenhauer, H.-J., Hromkovič, J., Klasing, R., Seibert, S., Unger, W.: Towards the notion of stability of approximation for hard optimization tasks and the traveling salesman problem. In: Bongiovanni, G., Petreschi, R., Gambosi, G. (eds.) CIAC 2000. LNCS, vol. 1767, pp. 72–86. Springer, Heidelberg (2000). doi:10.1007/3-540-46521-9_7
10. Böckenhauer, H.-J., Hromkovič, J., Klasing, R., Seibert, S., Unger, W.: An improved lower bound on the approximability of metric TSP and approximation algorithms for the TSP with sharpened triangle inequality. In: Reichel, H., Tison, S. (eds.) STACS 2000. LNCS, vol. 1770, pp. 382–394. Springer, Heidelberg (2000). doi:10.1007/3-540-46541-3_32
11. Böckenhauer, H.-J., Hromkovič, J., Seibert, S.: Stability of approximation. In: Gonzalez, T.F. (ed.) Handbook of Approximation Algorithms and Metaheuristics, Chapman & Hall/CRC, Chap. 31 (2007)
12. Böckenhauer, H.-J., Seibert, S.: Improved lower bounds on the approximability of the traveling salesman problem. RAIRO Theor. Inf. Appl. **34**, 213–255 (2000)
13. Brimberg, J., Mladenović, N., Todosijević, R., Urošević, D.: General variable neighborhood search for the uncapacitated single allocation p-hub center problem, Optimization Letters. doi:10.1007/s11590-016-1004-x
14. Campbell, J.F.: Integer programming formulations of discrete hub location problems. Eur. J. Oper. Res. **72**, 387–405 (1994)
15. Campbell, J.F., Ernst, A.T.: Hub location problems. In: Drezner, Z., Hamacher, H.W. (eds.) Facility Location: Applications and Theory, pp. 373–407. Springer, Berlin (2002)
16. Chen, L.-H., Cheng, D.-W., Hsieh, S.-Y., Hung, L.-J., Lee, C.-W., Wu, B.Y.: Approximation algorithms for single allocation k-hub center problem. In: Proceedings of the 33rd Workshop on Combinatorial Mathematics and Computation Theory (CMCT 2016), pp. 13–18 (2016)
17. Chen, L.-H., Hsieh, S.-Y., Hung, L.-J., Klasing, R., Lee, C.-W., Wu, B.Y.: On the complexity of the star p-hub center problem with parameterized triangle inequality. In: Fotakis, D., Pagourtzis, A., Paschos, V.T. (eds.) CIAC 2017. LNCS, vol. 10236, pp. 152–163. Springer, Cham (2017). doi:10.1007/978-3-319-57586-5_14

18. Chen, L.-H., Cheng, D.-W., Hsieh, S.-Y., Hung, L.-J., Lee, C.-W., Wu, B.Y.: Approximation algorithms for the star k-hub center problem in metric graphs. In: Dinh, T.N., Thai, M.T. (eds.) COCOON 2016. LNCS, vol. 9797, pp. 222–234. Springer, Cham (2016). doi:10.1007/978-3-319-42634-1_18

19. Ernst, A.T., Hamacher, H., Jiang, H., Krishnamoorthy, M., Woeginger, G.: Uncapacitated single and multiple allocation p-hub center problem. Comput. Oper. Res. **36**, 2230–2241 (2009)

20. Garey, M.R., Johnson, D.S.: Computers and Intractability: a guide to the theory of NP-completeness. W. H. Freeman and Company, San Francisco (1979)

21. Hromkovič, J.: Stability of approximation algorithms and the knapsack problem. In: Karhumäki, J., Maurer, H., Paun, G., Rozenberg, G. (eds.) Jewels are Forever, pp. 238–249. Springer, Heidelberg (1999)

22. Hromkovič, J.: Algorithmics for Hard Problems - Introduction to Combinatorial Optimization, Randomization, Approximation, and Heuristics, 2nd edn. Springer, Heidelberg (2003)

23. Kara, B.Y., Tansel, B.Ç.: On the single-assignment p-hub center problem. Eur. J. Oper. Res. **125**, 648–655 (2000)

24. Liang, H.: The hardness and approximation of the star p-hub center problem. Oper. Res. Lett. **41**, 138–141 (2013)

25. Meyer, T., Ernst, A., Krishnamoorthy, M.: A 2-phase algorithm for solving the single allocation p-hub center problem. Comput. Oper. Res. **36**, 3143–3151 (2009)

26. Pamuk, F.S., Sepil, C.: A solution to the hub center problem via a single-relocation algorithm with tabu search. IIE Trans. **33**, 399–411 (2001)

27. Rabbani, M., Kazemi, S.M.: Solving uncapacitated multiple allocation p-hub center problem by Dijkstra's algorithm-based genetic algorithm and simulated annealing. Int. J. Ind. Eng. Comput. **6**, 405–418 (2015)

28. Todosijević, R., Urošević, D., Mladenović, N., Hanafi, S.: A general variable neighborhood search for solving the uncapacitated r-allocation p-hub median problem. Optim. Lett. doi:10.1007/s11590-015-0867-6

29. Yaman, H., Elloumi, S.: Star p-hub center problem and star p-hub median problem with bounded path lengths. Comput. Oper. Res. **39**, 2725–2732 (2012)

30. Yang, K., Liu, Y., Yang, G.: An improved hybrid particle swarm optimization algorithm for fuzzy p-hub center problem. Comput. Ind. Eng. **64**, 133–142 (2013)

31. Yang, K., Liu, Y., Yang, G.: Solving fuzzy p-hub center problem by genetic algorithm incorporating local search. Appl. Soft Comput. **13**, 2624–2632 (2013)

32. Yang, K., Liu, Y., Yang, G.: Optimizing fuzzy p-hub center problem with generalized value-at-risk criterion. Appl. Math. Model. **38**, 3987–4005 (2014)

Approximation Algorithms for the Maximum Weight Internal Spanning Tree Problem

Zhi-Zhong Chen[1(✉)], Guohui Lin[2(✉)], Lusheng Wang[3], Yong Chen[2], and Dan Wang[3]

[1] Division of Information System Design, Tokyo Denki University, Hatoyama, Saitama 350-0394, Japan
`zzchen@mail.dendai.ac.jp`
[2] Department of Computing Science, University of Alberta, Edmonton, AB T6G 2E8, Canada
`{guohui,yong5}@ualberta.ca`
[3] Department of Computer Science, City University of Hong Kong, Tat Chee Avenue, Kowloon, Hong Kong, China
`cswangl@cityu.edu.hk`

Abstract. Given a vertex-weighted connected graph $G = (V, E)$, the maximum weight internal spanning tree (MwIST for short) problem asks for a spanning tree T of G such that the total weight of the internal vertices in T is maximized. The unweighted variant, denoted as MIST, is NP-hard and APX-hard, and the currently best approximation algorithm has a proven performance ratio 13/17. The currently best approximation algorithm for MwIST only has a performance ratio $1/3 - \epsilon$, for any $\epsilon > 0$. In this paper, we present a simple algorithm based on a novel relationship between MwIST and the maximum weight matching, and show that it achieves a better approximation ratio of 1/2. When restricted to claw-free graphs, a special case been previously studied, we design a 7/12-approximation algorithm.

Keywords: Maximum weight internal spanning tree · Maximum weight matching · Approximation algorithm · Performance analysis

1 Introduction

In the *maximum weight internal spanning tree* (MwIST for short) problem, we are given a vertex-weighted connected graph $G = (V, E)$, where each vertex v of V has a nonnegative weight $w(v)$, with the objective to compute a spanning tree T of G such that the total weight of the internal vertices in T, denoted as $w(T)$, is maximized. MwIST has applications in the network design for cost-efficient communication [19] and water supply [1].

This work was supported by KAKENHI Japan Grant No. 24500023 (ZZC), NSERC Canada (GL), GRF Hong Kong Grants CityU 114012 and CityU 123013 (LW), NSFC Grant No. 61672323 (GL), CSC Grant No. 201508330054 (YC).

Y. Cao and J. Chen (Eds.): COCOON 2017, LNCS 10392, pp. 124–136, 2017.
DOI: 10.1007/978-3-319-62389-4_11

When the vertex weights are uniform, or simply vertex-unweighted, the problem is referred to as the *maximum internal spanning tree* (MIST for short) problem. MIST is clearly NP-hard because it includes the NP-hard Hamiltonian-path [7] problem as a special case. Furthermore, MIST has been proven APX-hard [13], suggesting that it does not admit a polynomial-time approximation scheme (PTAS). In the literature, much research is done on designing (polynomial-time, if not specified) approximation algorithms for MIST to achieve the worst-case performance ratio as close to 1 as possible.

The probably first approximation for MIST is a local search algorithm, which achieves a ratio of $1/2$ and is due to Prieto and Sliper [15]. Salamon and Wiener [19] later modified slightly Prieto and Sliper's algorithm to make it run faster (in linear-time) while achieving the same ratio of $1/2$. Besides, two special cases of MIST were considered by Salamon and Wiener [19]: when restricted to claw-free graphs, they designed a $2/3$-approximation algorithm; when restricted to cubic graphs, they designed a $5/6$-approximation algorithm. Later, Salamon [18] proved that the $1/2$-approximation algorithm in [19] actually achieves a performance ratio of $3/(r+1)$ for the MIST problem on r-regular graphs ($r \geq 3$). Based on local optimization, Salamon [17] presented an $O(n^4)$-time $4/7$-approximation algorithm for MIST restricted to graphs without leaves. The algorithm was subsequently simplified and re-analyzed by Knauer and Spoerhase [8] to run faster (in cubic time), and it becomes the first improved $3/5$-approximation for the general MIST. Via a deeper local search strategy than those in [8,17], Li *et al.* [9] presented a further improved approximation algorithm for MIST with ratio $2/3$. At the same time, Li and Zhu [13] presented another $2/3$-approximation algorithm for MIST.

Unlike the other previously known approximation algorithms for MIST, the $2/3$-approximation by Li and Zhu [13] is based on a simple but crucial observation that the maximum number of internal vertices in a spanning tree of a graph G can be upper bounded by the maximum number of edges in a triangle-free 2-matching (a.k.a. path-cycle cover) of G. The time complexity of this approximation algorithm is dominated by computing the maximum triangle-free 2-matching, $O(nm^{1.5} \log n)$, where n and m are the numbers of vertices and edges in G, respectively. Li and Zhu [12] claimed that they are able to further improve their design to achieve a $3/4$-approximation algorithm for MIST, of the same time complexity. Recently, Chen *et al.* [2] gave another $3/4$-approximation algorithm for MIST, which is simpler than the one in [12]; and they showed that by applying three more new ideas, the algorithm can be refined into a $13/17$-approximation algorithm for MIST of the same time complexity. This is currently the best approximation algorithm for MIST.

The parameterized MIST by the number of internal vertices k, and its special cases and variants, have also been extensively studied in the literature [1,4–6,10,11,14–16]. The best known kernel for the general problem has a size $2k$, which leads to the fastest known algorithm with running time $O(4^k n^{O(1)})$ [10].

For the vertex-weighted version, MwIST, Salamon [17] designed the first $O(n^4)$-time $1/(2\Delta - 3)$-approximation algorithm, based on local search, where Δ is the maximum degree of a vertex in the input graph. For MwIST on claw-free graphs without leaves, Salamon [17] also designed an $O(n^4)$-time 1/2-approximation algorithm. Subsequently, Knauer and Spoerhase [8] proposed the first constant-ratio $1/(3 + \epsilon)$-approximation algorithm for the general MwIST, for any constant $\epsilon > 0$. The algorithm is based on a new pseudo-polynomial time local search algorithm, that starts with a depth-first-search tree and applies six rules to reach a local optimum. It yields a 1/3-approximation for MwIST and then is extended to a polynomial time $1/(3 + \epsilon)$-approximation scheme. The authors also showed that the ratio of 1/3 is asymptotically tight.

In this paper, we deal with the MwIST problem. We first prove a novel relationship between the total weight of the internal vertices in a spanning tree of the given vertex-weighted graph and the maximum weight matching of an edge-weighted graph, that is constructed out of the given vertex-weighted graph. Based on this relationship, we present a simple 1/2-approximation algorithm for MwIST; this ratio 1/2 significantly improves upon the previous known ratio of 1/3. When restricted to claw-free graphs, a special case previously studied in [17,19], we design a 7/12-approximation algorithm, improving the previous best ratio of 1/2.

2   The 1/2-approximation Algorithm

Recall that in the MwIST problem, we are given a connected graph $G = (V, E)$, where each vertex v of V has a nonnegative weight $w(v)$, with the objective to compute a spanning tree T of G such that the total weight of the internal vertices in T, denoted as $w(T)$, is maximized. We note that for such an objective function, we may assume without loss of generality that every leaf in the given graph G has weight 0.

We construct an edge-weighted graph based on $G = (V, E)$. In fact, the structure of the new graph is identical to that of G: the vertex set is still V, but instead the vertices have no weights; the edge set is still E, where the weight of each edge $e = \{u, v\}$ is $w(e) = w(u) + w(v)$, i.e., the weight of an edge is the total weight of its two ending vertices in the original graph. Since there is no ambiguity when we discuss the edge weights or the vertex weights, the new edge-weighted graph is still referred to as G. The weight of an edge subset refers to the total weight of the edges therein; while the weight of an acyclic subgraph refers to the total weight of the internal (and those surely will become internal) vertices therein.

Let M^* denote the maximum weight matching of (the edge-weighted graph) G, which can be computed in $O(n \min\{m \log n, n^2\})$-time, where $n = |V|$ and $m = |E|$.

Lemma 1. *Given a spanning tree T of G, we can construct a matching M of G such that $w(T) \leq w(M)$.*

Proof. We construct M iteratively. Firstly, we root the tree T at an internal vertex r, and all the edges of T are *unmarked*; then in every iteration we include into M an unmarked edge $e = \{u, v\}$ of T such that (1) both u and v are internal and (2) e is the closest to the root r measured by the number of edges on the path from r to e, followed by marking all the edges incident at u or v. This way, the total weight of the two internal vertices u and v in the tree T is transferred to M by adding the edge e to M. At the time this iterative procedure stops, there is no unmarked edge of T connecting two internal vertices, and thus every internal vertex whose weight has not been transferred to M must be adjacent to at least a leaf each via an unmarked edge.

Next, we iteratively include into M a remaining unmarked edge $e = \{u, v\}$ of T, followed by marking all the edges incident at u or v. This way, the total weight of u and v, which is greater than or equal to the weight of the internal vertex between u and v, is transferred to M by adding the edge e to M. At the end of this procedure, T contains no more unmarked edges. Since leaves in the tree T count nothing towards $w(T)$, we conclude that $w(T) \leq w(M)$. This proves the lemma.   □

The following corollary directly follows from Lemma 1, stating an upper bound on the total weight of an optimal solution to the MwIST problem.

Corollary 1. *Let T^* denote an optimal (maximum weight internal) spanning tree of G. Then, $w(T^*) \leq w(M^*)$.*

We next start with M^* to construct a spanning tree T. Let the edges of M^* be $e_1, e_2, \ldots, e_k$; let $e_j = \{a_j, b_j\}$, such that $w(a_j) \geq w(b_j)$, for all $j = 1, 2, \ldots, k$. Note that there could be vertices of degree 0 in the spanning subgraph $G[V, M^*]$ with the edge set M^*, and there could be edges of weight 0 in M^*; let X denote the set of such degree-0 vertices and the end-vertices of such weight-0 edges. Essentially we do not worry about the degree of any vertex of X in our final tree T, since their weights (if any) are not counted towards $w(M^*)$. This way, we assume without loss of generality that $w(a_j) > 0$ for each edge e_j of M^*, and consequently the degree of a_j is $d_G(a_j) \geq 2$, that is, a_j is adjacent to at least one other vertex than b_j in the graph G. Let $A = \{a_j \mid j = 1, 2, \ldots, k\}$, and $B = \{b_j \mid j = 1, 2, \ldots, k\}$; note that $V = A \cup B \cup X$.

Let $E^{aa} = E(A, A \cup X)$, i.e., the set of edges each connecting a vertex of A and a vertex of $A \cup X$, and $E^{ab} = E(A, B)$, i.e., the set of edges each connecting a vertex of A and a vertex of B. Our construction algorithm first computes a maximal acyclic subgraph of G, denoted as H_0, by adding a subset of edges of E^{aa} to M^*. This subset of edges is a maximum weight spanning forest on $A \cup X$, and it can be computed in $O(|E^{aa}| \log n)$-time via a linear scan. In the achieved subgraph H_0, if one connected component C contains more than one edge, then the vertex a_j of each edge $e_j = \{a_j, b_j\}$ in C has degree at least 2, i.e. is internal. Therefore, the total weight of the internal vertices in the component C is at least half of $w(C \cap M^*)$, and C is called *settled* and left alone by the algorithm.

Our algorithm next considers an arbitrary edge of M^* that is not yet in any settled component, say $e_j = \{a_j, b_j\}$. In other words, the edge e_j is an *isolated* component in the subgraph H_0. This implies that the vertex a_j is not incident to any edge of E^{aa}, and thus it has to be adjacent to some vertex in $B - \{b_j\}$. If a_j is adjacent to some vertex b_i in a settled component, then this edge (a_j, b_i) is added to the subgraph H_0 (the edge e_j is said *merged* into a settled component) and the iteration ends. The updated component remains settled, as $w(a_j) \geq w(e_j)/2$ is saved towards the weight of the final tree T.

In the other case, the vertex a_j is adjacent to a vertex b_i, such that the edge $e_i = \{a_i, b_i\}$ is also an isolated component in the current subgraph. After adding the edge (a_j, b_i) to the subgraph, the algorithm works with the vertex a_i exactly the same as with a_j at the beginning. That is, if a_i is adjacent to some vertex b_ℓ in a settled component, then this edge (a_i, b_ℓ) is added to the subgraph (the component that a_i belongs to is *merged* into a settled component) and the iteration ends; if a_i is adjacent to a vertex b_ℓ, such that the edge $e_\ell = \{a_\ell, b_\ell\}$ is also an isolated component in the current subgraph, then the edge (a_i, b_ℓ) is added to the subgraph, the algorithm works with the vertex a_ℓ exactly the same as with a_j at the beginning; in the last case, a_i is adjacent to a vertex b_ℓ inside the current component that a_i belongs to, then the edge (a_i, b_ℓ) is added to the current component to create a cycle, subsequently the lightest edge of M^* in the cycle is removed, the iteration ends, and the current component becomes settled. We note that in the above last case, the formed cycle in the current component contains at least 2 edges of M^*; breaking the cycle by removing the lightest edge ensures that at least half of the total weight of the edges of M^* in this cycle (and thus in this component) is saved towards the weight of the final tree T. Therefore, when the iteration ends, the resulting component is settled.

When the second step of the algorithm terminates, there is no isolated edge of M^* in the current subgraph, denoted as H_1, and each component is acyclic and settled. In the last step, the algorithm connects the components of H_1 into a spanning tree using any possible edges of E. We denote the entire algorithm as APPROX.

Lemma 2. *At the end of the second step of the algorithm* APPROX, *every component C of the achieved subgraph H_1 is acyclic and settled (i.e., $w(C) \geq w(C \cap M^*)/2$).*

Proof. Let C denote a component; $C \cap M^*$ is the subset of M^*, each edge of which has both end-vertices in C.

If C is obtained at the end of the first step, then C is acyclic and for every edge $e_j \in C \cap M^*$, the vertex a_j has degree at least 2, and thus $w(C) \geq w(C \cap M^*)/2$.

If a subgraph of C is obtained at the end of the first step but C is finalized in the second step, then C is also acyclic and for every edge $e_j \in C \cap M^*$, the vertex a_j has degree at least 2, and thus $w(C) \geq w(C \cap M^*)/2$.

If C is newly formed and finalized in the second step, then at the time C was formed, there was a cycle containing at least 2 edges of M^* of which the lightest one is removed to ensure the acyclicity, and thus the total weight of the internal vertices on this path is at least half of the total weight of the edges of M^* on this cycle. Also, the vertex a_j of every edge e_j not on the cycle has degree at least 2. Thus, $w(C) \geq w(C \cap M^*)/2$. □

Theorem 1. APPROX *is a 1/2-approximation for the MwIST problem.*

Proof. One clearly sees that APPROX runs in polynomial time, and in fact the running time is dominated by computing the maximum weight matching M^*.

From Lemma 2, at the end of the second step of the algorithm APPROX, every component C of the achieved subgraph H_1 is acyclic and satisfies $w(C) \geq w(C \cap M^*)/2$. Since there is no edge of M^* connecting different components of the subgraph H_1, the total weight of the internal vertices in H_1 is already at least $w(M^*)/2$, i.e. $w(H_1) \geq w(M^*)/2$. The last step of the algorithm may only increase the total weight. This proves that the total weight of the internal vertices of the tree T produced by APPROX is

$$w(T) \geq w(H_1) \geq w(M^*)/2 \geq w(T^*)/2,$$

where the last inequality is by Corollary 1, which states that $w(M^*)$ is an upper bound on the optimum. Thus, APPROX is a 1/2-approximation for the MwIST problem.   □

3   A 7/12-approximation Algorithm for Claw-Free Graphs

We present a better approximation algorithm for the MwIST problem on claw-free graphs. A graph $G = (V, E)$ is called *claw-free* if, for every vertex, at least two of its arbitrary three neighbors are adjacent. We again assume without loss of generality that every leaf in the graph G has weight 0. Besides, we also assume that $|V| \geq 5$.

We first present a reduction rule, which is a subcase of Operation 4 in [2], that excludes certain induced subgraphs of the given graph G from consideration. Each of these subgraphs is *hanging* at a cut-vertex of G (full details in [3]) and can be dealt with separately from the other part of G.

Operation 1. *If G has a cut-vertex v such that one connected component C of $G - v$ has two, three or four vertices, then obtain G_1 from $G - V(C)$ by adding a new leaf u of weight 0 and a new edge $\{v, u\}$.*

Let $tw(C)$ denote the maximum total weight of the internal vertices in a spanning tree of the subgraph induced on $V(C) \cup v$, in which $w(v)$ is revised to 0. Then there is an optimal spanning tree T_1 of G_1 of weight $w(T_1)$ if and only if there is an optimal spanning tree T of G of weight $w(T) = w(T_1) + tw(C)$.

Proof. Let G_c denote the subgraph induced on $V(C) \cup v$, that is, $G_c = G[V(C) \cup v]$; and let T_c denote the spanning tree of G_c achieving the maximum total weight of the internal vertices, that is, $w(T_c) = tw(C)$ (T_c can be computed in $O(1)$-time).

Note that in T_1, the leaf u must be adjacent to v and thus $w(v)$ is counted towards $w(T_1)$. We can remove the edge $\{v, u\}$ and u from T_1 while attach the tree T_c to T_1 by collapsing the two copies of v. This way, we obtain a spanning tree T of G, of weight $w(T) = w(T_1) + w(T_c)$ since $w(v)$ is not counted towards $w(T_c)$.

Conversely, for any spanning tree T of G, the vertex v is internal due to the existence of C. We may duplicate v and separate out a subtree T_c on the set of vertices $V(C) \cup v$, in which the weight of v is revised to 0. This subtree T_c is thus a spanning tree of G_c, and every vertex of $V(C)$ is internal in T if and only if it is internal in T_c. We attach the 0-weight vertex u to the vertex v in the remainder tree via the edge $\{v, u\}$, which is denoted as T_1 and becomes a spanning tree of G_1; note that the vertex v is internal in T_1. It follows that $w(T) = w(T_c) + w(T_1)$.   □

Let M^* denote the maximum weight matching of G. Let the edges of M^* be $e_1, e_2, \ldots, e_k$; let $e_j = \{a_j, b_j\}$, such that $w(a_j) \geq w(b_j)$, for all $j = 1, 2, \ldots, k$. For convenience, a_j and b_j are referred to as the *head* and the *tail* vertices of the edge e_j, respectively. The same as in the last section, we assume without loss of generality that $w(a_j) > 0$ for each j, and consequently the degree of a_j is $d_G(a_j) \geq 2$, that is, a_j is adjacent to at least one vertex other than b_j in the graph G. Let $A = \{a_j \mid j = 1, 2, \ldots, k\}$, $B = \{b_j \mid j = 1, 2, \ldots, k\}$, and $X = V - (A \cup B)$.

Let $E^{aa} = E(A, A)$, i.e., the set of edges each connecting two vertices of A, $E^{ax} = E(A, X)$, i.e., the set of edges each connecting a vertex of A and a vertex of X, and $E^{ab} = E(A, B)$, i.e., the set of edges each connecting a vertex of A and a vertex of B, respectively.

Let $M^{aa} \subseteq E^{aa}$ be a maximum cardinality matching within the edge set E^{aa}. We next prove a structure property of the spanning subgraph $G[V, M^* \cup M^{aa}]$, which has the edge set $M^* \cup M^{aa}$. For an edge $e_j = \{a_j, b_j\}$ of M^*, if a_j is not incident to any edge of M^{aa}, then e_j is called *isolated* in $G[V, M^* \cup M^{aa}]$.

Lemma 3. *Assume that two edges $e_{j_1} = \{a_{j_1}, b_{j_1}\}$ and $e_{j_2} = \{a_{j_2}, b_{j_2}\}$ of M^* are connected by the edge $\{a_{j_1}, a_{j_2}\} \in M^{aa}$ in $G[V, M^* \cup M^{aa}]$. Then there is at most one isolated edge $e_{j_3} = \{a_{j_3}, b_{j_3}\}$ whose head a_{j_3} can be adjacent to a_{j_1} or a_{j_2}.*

Proof. By contradiction, assume that there are two isolated edges $e_{j_3} = \{a_{j_3}, b_{j_3}\}$ and $e_{j_4} = \{a_{j_4}, b_{j_4}\}$ such that both the vertices a_{j_3} and a_{j_4} are adjacent to a_{j_1} or a_{j_2}. Then from the maximum cardinality of M^{aa}, a_{j_3} and a_{j_4} must be both adjacent to a_{j_1} or both adjacent to a_{j_2}. Suppose they are both adjacent to a_{j_1}; from the claw-free property, at least two of a_{j_2}, a_{j_3} and a_{j_4} are adjacent, which contradicts the maximum cardinality of M^{aa}. This proves the lemma.   □

For an isolated edge $e_{j_3} = \{a_{j_3}, b_{j_3}\}$ whose head is adjacent to an edge $\{a_{j_1}, a_{j_2}\} \in M^{aa}$ (i.e., satisfying Lemma 3), and assuming that $\{a_{j_2}, a_{j_3}\} \in E^{aa}$, we add the edge $\{a_{j_2}, a_{j_3}\}$ to $G[V, M^* \cup M^{aa}]$; consequently the edge e_{j_3} is no longer isolated. Let N^{aa} denote the set of such added edges associated with M^{aa}. At the end, the achieved subgraph is denoted as $H_0 = G[V, M^* \cup M^{aa} \cup N^{aa}]$.

Lemma 4. *In the subgraph $H_0 = G[V, M^* \cup M^{aa} \cup N^{aa}]$,*

- *every connected component containing more than one edge has either two or three edges from M^*, with their head vertices connected (by the edges of $M^{aa} \cup N^{aa}$) into a path; it is called a* type-I *component (see Fig. 1a) and a* type-II *component (see Fig. 1b), respectively;*

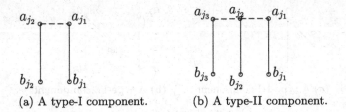

(a) A type-I component.   (b) A type-II component.

Fig. 1. The configurations of a type-I component and a type-II component.

– *for every isolated edge $e_j = \{a_j, b_j\}$, the head vertex is incident with at least one edge of $E^{ax} \cup E^{ab}$, but with no edge of E^{aa}.*

Proof. The proof directly follows the definition of H_0 and Lemma 3.   □

The following lemma is analogous to Lemma 3.

Lemma 5. *Any vertex of X can be adjacent to the head vertices of at most two isolated edges in the subgraph $H_0 = G[V, M^* \cup M^{aa} \cup N^{aa}]$.*

Proof. By contradiction, assume that $x \in X$ and there are three isolated edges $e_{j_k} = \{a_{j_k}, b_{j_k}\}$, $k = 1, 2, 3$, in the subgraph $H_0 = G[V, M^* \cup M^{aa} \cup N^{aa}]$, such that the edge $\{a_{j_k}, x\} \in E^{ax}$. From the claw-free property, at least two of a_{j_1}, a_{j_2} and a_{j_3} are adjacent, which contradicts Lemma 4. This proves the lemma. □

For an isolated edge $e_j = \{a_j, b_j\}$ in the subgraph $H_0 = G[V, M^* \cup M^{aa} \cup N^{aa}]$ whose head is adjacent to a vertex $x \in X$ (i.e., satisfying Lemma 5), we add the edge $\{a_j, x\}$ to H_0; consequently the edge e_j is no longer isolated. Let N^{ax} denote the set of such added edges associated with X. At the end, the achieved subgraph is denoted as $H_1 = G[V, M^* \cup M^{aa} \cup N^{aa} \cup N^{ax}]$.

Lemma 6. *In the subgraph $H_1 = G[V, M^* \cup M^{aa} \cup N^{aa} \cup N^{ax}]$,*

– *every connected component of H_0 containing more than one edge remains unchanged in H_1;*
– *every connected component containing a vertex x of X and some other vertex has either one or two edges from M^*, with their head vertices connected (by the edges of N^{ax}) to the vertex x; it is called a type-III component (see Fig. 2a) and a type-IV component (see Fig. 2b), respectively;*
– *for every isolated edge $e_j = \{a_j, b_j\}$, the head vertex is incident with at least one edge of E^{ab}, but with no edge of $E^{aa} \cup E^{ax}$.*

Proof. The proof directly follows the definition of H_1 and Lemmas 4 and 5.   □

Let E_0^{ab} denote the subset of E^{ab}, to include all the edges $\{a_j, b_\ell\}$ where both the edges $e_j = \{a_j, b_j\}$ and $e_\ell = \{a_\ell, b_\ell\}$ are isolated in the subgraph $H_1 = G[V, M^* \cup M^{aa} \cup N^{aa} \cup N^{ax}]$. Let $M^{ab} \subseteq E_0^{ab}$ be a maximum cardinality matching within the edge set E_0^{ab}. Let $H_2 = G[V, M^* \cup M^{aa} \cup N^{aa} \cup N^{ax} \cup M^{ab}]$ be the subgraph obtained from H_1 by adding all the edges of M^{ab}. One clearly

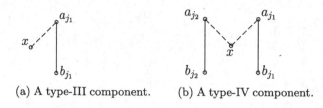

(a) A type-III component.   (b) A type-IV component.

Fig. 2. The configurations of a type-III component and a type-IV component.

sees that all the isolated edges in the subgraph H_1 are connected by the edges of M^{ab} into disjoint paths and cycles; while a path may contain any number of isolated edges, a cycle contains at least two isolated edges. Such a path and a cycle component are called a *type-V* component (see Fig. 3a) and a *type-VI* component (see Fig. 3b), respectively.

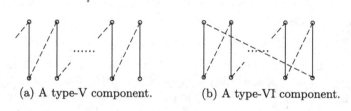

(a) A type-V component.   (b) A type-VI component.

Fig. 3. The configurations of a type-V component and a type-VI component.

Note that in a type-V component, there is exactly one head vertex of degree 1 and there is exactly one tail vertex of degree 1. We assume that for the tail vertex in a type-V component, it is not adjacent to the head of any other edge (via an edge of E^{ab}) in the same component; otherwise, through an edge exchange, the component is decomposed into a smaller type-V component and a new type-VI component.

Lemma 7. *In the subgraph $H_2 = G[V, M^* \cup M^{aa} \cup N^{aa} \cup N^{ax} \cup M^{ab}]$, for every type-V component, the degree-1 head vertex is adjacent (via an edge of E^{ab}) to the tail vertex of an edge in a type-I, -II, -III, or -IV component; on the other hand, the tail vertex of every edge in a type-I, -II, -III, or -IV component is adjacent to at most one such head vertex.*

Proof. We first show that the degree-1 head vertex in a type-V component C, denoted as a_j, cannot be adjacent to the tail of any edge in another type-V or a type-VI component C'. By contradiction, assume $\{a_j, b_\ell\} \in E^{ab}$ and e_ℓ is in C'. If the tail b_ℓ is already incident to some edge of M^{ab}, say $\{a_i, b_\ell\}$, then by the claw-free property at least two of a_i, a_j, a_ℓ must be adjacent, contradicting the fact that they are all isolated in the subgraph H_1. In the other case, the tail b_ℓ is the tail vertex of C' (which is a type-V component too), then it violates the maximum cardinality of M^{ab} since $\{a_j, b_\ell\} \in E^{ab}$ can be added to increase the size of M^{ab}. This proves the first half of the lemma.

The second half can be proven by a simple contradiction using the claw-free property of the graph. □

Subsequently, every type-V component C is connected to a type-I, -II, -III, or -IV component C', via the edge between the degree-1 head vertex of C and the tail vertex of an edge in $C' \cap M^*$. This way, the degree-1 tail vertex of C takes up the role of "the tail vertex" of the edge in $C' \cap M^*$, to become a tail vertex in the newly formed bigger component. For simplicity, the type of the component C' is passed to the newly formed bigger component. Denote this set of newly added edges as N^{ab}, which is a subset of $E^{ab} - M^{ab}$. The achieved subgraph is denoted as $H_3 = G[V, M^* \cup M^{aa} \cup N^{aa} \cup N^{ax} \cup M^{ab} \cup N^{ab}]$.

Lemma 8. *In the subgraph* $H_3 = G[V, M^* \cup M^{aa} \cup N^{aa} \cup N^{ax} \cup M^{ab} \cup N^{ab}]$,

- *there is no isolated type-V component;*
- *the head vertex of every edge of* M^* *has degree at least 2.*

Proof. The first half of the lemma follows from Lemma 7; the second half holds since there is no more isolated type-V component, which is the only type of component containing a degree-1 head vertex. □

We next create a set F of edges that are used to interconnect the components in the subgraph H_3. F is initialized to be empty. By Lemma 8, for every type-I, -II, -III, or -IV component C in the subgraph H_3, of weight $w(C \cap M^*)$, it is a tree and the total weight of the internal vertices therein is at least $\frac{1}{2}w(C \cap M^*)$; for every type-VI component C, which is a cycle, by deleting the lightest edge of $C \cap M^*$ from C we obtain a path and the total weight of the internal vertices in this path is also at least $\frac{1}{2}w(C \cap M^*)$. In the next theorem, we show that every component C in the subgraph H_3 can be converted into a tree on the same set of vertices, possibly with one edge specified for connecting a leaf of this tree outwards, such that the total weight of the internal vertices (and the leaf, if specified, which will become internal) in the tree is at least $\frac{2}{3}w(C \cap M^*)$. The specified edge for the interconnection purpose, is added to F. At the end of the process, the component C is called *settled*. A settled component C can be expressed in multiple equivalent ways, for example, that the total weight of the internal vertices (and the leaf, if specified, which will become internal) in the resulting tree is at least $\frac{2}{3}w(C \cap M^*)$, or that the total weight of the internal (and the leaf, if specified, which will become internal) vertices in the resulting tree is at least twice the total weight of the leaves (excluding the specified leaf, if any). We may use any of these ways for convenience of presentation.

In the sequel, we abuse the vertex notation to also denote its weight in math formulae; this simplifies the presentation and the meaning of the notation is easily distinguishable. For estimating the total weight of the internal vertices in a tree in the sequel, we frequently use the following inequality:

$$\forall w_1, w_2, w_3 \in \mathbb{R}, \ w_1 + w_2 + w_3 - \min\{w_1, w_2, w_3\} \geq 2\min\{w_1, w_2, w_3\}.$$

The following theorem summarizes the five lemmas and their technical proofs in [3] that every component in the subgraph H_3 can be settled.

Theorem 2. *Every component in the subgraph H_3 can be settled.*

Theorem 3. *The MwIST problem on claw-free graphs admits a 7/12-approximation algorithm.*

Proof. The above Theorem 2 states that every component of the subgraph $H_3 = G[V, M^* \cup M^{aa} \cup N^{aa} \cup N^{ax} \cup M^{ab} \cup N^{ab}]$ can be settled, without affecting any other components. Also, such a settling process for a component takes only linear time, by scanning once the edges in the subgraph induced on the set of vertices of the component. By settling, essentially the component is converted into a tree, possibly with one edge of F specified for connecting a leaf of the tree outwards.

In the next step of the algorithm, it iteratively processes the heaviest component C, i.e. with the largest $w(C \cap M^*)$. If the component C has been associated with an edge e of F, and using the edge e to connect a leaf of the resulting tree for C outwards does not create a cycle, then the algorithm does this and C is processed. This guarantees that the total weight of the internal vertices in $V(C)$ is at least $2w(C \cap M^*)/3$. If using the edge e to connect a leaf of the resulting tree for C outwards would create a cycle, the algorithm processes C by replacing C with another tree that guarantees that the total weight of the internal vertices in $V(C)$ is at least $\frac{1}{2}w(C \cap M^*)$. Notice that the latter case happens only because of (at least) one edge of F in an earlier iteration where a distinct component C' was processed, which connects a vertex of C' into a vertex of C. Therefore, every such C is associated with a distinct component C' processed by the algorithm in an earlier iteration, and thus $w(C') \geq w(C)$. On the other hand, every such component C' is associated to one C only, due to its edge in F connecting a leaf outwards into a vertex of C. It follows that for this pair of components C and C', the total weight of the internal vertices in $V(C) \cup V(C')$ is at least

$$w(C)/2 + 2w(C')/3 \geq 7(w(C) + w(C'))/12.$$

After all components of H_3 are processed, we obtain a forest for which the total weight of the internal vertices therein is at least $7w(M^*)/12$. The algorithm lastly uses any other available edges of E to interconnect the forest into a final tree, denoted as T; clearly $w(T) \geq 7w(M^*)/12$.

The time for the interconnecting purpose is at most $O(m \log n)$. Therefore, by Corollary 1 we have a 7/12-approximation algorithm for the MwIST problem on claw-free graphs.   □

4   Concluding Remarks

We presented a 1/2-approximation algorithm for the vertex weighted MIST problem, improving the previous best ratio of $1/(3 + \epsilon)$. The key ingredient in the design and analysis of our algorithm is a novel relationship between MwIST and the maximum weight matching. When restricted to claw-free graphs, we presented a 7/12-approximation algorithm, improving the previous best ratio of 1/2. It would be interesting to see whether this newly uncovered relationship, possibly combined with other new ideas, can be explored further.

The authors are grateful to two reviewers for their insightful comments and changes that improve the presentation greatly.

References

1. Binkele-Raible, D., Fernau, H., Gaspers, S., Liedloff, M.: Exact and parameterized algorithms for max internal spanning tree. Algorithmica **65**, 95–128 (2013)
2. Chen, Z.-Z., Harada, Y., Wang, L.: An approximation algorithm for maximum internal spanning tree. CoRR, abs/1608.00196 (2016)
3. Chen, Z.-Z., Lin, G., Wang, L., Chen, Y., Wang, D.: Approximation algorithms for the maximum weight internal spanning tree problem. arXiv, 1608.03299 (2016)
4. Coben, N., Fomin, F.V., Gutin, G., Kim, E.J., Saurabh, S., Yeo, A.: Algorithm for finding k-vertex out-trees and its application to k-internal out-branching problem. J. Comput. Syst. Sci. **76**, 650–662 (2010)
5. Fomin, F.V., Gaspers, S., Saurabh, S., Thomasse, S.: A linear vertex kernel for maximum internal spanning tree. J. Comput. Syst. Sci. **79**, 1–6 (2013)
6. Fomin, F.V., Lokshtanov, D., Grandoni, F., Saurabh, S.: Sharp separation and applications to exact and parameterized algorithms. Algorithmica **63**, 692–706 (2012)
7. Garey, M.R., Johnson, D.S.: Computers and Intractability: A Guide to the Theory of NP-completeness. W. H. Freeman and Company, San Francisco (1979)
8. Knauer, M., Spoerhase, J.: Better approximation algorithms for the maximum internal spanning tree problem. In: Dehne, F., Gavrilova, M., Sack, J.-R., Tóth, C.D. (eds.) WADS 2009. LNCS, vol. 5664, pp. 459–470. Springer, Heidelberg (2009). doi:10.1007/978-3-642-03367-4_40
9. Li, W., Chen, J., Wang, J.: Deeper local search for better approximation on maximum internal spanning trees. In: Schulz, A.S., Wagner, D. (eds.) ESA 2014. LNCS, vol. 8737, pp. 642–653. Springer, Heidelberg (2014). doi:10.1007/978-3-662-44777-2_53
10. Li, W., Wang, J., Chen, J., Cao, Y.: A $2k$-vertex kernel for maximum internal spanning tree. In: Dehne, F., Sack, J.-R., Stege, U. (eds.) WADS 2015. LNCS, vol. 9214, pp. 495–505. Springer, Cham (2015). doi:10.1007/978-3-319-21840-3_41
11. Li, X., Jiang, H., Feng, H.: Polynomial time for finding a spanning tree with maximum number of internal vertices on interval graphs. In: Zhu, D., Bereg, S. (eds.) FAW 2016. LNCS, vol. 9711, pp. 92–101. Springer, Cham (2016). doi:10.1007/978-3-319-39817-4_10
12. Li, X., Zhu, D.: A 4/3-approximation algorithm for the maximum internal spanning tree problem. CoRR, abs/1409.3700 (2014)
13. Li, X., Zhu, D.: Approximating the maximum internal spanning tree problem via a maximum path-cycle cover. In: Ahn, H.-K., Shin, C.-S. (eds.) ISAAC 2014. LNCS, vol. 8889, pp. 467–478. Springer, Cham (2014). doi:10.1007/978-3-319-13075-0_37
14. Prieto, E.: Systematic kernelization in FPT algorithm design. Ph.D. thesis. The University of Newcastle, Australia (2005)
15. Prieto, E., Sloper, C.: Either/or: using VERTEX COVER structure in designing FPT-algorithms — the case of k-INTERNAL SPANNING TREE. In: Dehne, F., Sack, J.-R., Smid, M. (eds.) WADS 2003. LNCS, vol. 2748, pp. 474–483. Springer, Heidelberg (2003). doi:10.1007/978-3-540-45078-8_41
16. Prieto, E., Sloper, C.: Reducing to independent set structure - the case of k-internal spanning tree. Nordic J. Comput. **12**, 308–318 (2005)

17. Salamon, G.: Approximating the maximum internal spanning tree problem. Theoret. Comput. Sci. **410**, 5273–5284 (2009)
18. Salamon, G.: Degree-based spanning tree optimization. Ph.D. thesis. Budapest University of Technology and Economics, Hungary (2009)
19. Salamon, G., Wiener, G.: On finding spanning trees with few leaves. Inf. Process. Lett. **105**, 164–169 (2008)

Incentive Ratios of a Proportional Sharing Mechanism in Resource Sharing

Zhou Chen[1]([✉]), Yukun Cheng[2], Qi Qi[1], and Xiang Yan[3]

[1] The Hong Kong University of Science and Technology, Sai Kung, Hong Kong
zchenaq@connect.ust.hk, kaylaqi@ust.hk
[2] Suzhou University of Science and Technology, Suzhou, China
ykcheng@amss.ac.cn
[3] Shanghai Jiaotong University, Shanghai, China
xyansjtu@163.com

Abstract. In a resource sharing system, resources are shared among multiple interconnected peers. Peers act as both suppliers and customers of resources by making a certain amount of their resource directly available to others. In this paper we focus on a proportional sharing mechanism, which is fair, efficient and guarantees a market equilibrium in the resource sharing system. We study the incentives an agent may manipulate such a mechanism, by the *vertex splitting strategy*, for personal gains and adopt a concept called *incentive ratio* to quantify the amount of gains. For the resource sharing system where the underlying network ia a cycle, we prove that the incentive ratio on this kind of network is bounded by $2 \leq \zeta \leq 4$. Furthermore, the incentive ratio on an even cycle, a cycle with even number of vertices, is proved to be exactly 2.

Keywords: Mechanism design · Market equilibrium · Combinatorial optimization · Incentive ratio · Resource sharing

1 Introduction

With the rapid growth of wireless and mobile Internet, the resource exchange or sharing over networks becomes widely applied, which goes beyond the peer-to-peer (P2P) bandwidth sharing idea [14]. Peers in networks act as both suppliers and customers of resources (such as processing power, disk storage or network bandwidth), and make their resources directly available to others according to preset rules [12]. The problem we considered here can be modeled as an undirected graph G, where each vertex v represents an agent with w_v units of divisible resources (or weight) to be distributed among its neighbors. The utility U_v is determined by the total amount of resources obtained from its neighbors. The resource exchange system can be viewed as a pure exchange economy, in which each agent owns some divisible resource and trades its resource to its neighbors in the network. An efficient allocation in a pure exchange economy can be characterized by the market equilibrium. However, the existence of a market

© Springer International Publishing AG 2017
Y. Cao and J. Chen (Eds.): COCOON 2017, LNCS 10392, pp. 137–149, 2017.
DOI: 10.1007/978-3-319-62389-4_12

equilibrium does not imply that it can be achieved easily. Wu and Zhang [14] have proposed the proportional response protocol (also called proportional sharing mechanism) for this model, under which each peer contributes its resource to each of its neighbors in proportion to the amount received from its neighbors. They showed that this distributed proportional dynamics converged to a proportional sharing equilibrium and surprisingly this equilibrium is equivalent to a market equilibrium. Their results were obtained by a combinatorial proof, based on the *bottleneck decomposition* method.

Wu and Zhang's results implied that the proportional sharing mechanism is fair and efficient. But agents are strategic in reality. The market equilibrium is determined by agents' reported information, which may be misreported by the agents for gaining more utility. It has been known that in the linear Fisher market an agent may lie to improve its utility [1,4]. An immediate question is: is it possible for an agent to improve its utility via strategic behaviors in our setting? It is known that an agent can not benefit from misreporting its weight or neighborhood information in the proportional sharing mechanism [6,7]. In this paper, we consider another commonly used strategic behavior where an agent v splits itself into a set of nodes, such as taking new IP addresses in the network environment. Each copied node is connected to some of its original neighbors and is assigned a certain amount of resource (summing up to its original amount of resource) to manipulate the proportional sharing mechanism. The new utility of this agent is the sum of the utility from all its copied nodes. We call this strategic behavior as *vertex splitting*.

Example 1. Consider a cycle G containing three vertices $\{x, y, v\}$ as shown in Fig. 1-(a). In the proportional sharing mechanism, agent v gets utility 10 from y. In Fig. 1-(b), v splits itself into two copied nodes v^l and v^r, and assigns 5 units of resource to each node respectively. Then the utilities of v^l and v^r are 2 and 12, respectively, which means the total utility of v increases to 14 after strategic manipulation.

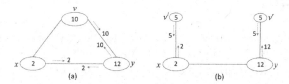

Fig. 1. An example that agent v can improve its utility by vertex splitting strategy. The numbers in the circles represent weights and dashed arrows indicate resource exchange.

Example 1 shows that the proportional sharing mechanism is not incentive compatible to the vertex splitting strategy. Thus there is a challenge that characterize how much one can improve its utility. In this paper, we will quantify the amount of improvement by a concept called *incentive ratio* [5]. Incentive ratio is defined as the ratio of the largest possible utility by manipulation over its original utility. We prefer a smaller incentive ratio because it implies that the mechanism is more robust to strategic behaviors.

Related Works. The voluntary cooperation of participating agents have been a key to the successes of automated Internet application through information and communication technology. [8] pioneered the study of incentive techniques for peer-to-peer resource sharing systems in terms of mechanism design and performance analysis. After that, [14] showed that a proportional response protocol is equivalent to the market equilibrium in the resource exchange system and thus considered to be a fair and efficient solution. On the other hand, market equilibrium may not be truthful. Agents' strategic behaviors are studied in the Fisher market equilibrium for linear markets [1] and for constant elasticity of substitution markets [3]. Then the concept of *incentive ratio* [5] is introduced to measure such strategic behaviors. The incentive ratio is similar to the ratio of *price of anarchy* [10,13], the former one measures individual gains one may acquire in deviation from truthful behavior, while the latter characterizes the loss of social efficiency in a selfish Nash equilibrium. [4,5] proved that the incentive ratio is exactly 2 in a Fisher market with linear utilities and Leontief utilities, and 1.445 with Cobb-Douglas utilities. No constant incentive ratio is known except Fisher market. The resource exchange model considered in this paper is a special case of the Arrow-Debreu market. It is proved that the proportional sharing mechanism is truthful against two kinds of strategic behaviors: misreporting on its connectivity or its own resource weight [6,7]. The vertex splitting behavior, studied in this paper, is motivated by the strategic behavior of false-name bidding, introduced by [16] in the auction design. Agents can play this strategy by creating multiple fictitious identities. It has been proved that the well known VCG mechanism is not robust to the false-name bidding strategy, which is followed by the study on the false-name-proof mechanism design [9,15], and on the efficiency guarantee of the VCG mechanism [2].

Main Results. In this paper, we consider resource sharing on a network with cycles only. A cycle network is the simplest graph that is still connected after an agent playing vertex splitting strategy. Such connectivity poses a big challenge for the analysis of incentive ratio, because each copied node's utility depends on the initial assigned weights of all copied nodes as well as the weights of all other nodes. The particular interesting techniques we adopt here are to identify the worst case of the splitting strategy (Proposition 6) and design auxiliary graphs in the proof of general cycles (Lemma 2). Furthermore, edge classification analysis is introduced for cycles with even number of vertices. Compared with a recent work [11] proving that the incentive ratio for the linear exchange economy is unbounded, we show a bounded incentive ratio of $2 \leq \zeta \leq 4$ on a general cycle and tight bound of 2 on a cycle with even number of vertices, for a resource sharing system. Our results fill in the blank of bounded incentive ratios in Arrow-Debreu markets.

2   Preliminary

The resource exchange system studied in this paper is modeled as an undirected and weighted network $G = (V, E; w)$, where each vertex $v \in V$ represents an agent with an amount of resource (weight) $w_v > 0$ exchanging with its neighbors. W.l.o.g, we assume G is connected. $\Gamma(v) = \{u : (v,u) \in E\}$ is the set of neighbors of v, and x_{vu} denotes the amount of resource v allocates to its neighbor u ($0 \leq x_{vu} \leq w_v$). We call a vector $X = (x_{vu})$ a *feasible allocation* if $\forall v \in V, \sum_{u \in \Gamma(v)} x_{vu} = w_v$, i.e. v allocates all its resource out. The utility of agent v is defined as $U_v(X) = \sum_{u \in \Gamma(v)} x_{uv}$, which is the total amount of resource received from all its neighbors. Given a set $S \subseteq V$, let $w(S) = \sum_{v \in S} w_v$ and $\Gamma(S) = \cup_{v \in S} \Gamma(v)$. It is possible that $S \cap \Gamma(S) \neq \emptyset$. Define $\alpha(S) = \frac{w(\Gamma(S))}{w(S)}$ as an inclusive expansion ratio of S, named by α-*ratio*. A set B is called a *bottleneck* of G if $\alpha(B) = \min_{S \subseteq V} \alpha(S)$, and a bottleneck with maximal size is called the *maximal bottleneck*. The bottleneck decomposition of a graph G is defined as follows.

Bottleneck Decomposition. Given network $G = (V, E; w)$, start with $V_1 = V$, $G_1 = G$ and $i = 1$. Find the maximal bottleneck B_i of G_i and let G_{i+1} be the induced subgraph on the vertex set $V_{i+1} = V_i - (B_i \cup C_i)$, where $C_i = \Gamma(B_i) \cap V_i$ is the neighbor set of B_i in G_i. Repeat until $G_{i+1} = \emptyset$ and set $k = i$ where $G_{i+1} = \emptyset$. Finally, $\mathcal{B} = \{(B_1, C_1), \cdots, (B_k, C_k)\}$ is the bottleneck decomposition of G. (B_i, C_i) is the i-th bottleneck pair and $\alpha_i = \frac{w(C_i)}{w(B_i)}$ is the α-ratio of (B_i, C_i).

Proposition 1 ([14]). *Given a graph G, the bottleneck decomposition of G is unique and*

(1) $0 < \alpha_1 < \alpha_2 < \cdots < \alpha_k \leq 1$;
(2) if $\alpha_i = 1$, then $i = k$ and $B_i = C_i$; otherwise B_i is an independent set and $B_i \cap C_i = \emptyset$;
(3) there is no edge between B_i and B_j, $i \neq j \in \{1, \cdots, k\}$;
(4) if there is an edge between B_i and C_j, then $j \leq i$.

B-class and C-class. Given $\mathcal{B} = \{(B_1, C_1), \cdots, (B_k, C_k)\}$ of G, each vertex in B_i (or C_i) with $\alpha_i < 1$ is called a B-class (or C-class) vertex. For the special case $B_k = C_k$, i.e. $\alpha_k = 1$, vertices in B_k are both in B-class and C-class.

BD Mechanism. Given the bottleneck decomposition $\mathcal{B}$ of G, an allocation is determined by the following algorithm [14], which is named as BD Mechanism.

- For (B_i, C_i) with $\alpha_i < 1$, consider the bipartite graph $\widehat{G} = (B_i, C_i; E_i)$ with $E_i = (B_i \times C_i) \cap E$. Construct a network $N = (V_N, E_N)$ with $V_N = \{s, t\} \cup B_i \cup C_i$ and directed edges (s, u) with capacity w_u for $u \in B_i$, (v, t) with capacity $\frac{w_v}{\alpha_i}$ for $v \in C_i$ and (u, v) with capacity $+\infty$ for $(u, v) \in E_i$. The max-flow min-cut theorem ensures a maximal flow $\{f_{uv}\}$, $u \in B_i$ and $v \in C_i$, such that $\sum_{v \in \Gamma(u) \cap C_i} f_{uv} = w_u$ and $\sum_{u \in \Gamma(v) \cap B_i} f_{uv} = \frac{w_v}{\alpha_i}$. Let the allocation be $x_{uv} = f_{uv}$ and $x_{vu} = \alpha_i f_{uv}$. Then we have $\sum_{u \in \Gamma(v) \cap B_i} x_{vu} = \sum_{u \in \Gamma(v) \cap B_i} \alpha_i \cdot f_{vu} = w_v$.

- For $\alpha_k = 1$ (i.e., $B_k = C_k$), construct a bipartite graph $\widehat{G} = (B_k, B'_k; E'_k)$ where B'_k is a copy of B_k. There is an edge $(u, v') \in E'_k$ if and only if $(u, v) \in$

$E[B_k]$. Construct a network by the above method. For any edge $(u, v') \in E'_k$, there exists flow $f_{uv'}$ such that $\sum_{v' \in \Gamma(u) \cap B'_k} f_{uv'} = w_u$. Let the allocation be $x_{uv} = f_{uv'}$.

- For any other edge $(u, v) \notin (B_i \times C_i) \cap E$, $i = 1, 2, \cdots, k$, define $x_{uv} = 0$.

BD Mechanism assigns all resource of each agent to its neighbors in the same pair. Therefore, all available resources exchange along edges in $(B_i \times C_i) \cap E$, $i = 1, \cdots, k$. Next we show that BD mechanism is a proportional sharing mechanism.

Proportional Sharing. A mechanism is called *proportional sharing* if the allocation from this mechanism satisfies $x_{vu} = \frac{x_{uv}}{\sum_{k \in \Gamma(v)} x_{kv}} w_v$, i.e., each agent v's resource is allocated proportionally to what it receives from its neighbors.

Proposition 2 ([14]). *BD Mechanism is a proportional sharing mechanism.*

As stated before, the resource sharing system can be modeled as an exchange economy where each agent u sells its own divisible resource and uses the money earned through trading to buy its neighbors' resource. A proportional sharing equilibrium is a market equilibrium, which is considered as an efficient allocation in the exchange economy. Given a bottleneck decomposition, if a price vector P is well defined as: $p_u = \alpha_i w_u$ if $u \in B_i$; and $p_u = w_u$ otherwise, then such a price vector together with the allocation X obtained from BD Mechanism is a market equilibirum.

Proposition 3 ([14]). *(p, X) is a market equilibrium and each agent u's utility is $U_u = w_u \cdot \alpha_i$ if $u \in B_i$; and $U_u = \frac{w_u}{\alpha_i}$ if $u \in C_i$.*

Note that $U_u \geq w_u$ if u is in C_i and $U_u \leq w_u$ if u is in B_i because $\alpha_i \leq 1$ by Proposition 1-(1). Let α_u be the α-ratio of u, i.e. $\alpha_u = \alpha_i$, if $u \in B_i \cup C_i$.

Resource exchange game. Although BD Mechanism ensures a fair and efficient allocation as network mechanism, an agent may not follow BD Mechanism at the execution level. Can agents play strategically to increase their utilities? We call such a problem with incentive consideration as the *resource exchange game*.

Recently, [6,7] studied two kinds of manipulative moves: cheating on its resource amount and its connectivity. They proved that BD Mechanism is truthful against these two strategies.

Proposition 4 ([7]). *For any agent u in a resource exchange game, the utility of u is continuous and monotonically nondecreasing on its reported value x. As $x \leq w_u$, the dominant strategy of agent u is to report its true weight.*

Proposition 5 ([6]). *For any agent u in a resource exchange game, the utility of u cannot be increased by cutting any incident edges connecting to it.*

In this paper, we consider another strategic move, called *vertex splitting strategy*, where an agent splits itself into several copied nodes, and assigns a weight to each copied node. Formally, in a resource sharing network G, the collection $\mathbf{w} = (w_1, w_2, \cdots, w_n) \in R^n$ is referred as the *weight profile*, and $\mathbf{w}_{-v}$ is the weight profile without v for any vertex v. Then the utility of agent v can be written as $U_v(G; \mathbf{w})$. After v splits itself into m nodes, $1 \leq m \leq d_v$ (d_v is the degree of v), and assigns an amount of resource to each node respectively, its new utility is denoted by $U_v'(G'; w_{v^1}, \cdots, w_{v^m}, \mathbf{w}_{-v})$. Here G' is the resulting graph and $w_{v^i} \in [0, w_v]$ is the amount of resource assigned to node v^i with the constraint $\sum_{i=1}^{m} w_{v^i} = w_v$.

As shown in Example 1, BD mechanism is not robust against vertex splitting strategy. To quantify the incentives an agent may manipulate the BD mechanism, we will redefine the incentive ratio, originally defined by [5], for the resource exchange game.

Definition 1 (Incentive Ratio). *The incentive ratio of agent v under BD Mechanism for the vertex splitting strategy is defined as*

$$\zeta_v = \max_{1 \leq m \leq d_v} \max_{w_{v^i} \in [0, w_v], \sum_{i=1}^{m} w_{v^i} = w_v; \mathbf{w}_{-v}; G'} \frac{U_v'(G'; w_{v^1}, \cdots, w_{v^m}, \mathbf{w}_{-v})}{U_v(G; \mathbf{w}_v)}.$$

The incentive ratio of BD mechanism in the resource exchange game is defined to be $\zeta = \max_{v \in V} \zeta_v$.

Note that we only need to consider a special case for vertex splitting strategy where agent v splits itself into d_v nodes and each node is connected to one of its neighbors. Proposition 6 below shows this special case will not reduce the generality of the result. The proof will be provided in our full version.

Proposition 6. *In a resource exchange game, the incentive ratio of BD mechanism with respect to vertex splitting strategy can be achieved by splitting d_v nodes and each node is connected to one neighbor, where d_v is the degree of agent v.*

3   Incentive Ratio of BD Mechanism on Cycles

In this section, we study the incentive ratio of BD Mechanism with respect to the vertex splitting strategy on cycles. Suppose the strategic vertex u splits itself into two copied nodes u^l and u^r with weight assignment (w_{u^l}, w_{u^r}). Since the underlying network is a cycle, the resulting graph G' shall be a path after the manipulation, which is still connected. Therefore, the utilities of copied nodes depend on both of their weights, written as $U_{u^l}'(w_{u^l}, w_{u^r})$ and $U_{u^r}'(w_{u^l}, w_{u^r})$. Let $\mathcal{B}' = \{(B_1', C_1'), \cdots, (B_{k'}', C_{k'}')\}$ be the bottleneck decomposition of G'. The α-ratio of pair (B_i', C_i') is denoted by α_i', $i = 1, \cdots, k'$.

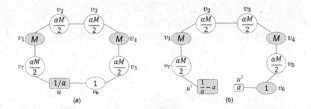

Fig. 2. An example that shows the lower bound of incentive ratio on cycles, where the gray or white vertices represent the B-class or C-class vertices respectively.

3.1   Incentive Ratio Bounds on Cycles

We first use the following example (Example 2) with 8 vertices to show that the lower bound of incentive ratio on cycles is 2.

Example 2. G is a cycle with eight vertices shown in Fig. 2-(a). Vertices' weights are $w_u = \frac{1}{\alpha}$, $w_{v_1} = w_{v_4} = M$, $w_{v_2} = w_{v_3} = w_{v_5} = w_{v_7} = \alpha M/2$ and $w_{v_6} = 1$, where $\alpha \in (0, 1)$ is a preset value and M is a large enough number satisfying $M > \frac{1}{\alpha} - \alpha$. The bottleneck decomposition of G is $\{(B_1, C_1)\}$, where $B_1 = \{u, v_1, v_4\}$ and $C_1 = \{v_2, v_3, v_5, v_6, v_7\}$ with $\alpha_1 = \alpha$. The utility of u is $U_u = w_u \cdot \alpha_1 = 1$.

Now vertex u splits itself into two copied nodes with weights α and $\frac{1}{\alpha} - \alpha$ respectively in Fig. 2-(b). The condition of $M > \frac{1}{\alpha} - \alpha$ promises that $(B_1', C_1') = \{\{u^1, v_1\}, \{v_2, v_7\}\}$, and $(B_2', C_2') = \{\{v_4, v_6\}, \{v_3, v_5, u^2\}\}$. The α-ratios are $\alpha_1' = \frac{\alpha M}{M+(\frac{1}{\alpha}-\alpha)}$ and $\alpha_2' = \alpha$. So $\lim_{M\to\infty} U_{u^1}' = \lim_{M\to\infty}(\alpha - \frac{1}{\alpha}) \cdot \alpha_1' = 1 - \alpha^2$, $U_{u^2}' = \alpha \cdot \frac{1}{\alpha_2'} = 1$ and $\lim_{M\to\infty} \zeta_u = \frac{\lim_{M\to\infty} U_u'}{U_u} = \frac{\lim_{M\to\infty} U_{u^1}' + U_{u^2}'}{U_u} = 2 - \alpha^2$. So if α is fairly small, the incentive ratio is close to 2 as much as desired.

In the following, we discuss the upper bound of incentive ratio. As stated before, the utilities of copied nodes depend on weights w_{u^l} and w_{u^r} simultaneously. This brings us a big challenge to compute U_{u^l}' and U_{u^r}' with two variables of w_{u^l} and w_{u^r}, and it's also hard to characterize the connections between U_{u^l}' (or U_{u^r}') and U_u. The vertex splitting strategy can be decomposed into two steps:

(1) Fix vertex u's weight w_u and add an additional node x with weight $w_{u^r} \in [0, w_u]$, which is adjacent to u's right neighbor v;
(2) Remove edge (v, u) from the network obtained in step 1, and decrease u's weight to $w_{u^l} = w_u - w_{u^r}$.

Then the utilities of node x and vertex u are $U_{u^r}'(w_{u^r}, w_{u^l})$ and $U_{u^l}'(w_{u^r}, w_{u^l})$ respectively. Based on the construction of the network in the first step, we define an auxiliary graph $\widehat{G}$ as follows: given a vertex u and its neighbor v on a cycle G, add a node x with weight $x \in [0, w_u]$ adjacent to v. For easy of presentation, we use x to represent the new node and its weight. So $\widehat{V} = V(G) \cup \{x\}$ and $\widehat{E} = E(G) \cup (x, v)$. As utility $\widehat{U}_u$ of u in $\widehat{G}$ shall change with the weight of the new node x, given the weights of all vertices, it can be described as a function of $x : \widehat{U}_u(x)$. Thus our proof is constructed based on three networks as shown in Fig. 3: the original cycle G, the split network G' and the auxiliary network $\widehat{G}$.

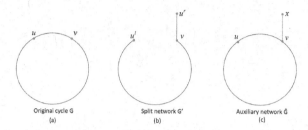

Fig. 3. The original cycle G, the split cycle G' and the auxiliary cycle $\widehat{G}$.

Lemma 1. $U'_{u^l}(w_{u^l}, w_{u^r}) \leq 2U_u$ and $U'_{u^r}(w_{u^l}, w_{u^r}) \leq 2U_u$.

Proof. Here we only give the proof for the result $U'_{u^l}(w_{u^l}, w_{u^r}) \leq 2U_u$, the upper bound of $U'_{u^r}(w_{u^l}, w_{u^r})$ can be deduced similarly. We focus on G' and $\widehat{G}$ first. If other vertices' weights are fixed and only the weight w_{u^l} of u^l increases up to w_u in G', then the monotonicity of utility on weight by Proposition 4 promises $U'_{u^l}(w_{u^l}, w_{u^r}) \leq U'_{u^l}(w_u, w_{u^r})$. From the structures of G' and $\widehat{G}$, we observe that G' can be obtained by deleting the edge (u, v) from $\widehat{G}$ and replacing u and x with u^l and u^r respectively. So the incentive compatibility of BD Mechanism for deleting edge strategy by Proposition 5 guarantees $U'_{u^l}(w_u, w_{u^r}) \leq \widehat{U}_u(w_{u^r})$, where $x = w_{u^r}$. We can further prove $\widehat{U}_u(x) \leq 2U_u, \forall x \in [0, w_u]$ in Lemma 2 below (The proof is in Subsect. 3.2).

Lemma 2. In the auxiliary network $\widehat{G}$, the utility $\widehat{U}_u(x) \leq 2U_u, \forall x \in [0, w_u]$.

Combining the above two inequalities and Lemma 2, we have $U'_{u^l}(w_{u^l}, w_{u^r}) \leq U'_{u^l}(w_u, w_{u^r}) \leq \widehat{U}_u(w_{u^r}) \leq 2U_u$.   □
Therefore, we have

Theorem 1. For the resource exchange system, the incentive ratio of BD Mechanism for the vertex splitting strategy on cycles is $2 \leq \zeta \leq 4$.

Proof. From Lemma 1, we have $U'_{u^l}(w_{u^l}, w_{u^r}) \leq 2U_u$ and $U'_{u^r}(w_{u^l}, w_{u^r}) \leq 2U_u$. Thus it is easy to compute $U'_u = U'_{u^r}(w_{u^l}, w_{u^r}) + U'_{u^l}(w_{u^l}, w_{u^r}) \leq 4U_u$, which means $\zeta \leq 4$.   □

3.2   Proof of Lemma 2

Lemma 2 characterizes the utility change of u caused by introducing another node x with weight $x \in [0, w_u]$. To obtain $\widehat{U}_u(x) \leq 2U_u, \forall x \in [0, w_u]$, the structure of $\widehat{G}$ and the new bottleneck decomposition $\widehat{\mathcal{B}}$ are the keys.

Definition 2 (Bottleneck Decomposition). Let the auxiliary graph be $\widehat{G}$ by adding one node x with weight $x \in [0, w_u]$ adjacent to one of u's neighbors v. Denote its bottleneck decomposition as $\widehat{\mathcal{B}} = \{(\widehat{B}_1, \widehat{C}_1), \cdots, (\widehat{B}_{\widehat{k}}, \widehat{C}_{\widehat{k}})\}$ with α-ratio $\widehat{\alpha}_j$, $j = 1, \cdots, \widehat{k}$. Similarly, $\widehat{V}_1 = V \cup \{x\}$, $\widehat{V}_{j+1} = \widehat{V}_j - (\widehat{B}_j \cup \widehat{C}_j)$ for $j = 1, \cdots, \widehat{k}-1$. We call vertices in $\widehat{B}_j$ (or $\widehat{C}_j$), $1 \leq j \leq \widehat{k}$, $\widehat{B}$-class (or $\widehat{C}$-class).

Recall vertex v is the neighbor of u adjacent to x in $\widehat{G}$. Let $(\widehat{B}_s, \widehat{C}_s)$ be the first pair in $\widehat{B}$ which differs from the one in $\mathcal{B}$. As the difference is due to the appearance of x, x and v are in $\widehat{B}_s \cup \widehat{C}_s$. Before we give the formal proof for Lemma 2, we propose a useful characterization of $\widehat{B}$ below, its proof will be provided in our full version.

Proposition 7. *If $v \in C_j$, $1 \le j \le k$, then the bottleneck decomposition $\widehat{B}$ satisfies: there exist two indexes $s, t \in \{1, 2, \cdots, \widehat{k}\}$ with $s \le t$, such that*

(1) $(\widehat{B}_1, \widehat{C}_1) = (B_1, C_1), \cdots, (\widehat{B}_{s-1}, \widehat{C}_{s-1}) = (B_{s-1}, C_{s-1})$;

(2) $(\widehat{B}_{t+1}, \widehat{C}_{t+1}) = (B_{j+1}, C_{j+1}), \cdots, (\widehat{B}_{\widehat{k}}, \widehat{C}_{\widehat{k}}) = (B_k, C_k)$;

(3) $\bigcup_{h=s}^{t} \widehat{B}_h = \left(\bigcup_{h=s}^{j} B_h \right) \cup \{x\}$ and $\bigcup_{h=s}^{t} \widehat{C}_h = \left(\bigcup_{h=s}^{j} C_h \right)$;

(4) for each vertex $y \in \widehat{B}_h \cup \widehat{C}_h$, $h = s, \cdots, t$, $\widehat{\alpha}_y \le \alpha_y$.

Proof of Lemma 2

- Case 1. $v \in C_j$, $1 \le j \le k$. Based on Proposition 7, we conclude:

  (a) if $u \in B_h \cup C_h$, $h = 1, \cdots, s-1$ or $h = j+1, \cdots, k$, then Proposition 7-(1) and (2) guarantee that the pairs keep the same which leads to $\widehat{U}_u(x) = U_u$.
  (b) if $u \in B_h$, $h = s, \cdots, j$, then u is still a $\widehat{B}$-class vertex by Proposition 7-(3). From Proposition 7-(4), we know that $\widehat{U}_u(x) = w_u \cdot \widehat{\alpha}_u \le w_u \cdot \alpha_u = U_u$.
  (c) if $u \in C_h$, $h = s, \cdots, j$, then $U_u \ge w_u$ and u is still a $\widehat{C}$-class vertex by Proposition 7-(3). $\widehat{U}_u(x) = w_u / \widehat{\alpha}_u \ge w_u / \alpha_u = U_u$. This result holds for each vertex $y \in C_h$, $h = s, \cdots, j$. W.l.o.g, let $\widehat{U}_y = U_y + \delta_y$, where $\delta_y \ge 0$. By the allocation rule from BD Mechanism, we know all resources owned by vertices in $\bigcup_{h=s}^{j} B_h \cup \{x\} = \bigcup_{h=s}^{t} \widehat{B}_h$ are assigned to $\bigcup_{h=s}^{j} C_h = \bigcup_{h=s}^{t} \widehat{C}_h$, and vice versa.

$$\sum_{h=s}^{j} w(B_h) + x = \sum_{h=s}^{t} w(\widehat{B}_h) = \sum_{h=s}^{t} \sum_{y \in \widehat{C}_h} \widehat{U}_y = \sum_{h=s}^{j} \sum_{y \in C_h} (U_y + \delta_y) = \sum_{h=s}^{j} w(B_h) + \delta$$

Then $\sum_{h=s}^{j} \sum_{y \in C_h} \delta_y = \delta = x$ and $\delta_y \le x$ for each $y \in C_h$, $s \le h \le j$. Especially for vertex $u \in C_h$, we have $\widehat{U}_u(x) = U_u + \delta_u \le U_u + x \le U_u + w_u \le 2U_u$, where $w_u \le U_u$ because u is a C-class vertex.

- Case 2. $v \in B_j$, $1 \le j \le k$. Since v is a B-class vertex, u must be in C_i with index $i \le j$ by Proposition 1. Therefore, $\alpha_u \le \alpha_j$ and $U_u \ge w_u$. So if u is a $\widehat{B}$-class vertex in $\widehat{G}$, its utility $\widehat{U}_u(x) \le w_u \le U_u$. In the following we focus on the case that both u and x are $\widehat{C}$-class vertices. The proof of $\widehat{U}_u(x) \le 2U_u$ when x is a $\widehat{B}$-class vertex is left in the full version.

  x is in $\widehat{C}$-class implies that v is in $\widehat{B}$-class because v is the unique neighbor of x in $\widehat{G}$. Then $(\widehat{B}_h, \widehat{C}_h) = (B_h, C_h)$, $h = 1, \cdots, j-1$ and $\widehat{V}_j = V_j \cup \{x\}$. Since $(\widehat{B}_j, \widehat{C}_j - \{x\})$ is a candidate pair in V_j (i.e., $\widehat{C}_j - \{x\} = \Gamma(\widehat{B}_j) \cap V_j$), $\alpha_j \le \frac{w(\widehat{C}_j - \{x\})}{w(\widehat{B}_j)} \le \frac{w(\widehat{C}_j)}{w(\widehat{B}_j)} = \widehat{\alpha}_j$ by the definition of bottleneck. So

(a) if $u \in C_i$, $1 \leq i \leq j - 1$, then u must be in $\widehat{C}_i$ with $\widehat{\alpha}_u = \alpha_u$ and $\widehat{U}_u(x) = U_u$.

(b) if $u \in C_j$ and $u \in \widehat{C}_h \neq C_j$ (Note here we assume u is in $\widehat{C}$-class), then $h \geq j$. Therefore, $\widehat{\alpha}_u = \widehat{\alpha}_h \geq \widehat{\alpha}_j \geq \alpha_j = \alpha_u$ and $\widehat{U}_u(x) = w_u/\widehat{\alpha}_u \leq w_u/\alpha_u = U_u$. □

4   Incentive Ratio on Even Cycles

In this section, we focus on the incentive ratio of BD Mechanism on a cycle with an even number of vertices, called *even cycle* for short.

Theorem 2. *For the resource sharing system, the incentive ratio of BD Mechanism for the vertex splitting strategy on even cycles is exactly 2.*

Example 2 tells us the lower bound of the incentive ratio on even cycles is 2. To obtain the tight bound of 2, we shall prove that the upper bound is also 2.

Lemma 3. *If the network G of resource exchange system is an even cycle and $\widehat{G}$ is the auxiliary graph of G, then $\widehat{U}_u(x) \leq U_u$, for any $x \in [0, w_u]$.*

To simplify our discussion, let the vertex set of even cycle G be $\{v_1, \cdots, v_{2n}\}$. Given the bottleneck decomposition $\mathcal{B} = \{(B_1, C_1) \cdots, (B_k, C_k)\}$, all vertices are categorized as B-class or C-class. For the pair (B_i, C_i) with $\alpha_1 < 1$, we know the B_i is independent by Proposition 1. But for (B_k, C_k) with $\alpha_k = 1$ (i.e. $B_k = C_k$), we redefine B_k and C_k as follows to obtain two disjoint vertex subsets: for vertex v_i in $B_k = C_k$ with an odd index i, let v_i be in B_k; otherwise let v_i be in C_k. So after such preprocessing, there is no edge between any two B-class vertices. According to the classes that the endpoints u and v belong to, we categorize each edge (u, v) as B-C edge or C-C edge. There is no B-B edge in G and

Proposition 8. *On an even cycle G, the number of C-C edges is even.*

Definition 3. *Two vertices u and v are called B-C connected in G, if there exists a path in G connecting them and all edges on this path are B-C edges. Such a path is called a B-C connected path.*

W.l.o.g. we assume each $(\widehat{B}_h, \widehat{C}_h)$ is connected, that means the edges from $(\widehat{B}_h \times \widehat{C}_h) \cap \widehat{E}$ construct a path and any two vertices in $\widehat{B}_h \cup \widehat{C}_h$ are connected by it. If not, $(\widehat{B}_h, \widehat{C}_h)$ is composed of several connected subpairs and each subpair's α-ratio is equal to $\widehat{\alpha}_h$. Therefore, we only need to consider connected subparis.

Proof of Lemma 3 (Sketch). Recall the proof in Lemma 2. We know $\widehat{U}_u(x) \leq U_u$ does not hold for two cases: (1) $v \in C_j$ and $u \in C_i \cap \widehat{C}_l$, $i = s, \cdots, j$; (2) $v \in B_j$, x is in $\widehat{B}$-class and $u \in C_i \cap \widehat{C}_l$, $i \leq j$. So it is enough for us to show $\widehat{U}_u(x) \leq U_u$ for the above two cases on an even cycle.

As noted before, $(\widehat{B}_s, \widehat{C}_s)$ is the first bottleneck pair in $\widehat{\mathcal{B}}$ which differs from $\mathcal{B}$. We must have $x, v \in \widehat{B}_s \cup \widehat{C}_s$. Suppose on the contrary that $\widehat{U}_u(x) > U_u$, meaning $\exists x \in [0, w_u]$, $\widehat{\alpha}_l(x) = \widehat{\alpha}_u(x) < \alpha_u$. Then we next show that G must have an odd number of C-C edges, which is a contradiction to Proposition 8.

(1) $v \in C_j$ and $u \in C_i \cap \widehat{C}_l$ with assumption $\widehat{\alpha}_l < \alpha_i$. So (u, v) is a C-C edge in G. Firstly, we start from the initial condition that $u \in C_i \cap \widehat{C}_l$ with $\widehat{\alpha}_l < \alpha_i$, and prove that there exists another vertex $u' \in C_{i'} \cap \widehat{C}_{l'}$ with $\widehat{\alpha}_{l'} < \alpha_{i'}$, $l' < l$, and there is a B-C connected path $P[u, u']$ between u and u'. Secondly, we continue to use the same analysis for $u' \in C_{i'} \cap \widehat{C}_{l'}$ with $\widehat{\alpha}_{l'} < \alpha_{i'}$. Because in each step of analysis, we could always find a pair $(\widehat{B}_{l'}, \widehat{C}_{l'})$ whose index is smaller than the previous one. The finiteness of network makes us reach at pair $(\widehat{B}_s, \widehat{C}_s)$. For pair $(\widehat{B}_s, \widehat{C}_s)$, there exists a vertex $u'' \in C_{i''} \cap \widehat{C}_s$ and there is a B-C connected path $P[u, u'']$ by the previous analysis. On the other hand, we know there is a path $P[u'', v]$ along the edges in $(\widehat{B}_s \times \widehat{C}_s) \cap \widehat{E}$ connecting v and u'' based on our assumption. Proposition 7 shows that $\widehat{B}_s$ or $\widehat{C}_s$ only have B-class or C-class vertices respectively. So path $P[u'', v]$, on which the vertices in $\widehat{B}_s$ and $\widehat{C}_s$ alternate, is B-C connected. By merging $P[u, u'']$ and $P[u'', v]$, we find a B-C connected path between u and v. Since there are two paths between any two vertices on a cycle, for u and v, there is a B-C connected path between them and another one is a C-C edge (u, v). So the whole cycle G contains only one C-C edge (u, v), contradicts to Proposition 8.

(2) $v \in B_j$, x is in $\widehat{B}$-class and $u \in C_i \cap \widehat{C}_l$, $i \leq j$, with $\widehat{\alpha}_l < \alpha_i$. So (u, v) is a B-C edge. This case is more complicated since Proposition 7 no longer holds and a $\widehat{B}$ set may have some C-class vertices. Our objective is also to derive a contradiction that even cycle G contains an odd number of C-C edges. The following claim is essential to our analysis, its proof is provided in the full version.

Proposition 9. *If $\widehat{B}_h$, $s \leq h \leq l$, has a C-class vertex a, then there is a path $P[v, a]$ from v to a which contains an even number of C-C edges.*

Firstly we deal with pair $(\widehat{B}_l, \widehat{C}_l)$ where u is in. We discuss two cases that whether $\widehat{B}_l$ contains C-class vertices or not. If $\widehat{B}_l$ has some C-class vertices (Case A), we care for the special one, say a, such that the path $P[u, a]$ along the edges in $(\widehat{B}_l \times \widehat{C}_l) \cap \widehat{E}$ does not contain any other vertices both in C-class and $\widehat{B}_l$, except for a. On one hand, we prove path $P[u, a]$ only has one C-C edge. On the other hand, Proposition 9 ensures there is a path $P[a, v]$ containing an even number of C-C edges. So by merging them, we get a path $P[u, v]$ having an odd number of C-C edges from u to v. In addition, cycle G is the union of path $P[u, v]$ and edge (u, v). Thus G has an odd number of C-C edges as (u, v) is a B-C edge. It is a contradiction. If $\widehat{B}_l$ dose not have C-class vertices (Case B), then $\widehat{B}_l$ and $\widehat{C}_l$ only contain B-class and C-class vertices respectively. The analysis for (1) can be utilized here. Thus there is a vertex $u' \in C_{i'} \cap \widehat{C}_{l'}$ with $\widehat{\alpha}_{l'} < \alpha_{i'}$, $l' < l$ and u and u' are B-C connected. Secondly, we continue to use the same analysis for u' and Case A or Case B may arise. Once Case A happens, a contradiction is derived. Otherwise, we continue to study Case B until we reach pair $(\widehat{B}_s, \widehat{C}_s)$. For pair $(\widehat{B}_s, \widehat{C}_s)$, there is a vertex $u'' \in C_{i''} \cap \widehat{C}_s$ which is B-C connected to u, as proved before. We also know there must be a path $P[v, u'']$ along edges in

$(\widehat{B}_s \times \widehat{C}_s) \cap \widehat{E}$ to connect v and u''. The vertex b on $P[v, u'']$ adjacent v must be in $\widehat{B}_s$ and is a C-class vertex since v is in $\widehat{C}_s$ and B_j. Thus $\widehat{B}_s$ at least has one C-class vertex and Case A appears, which derives a contradiction.   □

5   Conclusion

Our paper investigates the possible strategic behaviors of agents with respect to a proportional sharing mechanism, i.e. BD Mechanism, in a resource sharing system. We discuss the incentives of agents to play the vertex splitting strategy, and characterize how much utility can be improved by incentive ratio. Our main results are to prove that the incentive ratios are bounded by $2 \le \zeta \le 4$ on general cycles and are tight of 2 on even cycles. There is a space left that how to narrow the gap of upper and lower bounds of incentive ratio on general cycles and further how to explore the incentive ratio of BD Mechanism on any other graphs. We have done a lot of numerical experiments on cycles containing different number of vertices, and on random graphs in which each edge is generated with probability $p \in \{0.1, 0.2, \cdots, 0.9\}$. All simulation results demonstrate that the incentive ratio is no more than 2 as shown in Fig. 4. Thus, two questions are raised from our work: 1. Is it possible to prove the incentive ratio is 2 for any cycle? 2. Is it possible to prove the incentive ratio is no more than 2 for any graph? These will be interesting open problems for future work.

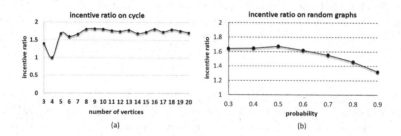

Fig. 4. The numerical experiment results.

Acknowledgments. This research was partially supported by the National Nature Science Foundation of China (No. 11301475, 11426026, 61632017, 61173011), by a Project 985 grant of Shanghai Jiao Tong University, and by the Research Grant Council of Hong Kong (ECS Project No. 26200314, GRF Project No. 16213115 and GRF Project No.16243516).

References

1. Adsul, B., Babu, C.S., Garg, J., Mehta, R., Sohoni, M.: Nash equilibria in fisher market. In: Kontogiannis, S., Koutsoupias, E., Spirakis, P.G. (eds.) SAGT 2010. LNCS, vol. 6386, pp. 30–41. Springer, Heidelberg (2010). doi:10.1007/978-3-642-16170-4_4
2. Alkalay-Houlihan, C., Vetta, A.: False-name bidding and economic efficiency in combinatorial auctions. In: AAAI, pp. 538–544 (2014)
3. Brânzei, S., Chen, Y., Deng, X., Aris, F., Frederiksen, S., Zhang, J.: The fisher market game: equilibrium and welfare. In: AAAI (2014)
4. Chen, N., Deng, X., Zhang, H., Zhang, J.: Incentive ratios of fisher markets. In: Czumaj, A., Mehlhorn, K., Pitts, A., Wattenhofer, R. (eds.) ICALP 2012. LNCS, vol. 7392, pp. 464–475. Springer, Heidelberg (2012). doi:10.1007/978-3-642-31585-5_42
5. Chen, N., Deng, X., Zhang, J.: How profitable are strategic behaviors in a market? In: Demetrescu, C., Halldórsson, M.M. (eds.) ESA 2011. LNCS, vol. 6942, pp. 106–118. Springer, Heidelberg (2011). doi:10.1007/978-3-642-23719-5_10
6. Cheng, Y., Deng, X., Pi, Y., Yan, X.: Can bandwidth sharing be truthful? In: Hoefer, M. (ed.) SAGT 2015. LNCS, vol. 9347, pp. 190–202. Springer, Heidelberg (2015). doi:10.1007/978-3-662-48433-3_15
7. Cheng, Y., Deng, X., Qi, Q., Yan, X.: Truthfulness of a proportional sharing mechanism in resource exchange. In: IJCAI, pp. 187–193 (2016)
8. Feldman, M., Lai, K., Stoica, I., Chuang, J.: Robust incentive techniques for peer-to-peer networks. In: EC (2004)
9. Iwasaki, A., Conitzer, V., Omori, Y., Sakurai, Y., Todo, T., Guo, M., Yokoo, M.: Worst-case efficiency ratio in false-name-proof combinatorial auction mechanisms. In: AAMAS, pp. 633–640 (2010)
10. Koutsoupias, E., Papadimitriou, C.: Worst-case equilibria. In: STACS, pp. 404–413 (1999)
11. Polak, I.: The incentive ratio in exchange economies. In: Chan, T.-H.H., Li, M., Wang, L. (eds.) COCOA 2016. LNCS, vol. 10043, pp. 685–692. Springer, Cham (2016). doi:10.1007/978-3-319-48749-6_49
12. Schollmeier, R.: A definition of peer-to-peer networking for the classification of peer-to-peer architectures and applications. In: P2P, pp. 101–102 (2001)
13. Roughgarden, T., Tardos, É.: How bad is selfish routing? J. ACM 49(2), 236–259 (2002)
14. Wu, F., Zhang, L.: Proportional response dynamics leads to market equilibrium. In: STOC, pp. 354–363 (2007)
15. Yokoo, M.: The characterization of strategy/false-name proof combinatorial auction protocols: Price-oriented, rationing-free protocol. In: IJCAI, pp. 733–739 (2003)
16. Yokoo, M., Sakurai, Y., Matsubara, S.: Robust combinatorial auction protocol against false-name bids. Artif. Intell. 130, 167–181 (2001)

Efficient Enumeration of Maximal k-Degenerate Subgraphs in a Chordal Graph

Alessio Conte[1]($\boxtimes$), Mamadou Moustapha Kanté[2], Yota Otachi[3], Takeaki Uno[4], and Kunihiro Wasa[4]

[1] Università di Pisa, Pisa, Italy
conte@di.unipi.it
[2] Université Clermont Auvergne, LIMOS, CNRS, Aubière, France
mamadou.kante@uca.fr
[3] Kumamoto University, Kumamoto, Japan
otachi@cs.kumamoto-u.ac.jp
[4] National Institute of Informatics, Tokyo, Japan
{uno,wasa}@nii.ac.jp

Abstract. In this paper, we consider the problem of listing the maximal k-degenerate induced subgraphs of a chordal graph, and propose an output-sensitive algorithm using delay $O(m \cdot \omega(G))$ for any n-vertex chordal graph with m edges, where $\omega(G) \le n$ is the maximum size of a clique in G. The problem generalizes that of enumerating maximal independent sets and maximal induced forests, which correspond to respectively 0-degenerate and 1-degenerate subgraphs.

1 Introduction

One of the fundamental problems in network analysis is finding subgraphs with some desired properties. A great body of literature has been devoted to develop efficient algorithms for many different types of subgraphs, such as frequent subgraphs [12], dense subgraphs [13] or complete subgraphs [5,9]. A more comprehensive list can be found in [20].

Dense subgraphs are object of extensive research, especially due to their close relationship to community detection; however, one may be interested in finding *sparse* graphs as many networks are sparse even if locally dense. For instance, [21] addresses the enumeration of induced trees in k-degenerate graphs.

The *degeneracy* of a graph is the smallest value k for which every subgraph of the graph has a vertex of degree *at most* k. A graph is said to be k-degenerate if its degeneracy is k or less. Degeneracy is also referred to as the coloring number or k-core number, as a k-degenerate graph may contain a k-core but not a $k+1$-core, and is a widely used sparsity measure [5,9,15,19,21]. Several studies tend to take into account the degeneracy of graphs, as it tends to be very small

M.M. Kanté is supported by French Agency for Research under the GraphEN project (ANR-15-CE-0009).

Y. Cao and J. Chen (Eds.): COCOON 2017, LNCS 10392, pp. 150–161, 2017.
DOI: 10.1007/978-3-319-62389-4_13

in real-world networks [19], many important graph classes in structural graph theory are degenerate [15]. Furthermore, it is straightforward to see that k-degenrate subgraphs generalize well known structures, as 0-degenerate subgraphs correspond to independent sets, while 1-degenerate subgraphs correspond to induced forests.

Alon *et al.* [1] investigated the size of the largest k-degenerate induced subgraph in a graph, giving tight lower bounds in relation to the degree sequence of the graph. Whilst Pilipczuk *et al.* [16] showed that a maximum k-degenerate induced subgraph can be found in randomized time $O((2 - \epsilon_k)^n n^{O(1)})$, for some $\epsilon_k > 0$ depending only on k, and moreover showed that there are at most $(2 - \epsilon_k)^n$ such subgraphs. See [2, 14] for other recent studies on degeneracy.

In this paper we address the enumeration of maximal k-degenerate induced subgraphs, and provide an efficient polynomial delay algorithm for chordal input graphs. An enumeration algorithm is of polynomial delay if the maximum computation time between two outputs is bounded by a polynomial in the size of the input. Enumeration algorithms are of high importance in several areas such as data-mining, biology, artificial intelligence, or databases. (see for instance [7, 20]).

Chordal graphs (also known as triangulated graphs) have been a topic of intensive study in computer science due to the applications in phylogenetic networks and also many NP-complete problems become tractable when the inputs are chordal graphs [3, 8, 11, 17, 18]. A graph is chordal if and only if every cycle of length 4 or more has a chord, i.e. an edge joining two non-consecutive vertices. Chordal graphs have been equivalently characterized in different ways: they are the graphs that allow a *perfect elimination ordering*, that is an elimination ordering in which every eliminated vertex is *simplicial* (its neighbors form a clique) [17, 18]; the graphs that allow a *clique tree* [3] (see Sect. 2.1); the intersection graphs of subtree families in trees [11]. In our case, we will consider the characterization by clique-trees. It is well-known that n-vertex chordal graphs have at most n maximal cliques. A clique-tree of a chordal graph G is a tree T whose nodes are in bijection with the set of maximal cliques, and such that for each vertex x the set of maximal cliques containing x form a subtree of T.

Our algorithm is based on the well-known *Extension Problem* (also known as *backtracking* or *flashlight* or *binary partition*) and uses the clique-tree. The enumeration can be reduced to the following question: Given two sets of vertices S and X, decide whether there is a maximal k-degenerate graph which contains S and does not intersect X. Indeed, if we can answer this question in polynomial time, the algorithm can be summarized as follows: start from the empty set, and in each iteration with given sets (S, X) pick a vertex v and partition the problem into those containing v (a call to the iteration $(S \cup \{v\}, X)$) or those not containing v (a call to the iteration $(S, X \cup \{v\})$), both calls depending on the answer given by the Extension problem. The delay of such algorithms is usually $O(n \cdot poly(n))$ with $poly(n)$ being the time to decide the Extension problem. This problem, however, can be shown to be NP-Complete for generic graphs (the proof is omitted for space reasons), and even on chordal graphs its complexity is not clear. We thus need some additional techniques: for our algorithm we do not consider all possible sets (S, X) for the Extension problem, but only some special cases driven by the clique-tree. Our special case of the Extension problem is the following (we consider the clique-tree T to be rooted):

Input. A node C of T, a partition (S, X) of the set of vertices in all the cliques preceding C in a pre-order traversal of T and a partition (S', X') of $C \backslash (S \cup X)$.
Output. Decide whether there is a maximal solution containing $S \cup S'$ and avoiding $X \cup X'$.

We propose a notion of *greedy solution* and show that this special case of the Extension problem is a Yes-instance if and only if a greedy solution exists; we also propose an $O(m)$-time algorithm to compute the greedy solution.

2 Preliminaries

An algorithm is said to be *output-polynomial* if the running time is bounded by a polynomial in the input and the output sizes. The delay is the maximum computation time between two outputs, pre-processing, and post-processing. If the delay is polynomial in the input size, the algorithm is called *polynomial delay*.

For two sets A and B we denote by $A \backslash B$ the set $\{x \in A \mid x \notin B\}$. Our graph terminology is standard, we refer to the book [6]. In this paper, we assume that graphs are simple, finite, loopless, and each graph is given with a linear ordering of its vertices. We can further assume graphs to be connected as the solutions of a non-connected graph are obtained by combining those of its connected components. We use n and m to denote respectively the numbers of vertices and edges in any graph. The vertex set of a graph G is denoted by $V(G)$ and its edge set by $E(G)$. The subgraph of G induced by $X \subseteq V(G)$, denoted by $G[X]$, is the graph $(X, (X \times X) \cap E(G))$; and we write $G \backslash X$ to denote $G[V(G) \backslash X]$. For a vertex x of G we denote by $N_G(x)$ the set of neighbors of x, i.e., the set $\{y \in V(G) \mid xy \in E(G)\}$, and we let $N_G[x]$, the *closed neighborhood of x*, be $N_G(x) \cup \{x\}$; the degree of a vertex x, $d_G(x)$, is defined as the size of $N_G(x)$.

A tree is an acyclic connected graph. A *clique* of a graph G is a subset C of G that induces a complete graph, and a *maximal clique* is a clique C of G such that $C \cup \{x\}$ is not a clique for all $x \in V(G) \backslash C$. Depending on the context, C may refer to its set of vertices, the subgraph induced by C or the corresponding node in the clique tree. We denote by $\mathcal{Q}(G)$ the set of maximal cliques of G, and by $\omega(G)$ the maximum number of vertices in a clique in $\mathcal{Q}(G)$. For a vertex x, we denote by $\mathcal{Q}(G, v)$ the set of maximal cliques containing x.

For a *rooted tree* T and two nodes u and v of T, we call v an *ancestor* of u, and u a *descendant* of v, if v is on the unique path from the root to u; u and v are *incomparable* if v is neither an ancestor or descendant of u. In what follows, we omit the subscript G and fix a graph $G = (V(G), E(G))$ if it is clear from the context.

2.1   Chordal Graphs and Clique Trees

A graph G is a *chordal graph* if it does not contain an induced cycle of length more than three. It is well-known that a chordal graph G has at most n maximal cliques, and they can be enumerated in linear time [4]. With every chordal graph G, one can associate a tree that we denote by $\mathcal{QT}(G)$, called *clique tree*, whose nodes are the maximal cliques of G and such that for every vertex $x \in V(G)$ the set $\mathcal{Q}(G, x)$ is a subtree of $\mathcal{QT}(G)$ [11]. Moreover, for every chordal graph G, one can compute a clique tree in linear time (see for instance [10]). In the rest of the paper all clique trees are considered rooted.

2.2 K-degenerate Graphs

A graph G is a k-*degenerate graph* if for any induced subgraph H in G, H has a vertex whose degree is at most k. The degeneracy of a graph is the minimum value k for which the graph is k-degenerate, and is a well known sparsity measure [5,9,15,19,21]. We consider the following question.

Problem 1. Given a chordal graph G and a positive integer k, enumerate all maximal k-degenerate induced subgraphs in G, with polynomial delay.

Note that a complete graph K_n is an $(n-1)$-degenerate graph, as all its vertices have degree $n-1$. Therefore, for any clique C of a graph G, any k-degenerate induced subgraph of G may have no more than $k+1$ vertices belonging to C. Chordal graphs have the following property.

Theorem 1. *The degeneracy of a chordal graph is exactly $\omega(G) - 1$.*

Proof. Since the degeneracy is a hereditary property (i.e., any subgraph of a k-degenrate graph is k-degenerate), and the complete graph K_n has degeneracy $n-1$, $\omega(G) - 1$ is a lower bound for the degeneracy of any graph. The fact that $\omega(G) - 1$ is an upper bound on chordal graphs relies on the fact that every chordal graph has at least a vertex whose neighbour is a clique [17]. Therefore, in any chordal graph we can find a vertex of degree at most $\omega(G) - 1$. $\square$

3   Enumeration Algorithm

This section describes our algorithm for enumerating all maximal k-degenerate induced subgraphs of a given chordal graph $G = (V, E)$. In the following, we sometimes refer to maximal k-degenerate induced subgraphs as *solutions*, and we denote them by their vertex set as for cliques, to ease the reading.

Our proposed algorithm is based on the binary partition method. The outline of our algorithm is as follows: We start with an empty induced subgraph S. Then we pick a vertex v from G and add v to S. If $S + v$ is a maximal k-degenerate subgraph, then we output $S + v$, otherwise we choose another vertex and add it to $S + v$. After that we backtrack and add v to an excluded set X, to generate all solutions that contain S and not v. By recursively applying the above operation to G we can enumerate all solutions. However, certain pairs (S, X)

may not generate a solution, as there may be no maximal k-degenerate induced subgraph containing S but no vertex in X (e.g., if $S = \emptyset, X = V$). If we test all the possibilities that will not lead to a solution, the cost of this process is not output sensitive, i.e. not bounded by a polynomial in the number of solutions. To develop efficient enumeration algorithm, we have to limit such redundant testing as much as possible. To achieve this we focus on the rooted clique tree $\mathcal{QT}(G)$ and introduce the concepts of *greedy filling* and *partial solution*. In what follows we let G be a fixed chordal graph.

3.1 Greedy Filling Strategy

Let R be a fixed maximal clique, called the *root* of $\mathcal{QT}(G)$, and let us root $\mathcal{QT}(G)$ at R. For a maximal clique C of G, whose parent in $\mathcal{QT}(G)$ is the clique P, we call *private vertices of C* be the set of vertices in $C \setminus P$. Because all cliques in $\mathcal{QT}(G)$ are different and inclusion-maximal, and by the properties of the clique tree, one can deduce the following.

Lemma 1. *Given a clique tree $\mathcal{QT}(G)$, every clique in $\mathcal{QT}(G)$ contains at least one private vertex, and every vertex v is private in exactly one clique in $\mathcal{QT}(G)$.*

Let C be a maximal clique of G. For $X \subseteq V(G)$, let $A(C, X) = 1$ if $|C \setminus X| \geq k + 1$, and $A(C, X) = 0$ otherwise. For any vertex $v \in X$, $A(v, X) = \sum_{C \in \mathcal{Q}(G,v)} A(C, X)$, i.e. the number of maximal cliques containing v for which $|C \setminus X| \geq k + 1$. As adding more than $k + 1$ vertices from the same clique to any solution S would cause S to not be k-degenerate anymore, we say that C is **saturated** in S if $|C \cap S| = k + 1$.

The function A allows us to check the maximality of a k-degenerate subgraph, thanks to the following lemma.

Lemma 2. *Let $G = (V, E)$ be a chordal graph and $M \subseteq V$ be a k-degenerate subgraph of G, with $X = V \setminus M$. M is maximal if and only if $A(x, X) \geq 1$ for each $x \in X$.*

Proof. Assume $A(x, X) \geq 1$ for each $x \in X$ and there exists a k-degenerate subgraph $M' \supset M$, with $v \in M' \setminus M$. As $v \in X$ we have $A(v, X) \geq 1$, thus there exists a clique C containing v s.t. $|C \setminus X| \geq k + 1$. As $M = V \setminus X$, we have $|C \setminus X| = |C \cap M| \geq k + 1$. As $M \cup \{v\} \subseteq M'$ we have $|C \cap M'| \geq k + 2$, thus M' contains a complete subgraph with $k + 2$ vertices and is not k-degenerate, which contradicts the hypothesis.

On the other hand, if for a vertex $x \in X$ we have $A(x, X) = 0$, then for any clique C containing x we have $|(M \cup \{x\}) \cap C| \leq k + 1$, since $|C \setminus X| = |C \cap M| < k + 1$. Thus the largest clique in $M \cup \{x\}$ has size at most $k + 1$, and as $M \cup \{x\}$ is a chordal graph (it is an induced subgraph of G) it is k-degenerate by Theorem 1. Thus M is not maximal, which contradicts the hypothesis. □

We now define the notion of *partial solution* as a pair of disjoint vertex subsets (S, X), where S contains vertices (to include) in the k-degenerate induced subgraph, and X is a set of vertices that must be excluded from the solution, with some additional properties:

Definition 1 (partial solution). *A pair (S, X) of subsets of $V(G)$ with $S \cap X = \emptyset$ is a* partial solution *if*

1. *$|S \cap C| \leq k + 1$ for any maximal clique C,*
2. *$\forall x \in X, A(x, X) \geq 1$,*
3. *for each maximal clique C, if $Pv(C) \cap (S \cup X) \neq \emptyset$, then $C' \subseteq S \cup X$ for all ancestors C' of C.*

Given a pair (S, X) of disjoint subsets of $V(G)$, it is not trivial to decide whether there exists a solution $M \supseteq S$ with $M \cap X = \emptyset$. However, as we will later demonstrate, this is always true if (S, X) is a partial solution. Next, we introduce the strategy that will be used by our algorithm to guarantee the existence of solutions. Let $\pi : \{1, \ldots, |\mathcal{Q}(G)|\} \to \mathcal{Q}(G)$ be a fixed linear ordering of $\mathcal{Q}(G)$ obtained from a pre-order traversal of $\mathcal{QT}(G)$, and let us call $\pi^{-1}(C)$ the *rank* of $C \in \mathcal{Q}(G)$. We use the rank of the cliques to define the order in which they are considered by the following procedure.

Definition 2 (Greedy filling). *The* greedy filling *of a partial solution (S, X) consists in the following. Let C be the maximal clique with the smallest rank for which $C \setminus (S \cup X) \neq \emptyset$. Add vertices one by one from C to S until C is saturated for S or $C \setminus (S \cup X) = \emptyset$. Then add the remaining vertices in $C \setminus (S \cup X)$ to X, if any, and repeat the process until no such clique C exists.*

Finally, we can now show that a partial solution can always be extended into a maximal one by means of a greedy filling.

Lemma 3. *For any partial solution (S, X), the greedy filling yields a maximal k-degenerate subgraph M of G such that $S \subseteq M$ and $M \cap X = \emptyset$.*

Proof. Let M be the greedy filling of (S, X). By definition, $S \subseteq M$ and $X \cap M = \emptyset$. Let $X_M = V \setminus M$.

We prove the statement by showing that at all times during the greedy filling (S, X) maintains the property of being a partial solution (see Definition 1), so in the end we have $A(x, X_M) \geq 1$ for each $x \in X_M$, making M a maximal k-degenerate subgraph by Lemma 2. Let Q be the maximal clique of the smallest *rank* for which $Q \setminus (S \cup X) \neq \emptyset$. Let (S', X') be the new pair constructed from Q by the greedy filling, and let (S_Q, X_Q) be the partition of $Q \setminus (S \cup X)$ such that $S' = S \cup S_Q$ and $X' = X \cup X_Q$. First notice that for all the ancestors Q' of Q we have $Q' \setminus (S \cup X) = \emptyset$ as their rank is smaller than the one of Q.

By definition of greedy filling, $|Q \cap S'| = |(Q \cap S) \cup S_Q| \leq k+1$. If $X_Q = \emptyset$, then $X' = X$ and $A(x, X') = A(x, X) \geq 1$ for each $x \in X'$. Otherwise, by definition of greedy filling, Q is saturated in S' ($|Q \cap S'| = k + 1$). Hence $A(Q, X') = 1$, and for each $x \in Q$ $A(x, X') \geq 1$, while for each $x \in X' \setminus Q = X \setminus Q$ $A(x, X') = A(x, X) \geq 1$. Thus, (S', X') is a partial solution, which completes the proof. $\square$

3.2 Binary Partition Method

We are now ready to describe our algorithm $k\text{MIG}(G, k)$, whose pseudo-code is given in Algorithm 1.

The principle is to start from the partial solution $S = \emptyset, X = \emptyset$, where S represent the vertices that will be in the solution, and X the vertices that are excluded from the solution, and proceed with binary partition: in each recursive call we consider a vertex $v \in Q$, initially from the clique Q with the smallest rank, i.e. the root of $\mathcal{QT}(G)$; we will first add v to S and find all the solutions containing $S \cup \{v\}$ and nothing in X; then add v to X and find all the solutions containing S and nothing in $X \cup \{v\}$, if any exists. At any step, we keep the invariant that (S, X) is a partial solution: If we add v to S (Line 12), this is equivalent to performing a step of the greedy filling, thus we know that $(S \cup \{v\}, X)$ is still a partial solution (see proof of Lemma 3). When, on the other hand, we try to add v to X (Line 14), we only explore this road if there exists a solution that contains all the vertices in S and no vertex in $X \cup \{v\}$. Thanks to (S, X) being a partial solution we will be able to discover this efficiently, and we will demonstrate (Lemma 4 in Sect. 3.3) that this is true if and only if $(S, X \cup \{v\})$ is still a partial solution. Only once $Q \setminus (S \cup X)$ is empty, we then proceed to the clique Q' next in the ranking (Lines 16–17). This guarantees that Q is always the clique of smallest rank such that $Q \setminus (S \cup X) \neq \emptyset$, thus condition 1 of Definition 1 still holds, and so (S, X) is still a partial solution. It is important

---

Algorithm 1. $k\text{MIG}$: Enumerating all maximal k-degenerate induced subgraphs in a chordal graph $G = (V, E)$

---

1 **Procedure** $k\text{MIG}(G, k)$
2   Compute $\mathcal{QT}(G)$ of G;
3     $R \leftarrow$ the root clique of $\mathcal{QT}(G)$;
4     $\pi : \{1, \ldots |\mathcal{Q}(G)|\} \to \mathcal{Q}(G) \leftarrow$ the pre-order traversal of $\mathcal{QT}(G)$;
5   Call Sub$k\text{MIG}(G, R, \emptyset, \emptyset, k)$;

6 **Procedure** Sub$k\text{MIG}(G, Q, S, X, k)$
7   **if** $V = S \cup X$ **then**
8   Output S
9   **if** $Q \setminus (S \cup X) \neq \emptyset$ **then**
10         $v \leftarrow$ the smallest vertex in $Q \setminus (S \cup X)$;
11   **if** $|Q \cap S| < k + 1$ **then**
12   Sub$k\text{MIG}(G, Q, S \cup \{v\}, X, k)$
13   **if** *there exists a solution* S^* *s.t.* $S \subseteq S^* \wedge S^* \cap (X \cup \{v\}) = \emptyset$ **then**
14   Sub$k\text{MIG}(G, Q, S, X \cup \{v\}, k)$
15   **else**
16         $Q' \leftarrow \pi(\pi^{-1}(Q) + 1)$;
17   Sub$k\text{MIG}(G, Q', S, X, k)$

to remark that, as all ancestors of Q are fully contained in $S \cup X$, and $v \notin S \cup X$, then v is always a *private vertex* of Q, not contained in the ancestors of Q.

Finally, if $S \cup X = V$ we can output S as a solution: by keeping the invariant that (S, X) is a partial solution, we know by Lemma 2 that S is a maximal k-degenerate induced subgraph of G.

3.3 Correctness

In this section we show the following theorem, that is the correctness of our algorithm.

Theorem 2. *Let G be a chordal graph and k be a non-negative integer. Then $k\mathit{MIG}(G, k)$ outputs all and only maximal k-degenerate induced subgraphs of G without duplicates.*

As mentioned in the description, $k\mathtt{MIG}(G, k)$ uses binary partition, thus every recursive call has either a single child (Line 17) which will simply extend the current solution, or will produce two recursive calls (Lines 12 and 14) that will lead to different solutions, as the first one considers only solutions for which $v \in S$, and the second only solutions for which $v \notin S$ (if any). Thus the same solution cannot be found more than once.

Furthermore, as we keep the invariant that (S, X) is a partial solution, by Lemma 2 we know that when $V = S \cup X$ then S is a maximal k-degenerate induced subgraph, thus $k\mathtt{MIG}(G, k)$ outputs only solutions.

Finally, any solution, i.e. maximal k-degenerate induced subgraph M is found by the algorithm, and we can prove this by induction: consider the set of cliques $Q_1, Q_2, \ldots$ in $\mathcal{QT}(G)$, ordered by ranking. As base condition assume that (S, X) is a partial solution such that $S \subseteq M, X \cap M = \emptyset$; this is always true in the beginning, when $(S = \emptyset, X = \emptyset)$. Let Q_i be the clique that we are considering, i.e. the one of smallest rank such that $Q_i \setminus (S \cup X) \neq \emptyset$, and v be the smallest vertex in $Q_i \setminus (S \cup X)$. If $v \in M$, then the recursive call in Line 12 will consider a partial solution which has one more vertex in common with M, i.e. $(S \cup \{v\}, X)$. Otherwise, $v \notin M$, that is, there exists a solution S^* such that $S \subseteq S^*$ and $S^* \cap (X \cup \{v\}) = \emptyset$, thus the recursive call in Line 14 is executed; this recursive call will consider a partial solution that has one more vertex in common with $V \setminus M$, i.e. $(S, X \cup \{v\})$. In both cases the base condition is still true, thus by induction $k\mathtt{MIG}(G, k)$ will find M. In order to prove Theorem 2, it only remains to show how to decide whether, given (S, X), there is a solution containing S but nothing in $X \cup \{v\}$, i.e., how to compute Line 13. This is shown in the following lemma.

Lemma 4. *Let (S, X) be any partial solution of G, Q be a clique such that its ancestor cliques are fully contained in $S \cup X$, and $v \notin S \cup X$ be a private vertex of Q. Then, there exists a solution S^* such that $S \subseteq S^*$ and $S^* \cap (X \cup \{v\}) = \emptyset$, if and only if $A(x, X \cup \{v\}) \geq 1$ for each vertex $x \in N[v] \cap (X \cup \{v\})$.*

Proof. Let $X' = X \cup \{v\}$. If for each vertex $x \in N[v] \cap X'$, $A(x, X') \geq 1$, then (S, X') still satisfies all the properties in Definition 1, as $A(w, X)$ is unchanged for any vertex $w \in X \setminus N(v)$. Thus (S, X') is a partial solution, and a solution S^* is given by Lemma 3.

Otherwise, there is a vertex $x \in X'$ such that $A(x, X') = 0$, i.e., there is no clique Q containing x such that $|Q \setminus X'| \geq k + 1$. As $X' \subseteq V \setminus S^*$ for any solution S^* disjoint from X', there is no clique Q containing x such that $|Q \setminus (V \setminus S^*)| \geq k + 1$, thus $A(x, V \setminus S^*) = 0$, and there is no maximal solution S^* by Lemma 2. □

Thus Theorem 2 is true, and $k\mathtt{MIG}(G, k)$ finds all and only maximal k-degenerate induced subgraphs of the chordal graph G exactly once.

4   Complexity Analysis

In this section we analyze the cost of our algorithm, and prove that it can enumerate all maximal k-degenerate subgraphs of G in $O(m \cdot \omega(G))$ time per solution. First, we recall some important properties of cliques in chordal graphs.

Remark 1 (From [3] and [10]). Let G be a connected chordal graph with $n > 1$ vertices and m edges. Then the number of maximal cliques in G is at most $n - 1$, and the sum of their sizes is $\sum_{C \in \mathcal{Q}(G)} |C| = O(m)$.

And regarding the cliques in G containing a specific node, we can state the following.

Lemma 5. *In a chordal graph G, the number of cliques containing a vertex v is at most $|N(v)|$.*

Proof. Consider $G[N[v]]$, the subgraph of G induced by vertices of $N[v]$. $G[N[v]]$ is chordal as it is an induced subgraph of a chordal graph, it has $|N[v]|$ vertices, and at most $|N[v]| - 1 = |N(v)|$ maximal cliques, which exactly correspond to the maximal cliques in G containing v. □

Now, consider the cost of executing Line 13, which dominates the cost of each iteration of the algorithm. We show in the next lemma that it can be done efficiently by exploiting Lemma 4. We recall that $\omega(G)$ denotes the maximum size of a clique in G.

Lemma 6. *Line 13 can be executed in time $O(\omega(G) \cdot |N(v)|)$.*

Proof. By Lemma 4 it is sufficient to check, for every vertex in $x \in N[v]$, whether there must be a clique Q' containing x such that $|Q' \setminus (X \cup \{v\})| \geq k + 1$. As (S, X) is a partial solution, if a vertex x is not contained in any clique such that $|Q' \setminus (X \cup \{v\})| \geq k + 1$, there exists a clique Q' such that $|Q' \setminus X| \geq k + 1 > |Q' \setminus (X \cup \{v\})| = k$, thus x is contained in one of the cliques containing v.

Assume we have a table that keeps track of the value $B(Q) = |Q \setminus (X \cup \{v\})|$ for every clique Q, and one that keeps the value $A(x) = |\{Q \mid x \in Q \text{ and } B(Q) \geq k+1\}|$. When adding v to X, we can update the B table by decrementing $B(Q)$ by 1 for every clique containing v. The number of such cliques in a chordal graph is at most $|N(v)|$ by Lemma 5. Every time the value of $B(Q)$ is decremented to less than $k + 1$, we can update the A table by decrementing $A(x)$ by 1 for each vertex x in Q. During this process, the check fails if and only if $A(x)$ is decremented to 0 for any x. The time required is $|Q| \leq \omega(G)$ for each considered clique, for a total cost of $O(\omega(G) \cdot |N(v)|)$. $\square$

Finally, we are ready to prove the complexity bound for $k\mathtt{MIG}(G, k)$.

Theorem 3. $k\mathtt{MIG}(G, k)$ *runs with delay* $O(m \cdot \omega(G))$.

Proof. First, we need to compute $\mathcal{QT}(G)$, which takes $O(n+m)$ time [10]. Note that $O(m + n) = O(m)$ as G is connected. Computing a pre-order traversal of $\mathcal{QT}(G)$ takes $O(n)$ time as $\mathcal{QT}(G)$ has at most n nodes.

In each recursive call we add a vertex either in S or in X or consider a next maximal clique. Hence, the depth of the tree of recursive calls is bounded by $2n$. To bound the delay between two solutions M and M', it is enough to bound the sum of the cost of all recursive calls in the path from the recursive call outputting M to the one that outputs M'. For clarity, let us use the term *recursive node* to refer a node in the tree of the recursive calls. Note that the recursive nodes that output a solution are exactly the leaves of this tree, thus the path between M and M' is bounded by the cost of a root-to-leaf and a leaf-to-root path.

As to execute Line 13 we use tables A and B (see Lemma 6), let us explain how to initialise them (we already explain in Lemma 6 how to update them). For each vertex x, we set $A(x) = |\{Q \in \mathcal{Q}(G, x) \mid |Q| \geq k+1\}|$, and set $B(Q) = |Q|$ for each $Q \in \mathcal{Q}(G)$. In order to set these values we can simply iterate over all maximal cliques in $\mathcal{QT}(G)$: initialising $B(Q)$ takes $O(1)$ time, and if $|Q| \geq k+1$ we increment $A(x)$ by 1 for each $x \in Q$, which takes $O(|Q|)$ time. The total running time for initialising the tables A and B take thus $O(n + m) = O(m)$ time (see Remark 1).

Let $v_1, \ldots v_t$ be the recursive nodes in the path from the root to the node that outputs M'. First, $t \leq 2n$ as in each step either we add v to S or to X or we take another Q. The delay now is the sum of the cost of each v_i. Lines 9–14 can be done in time $O(|N(x)| \cdot \omega(G))$ by Lemma 6. The cost for Lines 16–17 is $O(1)$. By summing, we have the upper bound $\sum_{Q \in \mathcal{Q}(G)} O(1) + \sum_{x \in V(G)} O(|N(x)| \cdot \omega(G)) = O(m \cdot \omega(G))$. The $O(m)$ preprocessing cost is negligible as there always exists at least one solution. $\square$

Note that this holds for any value of k: indeed, by Theorem 1 we know that chordal graphs are $\omega(G) - 1$-degenerate, thus for any $k \geq \omega(G)$, the problem is trivial as the only maximal solution is G itself.

5   Conclusion

We presented the first output-polynomial algorithm for enumerating maximal k-degenerate induced subgraphs in a chordal graph. The algorithm runs in $O(m \cdot \omega(G))$ time per solution for any given k. It would be interesting for future work to investigate the feasibility of an output-polynomial algorithm for general graphs. It is worth noticing that the enumeration of maximal independent sets in graphs is a special case as X is an independent set in G if and only if $G[X]$ is 0-degenerate.

References

1. Alon, N., Kahn, J., Seymour, P.D.: Large induced degenerate subgraphs. Graph. Combin. **3**(1), 203–211 (1987)
2. Bauer, R., Krug, M., Wagner, D.: Enumerating and generating labeled k-degenerate graphs. In: 2010 Proceedings of the Seventh Workshop on Analytic Algorithmics and Combinatorics, pp. 90–98. SIAM, Philadelphia (2010)
3. Blair, J.R., Peyton, B.: An introduction to chordal graphs and clique trees. In: George, A., Gilbert, J.R., Liu, J.W.H. (eds.) Graph Theory and Sparse Matrix Computation, pp. 1–29. Springer, New York (1993)
4. Chandran, L.S.: A linear time algorithm for enumerating all the minimum and minimal separators of a chordal graph. In: Wang, J. (ed.) COCOON 2001. LNCS, vol. 2108, pp. 308–317. Springer, Heidelberg (2001). doi:10.1007/3-540-44679-6_34
5. Conte, A., Grossi, R., Marino, A., Versari, L.: Sublinear-space bounded-delay enumeration for massive network analytics: Maximal cliques. In: ICALP (2016)
6. Diestel, R.: Graph Theory (Graduate Texts in Mathematics). Springer, New York (2005)
7. Eiter, T., Makino, K., Gottlob, G.: Computational aspects of monotone dualization: a brief survey. Discrete Appl. Math. **156**(11), 2035–2049 (2008)
8. Enright, J., Kondrak, G.: The application of chordal graphs to inferring phylogenetic trees of languages. In: Fifth International Joint Conference on Natural Language Processing, IJCNLP, pp. 545–552 (2011)
9. Eppstein, D., Löffler, M., Strash, D.: Listing all maximal cliques in sparse graphs in near-optimal time. In: Cheong, O., Chwa, K.-Y., Park, K. (eds.) ISAAC 2010. LNCS, vol. 6506, pp. 403–414. Springer, Heidelberg (2010). doi:10.1007/978-3-642-17517-6_36
10. Galinier, P., Habib, M., Paul, C.: Chordal graphs and their clique graphs. In: Nagl, M. (ed.) WG 1995. LNCS, vol. 1017, pp. 358–371. Springer, Heidelberg (1995). doi:10.1007/3-540-60618-1_88
11. Gavril, F.: The intersection graphs of subtrees in trees are exactly the chordal graphs. J. Comb. Theor. Ser. B **16**(1), 47–56 (1974)
12. Kuramochi, M., Karypis, G.: Frequent subgraph discovery. In: Proceedings IEEE International Conference on Data Mining, pp. 313–320. IEEE (2001)
13. Lee, V.E., Ruan, N., Jin, R., Aggarwal, C.: A survey of algorithms for dense subgraph discovery. In: Aggarwal, C.C., Wang, H. (eds.) Managing and Mining Graph Data, pp. 303–336. Springer, Heidelberg (2010)
14. Lukot'ka, R., Mazák, J., Zhu, X.: Maximum 4-degenerate subgraph of a planar graph. Electron. J. Comb. **22**(1), P1–11 (2015)

15. Nešetřil, J., Ossona de Mendez, P.: Sparsity: Graphs, Structures, and Algorithms. Algorithms and Combinatorics, vol. 28. Springer, Heidelberg (2012)
16. Pilipczuk, M., Pilipczuk, M.: Finding a maximum induced degenerate subgraph faster than 2^n. In: Thilikos, D.M., Woeginger, G.J. (eds.) IPEC 2012. LNCS, vol. 7535, pp. 3–12. Springer, Heidelberg (2012). doi:10.1007/978-3-642-33293-7_3
17. Rose, D.J., Tarjan, R.E., Lueker, G.S.: Algorithmic aspects of vertex elimination on graphs. SIAM J. Comput. **5**(2), 266–283 (1976)
18. Tarjan, R.E., Yannakakis, M.: Simple linear-time algorithms to test chordality of graphs, test acyclicity of hypergraphs, and selectively reduce acyclic hypergraphs. SIAM J. Comput. **13**(3), 566–579 (1984)
19. Ugander, J., Karrer, B., Backstrom, L., Marlow, C.: The anatomy of the facebook social graph. arXiv preprint arXiv: 1111.4503 (2011)
20. Wasa, K.: Enumeration of enumeration algorithms. arXiv preprint arXiv:1605.05102 (2016)
21. Wasa, K., Arimura, H., Uno, T.: Efficient enumeration of induced subtrees in a K-degenerate graph. In: Ahn, H.-K., Shin, C.-S. (eds.) ISAAC 2014. LNCS, vol. 8889, pp. 94–102. Springer, Cham (2014). doi:10.1007/978-3-319-13075-0_8

Reoptimization of Minimum Latency Problem

Wenkai Dai$^{(\boxtimes)}$

Saarland University, Saarbrücken, Germany
wenkai.dai@gmail.com

Abstract. The minimum latency problem (MLP) is a prominent variant
of the traveling salesman problem, which is APX-hard and currently has
the best approximation ratio about 3.59 on metric graphs [14]. In this
paper, we consider several reoptimization variants of the metric MLP. In
a reoptimization problem, an optimal solution to an instance is given,
the goal is to obtain a good solution for a locally modified instance. Here,
we consider four common local modifications: adding (resp., removing)
a vertex and increasing (resp., decreasing) the cost of an edge. First, we
prove that these four reoptimization problems are NP-hard. Then, we
provide 7/3-approximation and 3-approximation algorithms for adding
and removing a vertex respectively. As for changing the cost of an edge
e^*, we study them by parameterizing the position of e^*. For increasing
the cost, our approximation ratios range from 2.1286 to 4/3 during e^*
moving from the first edge of the given optimal tour to the last edge.
About decreasing the cost, we show that if the given optimal tour visits
e^* as the i-th edge then the problem is NP-hard and 2-approximable for
$i \geq 3$, while it has a PTAS but not FPTAS for a constant i. However, if
e^* is not in the given optimal solution, the problem for decreasing the
edge cost is approximable by at least $2.1286 + \mathcal{O}(1/n)$, where n is the
number of vertices in the given optimal solution. Moreover, we show that
relaxing the optimality of the given solution causes the approximability
of the problem to remove a vertex to be as hard as the metric MLP itself.

1 Introduction

The *minimum latency problem* (MLP) is a classic variant of the *traveling salesman problem* (TSP). Similar to TSP, it is not efficiently approximable in general
weighted graphs. Thus, we are only interested in its metric version. MLP seeks
for a tour starting from a fixed vertex which minimizes the sum of distances
from each vertex to this starting vertex.

This problem is NP-hard even when the metric space is induced by a weighted
tree [23], or when the edge costs are restricted into 1 and 2 [10,21]. Although it
is also APX-hard, the former studies reveal it is much harder to be approximated
than TSP. At present, the best approximation ratio of MLP is about 3.59 by
Chaudhuri et al. [14], while the metric TSP is 1.5-approximable [15].

In this paper, we study reoptimization aspects of MLP. Reoptimization is a
framework in the context of NP-hard optimization problems: given an optimal
solution to an *initial instance* I', then a *local modification* turns I' into a similar

© Springer International Publishing AG 2017
Y. Cao and J. Chen (Eds.): COCOON 2017, LNCS 10392, pp. 162–174, 2017.
DOI: 10.1007/978-3-319-62389-4_14

instance I, the goal is to compute a solution to the *modified instance I*. Actually, the scenario of reoptimization is motivated by real-world applications, e.g., a TSP example in practice: the salesman has known an optimal tour on some cities after working for some years, but on one day he is required to visit an extra city by his company; instead of computing a new tour from scratch, he may prefer to extend the former tour locally to integrate this new city.

The concept of reoptimization was first mentioned in [22]. Since then, the concept of reoptimization has been applied to various problems, including the traveling salesman problem [1,3–5,20], the Steiner tree problem [6,9,17], the knapsack problem [2], maximum weight induced heredity problems [12], covering problems [8], scheduling problems [11] and the shortest common superstring problem [7]. There are some surveys on reoptimization [13,16,18].

In this paper, we study the reoptimization of MLP for four typical local modifications: adding (resp., removing) a vertex denoted by MLP^{V+} (resp., MLP^{V-}) and increasing (resp., decreasing) the cost of an edge denoted by MLP^{E+} (resp., MLP^{E-}). Surprisingly, these four problems have not received much attention in previous researches. To the best of our knowledge, only MLP^{V+} was given a 3-approximation algorithm by Ausiello et al. [16]. Firstly, we show that these four problems are NP-hard. By a general method in [18], the hardness of MLP^{V+}, MLP^{E+} and MLP^{E-} can be easily established by polynomial-time Turing reductions; but for MLP^{V-}, we give a Karp reduction instead since a trivial instance extensible to a general instance of MLP^{V-} is hard to be detected. For MLP^{V+}, we provide a 7/3-approximation algorithm to improve the 3-approximation result by Ausiello et al. [16]. We also show that MLP^{V-} is at least 3-approximable, which outperforms computing from scratch for MLP. Furthermore, we provide a new idea to study the reoptimization of altering the cost of an edge. By parameterizing the position of the *modified* edge e^*, we can exhibit a more fine-grained analysis of how the approximability of MLP^{E+} and MLP^{E-} varies in Table 1. In particular, for the MLP^{E-} where the given optimal tour $\mathcal{OPT}'$ contains e^* as the i-th edge, unlike reoptimizing TSP, $\mathcal{OPT}'$ cannot remain optimal anymore in the modified instance, while this problem is NP-hard and has a PTAS, but no FPTAS unless P=NP, for every constant $i \geq 3$.

Goyal and Mömke [17] proposed a different reoptimization model which slacks the optimality of the given solution. They were concerned with how the approximability of reoptimization varies when the quality of the given solution declines. Following their way, we prove that MLP^{V-} cannot be approximated better than MLP itself if the given solution of the initial instance is not optimal anymore.

2 Preliminaries

For a graph G, let $V(G)$ and $E(G)$ denote the set of vertices and the set of edges of G respectively. A *Hamiltonian path (cycle also called tour)* of a graph is a path (cycle) that visits each vertex exactly once. A *complete, edge-weighted* graph $G = (V, E, c)$, where $c : E \rightarrow \mathbb{R}_{\geq 0}$, is called *metric* if the *triangle inequality:* $c(\{a, b\}) \leq c(\{a, c\}) + c(\{b, c\})$ is satisfied for all vertices $a, b, c \in V$.

Table 1. Approximation results about MLP^{E+} and MLP^{E-}, where $E(OPT')$ denotes the set of edges in the given optimal solution OPT', the parameter i indicates that e^* is the i-th edge in OPT' for $1 \leq i \leq n-1$, and n is the number of vertices of OPT'.

Problem	Approximation results			Ref.
MLP^{E+}	$i = 1$	$i \neq 1, n-1$	$i = n-1$	
$e^* \in E(OPT')$	2.1286	2	$4/3$	Theorem 5
$e^* \notin E(OPT')$	OPT' remains optimal after increasing			
MLP^{E-}	$i = 1, 2$	constant i	$i \geq 3$	
$e^* \in E(OPT')$	P	PTAS, no FPTAS	2	Theorems 6–8
$e^* \notin E(OPT')$	$2.1286 + \mathcal{O}(1/n)$			Theorem 9

The minimum latency problem (MLP) has an input instance (G, s), where $G = (V, E, c)$ is a metric graph with $V = \{v_1, \ldots, v_n\}$ and a fixed vertex $s \in V$ which is called *source*. Let a tour $P = (v_1, \ldots, v_n)$ starting at v_1 be a *subgraph (path)* in G consisting of n vertices and a sequence of $n-1$ edges, then P is a *feasible tour* for the MLP instance (G, v_1). Please note that the feedback edge (v_n, v_1) is not included in P. The cost of P is defined as $c(P) = \sum_{e \in E(P)} c(e)$. A *subtour (subpath)* of P from v_i to v_j is denoted by $v_i P v_j$. The *latency* $l(v_i, P, G)$ of a vertex $v_i \in V$ along the tour P is the cost of the subtour $v_1 P v_i$, i.e., $l(v_i, P, G) = c(v_1 P v_i)$. The starting vertex (source) $v_1 \in V$ has $l(v_1, P, G) = 0$. The *(total) latency* $L(P, G)$ of P is the sum of latencies for all vertices in P. Alternatively, Definition 1 expresses the latency of P in the form of edges.

Definition 1. *Let* $P = (e_1, \ldots, e_{n-1})$ *be a feasible tour for the MLP instance* (G, v_1). *The (total) latency* $L(P, G)$ *of* P *can be expressed by* (1).

$$L(P, G) = \sum_{i=1}^{n-1} (n - i) \cdot c(e_i) \tag{1}$$

Definition 2 (Metric Minimum Latency Problem (MLP)). *Given an MLP instance* (G, s), *the problem is to find a feasible tour* P *such that* P *has the* minimum *total latency for all feasible tours starting at* s.

For a metric graph $G = (V, E, c)$ and a vertex $t \in V$, we define $G - \{t\}$ to be a metric graph $G' = (V', E', c')$ s.t., $V = V' \cup \{t\}$ and $\forall e \in E' : c'(e) = c(e)$.

Definition 3. MLP^{V+} *(resp.,* MLP^{V-}) *is the following optimization problem:*

Input: *A vertex* t *(also called the modified vertex), two MLP instances* (G, s) *and* (G', s) *on metric graphs* $G = (V, E, c)$ *and* $G' = (V', E', c')$ *respectively, where* $G' = G - \{t\}$ *(resp.,* $G = G' - \{t\}$), *and an optimal tour* OPT' *for* (G', s);
Question: *Find an optimal tour* OPT *for the MLP instance* (G, s).

Definition 4. MLP^{E+} *(resp.,* MLP^{E-}) *is the following optimization problem:*
Input: *A fixed edge* e^* *(also called the modified edge), two MLP instances* (G, s)

and (G', s) on metric graphs $G = (V, E, c)$ and $G' = (V, E, c')$ respectively, where the edge $e^* \in E$ satisfies $c'(e^*) < c(e^*)$ (resp., $c'(e^*) > c(e^*)$) and $\forall e \in E \setminus \{e^*\}$: $c(e) = c'(e)$, and an optimal tour OPT' for (G', s).
Question: Find an optimal tour OPT for the MLP instance (G, s).

For reoptimization problems defined by Definitions 3–4, from now on, we implicitly use G' and OPT' (resp., G and OPT) to denote the input metric graph and the given optimal tour (resp., an arbitrary optimal tour) of the initial (resp., modified) MLP instance. We will use opt' (resp., opt) to denote the objective value of an optimal solution for the initial (resp., modified) instance throughout this paper. Since OPT' is also a graph, $V(OPT')$ and $E(OPT')$ denote vertices and edges of OPT' respectively, while these notations are applied to OPT similarly.

Particularly, we will implicitly assume that $V(G') = \{v_1, \ldots, v_n\}$ of n vertices and $OPT' = (v_1, \ldots, v_n)$ in our discussions, unless otherwise explicitly stated.

Let MLP_ϵ^{V-} denote a variant of MLP^{V-} such that the given solution is not optimal but a $(1+\epsilon)$-approximation solution for the initial instance, where $\epsilon > 0$.

In Definition 5, we introduce a useful NP-complete problem for later proofs.

Definition 5 (Restricted Hamiltonian Cycle Problem (RHC) [19]). *Given a general (unweighted) graph G, and a Hamiltonian path P in G where two endpoints of P are not adjacent in G, the problem is to decide whether G contains a Hamiltonian cycle.*

3   Hardness Results

In this section, we show that for the reoptimization problems defined in Sect. 2 there is no hope for polynomial-time exact algorithms.

Theorem 1. MLP^{V+}, MLP^{E+} *and* MLP^{E-} *are NP-hard.*

The proof of Theorem 1 can be easily established via a general framework introduced in [18]. The basic idea is to employ a polynomial-time Turing reduction, which tries to find a trivial MLP instance I_0 such that a general MLP instance I can be obtained by modifying I_0 polynomially many times, then I is solved by using an oracle of the MLP reoptimization problem polynomially many times.

However, this framework cannot be applied to MLP^{V-} directly. Instead, for Theorem 2, we take the Karp reduction to show the hardness of MLP^{V-}.

Theorem 2. MLP^{V-} *is NP-hard.*

Proof. The reduction is from RHC (Definition 5). Given an instance of RHC $(G^\star, P^\star)$, where $G^\star = (V^\star, E^\star)$, $V^\star = \{v_1, \ldots, v_n\}$ and a Hamiltonian path (HP) $P^\star = (v_1, \ldots, v_n)$, we construct an instance of MLP^{V-} as follows.

The initial instance (G', s) has the source s and the metric graph $G' = (V', E', c')$, where $V' = V^\star \cup \{s, t, d\}$ and $c' : E \to \{1, 2\}$. The modified instance (G, s) has the metric graph $G = (V, E, c)$, where $V = V' \setminus \{t\}$ and $\forall e \in E$: $c(e) = c'(e)$. The cost function c' assigns the cost one to edges $\{s, t\}$, $\{t, d\}$ and

$\{d, v_1\}$. For each $v \in V^* \setminus \{v_1\}$, if $\{v_1, v\} \in E^*$ then $c'(\{s, v\}) = 1$. Moreover, for any two vertices $v_i, v_j \in V^*$, if $\{v_i, v_j\} \in E^*$ then $c'(\{v_i, v_j\}) = 1$. For all unmentioned edges in G', c' gives each edge the cost two. The construction is completed by setting the given optimal tour $\mathcal{OPT}' = (s, t, d, v_1, v_2, \ldots, v_n)$.

It is easy to verify that G and G' are metric. Clearly, $\mathcal{OPT}'$ is optimal to (G', s) since each edge of $\mathcal{OPT}'$ has the cost one in G'. Hence, we have constructed a valid instance of MLP^{V-} in polynomial time. Now, we claim that G^* has a Hamiltonian cycle (HC) if and only if $\mathsf{opt} \leq \mathsf{opt}' - (n+2)$.

If there is a HC T^* in G^*, then there must be a vertex $v^* \in V^*$ such that $\{v_1, v^*\} \in E(T^*)$. Now, we can obtain one tour $P = (s, v^*T^*v_1, d)$ in G, where $v^*T^*v_1$ is a HP in G^* from v^* to v_1. Clearly P has $L(P, G) = \mathsf{opt}' - (n+2)$.

Conversely, if a tour P for (G, s) has the latency $\mathsf{opt}' - (n+2)$ then each edge of P has the cost one. This also implies that $\{v_1, d\}$ is the last edge of P and d is the last vertex of P. Let $\{s, v^*\}$ be the first edge of P, then $\{s, v^*\}$ has the cost one and the vertex v^* must be adjacent to v_1 in G^*. If $\{v_i, v_j\} \in E(P)$, where $v_i, v_j \in V^*$, then $\{v_i, v_j\} \in E^*$. Therefore, removing vertices d and s from P and rejoining v^* and v_1 in P must produce a HC for the graph G^*.

Finally, since RHC is NP-hard [19], this implies that MLP^{V-} is NP-hard. $\square$

Algorithm 1.1. The simple approximation algorithm for MLP^{V+}.

Input   : An instance $(G, G', \mathcal{OPT}', s, t)$ of MLP^{V+};
Output: A feasible tour T_{out} of (G, s) starting at the source s;

1 Let the given optimal tour $\mathcal{OPT}' = (v_1, \ldots, v_n)$ and the source $s = v_1$;
2 The first feasible tour of (G, s): $T_0 := (v_1, \ldots, v_n, t)$;
3 The second feasible tour of (G, s): $T_1 := (v_1, \ldots, v_{n-1}, t, v_n)$;
4 **return** a tour T_{out} in $\{T_0, T_1\}$ with the least total latency;

---

4   Approximation Results of MLP Reoptimization

4.1   Locally Modifying a Single Vertex

Theorem 3. *Algorithm 1.1 is a 7/3-approximation algorithm for* MLP^{V+}.

Proof. The input instance is $(G, G', \mathcal{OPT}', v_1, t)$, where $G' = (V', E', c')$, $G = (V, E, c)$, the modified vertex $t \in V$, $G' = G - \{t\}$, v_1 is the source. Recall that $\forall e \in E' : c'(e) = c(e)$ and $c(\mathcal{OPT}') = l(v_n, \mathcal{OPT}', G')$ is the cost of $\mathcal{OPT}'$ in G'.

In G, let $c_{\min}(t)$ denote the minimum edge cost for all edges incident on t, then there must be $v^* \in V'$ s.t. $c(\{v^*, t\}) = c_{\min}(t)$. For MLP^{V+}, it is obvious that $\mathsf{opt}' < \mathsf{opt}$. Due to $l(t, \mathcal{OPT}, G) \geq c_{\min}(t)$, we know that $\mathsf{opt}' + c_{\min}(t) \leq \mathsf{opt}$.

Since G is metric, the triangle inequality implies that

$$c(\{v_n, t\}) \leq c(\{v^*, t\}) + c(\{v^*, v_n\}) \leq c_{\min}(t) + c(\mathcal{OPT}'). \qquad (2)$$

Recalling Definition 1, by the factors given in (1), we know that

$$2 \cdot c(\mathcal{OPT}') \leq \text{opt}' + c(\{v_{n-1}, v_n\}). \tag{3}$$

According to Algorithm 1.1, we can bound the total latency of T_0 in G by (4).

$$\begin{aligned} L(T_0, G) &\leq \text{opt}' + c(\mathcal{OPT}') + c(\{v_n, t\}) \\ &\leq \text{opt}' + 2 \cdot c(\mathcal{OPT}') + c_{\min}(t) \\ &\leq 2 \cdot \text{opt} + c(\{v_{n-1}, v_n\}) \end{aligned} \tag{4}$$

Particularly, if $c(\{v_n, t\}) = c_{\min}(t)$ then T_0 directly leads to 2-approximation. Hence, we will simply assume $v^* \neq v_n$ in the remainder of this proof.

Now, we discuss the tour T_1, which visits t right after v_{n-1} but before v_n. Firstly, for each $v_i \in \{v_1, \ldots, v_{n-1}\}$, it has $l(v_i, \mathcal{OPT}', G') = l(v_i, T_1, G)$. The latency of v_n in T_1 is increased compared to its latency in $\mathcal{OPT}'$. By the triangle inequality, we can bound this increased value as follows:

$$l(v_n, T_1, G) - l(v_n, \mathcal{OPT}', G') \leq 2 \cdot c(\{v_{n-1}, t\}). \tag{5}$$

The latency of t in T_1, i.e., $l(t, T_1, G)$, is $c(v_1 \mathcal{OPT}' v_{n-1}) + c(\{v_{n-1}, t\})$. The cost $c(\{v_{n-1}, t\})$ is also bounded by $c(v_1 \mathcal{OPT}' v_{n-1}) + c_{\min}(t)$ if $v^* \neq v_n$. According to the above inequalities and Definition 1, the latency of T_1 in G is bounded by the following equality.

$$\begin{aligned} L(T_1, G) &\leq \text{opt}' + c(v_1 \mathcal{OPT}' v_{n-1}) + 3 \cdot c(\{v_{n-1}, t\}) \\ &\leq \text{opt}' + 4 \cdot c(v_1 \mathcal{OPT}' v_{n-1}) + 3 \cdot c_{\min}(t) \\ &\leq 3 \cdot \text{opt}' - 2 \cdot c(\{v_{n-1}, v_n\}) + 3 \cdot c_{\min}(t) \\ &\leq 3 \cdot \text{opt} - 2 \cdot c(\{v_{n-1}, v_n\}) \end{aligned} \tag{6}$$

Since Algorithm 1.1 outputs T_{out} as the better one in $\{T_0, T_1\}$, then we can combine (4) and (6) to finally show that $L(T_{\text{out}}, G) \leq (7/3) \cdot \text{opt}$.   □

Due to the limited space, the proof of Theorem 4 is not provided.

Theorem 4. MLP^{V-} *is 3-approximable by short-cutting the vertex t of $\mathcal{OPT}'$.*

4.2 Altering the Cost of an Edge

Definition 4 specially requires the graph after changing the edge cost to be metric. Then this constraint leads to a useful observation, which is given in Lemma 1.

Lemma 1 ([4]). *Let $G = (V, E, c)$ and $G = (V, E, c')$ be two metric graphs such that the cost functions c and c' coincide except for one edge $e^* \in E$. Then every edge adjacent to e^* has the cost at least $|c(e^*) - c'(e^*)|/2$.*

For MLP^{E+}, if $e^* \notin E(\mathcal{OPT}')$ then $\mathcal{OPT}'$ remains optimal after increasing the cost of e^*. Thus, we only need to consider the restricted MLP^{E+} where $e^* \in E(\mathcal{OPT}')$.

Theorem 5. *The restricted* MLP^{E+} *where* e^* *is the i-th edge of the given optimal tour* $\mathcal{OPT}'$ *for* $1 \le i \le n-1$, *can gain an approximation ratio* α *as follows: if* $i \neq 1, n-1$ *then* $\alpha = 2$; *if* $i = n-1$ *then* $\alpha = 4/3$; *and if* $i = 1$ *then* $\alpha = 2.1286$.

Proof. Let an instance of this restricted MLP^{E+} be $(G, G', \mathcal{OPT}', v_1, e^*)$, where $G = (V, E, c)$, $G' = (V, E, c')$, $\mathcal{OPT}' = (v_1, \ldots, v_n)$ and v_1 is the source. We note that $|E(\mathcal{OPT})| = n - 1$, $\forall e \in E \setminus \{e^*\} : c'(e) = c(e)$ and e^* is the i-th edge of $\mathcal{OPT}'$. Define $\delta = c(e^*) - c'(e^*)$ for $\delta > 0$. Clearly, MLP^{E+} has $\text{opt}' \le \text{opt}$.

For each $i \in \{1, \ldots, n-1\}$, the latency of $\mathcal{OPT}'$ after increasing is given as

$$L(\mathcal{OPT}', G) = \text{opt}' + (n - i) \cdot \delta. \tag{7}$$

Firstly, we show that $\mathcal{OPT}'$ is still a 2-approximation tour for the modified instance (G, v_1) when $i \notin \{1, n-1\}$. If $i \notin \{1, n-1\}$, the i-th edge e^* of $\mathcal{OPT}'$ implies that $\mathcal{OPT}'$ must contain the $(i-1)$-th edge e_l^* and the $(i+1)$-th edge e_r^*. Since these two edges are adjacent to e^*, Lemma 1 directly shows that

$$\delta/2 \le \min\{c'(e_l^*), c'(e_r^*)\}. \tag{8}$$

Recall Definition 1, we can use (8) to further derive that

$$(n-i) \cdot \delta \le (n - (i-1)) \cdot c'(e_l^*) + (n - (i+1)) \cdot c'(e_r^*) \le \text{opt}'. \tag{9}$$

Bounding the term $(n - i) \cdot \delta$ in (7) with (9), it directly implies that

$$L(\mathcal{OPT}', G) \le 2 \cdot \text{opt}' \le 2 \cdot \text{opt}. \tag{10}$$

Secondly, we prove that $\mathcal{OPT}'$ is a 4/3-approximation tour for the (G, v_1) when $i = n - 1$. If $i = n - 1$, $e^* = \{v_{n-1}, v_n\}$ is the last edge of $\mathcal{OPT}'$. Thus, e^* cannot be in any optimal tour of (G, v_1), otherwise $\mathcal{OPT}'$ remains optimal for (G, v_1).

Let $\mathcal{OPT}$ be an arbitrary optimal tour of (G, v_1). Firstly, we assume v_{n-1} is visited before v_n in $\mathcal{OPT}$. Now, we consider a special case where v_n is the n-th vertex and v_{n-1} is the $(n-2)$-th vertex visited in $\mathcal{OPT}$. In such a case, let e_k^n denote the $(n - k)$-th edge traversed in $\mathcal{OPT}$ for $k \in \{1, 2, 3\}$ respectively. Then, for each $k = 1, 2, 3$, the edge e_k^n is adjacent to e^*, which implies $c(e_k^n) \ge \delta/2$.

$$\begin{aligned} \text{opt} &\ge 3 \cdot c(e_3^n) + 2 \cdot c(e_2^n) + c(e_1^n) \\ &\ge (3 + 2 + 1) \cdot (\delta/2) \ge 3 \cdot \delta \end{aligned} \tag{11}$$

By Definition 1, we can further infer (11). By using $\delta \le \text{opt}/3$ in (7), we show the conclusion $L(\mathcal{OPT}', G) \le \frac{4}{3} \cdot \text{opt}$ if $i = n - 1$. For other different cases, the sum of factors in (11) should be larger than 6, then approximation ratios cannot be worse than 4/3. On the other hand, a similar discussion can be given if v_{n-1} is visited after v_n in $\mathcal{OPT}$.

Finally, if $i = 1$, we will show that $\mathcal{OPT}'$ cannot directly provide a good approximation for (G, v_1). Thus, we will find the other tour to complement $\mathcal{OPT}'$ to achieve a 2.1286-approximation. If $i = 1$ then $e^* = \{v_1, v_2\}$ is the first edge of $\mathcal{OPT}'$. Let $\mathcal{OPT}$ be an arbitrary optimal tour for (G, v_1). Note that e^* cannot be the first edge of $\mathcal{OPT}$, otherwise $\mathcal{OPT}'$ remains optimal after increasing. The first edge of $\mathcal{OPT}$, denoted by $e_f = \{v_1, v_f^*\}$, must be adjacent to e^*, which also implies $\delta \leq 2 \cdot c(e_f)$.

$$L(\mathcal{OPT}', G) \leq (1 + 2 \cdot \rho) \cdot \mathsf{opt} \tag{12}$$

By Definition 1, we can set $\rho \cdot \mathsf{opt} := (n-1) \cdot c(e_f)$, where $0 < \rho < 1$. Then we can rewrite (7) to (12). When $\rho \to 1$, the approximation ratio $(1 + 2 \cdot \rho)$ given by $\mathcal{OPT}'$ gets close to three, which is not good enough. Hence, we will obtain an alternative tour T_{alt} for $\rho \to 1$.

Clearly, the edge $e_f = \{v_1, v_f^*\}$ can be found by the exhaustive search. After removing v_1 from $\mathcal{OPT}$, the remaining part of $\mathcal{OPT}$ is still an optimal tour for the MLP instance $(G - \{v_1\}, v_f^*)$. By the algorithm of Chaudhuri et al. [14], we can compute a 3.59-approximation tour T for the instance $(G - \{v_1\}, v_f^*)$. Let the tour T_{alt} traverse e_f at first and then T. The latency of T_{alt} is bounded by

$$\begin{aligned} L(T_{\mathrm{alt}}, G) &\leq \rho \cdot \mathsf{opt} + 3.59 \cdot (1 - \rho) \cdot \mathsf{opt} \\ &\leq 3.59 \cdot \mathsf{opt} - 2.59 \cdot \rho \cdot \mathsf{opt}. \end{aligned} \tag{13}$$

The better tour of $\mathcal{OPT}'$ and T_{alt} will be the final output T_{out}. Thus, by combining (12) and (13), we obtain $L(T_{\mathrm{out}}, G) \leq 2.1286 \cdot \mathsf{opt}$. $\square$

For $\mathrm{MLP}^{\mathrm{E}-}$, we firstly study the restricted version that has $e^* \in E(\mathcal{OPT}')$. By Theorems 6–7, we show the complexity dichotomy for this restricted $\mathrm{MLP}^{\mathrm{E}-}$.

Theorem 6. *For $\mathrm{MLP}^{\mathrm{E}-}$, if the modified edge e^* is the first or second edge of the given optimal tour $\mathcal{OPT}'$, then $\mathcal{OPT}'$ remains optimal in the modified instance.*

Proof. For $\mathrm{MLP}^{\mathrm{E}-}$, if e^* is the first edge of $\mathcal{OPT}'$, then no tour can lose more latency than $\mathcal{OPT}'$ after decreasing the cost of e^* by Definition 1. Moreover, if e^* is the second edge of $\mathcal{OPT}'$ then e^* cannot contain the source. It means no feasible tour can visit e^* as the first edge. Thus, no feasible tour can lose more latency than $\mathcal{OPT}'$ after decreasing. Hence, $\mathcal{OPT}'$ remains optimal for above cases. $\square$

Theorem 7. *The restricted $\mathrm{MLP}^{\mathrm{E}-}$ where e^* is the i-th edge of the given optimal tour $\mathcal{OPT}'$ of the initial instance is NP-hard for a constant $i \geq 3$ and has no FPTAS unless P=NP.*

Proof. To show NP-hardness, we reduce from RHC (Definition 5). Given an instance (G^*, P^*) of RHC, where $G^* = (V^*, E^*)$, $V^* = \{v_1, \ldots, v_n\}$, and Hamiltonian path (HP) $P^* = (v_1, \ldots, v_n)$, we will construct an instance of $\mathrm{MLP}^{\mathrm{E}-}$ where the modified edge e^* is visited as the third edge of $\mathcal{OPT}'$.

We construct the initial instance (G', s), where $G' = (V, E, c')$, $V = V^* \cup \{s, a, b\}$, $c' : E \to \{2, 2.5, 3\}$ and s is the source. The cost function c' gives the cost two to each of edges $\{s, a\}$, $\{s, b\}$, $\{b, v_1\}$ and $\{a, b\}$. For any $v_i, v_j \in V^*$, if $\{v_i, v_j\} \in E^*$ then $c'(\{v_i, v_j\}) = 2$. For each $v \in V^* \setminus \{v_1\}$, if $\{v_1, v\} \in E^*$ then $c'(\{a, v\}) = 2.5$. Each unmentioned edge of E has cost three in c'. The modified instance (G, s) has $G = (V, E, c)$, where $\forall e \in E \setminus \{e^*\} : c(e) = c'(e)$. The given optimal tour $\mathcal{OPT}'$ of (G', s) is $(s, a, b, v_1, v_2 \ldots, v_n)$. The modified edge e^* is $\{b, v_1\}$, which has the cost $c(e^*) = 1$ after decreasing the cost.

It is easy to see that G' is metric due to $c' : E' \rightarrow \{2, 2.5, 3\}$. The graph G is also metric since $c : E' \rightarrow \{1, 2, 2.5, 3\}$ and only one edge e^* has the cost one. The constructed $\mathcal{OPT}'$ is optimal to (G', s) since $\forall e \in E(\mathcal{OPT}') : c'(e) = 2$. Clearly, we constructed a valid instance of $\mathrm{MLP}^{\mathrm{E}-}$ in polynomial time. Now, we claim that G^* has a Hamiltonian cycle (HC) if and only if opt $\leq$ opt$' - n - 1/2$.

Given a HC T^* in G^*, there is a vertex $v^* \in V^*$ such that $\{v_1, v^*\} \in E(T^*)$. A HP from v_1 to v^* in G^*, denoted by $v_1 P^* v^*$, is generated by removing $\{v_1, v^*\}$ from T^*. Then a tour $P = (s, b, v_1 P^* v^*, a)$ is a feasible tour for (G, s). The latency of P in G is $L(P, G) = $ opt$' - n - 1/2$.

Conversely, let P be a feasible tour to (G, s) such that $L(P, G) = $ opt$' - n - \frac{1}{2}$. By this latency, we can infer as follows: e^* is the second edge of P (v_1 is visited after b); a is the last vertex of P; and the last edge of P, denoted by $\{v^*, a\}$, has $c(\{v^*, a\}) = 2.5$. Then $c(\{v^*, a\}) = 2.5$ implies $\{v^*, v_1\} \in E^*$. Thus, removing vertices a, b and s from P and rejoining v^* and v_1 must form one HC for G^*.

Due to NP-hardness of RHC [19], our constructed instance is also NP-hard.

Now, we suppose there is an FPTAS $\mathcal{A}$ that returns a tour $T_{\mathcal{A}}$ with $L(T_{\mathcal{A}}, G) \leq (1 + \theta) \cdot$ opt in the polynomial time poly$(1/\theta, |G|)$. The construction is the same as above. For the modified instance (G, v_1), every non-optimal tour T has

$$L(T, G) \geq \text{opt} + 0.5 \geq \left(1 + \frac{1}{2 \cdot \text{opt}}\right) \cdot \text{opt} .$$

By setting $\theta < 1/(2 \cdot \text{opt})$, $\mathcal{A}$ must return an optimal tour to (G, v_1) in polynomial time, which decides whether G^* has a HC. Hence, no FPTAS unless P = NP. □

Theorem 8. *For* $\mathrm{MLP}^{\mathrm{E}-}$, *if the modified edge* e^* *is the i-th edge of the given optimal tour* $\mathcal{OPT}'$ *for* $3 \leq i \leq n - 1$, *then* $\mathcal{OPT}'$ *is 2-approximation tour after decreasing the cost of* e^*, *and a* PTAS *is admitted if i is a constant.*

Proof. Let $(G, G', \mathcal{OPT}', v_1, e^*)$ be an instance of $\mathrm{MLP}^{\mathrm{E}-}$, where $G = (V, E, c)$, $G' = (V, E, c')$, $V = \{v_1, \ldots, v_n\}$, $\mathcal{OPT}' = (v_1, \ldots, v_n)$ and v_1 is the source. We note that $|E(\mathcal{OPT})| = n - 1$ and $\forall e \in E \setminus \{e^*\} : c'(e) = c(e)$. Let $\mathcal{OPT}$ denote an arbitrary optimal tour to (G, v_1) s.t. $\mathcal{OPT} \neq \mathcal{OPT}'$. Since the cost of e^* is decreased, it holds that opt$' \geq$ opt. We also define $\delta = c'(e^*) - c(e^*) > 0$.

Clearly, $\mathcal{OPT}$ must take e^* as the j-th edge for $1 < j < i$, otherwise $\mathcal{OPT}'$ remains optimal after decreasing the cost of e^*. By Definition 1, we can express the latency of $\mathcal{OPT}'$ in G as follows:

$$\begin{aligned} L(\mathcal{OPT}', G) &= \text{opt}' - (n - i) \cdot \delta \\ &\leq \text{opt} + (n - j) \cdot \delta - (n - i) \cdot \delta \\ &\leq \text{opt} + (i - j) \cdot \delta . \end{aligned} \tag{14}$$

Now, we try to bound $(i - j) \cdot \delta$. We note that e^* cannot be the last edge for $\mathcal{OPT}$, otherwise $j < i$ is violated and $\mathcal{OPT}'$ remains optimal in (G, v_1). Thus, there must be two edges $e_{j-1}, e_{j+1} \in E(\mathcal{OPT})$ such that e_{j-1} and e_{j+1} are traversed in $\mathcal{OPT}$ exactly before and after the edge e^* respectively. Moreover, Lemma 1 indicates that $c(e_{j-1}) \geq \delta/2$ and $c(e_{j+1}) \geq \delta/2$. Hence, by Definition 1, we can bound δ via (15).

$$(n - (j - 1)) \cdot c(e_{j-1}) + (n - (j + 1)) \cdot c(e_{j+1}) \leq \mathsf{opt}$$
$$\implies \delta \leq \frac{\mathsf{opt}}{(n - j)} \tag{15}$$

By bounding δ in (14) with (15), and due to $n > i$, we can conclude via (16).

$$L(\mathcal{OPT}', G) \leq \left(1 + \frac{(i - j)}{(n - j)}\right) \cdot \mathsf{opt} \leq 2 \cdot \mathsf{opt} \tag{16}$$

For a constant i, $i-j$ is also constant. Then, given a constant i, the approximation ratio $(i - j)/(n - j)$ shown in (16) can be arbitrarily close to one if $n \to \infty$. By this observation, a PTAS can be designed if i is constant. Given an error parameter θ, the PTAS $\mathcal{A}$ will output the given optimal tour $\mathcal{OPT}'$ directly if $(i - j)/(n - j) \leq \theta$, otherwise an optimal tour $\mathcal{OPT}$ of (G, v_1) will be searched exhaustively. The run-time of the exhaustive search is bounded within $\mathcal{O}(n^{1/\theta})$. Thus, $\mathrm{MLP}^{\mathrm{E}-}$ has a PTAS if e^* is the i-th edge of $\mathcal{OPT}'$ for a constant i.   □

For $\mathrm{MLP}^{\mathrm{E}-}$, if $e^* \notin E(\mathcal{OPT}')$ then each optimal tour $\mathcal{OPT}$ of the modified instance must contain e^*. Thus, similar to Theorem 5, we introduce Theorem 9. Unfortunately, we currently have no idea about how to determine the position of e^* in $\mathcal{OPT}$ except that e^* contains the source. If e^* includes the source, this case results in the worst ratio $2.1286 + \mathcal{O}(1/n)$. Since Theorem 9 can be proved in a similar way of proving Theorem 5, the detailed proof is omitted to save space.

Theorem 9. *For the restricted* $\mathrm{MLP}^{\mathrm{E}-}$ *where* $e^* \notin E(\mathcal{OPT}')$ *and* e^* *is the* j-th *edge in an optimal tour* $\mathcal{OPT}$ *of the modified instance for* $1 \leq j \leq n - 1$, *it can be approximated by a ratio* α *as follows: if* $j = 1$ *then* $\alpha = 2.1286 + \mathcal{O}(1/n)$; *if* $j = n - 1$ *then* $\alpha = 1.5$; *otherwise* $\alpha = 2$.

5   Relaxing the Optimality of the Given Solution

In this section, we study $\mathrm{MLP}^{\mathrm{V}-}_{\epsilon}$, which is a variant of $\mathrm{MLP}^{\mathrm{V}-}$ by relaxing the optimality of the given solution. We will understand the optimality of the given solution is critical for $\mathrm{MLP}^{\mathrm{V}-}$ to outperform MLP since Theorem 10 shows that $\mathrm{MLP}^{\mathrm{V}-}_{\epsilon}$ cannot be approximated better than MLP anymore for a small $\epsilon > 0$.

Theorem 10. *If* $\mathrm{MLP}^{\mathrm{V}-}_{\epsilon}$ *is* α-*approximable then MLP is also* α-*approximable.*

Proof (Sketch). We give an approximation factor preserving reduction from MLP. Given an instance (G^*, s) of MLP, where $G^* = (V^*, E^*, c^*)$ and $s \in V^*$ is the source, we construct a $\mathrm{MLP}^{\mathrm{V}-}_{\epsilon}$ instance $(G, G', T_\epsilon, s, t)$ with $\epsilon > 0$ as follows.

Let $G = (V, E, c)$ be the metric graph generated by removing a vertex t from a metric graph $G' = (V'E', c')$. Firstly, we set $G = G^*$. Then the graph G' has the following settings: $V' = V \cup \{t\}$, $E' = E \cup E_t$ and $\forall e \in E : c'(e) = c(e)$, where E_t is the set of edges incident with t. Since $\forall e \in E : c(e) = c^\star(e)$, we only need to consider the costs for the edges in E_t.

We assume there is a γ-approximation algorithm $\mathcal{A}$ for the MLP, e.g., the 3.59-approximation algorithm by Chaudhuri et al. [14]. By applying $\mathcal{A}$ to (G, s), we obtain a γ-approximation tour $\mathcal{T}$. For each vertex $v \in V$, we set

$$c'(\{v, t\}) = L(\mathcal{T}, G) \cdot \frac{2 \cdot \gamma - 1}{\epsilon}.$$

Finally, let the given feasible tour T_ϵ be a tour of $\mathcal{T}$ followed by the vertex t.

At first, we show that this construction is a valid instance of $\mathrm{MLP}_\epsilon^{V-}$. It is easy to verify that G' is also metric if G is metric. Then we show that T_ϵ is a valid given solution. Clearly, our settings imply that $\forall e \in E_t : c'(e) > L(\mathcal{T}, G)$. Thus, it is better to visit t at the end for any optimal tour of (G', s). Now, we can prove that T_ϵ is a $(1 + \epsilon)$-approximation tour for the initial instance (G', s).

$$
\begin{aligned}
\frac{L(T_\epsilon, G')}{\mathrm{opt}'} &\leq \frac{\gamma \cdot \mathrm{opt} + c(\mathcal{T}) + L(\mathcal{T}, G) \cdot \frac{2\gamma-1}{\epsilon}}{\mathrm{opt} + L(\mathcal{T}, G) \cdot \frac{2\gamma-1}{\epsilon}} \\
&\leq 1 + \frac{(2\gamma - 1) \cdot \mathrm{opt}}{\mathrm{opt} + L(\mathcal{T}, G) \cdot \frac{2\gamma-1}{\epsilon}} \\
&\leq 1 + \epsilon
\end{aligned}
\tag{17}
$$

After removing t, the modified instance (G, s) is the same as the given MLP instance $(G^\star, s)$. Thus, $\mathrm{MLP}_\epsilon^{V-}$ cannot be approximated better than MLP. $\square$

Acknowledgement. This is part of the author's master thesis, written under the supervision of Tobias Mömke at Saarland University.

References

1. Archetti, C., Bertazzi, L., Speranza, M.G.: Reoptimizing the traveling salesman problem. Networks **42**(3), 154–159 (2003)
2. Archetti, C., Bertazzi, L., Speranza, M.G.: Reoptimizing the 0–1 knapsack problem. Discret. Appl. Math. **158**(17), 1879–1887 (2010)
3. Ausiello, G., Escoffier, B., Monnot, J., Paschos, V.: Reoptimization of minimum and maximum traveling salesman's tours. J. Discret. Algorithms **7**(4), 453–463 (2009). http://www.sciencedirect.com/science/article/pii/S1570866708001019
4. Böckenhauer, H.J., Forlizzi, L., Hromkovic, J., Kneis, J., Kupke, J., Proietti, G., Widmayer, P.: On the approximability of TSP on local modifications of optimally solved instances. Algorithmic Oper. Res. **2**(2), 83 (2007). https://journals.lib.unb.ca/index.php/AOR/article/view/2803
5. Berg, T., Hempel, H.: Reoptimization of traveling salesperson problems: changing single edge-weights. In: Dediu, A.H., Ionescu, A.M., Martín-Vide, C. (eds.) LATA 2009. LNCS, vol. 5457, pp. 141–151. Springer, Heidelberg (2009). doi:10.1007/978-3-642-00982-2_12

6. Bilò, D., Böckenhauer, H.-J., Hromkovič, J., Královič, R., Mömke, T., Widmayer, P., Zych, A.: Reoptimization of steiner trees. In: Gudmundsson, J. (ed.) SWAT 2008. LNCS, vol. 5124, pp. 258–269. Springer, Heidelberg (2008). doi:10.1007/978-3-540-69903-3_24

7. Bilò, D., Böckenhauer, H.J., Komm, D., Královič, R., Mömke, T., Seibert, S., Zych, A.: Reoptimization of the shortest common superstring problem. Algorithmica **61**(2), 227–251 (2011). http://dx.doi.org/10.1007/s00453-010-9419-8

8. Bilò, D., Widmayer, P., Zych, A.: Reoptimization of weighted graph and covering problems. In: Bampis, E., Skutella, M. (eds.) WAOA 2008. LNCS, vol. 5426, pp. 201–213. Springer, Heidelberg (2009). doi:10.1007/978-3-540-93980-1_16

9. Bilò, D., Zych, A.: New advances in reoptimizing the minimum steiner tree problem. In: Rovan, B., Sassone, V., Widmayer, P. (eds.) MFCS 2012. LNCS, vol. 7464, pp. 184–197. Springer, Heidelberg (2012). doi:10.1007/978-3-642-32589-2_19

10. Blum, A., Chalasani, P., Coppersmith, D., Pulleyblank, B., Raghavan, P., Sudan, M.: The minimum latency problem. In: Proceedings of the Twenty-sixth Annual ACM Symposium on Theory of Computing, STOC 1994, NY, USA, pp. 163–171 (1994). http://doi.acm.org/10.1145/195058.195125

11. Boria, N., Croce, F.D.: Reoptimization in machine scheduling. Theor. Comput. Sci. **540**, 13–26 (2014). http://dx.doi.org/10.1016/j.tcs.2014.04.004

12. Boria, N., Monnot, J., Paschos, V.T.: Reoptimization of maximum weight induced hereditary subgraph problems. Theor. Comput. Sci. **514**, 61–74 (2013). http://dx.doi.org/10.1016/j.tcs.2012.10.037

13. Boria, N., Paschos, V.T.: A survey on combinatorial optimization in dynamic environments. RAIRO - Oper. Res. **45**(3), 241–294 (2011). http://dx.doi.org/10.1051/ro/2011114

14. Chaudhuri, K., Godfrey, B., Rao, S., Talwar, K.: Paths, trees, and minimum latency tours. In: 44th Annual IEEE Symposium on Foundations of Computer Science, 2003, Proceedings, pp. 36–45, October 2003

15. Christofides, N.: Worst-case analysis of a new heuristic for the travelling salesman problem. Technical report 388, Graduate School of Industrial Administration, Carnegie Mellon University (1976)

16. Giorgio Ausiello, V.B., Escoffier, B.: Complexity and approximation in reoptimization. In: Cooper, S.B., Sorbi, A. (eds.) Computability in Context: Computation and Logic in the Real World, pp. 101–129. World Scientific, Singapore (2011)

17. Goyal, K., Mömke, T.: Robust reoptimization of Steiner trees. In: 35th IARCS Annual Conference on Foundation of Software Technology and Theoretical Computer Science, FSTTCS 16–18, 2015, Bangalore, India, pp. 10–24 (2015). http://dx.doi.org/10.4230/LIPIcs.FSTTCS.2015.10

18. Böckenhauer, H.-J., Hromkovič, J., Mömke, T., Widmayer, P.: On the hardness of reoptimization. In: Geffert, V., Karhumäki, J., Bertoni, A., Preneel, B., Návrat, P., Bieliková, M. (eds.) SOFSEM 2008. LNCS, vol. 4910, pp. 50–65. Springer, Heidelberg (2008). doi:10.1007/978-3-540-77566-9_5

19. Hromkovič, J.: Algorithmics for Hard Problems: Introduction to Combinatorial Optimization, Randomization, Approximation, and Heuristics. Springer-Verlag New York Inc., New York (2001)

20. Monnot, J.: A note on the traveling salesman reoptimization problem under vertex insertion. Inf. Process. Lett. **115**(3), 435–438 (2015). http://dx.doi.org/10.1016/j.ipl.2014.11.003

21. Papadimitriou, C.H., Yannakakis, M.: The traveling salesman problem with distances one and two. Math. Oper. Res. **18**(1), 1–11 (1993). http://dx.doi.org/10.1287/moor.18.1.1

22. Schäffter, M.W.: Scheduling with forbidden sets. Discrete Appl. Math. **72**(1–2), 155–166 (1997). http://dx.doi.org/10.1016/S0166-218X(96)00042-X
23. Sitters, R.: The minimum latency problem is NP-hard for weighted trees. In: Proceedings of the 9th Integer Programming and Combinatorial Optimization Conference, pp. 230–239 (2002)

Pure Nash Equilibria in Restricted Budget Games

Maximilian Drees[1]($\boxtimes$), Matthias Feldotto[2], Sören Riechers[2],
and Alexander Skopalik[2]

[1] Departement of Applied Mathematics, University of Twente,
Enschede, The Netherlands
m.w.drees@utwente.nl

[2] Department of Computer Science and Heinz Nixdorf Institute,
Paderborn University, Paderborn, Germany
{feldi,soerenri,skopalik}@mail.upb.de

Abstract. In budget games, players compete over resources with finite budgets. For every resource, a player has a specific demand and as a strategy, he chooses a subset of resources. If the total demand on a resource does not exceed its budget, the utility of each player who chose that resource equals his demand. Otherwise, the budget is shared proportionally. In the general case, pure Nash equilibria (NE) do not exist for such games. In this paper, we consider the natural classes of singleton and matroid budget games with additional constraints and show that for each, pure NE can be guaranteed. In addition, we introduce a lexicographical potential function to prove that every matroid budget game has an approximate pure NE which depends on the largest ratio between the different demands of each individual player.

1 Introduction

Resource allocation problems are widely considered in theory and practice. In computing centers, for example, resources such as processing power and available data rate have to be divided such that the overall performance is optimized. In our paper, we consider the problem that service providers often cannot satisfy the needs of all clients. Here, the total payoff obtainable from a system is often independent of the number of its participants. For example, the computational capacity of a server is usually fixed and does not grow with the number of requests. In a different use case, the overall size of connections between a service provider and all clients may be limited by the amount of data the provider can process. In our model, this is reflected by a limited budget for each resource. Now, different clients may have different agreed target uses with a provider, which we model by different weights, also called demands throughout the paper. In case a provider cannot fulfill the requirements of all clients, the available resource

---

This work was partially supported by the German Research Foundation (DFG) within the Collaborative Research Centre "On-The-Fly Computing" (SFB 901).

Y. Cao and J. Chen (Eds.): COCOON 2017, LNCS 10392, pp. 175–187, 2017.
DOI: 10.1007/978-3-319-62389-4_15

needs to be split, resulting in clients not being supplied with their full demand. In video streaming, for example, this may lead to a lower quality stream for certain clients. Additionally, we allow part of a resource to be reserved by some external party, which we model as offsets in our setting.

We consider this model in a game theoretic setting called budget games. Here, we are interested in the effects of rational decision making by individuals. In our context, the clients act as the players, who compete over resources with a finite budget. We assume that clients can choose freely among different strategies, with each available strategy being a subset of resources. A player has a specific demand on every resource. For example, in cloud computing, we view each strategy as a distribution of the necessary computing power on different computing centers. Now, each player strives to maximize the overall amount of resource capacities that is supplied to him. Our main interest lies in states in which no client wants to deviate from his current strategy, as this would yield no or only a marginal benefit for him in the given situation. These states are called pure Nash equilibria, or approximate pure Nash equilibria, respectively. Instead of a global instance enforcing such stable states, they occur as the result of player-induced dynamics. At every point in time, exactly one player changes his strategy such that the amount of received demand is maximized, assuming the strategies of the other players are fixed. It is known that in general, pure Nash equilibria do not exist in budget games. In our earlier research, we considered pure Nash equilibria in ordered budget games [7], where the order of the players arriving at a resource influences the distribution of its budget. In [6], we further discussed approximate pure Nash equilibria in standard budget games, where the resource is distributed proportionally between the players on the basis of their demands. However, the question whether there are pure Nash equilibria for certain restricted instances of standard budget games remained open. In this paper, we focus on budget games with restrictions on the strategies of the players and show that there are indeed certain properties under which pure Nash equilibria always exist. Matroid budget games capture the natural assumption that for any player, the value of a resource is independent of the other chosen resources. A special case are singleton budget games in which each player can only choose one resource at a time.

Our contribution. For matroid budget games, we show that under the restriction of fixed demands per player, they possess the finite improvement property. This implies that the player-induced dynamic mentioned above always leads to a pure Nash equilibrium. On the other hand, we also show that even under this restriction, the matroid property is still required for the existence of pure Nash equilibria. Without any extra conditions on the demands, we can guarantee approximate pure Nash equilibria with a small approximation ratio depending on the maximum ratio between the demands of a single player. By further limiting the structure of the strategies to singleton, we can loosen the restriction on the demands and still obtain positive results regarding equilibria. In some cases, singleton budget games are weakly acyclic, i.e., there is an improving path from each initial state to a pure Nash equilibrium. For the additional class of offset budget games we can guarantee the existence of pure Nash equilibria under some additional restrictions.

Related Work. Budget games share many properties with congestion games. Although the specific structure of the utility functions makes budget games a special case, the fact that the demand of a player can vary between resources also qualifies them as a more general model for representing different impacts of players on resources. In congestion games, players choose among subsets of resources while trying to minimize personal costs. In the initial (unweighted) version [17], the cost of each resource depends only on the number of players choosing it and it is the same for each player using that resource. They are exact potential games [16] and therefore always possess pure Nash equilibria. In the weighted version [15], each player has a fixed weight and the cost of a resource depends on the sum of weights. For this larger class of games, pure Nash equilibria can no longer be guaranteed. Ackermann et al. [1] determined that the structure of the strategy spaces is a crucial property in this matter. While a matroid congestion game always has a pure Nash equilibrium, every non-matroid set system induces a game without it. Harks and Klimm [10] gave a complete characterization of the class of cost functions for which every weighted congestion game possesses a pure Nash equilibrium. The cost functions have to be affine transformations of each other as well as be affine or exponential. Another extension considers player-specific payoff functions for the resources, which depend only on the number of players using a resource, but are different for each player [15]. For singleton strategy spaces, these games maintain pure Nash equilibria. Ackerman et al. [1] showed that, again, every player-specific matroid congestion game has a pure Nash equilibrium, while this is also a maximal property.

In a model similar to ours, each player does not choose only his resources, but also his demand on them [11]. In contrast, the players in our model cannot influence their demands. These games have pure Nash equilibria if the cost functions are either exponential or affine. Mavronicolas et al. [14] combined the concepts of weighted and player-specific congestion games and gave a detailed overview of the existence of pure Nash equilibria. In these games, the cost function $c_{i,r}$ of player i for resource r consists of a base function c_r, which depends on the weights of all players using r, as well as a constant $k_{i,r}$, both connected by abelian group operations. Later, Gairing and Klimm [8] characterized the conditions for pure Nash equilibria in general player-specific congestion games with weighted players. Pure Nash equilibria exist, if, and only if, the cost functions of the resources are affine transformations of each other as well as affine or exponential. Another generalization of congestion games is given by Byde et al. [3] and Voice et al. [19]. They introduce the model of games with congestion-averse utility functions. They show under which properties pure Nash equilibria exist and identify the matroid as a required property for the existence in most cases. Although they consider more general utility functions than standard congestion games, their model does not consider players' weights or demands.

Instead of assigning the whole cost of a resource to each player using it, the cost can also be shared between those players, so that everyone only pays a part of it. Such games are known as cost sharing games [12]. One method to determine the share of each player is proportional cost sharing, in which the share increases with the weight of a player. This is exactly what we are doing with budget games, but with utilities instead of costs. Under proportional cost sharing, which corresponds to our utility functions, pure Nash equilibria again do not exist in general [2]. Kollias and Roughgarden [13] took a different approach by considering weighted games in which the share of each player is identical to his Shapley value [18]. Using this method, every weighted congestion game yields a weighted potential function. However, we do not approach this from a mechanism design angle. Instead, we consider this system and especially the structure of the utility functions as given by the scenarios we introduced. Negative results regarding both existence and complexity of pure Nash equilibria lead to the study of approximate pure Nash equilibria [5]. Caragiannis et al. [4] and Hansknecht et al. [9] showed the existence of approximate pure Nash equilibria for weighted congestion games.

Model. A budget game $\mathcal{B}$ is a tuple $(\mathcal{N}, \mathcal{R}, (b_r)_{r \in \mathcal{R}}, (\mathcal{S}_i)_{i \in \mathcal{N}}, (\mathcal{D}_i)_{i \in \mathcal{N}})$ where the set of players is denoted by $\mathcal{N} = \{1, \ldots, n\}$, the set of resources by $\mathcal{R} = \{r_1, \ldots, r_m\}$, the budget of resource r by $b_r \in \mathbb{R}_{>0}$, the strategy space of player i by $\mathcal{S}_i$ and the demands of player i by $\mathcal{D}_i = (d_i(r_1), \ldots, d_i(r_m))$. Each strategy $s_i \subseteq 2^{\mathcal{R}}$ is a subset of resources. We call $d_i(r_j) > 0$ the demand of i on r_j and say that a strategy s_i uses a resource r_j if $r_j \in s_i$. The set of strategy profiles is denoted by $\mathcal{S} := \mathcal{S}_1 \times \ldots \times \mathcal{S}_n$. Each player i has a private utility function $u_i : \mathcal{S} \to \mathbb{R}_{\geq 0}$, which he strives to maximize. For a strategy profile $\mathsf{s} = (s_1, \ldots, s_n)$, let $T_r(\mathsf{s}) := \sum_{i \in \mathcal{N}: r \in s_i} d_i(r)$ be the total demand on resource r. The utility of player i from resource r is denoted by $u_{i,r}(\mathsf{s}) \in \mathbb{R}_{\geq 0}$ and defined as $u_{i,r}(\mathsf{s}) = 0$ if $r \notin s_i$ and $u_{i,r}(\mathsf{s}) := d_i(r) \cdot c_r(\mathsf{s})$ if $r \in s_i$, where $c_r(\mathsf{s}) := min(1, {}^{b_r}/_{T_r(\mathsf{s})})$ denotes the utility per unit demand. The total utility of i is $u_i(\mathsf{s}) := \sum_{r \in \mathcal{R}} u_{i,r}(\mathsf{s})$. When increasing the demand on a resource by some value d, we write $c_r(\mathsf{s}) \oplus d := min(1, {}^{b_r}/_{(T_r(\mathsf{s})+d)})$. If $\mathcal{M}_i = (\mathcal{R}, \mathcal{I}_i)$ is a matroid with $\mathcal{I}_i = \{x \subseteq s \mid s \in \mathcal{S}_i\}$ for every player i, we call $\mathcal{B}$ a matroid budget game. A matroid budget game is called a singleton budget game if every strategy uses exactly one resource. Let $\mathsf{s} \in \mathcal{S}$ and $i \in \mathcal{N}$. We denote with $\mathsf{s}_{-i} := (s_1, \ldots, s_{i-1}, s_{i+1}, \ldots, s_n)$ the strategy profile excluding i. For any $s_i' \in \mathcal{S}_i$, we can extend this to $(\mathsf{s}_{-i}, s_i') := (s_1, \ldots, s_{i-1}, s_i', s_{i+1}, \ldots, s_n) \in \mathcal{S}$. The best response of i to s_{-i} is denoted by $s_i^b \in \mathcal{S}_i$: i.e., $u_i(\mathsf{s}_{-i}, s_i^b) \geq u_i(\mathsf{s}_{-i}, s_i)$ for all $s_i \in \mathcal{S}_i$. We call the switch from s_i to s_i' with $u_i(\mathsf{s}_{-i}, s_i) < u_i(\mathsf{s}_{-i}, s_i')$ an improving move for player i. Sequential execution of best response improving moves creates a best response dynamic. A strategy profile s is called an α-approximate pure Nash equilibrium if $\alpha \cdot u_i(\mathsf{s}) \geq u_i(\mathsf{s}_{-i}, s_i')$ for every $i \in \mathcal{N}$ and $s_i' \in \mathcal{S}_i$. For $\alpha = 1$, s is simply called a pure Nash equilibrium. For the rest of this paper, we mostly omit the prefix *pure*. If from any initial strategy profile there is a path of improving moves which reaches an (α-approximate) Nash equilibrium, then the game is said to be weakly acyclic. If from any initial strategy profile each path of improving moves reaches an (α-approximate) Nash equilibrium, then the game possesses the finite improvement property. For a strategy profile s, the lexicographical potential function $\phi : \mathcal{S} \to \mathbb{R}_{>0}^m$ is defined as $\phi(\mathsf{s}) := (c_{r_1}(\mathsf{s}), \ldots, c_{r_m}(\mathsf{s}))$ with the entries $c_{r_k}(\mathsf{s})$ being sorted in ascending order. The augmented lexicographical potential function $\phi^* : \mathcal{S} \to$

$\mathbb{R}_{>0}^{m+1}$ extends this definition with $\phi^*(\mathbf{s}) := (T(\mathbf{s}), c_{r_1}(\mathbf{s}), \ldots, c_{r_m}(\mathbf{s}))$, whereas $T(\mathbf{s}) := \sum_{i \in \mathcal{N}} \sum_{r \in s_i} d_i(r)$ is the total demand by all players under $\mathbf{s}$.

2   Matroid Budget Games

By definition, all strategies of player i in a matroid budget game have the same size m_i. In addition, for any strategy profile $\mathbf{s}$, a strategy change from s_i to s_i' can be decomposed into a sequence $s_i = s_i^0, s_i^1, \ldots, s_i^{m_i} = s_i'$ of *lazy* moves which satisfy $s_i^k \in \mathcal{S}_i$, $|s_i^k \setminus s_i^{k+1}| \le 1$ and $u_i(\mathbf{s}^{-i}, s_i^k) < u_i(\mathbf{s}_{-i}, s_i^{k+1})$ for $0 \le k \le m_i$ (see [1]). A matroid budget game has fixed demands if there exists a constant $d_i \in \mathbb{R}_{>0}$ for every player i such that $d_i(r) = d_i$ for all $r \in \mathcal{R}$.

Theorem 1. *A matroid budget game with fixed demands reaches a pure Nash equilibrium after a finite number of improving moves.*

Proof. We show that a single lazy move already increases the lexicographical potential function ϕ. Let player i perform a lazy move in strategy profile $\mathbf{s}$, switching resource r_1 for r_2. Let $\mathbf{s}'$ be the resulting strategy profile. We get $u_{i,r_1}(\mathbf{s}) = d_i \cdot c_{r_1}(\mathbf{s}) < d_i \cdot c_{r_2}(\mathbf{s}') = u_{i,r_2}(\mathbf{s}')$ or simply $c_{r_1}(\mathbf{s}) < c_{r_2}(\mathbf{s}')$. Since $c_{r_1}(\mathbf{s}) < c_{r_1}(\mathbf{s}')$ also holds due to $T_{r_1}(\mathbf{s}') = T_{r_1}(\mathbf{s}) - d_i$, we get $\phi(\mathbf{s}) <_{\text{lex}} \phi(\mathbf{s}')$ and see that ϕ is strictly increasing regarding the lexicographical order for every improving lazy move. Since the number of different values of ϕ is finite, the best response eventually reaches a strategy profile without any further improving move. By definition, this is a pure Nash equilibrium.   □

For this result, the structure of the strategy spaces is a crucial property. Consider the budget game $\mathcal{B}_0$ shown in Fig. 1, which is defined as follows: $\mathcal{N} = \{1, 2, 3\}$, $\mathcal{R} = \{r_1, r_2, r_3, r_4\}$, $b_r = 2$ for all r, $\mathcal{S}_1 = \{s_1^1 = \{r_1, r_2\}, s_1^2 = \{r_3, r_4\}\}$, $\mathcal{S}_2 = \{s_2^1 = \{r_1, r_3\}, s_2^2 = \{r_2, r_4\}\}$, $\mathcal{S}_3 = \{s_3 = \{r_1, r_4\}\}$ and $d_1 = 2, d_2 = d_3 = 1$.

Theorem 2. *There is a budget game with fixed demands which is not a matroid budget game and does not have a pure Nash equilibrium.*

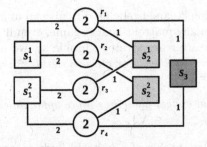

Fig. 1. The budget game $\mathcal{B}_0$ with fixed demands and no pure Nash equilibrium.

Proof. We analyze the game $\mathcal{B}_0$. Player 3 has only one strategy, so we focus on the four different strategy profiles resulting from the choices of player 1 and 2 (see Table 1). In each strategy profile, one player is able to increase his utility through a unilateral strategy change, so no pure Nash equilibrium exists.   □

Table 1. Overview of the different strategy profiles (restricted to players 1 and 2) and the corresponding utilities of the budget game $\mathcal{B}_0$.

players	strategy profiles			
	(s_1^1, s_2^1)	(s_1^1, s_2^2)	(s_1^2, s_2^1)	(s_1^2, s_2^2)
1	$2 + 1 = 3$	$\frac{4}{3} + \frac{2}{3} = 2$	$\frac{4}{3} + \frac{2}{3} = 2$	$2 + 1 = 3$
2	$1 + \frac{1}{2} = \frac{3}{2}$	$\frac{2}{3} + 1 = \frac{5}{3}$	$\frac{2}{3} + 1 = \frac{5}{3}$	$1 + \frac{1}{2} = \frac{3}{2}$

When considering singleton budget games with fixed demands, a Nash equilibrium can also be computed efficiently. Before proving this, we introduce a technical result.

Lemma 1. *Let $d_1, d_2 \in \mathbb{R}_{>0}$ with $d_1 \leq d_2$ and $b_r, T_r(s) \in \mathbb{R}_{\geq 0}$ with $T_r(s) + d_1 \geq b_r$. Then $d_1 \cdot \min\left(1, \frac{b_r}{T_r(s)+d_1}\right) \leq d_2 \cdot \min\left(1, \frac{b_r}{T_r(s)+d_2}\right)$.*

We omit the proof of this lemma. In the context of budget games, it implies that a player with higher demand always receives a higher utility from the same resource than another player with lower demand.

Theorem 3. *For a singleton budget game with fixed demands, pure Nash equilibria can be computed in time $\mathcal{O}(n)$.*

Proof. We start with an *empty* strategy profile where $s_i = \emptyset$ for every player i. The players then choose their actual strategy sequentially in ascending order of their demands. We show that a strategy choice made by player j does not change the best response of any player i with $d_i \leq d_j$. Let s be the strategy profile the moment before j chooses his strategy. If j picks the same resource r as i, then $d_j \cdot (c_r(s) \oplus d_j) \geq d_j \cdot (c_{r'}(s) \oplus d_j) \geq d_i \cdot (c_{r'}(s) \oplus d_i)$ for all $r' \in \mathcal{R}$ due to Lemma 1, meaning that r is still the best response for i.   □

The potential function ϕ can also be used to give an upper bound on approximate Nash equilibria in any matroid budget game, even if the demands are not fixed. Starting with an arbitrary strategy profile s_0, we only allow improving moves that also strictly increase ϕ. For player i, let $d_i^{max} := \max\{d_i(r) \mid r \in \mathcal{R}\}$ and $d_i^{min} := \min\{d_i(r) \mid r \in \mathcal{R}\}$.

Theorem 4. *A matroid budget game has an α-approximate pure Nash equilibrium for $\alpha = \max\{d_i^{max}/d_i^{min} \mid i \in \mathcal{N}\}$.*

Proof. Let s be a strategy profile of a matroid budget game $\mathcal{B}$ in which player i can switch resource r_1 for r_2 to increase his utility. We restrict the best response dynamic such that we only allow this lazy move if it also satisfies

$d_i(r_1) \cdot c_{r_1}(\mathsf{s}) < d_i(r_1) \cdot (c_{r_2}(\mathsf{s}) \oplus d_i(r_1))$. If this condition holds, player i would still profit from the lazy move if his demands on both r_1 and r_2 were the same. Such a lazy move would also increase ϕ as shown in the proof of Theorem 1. Therefore, the number of such improving moves is finite and this restricted best response dynamic arrives at a strategy profile s^α. Let s be a strategy profile which originates from s^α through a unilateral improving move by player i to s_i and let $\Delta_\alpha = |s_i \setminus s_i^\alpha|$. We assign an index k to every $r_k^\alpha \in s_i^\alpha$ and every $r_k \in s_i$. If a resource r is used by both s_i^α and s_i, then it has the same index ℓ for both strategies, where $\ell \geq \Delta_\alpha$. The improving move from s_i^α to s_i consists only of lazy moves with $d_i(r_k^\alpha) < d_i(r_k)$ and $c_{r_k^\alpha}(\mathsf{s}^\alpha) \geq (c_{r_k}(\mathsf{s}^\alpha) \oplus d_i(r_k))$. Since $\frac{d_i(r_k)}{d_i(r_k^\alpha)} \leq \frac{d_i^{\max}}{d_i^{\min}}$ holds for all resources, we get

$$u_i(\mathsf{s}) = \sum_{r \in s_i} u_{i,r}(\mathsf{s}) = \sum_{k=1}^{\Delta_\alpha} d_i(r_k) \cdot (c_{r_k}(\mathsf{s}^\alpha) \oplus d_i(r_k)) + \sum_{k=\Delta_\alpha+1}^{m_i} d_i(r_k^\alpha) \cdot c_{r_k^\alpha}(\mathsf{s}^\alpha)$$

$$\leq \sum_{k=1}^{\Delta_\alpha} \frac{d_i^{\max}}{d_i^{\min}} \cdot d_i(r_k^\alpha) \cdot c_{r_k^\alpha}(\mathsf{s}^\alpha) + \sum_{k=\Delta_\alpha+1}^{m_i} \frac{d_i^{\max}}{d_i^{\min}} \cdot d_i(r_k^\alpha) \cdot c_{r_k^\alpha}(\mathsf{s}^\alpha)$$

$$= \frac{d_i^{\max}}{d_i^{\min}} \cdot \sum_{k=1}^{m_i} d_i(r_k^\alpha) \cdot c_{r_k^\alpha}(\mathsf{s}^\alpha) = \frac{d_i^{\max}}{d_i^{\min}} \cdot u_i(\mathsf{s}^\alpha) \qquad \square$$

3   Singleton Budget Games with Two Demands

We now consider singleton budget games with two demands: i.e., $d_i(r) \in \{d^-, d^+\}$ for every demand of a player i on a resource r. We assume $d^- < d^+$. Also, all budgets are uniform: i.e., $b_r = b_{r'}$ for all resources r, r'. Finally, every resource r is available to every player i: i.e., there is a strategy $s_i \in \mathcal{S}_i$ using r. This models situations in which each player partitions the resources into two sets such that he prefers the resources from the first set over those in the second and he regards all resources from the same set as equally good. In our model, a more prefered resource is identified by a higher demand. Note that the preferences of two different players do not have to be the same. We show that Algorithm 1 always computes a Nash equilibrium by using the best response dynamic, which proves Theorem 5. The algorithm utilizes the best response dynamic and only controls the order of the improving moves. For the following discussion, we separate the improving moves into different types. The type depends on the demand of the corresponding player before and after his strategy change. Since we consider only two demands, there are only four different types: $d^+ \to d^+$, $d^+ \to d^-$, $d^- \to d^+$ and $d^- \to d^-$. We immediately see that in the intermediate strategy profile right after Phase 1 of the algorithm, no improving move of type $d^+ \to d^-$ exists. In addition, we now introduce the concepts of pushing and pulling strategy changes.

Let $\mathcal{B}$ be a singleton budget game with players i, j, resources r_1, r_2 and strategy profile s. In s, let $s_i = \{r_1\}$ and $s_j = \{r_2\}$ with $u_i(\mathsf{s}) < u_i(\mathsf{s}_{-i}, r_2)$ and $u_j(\mathsf{s}) \geq u_j(\mathsf{s}_{-j}, r)$ for all $r \in \mathcal{R}$. Denote $\mathsf{s}' = (\mathsf{s}_{-i}, r_2)$. If $u_j(\mathsf{s}') < u_j(\mathsf{s}'_{-j}, r_3)$ for

Algorithm 1. ComputeNE

$s \leftarrow$ arbitrary initial strategy profile

Phase 1:

while there is a player in s with best resp. improving move of type $d^+ \rightarrow d^-$ **do**
     perform best response improving move of type $d^+ \rightarrow d^-$
     $s \leftarrow$ resulting strategy profile

Phase 2:

while current strategy profile s is not a pure Nash equilibrium **do**
     if there is a player with best resp. improving move of type $d^+ \rightarrow d^-$ **then**
         perform best response improving move of type $d^+ \rightarrow d^-$
     else if there is a player i with b.-r. improving move of type $d^+ \rightarrow d^+$ **then**
         $\mathcal{N}' \leftarrow \{j \in \mathcal{N} \mid j$ has best response improving move of type $d^+ \rightarrow d^+\}$
         choose $i \in \mathcal{N}'$ such that $T_{s_i}(s) \geq T_{s_j}(s)$ for all $j \in \mathcal{N}'$
         perform best response improving move of i
     else
         perform any best response improving move $\triangleright\ d^- \rightarrow d^-$ or $d^- \rightarrow d^+$
     $s \leftarrow$ resulting strategy profile
     return s $\triangleright\ s$ is pure Nash equilibrium

---

some $r_3 \in \mathcal{R}$, then the strategy change by i from r_1 to r_2 is called a *pushing strategy change* for j. In the same scenario, let $u_i(s) \geq u_i(s_{-i}, r)$ for all $r \in \mathcal{R}$ and $u_j(s) < u_j(s_{-j}, r_3)$ for some $r_3 \in \mathcal{R}$. Denote $s^* = (s_{-j}, r_3)$. If $u_i(s^*) < u_i(s^*_{-i}, r_2)$, then the strategy change by j from r_2 to r_3 is called a *pulling strategy change* for i (see Fig. 2). If a strategy change is both pushing and pulling for the same player, we always regard it as the former. On the basis of these characterizations, we analyze the effects of strategy changes between the players. The proofs of the following three lemmata are done by case distinction and omitted here.

Lemma 2. *Let s be a strategy profile during Phase 2 of Algorithm 1. In s, no best response improving move of type $d^+ \rightarrow d^-$ is created by a pushing strategy change.*

Lemma 3. *Let s be a strategy profile during Phase 2 of Algorithm 1. In s, no best response improving move of type $d^+ \rightarrow d^-$ is created by a pulling strategy change of type $d^- \rightarrow d^-$.*

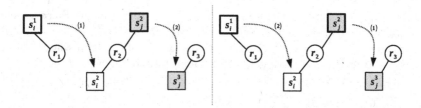

Fig. 2. Examples for pushing (left) and pulling (right) strategy changes.

Lemma 4. *Let s be a strategy profile during Phase 2 of Algorithm 1. In s, no best response improving move of type $d^+ \to d^-$ is created by a pulling strategy change of type $d^+ \to d^+$.*

We use the augmented lexicographical potential function ϕ^* to show that our algorithm actually terminates. With the three lemmata above, we see that during Phase 2, any improving move of type $d^+ \to d^-$ has to be created by a pulling strategy change of type $d^- \to d^+$. By executing both strategy changes right after another, we can combine them into a so-called macro strategy change. In a macro strategy change, two players i, j change their resources, with r being both the old resource of i and the new resource of j and $d_i(r) = d_j(r)$. As a result, the total demand on r does not change during a macro strategy change. An example can be seen in Fig. 3. Although not associated with an actual player, we say that a macro strategy change is performed by a virtual player. The following lemma shows that in our case, this virtual player would actually benefit from his strategy change. Again, we omit a proof.

Lemma 5. *Let s be a strategy profile in which a macro strategy change of type $d^+ \to d^+$ from r_1 to r_3 is executed. Then $d^+ \cdot c_{r_1}(s) < d^+ \cdot (c_{r_3}(s) \oplus d^+)$.*

Using this lemma, we conclude that a macro strategy change strictly increases ϕ^*. Its type is $d^+ \to d^+$ and from the results in the previous section, we know that such a strategy change strictly increases ϕ. Since the total demand of all players does not change, this holds for ϕ^* as well.

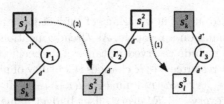

Fig. 3. Example for a macro strategy change. This sequence of strategy changes is equivalent to the strategy change of a virtual player k from r_1 to r_3.

Theorem 5. *A singleton budget game with two demands and uniform budgets is weakly acyclic.*

Proof. By construction, the output of Algorithm 1 is a Nash equilibrium. It remains to show that it actually terminates at some point. The number of improving moves in the first phase is at most n, as every player changes his strategy at most once. For the second phase, we use the augmented lexicographical potential function ϕ^*. This function is strictly increasing regarding $<_{\text{lex}}$ for all strategy changes of type $d^- \to d^-$ and $d^+ \to d^+$, since ϕ is strictly increasing for these types and the total demand of all players does not change. For strategy changes of type $d^- \to d^+$, ϕ^* is also strictly increasing because the total demand is always the first entry in $\phi^*(s)$ and it increases. ϕ^* can only decrease for improving moves of type $d^+ \to d^-$.

Let s_1 be the strategy profile right after Phase 1 has terminated. Then s_1 contains no best response improving move of type $d^+ \to d^-$. According to Lemmata 2, 3 and 4, such moves can appear only as the result of a pulling strategy change of type $d^- \to d^+$. In this case, both can be regarded as a single macro strategy change of type $d^+ \to d^+$. Because of Lemma 5 and the fact that such a macro strategy change does not change the total demand, ϕ^* strictly increases for such a macro strategy change, too. If a pulling strategy change creates multiple best response improving moves of type $d^+ \to d^-$ to a resource r, then the algorithm executes one of them, chosen by some arbitrary rule. Afterwards, the total demand on r is the same as it was before the pulling strategy change. Hence, the other best response moves of type $d^+ \to d^-$ cease to exist.

ϕ^* strictly increases after at least every second strategy change, so our algorithm has to terminate at some point. The resulting strategy profile is a Nash equilibrium. During its execution, the algorithm performs only best response improving moves and decides only the order in which they are completed. Therefore, the game is weakly acyclic.   □

We do not know if this result carries over to matroid budget games under the same restrictions. Regarding budget games with two demands and uniform budgets, but arbitrary strategy space structures, we already know that Nash equilibria do not exist in general [6].

4   Singleton Offset Budget Games

In this section, we introduce a new variant of budget games that allows a fixed offset to the total demand on a resource. As already mentioned in our introduction, this allows us to include reserved instances for specific users in our games. An offset $\sigma_r \in \mathbb{R}_{\geq 0}$ for resource $r \in \mathcal{R}$ changes the utility of player i from r under strategy profile s to $u_{i,r}(s) = d_i(r) \cdot \min\left(1, b_r / (T_r(s) + \sigma_r)\right)$. It is easy to see that by setting $\sigma_r = 0$ for every $r \in \mathcal{R}$, every offset budget game becomes a regular budget game. We now consider budget games with two additional restrictions: a total order on the players (based on their demands) and increasing demand ratios. Let i, j be players with $d_i(r) \leq d_j(r)$ for some resource r. The first restriction states that although the demands of the players are no longer fixed, their order is the same for every resource, so $d_i(r') \leq d_j(r')$ for all resources r'. This is a quite natural assumption, as bigger players (like global companies) normally have high demands on all resources. The second restriction requires larger players to have bigger deviations between their demands: i.e., $d_i(r')/d_i(r) \leq d_j(r')/d_j(r)$ (assuming $d_i(r) \leq d_i(r')$). Again, this assumption is only natural. Larger players (e.g., jobs on servers) offer more room for optimization and are more influenced by their choice of resource (e.g., servers which better support certain kinds of operations) than smaller ones, which are already quite compact. For this class we can guarantee the existence of pure Nash equilibria.

Theorem 6. *Singleton offset budget games with ordered players and increasing demand ratios always have a pure Nash equilibrium.*

Proof. Proof by induction over the number of players. For a game with n players, we denote the offset of resource r by $\sigma_n(r)$. For $n = 2$, the statement becomes trivial. For $n > 2$, we assume without loss of generality that $d_n(r_1) \geq d_i(r_1)$ for all $i \in \mathcal{N}$ and $d_n(r_1) \geq d_n(r)$ for all $r \in \mathcal{R}$. Fix the strategy of n to $\{r_1\}$. The resulting game is identical to one with $n - 1$ players and $\sigma_{r_1}^{n-1} = \sigma_{r_1}^n + d_n(r_1)$ and $\sigma_r^{n-1} = \sigma_r^n$ for all $r \in \mathcal{R} \setminus \{r_1\}$. By induction hypothesis, this game has a Nash equilibrium $\mathbf{s}'$. Let $\mathbf{s} = (\mathbf{s}', r_1)$ and assume that $\mathbf{s}$ is not a Nash equilibrium for n. We get $u_n(\mathbf{s}) = d_n(r_1) \cdot (c_{r_1}(\mathbf{s}) \oplus \sigma_{r_1}^n) < d_n(r_2) \cdot (c_{r_2}(\mathbf{s}) \oplus (\sigma_{r_2}^n + d_n(r_2))) = u_n(\mathbf{s}_{-n}, r_2)$ for some $r_2 \in R$. Let i be another player on r_1, i.e. $s_i = \{r_1\}$. Since r_1 is the best response of i, the following has to hold: $u_i(\mathbf{s}) = d_i(r_1) \cdot (c_{r_1}(\mathbf{s}) \oplus \sigma_{r_1}^n) \geq d_i(r_2) \cdot (c_{r_2}(\mathbf{s}) \oplus (\sigma_{r_2}^n + d_i(r_2))) = u_i(\mathbf{s}_{-i}, r_2)$. By combining these two, we get

$$\frac{d_i(r_2)}{d_i(r_1)} \cdot (c_{r_2}(\mathbf{s}) \oplus (\sigma_{r_2}^n + d_i(r_2))) < \frac{d_n(r_2)}{d_n(r_1)} \cdot (c_{r_2}(\mathbf{s}) \oplus (\sigma_{r_2}^n + d_n(r_2))). \qquad (1)$$

Since the players are ordered, we know that $d_n(r_2) \geq d_i(r_2)$ and therefore $(c_{r_2}(\mathbf{s}) \oplus (\sigma_{r_2}^n + d_i(r_2))) \geq (c_{r_2}(\mathbf{s}) \oplus (\sigma_{r_2}^n + d_n(r_2)))$. Equation 1 can thus be simplified to $\frac{d_i(r_1)}{d_i(r_2)} > \frac{d_n(r_1)}{d_n(r_2)}$. This contradicts our restriction that the demand ratios are increasing and implies that i cannot exist. n is the only player on resource r_1 and since this is his preferred resource, it is also his best response. We conclude that $\mathbf{s}$ has to be a Nash equilibrium for all n players. $\qquad \square$

This result holds for regular budget games, in particular.

Corollary 1. *Singleton budget games with ordered players and increasing demand ratios always have a pure Nash equilibrium.*

For matroid (offset) budget games with only two resources, this result can be improved even more. In this case, we can drop both restrictions regarding ordered players and increasing demand ratios. In addition, besides the pure existence of equilibria, such games are also weakly acyclic. The proof of the following result is similar to the one of Theorem 6, so we omit it here.

Theorem 7. *Every matroid (offset) budget game with two resources is weakly acyclic.*

5 Conclusion

The model of budget games enables us to analyze different effects which appear specifically in, but are not limited to, cloud computing. In emerging markets with shared resources, the question of resource allocation becomes more and more important. We focus on a specific method of distributing resources among the market participants. On the one hand, this model can be extended with additional properties of the cloud computing scenario (e.g., prices). On the other hand, other allocation mechanisms can be investigated.

References

1. Ackermann, H., Röglin, H., Vöcking, B.: Pure Nash equilibria in player-specific and weighted congestion games. Theor. Comput. Sci. **410**(17), 1552–1563 (2009)
2. Anshelevich, E., Dasgupta, A., Kleinberg, J., Tardos, E., Wexler, T., Roughgarden, T.: The price of stability for network design with fair cost allocation. SIAM J. Comput. **38**(4), 1602–1623 (2008)
3. Byde, A., Polukarov, M., Jennings, N.R.: Games with congestion-averse utilities. In: Mavronicolas, M., Papadopoulou, V.G. (eds.) SAGT 2009. LNCS, vol. 5814, pp. 220–232. Springer, Heidelberg (2009). doi:10.1007/978-3-642-04645-2_20
4. Caragiannis, I., Fanelli, A., Gravin, N., Skopalik, A.: Approximate pure Nash equilibria in weighted congestion games: existence, efficient computation, and structure. ACM Trans. Econ. Comput. **3**(1), 2 (2015)
5. Chien, S., Sinclair, A.: Convergence to approximate Nash equilibria in congestion games. Games Econ. Behav. **71**(2), 315–327 (2011)
6. Drees, M., Feldotto, M., Riechers, S., Skopalik, A.: On existence and properties of approximate pure Nash equilibria in bandwidth allocation games. In: Proceedings of the 8th International Symposium on Algorithmic Game Theory, pp. 178–189 (2015)
7. Drees, M., Riechers, S., Skopalik, A.: Budget-restricted utility games with ordered strategic decisions. In: Proceedings of the 7th International Symposium on Algorithmic Game Theory, pp. 110–121 (2014)
8. Gairing, M., Klimm, M.: Congestion games with player-specific costs revisited. In: Vöcking, B. (ed.) SAGT 2013. LNCS, vol. 8146, pp. 98–109. Springer, Heidelberg (2013). doi:10.1007/978-3-642-41392-6_9
9. Hansknecht, C., Klimm, M., Skopalik, A.: Approximate pure Nash equilibria in weighted congestion games. In: Proceedings of the 17th International Workshop on Approximation Algorithms for Combinatorial Optimization Problems, pp. 242–257. Schloss Dagstuhl-Leibniz-Zentrum fuer Informatik (2014)
10. Harks, T., Klimm, M.: On the existence of pure Nash equilibria in weighted congestion games. In: Abramsky, S., Gavoille, C., Kirchner, C., Meyer auf der Heide, F., Spirakis, P.G. (eds.) ICALP 2010. LNCS, vol. 6198, pp. 79–89. Springer, Heidelberg (2010). doi:10.1007/978-3-642-14165-2_8
11. Harks, T., Klimm, M.: Congestion games with variable demands. Mathemat. Oper. Res. **41**(1), 255–277 (2015)
12. Jain, K., Mahdian, M.: Cost sharing. In: Algorithmic Game Theory, pp. 385–410 (2007)
13. Kollias, K., Roughgarden, T.: Restoring pure equilibria to weighted congestion games. ACM Trans. Econ. Comput. **3**(4), 21 (2015)
14. Mavronicolas, M., Milchtaich, I., Monien, B., Tiemann, K.: Congestion games with player-specific constants. In: Kučera, L., Kučera, A. (eds.) MFCS 2007. LNCS, vol. 4708, pp. 633–644. Springer, Heidelberg (2007). doi:10.1007/978-3-540-74456-6_56
15. Milchtaich, I.: Congestion games with player-specific payoff functions. Games Econ. Behav. **13**(1), 111–124 (1996)
16. Monderer, D., Shapley, L.S.: Potential games. Games Econ. Behav. **14**(1), 124–143 (1996)
17. Rosenthal, R.W.: A class of games possessing pure-strategy Nash equilibria. Int. J. Game Theory **2**(1), 65–67 (1973)

18. Shapley, L.S.: A Value for n-Person Games. Technical report, DTIC Document (1952)
19. Voice, T., Polukarov, M., Byde, A., Jennings, N.R.: On the impact of strategy and utility structures on congestion-averse games. In: Leonardi, S. (ed.) WINE 2009. LNCS, vol. 5929, pp. 600–607. Springer, Heidelberg (2009). doi:10.1007/978-3-642-10841-9_61

A New Kernel for Parameterized Max-Bisection Above Tight Lower Bound

Qilong Feng, Senmin Zhu, and Jianxin Wang$^{(\boxtimes)}$

School of Information Science and Engineering,
Central South University, Changsha 410083, People's Republic of China
jxwang@csu.edu.cn

Abstract. In this paper, we study kernelization of Parameterized Max-Bisection above Tight Lower Bound problem, which is to find a bisection (V_1, V_2) of G with at least $\lceil |E|/2 \rceil + k$ crossing edges for a given graph $G = (V, E)$. The current best vertex kernel result for the problem is of size $16k$. Based on analysis of the relation between maximum matching and vertices in Gallai-Edmonds decomposition of G, we divide graph G into a set of blocks, and each block in G is closely related to the number of crossing edges of bisection of G. By analyzing the number of crossing edges in all blocks, an improved vertex kernel of size $8k$ is presented.

1 Introduction

Given a graph $G = (V, E)$, for two subsets V_1, V_2 of V, if $V_1 \cup V_2 = V$, $V_1 \cap V_2 = \emptyset$, and $||V_1| - |V_2|| \leq 1$, then (V_1, V_2) is called a *bisection* of G. An edge of G with one endpoint in V_1 and the other endpoint in V_2 is called a *crossing edge* of (V_1, V_2). The Maximum Bisection problem is to find a bisection (V_1, V_2) of G with maximum number of crossing edges. Jansen et al. [8] proved that the Maximum Bisection problem is NP-hard on planar graph. Díaz and Kamiński [1] proved that the Maximum Bisection is NP-hard on unit disk graphs.

Frieze and Jerrum [4] gave an approximation algorithm for the Maximum Bisection problem with ratio 0.651. Ye [13] presented an improved approximation algorithm with ratio 0.699. Halperin and Zwick [7] gave an approximation algorithm with ratio 0.701. Feige et al. [3] studied the Maximum Bisection problem on regular graphs, and presented an approximation algorithm with ratio 0.795. Karpiński et al. [9] studied approximation algorithms for Maximum Bisection problem on low degree regular graphs and planar graphs. For three regular graphs, an approximation algorithm of ratio 0.847 was presented in [9]. For four and five regular graphs, two approximation algorithms with ratios 0.805, 0.812 were presented in [9], respectively. For planar graph of a sublinear degree, a polynomial time approximation scheme was presented in [9]. Jansen et al. [8] studied Maximum Bisection problem on planar graphs, and gave the first polynomial time approximation scheme for the problem.

This work is supported by the National Natural Science Foundation of China under Grants (61420106009, 61232001, 61472449, 61672536, 61572414).

Y. Cao and J. Chen (Eds.): COCOON 2017, LNCS 10392, pp. 188–199, 2017.
DOI: 10.1007/978-3-319-62389-4_16

For a given graph G, it is easy to find a bisection with $\lceil |E|/2 \rceil$ crossing edges by probabilistic method. In this paper, we study the following problem.

Parameterized Max-Bisection above Tight Lower Bound (PMBTLB): Given a graph $G = (V, E)$ and non-negative integer k, find a bisection of G with at least $\lceil |E|/2 \rceil + k$ crossing edges, or report that no such bisection exists.

Gutin and Yeo [5] gave a vertex kernel of size $O(k^2)$ for the PMBTLB problem. Based on the relation between edges in maximum matching and crossing edges, a parameterized algorithm of running time $O^*(16^k)$ was presented in [5]. Mnich and Zenklusen [11] presented a vertex kernel of size $16k$ for PMBTLB problem based on Gallai-Edmonds decomposition of the given graph.

In this paper, we further analyze the relation between maximum matching and vertices in Gallai-Edmonds decomposition for a given graph G. The vertices in Gallai-Edmonds decomposition are divided into several categories, which play important role in getting improved kernel. Based on the categories of vertices, we divide graph G into a set of blocks, where each block is closely related to the number of crossing edges of bisection of G. By analyzing the number of crossing edges in all blocks, a vertex kernel of size $8k$ is presented.

2   Preliminaries

For a given graph $G = (V, E)$, we use n, m to denote the number of vertices in V and the number of edges in E, respectively. Assume that all the graphs discussed in the paper are loopless undirected graph with possible parallel edges. For a graph $G = (V, E)$, if $|V|$ is a odd number, we can add an isolated vertex into G such that the number of crossing edges in each bisection of G is not changed. For simplicity, we assume that all the graphs in the paper have even number of vertices.

For two subsets $A, B \subseteq V$, let $E(A)$ be the set of edges in $G[A]$, and let $E(A, B)$ be the set of edges with one endpoint in A and the other endpoint in B. For two vertices u and v in G, for simplicity, let uv denote an edge between u and v, and let $E(u, v)$ denote the set of edges between u and v. For a vertex v in G, let $d(v)$ denote the degree of v in G. For a subset $X \subseteq V$ and a vertex v in X, let $d_X(v)$ denote the degree of v in induced subgraph $G[X]$, and let $\delta(G[X])$ be the number of connected components in $G[X]$. For a subgraph H of G, let $V(H)$ be the set of vertices contained in H. Let $N[V(H)]$ denote the set of neighbors of vertices in $V(H)$, where $V(H)$ is contained in $N[V(H)]$, and let $N(V(H)) = N[V(H)] - V(H)$.

Given a matching M in G, let $V(M)$ denote the set of vertices in M. If a vertex u in G is not contained in $V(M)$, then u is called an *unmatched vertex*. Matching M is called a *near-perfect matching* of G if there is exactly one unmatched vertex in G. For a connected graph G, and any vertex u in G, if the size of maximum matching in $G \backslash \{u\}$ is equal to the size of maximum matching in G, then G is called a *factor-critical* graph. A Gallai-Edmonds decomposition of graph G is a tuple (X, Y, Z), where X is the set of vertices in G

which are not covered by at least one maximum matching of G, Y is $N(X)$, and $Z = V(G)\backslash(X \cup Y)$. The Gallai-Edmonds decomposition of G can be obtained in polynomial time [10].

Lemma 1 ([2,10]). *For a given graph G, a Gallai-Edmonds decomposition (X, Y, Z) of G has the following properties:*

1. *the components of the subgraph induced by X are factor-critical,*
2. *the subgraph induced by Z has a perfect matching,*
3. *if M is any maximum matching of G, it contains a near-perfect matching of each component of $G[X]$, a perfect matching of each component of $G[Z]$, and matches all vertices of Y with vertices in distinct components of $G[X]$,*
4. *the size of the maximum matching is $\frac{1}{2}(|V| - \delta(G[X]) + |Y|)$.*

For two subsets A, B of V, if $A \cap B = \emptyset$ and $|A| = |B|$, then (A, B) is called a *basic block* of graph G. Let $\mathcal{C} = \{C_1, \ldots, C_h\}$ be the set of basic blocks of G, where $C_i = (A_i, B_i)$. For a basic block $C_i = (A_i, B_i)$, let $V(C_i)$ denote the set of vertices in $A_i \cup B_i$. Given two basic blocks $C_i, C_j \in \mathcal{C}$, for simplicity, let $E(C_i, C_j) = E(V(C_i), V(C_j))$. For all basic blocks in $\mathcal{C}$, if $V(C_i) \cap V(C_j) = \emptyset$ $(i \neq j)$ and $\bigcup_{i=1}^{h} V(C_i) = V$, then $\mathcal{C}$ is called a *block cluster* of G. For a basic block $C \in \mathcal{C}$, we use $\mathcal{C} - C$ to denote $\mathcal{C}\backslash\{C\}$.

Based on the block cluster $\mathcal{C}$ and V, a bisection (V_1, V_2) of G can be constructed in the following way: for each basic block $C_i = (A_i, B_i)$ in $\mathcal{C}$, put all vertices in A_i into V_1 and V_2 with probability $1/2, 1/2$, respectively; if A_i is put into V_1, then B_i will be put into V_2, and if A_i is put into V_2, then B_i will be put into V_1.

Let $r_1 = \sum_{i=1}^{h} |E(A_i, B_i)|$, $r_2 = \sum_{i=1}^{h-1} \sum_{j=i+1}^{h} |E(C_i, C_j)|$, and $r_3 = \sum_{i=1}^{h} (|E(A_i)| + |E(B_i)|)$. For a basic block $C_i = (A_i, B_i)$ in $\mathcal{C}$, let $r(C_i) = |E(A_i, B_i)| - |E(A_i)| - |E(B_i)|$. Let $r(\mathcal{C}) = \sum_{i=1}^{h} r(C_i)$.

Lemma 2. *For any block cluster $\mathcal{C}$ of graph G, there exists a bisection (V_1', V_2') of G obtained from $\mathcal{C}$ such that $|E(V_1', V_2')|$ is at least $\lceil m/2 \rceil + r(\mathcal{C})/2$.*

Proof. For any two basic blocks C_i, C_j $(i \neq j)$ in $\mathcal{C}$, we now analyze the expected number of crossing edges from $E(C_i, C_j)$ for bisection (V_1, V_2). Assume that A_i is in V_1, and B_i is in V_2. In the process of constructing (V_1, V_2), A_j is put into V_1 and V_2 with probability $1/2, 1/2$, respectively. Therefore, the expected number of crossing edges from $E(C_i, C_j)$ is $|E(C_i, C_j)|/2$. Moreover, for a basic block C_i in $\mathcal{C}$, if $E(A_i, B_i) \neq \emptyset$, then the edges in $E(A_i, B_i)$ are all crossing edges, and edges in $E(A_i) \cup E(B_i)$ are not crossing edges. Therefore, the expected number of crossing edges in (V_1, V_2) is $r_1 + r_2/2 = (2r_1 + r_2)/2 = (r(\mathcal{C}) + r_3 + r_1 + r_2)/2$. Since $r_1 + r_2 + r_3 = m$, $(r(\mathcal{C}) + r_3 + r_1 + r_2)/2 = m/2 + r(\mathcal{C})/2$. Therefore, there must exist a bisection (V_1', V_2') of G with $|E(V_1', V_2')| \geq \lceil m/2 \rceil + r(\mathcal{C})/2$. $\square$

Lemma 3. *For a given instance (G, k) of PMBTLB problem and any block cluster $\mathcal{C}$ of G, if $r(\mathcal{C}) \geq 2k$, then G has a a bisection of size at least $\lceil m/2 \rceil + k$ based a standard derandomization as given by Ries and Zenklusen [12].*

3 Kernelization for PMBTLB Problem

For a given instance (G, k) of PMBTLB, assume that (X, Y, Z) is a Gallai-Edmonds decomposition of G. Let MM be a maximum matching of G. Based on the degree of vertices in X and the maximum matching MM, we divide X into following subsets:

$X_0 = \{v | v \in X, d(v) = 0\}$,
$X_1 = \{v | v \in X, d_X(v) = 0, v \in V(MM)\}$,
$X_2 = \{v | v \in X, d_X(v) = 0, v \notin V(MM)\}$,
$X_3 = \{v | v \in X, d_X(v) \geq 1, \exists u \in Y, uv \in MM\}$,
$X_4 = \{v | v \in X, \exists u \in X, uv \in MM\}$,
$X_5 = \{v | v \in X, d_X(v) \geq 1, v \notin V(MM)\}$.

We now give the process to construct a block cluster C of graph G, as given in Fig. 1. Assume that $C = \{C_1, \ldots, C_h\}$ is the block cluster of G obtained by algorithm BBDA1 in Fig. 1.

BBDA1(G, MM)
Input: a graph $G = (V, E)$, and a maximum matching MM in G.
Output: a block cluster C of G.
1. $C = \emptyset$;
2.   **for** each edge uv in MM **do**
2.1   let $A = \{u\}$, $B = \{v\}$;
2.2   construct a basic block $C = (A, B)$, and add it into C;
3.   let $V' = V \backslash V(MM)$;
4.   **if** V' is not empty **then**
4.1   randomly choose $|V'|/2$ vertices to put into A, and put the remaining vertices into B;
4.2   construct a basic block $C = (A, B)$, and add it into C;
5.   return C.

Fig. 1. Algorithm for constructing block cluster C

Lemma 4 ([6]). *If M is a matching in a graph G, then G has a bisection of size at least $\lceil m/2 \rceil + \lfloor |M|/2 \rfloor$, which can be found in $O(m + n)$ time.*

By Lemma 4, we can get that the size of matching M is less than $2k$, otherwise a bisection with at least $\lceil m/2 \rceil + k$ crossing edges can be found in polynomial time. For a maximum matching MM of G, if $V' = V \backslash V(MM)$ is empty, then G is a graph with perfect matching. Since the size of matching MM is less than $2k$, the number of vertices in G is bounded by $4k$.

In the following, assume that V' is not empty. Since $V' = V \backslash V(MM)$, V' is an independent set. Assume that C_h is the basic block constructed by step 4 of algorithm BBDA1. According to the construction process of C, for each basic block C_i in $C - C_h$, $r(C_i) \geq 1$. Especially, $r(C_h) = 0$. We now construct a new block cluster based on $\mathcal{C}$. The general idea is to move vertices of $V(C_h)$ to the basic blocks in $\mathcal{C} - C_h$ to get new basic blocks. In the construction process, if no vertex in added into a basic block $C = (A, B)$, then the value $r(C)$ is not changed. Since vertices in $V(C_h)$ form an independent set, after removing some vertices of C_h, the vertices in the remaining basic block C_h still form an independent set, and $r(C_h)$ is still zero. In the following, we give the process to get a new block cluster $\mathcal{C}' = \{C_1', \ldots, C_h'\}$ based on $\mathcal{C}$, which is given in Fig. 2.

BBDA2$((X, Y, Z), \mathcal{C})$
Input: a Gallai-Edmonds decomposition (X, Y, Z) of G, and a block cluster $\mathcal{C}$
of G returned by algorithm BBDA1.
Output: a new block cluster $\mathcal{C}'$ of G and a vertex set S.
1.   let $\mathcal{C}_Y = \{C_i, \ldots, C_j\}$ be the subset of $\mathcal{C}$ such that for each C_l in $\mathcal{C}_Y$,
      $V(C_l)$ contains one vertex from Y;
2. $\mathcal{C}' = \emptyset$; $C_h' = C_h$; $\mathcal{C}_Y' = \mathcal{C}_Y$; $S = X_2$;
3.   **for** each C_l in $\mathcal{C}_Y'$ **do**
3.1   assume A_l of C_l contains one vertex from Y, denoted by u_l;
3.2   **for** each vertex v in S **do**
3.3   **if** there exists a vertex w in S with $|E(v, u_l)| > |E(w, u_l)|$ **then**
3.4 $B_l = B_l \cup \{v\}$; $A_l = A_l \cup \{w\}$; $S = S \backslash \{v, w\}$; $C_h' = C_h' \backslash \{v, w\}$;
3.5 $C_l = (A_l, B_l)$;
4. $\mathcal{C}' = (\mathcal{C} - \mathcal{C}_Y - C_h) \cup \mathcal{C}_Y' \cup C_h'$;
5.   **return** $\mathcal{C}'$ and S.

Fig. 2. Algorithm for constructing block cluster $\mathcal{C}'$

Let $\mathcal{C}'$ be the block cluster returned by algorithm BBDA2. We now analyze the difference between $r(\mathcal{C})$ and $r(\mathcal{C}')$. For two vertices v, w in X_2 that are added into C_l in step 3 of algorithm BBDA2, $r(C_l)$ is increased by at least one. Assume that S is returned by algorithm BBDA2. For any two vertices w, v in S, it is easy to see that for each vertex u in G, $|E(w, u)| = |E(v, u)|$. Since all vertices in $|X_2 \backslash S|$ are moved to $\mathcal{C}_Y'$ in algorithm BBDA2, $r(\mathcal{C}_Y') - r(\mathcal{C}_Y)$ is at least $|X_2 \backslash S|/2$, and $r(\mathcal{C}') - r(\mathcal{C})$ is at least $|X_2 \backslash S|/2$. In algorithm BBDA1, each edge in MM is chosen to construct a basic block. Therefore, the value $r(\mathcal{C})$ is at least $|MM|$. Since $\mathcal{C}'$ is constructed based on $\mathcal{C}$, we have

$$r(\mathcal{C}') \geq |MM| + |X_2 \backslash S|/2. \tag{1}$$

Since the vertices in X_0, X_2 and X_5 are not in $V(MM)$, in algorithm BBDA1, $V(C_h) = X_0 \cup X_2 \cup X_5$. In algorithm BBDA2, the vertices in $X_2 \backslash S$ are moved from C_h to $\mathcal{C}_Y'$. Therefore, $V(C_h') = X_0 \cup S \cup X_5$.

Lemma 5. *For any basic block $C_l' = (A_l', B_l')$ in $\mathcal{C}'$, where $V(C_l')$ contains one vertex u of Y and $u \in N(S)$, assume that $u \in A_l'$. Then, S cannot be connected to any vertex in B_l', and the number of basic blocks in $\mathcal{C}'$ containing one vertex in $N(S)$ is $|N(S)|$.*

For any connected component H in $G[X]$ with at least three vertices, there is exactly one vertex v in $V(H)$ such that v is in either X_3 or X_5, and other vertices in $V(H)\backslash\{v\}$ are in X_4.

Lemma 6. $|X_5| < 2k$.

Proof. If X_5 is empty, then this lemma is correct. Let $\mathcal{H} = \{H_1, \ldots, H_l\}$ be the set of connected components in $G[X]$, each of which has size at least three and contains one vertex in X_5. For any H_i $(1 \leq i \leq l)$ in $\mathcal{H}$, there exists a perfect matching in $G[V(H_i)\backslash\{v\}]$, and the number of edges from $E(H_i)$ in MM is $(|V(H_i)| - 1)/2$. By above discussion, the number of edges in MM is less than $2k$. Therefore, $\sum_{i=1}^{l}(|V(H_i)| - 1)/2 < 2k$. Thus, $|X_5| < 2k$.   □

In the following, we will construct two new block clusters $\mathcal{C}''$ and $\mathcal{C}'''$ based on $\mathcal{C}'$ by adding vertices in X_0, X_5 and S into basic blocks of $\mathcal{C}' - C_h'$.

Case 1. $|X_0| \geq |S|$. Under this case, the general idea to construct $\mathcal{C}''$ is to use the vertices in S and X_0 firstly. When all vertices in S are added into basic blocks in $\mathcal{C}' - C_h'$ to get new basic blocks in $\mathcal{C}''$, we consider the vertices X_5 and the remaining vertices in X_0 to construct basic blocks in $\mathcal{C}'''$ based on $\mathcal{C}''$.

By Lemma 5, find a basic block $C_i' = (A_i', B_i')$ in $\mathcal{C}'$ such that $V(C_i') \cap N(S) = \{u\}$, and assume that u is contained in A_i'. A new basic block $C_i'' = (A_i'', B_i'')$ of $\mathcal{C}''$ can be constructed from C_i' by the following steps: $B_i'' = B_i' \cup S$; arbitrarily choose $|S|$ vertices from X_0, denoted by P; $A_i'' = A_i' \cup P$. Let $C_h'' = C_h'\backslash(S \cup P)$, $\mathcal{C}'' = (\mathcal{C}' - C_i' - C_h') \cup C_i'' \cup C_h''$. Since no vertex in S is connected to any vertex in B_i' and no vertex in P is connected to any vertex in $V(C_i')$, $r(C_i'') - r(C_i')$ is at least $|S|$. Thus,

$$r(\mathcal{C}'') - r(\mathcal{C}') \geq |S|. \tag{2}$$

Let $\mathcal{H}$ be the set of connected components in $G[X]$ such that for each connected component H in $\mathcal{H}$, $V(H)$ contains one vertex of X_5, and $N(V(H)) \cap Y \neq \emptyset$. For each connected component H in $\mathcal{H}$, assume that E' is the set of edges in H that are contained in MM. Let $\mathcal{C}_H$ be a subset of $\mathcal{C}''$, where $\mathcal{C}_H$ can be constructed by the edges in E'. Since H is factor-critical, there exists a vertex v in $V(H)$ such that v is an unmatched vertex. If v is not connected to any vertex in Y, then find a vertex u in $V(H)$ that is connected to some vertices in Y, and find a perfect matching M' in $G[V(H)\backslash\{u\}]$. Construct a set $\mathcal{F}$ of basic blocks by the edges in M', and let $\mathcal{C}'' = (\mathcal{C}'' - \mathcal{C}_H) \cup \mathcal{F}$. After dealing with all connected components in $\mathcal{H}$ by the above process, a new maximum matching MM' can be obtained, and for each connected component H in $G[X]$ satisfying that $V(H)$ has at least three vertices and $N(H) \cap Y \neq \emptyset$, if v is an unmatched vertex in H, then v is connected to at least one vertex in Y.

The vertices in X_5 are divided into the following two types. Let X_5^y be a subset of X_5 such that each vertex v in X_5^y is connected to at least one vertex in Y, and let X_5^i be a subset of X_5 such that each vertex u in X_5^i is not connected to any vertex in Y.

Lemma 7. *Given a vertex $v \in X_5^y$, for any basic block $C_l'' = (A_l'', B_l'')$ in C'', where $V(C_l'')$ contains one vertex u of Y and $u \in N(v)$, assume that $u \in A_l''$. Then, v cannot be connected to any vertex in B_l''.*

Let $X_0' = X_0 \backslash P$. We now construct basic blocks of C''' by vertices in X_0' and X_5^y. Assume that $|X_0'| \geq |X_5^y|$. For a vertex v in X_5^y, by Lemma 7, there exists a basic block $C_i'' = (A_i'', B_i'')$ in C'' containing u such that $u \in N(v)$. Without loss of generality, assume that u is contained in A_i''. A new basic block $C_i''' = (A_i''', B_i''')$ of C''' can be constructed from C_i'' by the following process: $B_i''' = B_i'' \cup \{v\}$; arbitrarily choose a vertex w in X_0', let $A_i''' = A_i'' \cup \{w\}$. Let $C_h''' = C_h'' \backslash \{v, w\}$, $C''' = (C'' - C_i'' - C_h'') \cup \{C_i'''\} \cup \{C_h'''\}$. Since v is not connected to any vertex in B_i'' and w is not connected to any vertex in $V(C_i'')$, it is easy to see that $r(C_i''') - r(C_i'') \geq 1$. By considering all vertices in X_5^y, we have

$$r(C''') - r(C'') \geq |X_5^y|. \tag{3}$$

For the case when $|X_0'| < |X_5^y|$, a similar construction process can be obtained as above. By considering all vertices in X_0', we can get that

$$r(C''') - r(C'') \geq |X_0'|. \tag{4}$$

Lemma 8. *Based on MM', $|X_5^i| \leq \delta(G \backslash X_0)$.*

Proof. We prove this lemma by discussing the types of connected components in $G \backslash X_0$. For a connected component H in $G \backslash X_0$, if $V(H)$ contains no vertex of X, then no vertex in $V(H)$ is contained in X_5^i. If all vertices in $V(H)$ are from X, then H is a factor-critical connected component, and no vertex in $V(H)$ is connected to Y. Under this case, one vertex from $V(H)$ is contained in X_5^i. Assume that $V(H) \backslash X$ is not empty. By the construction process of MM', if a vertex v in $N(Y) \cap V(H)$ is contained in X_5, then v is in X_5^y, and no vertex in $V(H)$ is in X_5^i. Therefore, $|X_5^i| \leq \delta(G \backslash X_0)$. □

Rule 1. For a given instance (G, k) of PMBTLB, if $|X_0| + \delta(G \backslash X_0) > \frac{n}{2}$, then arbitrarily delete $|X_0| + \delta(G \backslash X_0) - \frac{n}{2}$ vertices from X_0.

Lemma 9. *Rule 1 is safe.*

Proof. Assume that $|X_0| + \delta(G \backslash X_0) > \frac{n}{2}$, and assume that (V_1, V_2) is a maximum bisection of G. If $X_0 = \emptyset$, then the number of connected components in $G \backslash X_0$ is at most $n/2$, because each connected component in $G \backslash X_0$ has at least two vertices. Assume that the number of connected components in $G \backslash X_0$ is at least one, otherwise, the PMBTLB problem can be trivially solved. Assume that $\mathcal{H} = \{H_1, \ldots, H_l\}$ is the set of connected components in $G \backslash X_0$. We now prove that all vertices in X_0 cannot be contained totally in V_1 or V_2. Assume that all vertices of X_0 are contained in V_1. Since $|X_0| + \delta(G \backslash X_0) > \frac{n}{2}$, there must exist a connected component H_i in $\mathcal{H}$ such that all vertices in $V(H_i)$ are contained in V_2. Choose a vertex v in $V(H_i)$, and find a vertex u of X_0 in V_1. Let $V_1' = (V_1 \backslash \{u\}) \cup \{v\}$, $V_2' = (V_2 \backslash \{v\}) \cup \{u\}$. Then, $|E(V_1', V_2')| - |E(V_1, V_2)| \geq 1$, contradicting the fact that

(V_1, V_2) is a maximum bisection of G. Therefore, $V_1 \cap X_0 \neq \emptyset$ and $V_2 \cap X_0 \neq \emptyset$. Assume that X_0' and X_0'' are two subsets of X_0 such that X_0' is contained in V_1, and X_0'' is contained in V_2. Without loss of generality, assume that $|X_0'| \leq |X_0''|$. Let R be any subset of X_0'' of size $|X_0'|$. Let $V_1'' = V_1 - X_0'$, $V_2'' = V_2 - R$. Then, $|V_1''| = |V_2''|$. Denote the new graph with bisection (V_1'', V_2'') by G'. It is easy to see that $|E(V_1, V_2)|$ is bounded by $\lceil m/2 \rceil + k$ if and only if $|E(V_1'', V_2'')|$ is bounded by $\lceil m/2 \rceil + k$. Repeat the above process until $|X_0| + \delta(G \backslash X_0) \leq \frac{n}{2}$. $\square$

For a given instance (G, k) of PMBTLB, by applying Rule 1 on G exhaustively, we can get the following result.

Lemma 10. *For a given instance (G, k) of PMBTLB, if $|X_0| \geq |S|$, then the number of vertices in G is bounded by $8k$.*

Proof. Based on block cluster $\mathcal{C}'''$, we prove this lemma by analyzing the sizes of X_0' and X_5^y.

(1) $|X_0'| \geq |X_5^y|$. By Lemma 3 and inequalities (1), (2) and (3), we can get that $r(\mathcal{C}''') = |MM| + \frac{|X_2 \backslash S|}{2} + |S| + |X_5^y| < 2k$. Based on Gallai-Edmonds decomposition (X, Y, Z) and MM, we know that $|MM| = (|Z| + |Y| + |X \backslash (X_0 \cup X_2 \cup X_5)|)/2$. Then, $|Z| + |Y| + |X \backslash (X_0 \cup S \cup X_5)| + 2|S| + 2|X_5^y| < 4k$. Since $X_5 = X_5^y \cup X_5^i$, and S, X_5^i, X_5^y, X_0 are disjoint, we can get that

$$|Z| + |Y| + |X| - |X_0| - |X_5^i| + |S| + |X_5^y| < 4k. \tag{5}$$

By Lemmas 8 and 9, $|X_0| + |X_5^i| \leq \frac{n}{2}$. Then, we can get that

$$|V| = |Z| + |Y| + |X| = |Z| + |Y| + |X \backslash (X_0 \cup X_{5i})| + |X_0| + |X_5^i|$$
$$= \underbrace{|Z| + |Y| + |X| - |X_0| - |X_5^i|}_{< 4k \text{ by } (5)} + |X_0| + |X_5^i|$$
$$< 4k + |V|/2.$$

Therefore, $|V| < 8k$.

(2) $|X_0'| < |X_5^y|$. By Lemma 3 and inequalities (1), (2) and (4), we can get that $r(\mathcal{C}''') = |MM| + \frac{|X_2 \backslash S|}{2} + |S| + |X_0'| < 2k$. Since $P \subseteq X_0$, $X_0' = X_0 \backslash P$ and $|S| = |P|$, we can get that $r(\mathcal{C}''') = |MM| + \frac{|X_2 \backslash S|}{2} + |X_0| < 2k$. Similarly, $|MM| = (|Z| + |Y| + |X \backslash (X_0 \cup X_2 \cup X_5)|)/2$. Then, $|Z| + |Y| + |X \backslash (X_0 \cup S \cup X_5)| + 2|X_0| < 4k$. Since S, X_5, X_0 are disjoint and $|X_0| \geq |S|$, we can get that

$$|Z| + |Y| + |X| - |X_5| < 4k. \tag{6}$$

By Lemma 6, $|X_5| < 2k$. Then, we can get that

$$|V| = |Z| + |Y| + |X| = |Z| + |Y| + |X \backslash X_5| + |X_5|$$
$$= \underbrace{|Z| + |Y| + |X| - |X_5|}_{< 4k \text{ by } (6)} + |X_5|$$
$$< 4k + 2k = 6k.$$

Therefore, if $|X_0| \geq |S|$, the number of vertices in G is bounded by $8k$. $\square$

Case 2. $|X_0| < |S|$. Under this case, the general idea to construct C'' is to use the vertices in S and X_0 firstly. When all vertices in X_0 are added into basic blocks in $C' - C'_h$ to get new basic blocks in C'', we consider vertices in X_5 and the remaining vertices in S to construct basic blocks in C''' based on C''.

By Lemma 5, find a basic block $C'_i = (A'_i, B'_i)$ in C' such that $V(C'_i) \cap N(S) = \{u\}$, and assume that u is contained in A'_i. A new basic block $C''_i = (A''_i, B''_i)$ of C'' can be constructed from C'_i by the following steps: $A''_i = A'_i \cup X_0$; arbitrarily choose $|X_0|$ vertices from S, denoted by Q; $B''_i = B'_i \cup Q$. Let $C''_h = C'_h \setminus (X_0 \cup Q)$, $C'' = (C' - C'_i - C'_h) \cup \{C''_i\} \cup \{C''_h\}$. Since no vertex in Q is connected to any vertex in B'_i and no vertex in X_0 is connected to any vertex in $V(C'_i)$, $r(C''_i) - r(C'_i)$ is at least $|X_0|$. Thus,

$$r(C'') - r(C') \geq |X_0|. \tag{7}$$

Since $|X_0| < |S|$, S is not empty. For a connected component H in $G[X]$, and for any three vertices w, v and u, where $w \in S$, $v \in V(H)$ and $u \in N(S)$, if $N(V(H)) = N(S)$ and $|E(w, u)| = |E(v, u)|$, then H is called a *special component* in $G[X]$.

Let $\mathcal{H}$ be the set of connected components in $G[X]$ such that for each connected component H in $\mathcal{H}$, $V(H)$ contains one vertex of X_5 and H is not a special component. For each connected component H in $\mathcal{H}$, assume that E' is the set of edges in H that are contained in MM. Let $\mathcal{C}_H$ be a subset of C'', where $\mathcal{C}_H$ can be constructed by the edges in E'. Since H is factor-critical, there exists a vertex v in $V(H)$ such that v is an unmatched vertex. Based on $N(S)$ and $N(v)$, we give following five conditions: (1) $N(v) \cap N(S) = \emptyset$; (2) $N(v) \setminus N(S) \neq \emptyset$; (3) $N(v) \subset N(S)$; (4) $N(v) = N(S)$, and for any two vertices $w \in S$ and $u \in N(S)$, $|E(v, u)| > |E(w, u)|$; (5) $N(v) = N(S)$, and for any two vertices $w \in S$ and $u \in N(S)$, $|E(w, u)| < |E(v, v)|$.

We now introduce how to get a new maximum matching based on the above five conditions. Since $V(H)$ is not a special component, if v does not satisfy any condition from conditions (1)–(5), then a vertex u in H can be found such that u satisfies one of the above conditions. Then, find a perfect matching M' in $G[V(H) \setminus \{u\}]$, and construct a set $\mathcal{F}$ of basic blocks by the edges in M'. Let $C'' = (C'' - \mathcal{C}_H) \cup \mathcal{F}$. After dealing with all connected components in $\mathcal{H}$ by above process, a new maximum matching MM' can be obtained.

The vertices in X_5 are divided into the following three types. Let X_5^1 be a subset of X_5 such that for each vertex v in X_5^1, the connected component in $G[X]$ containing v is a special component. Let X_5^2 be a subset of X_5 such that for each vertex v in X_5^2, v satisfies one of conditions (1), (2), and (4). Let X_5^3 be a subset of X_5 such that for each vertex v in X_5^3, v satisfies one of conditions (3) and (5).

Lemma 11. *For any vertex v in X_5^2 and any vertex w in S, there exists a vertex u in $N(v)$ with $|E(v, u)| > |E(w, u)|$ such that a basic block $C''_i = (A''_i, B''_i)$ in C'' can be found with $V(C''_i) \cap Y = \{u\}$. Assume that $A''_i \cap Y = \{u\}$. Then, no vertex in B''_i is connected to $\{v, w\}$.*

Lemma 12. *For any vertex v in X_5^3 and any vertex w in S, there exists a vertex u in $N(w)$ with $|E(w, u)| > |E(v, u)|$ such that a basic block $C_i'' = (A_i'', B_i'')$ in C'' can be found with $V(C_i'') \cap Y = \{u\}$. Assume that $A_i'' \cap Y = \{u\}$. Then, no vertex in B_i'' is connected to $\{v, w\}$.*

Let $S' = S \backslash Q$. We now construct basic blocks of C''' by vertices in S', X_5^2 and X_5^3. Assume that $|S'| \geq |X_5^2| + |X_5^3|$. For a vertex v in X_5^2 and any vertex w in S', by Lemma 11, there exists a vertex u in $N(v)$ with $|E(v, u)| > |E(w, u)|$ such that a basic block $C_i'' = (A_i'', B_i'')$ in C'' can be found with $V(C_i'') \cap Y = \{u\}$. Without loss of generality, assume that u is contained in A_i''. A new basic block $C_i''' = (A_i''', B_i''')$ can be constructed from C_i'' by the following process: $B_i''' = B'' \cup \{v\}$ and $A_i''' = A'' \cup \{w\}$. Let $C_h''' = C_h'' \backslash \{v, w\}$, $C''' = (C'' - C_i'' - C_h'') \cup \{C_i'''\} \cup \{C_h'''\}$. By Lemma 11, no vertex in B_i'' is connected to $\{v, w\}$. It is easy to see that $r(C_i''') - r(C_i'') \geq 1$.

For a vertex v in X_5^3 and any vertex w in S', by Lemma 12, there exists a vertex u in $N(w)$ with $|E(w, u)| > |E(v, u)|$ such that a basic block $C_j'' = (A_j'', B_j'')$ in C'' can be found with $V(C_j'') \cap Y = \{u\}$. Without loss of generality, assume that u is contained in A_j''. A new basic block C_j''' can be constructed from C_j'' by the following process: $B_j''' = B'' \cup \{w\}$ and $A_j''' = A'' \cup \{v\}$. Let $C_h''' = C_h'' \backslash \{v, w\}$, $C''' = (C'' - C_j'' - C_h'') \cup \{C_j'''\} \cup \{C_h'''\}$. By Lemma 12, no vertex in B_j'' is connected to $\{v, w\}$. It is easy to see that $r(C_j''') - r(C_j'') \geq 1$.

By considering all vertices in $X_5^2 \cup X_5^3$, we can get that

$$r(C''') - r(C'') \geq |X_5^2| + |X_5^3|. \tag{8}$$

For the case when $|S'| < |X_5^2| + |X_5^3|$, a similar construction process can be obtained as above. By considering all vertices in S', we can get that

$$r(C''') - r(C'') \geq |S'|. \tag{9}$$

Rule 2. For a given instance (G, k) of PMBTLB, if $|S| + |X_5^1| > \frac{n}{2}$, then arbitrarily delete $|S| + |X_5^1| - \frac{n}{2}$ vertices from S.

Lemma 13. *Rule 2 is safe.*

Proof. Assume that $|S| + |X_5^1| > \frac{n}{2}$, and assume that (V_1, V_2) is a maximum bisection of G. Since $|S| > |X_0|$, S is not empty. We prove this lemma by discussing whether X_5^1 is empty or not.

(a) $X_5^1 \neq \emptyset$. Assume that $\mathcal{H} = \{H_1, \ldots, H_l\}$ is the set of connected components in $G[X]$ containing one vertex of X_5^1, and assume that $S = \{v_1, \ldots, v_j\}$. We now prove that all vertices in S cannot be contained totally in V_1 or V_2. Assume that all vertices of S are contained in V_1. Since $|S| + |X_5^1| > \frac{n}{2}$, there must exist a connected component H_i in $\mathcal{H}$ such that all vertices in $V(H)$ are contained in V_2. Choose a vertex v in H_i, and find a vertex u of S in V_1. Let $V_1' = (V_1 - \{u\}) \cup \{v\}$, $V_2' = (V_2 - \{v\}) \cup \{u\}$. Then, $|E(V_1', V_2')| - |E(V_1, V_2)| \geq 1$, contradicting the fact that (V_1, V_2) is a maximum bisection of G. Therefore, $V_1 \cap S \neq \emptyset$ and $V_2 \cap S \neq \emptyset$.

(b) $X_5^1 = \emptyset$. Since $|S| > \frac{n}{2}$, $V_1 \cap S \neq \emptyset$ and $V_2 \cap S \neq \emptyset$.

Assume that S' and S'' are two subsets of S such that S' is contained in V_1, and S'' is contained in V_2. Without loss of generality, assume that $|S'| \leq |S''|$. Let R be any subset of S'' of size $|S'|$. Let $V_1'' = V_1 - S'$, $V_2'' = V_2 - R$. Then, $|V_1''| = |V_2''|$. Denote the new graph with bisection (V_1'', V_2'') by G'. It is easy to see that $|E(V_1, V_2)|$ is bounded by $\lceil m/2 \rceil + k$ if and only if $|E(V_1'', V_2'')|$ is bounded by $\lceil m'/2 \rceil + k$, where m' is the number of edges in G'. Repeat the above process until $|S| + |X_5^1| \leq \frac{n}{2}$.   □

For a given instance (G, k) of PMBTLB, by applying Rule 2 on G exhaustively, we can get following result.

Lemma 14. *For a given instance (G, k) of PMBTLB, if $|S| > |X_0|$, then the number of vertices in G is bounded by 8k.*

Proof. Based on block cluster $\mathcal{C}'''$, we prove this lemma by analyzing the sizes of S', X_5^2 and X_5^3.

(1) $|S'| \geq |X_5^2| + |X_5^3|$. By Lemma 3 and inequalities (1), (7) and (8), we can get that $r(\mathcal{C}''') = |MM| + \frac{|X_2 \setminus S|}{2} + |X_0| + |X_5^2| + |X_5^3| < 2k$. Since $|MM| = (|Z| + |Y| + |X \setminus (X_0 \cup X_2 \cup X_5)|)/2$, we can get that $|Z| + |Y| + |X \setminus (X_0 \cup S \cup X_5)| + 2(|X_0| + |X_5^2| + |X_5^3|) < 4k$. Since $X_5 = X_5^1 \cup X_5^2 \cup X_5^3$, and S, X_5^1, X_5^2, X_5^3, X_0 are disjoint, we have

$$|Z| + |Y| + |X| - |S| - |X_5^1| + |X_0| + |X_5^2| + |X_5^3| < 4k. \tag{10}$$

By Lemma 13, $|S| + |X_5^1| < n/2$. Then, we can get that

$$\begin{aligned} |V| = |Z| + |Y| + |X| &= |Z| + |Y| + |X \setminus (S \cup X_5^1)| + |S| + |X_5^1| \\ &= \underbrace{|Z| + |Y| + |X| - |S| - |X_5^1|}_{<4k \text{ by } (10)} + |S| + |X_5^1| \\ &< 4k + |V|/2. \end{aligned}$$

Therefore, $|V| < 8k$.

(2) $|S| < |X_5^2| + |X_5^3|$. By Lemma 3 and inequalities (1), (7) and (9), we have $r(\mathcal{C}''') = |MM| + \frac{|X_2 \setminus S|}{2} + |X_0| + |S'| < 2k$. Since $Q \subseteq S$, $S' = S \setminus Q$ and $|X_0| = |Q|$, we can get that $r(\mathcal{C}''') = |MM| + \frac{|X_2 \setminus S|}{2} + |S| < 2k$. Similarly, $|MM| = (|Z| + |Y| + |X \setminus (X_0 \cup X_2 \cup X_5)|)/2$. Then, $|Z| + |Y| + |X \setminus (X_0 \cup S \cup X_5)| + 2|S| < 4k$. Since S, X_5, X_0 are disjoint and $|X_0| < |S|$, we can get that

$$|Z| + |Y| + |X| - |X_5| < 4k. \tag{11}$$

By Lemma 6, $|X_5| < 2k$. Then, we can get that

$$\begin{aligned} |V| = |Z| + |Y| + |X| &= |Z| + |Y| + |X \setminus X_5| + |X_5| \\ &= \underbrace{|Z| + |Y| + |X| - |X_5|}_{<4k \text{ by } (11)} + |X_5| \\ &< 4k + 2k = 6k. \end{aligned}$$

Therefore, if $|X_0| < |S|$, the number of vertices in G is bounded by 8k.   □

For a given instance (G, k) of PMBTLB, our kernelization algorithm is to apply Rule 1 and Rule 2 on G exhaustively. By Lemmas 10 and 14, we can get the following result.

Theorem 1. *The PMBTLB problem admits a vertex kernel of size $8k$.*

References

1. Díaz, J., Kamiński, M.: MAX-CUT and MAX-BISECTION are NP-hard on unit disk graphs. Theor. Comput. Sci. **377**(1–3), 271–276 (2007)
2. Edmonds, J.: Paths, trees, and flowers. Can. J. Math. **17**(3), 449–467 (1965)
3. Feige, U., Karpiński, M., Langberg, M.: A note on approximating Max-Bisection on regular graphs. Inf. Process. Lett. **79**(4), 181–188 (2001)
4. Frieze, A., Jerrum, M.: Improved approximation algorithms for max k-cut and max bisection. Algorithmica **18**(1), 67–81 (1997)
5. Gutin, G., Yeo, A.: Note on maximal bisection above tight lower bound. Inf. Process. Lett. **110**(21), 966–969 (2010)
6. Haglin, D.J., Venkatesan, S.M.: Approximation and intractability results for the maximum cut problem and its variants. IEEE Trans. Comput. **40**(1), 110–113 (1991)
7. Halperin, E., Zwick, U.: A unified framework for obtaining improved approximation algorithms for maximum graph bisection problems. Random Struct. Algorithms **20**(3), 382–402 (2002)
8. Jansen, K., Karpiński, M., Lingas, A., Seidel, E.: Polynomial time approximation schemes for max-bisection on planar and geometric graph. SIAM J. Comput. **35**(1), 163–178 (2000)
9. Karpiński, M., Kowaluk, M., Lingas, A.: Approximation algorithms for max bisection on low degree regular graphs and planar graphs. Electron. Colloq. Comput. Complex. **7**(7), 369–375 (2000)
10. Lovász, L., Plummer, M.D.: Matching Theory. NorthHolland, Amsterdam (1986)
11. Mnich, M., Zenklusen, R.: Bisections above tight lower bounds. In: Golumbic, M.C., Stern, M., Levy, A., Morgenstern, G. (eds.) WG 2012. LNCS, vol. 7551, pp. 184–193. Springer, Heidelberg (2012). doi:10.1007/978-3-642-34611-8_20
12. Ries, B., Zenklusen, R.: A 2-approximation for the maximum satisfying bisection problem. Eur. J. Oper. Res. **210**(2), 169–175 (2011)
13. Ye, Y.: A 0.699-approximation algorithm for max-bisection. Math. Program. **90**(1), 101–111 (2001)

Information Complexity of the AND Function in the Two-Party and Multi-party Settings

Yuval Filmus[1], Hamed Hatami[2], Yaqiao Li[2(✉)], and Suzin You[2]

[1] Technion — Israel Institute of Technology, Haifa, Israel
yuvalfi@cs.technion.ac.il
[2] McGill University, Montreal, Canada
hatami@cs.mcgill.ca, yaqiao.li@mail.mcgill.ca, suzinyou.sy@gmail.com

Abstract. In a recent breakthrough paper [M. Braverman, A. Garg, D. Pankratov, and O. Weinstein, From information to exact communication, STOC'13] Braverman *et al.* developed a local characterization for the zero-error information complexity in the two-party model, and used it to compute the exact internal and external information complexity of the 2-bit AND function. In this article, we extend their results on AND function to the multi-party number-in-hand model by proving that the generalization of their protocol has optimal internal and external information cost for certain distributions. Our proof has new components, and in particular it fixes a minor gap in the proof of Braverman *et al.*

1 Introduction

Although communication complexity has since its birth been witnessing steady and rapid progress, it was not until recently that a focus on an information-theoretic approach resulted in new and deeper understanding of some of the classical problems in the area. This gave birth to a new area of complexity theory called *information complexity*. Recall that communication complexity is concerned with minimizing the amount of communication required for players who wish to evaluate a function that depends on their private inputs. Information complexity, on the other hand, is concerned with the amount of information that the communicated bits reveal about the inputs of the players to each other, or to an external observer.

One of the important achievements of information complexity is the recent breakthrough of [BGPW13] that determines the exact asymptotics of the randomized communication complexity of one of the oldest and most studied problems in communication complexity, set disjointness:

$$\lim_{\varepsilon \to 0} \lim_{n \to \infty} \frac{R_\varepsilon(\text{DISJ}_n)}{n} \approx 0.4827. \tag{1}$$

Here $R_\varepsilon(\cdot)$ denotes the randomized communication complexity with an error of at most ε on every input, and DISJ_n denotes the set disjointness problem.

H. Hatami—Supported by an NSERC grant.

© Springer International Publishing AG 2017
Y. Cao and J. Chen (Eds.): COCOON 2017, LNCS 10392, pp. 200–211, 2017.
DOI: 10.1007/978-3-319-62389-4_17

Prior to the discovery of these information-theoretic techniques, proving the lower bound $R_\varepsilon(\text{DISJ}_n) = \Omega(n)$ had already been a challenging problem, and even Razborov's [Raz92] short proof of that fact is intricate and sophisticated.

Note that the set disjointness function is nothing but an OR of AND functions. More precisely, for $i = 1, \ldots, n$, if x_i is the Boolean variable which represents whether i belongs to Alice's set or not, and y_i is the corresponding variable for Bob, then $\bigvee_{i=1}^{n}(x_i \wedge y_i)$ is true if and only if Alice's input intersects Bob's input. Braverman *et al.* [BGPW13] exploited this fact to prove (1). Roughly speaking, they first determined the exact information complexity of the 2-bit AND function for any underlying distribution μ on the set of inputs $\{0,1\} \times \{0,1\}$, and then used the fact that amortized communication equals information complexity [BR14] to relate this to the communication complexity of the set disjointness problem. The constant 0.4827 in (1) is indeed the maximum of the information complexity of the 2-bit AND function over all measures μ that assign a zero mass to $(1,1) \in \{0,1\} \times \{0,1\}$. That is

$$\max_{\mu:\mu(11)=0} \text{IC}_\mu(\text{AND}) \approx 0.4827,$$

where $\text{IC}_\mu(\text{AND})$ denotes the information complexity of the 2-bit AND function with respect to the distribution μ with no error (see Definition 1 below). These results show the importance of knowing the exact information complexity of simple functions such as the AND function.

Although obtaining the asymptotics of $R_\varepsilon(\text{DISJ}_n)$ from the information complexity of the AND function is not straightforward and a formal proof requires overcoming some technical difficulties, the bulk of [BGPW13] is dedicated to computing the exact information complexity of the 2-bit AND function. This rather simple-looking problem had been studied previously by Ma and Ishwar [MI11, MI13], and some of the key ideas of [BGPW13] originate from their work. In [BGPW13] Braverman *et al.* introduced a protocol to solve the AND function, and proved that it has optimal internal and external information cost. Interestingly this protocol is not a conventional communication protocol as it has access to a continuous clock, and the players are allowed to "buzz" at randomly chosen times. Indeed, it is known [BGPW13] that no protocol with a bounded number of rounds can have optimal information cost for the AND function, and hence the infinite number of rounds, implicit in the continuous clock, is essential. We shall refer to this protocol as the *buzzers* protocol.

1.1   Our Contributions

Fixing the Argument of [BGPW13]: In order to show that the *buzzers* protocol has optimal information cost, inspired by the work of Ma and Ishwar [MI11, MI13], Braverman *et al.* came up with a local concavity condition, and showed that if a protocol satisfies this condition, then it has optimal information cost. This condition, roughly speaking, says that it suffices to verify that one does not gain any advantage over the conjectured optimal protocol if one of the players starts by sending a bit B. In the original paper [BGPW13], it is claimed that it suffices to verify this condition only for signals B that reveal arbitrarily small information about the inputs. As we shall see, however, this is *not* true, and one can construct counter-examples to this statement.

In Theorem 3 we prove a variant of the local concavity condition that allows one to consider only signals B with small information leakage, and then apply it to fix the argument in [BGPW13]. We have been informed through private communication that Braverman *et al.* have also independently fixed this error.

Extension of [BGPW13] to the Multi-party Setting: We then apply Theorem 3 to extend the result of [BGPW13] to the multi-party number-in-hand model by defining a generalization of the *buzzers* protocol, and then prove in Theorem 4 that it has optimal internal and external information cost when the underlying distribution satisfies the following assumption:

Assumption 1. *The support of μ is a subset of $\{0, 1, e_1, \ldots, e_k\}$, where e_i is the usual i^{th} basis vector $(0, \ldots, 0, 1, 0, \ldots, 0)$, and k is the number of parties.*

Note that in the two-party setting, every distribution satisfies this assumption and thus our results are complete generalizations of the results of [BGPW13] of the two-party setting. The distributions in Assumption 1 arise naturally in the study of the set disjointness problem, and as a result they have been considered previously in [Gro09].

This extension is not straightforward since in [BGPW13], a large part of the calculations for verifying the local concavity conditions are carried out by the software Mathematica, however, in the number-in-hand model, the number of players k can be arbitrary, one cannot simply rely on a computer for those calculations. Indeed, we had to look into the protocol and show that it suffices to analyze the behaviour of the information cost in a small time interval that is determined by the distribution of the inputs, and furthermore one can reduce all distributions into a class of essentially two-parameter distributions. Hence the problem can be reduced to one that has only a constant number of variables, thus allowing us to use Mathematica to verify the concavity condition. We believe our analysis provides new insights even for the two-party setting.

2   Preliminaries

2.1   Notation

We typically denote the random variables by capital letters (e.g. $A, B, C,$ X, Y, Π), and write $A_1 \ldots A_n$ to denote the random variable $(A_1, \ldots, A_n)$. Let $[k] := \{1, \ldots, k\}$, and $\text{supp}(\mu)$ denote the support of a measure μ. We denote the statistical distance (a.k.a. total variation distance) of two measures μ and ν on a sample space Ω by $|\mu - \nu| := \frac{1}{2} \sum_{a \in \Omega} |\mu(a) - \nu(a)|$. For every $\varepsilon \in [0, 1]$, $\text{H}(\varepsilon) = -\varepsilon \log \varepsilon - (1 - \varepsilon) \log(1 - \varepsilon)$ denotes the binary entropy, where here and throughout the paper $\log(\cdot)$ is in base 2, and $0 \log 0 = 0$. We also use $\text{H}(X)$ to denote the entropy of a random variable X.

Recall $D(\mu||\nu) := \sum_{a\in\Omega}\mu(a)\log\frac{\mu(a)}{\nu(a)}$ denotes the Kullback-Leibler divergence (a.k.a. relative entropy) between two distributions μ and ν. Let X and Y be two random variables, the standard notation $I(X,Y) := H(X) - H(X|Y)$ means the mutual information between X and Y, we also use $D(X||Y)$ to denote the divergence between the distributions of X and Y. For more on divergence and mutual information, see [CT12].

2.2   Communication Complexity and Information Complexity

We briefly review the notion of two-party communication complexity which was introduced by Yao [Yao79], see [KN97] for detailed definitions. In this model there are two players (with unlimited computational power), often called Alice and Bob, who wish to collaboratively compute a given function $f\colon \mathcal{X}\times\mathcal{Y}\to\mathcal{Z}$. Alice receives an input $x\in\mathcal{X}$ and Bob receives $y\in\mathcal{Y}$. Neither of them knows the other player's input, and they wish to communicate in accordance with an agreed-upon protocol π to compute $f(x,y)$. The protocol π specifies as a function of (only) the transmitted bits whether the communication is over, and if not, who sends the next bit. Furthermore π specifies what the next bit must be as a function of the transmitted bits and the input of the player who sends the bit. The *transcript Π* of a protocol π is the list of all the transmitted bits during the execution of the protocol. In the randomized communication model, the players have access to private random strings. These random strings are independent, and they can have any desired distributions individually.

While communication complexity cares about the number of transmitted bits, the information complexity cares about the *information* revealed by the communication, see [BBCR10] for a detailed discussion. To be able to measure information, we also need to assume that there is a prior distribution μ on $\mathcal{X}\times\mathcal{Y}$.

Definition 1. *The* external information cost *and the* internal information cost *of a protocol π with respect to a distribution μ on inputs from $\mathcal{X}\times\mathcal{Y}$ are defined as* $\mathrm{IC}^{ext}_\mu(\pi) = I(\Pi;XY)$ *and* $\mathrm{IC}_\mu(\pi) = I(\Pi;X|Y) + I(\Pi;Y|X)$, *respectively, where* $\Pi = \Pi_{XY}$ *is the transcript of the protocol when it is executed on the input XY.*

Given a function f and $\varepsilon \geqslant 0$, we define

$$\mathrm{IC}_\mu(f,\varepsilon) = \inf_\pi \mathrm{IC}_\mu(\pi), \qquad\qquad (2)$$

where π computes the value of $f(x,y)$ with error at most ε for *every* input (x,y), i.e., $\mathbf{Pr}[\pi(x,y)\neq f(x,y)] \leqslant \varepsilon$, for all $(x,y)\in\mathcal{X}\times\mathcal{Y}$. Given another distribution ν on $\mathcal{X}\times\mathcal{Y}$, define

$$\mathrm{IC}_\mu(f,\nu,\varepsilon) = \inf_\pi \mathrm{IC}_\mu(\pi), \qquad\qquad (3)$$

where π computes the value of $f(x,y)$ with error at most ε if the input (x,y) is sampled according to ν, i.e., $\mathbf{Pr}_{(x,y)\sim\nu}[\pi(x,y)\neq f(x,y)] \leqslant \varepsilon$.

In particular, $\mathrm{IC}_\mu(f,0)$ is the minimal information cost of all protocols that compute f correctly on *every* input, while $\mathrm{IC}_\mu(f,\mu,0)$ is the minimal information cost of all protocols that compute f correctly on the support of μ. Hence $\mathrm{IC}_\mu(f,\mu,0)$ can be strictly smaller than $\mathrm{IC}_\mu(f,0)$. Denote $\mathrm{IC}_\mu(f) = \mathrm{IC}_\mu(f,0)$.

Remark 1 (A warning regarding our notation). In the literature of information complexity it is common to use "$\mathrm{IC}_\mu(f, \varepsilon)$" to denote the distributional error case, i.e. what we denote by $\mathrm{IC}_\mu(f, \mu, \varepsilon)$. Unfortunately this has become the source of some confusions in the past, as sometimes "$\mathrm{IC}_\mu(f, \varepsilon)$" is used to denote both of the distributional error $[f, \mu, \varepsilon]$ and the point-wise error $[f, \varepsilon]$. To avoid ambiguity we distinguish these two cases by using different notations $\mathrm{IC}_\mu(f, \mu, \varepsilon)$ and $\mathrm{IC}_\mu(f, \varepsilon)$.

It turns out that $\mathrm{IC}_\mu(f, \varepsilon)$ and $\mathrm{IC}_\mu(f, \nu, \varepsilon)$ are both continuous with respect to ε. For our purpose we need the uniform continuity.

Theorem 1 ([BGPW16, Lemma 4.4]). *$\mathrm{IC}_\mu(f)$ is uniformly continuous with respect to μ, under the statistical distance of distributions.*

2.3   The Multi-party Number-In-Hand Model

The number-in-hand model is the most straightforward generalization of Yao's two-party model to the settings where more than two players are present. In this model there are k players who wish to collaboratively compute a function $f\colon \mathcal{X}_1 \times \ldots \times \mathcal{X}_k \to \mathcal{Z}$. The communication is in the shared blackboard model, which means that all the communicated bits are visible to all the players. Let μ be a probability distribution on $\mathcal{X}_1 \times \ldots \times \mathcal{X}_k$, and let $X = (X_1, \ldots, X_k)$ be sampled from $\mathcal{X}_1 \times \ldots \times \mathcal{X}_k$ according to μ. Definition 1 generalizes in a straightforward manner to $\mathrm{IC}_\mu^{ext}(\pi) = \mathrm{I}(\Pi; X)$, and $\mathrm{IC}_\mu(\pi) = \sum_{i=1}^k \mathrm{I}(\Pi; X_{-i}|X_i)$ where $X_{-i} := (X_1, \ldots, X_{i-1}, X_{i+1}, \ldots, X_k)$. Note also that $\mathrm{I}(\Pi; X|X_i) = \mathrm{I}(\Pi; X_{-i}|X_i)$, and thus we have $\mathrm{IC}_\mu(\pi) = \sum_{i=1}^k \mathrm{I}(\Pi; X|X_i)$. The notations $\mathrm{IC}_\mu(f)$, $\mathrm{IC}_\mu(f, \varepsilon)$, and $\mathrm{IC}_\mu(f, \nu, \varepsilon)$, and the continuity results also generalize in a straightforward manner to this setting.

3   The Local Characterization of the Optimal Information Cost

Let B be a random bit sent by one of the players, and let $\mu_0 = \mu|_{B=0}$ and $\mu_1 = \mu|_{B=1}$, or in other words $\mu_b(xy) := \mathbf{Pr}[XY = xy|B = b]$ for $b = 0, 1$. Denote $\mathbf{Pr}[\cdot|xy] := \mathbf{Pr}[\cdot|XY = xy]$.

Definition 2. *Let μ be a distribution and B be a signal sent by one of the players.*

- *B is called* unbiased *with respect to μ if $\mathbf{Pr}[B = 0] = \mathbf{Pr}[B = 1] = \frac{1}{2}$.*
- *B is called* non-crossing *if $\mu(xy) < \mu(x'y')$ implies that $\mu_b(xy) \leqslant \mu_b(x'y')$ for $b = 0, 1$.*
- *B is called ε-weak if $|\mathbf{Pr}[B = 0|xy] - \mathbf{Pr}[B = 1|xy]| \leqslant \varepsilon$ for every input xy.*

A protocol is said to be in normal form *with respect to μ if all its signals are unbiased and non-crossing with respect to μ.*

Let $\Delta(\mathcal{X} \times \mathcal{Y})$ denote the set of distributions on $\mathcal{X} \times \mathcal{Y}$. A measure $\mu \in \Delta(\mathcal{X} \times \mathcal{Y})$ is said to be *internal-trivial* (resp. *external-trivial*) for f if $\mathrm{IC}_\mu(f) = 0$ (resp. $\mathrm{IC}_\mu^{ext}(f) = 0$). These measures are characterized in [DFHL16].

3.1   The Local Characterization

Suppose that after a random bit B is sent, if $B = 0$, the players continue by running a protocol that is (almost) optimal for μ_0, and if $B = 1$, they run a protocol that is (almost) optimal for μ_1. Note that the amount of information that B reveals about the inputs to an external observer is $I(B; XY)$. This shows

$$\mathrm{IC}_\mu^{ext}(f) \leqslant I(B; XY) + \mathbb{E}_B[\mathrm{IC}_{\mu_B}^{ext}(f)], \tag{4}$$

and similarly

$$\mathrm{IC}_\mu(f) \leqslant I(B; X|Y) + I(B; Y|X) + \mathbb{E}_B[\mathrm{IC}_{\mu_B}(f)]. \tag{5}$$

In [BGPW13] it is shown that (4) and (5) essentially characterize $\mathrm{IC}_\mu^{ext}(f, \mu, 0)$ and $\mathrm{IC}_\mu(f, \mu, 0)$. Denote $\mathrm{I}_B^{ext} := I(B; XY)$, and $\mathrm{I}_B := I(B; X|Y) + I(B; Y|X)$.

Theorem 2 ([BGPW13]).   *Suppose that* $C\colon \Delta(\mathcal{X} \times \mathcal{Y}) \to [0, \log(|\mathcal{X} \times \mathcal{Y}|)]$ *satisfies*

(i) $C(\mu) = 0$ *for every measure* μ *such that* $\mathrm{IC}_\mu(f, \mu, 0) = 0$, *and*
(ii) *for every signal* B *that can be sent by one of the players*

$$C(\mu) \leqslant \mathrm{I}_B + \mathbb{E}_B[C(\mu_B)].$$

Then $C(\mu) \leqslant \mathrm{IC}_\mu(f, \mu, 0)$. *Similarly if* $\mathrm{IC}_\mu(f, \mu, 0)$ *is replaced by* $\mathrm{IC}_\mu^{ext}(f, \mu, 0)$, *and* I_B *is replaced by* I_B^{ext}, *then* $C(\mu) \leqslant \mathrm{IC}_\mu^{ext}(f, \mu, 0)$. *Furthermore, in both of the external and the internal cases, it suffices to verify (ii) only for non-crossing unbiased signals* B.

In light of Theorem 2, in order to determine the values of $\mathrm{IC}_\mu(f, \mu, 0)$, one has to first prove an upper bound by constructing a protocol (or a sequence of protocols) for every measure μ. Then it suffices to verify that the bound satisfies the conditions of Theorem 2.

A Counterexample: In [BGPW13], $\mathrm{IC}_\mu(\mathrm{AND}_2)$ is determined using Theorem 2. However, the proof presented in [BGPW13] contains a small gap. It is claimed that it suffices to verify (2) for sufficiently weak signals. While it is not difficult to see that indeed it suffices to verify (2) for signals B which are ε-weak for an absolute constant $\varepsilon > 0$, in [BGPW13] the condition (ii) is only verified for ε that is smaller than a function of μ. This is not sufficient, and one can easily construct a counter-example by allowing the signal B to become increasingly weaker as μ moves closer to the boundary (by boundary we mean the set of measures μ that satisfy Theorem 2 (i)). Indeed, for example, set $C(\mu) = K$ for a very large constant K if μ does not satisfy Theorem 2 (i), and otherwise set $C(\mu) = 0$. Obviously (2) holds if μ is on the boundary. On the other hand, if μ is not on the boundary, then by taking B to be sufficiently weak as a function of μ, we can guarantee that μ_0 and μ_1 are not on the boundary either, and thus

(2) holds in this case as well. However, taking K to be sufficiently large violates the desired conclusion that $C(\mu) \leqslant \mathrm{IC}_\mu(f, \mu, 0)$.

To fix this error, we start by noting that one can obtain a similar characterization of $\mathrm{IC}_\mu(f)$ (see the full version paper). Using the uniform continuity of $\mathrm{IC}_\mu(f)$ with respect to μ, we prove that it suffices to verify Condition (ii) for signals B that are weaker than quantities that can depend on μ. This as we shall see suffices to fix the proof of [BGPW13]. The proof of Theorem 3 is presented in Sect. 3.3.

Theorem 3 (Main Theorem 1). *Let $w \colon (0,1] \to (0,1]$ be a non-decreasing function, $\Omega \subseteq \Delta(\mathcal{X} \times \mathcal{Y})$ be a subset of measures containing the internal trivial distributions for function f. Let $\delta(\mu)$ denote the distance of μ from Ω. Suppose that $C \colon \Delta(\mathcal{X} \times \mathcal{Y}) \to [0, \log(|\mathcal{X} \times \mathcal{Y}|)]$ satisfies*

(i) $C(\mu)$ is uniformly continuous with respect to μ;
(ii) $C(\mu) = \mathrm{IC}_\mu(f)$ if $\delta(\mu) = 0$, and
(iii) for every non-crossing unbiased $w(\delta(\mu))$-weak signal B that can be sent by one of the players,

$$C(\mu) \leqslant \mathrm{I}_B + \mathbb{E}_B[C(\mu_B)]. \tag{6}$$

Then $C(\mu) \leqslant \mathrm{IC}_\mu(f)$. Similarly, if we replace Ω by a subset containing the external trivial distributions for function f, in Condition (ii) replace $\mathrm{IC}_\mu(f)$ by $\mathrm{IC}_\mu^{ext}(f)$, and in Condition (iii) replace I_B by I_B^{ext}, then $C(\mu) \leqslant \mathrm{IC}_\mu^{ext}(f)$.

3.2   Communication Protocols as Random Walks on $\Delta(\mathcal{X} \times \mathcal{Y})$

Consider a protocol π and a prior distribution μ on $\mathcal{X} \times \mathcal{Y}$. Suppose that in the first round Alice sends a random signal B to Bob, we can interpret this as a random update of the prior distribution μ to a new distribution $\mu_0 := \mu|_{B=0}$ or $\mu_1 := \mu|_{B=1}$ depending on the value of B. Similarly if Bob is sending a signal. Therefore, we can think of a protocol as a random walk on $\Delta(\mathcal{X} \times \mathcal{Y})$ that starts at μ, and every time that a player sends a signal, it moves to a new distribution. The random walk terminates at $\mu_\Pi := \mu|_\Pi$. Since Π is random variable, μ_Π is also a random variable taking value in $\Delta(\mathcal{X} \times \mathcal{Y})$. Recall $\mathrm{I}(X; \Pi|Y) = \mathbb{E}_{\pi \sim \Pi, y \sim Y} \mathrm{D}(X|_{\Pi=\pi, Y=y} \| X|_{Y=y})$ and $\mathrm{I}(XY; \Pi) = \mathbb{E}_{\pi \sim \Pi} \mathrm{D}(XY|_{\Pi=\pi} \| XY)$, this shows, interestingly, the distribution of μ_Π (which is a distribution on the set $\Delta(\mathcal{X} \times \mathcal{Y})$) completely determines the information cost of the protocol π.

Proposition 1 [BS15]. *Let π and τ be two communication protocols with the same input set $\mathcal{X} \times \mathcal{Y}$ endowed with a probability measure μ. Let Π and T denote the transcripts of π and τ, respectively. If μ_Π has the same distribution as μ_T, then $\mathrm{IC}_\mu(\pi) = \mathrm{IC}_\mu(\tau)$ and $\mathrm{IC}_\mu^{ext}(\pi) = \mathrm{IC}_\mu^{ext}(\tau)$.*

Let $\mathcal{C}_\mu^T(\Delta(\mathcal{X} \times \mathcal{Y}))$ denote the set of all probability distributions on $\Delta(\mathcal{X} \times \mathcal{Y})$ that can be obtained, starting from the distribution μ, through communication protocols that perform a given communication task T. The information cost of performing the task T is the infimum of the information costs of the distributions in $\mathcal{C}_\mu^T(\Delta(\mathcal{X} \times \mathcal{Y}))$. Although this infimum is not always attained (see [BGPW13]), if one takes the closure of $\mathcal{C}_\mu^T(\Delta(\mathcal{X} \times \mathcal{Y}))$ (under weak convergence) then one can replace the infimum with minimum. For the 2-bit AND function, the buzzers protocol of [BGPW13] yields the distribution in the closure of $\mathcal{C}_\mu^T(\Delta(\mathcal{X} \times \mathcal{Y}))$ that achieves the minimum information cost. We believe that the following is an important open problem.

Problem 1 Define a paradigm such that for every communication task T and every measure μ on an input set $\mathcal{X} \times \mathcal{Y}$, the set of distributions on $\Delta(\mathcal{X} \times \mathcal{Y})$ resulting from the protocols performing the task T in this paradigm is exactly equal to the closure of $\mathcal{C}_\mu^T(\Delta(\mathcal{X} \times \mathcal{Y}))$.

Partial progress on this problem has been made in [DF16, DFHL16].

3.3   Proof of Theorem 3

We present the proof for the internal case only, as the external case is similar. We need two technical lemmas (see proofs in the full version paper). The signal simulation lemma says every signal can be perfectly simulated by a non-crossing unbiased ε-weak signal sequence, it generalizes [BGPW13, Lemma 5.2].

Lemma 1 (Signal Simulation). *Let $\varepsilon > 0$, and consider $\mu \in \Delta(\mathcal{X} \times \mathcal{Y})$ and a signal B sent by one of the players. There exists a sequence of non-crossing unbiased ε-weak random signals $\mathcal{B} = (B_1 B_2 \ldots)$ that with probability 1 terminates, and furthermore $\mu|_\mathcal{B}$ has the same distribution as $\mu|_B$.*

Lemma 2. *Let $w, \delta(\mu)$ and C be as in Theorem 3, and suppose C satisfies Conditions (i), (ii) and (iii). Let τ be a protocol that terminates with probability 1, and further assume τ is in normal form and every signal sent in τ is ε-weak. Given a probability distribution $\mu \in \Delta(\mathcal{X} \times \mathcal{Y})$, for every node u in the protocol tree of τ, let μ_u be the probability distribution conditioned on the event that the protocol reaches u. If μ satisfies $w(\delta(\mu_u)) \geqslant \varepsilon$ for every internal node u, then*

$$C(\mu) \leqslant \mathrm{IC}_\mu(\tau) + \underset{\ell}{\mathbb{E}}[C(\mu_\ell)],$$

where the expected value is over all leaves ℓ of τ chosen according to the distribution (on the leaves) when the inputs are sampled according to μ.

Proof of Theorem 3. Firstly by (ii), $\delta(\mu) = 0$ implies $C(\mu) = \mathrm{IC}_\mu(f) \leqslant \mathrm{IC}_\mu(f)$. Hence assume $\delta(\mu) > 0$. Consider an arbitrary signal B sent by Alice. As we discussed before, one can interpret B as a one step random walk that starts at μ and jumps either to μ_0 or to μ_1 with corresponding probabilities $\mathbf{Pr}[B = 0|X = x]$ and $\mathbf{Pr}[B = 1|X = x]$. The idea behind the proof is to use Lemma 1 to simulate this random jump with a random walk that has smaller steps so that we can apply the concavity assumption of the theorem to those steps.

Let π be a protocol that computes $[f, 0]$. For $0 < \eta < \delta(\mu)$, applying Lemma 1 one gets a new protocol $\tilde{\pi}$ by replacing every signal sent in π with a random walk consisting of $w(\eta)$-weak non-crossing unbiased signals. Note $\tilde{\pi}$ terminates with probability 1. Moreover, since $\tilde{\pi}$ is a perfect simulation of π, by Proposition 1 we have $\mathrm{IC}_\mu(\pi) = \mathrm{IC}_\mu(\tilde{\pi})$.

For every node v in the protocol tree of $\tilde{\pi}$, let μ_v be the measure μ conditioned on the event that the protocol reaches the node v. Obtain τ from $\tilde{\pi}$ by terminating at every node v that satisfies $\delta(\mu_v) \leqslant \eta$. Note that by the construction, Condition (iii) is satisfied on every *internal* node v of τ, as every such node satisfies $\eta < \delta(\mu_v)$, thus $w(\eta) \leqslant w(\delta(\mu_v))$ implying the signal sent on node v is $w(\delta(\mu_v))$-weak. Hence by Lemma 2,

$$C(\mu) \leqslant \mathrm{IC}_\mu(\tau) + \mathbb{E}_\ell[C(\mu_\ell)],$$

where the expected value is over all leaves of τ. For every μ_ℓ, let $\mu'_\ell \in \Omega$ be a distribution such that $\delta(\mu_\ell) = |\mu_\ell - \mu'_\ell|$. By Conditions (i) and (ii), and the uniform continuity of $\mathrm{IC}_\mu(f)$, we have that for every $\varepsilon > 0$ there exists $\eta > 0$, such that for all μ_ℓ, as long as $\delta(\mu_\ell) = |\mu_\ell - \mu'_\ell| \leqslant \eta$, then

$$C(\mu_\ell) \leqslant C(\mu'_\ell) + \varepsilon = \mathrm{IC}_{\mu'_\ell}(f) + \varepsilon \leqslant \mathrm{IC}_{\mu_\ell}(f) + \varepsilon + \varepsilon = \mathrm{IC}_{\mu_\ell}(f) + 2\varepsilon.$$

As a result, $C(\mu) \leqslant \mathrm{IC}_\mu(\tau) + \mathbb{E}_\ell[\mathrm{IC}_{\mu_\ell}(f) + 2\varepsilon] = \mathrm{IC}_\mu(\tau) + \mathbb{E}_\ell[\mathrm{IC}_{\mu_\ell}(f)] + 2\varepsilon$. Since μ_ℓ is generated by truncating $\tilde{\pi}$, we have $\mathrm{IC}_\mu(\tau) + \mathbb{E}_\ell[\mathrm{IC}_{\mu_\ell}(f)] \leqslant \mathrm{IC}_\mu(\tilde{\pi}) = \mathrm{IC}_\mu(\pi)$. Therefore $C(\mu) \leqslant \mathrm{IC}_\mu(\pi) + 2\varepsilon$. As this holds for arbitrary ε, we must have $C(\mu) \leqslant \mathrm{IC}_\mu(\pi)$. $\qquad\qquad\square$

4   The Multi-party AND Function in the Number-In-Hand Model

We now show that the buzzers protocol proposed in [BGPW13] can be generalized to the multi-party number-in-hand setting. Denote $\mu_x := \mu(\{x\})$ for every $x \in \{0,1\}^k$, and assume that $\mu_{e_1} \leqslant \ldots \leqslant \mu_{e_k}$.

- There is a clock whose time starts at 0 and increases continuously to $+\infty$.
- Let $t_i := \ln \frac{\mu_{e_i}}{\mu_{e_1}}$ for $i = 1, \ldots, k$, and let $t_{k+1} := \infty$.
- For every $i = 1, \ldots, k$, if $x_i = 0$ then the i-th player privately picks an independent random variable T_i with exponential distribution with parameter $\lambda = 1$, and if time reaches $t_i + T_i$, the player announces that his/her input is 0, and the protocol terminates immediately with all the players knowing that $\bigwedge_{i=1}^k x_i = 0$.
- If the clock reaches $+\infty$ without any player announcing their input, the players will know that $\bigwedge_{i=1}^k x_i = 1$.

Fig. 1. The protocol $\pi_\mu^\wedge$ for solving the multi-party AND function on a distribution μ.

For every $i \in [k]$, if $x_i = 0$ then the i-th player activates a buzzer at time t_i and becomes "active". The protocol terminates with $\bigwedge_{i=1}^k x_i = 0$ once the first buzz happens, otherwise the time reaches ∞ without anyone buzzing, and they decide $\bigwedge_{i=1}^k x_i = 1$.

Theorem 4 (Main Theorem 2). *For every μ satisfying Assumption 1, the protocol $\pi_\mu^\wedge$ has the smallest external and internal information cost.*

Observe that a direct calculation as in [BGPW13] is now simply impossible due to the nature of multi-party setting where the number of players k can be arbitrary. We have to reduce the problem to one that has only a constant number of variables before a computation can be performed.

4.1 Setting up

Let μ be a measure satisfying Assumption 1, and $X = (X_1, \ldots, X_k)$ be the random k-bit input. Let Π be the transcript of protocol $\pi_\mu^\wedge$, and $\Pi_x = \Pi|_{X=x}$.

To verify the concavity condition, consider an unbiased signal B with parameter ε sent by the player s. That is $\mathbf{Pr}[B = 0|X_s = 0] = \frac{1+\varepsilon\mathbf{Pr}[X_s=1]}{2}$ and $\mathbf{Pr}[B = 1|X_s = 1] = \frac{1+\varepsilon\mathbf{Pr}[X_s=0]}{2}$. Let μ^0 and μ^1 denote the distributions of $X^0 := X|_{B=0}$ and $X^1 := X|_{B=1}$, then $\mu = \frac{\mu^0+\mu^1}{2}$. Let Π^0 and Π^1 denote the random variables corresponding to the transcripts of $\pi_{\mu^0}^\wedge$ and $\pi_{\mu^1}^\wedge$, respectively.

Note Π contains the termination time t, and if $t < \infty$, also the name of the player who first buzzed. We denote by π_∞ the transcript corresponding to termination time $t = \infty$, and by π_t^m the termination time $t < \infty$ with the m-th player buzzing. For $t \in [0, \infty)$, let $\Phi_x(t)$ denote the total amount of active time spent by all players before time t if the input is x. For $t_r \leqslant t < t_{r+1}$, we have $\Phi_x(t) = \sum_{i:x_i=0} \max(t - t_i, 0) = \sum_{i\in[1,r],x_i=0}(t - t_i)$. The probability density function f_x of Π_x satisfies $f_x(\pi_t^m) = 0$ if $t_m > t$ or $x_m = 1$, and equals to $e^{-\Phi_x(t)}$ otherwise. Also $\mathbf{Pr}(\Pi_1 = \pi_\infty) = 1$. The distribution of the transcript Π is then $f(\pi_t^m) = \sum_x \mu_x f_x(\pi_t^m)$. Define f^0, f_x^0 and f^1, f_x^1 analogously for $\pi_{\mu^0}^\wedge$ and $\pi_{\mu^1}^\wedge$.

Denote $\beta_s := \mathbf{Pr}[X_s = 1]$, and $\zeta_s := \mathbf{Pr}[X_s = 0]$. For $B = 0$, the new starting times are $t_i^0 = t_i$ for $i \neq s$, and $t_s^0 = t_s - \gamma_0$ where $\gamma_0 = \ln\left(\frac{1+\varepsilon\beta_s}{1-\varepsilon\zeta_s}\right)$. On the other hand, for $B = 1$, we have the new starting times are $t_i^1 = t_i$ for $i \neq s$, and $t_s^1 = t_s + \gamma_1$ where $\gamma_1 = \ln\left(\frac{1+\varepsilon\zeta_s}{1-\varepsilon\beta_s}\right)$.

Let $\phi(x) := x\ln(x)$. The concavity conditions of Theorem 3 reduce to

$$\int_{-\infty}^{\infty} \sum_m \left(\phi(f(\pi_t^m)) - \frac{\phi(f^0(\pi_t^m)) + \phi(f^1(\pi_t^m))}{2}\right)$$
$$-\sum_m \sum_x \left(\phi(\mu_x f_x(\pi_t^m)) - \frac{\phi(\mu_x^0 f_x^0(\pi_t^m)) + \phi(\mu_x^1 f_x^1(\pi_t^m))}{2}\right) dt \geqslant 0, \quad (7)$$

for the external case, and

$$\sum_{j=1}^{k} \int_{-\infty}^{\infty} \sum_m \sum_{b=0}^{1} \left(\phi(f_{x_j=b}(\pi_t^m)) - \frac{\phi(f_{x_j=b}^0(\pi_t^m)) + \phi(f_{x_j=b}^1(\pi_t^m))}{2}\right)$$
$$-\sum_m \sum_x \left(\phi(\mu_x f_x(\pi_t^m)) - \frac{\phi(\mu_x^0 f_x^0(\pi_t^m)) + \phi(\mu_x^1 f_x^1(\pi_t^m))}{2}\right) dt \geqslant 0, \quad (8)$$

for the internal case, where $f_{x_j=b}(\pi_t^m) := \sum_{X:X_j=b} \mu_X f_X(\pi_t^m)$, and $f_{x_j=b}^0(\pi_t^m) := \sum_{X:X_j=b} \mu_X^0 f_X^0(\pi_t^m)$, $f_{x_j=b}^1(\pi_t^m) := \sum_{X:X_j=b} \mu_X^1 f_X^1(\pi_t^m)$.

Denote the function inside the integral of (8) by $\mathrm{concav}_\mu(t, j)$, and the function inside the integral of (7) by $\mathrm{concav}_\mu^{ext}(t)$. Hence to prove Theorem 4 it suffices to verify $\int_{-\infty}^{\infty} \mathrm{concav}_\mu^{ext}(t)dt \geqslant 0$ and $\sum_{j=1}^{k} \int_{-\infty}^{\infty} \mathrm{concav}_\mu(t, j)dt \geqslant 0$.

4.2 Reductions

Our first reduction is to deal with the uniform distribution on $e_1, \ldots, e_k$. In this measure all players are active at $t = 0$ and the protocol stops if one play buzzes or at infinity. For technical reasons, we need to analyze this particular measure separately. Later we will put this μ into Ω as in Theorem 3.

Statement 1. *Let μ be the measure $\mu_{e_1} = \ldots = \mu_{e_k} = 1/k$. The internal and external information cost of the protocol $\pi^{\wedge}$ is optimal with respect to μ.*

Observe that all the information is revealed if the input **1**, this degeneracy can be eliminated.

Statement 2. *It suffices to assume μ satisfying $\mu(\mathbf{1}) = 0$.*

Using the memoryless property of exponential distribution, one can shift the activation time of all the players by $-\ln(\mu_{e_s}/\mu_{e_1})$, and assume that $t_1 = -\ln(\mu_{e_s}/\mu_{e_1}), \ldots, t_s = 0, \ldots, t_k = \ln(\mu_{e_k}/\mu_{e_s})$. Our third major reduction says that it suffices to focus on the time interval $t \in [-\gamma_0, \gamma_1]$ instead of $(-\infty, \infty)$. This is achieved by showing that $\int_{-\infty}^{-\gamma_0} \mathrm{concav}_\mu^{ext}(t)dt \geqslant 0$ and $\int_{\gamma_1}^{\infty} \mathrm{concav}_\mu^{ext}(t)dt = 0$. Similarly for the internal case.

Statement 3. *It suffices to assume μ satisfies $\mu(\mathbf{1}) = 0$, and verify $\int_{-\gamma_0}^{\gamma_1} \mathrm{concav}_\mu^{ext}(t)dt \geqslant 0$ and $\sum_{j=1}^{k} \int_{-\gamma_0}^{\gamma_1} \mathrm{concav}_\mu(t, j)dt \geqslant 0$.*

Lastly we reduce all the distributions into a class of essentially two-parameter distributions. Firstly we observe that conditioned on the buzz time $t \in [-\gamma_0, \gamma_1]$, we have $\mu_{e_1}|_{t \geqslant t_s - \gamma_0} = \cdots = \mu_{e_{s-1}}|_{t \geqslant t_s - \gamma_0}$. Secondly, we show that one can *transfer* the mass on those e_j such that $\mu_{e_j} > \mu_{e_s}$ to μ_0, without harming the concavity condition we aimed to verify.

Statement 4. *It suffices to assume μ satisfies $\mu_{e_1} = \cdots = \mu_{e_{s-1}} = \beta$, $\mu_{e_s} = \cdots = \mu_{e_k} = e^{\gamma_0}\beta$, and $\mu_0 = 1 - (s-1)\beta - (k - s + 1)e^{\gamma_0}\beta$, where $0 < \beta < 1$.*

4.3 Proof of Theorem 4

Set Ω to be the set of all external (resp. internal) trivial measures together with the measure in Statement 1, obviously Condition (i) and (ii) are satisfied.

External Information Cost. Set $w(x) = ck^{-20}x^4$ for some fixed constant $c > 0$. Using Wolfram Mathematica, we obtain

$$(7) \geqslant \frac{(k + 5s - 6)(1 - 2\beta)\beta}{12(1 - \beta)\ln 2}\varepsilon^3 + O(\varepsilon^4). \tag{9}$$

Using the bound of the error term given in the full version paper, one can verify the concavity condition (7) is satisfied for all $w(\delta(\mu))$-weak signals as long as $\varepsilon < ck^{-20}\min\{\beta, 1 - k\beta\}^3$. Hence by Theorem 3, the protocol is optimal for external information cost. More details are presented in the full version paper.

Internal Information Cost. Similarly, using Wolfram Mathematica, we obtain

$$(8) \geqslant \begin{cases} \frac{(k+5s-6)(1-2\beta)\beta}{12(1-\beta)\ln 2}\varepsilon^3 + O(\varepsilon^4), & k = 2, \\ \frac{(k+5s-6)((3k-2)\beta^2-4(k-1)\beta+k-1)\beta}{12(1-\beta)(1-2\beta)\ln 2}\varepsilon^3 + O(\varepsilon^4), & k \geqslant 3. \end{cases} \tag{10}$$

Similarly one can pick an appropriate function w to verify that the concavity condition (8) is satisfied for all $w(\delta(\mu))$-weak signals when ε is sufficiently small. Hence by Theorem 3, the protocol is optimal for internal information cost.

More detailed analysis can be found in the full version of our paper.

References

[BBCR10] Barak, B., Braverman, M., Chen, X., Rao, A.: How to compress interactive communication [extended abstract]. In: STOC 2010, pp. 67–76. ACM, New York (2010)

[BGPW13] Braverman, M., Garg, A., Pankratov, D., Weinstein, O.: From information to exact communication (extended abstract). In: STOC 2013, pp. 151–160. ACM, New York (2013)

[BGPW16] Braverman, M., Garg, A., Pankratov, D., Weinstein, O.: Information lower bounds via self-reducibility. Theory Comput. Syst. **59**(2), 377–396 (2016)

[BR14] Braverman, M., Rao, A.: Information equals amortized communication. IEEE Trans. Inform. Theory **60**(10), 6058–6069 (2014)

[BS15] Braverman, M., Schneider, J.: Information complexity is computable, arXiv preprint arXiv:1502.02971 (2015)

[CT12] Cover, T.M., Thomas, J.A.: Elements of Information Theory. Wiley, New York (2012)

[DF16] Dagan, Y., Filmus, Y.: Grid protocols (2016, in preparation)

[DFHL16] Dagan, Y., Filmus, Y., Hatami, H., Li, Y.: Trading information complexity for error, arXiv preprint arXiv:1611.06650 (2016)

[Gro09] Gronemeier, A.: Asymptotically optimal lower bounds on the nih-multiparty information, arXiv preprint arXiv:0902.1609 (2009)

[KN97] Kushilevitz, E., Nisan, N.: Communication Complexity. Cambridge University Press, Cambridge (1997)

[MI11] Ma, N., Ishwar, P.: Some results on distributed source coding for interactive function computation. IEEE Trans. Inform. Theory **57**(9), 6180–6195 (2011)

[MI13] Ma, N., Ishwar, P.: The infinite-message limit of two-terminal interactive source coding. IEEE Trans. Inform. Theory **59**(7), 4071–4094 (2013)

[Raz92] Razborov, A.A.: On the distributional complexity of disjointness. Theoret. Comput. Sci. **106**(2), 385–390 (1992)

[Yao79] Yao, A.C.-C.: Some complexity questions related to distributive computing (preliminary report). In: STOC 1979, pp. 209–213. ACM (1979)

Optimal Online Two-Way Trading
with Bounded Number of Transactions

Stanley P.Y. Fung[(✉)]

Department of Informatics, University of Leicester, Leicester LE1 7RH, UK
pyf1@leicester.ac.uk

Abstract. We consider a two-way trading problem, where investors buy and sell a stock whose price moves within a certain range. Naturally they want to maximize their profit. Investors can perform up'to k trades, where each trade must involve the full amount. We give optimal algorithms for three different models which differ in the knowledge of how the price fluctuates. In the first model, there are global minimum and maximum bounds m and M. We first show an optimal lower bound of φ (where $\varphi = M/m$) on the competitive ratio for one trade, which is the bound achieved by trivial algorithms. Perhaps surprisingly, when we consider more than one trade, we can give a better algorithm that loses only a factor of $\varphi^{2/3}$ (rather than φ) per additional trade. Specifically, for k trades the algorithm has competitive ratio $\varphi^{(2k+1)/3}$. Furthermore we show that this ratio is the best possible by giving a matching lower bound. In the second model, m and M are not known in advance, and just φ is known. We show that this only costs us an extra factor of $\varphi^{1/3}$, i.e., both upper and lower bounds become $\varphi^{(2k+2)/3}$. Finally, we consider the bounded daily return model where instead of a global limit, the fluctuation from one day to the next is bounded, and again we give optimal algorithms, and interestingly one of them resembles common trading algorithms that involve stop loss limits.

1 Introduction

The model. We consider a scenario commonly faced by investors. The price of a stock varies over time. In this paper we use a 'day' as the smallest unit of time, so there is one new price each day. Let $p(i)$ be the price at day i. The investor has some initial amount of money. Over a time horizon of finite duration, the investor wants to make a bounded number of trades of this one stock. Each trade (b, s) consists of a buy transaction at day b, followed by a sell transaction at day s where $s > b$. (Thus one trade consists of two transactions.) Both transactions are 'all-in': when buying, the investor uses all the money available, and when selling all stock they currently own is sold. A sale must be made before the next purchase can take place. Also, no short selling is allowed, i.e., there can be no selling if the investor is not currently holding stock. When the end of the time horizon is reached, i.e., on the last day, no buying is allowed and the investor must sell off all the stocks that they still hold back to cash at the price of the day.

© Springer International Publishing AG 2017
Y. Cao and J. Chen (Eds.): COCOON 2017, LNCS 10392, pp. 212–223, 2017.
DOI: 10.1007/978-3-319-62389-4_18

There are a number of rationales for considering a bounded number of trades and/or that trades must involve all the money available. Individual, amateur investors typically do not want to make frequent transactions due to high transaction fees. Often transaction fees have a fixed component (i.e., a fixed amount or a minimum tariff per transaction, irrespective of the trading amount) which makes transaction fees disproportionally high for small trades. Frequent trading also requires constant monitoring of the markets which amateur investors may not have the time or resources for; often they only want to change their investment portfolios every now and then. Also, for investors with little money available, it is not feasible or sensible to divide them into smaller pots of money, in arbitrary fractions as required by some algorithms. The finiteness of the time horizon (and that its length is possibly unknown as well) corresponds to situations where an investor may be forced to sell and leave the market due to unexpected need for money elsewhere, for example.

Each trade with a buying price of $p(b)$ and a selling price of $p(s)$ gives a *gain* of $p(s)/p(b)$. This represents how much the investor has after the trade if they invested 1 dollar in the beginning. Note that this is a ratio, and can be less than 1, meaning there is a loss, but we will still refer to it as a 'gain'. If a series of trades are made, the overall gain or the *return* of the algorithm is the product of the gains of each of the individual trades. This correctly reflects the fact that all the money after each trade is re-invested in the next.

Since investors make decisions without knowing future stock prices, the problem is *online* in nature. We measure the performance of online algorithms with *competitive analysis*, i.e., by comparing it with the optimal offline algorithm OPT that knows the price sequence in advance and can therefore make optimal decisions. The *competitive ratio* of an online algorithm ONL is the worst possible ratio of the return of OPT to the return of ONL, over all possible input (price) sequences. The multiplicative nature of the definition of the return (instead of specifying a negative value for a loss) means that the competitive ratio can be computed in the normal way in the case of a loss: for example, if OPT makes a gain of 2 and ONL makes a 'gain' of $1/3$, then the competitive ratio is 6.

Three models on the knowledge of the online algorithm. We consider three different models on how the price changes, or equivalently, what knowledge the online algorithm has in advance. In the first model, the stock prices are always within a range $[m..M]$, i.e., m is the minimum possible price and M the maximum possible price. Both m and M are known to the online algorithm up front. In the second model, the prices still fluctuate within this range, but m and M are not (initially) known; instead only their ratio $\varphi = M/m$, called the *fluctuation ratio*, is known. In both these models the length of the time horizon (number of days) is unknown (until the final day arrives). In the third model, called the *bounded daily return model*, there is no global minimum or maximum price. Instead, the maximum fluctuation from day to day is bounded: namely, the price $p(i + 1)$ of the next day is bounded by the price $p(i)$ of the current day by $p(i)/\beta \leq p(i + 1) \leq \alpha p(i)$ for some $\alpha, \beta > 1$. This means the prices cannot suddenly change a lot. Many stock markets implement the so-called 'circuit breakers' where trading is stopped when such limits are reached. Here α, β and the trade duration T are known to the online algorithm. All three models are well-established in the algorithms literature; see e.g. [1,5].

Previous results and related work. Financial trading and related problems are obviously important topics and have been much studied from the online algorithms perspective. A comprehensive survey is given in [8]. Here we only sample some of the more important results and those closer to the problems we study here. In the *one-way search problem*, the online player chooses one moment of time to make a single transaction from one currency to another currency. Its return is simply the price at which the transaction takes place. A reservation price (RP) based policy is to buy as soon as the price reaches or goes above a pre-set *reservation price*. It is well-known that, if m and M are known, the RP policy with a reservation price of $\sqrt{Mm}$ is optimal and achieves a competitive ratio of $\sqrt{\varphi}$. If only φ is known, then no deterministic algorithm can achieve a ratio better than φ. With the help of randomization, however, a random mix of different RPs gives a competitive ratio of $O(\log \varphi)$ if φ is known. Even if φ is not known, a competitive ratio of $O(\log \varphi \cdot \log^{1+\epsilon}(\log \varphi))$ can be achieved. See [5] for all the above results and more discussions.

In the *one-way trading problem*, the objective is again to maximize the return in the other currency, but there can be multiple transactions, i.e., not all the money has to be traded in one go. (This distinction of terminology between search and trading is used in [5], but is called non-preemptive vs. preemptive in [8]. We prefer calling them *unsplittable* vs. *splittable* here.) The relation between one-way trading and randomized algorithms for one-way search is described in [5]. Many variations of one-way search or one-way trading problems have since been studied; some examples include the bounded daily return model [1,11], searching for k minima/maxima instead of one [7], time-varying bounds [4], unbounded prices [2], search with advice complexity [3], etc.

What we study here, however, is a *two-way* version of the unsplittable trading problem[1], which is far less studied. Here the online player has to first convert from one currency (say cash) to another (a stock), hopefully at a low price, and then convert back from the stock to cash at some later point, hopefully at a high price. All the money must be converted back to the first currency when or before the game ends. This model is relevant where investors are only interested in short term, speculative gains. For the models with known m, M or known φ and with one trade, Schmidt et al. [9] gave a φ-competitive algorithm; it uses the same RP for buying and selling. But consider the DO-NOTHING algorithm that makes no trades at all. Clearly it is also φ-competitive as ONL's gain is 1 and OPT's gain is at most φ (if the price goes from m to M). A number of common algorithms, such as those based on moving averages, were also studied in [8]. It was shown that they are φ^2-competitive (and not better), which are therefore even worse. It is easy to show that these algorithms have competitive

---

[1] In the terminology of [5] this should be called '*two-way search*', but we feel that the term does not convey its application in stock market trading.

ratios φ^k and φ^{2k} respectively when extended to k trades. Schroeder et al. [10] gave some algorithms for the bounded daily return model, without limits on the number of trades. However, most of these algorithms tend to make decisions that are clearly bad, have the worst possible performance (like losing by the largest possible factor every day throughout), and have competitive ratios no better than what is given by DO-NOTHING.

Our results. In this paper we consider the two-way unsplittable trading problem where a bounded number k of trades are permitted, and derive optimal algorithms. First we consider the model with known m and M. We begin by considering the case of $k = 1$. Although some naive algorithms are known to be φ-competitive and seemingly nothing better is possible, we are not aware of any matching general lower bound. We give a general lower bound of φ, showing that the naive algorithms cannot be improved. The result is also needed in subsequent lower bound proofs.

It may be tempting to believe that nothing can beat the naive algorithm also for more trades. Interestingly, we prove that for $k \geq 2$ this is not true. While naive algorithms like DO-NOTHING are no better than φ^k-competitive, we show that a reservation price-based algorithm is $\varphi^{(2k+1)/3}$-competitive. For example, when $k = 2$, it is $\varphi^{5/3}$-competitive instead of trivially φ^2-competitive. Furthermore, we prove a matching lower bound, showing that the algorithm is optimal.

Next, we consider the model where only φ is known, and give an algorithm with a competitive ratio of $\varphi^{(2k+2)/3}$, i.e., only a factor $\varphi^{1/3}$ worse than that of the preceding model. Again we show that this bound is optimal.

Finally we consider the bounded daily return model, and give two optimal algorithms where the competitive ratio depends on α, β and T. For example, with one trade and in the symmetric case $\alpha = \beta$, the competitive ratio is $\alpha^{2T/3}$. While this is exponential in T (which is unavoidable), naive algorithms could lose up to a factor of $\max(\alpha, \beta)$ every day, and DO-NOTHING has a competitive ratio of α^T. One of the algorithms uses the 'stop loss / lock profit' strategy commonly used in real trading; as far as we are aware, this is the first time where competitive analysis justifies this common stock market trading strategy, and in fact suggests what the stop loss limit should be.

Some proofs are omitted due to space constraints.

2   Known m and M

In this section, where we consider the model with known m and M, we can without loss of generality assume that $m = 1$. This is what we will do to simplify notations. It also means M and φ are equal and are sometimes used interchangeably.

Theorem 1. *For $k = 1$, no deterministic algorithm has a competitive ratio better than $\varphi^{1-\epsilon}$, for any $\epsilon > 0$.*

Proof. Choose $n = \lceil 1/\epsilon \rceil$ and define $v_i = M^{i/n}$ for $i = 0, 1, \ldots, n$. The following price sequence is released until ONL buys: $v_{n-1}, M, v_{n-2}, M, \ldots, v_i, M, \ldots, v_1$, M, v_0. If ONL does not buy at any point, or buys at price M, then its return is at most 1. Then OPT buys at v_1 and sells at M to get a return of $M^{1-1/n}$. So suppose ONL buys at v_i for some $1 \leq i \leq n-1$. (It cannot buy at v_0 as it is the last time step.) As soon as ONL bought, the rest of the sequence is not released; instead the price drops to m and the game ends. ONL's return is m/v_i. If $i = n-1$, then OPT makes no trade and its return is 1, so competitive ratio $= v_{n-1}/m = M^{1-1/n}$. Otherwise, if $i < n-1$, OPT buys at v_{i+1} (two days before ONL's purchase) and sells at the next day at price M, giving a return of M/v_{i+1}. The competitive ratio is therefore $Mv_i/(mv_{i+1}) = M^{1-1/n}$.

Thus in all cases the competitive ratio is at least $M^{1-1/n} \geq \varphi^{1-\epsilon}$. $\square$

Note that the proof does not require ONL to use only one trade: it cannot benefit even if it is allowed to use more trades. This fact will be used later in Theorems 3 and 5.

For $k > 1$, we analyze the following algorithm:

Algorithm 1. The reservation price algorithm, with known price range $[m..M]$

Upon release of the i-th price $p(i)$:
if $i = n$ **then**
  if currently holding stock **then**
    sell at price $p(n)$ and the game ends
else
  if $p(i) \leq M^{1/3}$ and currently not holding stock and not used up all trades **then**
    buy at price $p(i)$
  else if $p(i) \geq M^{2/3}$ and currently holding stock **then**
    sell at price $p(i)$

Theorem 2. *Algorithm 1 has competitive ratio $\varphi^{(2k+1)/3}$, for $k \geq 1$.*

Proof. First we make a few observations. We call a trade *winning* if its gain is higher than 1, and *losing* otherwise. Any winning trade made by ONL has a gain of at least $M^{1/3}$. If the algorithm makes a losing trade, it must be a forced sale at the end and the gain is not worse than $M^{-1/3}$. Moreover, it follows that the algorithm cannot trade anymore after it.

We consider a number of cases separately based on the sequence of win/loss trades. If we use W and L to denote a winning and a losing trade respectively, then following the above discussion, the possible cases are: nil (no trade), L, $W^j L$ for $1 \leq j \leq k-1$, and W^j for $1 \leq j \leq k$. (Here W^j denotes a sequence of j consecutive W's).

Case nil: Since ONL has never bought, the prices were never at or below $M^{1/3}$ (except possibly the last one, but neither OPT nor ONL can buy there) and hence OPT's return cannot be better than $(M/M^{1/3})^k = M^{2k/3}$. ONL's return is 1. So the competitive ratio is at most $M^{2k/3}$.

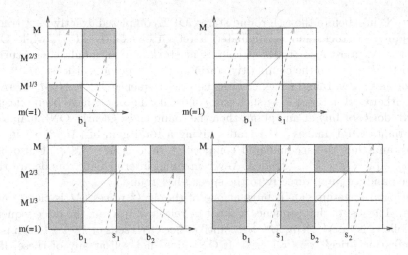

Fig. 1. Four cases illustrated, for $k = 2$. Horizontal axis is time, vertical axis is price. Shaded regions are the regions where the prices cannot fall into. Green solid arrows depict possible buying and selling actions of ONL, red dashed arrows for OPT. Top left: case L, Top right: case W, Bottom left: case WL, Bottom right: case WW. (Color figure online)

Case L: Suppose ONL buys at time b_1 and is forced to sell at the end. The prices before b_1 cannot be lower than $M^{1/3}$ (or else it would have bought) and the prices after b_1 cannot be higher than $M^{2/3}$ (or else it would have sold). Thus, it is easy to see (Fig. 1) that OPT cannot make any trade with gain higher than $M^{2/3}$. So the competitive ratio is at most $(M^{2/3})^k/M^{-1/3} = M^{(2k+1)/3}$.

Case W: Suppose ONL buys at time b_1 and sells at time s_1. Then before b_1, the prices cannot be below $M^{1/3}$; between b_1 and s_1, the prices cannot be higher than $M^{2/3}$; and after s_1, the prices cannot be lower than $M^{1/3}$. It can be seen from Fig. 1 that OPT can make at most one trade with gain M (crossing time s_1); any other trade it makes must be of gain at most $M^{2/3}$. So the competitive ratio is at most $M(M^{2/3})^{k-1}/M^{1/3} = M^{2k/3}$.

Case $W^j L$, $1 \leq j \leq k-1$: Similarly, we can partition the timeline into regions (Fig. 1), from which we can see that OPT can make at most j trades of gain M and the rest have gain at most $M^{2/3}$. Thus competitive ratio $= (M^j(M^{2/3})^{k-j})/((M^{1/3})^j M^{-1/3}) = M^{(2k+1)/3}$.

Case W^j, $1 < j \leq k-1$: This can only be better than the previous case, as OPT again can make at most j trades of gain M and the rest have gain at most $M^{2/3}$, but ONL's return is better than the previous case.

Case W^k: In this case the competitive ratio is simply $M^k/(M^{1/3})^k = M^{2k/3}$. $\square$

Theorem 3. *No deterministic algorithm has a competitive ratio better than $\varphi^{(2k+1)/3-\epsilon}$, for any $\epsilon > 0$ and $k \geq 1$.*

Proof. The prices are released in up to k rounds. The final round k is a special round. For all other rounds, we maintain the following invariants. For each $1 \leq$

$i \leq k - 1$, just before the i-th round starts, OPT completed exactly $i - 1$ trades, is holding no stock, and accumulated a return of exactly M^{i-1}, while ONL completed at most $i - 1$ trades, is holding no stock, and accumulated a return of at most $M^{(i-1)/3}$. So the competitive ratio up to this point is at least $M^{2(i-1)/3}$.

For any $i < k$, round i begins with the price sequence $M^{1/3}, M, M^{1/3}, M, \ldots$ until either ONL buys or $k - i$ such pairs of oscillating prices have been released. If ONL does not buy at any point, then the round ends. Clearly, ONL maintains its variants. OPT makes $k - i$ trades giving a total gain of $(M^{2/3})^{k-i}$ in this round, and thus the accumulated competitive ratio is $M^{2(k-1)/3}$. It also used $(i - 1) + (k - i) = k - 1$ trades. Any remaining intermediate rounds are then skipped and we jump directly to the special last round k.

Otherwise, assume ONL buys at one of the $M^{1/3}$ prices (M is clearly even worse). The rest of that sequence will not be released. Instead, the price sequence that follows is $m, M^{2/3}, m, M^{2/3}, \ldots$ until either ONL sells or $k - i + 1$ such pairs of oscillating prices were released. If ONL does not sell at any of these, then the price drops to m and the game ends (with no further rounds, not even the special round). ONL's gain in this round is $M^{-1/3}$. OPT uses all its remaining $k - i + 1$ trades and gains $(M^{2/3})^{k-i+1}$. Combining with the previous rounds, the competitive ratio is at most $M^{2(i-1)/3} M^{2(k-i+1)/3} / M^{-1/3} = M^{(2k+1)/3}$.

Otherwise ONL sells at one of the $M^{2/3}$ prices (m is even worse). The rest of that sequence will not be released; instead the price goes up to M and this round ends. OPT's gain in this round is M by making one trade from m to M; ONL gains $M^{1/3}$. Thus the invariants are maintained and we move on to the next round. (Regarding the invariant that ONL is not holding stock at the end of the round, we can assume w.l.o.g. that ONL does not buy at the last price M, since clearly it cannot make a profit doing so. In any case, even if it does buy, it can be treated as if it were buying at the beginning of the next round).

Finally, if we arrive at round k, then the same price sequence as in Theorem 1 is used to give an additional factor of $M^{1-\epsilon}$ to the competitive ratio. Note that at the start of this round, OPT has one trade left, and ONL has one or more trades left, but that will not help. Thus the competitive ratio is not better than $M^{2(k-1)/3+1-\epsilon} = M^{(2k+1)/3-\epsilon}$. □

3   Known φ only

For $k = 1$ DO-NOTHING is clearly still φ-competitive, and Theorem 1 still applies here, so we focus on $k > 1$. We adapt Algorithm 1 by buying only when it is certainly 'safe', i.e., when it is certain that the price is within the lowest $\varphi^{1/3}$ of the actual price range, and sells when it gains $\varphi^{1/3}$. The formal description is given in Algorithm 2. Let M_t be the maximum price observed up to and including day t. Note that M_t is a stepwise increasing function of t.

Theorem 4. *Algorithm 2 has competitive ratio $\varphi^{(2k+2)/3}$, for any $k \geq 2$.*

Proof. Clearly ONL gets the same as in Theorem 2: each winning trade has gain at least $\varphi^{1/3}$ and a losing trade, which can only appear as the last trade, has gain at least $\varphi^{-1/3}$. The difference is in how we bound OPT's gain.

Algorithm 2. The reservation price algorithm, with known φ

Upon release of the i-th price $p(i)$:
if $i = n$ **then**
   if currently holding stock **then**
     sell at price $p(n)$ and the game ends.
else
   if $i = 1$ **then**
     $M_1 := p(1)$
   else
     $M_i := \max(M_{i-1}, p(i))$
   if $p(i) \leq M_i/\varphi^{2/3}$ and currently not holding stock and not used up all trades **then**
     buy at price $p(i)$
   else if currently holding stock bought at price $p(b)$ and $p(i) \geq \varphi^{1/3}p(b)$ **then**
     sell at price $p(i)$

In the case of W^k (ONL makes k winning trades) then the same argument as Theorem 2 applies, so in the following we only consider the case where ONL did not use up all its trade, i.e., it is always able to buy if it is not holding.

A *sell event* happens at a day when ONL sells and makes a profit (i.e., excludes the forced sale at the end). An *M-change event* happens at day t when M_t changes. Each OPT trade (b^*, s^*) can be classified into one of the following types:

(1) There is at least one sell event during $[b^*, s^*]$. Clearly the number of such OPT trades is limited by the number of sell events. Each such trade can gain up to φ.

(2) There is no sell event during $[b^*, s^*]$, and at b^* ONL is holding or buying. Suppose ONL's most recent purchase is at time $b \leq b^*$. Then $p(b) \leq M_b/\varphi^{2/3} \leq M_{b^*}/\varphi^{2/3}$. It is holding stock throughout and still did not sell at s^* (or is forced to sell if s^* is the last day), hence $p(s^*) < p(b)\varphi^{1/3} \leq M_{b^*}/\varphi^{1/3}$. But clearly $p(b^*) \geq M_{b^*}/\varphi$, hence the gain of OPT is at most $\varphi^{2/3}$.

(3) There is no sell event during $[b^*, s^*]$, at b^* ONL is neither holding nor buying, and there is no M-change event in $(b^*, s^*]$. We have $p(b^*) > M_{b^*}/\varphi^{2/3}$ as otherwise ONL would have bought at b^*. Clearly $p(s^*) \leq M_{s^*} = M_{b^*}$. Hence the gain of OPT is $p(s^*)/p(b^*) < \varphi^{2/3}$.

(4) There is no sell event during $[b^*, s^*]$, at b^* ONL is neither holding nor buying, and there is/are M-change event(s) in $(b^*, s^*]$. Suppose there are a total of x such OPT trades, $(b_1^*, s_1^*), (b_2^*, s_2^*), \ldots, (b_x^*, s_x^*)$, in chronological order. Note that $p(b_i^*) > M_{b_i^*}/\varphi^{2/3}$ or else ONL would have bought at b_i^*. So for all i, $p(b_{i+1}^*) > M_{b_{i+1}^*}/\varphi^{2/3} \geq M_{s_i^*}/\varphi^{2/3} \geq p(s_i^*)/\varphi^{2/3}$. Thus the total gain of these x trades is

$$\prod_{i=1}^{x} \frac{p(s_i^*)}{p(b_i^*)} = \frac{1}{p(b_1^*)} \frac{p(s_1^*)}{p(b_2^*)} \cdots \frac{p(s_{x-1}^*)}{p(b_x^*)} \frac{p(s_x^*)}{1} < \frac{p(s_x^*)}{p(b_1^*)}(\varphi^{2/3})^{x-1} \leq \varphi^{(2x+1)/3}.$$

Suppose ONL makes y winning trades and one losing trade. Then OPT makes at most y trades of type (1), gaining at most φ^y from those. Then, if x of OPT's trades are of type (4), they in total gives another gain of at most $\varphi^{(2x+1)/3}$. The remaining trades are of types (2) and (3), gaining $\varphi^{2/3}$ each. The competitive ratio is therefore at most

$$\frac{\varphi^y \varphi^{(2x+1)/3} \varphi^{2(k-x-y)/3}}{\varphi^{y/3} \varphi^{-1/3}} = \varphi^{(2k+2)/3}.$$

If ONL makes $y < k$ winning trades and no losing trade, the competitive ratio can only be better, as OPT's return is as above but ONL's is $\varphi^{1/3}$ better.   □

Theorem 5. *No deterministic algorithm is better than $\varphi^{(2k+2)/3-\epsilon}$-competitive, for any $\epsilon > 0$ and $k \geq 2$.*

Proof. Again there will be a number of rounds. Round 1 is special, in that OPT will get a factor of φ better than ONL but will afterwards reveal knowledge of m and M. Rounds 2 to k are then similar to Theorem 3.

Round 1: The first price is 1. If ONL does not buy, then the price goes up to φ. OPT makes one trade and gains φ. Now we know the range is $[1..\varphi]$, and we can assume w.l.o.g. that ONL does not buy at φ. Then the round ends. At the end of this round, both OPT and ONL are not holding stock, OPT made one trade and ONL none, but ONL is a factor of φ behind in the return.

Otherwise, if ONL buys at 1, then the subsequent price sequence is $1/\varphi, 1$, $1/\varphi, 1, \ldots$ for up to k such pairs, until ONL sells. Now we know the range is $[1/\varphi..1]$. Without loss of generality we can assume ONL does not sell at $1/\varphi$ since it is clearly the lowest possible price. If ONL does not sell at any point, then the game ends with no further rounds. OPT makes k trades gaining φ^k, and ONL's gain is 1. The competitive ratio is φ^k, which is at least $\varphi^{(2k+2)/3}$. If ONL sells at some point with price 1, then the sequence stops and this round ends. OPT buys at $1/\varphi$ and sells at 1, getting a gain of φ. ONL's gain is 1. Both OPT and ONL used one trade, and OPT is a factor of φ ahead of ONL.

Each of rounds 2 to $k-1$ are the same as the intermediate rounds in Theorem 3, with OPT gaining a factor of $\varphi^{2/3}$ ahead of ONL in each round.

Finally, in round k we use the same price sequence in Theorem 1, which gives an extra factor of $\varphi^{1-\epsilon}$. Note that ONL may have more trades left then OPT (in addition to the same reason as in Theorem 3, in round 1 ONL may have done no trade), but again it is not useful for ONL.   □

4   Bounded Daily Return, Known Duration

Recall that in this model, the prices are bounded by $p(i)/\beta \leq p(i+1) \leq \alpha p(i)$ for some $\alpha, \beta > 1$. Trades can take place at days $0, 1, \ldots, T$.

Theorem 6. *No deterministic algorithm has a competitive ratio better than $\alpha^{T(2k \log \beta)/((k+1)\log \beta + k \log \alpha)}$.*

Proof. The adversary strategy is very simple and natural: whenever ONL is not holding stock, the price goes up by a factor of α every day, and while it is holding stock it goes down by β every day. Let the ONL trades be (b_i, s_i), $i = 1, \ldots, k$. (If there are fewer than k trades, auxillary ones with $b_i = s_i$ can be added.) Define $t_{2i-1} = b_i - s_{i-1}$ for $1 \leq i \leq k+1$, and $t_{2i} = s_i - b_i$ for $1 \leq i \leq k$. (For convenience define $s_0 = 0$ and $b_{k+1} = T$.) ONL's return is $1/(\beta^{t_2} \beta^{t_4} \cdots \beta^{t_{2k}})$. OPT's optimal actions, if allowed $k + 1$ trades, is to hold during the exact opposite intervals as ONL, i.e., buy at s_i and sell at b_{i+1} for $0 \leq i \leq k$. But since it can make at most k trades, its possible course of actions include skipping one of those trades, or making one of the trades 'span across two intervals', e.g., buying at s_i and selling at b_{i+2}. Thus OPT's return is one of

$$\alpha^{t_3} \alpha^{t_5} \cdots \alpha^{t_{2k+1}}, \alpha^{t_1} \alpha^{t_5} \cdots \alpha^{t_{2k+1}}, \ldots, \alpha^{t_1} \alpha^{t_3} \cdots \alpha^{t_{2k-1}},$$

$$\alpha^{t_1} \alpha^{t_3} \cdots \alpha^{t_{2k+1}}/\beta^{t_2}, \alpha^{t_1} \alpha^{t_3} \cdots \alpha^{t_{2k+1}}/\beta^{t_4}, \ldots, \alpha^{t_1} \alpha^{t_3} \cdots \alpha^{t_{2k+1}}/\beta^{t_{2k}}.$$

To attain the worst competitive ratio, these ratios should be equal, which means $t_1 = t_3 = \cdots = t_{2k+1}$ and $t_2 = t_4 = \cdots = t_{2k}$. This further implies $\alpha^{k t_1} = \alpha^{(k+1)t_1}/\beta^{t_2}$, which gives $\alpha^{t_1} = \beta^{t_2}$. Together with $t_1 + \cdots + t_{2k+1} = (k+1)t_1 + k t_2 = T$, this gives

$$t_1 = \frac{\log \beta}{(k+1) \log \beta + k \log \alpha} T, \quad t_2 = \frac{\log \alpha}{(k+1) \log \beta + k \log \alpha} T$$

The competitive ratio is then $\alpha^{k t_1}/(1/\beta^{k t_2}) = \alpha^{2k t_1} = \alpha^{T(2k \log \beta)/((k+1) \log \beta + k \log \alpha)}$. □

Theorem 7. *Algorithm 3 has competitive ratio $\alpha^{T(2k \log \beta)/((k+1) \log \beta + k \log \alpha)}$.*

Algorithm 3 may feel unnatural since it does not depend on the price sequence at all (this is called 'static' in [1]). But we prove that the following variation of the algorithm has the same competitive ratio: it sells only when the current price falls below h/β^{t_2} where h is the highest price seen since the last purchase. This coincides with the 'stop loss' strategy very common in real trading (more precisely 'trailing stop' [6] where the stop loss limit is not fixed but tracks the highest price seen thus far, to potentially capture the most profit).

Algorithm 3. Static algorithm for known α, β and T.

Set $t_1 = \frac{\log \beta}{(k+1) \log \beta + k \log \alpha} T$ and $t_2 = \frac{\log \alpha}{(k+1) \log \beta + k \log \alpha} T$, rounding to nearest integers.
Upon release of the i-th price $p(i)$:
if $i = T$ then
  if currently holding stock then
    sell at price $p(T)$ and the game ends
else
  if have not been holding stock for t_1 days and not used up all trades then
    buy at price $p(i)$
  else if have been holding stock for t_2 days then
    sell at price $p(i)$

Algorithm 4. Stop loss based algorithm for known α, β and T.

Set t_1 and t_2 as in Algorithm 3.
Upon release of the i-th price $p(i)$:
if $i = T$ **then**
    if currently holding stock **then**
        sell at price $p(T)$ and the game ends
else
    if have not been holding stock for t_1 days and not used up all trades **then**
        buy at price $p(i)$
        set $h = p(i)$
    else if currently holding stock **then**
        set $h = \max(h, p(i))$
        sell at price $p(i)$ if $p(i) < h/\beta^{t_2}$

Theorem 8. *Algorithm 4 has competitive ratio* $\alpha^{T(2k \log \beta)/((k+1)\log \beta + k \log \alpha)}$.

Proof. (Sketch) Recall that $\alpha^{t_1} = \beta^{t_2}$. Let r denote this common value, and the competitive ratio we want to prove is then r^{2k}. Our approach is to partition the time horizon so that in each partition x trades in OPT are associated to y trades in ONL such that the ratio between their gains is at most r^{x+y}. Since in total they make at most $2k$ trades, this proves the theorem.

Denote by (b_i, s_i) ONL's i-th trade. Let $H_i = [b_i, s_i]$ (a closed time interval) and $N_i = (s_{i-1}, b_i)$ (an open interval). Let h_i be the highest price during H_i. We can show the following properties:

(1) For any two days x and y in the same H_i, where $x < y$, we have $p(y) \geq p(x)/r$. As a direct consequence, $p(s_i) \geq h_i/r$.
(2) Without loss of generality we can assume $p(b_{i+1}) = p(s_i)r$.
(3) OPT would not buy or sell strictly within any N_i.

Consider an OPT trade (b^*, s^*). Suppose b^* falls within H_u and s^* falls within H_v, where $v \geq u$. Its gain is

$$\frac{p(s^*)}{p(b^*)} \leq \frac{h_v}{p(b_u)/r} \leq \frac{p(s_v)r}{p(b_u)/r} = \frac{p(s_v)}{p(b_u)}r^2$$

where the inequalities are due to (1). Then

$$\frac{p(s^*)}{p(b^*)} \leq \frac{p(s_v)}{p(b_v)}\frac{p(b_v)}{p(s_{v-1})}\frac{p(s_{v-1})}{p(b_{v-1})}\frac{p(b_{v-1})}{p(s_{v-2})} \cdots \frac{p(s_{u+1})}{p(b_{u+1})}\frac{p(b_{u+1})}{p(s_u)}\frac{p(s_u)}{p(b_u)}r^2$$

$$= \frac{p(s_v)}{p(b_v)}r\frac{p(s_{v-1})}{p(b_{v-1})}r \cdots r\frac{p(s_u)}{p(b_u)}r^2 = \frac{p(s_v)}{p(b_v)}\frac{p(s_{v-1})}{p(b_{v-1})} \cdots \frac{p(s_u)}{p(b_u)}r^{v-u+2}$$

due to (2). If no other OPT trade sells during H_u or buys during H_v, we can associate this $x = 1$ OPT trade with these $y = v - u + 1$ ONL trades $(b_u, s_u), \ldots, (b_v, s_v)$. The ratio between their gains is $r^{v-u+2} = r^{1+y}$. Now suppose there are two OPT trades (b^*, s^*) and (b^{**}, s^{**}) such that s^* and b^{**} are in the same H_i. Then $(p(s^{**})/p(b^{**}))(p(s^*)/p(b^*)) \le (p(s^{**})/p(b^*))r$ due to (1), and by losing a factor of r we can replace the two OPT trades with one (b^*, s^{**}). By repeatedly applying the argument, we can partition the time horizon into disjoint parts, where in each part x OPT trades have been merged into one, and this merged trade buys during some H_u and sells during some H_v, and no other OPT trade sells at H_u or buys at H_v. The ratio between the gains, after considering the loss due to merging the OPT trades, is $r^{1+y}r^{x-1} = r^{x+y}$.   □

References

1. Chen, G.-H., Kao, M.-Y., Lyuu, Y.-D., Wong, H.-K.: Optimal buy-and-hold strategies for financial markets with bounded daily returns. SIAM J. Comput. **31**(2), 447–459 (2001)
2. Chin, F.Y.L., Fu, B., Guo, J., Han, S., Hu, J., Jiang, M., Lin, G., Ting, H.F., Zhang, L., Zhang, Y., Zhou, D.: Competitive algorithms for unbounded one-way trading. Theoret. Comput. Sci. **607**(1), 35–48 (2015)
3. Clemente, J., Hromkovič, J., Komm, D., Kudahl, C.: Advice complexity of the online search problem. In: Mäkinen, V., Puglisi, S.J., Salmela, L. (eds.) IWOCA 2016. LNCS, vol. 9843, pp. 203–212. Springer, Cham (2016). doi:10.1007/978-3-319-44543-4_16
4. Damaschke, P., Ha, P.H., Tsigas, P.: Online search with time-varying price bounds. Algorithmica **55**, 619–642 (2009)
5. El-Yaniv, R., Fiat, A., Karp, R.M., Turpin, G.: Optimal search and one-way trading online algorithms. Algorithmica **30**(1), 101–139 (2001)
6. Glynn, P.W., Iglehart, D.L.: Trading securities using trailing stops. Manage. Sci. **41**(6), 1096–1106 (1995)
7. Lorenz, J., Panagiotou, K., Steger, A.: Optimal algorithms for k-search with application in option pricing. Algorithmica **55**(2), 311–328 (2009)
8. Mohr, E., Ahmed, I., Schmidt, G.: Online algorithms for conversion problems: a survey. Surv. Oper. Res. Manage. Sci. **19**, 87–104 (2014)
9. Schmidt, G., Mohr, E., Kersch, M.: Experimental analysis of an online trading algorithm. Electron. Notes Discrete Math. **36**, 519–526 (2010)
10. Schroeder, P., Schmidt, G., Kacem, I.: Optimal on-line algorithms for bi-directional non-preemptive conversion with interrelated conversion rates. In: Proceedings of the 4th IEEE Conference on Control, Decision and Information Technology, pp. 28–33 (2016)
11. Zhang, W., Xu, Y., Zheng, F., Dong, Y.: Optimal algorithms for online time series search and one-way trading with interrelated prices. J. Comb. Optim. **23**, 159–166 (2012)

Parameterized Shifted Combinatorial Optimization

Jakub Gajarský[1,4], Petr Hliněný[1(✉)], Martin Koutecký[2], and Shmuel Onn[3]

[1] Masaryk University, Brno, Czech Republic
{gajarsky,hlineny}@fi.muni.cz
[2] Charles University, Prague, Czech Republic
koutecky@kam.mff.cuni.cz
[3] Technion - Israel Institute of Technology, Haifa, Israel
onn@ie.technion.ac.il
[4] Technical University Berlin, Berlin, Germany
jakub.gajarsky@tu-berlin.de

Abstract. *Shifted combinatorial optimization* is a new nonlinear optimization framework which is a broad extension of standard combinatorial optimization, involving the choice of several feasible solutions at a time. This framework captures well studied and diverse problems ranging from so-called vulnerability problems to sharing and partitioning problems. In particular, every standard combinatorial optimization problem has its shifted counterpart, which is typically much harder. Already with explicitly given input set the shifted problem may be NP-hard. In this article we initiate a study of the parameterized complexity of this framework. First we show that shifting over an explicitly given set with its cardinality as the parameter may be in XP, FPT or P, depending on the objective function. Second, we study the shifted problem over sets definable in MSO logic (which includes, e.g., the well known MSO partitioning problems). Our main results here are that shifted combinatorial optimization over MSO definable sets is in XP with respect to the MSO formula and the treewidth (or more generally clique-width) of the input graph, and is W[1]-hard even under further severe restrictions.

Keywords: Combinatorial optimization · Shifted problem · Treewidth · MSO logic · MSO partitioning

J. Gajarský's research was partially supported by the European Research Council under the European Union's Horizon 2020 research and innovation programme (ERC Consolidator Grant DISTRUCT, grant agreement No 648527).

P. Hliněný, and partially J. Gajarský, were supported by the research centre Institute for Theoretical Computer Science (CE-ITI), project P202/12/G061 of the Czech Science Foundation.

M. Koutecký was partially supported by the project 17-09142S of the Czech Science Foundation.

Shmuel Onn was partially supported by the Dresner Chair at the Technion.

Y. Cao and J. Chen (Eds.): COCOON 2017, LNCS 10392, pp. 224–236, 2017.
DOI: 10.1007/978-3-319-62389-4_19

1   Introduction

The following optimization problem has been studied extensively in the literature.

(Standard) Combinatorial Optimization. Given $S \subseteq \{0,1\}^n$ and $\mathbf{w} \in \mathbb{Z}^n$, solve

$$\max\{\mathbf{ws} \mid \mathbf{s} \in S\} . \tag{1}$$

The complexity of the problem depends on $\mathbf{w}$ and the type and presentation of S. Often, S is the set of indicating (characteristic) vectors of members of a family of subsets over a ground set $[n] := \{1, \ldots, n\}$, such as the family of $s - t$ dipaths in a digraph with n arcs, the set of perfect matchings in a bipartite or arbitrary graph with n edges, or the set of bases in a matroid over $[n]$ given by an independence oracle.

Partly motivated by vulnerability problems studied recently in the literature (see a brief discussion below), in this article we study a broad nonlinear extension of CO, in which the optimization is over r choices of elements of S and which is defined as follows. For a set $S \subseteq \mathbb{R}^n$, let S^r denote the set of $n \times r$ matrices having each column in S,

$$S^r := \{\mathbf{x} \in \mathbb{R}^{n \times r} \mid \mathbf{x}^k \in S, \ k = 1, \ldots, r\} .$$

Call $\mathbf{x}, \mathbf{y} \in \mathbb{R}^{n \times r}$ equivalent and write $\mathbf{x} \sim \mathbf{y}$ if each row of $\mathbf{x}$ is a permutation of the corresponding row of $\mathbf{y}$. The *shift* of $\mathbf{x} \in \mathbb{R}^{n \times r}$ is the unique matrix $\overline{\mathbf{x}} \in \mathbb{R}^{n \times r}$ satisfying $\overline{\mathbf{x}} \sim \mathbf{x}$ and $\overline{\mathbf{x}}^1 \geq \cdots \geq \overline{\mathbf{x}}^r$, that is, the unique matrix equivalent to $\mathbf{x}$ with each row nonincreasing. Our nonlinear optimization problem follows:

Shifted Combinatorial Optimization (SCO). Given $S \subseteq \{0,1\}^n$ and $\mathbf{c} \in \mathbb{Z}^{n \times r}$, solve

$$\max\{\mathbf{c}\overline{\mathbf{x}} \mid \mathbf{x} \in S^r\} . \tag{2}$$

(Here $\mathbf{c}\overline{\mathbf{x}}$ is used to denote the ordinary scalar product of the vectors $\mathbf{c}$ and $\overline{\mathbf{x}}$.)

This problem easily captures many classical fundamental problems. For example, given a graph $G = (V, E)$ with n vertices, let $S := \{N[v] \mid v \in V\} \subseteq \{0,1\}^n$, where $N[v]$ is the characteristic vector of the closed neighborhood of v. Choose an integer parameter r and let $c_i^1 := 1$ for all i and $c_i^j := 0$ for all i and all $j \geq 2$. Then the optimal objective function value of (2) is n if and only if we can select a set D of r vertices in G such that every vertex belongs to the closed neighborhood of at least one of the selected vertices, that is, when D is a dominating set of G. Likewise, one can formulate the vertex cover and independent set problems in a similar way.

One specific motivation for the SCO problem is as follows. Suppose S is the set of indicators of members of a family over $[n]$. A feasible solution $\mathbf{x} \in S^r$ then represents a choice of r members of the given family such that the k-th column $\mathbf{x}^k$ is the indicator of the k-th member. Call element i in the ground set *k-vulnerable* in $\overline{\mathbf{x}}$ if it is used by at least k of the members represented by $\mathbf{x}$, that is, if the i-th row $\mathbf{x}_i$ of $\mathbf{x}$ has at least k ones. It is easy to see that the k-th column $\overline{\mathbf{x}}^k$ of the shift of $\mathbf{x}$ is precisely the indicator of the set of k-vulnerable elements in $\mathbf{x}$. So the shifted optimization problem is to maximize

$$\mathbf{c}\overline{\mathbf{x}} \;=\; \sum\{c_i^k \mid i \text{ is } k\text{-vulnerable in } \mathbf{x}, \; i = 1, \ldots, n, \; k = 1, \ldots, r\} \;.$$

Minimizing the numbers of k-vulnerable elements in $\mathbf{x}$ may be beneficial for survival of some family members under various attacks to vulnerable elements by an adversary, see e.g. [1,20] for more details. For example, to minimize the number of k-vulnerable elements for some k, we set $c_i^k := -1$ for all i and $c_i^j := 0$ for all i and all $j \neq k$. To *lexicographically* minimize the numbers of r-vulnerable elements, then of $(r-1)$-vulnerable elements, and so on, till that of 1-vulnerable elements, we can set $c_i^k := -(n+1)^{k-1}$ for all i, k.

As another natural example, consider $\mathbf{c}$ with $c_i^1 := 1$ and $c_i^j := -1$ for $1 < j \leq r$. Then $\mathbf{c}\overline{\mathbf{x}} = n$ if and only if the columns of $\mathbf{x}$ indicate a *partition* of S. This formulation hence allows us to optimize over partitions of the ground set (see Sect. 3). Or, consider $\mathbf{c}$ with $\mathbf{c}_i = (1, \ldots, 1, -1, \ldots, -1)$ of length $a > 0$ with $b \leq a$ ones, and let S be the family of independent sets of a graph G. Then $\max \mathbf{c}\overline{\mathbf{x}}$ relates to *fractional coloring of* G; it holds $\max \mathbf{c}\overline{\mathbf{x}} = bn$ if and only if G has a coloring by a colors in total such that every vertex receives b distinct colors – this is the so-called $(a : b)$-COLORING problem.

The complexity of the shifted combinatorial optimization (SCO) problem depends on $\mathbf{c}$ and on the presentation of S, and is often harder than the corresponding standard combinatorial optimization problem. Say, when S is the set of perfect matchings in a graph, the standard problem is polynomial time solvable, but the shifted problem is NP-hard even for $r = 2$ and cubic graphs, as the optimal value of the above 2-vulnerability problem is 0 if and only if the graph is 3-edge-colorable [17]. The minimization of 2-vulnerable arcs with S the set of s–t dipaths in a digraph, also called the MINIMUM SHARED EDGES problem, was recently shown to be NP-hard for r variable in [20], polynomial time solvable for fixed r in [1], and fixed-parameter tractable with r as a parameter in [5].

In the rest of this article we always assume that the number r of choices is variable. Call a matrix $\mathbf{c} \in \mathbb{Z}^{n \times r}$ *shifted* if $\mathbf{c} = \overline{\mathbf{c}}$, that is, if its rows are nonincreasing. In [13] it was shown that when $S = \{\mathbf{s} \in \{0,1\}^n \mid \mathbf{As} = \mathbf{b}\}$ where $\mathbf{A}$ is a totally unimodular matrix and $\mathbf{b}$ is an integer vector, the shifted problem with shifted $\mathbf{c}$, and hence in particular the above lexicographic vulnerability problem, can be solved in polynomial time. In particular this applies to the cases of S the set of s–t dipaths in a digraph and S the set of perfect matchings in a bipartite graph. In [17] it was shown that the shifted problem with shifted $\mathbf{c}$ is also solvable in polynomial time for S the set of bases of a matroid presented by an independence oracle (in particular, spanning trees in a graph), and even for the intersection of matroids of certain type.

Main Results and Paper Organization. In this article we continue on systematic study of shifted combinatorial optimization.

First, in Sect. 2, we consider the case when the set S is given explicitly. While the standard problem is always trivial in such case, the SCO problem can be NP-hard for explicit set S (Proposition 2.1). Our main results on this case can be briefly summarized as follows:

- **(Theorem 2.2)** The shifted combinatorial optimization problem, parameterized by $|S| = m$, is; (a) for general $\mathbf{c}$ in the complexity class XP and W[1]-hard w.r.t. m, (b) for shifted $\mathbf{c}$ in FPT, and (c) for shifted $-\mathbf{c}$ in P.
- **(Theorem 2.4)** The latter case (c) of shifted $-\mathbf{c}$ is in P even for sets S presented by a linear optimization oracle.

In Sect. 3, we study a more general framework of SCO for the set S definable in Monadic Second Order (MSO) logic. This rich framework includes, for instance, the well-studied case of so called MSO partitioning problems on graphs. We prove the following statement which generalizes known results about MSO partitioning:

- **(Theorem 3.5, Corollary 3.9)** The shifted combinatorial optimization problem, for (a) graphs of bounded treewidth and S defined in MSO_2 logic, or (b) graphs of bounded clique-width and S defined in MSO_1 logic, is in XP (parameterized by the width and the formula defining S).

In the course of proving this statement we also provide a connection of shifted optimization to separable optimization when the corresponding polyhedron is decomposable and 0/1 (Lemma 3.3).

To complement the previous tractability result, in Sect. 4 we prove the following negative result under much more restrictive parametrization.

- **(Theorem 4.2).** There exists a fixed First Order formula ϕ such that the associated MSO_1 partitioning problem, and hence also the SCO problem with S defined by ϕ, are W[1]-hard on graphs of bounded treedepth.

We use standard terminology of graph theory, integer programming and parameterized complexity. Due to space restrictions, the missing proofs of our statements are left for the full version [arXiv:1702.06844]. The statements with proofs presented only in the full version are marked with **(*)**.

2   Sets Given Explicitly

In this section we consider the shifted problem (2) over an explicitly given set $S = \{\mathbf{s}^1, \dots, \mathbf{s}^m\}$. We demonstrate that already this seemingly simple case is in fact nontrivial and interesting, and that the brute-force algorithm which tries all possible r-subsets of S is likely close to optimal:

Proposition 2.1 (*). *The SCO problem* (2) *is* NP-*hard for* 0/1 *shifted matrices* $\mathbf{c} = \overline{\mathbf{c}} \in \{0,1\}^{n \times r}$ *and explicitly given* 0/1 *sets* $S = \{\mathbf{s}^1, \dots, \mathbf{s}^m\} \subseteq \{0,1\}^n$. *Moreover, unless the* Exponential Time Hypothesis *(ETH) fails, it cannot be solved in time* $\mathcal{O}(n^{o(r)})$.

Note that the next results in this section concerning Shifted IP apply to the more general situation in which S may consist of arbitrary integer vectors, not necessarily 0/1. This is formulated as follows.

Shifted Integer Programming. Given $S \subseteq \mathbb{Z}^n$ and $\mathbf{c} \in \mathbb{Z}^{n \times r}$, similarly to (2), solve

$$\max\{\mathbf{c}\overline{\mathbf{x}} \mid \mathbf{x} \in S^r\}. \tag{3}$$

For $S = \{\mathbf{s}^1, \ldots, \mathbf{s}^m\}$ and nonnegative integers $r_1, \ldots, r_m$ with $\sum_{i=1}^m r_i = r$, let $\mathbf{x}(r_1, \ldots, r_m)$ be the matrix in S^r with first r_1 columns equal to $\mathbf{s}^1$, next r_2 columns equal to $\mathbf{s}^2$, and so on, with last r_m columns equal to $\mathbf{s}^m$, and define $f(r_1, \ldots, r_m) := \mathbf{c}\overline{\mathbf{x}}(r_1, \ldots, r_m)$.

We have got the following effective theorem in contrast with Proposition 2.1.

Theorem 2.2 (*). *The shifted integer programming problem* (3) *over an explicitly given set* $S = \{\mathbf{s}^1, \ldots, \mathbf{s}^m\} \subseteq \mathbb{Z}^n$ *reduces to the following nonlinear integer programming problem over a simplex,*

$$\max\left\{f(r_1, \ldots, r_m) \mid r_1, \ldots, r_m \in \mathbb{Z}_+, \; \sum_{k=1}^m r_k = r\right\}. \tag{4}$$

If $\mathbf{c} = \overline{\mathbf{c}}$ *is shifted then* f *is concave, and if* $-\mathbf{c}$ *is shifted then* f *is convex. Furthermore, the following cases hold:*

1. *With* m *parameter and* $\mathbf{c}$ *arbitrary, problem* (3) *is in* XP. *Moreover, the problem is* W[1]-*hard with parameter* m *even for 0/1 sets* S.
2. *With* m *parameter and* $\mathbf{c}$ *shifted, problem* (3) *is in* FPT.
3. *With* m *variable and* $-\mathbf{c}$ *shifted, problem* (3) *is in* P.

Let us first give an exemplary application of part 2 of Theorem 2.2 now. Bredereck et al. [3] study the WEIGHTED SET MULTICOVER (WSM) problem, which is as follows. Given a universe $U = \{u_1, \ldots, u_k\}$, integer demands $d_1, \ldots, d_k \in \mathbb{N}$ and a multiset $\mathcal{F} = \{F_1, \ldots, F_n\} \subseteq 2^U$ with weights $w_1, \ldots, w_n \in \mathbb{N}$, find a multiset $\mathcal{F}' \subseteq \mathcal{F}$ of smallest weight which satisfies all the demands – that is, for all $i = 1, \ldots, k$, $|\{F \in \mathcal{F}' \mid u_i \in F\}| \geq d_i$. It is shown [3] that this problem is FPT when the size of the universe is a parameter, and then several applications in computational social choice are given there.

Notice that $\mathcal{F}$ can be represented in a succinct way by viewing $\mathcal{F}$ as a set $\mathcal{F}_s = \{F_1, \ldots, F_K\}$ and representing the different copies of $F \in \mathcal{F}_s$ in $\mathcal{F}$ by defining K weight functions $w_1, \ldots, w_K$ such that, for each $i = 1, \ldots, K$, $w_i(j)$ returns the total weight of the first j lightest copies of F_i, or ∞ if there are less than j copies. We call this the *succinct variant*.

Bredereck et al. [3] use Lenstra's algorithm for their result, which only works when $\mathcal{F}$ is given explicitly. We note in passing that our approach allows us to extend their result to the succinct case.

Proposition 2.3 (*). *Weighted Set Multicover is in* FPT *with respect to universe size* k, *even in the succinct variant.*

Theorem 2.2, part 3, can be applied also to sets S presented implicitly by an oracle. A *linear optimization oracle* for $S \subseteq \mathbb{Z}^n$ is one that, queried on $w \in \mathbb{Z}^n$, solves the linear optimization problem $\max\{\mathbf{w}\mathbf{s} \mid \mathbf{s} \in S\}$. Namely, the oracle either asserts that the problem is infeasible, or unbounded, or provides an optimal solution. As mentioned before, even for $r = 2$, the shifted problem for perfect matchings is NP-hard, and hence for general $\mathbf{c}$ the shifted problem over S presented by a linear optimization oracle is also hard even for $r = 2$. In contrast, we have the following strengthening.

Theorem 2.4 (*). *The shifted problem* (3) *with* $\mathbf{c}$ *nondecreasing, over any set* $S \subset \mathbb{Z}^n$ *which is presented by a linear optimization oracle, can be solved in polynomial time.*

3   MSO-definable Sets: XP for Bounded Treewidth

In this section we study another tractable and rich case of shifted combinatorial optimization, namely that of the set S defined in the MSO logic of graphs. This case, in particular, includes well studied MSO partitioning framework of graphs (see below) which is tractable on graphs of bounded treewidth and clique-width. In the course of proving our results, it is useful to study a geometric connection of 0/1 SCO problems to separable optimization over decomposable polyhedra.

3.1   Relating SCO to Decomposable Polyhedra

The purpose of this subsection is to demonstrate how shifted optimization over 0/1 polytopes closely relates to an established concept of decomposable polyhedra. We refer to Ziegler [25] for definitions and terminology regarding polytopes.

Definition 3.1 (Decomposable polyhedron and Decomposition oracle). *A polyhedron* $P \subseteq \mathbb{R}^n$ *is* decomposable *if for every* $k \in \mathbb{N}$ *and every* $\mathbf{x} \in kP \cap \mathbb{Z}^n$, *there are* $\mathbf{x}^1, \ldots, \mathbf{x}^k \in P \cap \mathbb{Z}^n$ *with* $\mathbf{x} = \mathbf{x}^1 + \cdots + \mathbf{x}^k$, *where* $kP = \{ k\mathbf{y} \mid \mathbf{y} \in P \}$. *A* decomposition oracle *for a decomposable* P *is one that, queried on* $k \in \mathbb{N}$ *given in unary and on* $\mathbf{x} \in kP \cap \mathbb{Z}^n$, *returns* $\mathbf{x}^1, \ldots, \mathbf{x}^k \in P \cap \mathbb{Z}^n$ *with* $\mathbf{x} = \mathbf{x}^1 + \cdots + \mathbf{x}^k$.

This property is also called *integer decomposition property* or being *integrally closed* in the literature. The best known example are polyhedra given by totally unimodular matrices [2]. Furthermore, we will use the following notion.

Definition 3.2 (Integer separable (convex) minimization oracle). *Let* $P \subseteq \mathbb{R}^n$ *and let* $f(\mathbf{x}) = \sum_{i=1}^n f_i(x_i)$ *be a separable function on* $\mathbb{R}^n$. *An* integer separable minimization oracle *for* P *is one that, queried on this* f, *either reports that* $P \cap \mathbb{Z}^n$ *is empty, or that it is unbounded, or returns a point* $\mathbf{x} \in P \cap \mathbb{Z}^n$ *which minimizes* $f(\mathbf{x})$. *An* integer separable convex minimization oracle *for* P *is an integer separable minimization oracle for* P *which can only be queried on functions* f *as above with all* f_i *convex.*

We now formulate how these notions naturally connect with SCO.

Lemma 3.3 (*). *Let $(S, \mathbf{c}, r)$ be an instance of shifted combinatorial optimization, with $S \subseteq \{0,1\}^n$, $r \in \mathbb{N}$ and $\mathbf{c} \in \mathbb{Z}^{n \times r}$. Let $P \subseteq [0,1]^n$ be a polytope such that $S = P \cap \{0,1\}^n$ and let $Q \subseteq [0,1]^{n+n'}$ be some extension of P, that is, $P = \{\mathbf{x} \mid (\mathbf{x}, \mathbf{y}) \in Q\}$.*

Then, provided a decomposition oracle for Q and an integer separable minimization oracle for rQ, the shifted problem given by $(S, \mathbf{c}, r)$ can be solved with one call to the optimization oracle and one call to the decomposition oracle. Furthermore, if $\mathbf{c}$ is shifted, an integer separable convex minimization oracle suffices.

To demonstrate Lemma 3.3 we use it to give an alternative proof of the result of Kaibel et al. [13] that the shifted problem is polynomial when $S = \{\mathbf{x} \mid \mathbf{A}\mathbf{x} = \mathbf{b}, \mathbf{x} \in \{0,1\}^n\}$ and $\mathbf{A}$ is totally unimodular. It is known that $P = \{\mathbf{x} \mid \mathbf{A}\mathbf{x} = \mathbf{b}, 0 \le \mathbf{x} \le 1\}$ is decomposable and a decomposition oracle is realizable in polynomial time [23]. Moreover, it is known that an integer separable convex minimization oracle for rP is realizable in polynomial time [12]. Lemma 3.3 implies that the shifted problem is polynomial for this S when $\mathbf{c}$ is shifted.

The reason we have formulated Lemma 3.3 for S given by an extension Q of the polytope P corresponding to S, is the following: while P itself might not be decomposable, there always exists an extension of it which is decomposable. See the full version for more details.

Other potential candidates where Lemma 3.3 could be applied are classes of polytopes that are either decomposable, or allow efficient integer separable (convex) minimization. Some known decomposable polyhedra are stable set polytopes of perfect graphs, polyhedra defined by k-balanced matrices [24], polyhedra defined by nearly totally unimodular matrices [10], etc. Some known cases where integer separable convex minimization is polynomial are for $P = \{\mathbf{x} \mid \mathbf{A}\mathbf{x} = \mathbf{b}, \mathbf{x} \in \{0, 1, \ldots, r\}^n\}$ where the Graver basis of $\mathbf{A}$ has small size or when $\mathbf{A}$ is highly structured, namely when $\mathbf{A}$ is either an n-fold product, a transpose of it, or a 4-block n-fold product; see the books of Onn [21] and De Loera, Hemmecke and Köppe [18].

3.2 XP Algorithm for MSO-definable Set

We start with defining the necessary specialized terms. We assume that the reader is familiar with the standard *treewidth* of a graph (see also the full paper).

Given a matrix $\mathbf{A} \in \mathbb{Z}^{n \times m}$, we define the corresponding *Gaifman graph* $G = G(\mathbf{A})$ as follows. Let $V(G) = [m]$. We let $\{i, j\} \in E(G)$ if and only if there is an $r \in [n]$ with $\mathbf{A}[r, i] \ne 0$ and $\mathbf{A}[r, j] \ne 0$. Intuitively, two vertices of G are adjacent if the corresponding variables x_i, x_j occur together in some constraint of $\mathbf{A}\mathbf{x} \le \mathbf{b}$. The *(Gaifman) treewidth of a matrix* $\mathbf{A}$ is then the treewidth of its Gaifman graph, i.e., $tw(\mathbf{A}) := tw(G(\mathbf{A}))$.

The aforementioned MSO partitioning framework of graphs comes as follows.

MSO Partitioning Problem. Given a graph G, an MSO_2 formula φ with one free vertex-set variable and an integer r, the task is as follows;

- to find a partition $U_1 \dot\cup U_2 \ldots \dot\cup U_r = V(G)$ of the vertices of G such that $G \models \varphi(U_i)$ for all $i = 1, \ldots, r$, or
- to confirm that no such partition of $V(G)$ exists.

For example, if $\varphi(X)$ expresses that X is an independent set, then the φ-MSO PARTITIONING problem decides if G has an r-coloring, and thus, finding minimum feasible r (simply by trying $r = 1, 2, \ldots$) solves the CHROMATIC NUM-BER problem. Similarly, if $G \models \varphi(X)$ when X is a dominating set, minimizing r solves the DOMATIC NUMBER problem, and so on.

Rao [22] showed an algorithm for MSO PARTITIONING, for any MSO_2 formula φ, on a graph G with treewidth $tw(G) = \tau$ running in time $r^{f(\varphi,\tau)}|V(G)|$ (XP) for some computable function f. Our next result widely generalizes this to SCO over MSO-definable sets.

Definition 3.4 (MSO-definable sets). *For a graph G on $|V(G)| = n$ vertices, we interpret a $0/1$ vector $\mathbf{x} \in \{0,1\}^n$ as the set $X \subseteq V$ where $v \in X$ iff $x_v = 1$. We then say that $\mathbf{x}$ satisfies a formula φ if $G \models \varphi(X)$. Let*

$$S_\varphi(G) = \{\mathbf{x} \mid \mathbf{x} \text{ satisfies } \varphi \text{ in } G\}.$$

Let $\mathbf{c}$ be defined as $c_i^1 := 1$ for $1 \le i \le n$ and $c_i^j := -1$ for $2 \le j \le r$ and $1 \le i \le n$. Observe then the following: deciding whether the shifted problem with $S = S_\varphi(G)$, $\mathbf{c}$ and r, has an optimum of value n is equivalent to solving the MSO PARTITIONING problem for φ.

Theorem 3.5 (*). *Let G be a graph of treewidth $tw(G) = \tau$, let φ be an MSO_2 formula and $S_\varphi(G) = \{\mathbf{x} \mid \mathbf{x} \text{ satisfies } \varphi\}$. There is an algorithm solving the shifted problem with $S = S_\varphi(G)$ and any given $\mathbf{c}$ and r in time $r^{f(\varphi,\tau)} \cdot |V(G)|$ for some computable function f. In other words, for parameters φ and τ, the problem is in the complexity class XP.*

We will prove Theorem 3.5 using Lemma 3.3 on separable optimization over decomposable polyhedra. To that end we need to show the following two steps:

1. There is a $0/1$ extension Q of the polytope $P = \text{conv}(S_\varphi(G))$ which is decomposable and endowed with a decomposition oracle (Definition 3.1),
2. there is an integer separable minimization oracle (Definition 3.2) for the polytope rQ.

The first point is implied by a recent result of Kolman, Koutecký and Tiwary:

Proposition 3.6 ([15][1]). *Let G be a graph on n vertices of treewidth $tw(G) = \tau$, and φ be an MSO_2 formula with one free vertex-set variable. Then, for some computable functions f_1, f_2, f_3, there are matrices $\mathbf{A}, \mathbf{B}$ and a vector $\mathbf{b}$, computable in time $f_1(\varphi, \tau) \cdot n$, such that*

1. *the polytope $Q = \{(\mathbf{x}, \mathbf{y}) \mid \mathbf{A}\mathbf{x} + \mathbf{B}\mathbf{y} = \mathbf{b}, \mathbf{y} \geq \mathbf{0}\}$ is a 0/1 polytope which is an extension of the polytope $P = \mathrm{conv}(S_\varphi(G))$, and $Q \subseteq [0,1]^{f_2(\varphi,\tau)n}$,*
2. *Q is decomposable and endowed with a decomposition oracle, and*
3. *the (Gaifman) treewidth of the matrix $(\mathbf{A}\,\mathbf{B})$ is at most $f_3(\varphi, \tau)$.*

The second requirement of Lemma 3.3 follows from efficient solvability of the constraint satisfaction problem (CSP) of bounded treewidth, originally proven by Freuder [7]. We will use a natural weighted version of this folklore result.

For a CSP instance $I = (V, \mathcal{D}, \mathcal{H}, \mathcal{C})$ one can define the *constraint graph* of I as $G = (V, E)$ where $E = \{\{u, v\} \mid (\exists C_U \in \mathcal{H}) \lor (\exists w_U \in \mathcal{C})$ s.t. $\{u, v\} \subseteq U\}$. The *treewidth of a CSP instance* I is defined as the treewidth of the constraint graph of I.

Proposition 3.7 ([7]). *Given a CSP instance I of treewidth τ and maximum domain size $D = \max_{u \in V} |D_u|$, a minimum weight solution can be found in time $\mathcal{O}(D^\tau(n + |\mathcal{H}| + |\mathcal{C}|))$.*

Proposition 3.7 can be used to realize an integer separable minimization oracle for integer programs of bounded treewidth, as follows.

Lemma 3.8 (*). *Let $\mathbf{A} \in \mathbb{Z}^{n \times m}, \mathbf{b} \in \mathbb{Z}^m, \boldsymbol{\ell}, \mathbf{u} \in \mathbb{Z}^n$ be given s.t. $tw(\mathbf{A}) = \tau$, and let $D = \|\mathbf{u} - \boldsymbol{\ell}\|_\infty$. Then an integer separable minimization oracle over $P = \{\mathbf{x} \mid \mathbf{A}\mathbf{x} = \mathbf{b}, \boldsymbol{\ell} \leq \mathbf{x} \leq \mathbf{u}\}$ is realizable in time $D^\tau(n + m)$.*

The proof of Lemma 3.8 proceeds by constructing a CSP instance I based on the ILP $\mathbf{A}\mathbf{x} = \mathbf{b}, \boldsymbol{\ell} \leq \mathbf{x} \leq \mathbf{u}$, such that solving I corresponds to integer separable minimization over P. Since the treewidth of I is τ and the maximum domain size is D, Proposition 3.7 does the job.

Now, in our case of rQ where Q is a 0/1 polytope, we have got $D \leq r$. Consequently, we can finish the proof of Theorem 3.5 by using Lemma 3.3.

Besides treewidth, another useful width measure of graphs is the *clique-width* of a graph G which, briefly, means the minimum number k of labels needed to construct G along a so called k-expression. Rao's result [22] applies also to the MSO partitioning problem for MSO_1 formulas and graphs of bounded clique-width. We show the analogous extension of Theorem 3.5 next.

Corollary 3.9 (*). *Let G be a graph of clique-width $cw(G) = \gamma$ given along with a γ-expression, let ψ be an MSO_1 formula and $S_\psi(G) = \{\mathbf{x} \mid \mathbf{x}$ satisfies $\psi\}$. There is an algorithm solving the shifted problem with $S = S_\psi(G)$ and any given $\mathbf{c}$ and r in time $r^{f(\psi,\gamma)} \cdot |V(G)|$ for some computable f.*

---

[1] We remark that while [15, Theorem 4] explicitly speaks about generally weaker MSO_1 logic, it is folklore that in the realm of graphs of bounded treewidth the same can be equivalently stated with MSO_2 logic (as noted also in [15]).

While it is possible to prove Corollary 3.9 along the same lines as used above, we avoid repeating the previous arguments and, instead, apply the following technical tool which extends the known fact that a class of graphs is of bounded clique-width iff it has an MSO_1 interpretation in the class of rooted trees.

Lemma 3.10 (*). *Let G be a graph of clique-width $cw(G) = \gamma$ given along with a γ-expression Γ constructing G, and let ψ be an MSO_1 formula. One can, in time $\mathcal{O}(|V(G)| + |\Gamma| + |\psi|)$, compute a tree T and an MSO_1 formula φ such that $V(G) \subseteq V(T)$ and*

$$for\ every\ X\ it\ is\ T\ \models\ \varphi(X),\ iff\ X \subseteq V(G)\ and\ G \models \psi(X).$$

With Lemma 3.10 at hand, it is easy to derive Corollary 3.9 from previous Theorem 3.5 applied to the tree T.

Finally, we add a small remark regarding the input G in Corollary 3.9; we are for simplicity assuming that G comes along with its γ-expression since it is currently not known how to efficiently construct a γ-expression for an input graph of fixed clique-width γ. Though, one may instead use the result of [11] which constructs in FPT a so-called rank-decomposition of G which can be used as an approximation of a γ-expression for G (with up to an exponential jump, but this does not matter for a fixed parameter γ in theory).

4 MSO-definable Sets: W[1]-Hardness

Recall that natural hard graph problems such as CHROMATIC NUMBER are instances of MSO PARTITIONING and so also instances of shifted combinatorial optimization. While we have shown an XP algorithm for SCO with MSO-definable sets on graphs of bounded treewidth and clique-width in Theorem 3.5 and Corollary 3.9, it is a natural question whether an FPT algorithm could exist for this problem, perhaps under a more restrictive width measure.

Here we give a strong negative answer to this question. First, we note the result of Fomin et al. [6] proving W[1]-hardness of CHROMATIC NUMBER parameterized by the clique-width of the input graph. This immediately implies that an FPT algorithm in Corollary 3.9 would be very unlikely. Although, CHROMATIC NUMBER is special in the sense that it is solvable in FPT when parameterized by the treewith of the input. Here we prove that it is not the case of MSO PARTITIONING problems and SCO in general, even when considering restricted MSO_1 formulas and shifted **c**, and parameterizing by a much more restrictive *treedepth* parameter.

Definition 4.1 (Treedepth). *Let the* height *of a rooted tree or forest be the maximum root-to-leaf distance in it. The* closure *$cl(F)$ of a rooted forest F is the graph obtained from F by making every vertex adjacent to all of its ancestors. The* treedepth *$td(G)$ of a graph G is one more than the minimum height of a forest F such that $G \subseteq cl(F)$.*

Note that always $td(G) \geq tw(G) + 1$ since we can use the vertex sets of the root-to-leaf paths of the forest F (from Definition 4.1) in a proper order as the bags of a tree-decomposition of G.

Theorem 4.2 (*). *There exists a graph FO formula $\varphi(X)$ with a free set variable X, such that the instance of the* MSO PARTITIONING *problem given by φ, is* W*[1]-hard when parameterized by the treedepth of an input simple graph G.*

Consequently, the shifted problem with $S_\varphi(G)$ is also W*[1]-hard (for suitable **c**) when parameterized by the treedepth of G.*

The way we approach Theorem 4.2 is by a reduction from W[1]-hardness of CHROMATIC NUMBER with respect to clique-width [6], in which we exploit some special hidden properties of that reduction, as extracted in [8]. In a nutshell, the "difficult cases" of [6] can be interpreted in a special way into labeled rooted trees of small height, and here we further trade the labels (the number of which is the parameter) for increased height of a tree and certain additional edges belonging to the tree closure.

5   Conclusions and Open Problems

We close with several open problems we consider interesting and promising.

Parameterizing by r. It is interesting to consider taking r as a parameter. For example, Fluschnik et al. [5] prove that the MINIMUM SHARED EDGES problem is FPT parameterized by the number of paths. Omran et al. [20] prove that the MINIMUM VULNERABILITY problem is in XP with the same parameter. Since both problems are particular cases of the shifted problem, we ask whether the shifted problem with S being the set of $s - t$ paths of a (di)graph lies in XP or is NP-hard already for some constant r.

Further uses of Lemma 3.3. For example, which interesting combinatorial sets S can be represented as n-fold integer programs [18,21] such that the corresponding polyhedra are decomposable?

Approximation. The MINIMUM VULNERABILITY problem has also been studied from the perspective of approximation algorithms [20]. What can be said about the approximation of the shifted problem?

Going beyond 0/1. The results in Sect. 2 are the only known ones in which S does not have to be 0/1. What can be said about the shifted problem with such sets S that are not given explicitly, e.g., when S is given by a totally unimodular system?

References

1. Assadi, S., Emamjomeh-Zadeh, E., Norouzi-Fard, A., Yazdanbod, S., Zarrabi-Zadeh, H.: The minimum vulnerability problem. Algorithmica **70**(4), 718–731 (2014)
2. Baum, S., Trotter Jr., L.E.: Integer rounding and polyhedral decomposition for totally unimodular systems. In: Henn, R., Korte, B., Oettli, W. (eds.) Optimization and Operations Research, pp. 15–23. Springer, Heidelberg (1978)
3. Bredereck, R., Faliszewski, P., Niedermeier, R., Skowron, P., Talmon, N.: Elections with few candidates: prices, weights, and covering problems. In: Walsh, T. (ed.) ADT 2015. LNCS, vol. 9346, pp. 414–431. Springer, Cham (2015). doi:10.1007/978-3-319-23114-3_25
4. Chen, J., Huang, X., Kanj, I.A., Xia, G.: Strong computational lower bounds via parameterized complexity. J. Comput. Syst. Sci. **72**(8), 1346–1367 (2006)
5. Fluschnik, T., Kratsch, S., Niedermeier, R., Sorge, M.: The parameterized complexity of the minimum shared edges problem. In: FSTTCS 2015, vol. 45, pp. 448–462. LIPIcs, Schloss Dagstuhl (2015)
6. Fomin, F., Golovach, P., Lokshtanov, D., Saurab, S.: Clique-width: on the price of generality. In: SODA 2009, pp. 825–834. SIAM (2009)
7. Freuder, E.C.: Complexity of K-tree structured constraint satisfaction problems. In: Proceedings of the 8th National Conference on Artificial Intelligence, pp. 4–9 (1990)
8. Gajarský, J., Lampis, M., Ordyniak, S.: Parameterized algorithms for modular-width. In: Gutin, G., Szeider, S. (eds.) IPEC 2013. LNCS, vol. 8246, pp. 163–176. Springer, Cham (2013). doi:10.1007/978-3-319-03898-8_15
9. Ganian, R., Hliněný, P., Nešetřil, J., Obdržálek, J., Ossona de Mendez, P., Ramadurai, R.: When trees grow low: shrubs and fast MSO_1. In: Rovan, B., Sassone, V., Widmayer, P. (eds.) MFCS 2012. LNCS, vol. 7464, pp. 419–430. Springer, Heidelberg (2012). doi:10.1007/978-3-642-32589-2_38
10. Gijswijt, D.: Integer decomposition for polyhedra defined by nearly totally unimodular matrices. SIAM J. Discrete Math. **19**(3), 798–806 (2005)
11. Hliněný, P., Oum, S.: Finding branch-decompositions and rank-decompositions. SIAM J. Comput. **38**(3), 1012–1032 (2008)
12. Hochbaum, D.S., Shanthikumar, J.G.: Convex separable optimization is not much harder than linear optimization. J. ACM **37**(4), 843–862 (1990)
13. Kaibel, V., Onn, S., Sarrabezolles, P.: The unimodular intersection problem. Oper. Res. Lett. **43**(6), 592–594 (2015)
14. Knop, D., Koutecký, M., Masařík, T., Toufar, T.: Simplified algorithmic metatheorems beyond MSO: treewidth and neighborhood diversity, 1 March 2017. arXiv:1703.00544
15. Kolman, P., Koutecký, M., Tiwary, H.R.: Extension complexity, MSO logic, and treewidth, 28 February 2017. arXiv:1507.04907
16. Kreutzer, S.: Algorithmic meta-theorems. In: Electronic Colloquium on Computational Complexity (ECCC), vol. 16, p. 147 (2009)
17. Levin, A., Onn, S.: Shifted matroid optimization. Oper. Res. Lett. **44**, 535–539 (2016)
18. De Loera, J.A., Hemmecke, R., Köppe, M.: Algebraic and Geometric Ideas in the Theory of Discrete Optimization. MOS-SIAM Series on Optimization, vol. 14. SIAM, Philadelphia (2013)

19. Oertel, T., Wagner, C., Weismantel, R.: Integer convex minimization by mixed integer linear optimization. Oper. Res. Lett. **42**(6–7), 424–428 (2014)
20. Omran, M.T., Sack, J.-R., Zarrabi-Zadeh, H.: Finding paths with minimum shared edges. J. Comb. Optim. **26**(4), 709–722 (2013)
21. Onn, S.: Nonlinear discrete optimization. Zurich Lectures in Advanced Mathematics. European Mathematical Society. http://ie.technion.ac.il/~onn/Book/NDO.pdf
22. Rao, M.: MSOL partitioning problems on graphs of bounded treewidth and clique-width. Theor. Comput. Sci. **377**(1–3), 260–267 (2007)
23. Schrijver, A.: Combinatorial Optimization: Polyhedra and Effciency. Algorithms and Combinatorics, vol. 24. Springer, Heidelberg (2003)
24. Zambelli, G.: Colorings of k-balanced matrices and integer decomposition property of related polyhedra. Oper. Res. Lett. **35**(3), 353–356 (2007)
25. Ziegler, G.M.: Lectures on Polytopes. Graduate Texts in Mathematics, vol. 152. Springer, New York (1995)

Approximate Minimum Diameter

Mohammad Ghodsi[1,2], Hamid Homapour[1(✉)], and Masoud Seddighin[1(✉)]

[1] Sharif University of Technology, Tehran, Iran
ghodsi@sharif.edu, {homapour,mseddighin}@ce.sharif.edu
[2] IPM - Institute for Research in Fundamental Sciences, Tehran, Iran

Abstract. We study the minimum diameter problem for a set of inexact points. By inexact, we mean that the precise location of the points is not known. Instead, the location of each point is restricted to a continuous region (*Imprecise* model) or a finite set of points (*Indecisive* model). Given a set of inexact points in one of *Imprecise* or *Indecisive* models, we wish to provide a lower-bound on the diameter of the real points.

In the first part of the paper, we focus on *Indecisive* model. We present an $O(2^{\frac{1}{\epsilon^d}} \cdot \epsilon^{-2d} \cdot n^3)$ time approximation algorithm of factor $(1 + \epsilon)$ for finding minimum diameter of a set of points in d dimensions. This improves the previously proposed algorithms for this problem substantially.

Next, we consider the problem in *Imprecise* model. In d-dimensional space, we propose a polynomial time $\sqrt{d}$-approximation algorithm. In addition, for $d = 2$, we define the notion of α-separability and use our algorithm for *Indecisive* model to obtain $(1+\epsilon)$-approximation algorithm for a set of α-separable regions in time $O(2^{\frac{1}{\epsilon^2}} \cdot \frac{n^3}{\epsilon^{10} \cdot \sin(\alpha/2)^3})$.

Keywords: Indecisive · Imprecise · Computational Geometry · Approximation algorithms · Core-set

1 Introduction

Rapid growth in the computational technologies and the vast deployment of sensing and measurement tools turned Big Data to an interesting and interdisciplinary topic that attracted lots of attentions in the recent years.

One of the critical issues in Big Data is dealing with uncertainty. Computational Geometry (CG) is one of the fields, that is deeply concerned with large scale data and the issue of imprecision. The real data for a CG problem is often gathered by measurement and coordination instrument (GPS, Digital compass, etc.) which are inexact. In addition, the movement is an unavoidable source of inaccuracy. Hence, the data collected by these tools are often noisy and suffer from inaccuracies. In spite of this inaccuracy, most of the algorithms are based on the fact the input data are precise. These algorithms become useless in the presence of inaccurate data. Thus, it is reasonable to design algorithms that take this inaccuracy into account.

© Springer International Publishing AG 2017
Y. Cao and J. Chen (Eds.): COCOON 2017, LNCS 10392, pp. 237–249, 2017.
DOI: 10.1007/978-3-319-62389-4_20

Point is the primary object in a geometric problem. Roughly speaking, the input for most of the CG problems is a set of points in different locations of a metric space. The exact location of these points is one of the parameters that might be inaccurate. To capture this inaccuracy in geometric problems, several models have been proposed, e.g. Uncertain, Indecisive, Imprecise and Stochastic models [9]. Suppose that $\mathbb{P} = \{\rho_1, \rho_2, \ldots, \rho_n\}$ is the set of exact points. Due to the measurement inaccuracy, $\mathbb{P}$ is unknown to us. In the following, we give a short description of the way that each model tries to represent $\mathbb{P}$:

- **Uncertain (or locational uncertain) model**: in the *uncertain* model, the location of each point is determined by a probability distribution function. Formally, for every point $\rho_i \in \mathbb{P}$ a probability distribution $\mathcal{D}_i$ is given in input, that represent the probability that ρ_i appears in every location.
- **Indecisive model**: in the *indecisive* model, each input point can take one of k distinct possible locations. In other words, for each real point $\rho_i \in \mathbb{P}$, a set of k points is given, where the location of ρ_i equals to one of these points.
- **Imprecise model**: in the *imprecise* model, possible locations of a point is restricted to a region, i.e., for each point $\rho \in \mathbb{P}$, a finite region R_i is given in input. All we know is $\rho_i \in R_i$.
- **Stochastic model**: every point p in input has a deterministic location, but there is a probability that $p \notin \mathbb{P}$. *Stochastic* model is used for the case that false reading is possible, usually in database scenarios [2,4].

Let C be an input related to one of these models, that represents the exact point set $\mathbb{P}$. The question is, what information can we retrieve from C? For brevity, suppose that the objective is to compute a statistical value $D(\mathbb{P})$ (e.g., $D(\mathbb{P})$ can be the diameter or width of $\mathbb{P}$). Trivially, finding the exact value of $D(\mathbb{P})$ is not possible, since we don't have access to the real points $\mathbb{P}$. Instead, there may be several approaches, based on the input model. For example, one can calculate the distribution of $D(\mathbb{P})$ (for uncertain model), or can provide upper bound or lower bound on the value of $D(\mathbb{P})$ (for Indecisive and Imprecise models) [9,11].

In this paper, we investigate on *Indecisive* and *Imprecise* models and consider the diameter as the objective function. Furthermore, our approach is to compute bounds on the output. Thus, the general form of the problem we wish to observe is as follows: Given an input C corresponding to an *Indecisive* (*Imprecise*) data model of a set $\mathbb{P}$ of exact data. The goal is to lower-bound the value of the $D(\mathbb{P})$, i.e., finding a lower-bound on the minimum possible value for $D(\mathbb{P})$. In Sect. 1.1 you can find a formal definition of the problem. It is worth mentioning that this problem has many applications in computer networking and databases [6,7,14].

1.1   Model Definition and Problem Statement

Indecisive model. As mentioned, in *Indecisive* model, each actual point $\rho_i \in \mathbb{P}$ is represented via a finite set of points with different locations. For simplicity, suppose that we color the representatives of each point with a unique color. Thus, the input for a problem in *Indecisive* model is a set $\mathcal{P} = \{\mathcal{P}_1, \mathcal{P}_2, \ldots, \mathcal{P}_m\}$, where each $\mathcal{P}_i$ is a set of points representing alternatives of ρ_i, i.e., ρ_i is one of the elements in $\mathcal{P}_i$. All the points in $\mathcal{P}_i$ are colored with color $\mathcal{C}_i$. For Indecisive model, we assume that total number of the points is n, i.e. $\sum_i |\mathcal{P}_i| = n$.

In addition to uncertainty, this model can represent various situations. For example, consider an instance of a resource allocation problem where we have a set of resources and each resource has a number of copies (or alternatives) and the solutions are constrained to use exactly one of the alternatives. Indecisive model can be used to represent this kind of issues. To do this, each resource can be represented by a point in d dimensional Euclidean space. To represent the type of resources, each point is associated with a color indicating its type.

Imprecise model. In *Imprecise* model, the possible locations of a point is restricted by a finite and continuous area. The input for an imprecise problem instance is a set $\mathcal{R} = \{\mathcal{R}_1, \mathcal{R}_2, \ldots, \mathcal{R}_n\}$, where for every region $\mathcal{R}_i$, we know that $\rho_i \in \mathcal{R}_i$. Therefore, each point of $\mathcal{R}_i$ is a possible location for the actual point ρ_i.

This model can be applied in many situations. For example, we know that all measurements have a degree of uncertainty regardless of precision and accuracy. To represent an actual point, we can show the amount of inaccuracy with a circle (i.e. as a region) which is centered at the measured point with a radius equals to the size of tolerance interval in the measurement tool, which means that the actual point lies somewhere in this circle.

Problem Statement. For brevity, we use *MinDCS* and *MinDiam* to refer to the problem in models *Indecisive* and *Imprecise*, respectively. In the following, we formally define *MinDCS* and *MinDiam* problems.

Problem 1 (*MinDCS*). Given a set $\mathcal{P} = \{\mathcal{P}_1, \mathcal{P}_2, \ldots, \mathcal{P}_m\}$, with each $\mathcal{P}_i$ being a finite set of points in d-dimensional euclidean space with the same color $\mathcal{C}_i$ and $\sum_i |\mathcal{P}_i| = n$. A *color selection* of $\mathcal{P}$ is a set of m points, one from each $\mathcal{P}_i$. Find a color selection S of $\mathcal{P}$ so that the diameter of these points is the smallest among all options.

Denote by D_{min}, the diameter of the desired selection in *MinDCS*. Formally:

$$D_{min} = \min_{S = Sel(\mathcal{P})} \left(\max_{\forall p_i, p_j \in S} ||p_i - p_j|| \right),$$

where $Sel(\mathcal{P})$ is the set of all possible color selections of $\mathcal{P}$. Furthermore, we define OPT as the color selection that the diameter of its points is smallest among all possible color selections:

$$OPT = \arg \min_{S = Sel(\mathcal{P})} \left(\max_{\forall p_i, p_j \in S} ||p_i - p_j|| \right).$$

We also denote by $D(\mathcal{Y})$, the diameter of the point set $\mathcal{Y}$.

Problem 2 (*MinDiam*). In this problem, a set $\mathcal{R} = \{\mathcal{R}_1, \mathcal{R}_2, \ldots, \mathcal{R}_n\}$ is given, where each $\mathcal{R}_i$ is a bounded convex region in d-dimensional euclidean space.

We want to select one point from each region, such that the diameter of the selected points is minimized. Formally, we want to select a set $S = \{p_1, p_2, \ldots, p_n\}$ of points, where $p_i \in \mathcal{R}_i$ such that $D(S)$ is the smallest possible.

1.2   Related Works

For *Indecisive* model, Fan et al. [10] suggested a randomized algorithm with the time complexity $O(n^{1+\epsilon})$ for $(1 + \epsilon)$-approximation of the maximum diameter, where ϵ could be an arbitrarily small positive constant.

Zhang et al. [14] suggested an $O(n^k)$ time brute force algorithm to find minimum possible diameter in *Indecisive* model. Furthermore, Fleischer and Xu [6,7] showed that this problem can be solved in polynomial time for the L_1 and L_∞ metrics, while it is NP-hard for all other L_p metrics, even in two dimensions. They also gave an approximation algorithm with a constant approximation factor. By extending the definition of ϵ-kernels from Agarwal et al. [1] to avatar ϵ-kernels (the notion of avatar is the same as *Indecisive* model) in d dimensional Euclidean space, Consuegra et al. [3] proposed an $(1 + \epsilon)$-approximation algorithm with running time $O((n)^{(2d+3)} \cdot \frac{m}{\delta^d} \cdot (2d)^2 \cdot 2^{\frac{1}{\delta^d}} (\frac{2}{\delta^{d-1}})^{\lfloor \frac{d}{2} \rfloor} + (\frac{2}{\delta^{d-1}})^a)$ (where δ is the side length of the grid cells for constructing core-set, d is the dimension of the points, k is the maximum frequency of the alternatives of a points, and a is a small constant). Furthermore, for *Uncertain* model, Fan et al. [5] proposed an approximation algorithm to compute the expected diameter of a point set.

In *Imprecise* model, in the Euclidean plane, Löffler and van Kreveld [12] proposed an $O(n \log n)$ time algorithm for the minimum diameter of a set of imprecise points modeled as squares and a $(1 + \epsilon)$-approximation algorithm with running time $O(n^{c\epsilon^{-\frac{1}{2}}})$ for the points modeled as discs, where $c = 6.66$.

1.3   Our Results and Techniques

The first part of our work (Sect. 2) is devoted to the diameter problem in *Indecisive* model. In this case, we present an $O(2^{\frac{1}{\epsilon^d}} \cdot \frac{1}{\epsilon^{2d}} \cdot n^3)$ time approximation algorithm of factor $(1 + \epsilon)$. This improves the previous $O((n)^{(2d+3)} \cdot \frac{m}{\delta^d} \cdot (2d)^2 \cdot 2^{\frac{1}{\delta^d}} (\frac{2}{\delta^{d-1}})^{\lfloor \frac{d}{2} \rfloor} + (\frac{2}{\delta^{d-1}})^a)$ time algorithm substantially. The idea is to build several grids with different side-lengths and round the points in $\mathcal{P}$ to the grid points.

In the second part of our work (Sect. 3) we study the problem in *Imprecise* model. In d-dimensional space, we propose a polynomial time $\sqrt{d}$-approximation algorithm. We obtain this result by solving a linear program that formulates a relaxed version of the problem. Next, for $d = 2$, we define the notion of α-separability and use a combination of our previous methods to obtain a polynomial time $(1 + \epsilon)$-approximation algorithm for α-separable regions. For this purpose, we first determine a lower-bound on the minimum diameter by the $\sqrt{d}$ approximation algorithm, and convert the problem into an *Indecisive* instance. Next, we use the results for *Indecisive* model to obtain a $(1 + \epsilon)$-approximation.

2 Approximation Algorithm for *MinDCS*

We use a powerful technique that is widely used to design approximation algorithms for geometric problems. The technique is based on the following steps [1,8,13]:

1. Extract a small subset of data that inherited the properties of original data (i.e., coreset)
2. Run lazy algorithm on the coreset.

For a set P of points in R^d, and an optimization problem $\mathcal{X}$ (e.g., $\mathcal{X}(P)$ would be the diameter or width of P), a subset Q of the points of P is an ϵ-coreset, if

$$(1 - \epsilon)\mathcal{X}(P) \leq \mathcal{X}(Q) \leq (1 + \epsilon)\mathcal{X}(P).$$

We state this fact, by saying that Q is an ϵ-coreset of P for $\mathcal{X}(.)$. Considering the input data in indecisive model, we have the following definition of coreset: given a set P of n points colored with m colors. We say Q is an ϵ-coreset of P for $MinDCS$ iff: (i) Q contains at least one point of each color, (ii) $D(Q) \leq (1 + \epsilon)D_{min}$.

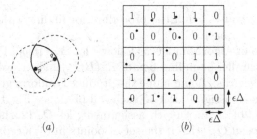

(a)   (b)

Fig. 1. (a) The possible area of two points p and q. (b) A binary assignment for a grid.

2.1 Approximate Minimum Diameter

Definition 1. *Given a set of colored points P and a set C of colors. P is C-legal iff for each $c \in C$ there exists a point $p \in P$ with color c.*

Definition 2 (Possible area). *Consider two points p and q. Draw two balls of radius $|pq|$, one of them centered at p and the other centered at q. Name the intersection area of these two balls as possible area (C_{pq}), see Fig. 1(a).*

Observation 1. *If p and q be the points that determine the diameter of a point set, all the points in the set must lay in C_{pq}.*

Regarding Observation 1, we compute an ϵ-approximation of $MinDCS$ by the process described in Algorithm 1. The algorithm operates as follows: let $S = \{S_1, S_2, \ldots\}$ be the set of all pairs of points in $\mathcal{P}$. For each $S_i = \{p_i, q_i\}$, let P_i be the set of points in $C_{p_i q_i}$. For each P_i which is C-legal, we compute an approximation of minimum diameter of P_i by Algorithm 2, as will be described further. Next, among all computations, we choose the one with the minimum value (D_{alg}).

Note that, since we consider all pairs of the points, for some C-legal pair S_i we have $D_{min} = |p_i q_i|$ and hence D_{alg} is an ϵ-approximation of D_{min}.

Algorithm 1. Minimum Diameter Approximation

1: **function** MINDIAMETERAPX($\mathcal{P}$)
2:     $D_{alg} = null$
3:     **for each** $S_i \in \mathcal{S}$ **do**
4:         **if** P_i is C-legal **then**
5:             $D_{apx}(P_i) = DiameterAPX(P_i, \{p_i, q_i\})$
6:             $D_{alg} = \min\{D_{alg}, D_{apx}(P_i)\}$
7:         **end if**
8:     **end for**
9:     **return** D_{alg}
10: **end function**

In Algorithm 2, you can find the procedure for finding an approximation of minimum diameter for each P_i.

The description of Algorithm 2 is as follows: let $\Delta = |p_i q_i|$. First, we compute the smallest axis parallel bounding box of P_i ($B(P_i)$). Next, we split $B(P_i)$ into the cells with side lengths $\epsilon \Delta$ and name the produced uniform grid as $\mathbb{G}$. A *binary assignment* of $\mathbb{G}$ is to assign 0 or 1 to each cell of $\mathbb{G}$, see Fig. 1(b). Consider all binary assignments of $\mathbb{G}$. In the jth assignment, let Q_j be the set of the cells with value '1'. We call Q_j *legal*, if the set of points in Q_j's cells is C-legal.

Number of the cells in $\mathbb{G}$ is $O(\frac{1}{\epsilon^d})$ and hence, there are at most $O(2^{\frac{1}{\epsilon^d}})$ legal assignments. For each cell of a legal Q_j, choose an arbitrary point in that cell as a representative. Next, we compute the diameter of the representatives in time $O((\frac{1}{\epsilon^d})^2)$. Regarding all the computations, we return the minimum of them as an approximation of $D(P_i)$.

Note that if S_i would be the optimum color selection of P_i, then obviously there exists an assignment j, such that Q_j is legal and includes the cells corresponding to the points in S_i.

In order to determine whether or not Q_j is legal, we can check in $O(n)$ time the existence of each color in at least one cell of Q_j.

Finally, for all C-legal P_i, we select the smallest among all approximations of P_i as an approximation of D_{min}.

Theorem 2. *The MinDCS problem can be approximated in time $O(2^{\frac{1}{\epsilon^d}} . \epsilon^{-2d} . n^3)$ of factor $(1 + \epsilon)$ for fixed dimensions.*

Proof. Let D_{alg} be the value returned by our algorithm, and $p, q \in OPT$ be the points with maximum distance in the optimal solution. Obviously, the set

Algorithm 2. Approximate D_{P_i} Respect to $p, q \in S_i$

1: **function** DIAMETERAPX(P_i, S_i)
2:     Let $S_i = \{p_i, q_i\}$
3:     $\Delta = |p_i q_i|$
4:     $D(P_i) = null$
5:     Let $\mathbb{G}$ be a uniform grid on $B(P_i)$ in d dimensional space with cells of size $\epsilon\Delta$
6:     **for each** binary assignment of cells of $\mathbb{G}$ **do**
7:         Let Q_j be the cells that is assigned a value of 1
8:         **if** Q_j is legal **then**
9:             Let Q'_j be the set of representative points of cells
10:            $D(P_i) = \min\{D(P_i), D(Q'_j)\}$
11:        **end if**
12:    **end for**
13:    **return** $D(P_i)$
14: **end function**

---

of points in C_{pq} is C-legal, and all the points in OPT are in C_{pq}. Consider the binary assignment corresponding to OPT, i.e., a cell is 1, iff it contains at least one point of OPT. Since this assignment is C-legal, it would be considered by our algorithm. Thus, $D_{alg} \leq (1+\epsilon)D(OPT)$. On the other hand, the assignment related to D_{alg} is also C-legal. Hence, $D_{alg} \geq (1-\epsilon)D(OPT)$.

There are n^2 different possible areas. For each of them we have $2^{\frac{1}{\epsilon^d}}$ different assignments. Checking the legality of each assignment takes $O(n)$ time. For a C-legal assignment we can find the minimum diameter in $O(\epsilon^{-2d})$ time since each assignment contains at most ϵ^{-d} cells. Thus, total running time would be $O(2^{\frac{1}{\epsilon^d}}.\epsilon^{-2d}.n^3)$.

It is worth mentioning that we can improve the running time to $O(2^{\frac{1}{\epsilon^d}}.(\epsilon^{-2d} + n^3))$ by a preprocessing phase that computes the diameter for every $2^{\frac{1}{\epsilon^d}}$ different binary assignments and uses these preprocessed values for every pair of S.

3   Approximation Algorithm for *MinDiam*

In this section, we consider the problem of finding minimum diameter in *Imprecise* model. As mentioned, in this model an imprecise point is represented by a continuous region. Given a set $\mathcal{R} = \{\mathcal{R}_1, \mathcal{R}_2, \ldots, \mathcal{R}_n\}$ where each $\mathcal{R}_i$ is a polygonal region. A **selection** $S = \{s_1, s_2, ..., s_n\}$ of $\mathcal{R}$'s is a set of points where $s_i \in \mathcal{R}_i$. In the minimum diameter problem, the goal is to find a selection S such that $D(S)$ is minimized among all options. We consider the problem for the case where every $\mathcal{R}_i$ is a polygonal convex region with constant complexity (i.e., number of edges).

In Sect. 3.1, we present a polynomial time $\sqrt{d}$-approximation algorithm for *MinDiam*.

3.1 $\sqrt{d}$-approximation

Our $\sqrt{d}$-approximation construction is based on LP-programming which uses the rectilinear distances (d_{l_1}) of the points. For now, suppose that d is constant and consider LP 1. In this LP, we want to select the points $s_1, s_2, \ldots, s_n$ such that $s_i \in \mathcal{R}_i$ and the rectilinear diameter of these points is minimized.

$$
\begin{aligned}
&\textbf{LP 1}: \\
&\text{minimize} \quad \ell \\
&\text{subject to} \quad d_{l_1}(s_i,\ s_j) \le \ell \qquad \forall i, j \\
&\qquad\qquad\qquad\ s_i \in \mathcal{R}_i \quad \forall i, j
\end{aligned}
\tag{1}
$$

We show that $d_{l_1}(s_i, s_j)$ can be expressed by 2^d linear constraints (note that $d_{l_1}(s_i, s_j) = |s_{i,1} - s_{j,1}| + \ldots + |s_{i,d} - s_{j,d}|$). Based on the relative value of $s_{i,k}, s_{j,k}$, there are 2^d different possible expressions for $d_{l_1}(s_i, s_j)$. For example, for $d = 2$, we have:

$$
d_{l_1}(s_i,\ s_j) = \begin{cases}
s_{i,1} - s_{j,1} + s_{i,2} - s_{j,2} & s_{i,1} \ge s_{j,1} \quad s_{i,2} \ge s_{j,2} \\
s_{j,1} - s_{i,1} + s_{i,2} - s_{j,2} & s_{i,1} < s_{j,1} \quad s_{i,2} \ge s_{j,2} \\
s_{i,1} - s_{j,1} + s_{j,2} - s_{i,2} & s_{i,1} \ge s_{j,1} \quad s_{i,2} < s_{j,2} \\
s_{j,1} - s_{i,1} + s_{j,2} - s_{i,2} & s_{i,1} < s_{j,1} \quad s_{i,2} < s_{j,2}
\end{cases}
$$

Call these expressions as $e_1, e_2, \ldots, e_{2^d}$. Note that if $d_{l_1}(s_i,\ s_j) \le \ell$, then for all $1 \le k \le 2^n$, $e_k \le \ell$. With this in mind, we can replace $d_{l_1}(s_i,\ s_j) \le \ell$ with these 2^d linear expressions. It is worth mentioning that we can reduce the number of expressions of this type in LP 1 using some additional variables, see Sect. 3.2 for more details.

Since the regions are convex, the constraint $s_i \in \mathcal{R}_i$ can be described by $|\mathcal{R}_i|$ linear inequalities where $|\mathcal{R}_i|$ is the complexity of region $\mathcal{R}_i$. Thus, the number of the constraints of this type is $|\mathcal{R}| = \sum_i |\mathcal{R}_i|$. Therefore, total number of the constraints would be $2^d \cdot n^2 + |\mathcal{R}|$ which is poly(n) for constant d.

Theorem 3. $D_{min} \le \ell \le \sqrt{d} D_{min}$.

Proof. Let S_{min} be the optimal selection of R so that $D_{min} = D(S_{min})$. Obviously, $\ell \ge D_{min}$. On the other hand, S_{min} is a feasible solution for LP 1 with rectilinear diameter at most $\sqrt{d} D_{min}$.

Theorem 3 shows that LP 1 results in a $\sqrt{d}$ approximation algorithm for $MinDiam$. We use this result to bound the optimal solution in Sect. 3.3.

3.2   Reducing the Number of Constraints

In this section, we describe how to reduce the number of constraints in LP 1. For this, we use Observation 4.

Observation 4. $|x|$ *is the smallest value* z *such that* $z \geq x$ *and* $z \geq -x$.

Thus, for the following LP, we have $z = |x|$:

$$
\begin{aligned}
&\textbf{LP 2}: \\
&\text{minimize} \quad z \\
&\text{subject to} \quad x \leq z \\
&\phantom{\text{subject to} \quad} -x \leq z
\end{aligned}
\tag{2}
$$

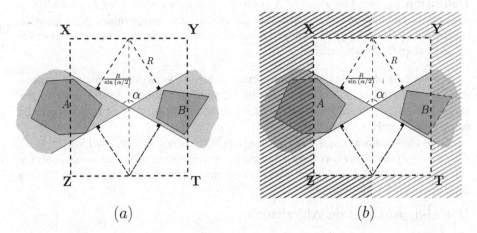

(a) $\qquad\qquad\qquad\qquad\qquad\qquad\qquad$ (b)

Fig. 2. (a) Regions are separated by two intersecting lines. The optimal solution S_{min}, must be entirely within the rectangle $XYZT$. (b) The points inside the green area are at least R distance away from the region A (similar fact is intact between the points inside the blue area and the region B). (Color figure online)

We use Observation 4 to modify LP 1 into LP 3. In LP 3, Q is a large enough integer, ensuring that minimizing ℓ is more important than $\sum d_{i,j,k}$.

$$
\begin{aligned}
&\textbf{LP 3}: \\
&\text{minimize} \quad Q\ell + \sum d_{i,j,k} \\
&\text{subject to} \quad \sum_k d_{i,j,k} \leq \ell && \forall i,j \\
&\phantom{\text{subject to} \quad} s_i \in \mathcal{R}_i && \forall i,j \\
&\phantom{\text{subject to} \quad} d_{i,j,k} \geq s_{i,k} - s_{j,k} && \forall i,j,k \\
&\phantom{\text{subject to} \quad} d_{i,j,k} \geq s_{j,k} - s_{i,k} && \forall i,j,k
\end{aligned}
\tag{3}
$$

Number of the constraints in LP 3 is $O(n^2 d + |\mathcal{R}|)$ which is $\mathsf{poly}(n)$.

3.3 $(1 + \epsilon)$-approximation for $d = 2$

In this section, we propose an ϵ-approximation algorithm for $MinDiam$ when $d = 2$ and the regions in $\mathcal{R}$ admit a special notion of separability, called α-separability.

α-separability

Definition 3. *Two intersecting lines, divide the plane into four areas. Two regions A and B are separated by these lines, if they completely belong to opposite areas, See Fig. 2(a). In Fig. 2(a), we name α as the degree of separation.*

Definition 4. *A set S of regions is α-separable, if **there exists** two regions X and Y in S, which are α-separable.*

Definition 5. *For two regions X and Y, maximum degree of separability is defined as the maximum α, such that X and Y are α-separable. Subsequently, maximum degree of separability for a set $\mathcal{R}$ of regions, is the maximum possible α, such that $\mathcal{R}$ is α-separable.*

It's easy to observe that the maximum degree of separability for two regions can be computed in polynomial time by computing the common tangent of the regions. Similarly, maximum degree of separability for a set $\mathcal{R}$ of regions, can be computed in polynomial time.

Note that the maximum degree of separability is not well defined for the case, where every pair of regions have a nonempty intersection. We postpone resolving this issue to Sect. 3.4. For now, we suppose that $\mathcal{R}$ contains at least two regions that are completely disjoint.

$(1 + \epsilon)$-approximation Algorithm

Theorem 5. *Let $\mathcal{R}$ be an α-separable set of regions. then there exists an ε-approximation algorithm for computing MinDiam of $\mathcal{R}$ with running time $O(2^{\frac{1}{\epsilon^2}} \cdot \frac{n^3}{\epsilon^{10} \cdot \sin(\alpha/2)^3})$.*

Proof. Assume that A, B are the regions in $\mathcal{R}$, that are α-separable. Figure 2(a) shows the regions and the separating lines. Let R be the value, with the property that $D_{min} \leq R \leq \sqrt{2}D_{min}$. Such R can be computed by the $\sqrt{d}$-approximation algorithm described in Sect. 3.1.

Argue that the points in the optimal solution S_{min} must be entirely within the rectangle $XYZT$ (See Fig. 2(a)), while every point out of the rectangle has distance more than R to at least one of the selected points from regions A or B (see Fig. 2(b)). The area of $XYZT$ is $\frac{4R^2}{\sin(\alpha/2)}$, that is, $O(\frac{R^2}{\sin(\alpha/2)})$.

Color each region R_i with color c_i. Now, construct a grid $\mathbb{G}$ on the rectangle $XYZT$ with cells of size ϵR. For each grid point p and region R_i containing p create a point with the same location as p and color c_i, see Fig. 3(a). Total number of the points is $O(\frac{n}{\epsilon^2 \sin(\alpha/2)})$.

Let P be the set of generated points. We give P as an input instance for $MinDCS$ problem. Next, using Algorithm 1 we find a color selection Q with $D(Q) \leq (1+\epsilon)D_{min}$ where D_{min} is the diameter of the optimal selection. While the diameter of each cell in $\mathbb{G}$ equals $\sqrt{2}\epsilon = O(\epsilon)$, total error of selection Q for $MinDiam$ is $O(\epsilon)$. Considering the time complexity obtained for $MinDCS$, total running time would be $O(2^{\frac{1}{\epsilon^2}} \cdot \frac{n^3}{\epsilon^{10} \cdot \sin(\alpha/2)^3})$.

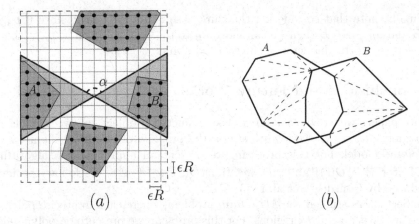

Fig. 3. (a) Constructing a grid on $XYZT$. (b) Special cases.

Note that $1/\sin(\alpha)$ is exponentially descending for $0 \leq \alpha \leq \pi/2$. As an example, for $\alpha > 20°$, $1/\sin(\alpha) < 3$ and for $\alpha > 45°$, $1/\sin(\alpha) < 1.5$. Thus, for large enough α (e.g. $\alpha > 10$), we can consider this value as a constant.

3.4   A Discussion on α-separability

As previously mentioned, α-separability is not well defined in the case that all pairs of the regions have a nonempty intersection. In this section, we wish to address this issue.

Note that by **Helly's theorem**, if every triple regions of the set $\mathcal{R}$ have a nonempty intersection, then the intersection of all the regions in $\mathcal{R}$ is nonempty, i.e., $\bigcap_i \mathcal{R}_i \neq \emptyset$. In this case, the optimal selection in $MinDiam$ is trivial: select the same point for all the regions. Thus, we can assume that there exists a set of three regions, namely, A, B, and C in $\mathcal{R}$ such that any pair of them has a nonempty intersection, but $A \cap B \cap C$ is empty.

Denote by p_A, p_B, and p_C the selected points from the regions A, B, and C, respectively. To solve the problem of $MinDCS$ on $\mathcal{R}$ we divide the problem into three sub problems: (i) $\{p_A, p_B\} \in A \cap B$, (ii) $p_A \in (A \setminus B)$ and $p_B \in (B \setminus A)$, and (iii) $p_A \in A \cap B$ and $p_B \in (A \setminus B)$.

For the first case, we can assume w.l.o.g that p_A and p_B are in the same location in $A \cap B$. Thus, we can remove $A \Delta B$ and only keep $A \cap B$. Since the intersection of $(A \cap B)$ and C is empty, they are separable.

For the second case, we partition $A \Delta B$ into the constant number of convex sub-regions (for example, by triangulating $A \Delta B$ as shown in Fig. 3(b)). Suppose that $A \setminus B$ is divided into r sub-regions and $B \setminus A$ is divided into s sub-regions. Then we solve $r \times s$ sub-problems, according to the sub-regions p_A and p_B belong. In all of these sub-problems, α-separability is well defined.

Finally, note that considering the third case is not necessary, since if $p_A \in A \cap B$ and $p_B \notin A \cap B$ then we can move p_B to the same location as p_A without any decrease in the diameter of the selected points.

4   Conclusions and Future Works

In this paper, we tried to address the diameter problem in two models of uncertainty. In Sect. 2, we investigate on the problem of minimum diameter in *Indecisive* model. For this problem, we present an approximation algorithm with factor $(1 + O(\epsilon))$ and running time $O(2^{\frac{1}{\epsilon^d}}.\epsilon^{-2d}.n^3)$. We follow the idea introduced by Consuegra et al. in [3].

In Sect. 3, we studied the $MinDiam$ problem, where each imprecise point is represented with a convex regions. For this problem, we presented a polynomial time $\sqrt{d}$-approximation algorithm for constant d. next, we used this result to give a $(1 + \epsilon)$-approximation for the case, where $d = 2$ and the regions are α-separate.

A future direction would be generalizing the $(1 + \epsilon)$-approximation algorithm proposed for $MinDiam$ to higher dimensions. In addition, one can think of removing the condition of α-separability for this problem.

References

1. Agarwal, P.K., Har-Peled, S., Varadarajan, K.R.: Approximating extent measures of points. J. ACM (JACM) **51**(4), 606–635 (2004)
2. Aggarwal, C.C.: Trio a system for data uncertainty and lineage. In: Aggarwal, C.C. (ed.) Managing and Mining Uncertain Data, pp. 1–35. Springer, Heidelberg (2009)
3. Consuegra, M.E., Narasimhan, G., Tanigawa, S.I.: Geometric avatar problems. In: IARCS Annual Conference on Foundations of Software Technology and Theoretical Computer Science (FSTTCS 2013), vol. 24, pp. 389–400. Schloss Dagstuhl-Leibniz-Zentrum fuer Informatik (2013)
4. Cormode, G., Li, F., Yi, K.: Semantics of ranking queries for probabilistic data and expected ranks. In: IEEE 25th International Conference on Data Engineering, ICDE 2009, pp. 305–316. IEEE (2009)
5. Fan, C., Luo, J., Zhong, F., Zhu, B.: Expected computations on color spanning sets. In: Fellows, M., Tan, X., Zhu, B. (eds.) AAIM/FAW -2013. LNCS, vol. 7924, pp. 130–141. Springer, Heidelberg (2013). doi:10.1007/978-3-642-38756-2_15
6. Fleischer, R., Xu, X.: Computing minimum diameter color-spanning sets. In: Lee, D.-T., Chen, D.Z., Ying, S. (eds.) FAW 2010. LNCS, vol. 6213, pp. 285–292. Springer, Heidelberg (2010). doi:10.1007/978-3-642-14553-7_27
7. Fleischer, R., Xu, X.: Computing minimum diameter color-spanning sets is hard. Inf. Process. Lett. **111**(21), 1054–1056 (2011)
8. Har-Peled, S.: Geometric Approximation Algorithms, vol. 173. American Mathematical Society, Boston (2011)
9. Jørgensen, A., Löffler, M., Phillips, J.M.: Geometric computations on indecisive points. In: Dehne, F., Iacono, J., Sack, J.-R. (eds.) WADS 2011. LNCS, vol. 6844, pp. 536–547. Springer, Heidelberg (2011). doi:10.1007/978-3-642-22300-6_45
10. Ju, W., Fan, C., Luo, J., Zhu, B., Daescu, O.: On some geometric problems of color-spanning sets. J. Comb. Optim. **26**(2), 266–283 (2013)

11. Löffler, M.: Data imprecision in computational geometry (2009)
12. Löffler, M., van Kreveld, M.: Largest bounding box, smallest diameter, and related problems on imprecise points. Computat. Geom. **43**(4), 419–433 (2010)
13. Zarrabi-Zadeh, H.: Geometric approximation algorithms in the online and data stream models. Ph.D. thesis, University of Waterloo (2008)
14. Zhang, D., Chee, Y.M., Mondal, A., Tung, A., Kitsuregawa, M.: Keyword search in spatial databases: towards searching by document. In: IEEE 25th International Conference on Data Engineering, ICDE 2009, pp. 688–699. IEEE (2009)

On Constant Depth Circuits Parameterized by Degree: Identity Testing and Depth Reduction

Purnata Ghosal, Om Prakash, and B.V. Raghavendra Rao$^{(\boxtimes)}$

Department of Computer Science and Engineering, IIT Madras, Chennai, India
purnatag@gmail.com, op708543@gmail.com, bvrr@cse.iitm.ac.in

Abstract. In this article we initiate the study of polynomials parameterized by degree by arithmetic circuits of small syntactic degree. We define the notion of fixed parameter tractability and show that there are families of polynomials of degree k that cannot be computed by homogeneous depth four $\Sigma\Pi^{\sqrt{k}}\Sigma\Pi^{\sqrt{k}}$ circuits. Our result implies that there is no parameterized depth reduction for circuits of size $f(k)n^{O(1)}$ such that the resulting depth four circuit is homogeneous.

We show that testing identity of depth three circuits with syntactic degree k is fixed parameter tractable with k as the parameter. Our algorithm involves an application of the hitting set generator given by Shpilka and Volkovich [APPROX-RANDOM 2009]. Further, we show that our techniques do not generalize to higher depth circuits by proving certain rank-preserving properties of the generator by Shpilka and Volkovich.

1 Introduction

Parameterized Complexity is the discipline where an additional parameter along with the input is used for measuring the complexity of computational problems. This leads to more fine-grained complexity classification of computational problems and a relaxed notion of tractability. Downey and Fellows [9] were the first to study complexity of problems with a parameter, and develop the area of parameterized complexity theory. Over the last two decades, parameterized complexity has played a pivotal role in algorithmic research [9].

Fixed Parameter Tractability (FPT) is the notion of tractability in Parameterized Complexity Theory. A decision problem with parameter k that is decidable in deterministic time $f(k)\mathsf{poly}(n)$ is said to be fixed parameter tractable (FPT for short). The whole area of parameterized complexity theory is centered around this definition of tractability. The parameterized intractable problems are based on the hierarchy of classes known as the W-hierarchy. The smallest member of W-hierarchy, W[1] consists of problems that are FPT equivalent to the clique problem with the size of the clique as the parameter.

Motivation: Parameterized complexity of problems based on graphs and other combinatorial structures played pivotal role in the development of Parameterized Algorithms and Complexity Theory. Many of the parameterized algorithms

© Springer International Publishing AG 2017
Y. Cao and J. Chen (Eds.): COCOON 2017, LNCS 10392, pp. 250–261, 2017.
DOI: 10.1007/978-3-319-62389-4_21

involve evaluation of polynomials of degree bounded by the parameter. For example, in [5], Björklund et. al., defined and used a degree k polynomial which is identically zero if an only if the given graph has no cycle of length k or smaller. Polynomials of degree bounded by the parameter has been extensively used to develop efficient randomized parameterized algorithms, see e.g., [2,4,13]. Further, obtaining a deterministic $f(k)n^{O(1)}$ time algorithm checking if a n variate polynomial of degree at most k is zero or not for some function f of k would lead to fast deterministic FPT algorithms for a wide variety of problems. The construction of representative sets in [12] can also be viewed a parameterized de-randomization of the polynomial identity testing problem for a special class of polynomials where the monomials have a matroidal structure.

Wide application of polynomials whose degree is bounded by the parameter in Parameterized Complexity Theory merits a study of such polynomials in algebraic models of computation. We initiate the study of polynomials with the degree as the parameter.

Our Model: Motivated by the applications of polynomials parameterized by degree, we study the parameterized complexity of polynomials parameterized by degree. More specifically, given a parameter $k = k(n)$, classify the families of polynomials of degree k based on the minimum size of the arithmetic circuit computing it. Here, the notion of efficiency is based on the fixed parameter tractability. A natural model of computation for polynomials with degree bounded by k would be arithmetic circuits where every gate computes a polynomial of degree at most k. However, such a circuit of size $f(k)n^{O(1)}$ for some function f of k can compute polynomials with coefficients as large as 2^{2^n} where n is the number of variables, and hence evaluation of such polynomials unlikely to be Fixed Parameter Tractable. Thus we need models were the coefficients computed can be represented in $f(k)n^{O(1)}$ many bits. Towards this, we consider the arithmetic circuits where the syntactic degree (see Sect. 2 for a definition) is bounded by the degree of the polynomial being computed as the computational model.

Further, we study the parameterized version of Arithmetic Circuit Identity Testing (ACIT) problem: testing if the given arithmetic circuit computes the zero polynomial or not. ACIT is one of the fundamental computational questions on polynomials. Schwartz [22] and Zippel [26] independently showed that there is a randomized polynomial time algorithm for ACIT. Their algorithm worked for a more general setting, where the polynomials are given in the black-box form. However, obtaining deterministic polynomial time algorithm for ACIT has been one of the prominent open questions for decades, playing a pivotal role in Algebraic Complexity Theory. Motivated by the application of ACIT in several parameterized algorithms [2,4,5,13] with degree as the parameter, we study the complexity of ACIT with degree as the parameter.

Our Results: We define the notion of fixed parameter tractability (FPT) of a family of polynomials parameterized by the degree. We show that there are polynomials that have FPT circuits but cannot be computed by depth four homogeneous $\Sigma\Pi^{\sqrt{k}}\Sigma\Pi^{\sqrt{k}}$ circuits of FPT size (Theorem 2). Also, we show a parameterized separation between depth three and four circuits (Theorem 4).

Then we study the parameterized complexity of the arithmetic circuit identity testing problem and show that non-black box identity testing of depth three circuits of syntactic degree at most k is fixed parameter tractable (Theorem 5). This result is obtained by the application of a hitting set generator defined by Shpilka and Volkovich [23]. Finally we show that the techniques used in Theorem 5 cannot be used to obtain efficient parameterized identity tests for depth four circuits by proving that the generator given by [23] preserves rank of certain matrix associated with polynomials (Theorem 6).

Related work: Though algebraic techniques have been well utilized in obtaining efficient parameterized algorithms, the focus on parameterized complexity of algebraic problems is very limited. Chen et. al. [8] studied the parameterized complexity of detecting and testing monomials in polynomials given as arithmetic circuits. Arvind et. al [3] obtained parameterized algorithms for solving systems of linear equations parameterized by the hamming weight. Engels [10] developed Parameterized Algebraic Complexity theory in analogy to Valiant's notion of Algebraic Complexity. Apart from these results, there have not been much attention on algebraic problems in the parameterized world.

Müller [16] studied several parameterized variants of ACIT and obtained randomized parameterized algorithms for those variants that use $O(k \log n)$ random bits where k is the parameter. Further, Chauhan and Rao [7] studied ACIT with the syntactic degree of the circuit as a parameter, and showed that the problem has a randomized algorithm that uses only $O(k \log n)$ random bits, where k is the syntactic degree. Finally, it can be seen easily from the observations in [7] that ACIT with syntactic degree as a parameter is equivalent to the same problem with the number of variables as a parameter.

2   Preliminaries

In this section we will introduce necessary notions on arithmetic circuits and parameterized complexity. For more details the reader is referred to [9].

An arithmetic circuit C over a field $\mathbb{F}$ and variables $X = \{x_1, \ldots, x_n\}$ is a labeled directed acyclic graph. The nodes in C are called gates. Gates of zero in-degree are called input gates and are labeled by either variables in X, or constants in $\mathbb{F}$. Other gates in C are labeled by either $\times$ or $+$. Gates with zero out-degree are called output gates. In our applications, an arithmetic circuit will have a unique output gate. Every gate in C naturally represents a polynomial in $\mathbb{F}[x_1, \ldots, x_n]$. The polynomial computed by C is the polynomial represented at the output gate.

Depth of an arithmetic circuit is the length of the longest path from an input gate to the output gate. In this paper, our focus is on constant depth arithmetic circuits. It should be noted that, constant depth circuits are interesting only when the fan-in of gates are allowed to be unbounded. $\Sigma\Pi\Sigma$ denotes the class of depth three circuits of the form $\sum_{i=1}^{r} \prod_{j=1}^{t} \ell_{i,j}$ for some $r, t \geq 0$ and $\ell_{i,j}$s are linear functions of the input variables. Similarly, depth four $\Sigma\Pi\Sigma\Pi$ circuits are defined. The fan-in restriction on the gates at a specific layer are denoted by superscripts, e.g., $\Sigma\Pi^k\Sigma$ denotes the sub class of $\Sigma\Pi\Sigma$ circuits where the middle layer of product gates have fan-in bounded by k.

Saxena [21] introduced the notion of *dual representation of polynomials*. Let $f \in \mathbb{F}[x_1,\ldots,x_n]$. Then f is said to have a dual representation of size t, if there are univariate polynomials $g_{ij}(x_j)$ such that $f = \sum_{i=1}^{t} g_{i1}(x_1)\cdots g_{in}(x_n)$.

The *syntactic degree* denoted by *syntdeg* for every gate v of an arithmetic circuit is defined as follows:

$$syntdeg(v) = \begin{cases} 1 & \text{if } v \text{ is an input gate} \\ \max\{syntdeg(v_1), syntdeg(v_2)\} & \text{if } v = v_1 + v_2 \\ syntdeg(v_1) + syntdeg(v_2) & \text{if } v = v_1 \times v_2 \end{cases}$$

We need the notion of *hitting sets* for arithmetic circuits. Let $\mathcal{C}_n$ be a class of polynomials in n variables. A set $\mathcal{H}_n \subseteq \mathbb{F}^n$ is called a hitting set the class $\mathcal{C}$ with n inputs, such that for all polynomials $f \in \mathcal{C}$, $f \not\equiv 0$, $\exists a \in \mathcal{H}$, $f(a) \neq 0$.

We also require the notion of *hitting set generators*. Consider a polynomial mapping $G = (G^1,\ldots,G^n) : \mathbb{F}^t \to \mathbb{F}^n$ where $G^1,\ldots,G^n$ are t variate polynomials. G is a generator for the circuit class $\mathcal{C}_n$ if for every polynomial $f \in \mathcal{C}$, $f \not\equiv 0$, it holds that $G(f) \not\equiv 0$ where $G(f) = f(G^1,\ldots,G^n)$.

It is known that the image of a generator for polynomials in $\mathcal{C}$ contains a hitting-set for all non-zero polynomials in $\mathcal{C}$ [23]. In this paper we will require a hitting set generator defined by Shpilka and Volkovich [23].

Definition 1 (S-V Generator [23]). *Let $a_1, a_2 \ldots a_n$ be distinct elements in the given field $\mathbb{F}$. Let $G_k^i \in \mathbb{F}[y_1, y_2 \ldots y_k, z_1, z_2 \ldots z_k]$ be the polynomial defined as follows:*

$$G_k^i(y_1, y_2 \ldots y_k, z_1, z_2 \ldots z_k) = \sum_{j=1}^{k} L_i(y_j)z_j, \text{ where } L_i(x) = \frac{\prod_{j\neq i}(x - a_j)}{\prod_{j\neq i}(a_i - a_j)}.$$

The generator G_k is defined as $G_k \triangleq (G_k^1, \ldots G_k^n)$.

For a polynomial f, let $G_k(f) \triangleq f(G_k^1,\ldots,G_k^n)$. In [23], Shpilka and Volkovich showed that G_k is a hitting set generator for sum of k read-once polynomials. Further, in [7] Chauhan and Rao showed that the generator G_k is also a hitting set generator for degree k polynomials:

Lemma 1. *For any n-variate polynomial f of degree k, $f \equiv 0 \iff G_k(f) \equiv 0$.*

A *parameterized problem* is a set $P \subseteq \Sigma^* \times \mathbb{N}$, where Σ is a finite set of alphabets. If $(x, k) \in \Sigma^* \times \mathbb{N}$ is an instance of a parameterized problem, we refer to x as the input and k as the parameter.

Definition 2 (Fixed Parameter Tractability). *A parameterized problem $P \subseteq \Sigma^* \times \mathbb{N}$ is fixed-parameter tractable (FPT) if there is a computable function $f : \mathbb{N} \to \mathbb{N}$, a constant $c \in \mathbb{N}$ and an algorithm that, given a pair $(x, k) \in \Sigma^* \times \mathbb{N}$, decides if $(x, k) \in P$ in at most $f(k)poly(n)$ steps, where n is the length of input.*

ACIT is the problem of testing whether a given arithmetic circuit C computes a polynomial p that is identically zero. There have been several parameterized variants of ACIT studied in the literature [16]. Following [7] we consider the syntactic degree of the arithmetic circuit as a parameter.

Problem para-ACIT
INPUT A polynomial p given as an arithmetic circuit C of syntactic degree k.
PARAMETER : k
OUTPUT YES iff $p \equiv 0$.

Chauhan and Rao [7] show that $\mathsf{para} - \overline{\mathsf{ACIT}} \in \mathsf{W[P]} - \mathsf{RFPT}$, a parameterized class analogous to RP. In this paper we consider the problem when restricted to depth three circuits.

We consider the notion of partial derivative matrix of a polynomial defined by Nisan [17]. Raz [18] used a variant of partial derivative matrix, which later were generalized by Kumar et al [15]. In this paper we consider yet another variant of partial derivative matrices:

Definition 3. *Let $f \in \mathbb{F}[x_1, \ldots, x_n]$ be a polynomial of degree d, $\varphi : X \to Y \cup Z$ be a partition of the input variables of f. Then the coefficient matrix $M_{f\varphi}$ has its rows indexed by monomials μ of degree at most d in variables in Y, and columns indexed by monomials ν of degree at most d in variables in Z. For monomials μ and ν respectively in variables Y and Z, the entry $M_{f\varphi}(\mu, \nu)$ is the coefficient of the monomial $\mu\nu$ in f.*

The definition above is an adaptation of [17] using a partition of variables and different from the notion of polynomial coefficient matrices in [15]. The coefficient matrix of a polynomial is well studied in the literature, however, the specific form used in the above definition has not been mentioned explicitly in the literature. The following fundamental properties of the rank of the coefficient matrix follow directly from [18].

Lemma 2. *If $f, g, h \in \mathbb{F}[X]$ such that $f = g + h$, then $\forall \varphi : X \to Y \cup Z$, $\mathsf{rank}(M_{f\varphi}) \leq \mathsf{rank}(M_{g\varphi}) + \mathsf{rank}(M_{h\varphi})$.*

Recall that an arithmetic circuit is a *formula* if the out-degree of every gate is either one or zero. Finally, we need the following result on testing identity of non-commutative formulas. Let $\mathbb{F}\{x_1, x_2 \ldots, x_n\}$ be the non-commutative polynomial ring over the field $\mathbb{F}$. The following theorem was proved by Raz and Shpilka [19].

Theorem 1 [19]. *Let $C \in \mathbb{F}\{X\}$ be a non-commutative arithmetic formula, then there is a white-box identity testing algorithm for C having time complexity $O(size(C))$.*

3   Parameterization by Degree

3.1   Fixed Parameter Tractability in Arithmetic Computation

In this section we define the notion of parameterized tractability of polynomials over $\mathbb{Z}$ parameterized by degree.

Definition 4. *Let $k = k(n)$. A family $(p_n)_{n \geq 0}$ of polynomials over $\mathbb{Z}$ is said to be degree k parameterized if*

- *There is a $c > 0$ such that p_n is an n^c variate polynomial for every $n \geq 0$;*
- *Degree of p_n is bounded by $k = k(n)$ for every $n \geq 0$; and*
- *The absolute value of the coefficients of p_n is bounded by $2^{g(k)n^c}$, for some function g that depends only on k.*

Since any arithmetic circuit can be homogenized efficiently, we have:

Proposition 1. [6] *For any parameterized polynomial family (p, k) if there is a family of arithmetic circuits $C = (C_n)_{n \geq 0}$ of size $f(k)n^c$ computing p, where $f(k)$ is a function of k and c is a constant, then there is a family of arithmetic circuits $C' = (C'_n)$ of size $f'(k)n^{c'}$ for p such that every gate in C'_n computes a polynomial of degree at most k.*

It can be seen that a circuit C_n of size $f(k)n^{O(1)}$ where every gate computes a polynomial of degree at most k can compute polynomials where the absolute value of the coefficient can be as large as $2^{2^{n^{O(1)}}}$ even when the only constants allowed in the circuit are $\{-1, 0, 1\}$. This makes the evaluation of such polynomial in FPT time unfeasible. A natural restriction solution would be to bound the syntactic degree of the circuit. An arithmetic circuit C_n is said to be of syntactic degree d if every gate in C_n has syntactic degree bounded by d.

A degree parameterized polynomial family $(p = (p_n)_{n \geq 0}, k)$ with k as the parameter is said to be *fixed parameter tractable* (FPT) if for every $n \geq 0$, there is an arithmetic circuit C_n of syntactic degree at most k and of size $f(k)n^c$ computing p_n where f is a function of k and c is a constant.

3.2   Parameterized Depth Reduction

Depth reduction is one of the most fundamental structural aspects in algebraic complexity theory: Given a polynomial family $p = (p_n)_{n \geq 0}$ and a size bound $s = s(n)$, what is the minimum depth of an arithmetic circuit of size s computing p? The parameterized depth reduction problem can be stated as:
Given a parameterized polynomial family (p, k) in FPT what is the minimum depth of a size $f(k)n^c$ circuit computing p where $f(k)$ is an arbitrary function of k and c is some constant.

By applying the well known depth reduction technique in [25], we have:

Proposition 2. [25] *Any parameterized polynomial family (p, k) in FPT can be computed by circuits of depth $f(k) \log n$ and size $f'(k)n^{O(1)}$, for some functions f and f' that depend only on the parameter.*

In a surprising result, Agrawal and Vinay [1] showed that any homogeneous polynomial p computed by polynomial size arithmetic circuits can be computed by depth four $\Sigma\Pi^{\sqrt{n}}\Sigma\Pi^{\sqrt{n}}$ homogeneous circuits of size $2^{O(\sqrt{n}\log^2 n)}$. Further, Tavenas [24] improved this bound to $2^{\sqrt{n}\log n}$. Over infinite fields, there is a depth three $\Sigma\Pi\Sigma$ circuit of size $2^{\sqrt{n}\log n}$ for p [14].

A parameterized counterpart of the depth reduction in [1] would be to transform a circuit C_n of size $f(k)n^{O(1)}$ and syntactic degree k to a depth four $\Sigma\Pi\Sigma\Pi$ circuit of syntactic degree k and size $f'(k)n^{O(1)}$ where f and f' are functions of k alone. Note that a $\Sigma\Pi\Sigma\Pi$ circuit C of syntactic degree k will have Π fan-in bounded by k at both of the Π layers and hence can assumed to be of the form $\Sigma\Pi^k\Sigma\Pi^k$. Further, if C is homogeneous with the bottom Π layer having syntactic degree t then C can be assumed to be a homogeneous $\Sigma\Pi^{k/t}\Sigma\Pi^t$ circuit. We first observe that we can replace Π gates with $\wedge$ (powering) gates in any depth four circuit with syntactic degree bounded by the parameter k. The proof is a direct application of Fischer's identity [11] twice and is omitted.

Lemma 3. *Let C be a $\Sigma\Pi^{k/t}\Sigma\Pi^t$ circuit of size s computing a polynomial p over $\mathbb{Z}$. Then there is a $\Sigma\wedge^{k/t}\Sigma\wedge^t\Sigma$ circuit C' of size $\max\{2^{k/t}, 2^t\}\cdot s$ computing p. Moreover, if C is homogeneous, so does C'.*

Thus a parameterized version of depth reduction in [1] would imply that every parameterized polynomial family (p, k) in FPT can be computed by a homogeneous $\Sigma\wedge^{O(\sqrt{k})}\Sigma\wedge^{O(\sqrt{k})}\Sigma$ circuit of size $f(k)n^{O(n)}$ for some function f of k. However, in the next section, we observe that this is not possible.

3.3   Parameterized Lower Bound for $\Sigma\wedge^{O(\sqrt{k})}\Sigma\wedge^{O(\sqrt{k})}\Sigma$ circuits

In this section we prove lower bounds against homogeneous $\Sigma\wedge^{O(\sqrt{k})}\Sigma\wedge^{O(\sqrt{k})}\Sigma$ circuits using the well known method of partial derivatives. To begin with, we obtain an upper bound on the dimension of partial derivatives of small powers of sum of powers of homogeneous linear forms:

Lemma 4. *Let $f = \ell_1^d + \ell_2^d + \cdots + \ell_t^d$ where $\ell_1, \ldots, \ell_t$ are linear forms in $\{x_1, \ldots, x_n\}$. Then, for any $r \leq k$, $\dim(\langle \partial^{=r} f^\alpha \rangle) \leq g(k)t^\alpha$ for $\alpha = o(k), d = o(k)$ and some computable function g that depends only on k.*

Now, we obtain a polynomial with large dimension of partial derivatives:

Lemma 5. *There is a polynomial $p \in \mathbb{F}[x_1, \ldots, x_n]$ of degree k that can be computed by polynomial size $\Pi\Sigma\wedge^2$ circuits with $\dim(\langle \partial^{\leq k/2} p \rangle) = n^{\Omega(k)}$.*

Combining Lemmas 4 and 5 (whose proofs we omit) along with the subadditivity property of the dimension of partial derivatives, we get:

Theorem 2. *There is a polynomial $p \in \Sigma\Pi^{k/2}\Sigma\wedge^2$ such that any homogeneous $\Sigma\wedge^{o(k)}\Sigma\wedge\Sigma$ circuit computing it will have top fan-in at least $n^{\Omega(k)}$.*

Proof. Consider $f \in \Sigma^s\wedge^{o(k)}\Sigma^t\wedge\Sigma$ such that $f = f_1 + \cdots + f_s$ where $f_i \in \wedge^{o(k)}\Sigma^t\wedge\Sigma$ and set $r = k/3$. By Lemma 4 we have $\dim(\langle \partial^{=r} f_i \rangle) \leq g(k)t^\alpha$ for some g. Then $\dim(\langle \partial^{=r} f \rangle) \leq \sum_i \dim(\langle \partial^{=r} f_i \rangle) \leq sg(k)t^\alpha$. Since $t = f(k)n^{O(1)}$ for some function f of k, $\alpha = o(k)$ and $\dim(\langle \partial^{=r} p \rangle) = n^{\Omega(k)}$, we have $s = n^{\Omega(k)}$.

It can be noted that the lower bound in Theorem 2 also holds for the degree k elementary symmetric polynomial $Sym_{n,k}(X) = \sum_{S\subseteq[n],|S|=k} \prod_{i\in S} x_i$:

Corollary 1. *Any homogeneous* $\Sigma \wedge^{o(k)} \Sigma \wedge \Sigma$ *circuit computing* $Sym_{n,k}$ *will have top fan-in at least* $n^{\Omega(k)}$.

However, the homogeneity condition in Corollary 1 is necessary due to the following result in [20]:

Proposition 3. $Sym_{n,k}$ *can be computed by* $2^{O(\sqrt{k})}n^{O(1)}$ *size* $\Sigma \wedge^{\sqrt{k}} \Sigma \wedge^{O(k)} \Sigma$ *circuit.*

3.4 Parameterized Lower Bounds for Depth Three Circuits

In this section, we prove parameterized lower bound against depth three $\Sigma\Pi\Sigma$ circuits of syntactic degree bounded by the parameter. We use the method of partial derivative matrix used in [18]. We begin with a lower bound on the measure for a polynomial computed by depth four circuits.

Lemma 6. *Let* $X = \{y_1, \ldots, y_m, z_1, \ldots, z_m\}$. *Let* $f = \prod_{i=1}^{k} Q_i$ *where* $Q_i = 1 + y_{\frac{(i-1)m}{k}+1}z_{\frac{(i-1)m}{k}+1} + y_{\frac{(i-1)m}{k}+2}z_{\frac{(i-1)m}{k}+2} + \ldots + y_{\frac{im}{k}}z_{\frac{im}{k}}$ *are multivariate quadratic polynomials in* $\mathbb{F}[X]$, *then* $\exists \varphi : X \to Y \cup Z$ *such that* $\mathsf{rank}(M_{f\varphi}) = \Omega((\frac{m+k}{k})^k)$.

Now, we need the folklore fact that for any partition of its variables, a polynomial with a 'small' dual representation will have rank of the coefficient matrix small.

Lemma 7. *(folklore) Suppose* $f = \sum_{i=1}^{t} g_{i,1}(x_1)g_{i,2}(x_2)\ldots g_{i,n}(x_n)$. *Then, for all partitions* $\varphi : X \to Y \cup Z$, $\mathsf{rank}(M_{f\varphi}) \leq t$ *where* $g_{i,1}, g_{i,2}, \ldots, g_{i,n}$ *are univariate polynomials in* $x_1, x_2, \ldots, x_n$ *respectively.*

We omit the proofs of the Lemmas 7 and 6. As an immediate consequence of these lemmas we have:

Theorem 3. *There exists a polynomial* $f \in \mathbb{F}[x_1, \ldots, x_n], f = \sum_{i=1}^{s}\prod_{j=1}^{d_i} Q_{i,j}$, *where* $s > 0$, $Q_{i,j}$ *is a multivariate quadratic polynomial with* $max_i\{d_i\} \leq k$, *such that* $f = \Sigma_{i=1}^{t}g_{i,1}(x_1)g_{i,2}(x_2)\ldots g_{i,n}(x_n) \implies t = n^{\Omega(k)}$ *where* $g_{i,1}g_{i,2}\ldots g_{i,n}$ *are univariate polynomials of syntactic degree* $k = O(\log n)$.

Further, we observe that depth three $\Sigma\Pi\Sigma$ circuits of syntactic degree k, compute polynomials of 'small' degree under every partition:

Lemma 8. *Let* $f \in \mathbb{F}[x_1, \ldots, x_n], f \in \Sigma\Pi^k\Sigma$ *and* $M_{f\varphi}$ *is the coefficient matrix corresponding to the partition of variables of* f, $\varphi : X \to Y \cup Z$. *Then* $\mathsf{rank}(M_{f\varphi}) \leq s \cdot 2^{O(k)}$, *where* s *is the smallest size of a* $\Sigma\Pi^k\Sigma$ *circuit for* f.

Proof. Let $f \in \mathbb{F}[x_1, \ldots, x_n], f \in \Sigma\Pi^k\Sigma$. Hence $f = \sum\prod^k\sum_i^n a_ix_i$, where $a_i \in \mathbb{F}$. Then, $\forall\varphi$, $\mathsf{rank}(M_{f\varphi}) \leq 2^{O(k)}s$ since $\mathsf{rank}(M_{\ell\varphi}) \leq 2$, for all linear functions $\ell = \sum_{i=1}^{n} a_ix_i$; and rank of the coefficient matrix is sub-multiplicative and sub-additive.

Combining Lemmas 6 and 8 we get the following immediately:

Theorem 4. $\Sigma\Pi^k\Sigma \subsetneq \Sigma\Pi^{k/2}\Sigma\Pi^2$.

In the next section, we study the parameterized complexity of the identity testing problem for the models considered in this section.

4   Para-ACIT for constant depth circuits

This section is devoted to the study of the ACIT problem for depth three and four circuits. We obtain a non-black box identity test for the class of $\Sigma\Pi^k\Sigma$ circuits and argue the limitations of the technique used in the test.

4.1   Depth Three Circuits Parameterized by Degree

In this section we show that the non-black box para-ACIT for depth three $\Sigma\Pi^k\Sigma$ circuits is FPT. The proof follows by the application of S-V generator and Theorem 1.

Theorem 5. *Let f be a polynomial of the form $\sum_{i=1}^{m}\prod_{j=1}^{d_i}\ell_{i,j}$ where $\ell_{i,j}s$ are linear forms. There is a white-box identity testing algorithm for C that runs in time $g(k).size(C)^{O(1)}$, where g is a computable function, i.e., the white-box ACIT for $\Sigma\Pi\Sigma$ circuits with syntactic degree as the parameter is in FPT.*

Theorem 5 can also be obtained by applying the Fischer's identity on $\Sigma\Pi^k\Sigma$ circuits to obtain equivalent $\Sigma\wedge^k\Sigma$ circuits and then applying the techniques in [21]. However, this works only over infinite fields, whereas our proof above works over any field, though the identity test needs a suitable extension of the underlying field.

A natural next step would be to obtain dual representations for parameterized families computed by $\Sigma\Pi^k\Sigma\Pi^k$ circuits. However, from Theorems 3 and 4 it follows that there are depth four $\Sigma\Pi^{k/t}\Sigma\Pi^t$ arithmetic circuits that cannot have dual representation of FPT size. This does not rule out the possibility of getting a dual representation via the application of S-V generator. However, in the following section, we show that every image of a polynomial f under the S-V generator has many partitions that preserve the rank of the polynomial coefficient matrix of f.

4.2   Rank Preserving Property of the S-V Generator

In this section, we show that the rank of the coefficient matrix of a polynomial is invariant under the S-V generator map G_k.

Theorem 6. *Let $f \in \mathbb{F}[x_1,\ldots,x_n]$ be a polynomial of degree $\leq k$. Let $g = G_{2k}(f)$. Then, $\exists\varphi, \mathrm{rank}(M_{f^\varphi}) \geq r \implies \mathbf{Pr}_{\varphi'}[\mathrm{rank}(M_{g^{\varphi'}}) = \Omega(r)] \geq \Omega(1/k^2)$ where the probability is over the uniform distribution over the set of all partitions φ' of a set of $4k$ variables into two equal parts.*

Proof. Fix $a_1,\ldots,a_n \in \mathbb{F}$ be distinct. Then,

$$G_k(x_i) = \sum_{p=1}^{k} z_p L_i(y_p) = \sum_{p=1}^{k} z_p \frac{\prod_{j\neq i}(y_p - a_j)}{\prod_{j\neq i}(a_i - a_j)}$$

$$= \sum_{p=1}^{k} z_p \frac{(y_p - a_1) \ldots (y_p - a_{i-1})(y_p - a_{i+1}) \ldots (y_p - a_n)}{(a_i - a_1) \ldots (a_i - a_{i-1})(a_i - a_{i+1}) \ldots (a_i - a_n)}$$

$$= \sum_{p=1}^{k} z_p (y_p^{n-1} - \sum_{j \neq i} a_j y_p^{n-2} + \ldots + (-1)^{n-1} \prod_{j \neq i} a_j).$$

Therefore, $G_k(x_i) = \sum_{p=1}^{k} \sum_{q=1}^{n} z_p y_p^{n-q} (-1)^q \prod_{\forall t j_t \neq i} a_{j_1} \ldots a_{j_q}$

$$= \sum_{\substack{p \in [k] \\ q \in [n]}} z_p y_p^{n-q} c_{pqi} \quad (\text{where } c_{pqi} = (-1)^q \prod_{\forall t j_t \neq i} a_{j_1} \ldots a_{j_q}).$$

Using the above for a monomial $m = x_{i_1} x_{i_2} \ldots x_{i_k}$, we get:

$$G_k(m) = \sum_{\substack{p_1, \ldots, p_k \in [k] \\ q_1 \ldots q_k \in [n-1]}} z_{p_1} \ldots z_{p_k} y_{p_1}^{n-q_1} \ldots y_{p_k}^{n-q_k} \prod_{j=1}^{k} c_{p_j q_j i_j}$$

Let $\mathcal{M}_k$ be the set of all degree k monomials in the variables $\{x_1, \ldots, x_n\}$. Let $\mathcal{S}_{nk}$ be the set of all monomials of the form $\prod_{i \in I} z_i y_i^{n-q_i}$, for all multi-sets $I \subseteq \{1, \ldots, k\}$ of size k and $1 \leq q_i \leq n - 1$ for every i.

Let $V = \mathrm{Span}(\mathcal{M}_k)$, and $W = \mathrm{Span}(\mathcal{S}_{nk})$ be the vector spaces spanned by the the sets. The vector space V contains all polynomials in $\mathbb{F}$ of degree k, and hence the dimension of V is $\binom{n+k}{k}$. Also, dimension of W is bounded by $\binom{2k}{k} n^k$. Note that G_k is indeed a linear map from V to W. Let C be the be the $\binom{n+k}{k} \times \binom{2k}{k} n^k$ matrix representing G_k. Then, $\forall w \in W$, $G_k(w) = Cw^T \in V$. Now, we argue that C has full row-rank.

Claim. C has full row-rank.

Proof (of the Claim). Suppose C is not of full row rank. Then $\exists \alpha_{i_1}, \ldots, \alpha_{i_r} \in \mathbb{R}$, such that $\sum_{j=1}^{r} \alpha_{i_j} C[i_j] = 0$ with $\alpha_{i_j} \neq 0$ for some j, where $C[i]$ represents the ith row of C, and $r \leq \dim(V)$. As G_k is linear, there must be some $v_{i_1}, \ldots, v_{i_r}$ such that $G_k(v_{i_j}) = C[i_j]$. Then we have:

$$\sum_{j=1}^{r} \alpha_{i_j} G_k(v_{i_j}) = \sum_{j=1}^{r} G_k(\alpha_{i_j} v_{i_j}) = 0 \implies G_k(\alpha_{i_1} v_{i_1} + \ldots + \alpha_{i_r} v_{i_r}) = 0.$$

We can see that $P \equiv \alpha_{i_1} v_{i_1} + \ldots + \alpha_{i_r} v_{i_r}$ is a polynomial of degree at most k in $\mathbb{F}[x_1, \ldots, x_n]$, such that $G_k(P) \equiv 0$, whereas $P \not\equiv 0$ since $\exists \alpha_{i_j} \neq 0$. This contradicts Lemma 1. Hence, the Claim is proved.

Consider a partition $\varphi : X \to A \cup B$ and suppose $\mathrm{rank}(M_{f\varphi}) \geq r$. Let $m_1, \ldots, m_r$ be r linearly independent rows of M_f (chosen arbitrarily). Let $p_1, \ldots, p_r$ be the polynomials representing these rows, i.e., $p_i = \sum_{S \subseteq B} M_f[m_i, m_S] m_S$. Then $p_1, \ldots, p_r$ are linearly independent, i.e., $\forall \alpha_1 \ldots \alpha_r \in \mathbb{F}, \sum_{i=1}^{r} \alpha_i p_i = 0 \implies \forall i, \alpha_i = 0$. Let $q_i = G_k(p_i), 1 \leq i \leq r$ then clearly, $\sum_{i=1}^{r} \alpha_i q_i = 0 \implies \forall i, \alpha_i = 0$. This however, is not sufficient, since the partition φ does not imply a partition on $Y \cup Z$. To overcome this difficulty we consider the generator G_{2k} rather than G_k. Note that for any degree k polynomial f, $G_{2k}(f) \equiv 0 \iff G_k(f) \equiv 0$. Suppose $G_{2k} : \mathbb{F}[x_1, \ldots, x_n] \to \mathbb{F}[Y' \cup Z']$ where $Y' = \{y_1, \ldots, y_{2k}\}$ and $Z' = \{z_1, \ldots, z_{2k}\}$. Consider arbitrary partitions: $Y' = Y_1 \cup Y_2$, $|Y_1| = |Y_2| = k$, and $Z' = Z_1 \cup Z_2$, $|Z_1| = |Z_2| = k$. Define the map $\widehat{G_{2k}} = (\widehat{G_{2k}^{(1)}}, \ldots, \widehat{G_{2k}^{(n)}})$, where

$$\widehat{G_{2k}^{(i)}} = \widehat{G_{2k}}(x_i) = \begin{cases} G_{2k}^{(i)}|_{\{w=0 \mid w \in Y_2 \cup Z_2\}} & \text{if } i \in A \\ G_{2k}^{(i)}|_{\{w=0 \mid w \in Y_1 \cup Z_1\}} & \text{if } i \in B \end{cases}$$

Note that the polynomial $G_{2k}^{(i)}|_{\{x=0 \mid x \in Y_2 \cup Z_2\}}$ is indeed a copy of G_k^i for every i, is defined over $Y_1 \cup Z_1$ for $i \in A$, and over $Y_2 \cup Z_2$ for $i \in B$. Now, the partition φ naturally induces a partition φ' of $Y' \cup Z'$. Let $q_i' = \widehat{G_{2k}}(p_i)$, then from the above observations, we have that the polynomials $q_1', \ldots, q_i'$ are linearly independent. Since these polynomials correspond to rows in the matrix $M_{g\varphi'}$, we have $\text{rank}(M_{g\varphi'}) \geq r$. Now, to prove the required probability bound, note that the choice of the partitions $Y' = Y_1 \cup Y_2$ and $Z' = Z_1 \cup Z_2$ was arbitrary, and the rank bound holds for every such partition. There are $\binom{2k}{k}^2$ such partitions. Thus, over a random choice of partition, $\mathbf{Pr}[\text{rank}(M_{g\varphi'}) \geq r] \geq \binom{2k}{k}^2 / \binom{4k}{2k} = \Omega(1/k^2)$.

Finally, we show that polynomials with high dimension of partial derivatives cannot have small rank polynomial coefficient matrices. Let $\langle \partial^{\leq k}(f) \rangle$ be the space spanned by k^{th} order partial derivatives of a polynomial f.

Then, we show that:

Lemma 9. *Let $f \in \mathbb{F}[X]$, where $X = \{x_1, \ldots, x_n\}$ be a polynomial of degree d. Then $\forall R \in \mathbb{R}, 0 \leq k \leq d, \dim(\langle \partial^{\leq k}(f) \rangle) \geq R \implies \exists \varphi : X \to Y \cup Z \cup \mathbb{F}$, such that $|Y| = |Z|$ and $\text{rank}(M_f^\varphi) = \Omega(R/2^k)$.*

References

1. Agrawal, M., Vinay, V.: Arithmetic circuits: a chasm at depth four. In: FOCS, pp. 67–75 (2008)
2. Amini, O., Fomin, F.V., Saurabh, S.: Counting subgraphs via homomorphisms. SIAM J. Discrete Math. **26**(2), 695–717 (2012)
3. Arvind, V., Köbler, J., Kuhnert, S., Torán, J.: Solving linear equations parameterized by hamming weight. Algorithmica **75**(2), 322–338 (2016)
4. Björklund, A.: Exact covers via determinants. In: STACS, pp. 95–106 (2010)
5. Björklund, A., Husfeldt, T., Taslaman, N.: Shortest cycle through specified elements. In: SODA, pp. 1747–1753 (2012)
6. Bürgisser, P.: Completeness and Reduction in Algebraic Complexity Theory, vol. 7. Springer Science & Business Media, Heidelberg (2013)
7. Chauhan, A., Rao, B.V.R.: Parameterized analogues of probabilistic computation. In: CALDAM, pp. 181–192 (2015)

8. Chen, Z., Fu, B., Liu, Y., Schweller, R.T.: On testing monomials in multivariate polynomials. Theor. Comput. Sci. **497**, 39–54 (2013)
9. Downey, R.G., Fellows, M.R.: Fundamentals of Parameterized Complexity. Texts in Computer Science. Springer, London (2013). http://dx.doi.org/10.1007/978-1-4471-5559-1
10. Engels, C.: Why are certain polynomials hard?: a look at non-commutative, parameterized and homomorphism polynomials. Ph.D. thesis, Saarland University (2016)
11. Fischer, I.: Sums of like powers of multivariate linear forms. Mathemat. Mag. **67**(1), 59–61 (1994)
12. Fomin, F.V., Lokshtanov, D., Panolan, F., Saurabh, S.: Efficient computation of representative families with applications in parameterized and exact algorithms. J. ACM **63**(4), 29:1–29:60 (2016)
13. Fomin, F.V., Lokshtanov, D., Raman, V., Saurabh, S., Rao, B.V.R.: Faster algorithms for finding and counting subgraphs. J. Comput. Syst. Sci. **78**(3), 698–706 (2012). http://dx.doi.org/10.1016/j.jcss.2011.10.001
14. Gupta, A., Kamath, P., Kayal, N., Saptharishi, R.: Arithmetic circuits: a chasm at depth three. In: FOCS 2013, pp. 578–587. IEEE (2013)
15. Kumar, M., Maheshwari, G., Sarma M.N., J.: Arithmetic circuit lower bounds via maxrank. In: Fomin, F.V., Freivalds, R., Kwiatkowska, M., Peleg, D. (eds.) ICALP 2013. LNCS, vol. 7965, pp. 661–672. Springer, Heidelberg (2013). doi:10. 1007/978-3-642-39206-1_56
16. Müller, M.: Parameterized randomization. Ph.D. thesis, Albert-Ludwigs-Universität Freiburg im Breisgau (2008)
17. Nisan, N.: Lower bounds for non-commutative computation. In: STOC, pp. 410–418. ACM (1991)
18. Raz, R.: Multi-linear formulas for permanent and determinant are of super-polynomial size. J. ACM **56**(2) (2009)
19. Raz, R., Shpilka, A.: Deterministic polynomial identity testing in non-commutative models. Comput. Complex. **14**(1), 1–19 (2005)
20. Saptharishi, R., Chillara, S., Kumar, M.: A survey of lower bounds in arithmetic circuit complexity. Technical report (2016). https://github.com/dasarpmar/lowerbounds-survey/releases
21. Saxena, N.: Diagonal circuit identity testing and lower bounds. In: Aceto, L., Damgård, I., Goldberg, L.A., Halldórsson, M.M., Ingólfsdóttir, A., Walukiewicz, I. (eds.) ICALP 2008. LNCS, vol. 5125, pp. 60–71. Springer, Heidelberg (2008). doi:10.1007/978-3-540-70575-8_6
22. Schwartz, J.T.: Fast probabilistic algorithms for verification of polynomial identities. J. ACM (JACM) **27**(4), 701–717 (1980)
23. Shpilka, A., Volkovich, I.: Improved polynomial identity testing for read-once formulas. In: Dinur, I., Jansen, K., Naor, J., Rolim, J. (eds.) APPROX/RANDOM-2009. LNCS, vol. 5687, pp. 700–713. Springer, Heidelberg (2009). doi:10.1007/978-3-642-03685-9_52
24. Tavenas, S.: Improved bounds for reduction to depth 4 and depth 3. Inf. Comput. **240**, 2–11 (2015)
25. Valiant, L.G., Skyum, S., Berkowitz, S., Rackoff, C.: Fast parallel computation of polynomials using few processors. SIAM J. Comput. **12**(4), 641–644 (1983)
26. Zippel, R.: Probabilistic algorithms for sparse polynomials. In: Ng, E.W. (ed.) Symbolic and Algebraic Computation. LNCS, vol. 72, pp. 216–226. Springer, Heidelberg (1979). doi:10.1007/3-540-09519-5_73

A Tighter Relation Between Sensitivity Complexity and Certificate Complexity

Kun He[1,2]($\boxtimes$), Qian Li[1,2], and Xiaoming Sun[1,2]

[1] CAS Key Lab of Network Data Science and Technology, Institute of Computing
Technology, Chinese Academy of Sciences, Beijing 100190, China
{hekun,liqian,sunxiaoming}@ict.ac.cn
[2] University of Chinese Academy of Sciences, Beijing 100049, China

Abstract. The *sensitivity conjecture*, proposed by Nisan and Szegedy in 1994 [20] which asserts that for any Boolean function, its sensitivity complexity is polynomially related to the block sensitivity complexity, is one of the most important and challenging problems in the study of decision tree complexity. Despite of a lot of efforts, the best known upper bounds of block sensitivity, as well as the certificate complexity, is still exponential in terms of sensitivity [1,5]. In this paper, we give a better upper bound for certificate complexity and block sensitivity, $bs(f) \leq C(f) \leq (\frac{8}{9} + o(1))s(f)2^{s(f)-1}$, where $bs(f), C(f)$ and $s(f)$ are the block sensitivity, certificate complexity and sensitivity, respectively. The proof is based on a deep investigation on the structure of the *sensitivity graph*. We also provide a tighter relationship between the 0-certificate complexity $C_0(f)$ and 0-sensitivity $s_0(f)$ for functions with small 1-sensitivity $s_1(f)$.

Keywords: Sensitivity conjecture · Sensitivity · Block sensitivity · Certificate complexity · Boolean functions

1 Introduction

The relation between sensitivity complexity and other decision tree complexity measures is one of the most important topics in Boolean function complexity theory. Sensitivity complexity is first introduced by Cook, Dwork and Reischuk [11,12] to study the time complexity of CREW-PRAMs. The sensitivity of a Boolean function, $s(f)$, is the maximum number of variables x_i in an input assignment $x = (x_1, \cdots, x_n)$ with the property that changing x_i changes the value of $f(x)$. Nisan [19] then introduced the concept of block sensitivity, and demonstrated the remarkable fact that block sensitivity can fully characterize the time complexity of CREW-PRAMs. Block sensitivity, $bs(f)$, is a generalization

This work was supported in part by the National Natural Science Foundation of China Grant 61433014, 61502449, 61602440, the 973 Program of China Grants No. 2016YFB1000201 and the China National Program for support of Top-notch Young Professionals.

© Springer International Publishing AG 2017
Y. Cao and J. Chen (Eds.): COCOON 2017, LNCS 10392, pp. 262–274, 2017.
DOI: 10.1007/978-3-319-62389-4_22

of sensitivity to the case when we are allowed to change disjoint blocks of variables. Block sensitivity turns out to be polynomially related to a number of other complexity measures for Boolean functions [9], such as decision tree complexity, certificate complexity, polynomial degree and quantum query complexity, etc. One exception is sensitivity. So far it is still not clear whether sensitivity complexity could be exponentially smaller than block sensitivity and other measures. The famous sensitivity conjecture, proposed by Nisan and Szegedy in 1994 [20], asserts that block sensitivity and sensitivity complexity are also polynomially related. According to the definition of sensitivity and block sensitivity, it is easy to see that $s(f) \leq bs(f)$ for any total Boolean function f. But in the other direction, it is much harder to prove an upper bound of block sensitivity in terms of sensitivity complexity.

We briefly recall some of the remarkable improvements about the upper bound of block sensitivity here. Kenyon and Kutin [17] proved that $bs(f) \leq (\frac{e}{\sqrt{2\pi}})e^{s(f)}\sqrt{s(f)}$. Then Ambainis et al. [1] proved that $bs(f) \leq C(f) \leq 2^{s(f)-1}s(f) - (s(f) - 1)$, where $C(f)$ is the certificate complexity, a complexity measure with clear combinational interpretation. Certificate complexity being at least c simply means that there is an input $x = (x_1, \cdots, x_n)$ that is not contained in an $(n - (c - 1))$-dimensional subcube of the Boolean hypercube on which f is constant. The last improvement is $bs(f) \leq C(f) \leq \max\{2^{s(f)-1}(s(f) - \frac{1}{3}), s(f)\}$ [3], only minus a constant $1/3$ from the term $s(f)$. Despite of a lot of efforts, the best known upper bound of block sensitivity is still exponential in terms of sensitivity. The best known separation between sensitivity and block sensitivity complexity is quadratic [4]: there exists a sequence of Boolean functions f with $bs(f) = \frac{2}{3}s(f)^2 - \frac{1}{3}s(f)$.

Recently, Tal [24] showed that any upper bound of the form $bs_l(f) \leq s(f)^{l-\varepsilon}$ for $\varepsilon > 0$ implies a subexponential upper bound on $bs(f)$ in terms of $s(f)$. Here $bs_l(f)$, the l-block sensitivity, defined by Kenyon and Kutin [17], is the block sensitivity with the size of each block at most l. There are also many works on the equivalent propositions of sensitivity conjecture, i.e., whether sensitivity complexity is polynomially related to Fourier degree, certificate complexity or quantum query complexity. Gopalan et al. [15] proved a 2-approximate version of the degree vs. sensitivity conjecture (the original one needs an ∞-approximation): every Boolean function with sensitivity s can be ϵ-approximated (in ℓ_2) by a polynomial whose degree is $O(s\log(1/\epsilon))$. Recently, Lovett et al. [22] proved a robust analog of the sensitivity conjecture, due to Gopalan et al., which relates the decay of the Fourier mass of a boolean function to moments of its sensitivity. Ben-David [7] provided a cubic separation between quantum query complexity and sensitivity, as well as a power of 2.1 separation between certificate complexity and sensitivity. While to solve the sensitivity conjecture seems very challenging for general Boolean functions, special classes of functions have been investigated, such as functions with graph properties [25], cyclically invariant functions [10], functions with small alternating number [18], disjunctive normal forms [21], constant depth regular read-k formulas [6], etc. We recommend readers [16] for an excellent survey about the sensitivity conjecture. For other recent progresses, see [1,2,5,8,13,14,23].

Our Results. In this paper, we give a better upper bound of block sensitivity in terms of sensitivity.

Theorem 1. *For any total Boolean function* $f : \{0,1\}^n \to \{0,1\}$,

$$bs(f) \le C(f) \le (\frac{8}{9} + o(1))s(f)2^{s(f)-1}.$$

Here $o(1)$ *denotes a term that vanishes as* $s(f) \to +\infty$.

Similar to the Ambainis et al. [3], our proof is also based on analysis of the structure of the *sensitivity graph*. However, their result, $max\{2^{s(f)-1}(s(f) - \frac{1}{3}), s(f)\}$, is about $(1-o(1))s(f)2^{s(f)-1}$. The improvement is multiplying by the factor $(1 - o(1))$ comparing to the previous result $s(f)2^{s(f)-1} - (s(f) - 1)$ [1], while our improvement is multiplying by factor $(\frac{8}{9}+o(1))$. Ambainis et al. [3] also investigated the Boolean functions with $s_1(f) = 2$, and showed that $C_0(f) \le \frac{9}{5}s_0(f)$ for this class of Boolean functions. In this paper, we also improve this bound.

Theorem 2. *Let* f *be a Boolean function with* $s_1(f) = 2$, *then*

$$C_0(f) \le \frac{37 + \sqrt{5}}{22}s_0(f) \approx 1.7835s_0(f).$$

Organization. We present preliminaries in Sect. 2. We give the overall structure of our proof for Theorem 1 in Sect. 3 and the detailed proofs for lemmas in Sect. 4. Finally, we conclude this paper in Sect. 5. The proof of Theorem 2 is given in Appendix.

2 Preliminaries

Let $f : \{0,1\}^n \to \{0,1\}$ be a Boolean function. For an input $x \in \{0,1\}^n$ and a subset $B \subseteq [n]$, x^B denotes the input obtained by flipping all the bits x_j such that $j \in B$.

Definition 1. *The* sensitivity *of* f *on input* x *is defined as* $s(f,x) := |\{i : f(x) \ne f(x^i)\}|$. *The sensitivity* $s(f)$ *of* f *is defined as* $s(f) := max_x s(f,x)$. *The* b-sensitivity $s_b(f)$ *of* f, *where* $b \in \{0,1\}$, *is defined as* $s_b(f) := max_{x \in f^{-1}(b)} s(f,x)$.

Definition 2. *The* block sensitivity $bs(f,x)$ *of* f *on input* x *is the maximum number of disjoint subsets* $B_1, B_2, \cdots, B_r$ *of* $[n]$ *such that for all* $j \in [r]$, $f(x) \ne f(x^{B_j})$. *The block sensitivity of* f *is defined as* $bs(f) := max_x bs(f,x)$. *The* b-block sensitivity $bs_b(f)$ *of* f, *where* $b \in \{0,1\}$, *is defined as* $bs_b(f) := max_{x \in f^{-1}(b)} bs(f,x)$.

Definition 3. *A partial assignment is a function* $p : [n] \to \{0, 1, *\}$. *We call* $S = \{i | p(i) \neq *\}$ *the support of this partial assignment. We define the co-dimension of* p, *denoted by co-dim(p), to be* $|S|$. *We say* x *is consistent with* p *if* $x_i = p_i$ *for every* $i \in S$. p *is called a* b-*certificate if* $f(x) = b$ *for any* x *consistent with* p, *where* $b \in \{0, 1\}$. *For* $B \subseteq S$, p^B *denotes the partial assignment obtained by flipping all the bits* p_j *such that* $j \in B$. *For* $i \in [n]/S$ *and* $b \in \{0, 1\}$, $p_{i=b}$ *denotes the partial assignment obtained by setting* $p_i = b$.[1]

Definition 4. *The* certificate complexity $C(f, x) \cdot$ *of* f *on* x *is the minimum co-dimension of* $f(x)$-*certificate that* x *is consistent with. The certificate complexity* $C(f)$ *of* f *is defined as* $C(f) := \max_x C(f, x)$. *The* b-*certificate complexity* $C_b(f)$ *of* f *is defined as* $C_b(f) := \max_{x \in f^{-1}(b)} C(f, x)$.

Definition 5. *For any set* $S \subseteq \{0, 1\}^n$, *let* $s(f, S)$ *to be* $\sum_{x \in S} s(f, x)$, *the average sensitivity of* S *is defined by* $s(f, S) / |S|$.

In this work we regard $\{0, 1\}^n$ as a set of vertices for an n-dimensional hypercube Q_n, where two nodes x and y have an edge if and only if the Hamming distance between them is 1. A Boolean function $f : \{0, 1\}^n \to \{0, 1\}$ can be regarded as a 2-coloring of the vertices of Q_n, where x is *black* if $f(x) = 1$ and x is *white* if $f(x) = 0$. Let $f^{-1}(1) = \{x | f(x) = 1\}$ be the set of all black vertices. If $f(x) \neq f(y)$, we call the edge (x, y) a sensitive edge and x is sensitive to y (y is also sensitive to x). We regard a subset $S \subseteq \{0, 1\}^n$ as the subgraph G induced by the vertices in S. Define the size of G, $|G|$, as the size of S. It is easy to see that $s(f, x)$ is the number of neighbors of x with a different color than x. A certificate is a monochromatic subcube, and $C(f, x)$ is the co-dimension of the largest monochromatic subcube which contains x.

There is a natural bijection between the partial assignments and the subcubes, where a partial assignment p corresponds to a subcube induced by the vertices consistent with p. Without ambiguity, we sometimes abuse these two concepts.

Definition 6. *Let* G *and* H *be two induced subgraphs of* Q_n. *Let* $G \cap H$ *denote the graph induced on* $V(G) \cap V(H)$. *For any two subcubes* G *and* H, *we call* H *a neighbor cube of* G *if their corresponding partial assignments* p_G *and* p_H *satisfy* $p_G = p_H^i$ *for some* i.

Our proofs rely on the following result proved by Ambainis and Vihrovs [5].

Lemma 1. *[5] Let* G *be a non-empty induced subgraph of* Q_k *induced by* $Q_k \cap f^{-1}(b)$ *where* $b \in \{0, 1\}$ *and satisfying that the sensitivity of every vertice in* G *is at most* s, *then either* $|G| \geq \frac{3}{2} \cdot 2^{k-s}$, *or* G *is a subcube of* Q_k *with co-dim(G)=s and* $|G| = 2^{k-s}$.

The following formulation of Turan's theorem [26] is also used in our proofs.

Lemma 2 (Turan's theorem [26]). *Every graph* G *with* n *vertices and average degree* Δ *has an independent set of size at least* $\frac{n}{\Delta + 1}$.

---

[1] The function p can be viewed as a vector, and we sometimes use p_i to represent $p(i)$.

3   The Sketch of Proof for Theorem 1

In this section, we give the sketch of the proof for Theorem 1. We first present some notations used in the proof. Let f be an n-input Boolean function. Let z be an input with $f(z) = 0$ and $C(f, z) = C_0(f) = m$. Then there exists a 0-certificate of co-dimension $C_0(f)$ consistent with z, and let H be the one with maximum average sensitivity if there are many such 0-certificates. W.l.o.g., we assume $z = 0^n$ and $H = 0^m *^{n-m}$. Among the m neighbor cubes of H, from Lemma 1 we have for any $i \in [m]$, either $|H^i \cap f^{-1}(1)| \geq \frac{3}{2} \cdot \frac{|H|}{2^{s_1(f)-1}}$ or $H^i \cap f^{-1}(1)$ is a subcube of H^i of size $\frac{|H|}{2^{s_1(f)-1}}$, which are called *heavy cube* and *light cube*, respectively. W.l.o.g., assume $H^1, H^2 \cdots, H^l$ are light cubes and $H^{l+1}, \cdots, H^m$ are heavy cubes, where $l \leq m$ is the number of light cubes. For $k > m$, let $N_k^0 = \{i \in [l] | (H^i \cap f^{-1}(1))_k = 0\}$. Intuitively, N_k^0 consists of the indexes of light cubes in which the kth bit of black subcube, $H^i \cap f^{-1}(1)$, is 0. Similarly, let $N_k^1 = \{i \in [l] | (H^i \cap f^{-1}(1))_k = 1\}$ and $N_k = N_k^0 \cup N_k^1$. In the following, we will assume $i \in [l]$ and $k > m$ while $i \in N_k$ is used.

For any subcube $H' \subseteq H$, we use $s_l(f, H')$ ($s_h(f, H')$ respectively) to denote the number of sensitive edges of H' adjacent to the light cubes (heavy cubes respectively). Similarly, for subcube $H' \subseteq H^i$ where $i \leq l$ and $H' \subseteq f^{-1}(0)$, we use $s_l(f, H')$ ($s_h(f, H')$ respectively) to denote the number of sensitive edges of H' adjacent to $H^{1,i}, \cdots, H^{i-1,i}, H^i, H^{i+1,i}, \cdots, H^{l,i}$ ($H^{l+1,i}, \cdots, H^{m,i}$ respectively). It is easy to see $s_l(f, H') + s_h(f, H') = s(f, H')$.

The overall structure of our proof is as follows.

$$
\left. \begin{array}{l}
\text{Lemma 1} \\
\text{Lemma 6} \Rightarrow \text{Lemma 3} \\
\left. \begin{array}{l} \text{Lemma 7} \\ \text{Lemma 8} \end{array} \right\} \Rightarrow \text{Lemma 4} \\
\text{Lemma 5}
\end{array} \right\} \Rightarrow \text{Theorem 1.}
$$

Lemmas 3, 4, 6, 7 and 8 are about the structures of light cubes while Lemma 5 gives a lower bound of the number of 1-inputs in heavy cubes. The main idea of our proof is to show that there are many 1-inputs in the heavy cubes. To see why it works, consider the extremal case where there are no light cubes (i.e. $l = 0$), then the average sensitivity of H is at least $m \cdot \frac{3}{2^{s_1(f)}}$. Because the average sensitivity of H can not exceed $s_0(f)$, we have $m \cdot \frac{3}{2^{s_1(f)}} \leq s_0(f)$ and $m \leq \frac{2}{3} s_0(f) 2^{s_1(f)-1}$.

More generally, the average sensitivity of H is at least $\frac{l}{2^{s_1(f)-1}} + \frac{3(m-l)}{2^{s_1(f)}}$. Let $L = s_0(f) 2^{s_1(f)-1} / l$. If $L \geq 2$, we have $l \leq s_0(f) 2^{s_1(f)-2}$ and $m \leq \frac{5}{6} s_0(f) 2^{s_1(f)-1}$. If $s_1(f) = 1$, it has already been shown that $C_0(f) \leq s_0(f)$ [3]. In the following proof, we assume $l > 0$, $L < 2$ and $s_1(f) \geq 2$. Note that if $i \in N_k^1$, then $H_{k=0}$ together with $H_{k=0}^i$ is another certificate of z of the same co-dimension with H. Thus according to the assumption that H is the one with maximum average sensitivity, we have

$$
s(f, H) - \big(s(f, H_{k=0}) + s(f, H_{k=0}^i)\big) = s(f, H_{k=1}) - s(f, H_{k=0}^i) \geq 0.
$$

By summing over different cubes and different bits, we get

$$
\sum_{k:|N_k^1|\geq s_1(f)-1}\ \sum_{i\in S_k^1}\Big(s_h(f,H_{k=1})-s_h(f,H_{k=0}^i)\Big)
$$

$$
=\sum_{k:|N_k^1|\geq s_1(f)-1}\ \sum_{i\in S_k^1}\left[\Big(s_l(f,H_{k=0}^i)-s_l(f,H_{k=1})\Big)-\Big(s(f,H_{k=0}^i)-s(f,H_{k=1})\Big)\right]
$$

$$
\geq\sum_{k:|N_k^1|\geq s_1(f)-1}\ \sum_{i\in S_k^1}\Big(s_l(f,H_{k=0}^i)-s_l(f,H_{k=1})\Big)
$$

$$
\geq\frac{(s_1(f)-1)|H|}{2^{s_1(f)-1}}\sum_{k:|N_k^1|\geq s_1(f)-1}|N_k^0|
$$

$$
\geq(\frac{1}{2}-o(1))\frac{(s_1(f)-1)^2|H|l}{2^{s_1(f)-1}}.
$$

$$(1)$$

Here $o(1)$ denotes a term that vanishes as $s_1(f)\rightarrow+\infty$, and S_k^1 is a subset of N_k^1 of size $s_1(f)-1$. The last but one inequality is due to the following lemma.

Lemma 3. *For any* $i\in N_k^1$, $s_l(f,H_{k=0}^i)-s_l(f,H_{k=1})\geq\frac{|N_k^0|\cdot|H|}{2^{s_1(f)-1}}$.

The last inequality is due to

Lemma 4. *If* $L<2$, *then* $\sum_{k:|N_k^1|\geq s_1(f)-1}|N_k^0|\geq(\frac{1}{2}-o(1))l(s_1(f)-1)$.

On the other side, we can show that

Lemma 5.

$$
\sum_{k:|N_k^1|\geq s_1(f)-1\ i\in S_k^1}\sum(s_h(f,H_{k=1})-s_h(f,H_{k=0}^i))\leq(s_1(f)-1)^2\sum_{l<t\leq m}|H^t\cap f^{-1}(1)|.
$$

The proofs of these three lemmas are postponed to the next section. We first finish the proof of Theorem 1 here. Equality 1 together with Lemma 5 demonstrates that there are many 1-inputs in the heavy cubes, i.e.

$$
\sum_{l<t\leq m}|H^t\cap f^{-1}(1)|\geq\frac{(\frac{1}{2}-o(1))l|H|}{2^{s_1(f)-1}}.
$$

This inequality gives a new lower bound of the number of 1-inputs in the heavy cubes neighboring H by relating it to the number of light cubes. This lower bound provides new structural insights of the sensitivity graph. Combining it with

$$
\frac{l}{2^{s_1(f)-1}}+\sum_{l<t\leq m}\frac{|H^t\cap f^{-1}(1)|}{|H|}\leq s_0(f),
$$

we get

$$
l\leq(\frac{2}{3}+o(1))2^{s_1(f)-1}s_0(f).
$$

Moreover, recall that $|H^t \cap f^{-1}(1)|/|H| \geq \frac{3}{2^{s_1(f)}}$, thus

$$\frac{l}{2^{s_1(f)-1}} + \frac{3}{2} \cdot \frac{m-l}{2^{s_1(f)-1}} \leq s_0(f).$$

Therefore,

$$C_0(f) = m \leq (\frac{8}{9} + o(1))s_0(f)2^{s_1(f)-1}.$$

Here, $o(1)$ denotes a term that vanishes as $s_1(f) \to +\infty$. Similarly, we can also obtain

$$C_1(f) \leq (\frac{8}{9} + o(1))s_1(f)2^{s_0(f)-1}.$$

Therefore,

$$C(f) \leq (\frac{8}{9} + o(1))s(f)2^{s(f)-1}.$$

where $o(1)$ denotes a term that vanishes as $s(f) \to +\infty$.

4   Proofs of the Lemmas

4.1   Proof of Lemma 3

Before giving the proof of Lemma 3, we first prove the following lemma which will be used.

Lemma 6. *If $i,j \in N_k$, then $|H^{i,j}_{k=1} \cap f^{-1}(1)| \geq \frac{|H|}{2^{s_1(f)-1}}$ and $|H^{i,j}_{k=0} \cap f^{-1}(1)| \geq \frac{|H|}{2^{s_1(f)-1}}$.*

Proof. W.l.o.g., assume $i \in N^1_k$. For any $x \in H^i \cap f^{-1}(1)$, there are $(s_1(f) - 1)$ vertices in H^i as well as $x^i \in H$ sensitive to x, thus $x^j \in H^{i,j}$ is in $f^{-1}(1)$, since otherwise x would be sensitive to $s_1(f)+1$ vertices. Therefore, $|H^{i,j}_{k=1} \cap f^{-1}(1)| \geq |H^i \cap f^{-1}(1)| = \frac{|H|}{2^{s_1(f)-1}}$. Similarly, if $j \in N^0_k$, we have $|H^{i,j}_{k=0} \cap f^{-1}(1)| \geq \frac{|H|}{2^{s_1(f)-1}}$.

If $j \in N^1_k$, note that $H^{i,j}_{k=0} \cap f^{-1}(1) \neq \emptyset$, since otherwise $H^{i,j}_{k=0}, H^i_{k=0}, H^j_{k=0}$ and $H_{k=0}$ would become a larger monochromatic subcube containing z, which is contradictory with the assumption of H, i.e., the maximum monochromatic subcube containing z. For any $y \in H^{i,j}_{k=0} \cap f^{-1}(1)$, y is sensitive to $y^i \in H^i$ and $y^j \in H^j$, thus y has at most $s_1(f) - 2$ sensitive edges in $H^{i,j}_{k=0}$. Therefore, $|H^{i,j}_{k=0} \cap f^{-1}(1)| \geq \frac{|H^{i,j}_{k=0}|}{2^{s_1(f)-2}} = \frac{|H|}{2^{s_1(f)-1}}$ according to Lemma 1.

Proof. (**Proof of Lemma 3**) Since $H_{k=1} \cap f^{-1}(1) = \emptyset$ and $H^i_{k=0} \cap f^{-1}(1) = \emptyset$, it is easy to see

$$s_l(f, H_{k=1}) = \sum_{j=1}^{l} |H^j_{k=1} \cap f^{-1}(1)| = \frac{|N^1_k| \cdot |H|}{2^{s_1(f)-1}} + \frac{(l - |N_k|)|H|}{2^{s_1(f)}} = \frac{(l + |N^1_k| - |N^0_k|)|H|}{2^{s_1(f)}}.$$

Similarly,

$$s_l(f, H^i_{k=0}) = \sum_{j=1, j \neq i}^{l} |H^{i,j}_{k=0} \cap f^{-1}(1)| + |H^i \cap f^{-1}(1)|.$$

If $j \notin N_k$, then for any $x \in H^j \cap f^{-1}(1)$, we have $x^i \in f^{-1}(1)$ since otherwise x would have $s_1(f) + 1$ sensitive edges, thus $|H^{i,j}_{k=0} \cap f^{-1}(1)| \geq |H^j_{k=0} \cap f^{-1}(1)| = \frac{|H^j \cap f^{-1}(1)|}{2} = \frac{|H|}{2^{s_1(f)}}$. If $j \in N_k$, $|H^{i,j}_{k=0} \cap f^{-1}(1)| \geq \frac{|H|}{2^{s_1(f)-1}}$ according to Lemma 6. Therefore, $s_l(f, H^i_{k=0}) \geq \frac{(l + |N^1_k| + |N^0_k|)|H|}{2^{s_1(f)}} = s_l(f, H_{k=1}) + \frac{|N^0_k| \cdot |H|}{2^{s_1(f)-1}}$.

4.2   Proof of Lemma 4

We first prove two lemmas. With these lemmas Lemma 4 follows. In our proofs, the logarithm uses base 2.

Lemma 7. *For any integer $c > 2$,*

$$\sum_{|N_k| \geq c} |N_k| \geq l\Big(\log l - \log \big(s_0(f)(s_1(f) - 1)(c - 2) + s_0(f)\big)\Big).$$

Proof. First note that for $i \leq l$, $H^i \cap f^{-1}(1)$ is a subcube and co-dim($H^i \cap f^{-1}(1)$) $= n - m - s_1(f) + 1$, which means $|\{k > m | (H^i \cap f^{-1}(1))_k \neq *\}| = s_1(f) - 1$. Let $w = |\{k > m | |N_k| \geq c\}|$. W.l.o.g., we assume that $|N_k| \geq c$ if and only if $k \in [m + 1, m + w]$. For any $y \in \{0, 1\}^w$, let $\mathbb{I}_{\overline{y}} = \{i \in [l] | \forall j \in [w] : (H^i \cap f^{-1}(1))_{j+m} \neq y_j\}$.

We claim that for any y, $|\mathbb{I}_{\overline{y}}|$ can not be "too large". Think about the graph $G = (V, E)$ where $V = \mathbb{I}_{\overline{y}}$ and $(i, j) \in E$ if $i, j \in N_k$ and $(H^i \cap f^{-1}(1))_k \neq (H^j \cap f^{-1}(1))_k$ for some $k \geq m + w + 1$. Note that $|N_k| \leq c - 1$ holds for any $k \geq m + w + 1$. For any $i \in N_k$ where $k \geq m + w + 1$, there are at most $|N_k| - 1 \leq c - 2$ different j, where $j \in N_k$, having an edge with i. Therefore for any $i \in \mathbb{I}_{\overline{y}}$,

$$\deg(i) \leq \sum_{k:i\in N_k, k \geq m+w+1} (|N_k| - 1) \leq \sum_{k:i\in N_k, k \geq m+w+1} (c - 2) \leq (s_1(f) - 1)(c - 2).$$

Thus according to Turan's theorem, there exists an independent set S of size

$$|S| = \frac{|V|}{\frac{\sum_i \deg(i)}{|V|} + 1} \geq \frac{|V|}{\max_i \deg(i) + 1} \geq \frac{|\mathbb{I}_{\overline{y}}|}{(s_1(f) - 1)(c - 2) + 1},$$

which means that there exists an input $x \in H$ such that $x^i \in f^{-1}(1)$ for any $i \in S$, therefore $|S| \leq s_0(f)$, implying

$$|\mathbb{I}_{\overline{y}}| \leq ((s_1(f) - 1)(c - 2) + 1)s_0(f).$$

Therefore, we have

$$\sum_{y\in\{0,1\}^w} |\mathbb{I}_{\overline{y}}| \le 2^w((s_1(f)-1)(c-2)+1)s_0(f). \tag{2}$$

On the other side, let $w_i = |\{k \in [m+1, m+w]|i \in N_k\}|$, then there are exact 2^{w-w_i} different ys such that $\mathbb{I}_{\overline{y}}$ contains i, thus

$$\sum_{y\in\{0,1\}^w} |\mathbb{I}_{\overline{y}}| = \sum_{i\le l} 2^{w-w_i} \ge l \cdot 2^{w - \sum_{i\le l} w_i/l} = l \cdot 2^{w - \sum_{k=m+1}^{m+w} |N_k|/l} = l \cdot 2^{w - \sum_{|N_k|\ge c} |N_k|/l}. \tag{3}$$

The inequality is due to the AM-GM inequality. Combining Inequality (2) and (3), we can finish the proof of the lemma.

Lemma 8. $\sum_{k>m} \big||N_k^0| - |N_k^1|\big| \le l\sqrt{2\ln L(s_1(f)-1)}.$

Proof. We sample an input $x \in H$ as $\Pr(x_k = 0) = p$ independently for each $k > m$. Here $p := \sum_{k>m}|N_k^0|/\sum_{k>m}|N_k|$. Recall that for $i \in [l]$, $|\{k > m : (H^i \cap f^{-1}(1))_k \ne *\}| = s_1(f) - 1$, then $\Pr(x^i \in f^{-1}(1)) = p^{d_i}(1-p)^{s_1(f)-1-d_i}$, where $d_i := |\{k > m : (H^i \cap f^{-1}(1))_k = 0\}|$. Therefore

$$\begin{aligned}
s_0(f) &\ge \mathbb{E}(s(f,x)) \\
&\ge \sum_{i\in[l]} \Pr(x^i \in f^{-1}(1)) \\
&= \sum_{i\in[l]} p^{d_i}(1-p)^{s_1(f)-1-d_i} \\
&\ge l p^{\frac{\sum_{i\in[l]} d_i}{l}} (1-p)^{\frac{\sum_{i\in[l]} s_1(f)-1-d_i}{l}} \\
&= l p^{\frac{\sum_{k>m} |N_k^0|}{l}} (1-p)^{\frac{\sum_{k>m} |N_k^1|}{l}} \\
&= l p^{\frac{p\sum_{k>m} |N_k|}{l}} (1-p)^{\frac{(1-p)\sum_{k>m} |N_k|}{l}} \\
&= l p^{p(s_1(f)-1)}(1-p)^{(1-p)(s_1(f)-1)}.
\end{aligned} \tag{4}$$

Step four is due to the AM-GM inequality. Step five is due to the fact $\sum_{i\in[l]} d_i = \sum_{k>m}|N_k^0|$ and $\sum_{i\in[l]} s_1(f) - 1 - d_i = \sum_{k>m}|N_k^1|$, which are immediate from the definition of d_i, N_k^0 and N_k^1. The last step is due to the fact that $\sum_{k>m}|N_k| = l(s_1(f)-1)$. By calculus, it is not hard to obtain $e^{2(p-1/2)^2} \le 2p^p(1-p)^{1-p}$ for $0 \le p \le 1$. Together with Inequality (4) and recall that $L = s_0(f)2^{s_1(f)-1}/l$, it implies $|p - \frac{1}{2}| \le \sqrt{\frac{\ln L}{2(s_1(f)-1)}}$. Therefore

$$\sum_{k>m} \big||N_k^1| - |N_k^0|\big| = |1 - 2p| \sum_{k>m} |N_k| \le l\sqrt{2\ln L(s_1(f)-1)}.$$

Now, we can prove Lemma 4. For any $c_2 > 2c_1$, first note that

$$\sum_{|N_k^1|<c_1,|N_k|\geq c_2} (|N_k^0|-|N_k^1|) = \sum_{|N_k^1|<c_1,|N_k|\geq c_2} |N_k|\left(1-\frac{2|N_k^1|}{|N_k|}\right) \geq \frac{c_2-2c_1}{c_2} \sum_{|N_k^1|<c_1,|N_k|\geq c_2} |N_k|.$$

Then we have

$$\sum_{|N_k^1|\geq c_1} 2|N_k^0| \geq \sum_{|N_k^1|\geq c_1,|N_k|\geq c_2} 2|N_k^0|$$

$$= \sum_{|N_k|\geq c_2} |N_k| - \sum_{|N_k^1|<c_1,|N_k|\geq c_2} |N_k| - \sum_{|N_k^1|\geq c_1,|N_k|\geq c_2} (|N_k^1|-|N_k^0|)$$

$$\geq \sum_{|N_k|\geq c_2} |N_k| - \frac{c_2}{c_2-2c_1} \sum_{|N_k^1|<c_1,|N_k|\geq c_2} (|N_k^0|-|N_k^1|) - \sum_{|N_k^1|\geq c_1,|N_k|\geq c_2} \left||N_k^1|-|N_k^0|\right|$$

$$\geq \sum_{|N_k|\geq c_2} |N_k| - \frac{c_2}{c_2-2c_1} \sum_{|N_k|\geq c_2} \left||N_k^0|-|N_k^1|\right|.$$

According to Lemmas 7 and 8, we have

$$\sum_{|N_k^1|\geq c_1} |N_k^0| \geq \frac{l(\log l - \log(s_0(f)(s_1(f)-1)(c_2-1)+s_0(f)))}{2} - \frac{lc_2\sqrt{2\ln L(s_1(f)-1)}}{2(c_2-2c_1)}$$

$$= \frac{l(s_1(f)-1-\log L - \log((s_1(f)-1)(c_2-2)+1))}{2} - \frac{lc_2\sqrt{2\ln L(s_1(f)-1)}}{2(c_2-2c_1)}.$$

Recall $L \leq 2$, and let $c_1 = s_1(f)-1$ and $c_2 = 3c_1$, thus

$$\sum_{|N_k^1|\geq s_1(f)-1} |N_k^0| \geq l(s_1(f)-1)(\frac{1}{2}-o(1)).$$

4.3   Proof of Lemma 5

Proof. Note that $H_{k=1} \cap f^{-1}(1) = \emptyset$ and $H_{k=0}^i \cap f^{-1}(1) = \emptyset$ for $i \in N_k^1$. Thus it is easy to see that

$$s_h(f, H_{k=1}) - s_h(f, H_{k=0}^i) = \sum_{l<t\leq m} \sum_{x\in H_{k=1}^t} (f(x) - f(x^{i,k})).$$

Therefore,

$$\sum_{k:|N_k^1|\geq s_1(f)-1}\sum_{i\in S_k^1}\left(s_h(f,H_{k=1})-s_h(f,H_{k=0}^i)\right)$$

$$=\sum_{k:|N_k^1|\geq s_1(f)-1}\sum_{i\in S_k^1}\sum_{l<t\leq m}\sum_{x\in H_{k=1}^t}\left(f(x)-f(x^{i,k})\right)$$

$$\leq\sum_{k:|N_k^1|\geq s_1(f)-1}\sum_{i\in S_k^1}\sum_{l<t\leq m}\sum_{\substack{x\in H^t,\\f(x)=1,f(x^{i,k})=0}}1$$

$$=\sum_{k:|N_k^1|\geq s_1(f)-1}\sum_{i\in S_k^1}\sum_{l<t\leq m}\left(\sum_{\substack{x\in H^t,f(x)=1,\\f(x^k)=0,f(x^{i,k})=0}}1+\sum_{\substack{x\in H^t,f(x)=1,\\f(x^k)=1,f(x^{i,k})=0}}1\right)$$

$$\leq\sum_{k:|N_k^1|\geq s_1(f)-1}\sum_{i\in S_k^1}\sum_{l<t\leq m}\left(\sum_{\substack{x\in H^t,f(x)=1,\\f(x^k)=0,f(x^{i,k})=0}}1+\sum_{\substack{x\in H^t,\\f(x)=1,f(x^i)=0}}1\right)$$

$$\leq\sum_{l<t\leq m}\left(\sum_{x\in H^t,f(x)=1}\sum_{\substack{k:|N_k^1|\geq s_1(f)-1,i\in S_k^1\\f(x^k)=0}}\sum 1+\sum_{x\in H^t,f(x)=1}\sum_{i\leq l:f(x^i)=0}\sum_{k:i\in S_k^1}1\right)$$

$$\leq\sum_{l<t\leq m}\left(\sum_{x\in H^t,f(x)=1}\sum_{\substack{k:|N_k^1|\geq s_1(f)-1,\\f(x^k)=0}}(s_1(f)-1)+\sum_{x\in H^t,f(x)=1}\sum_{i\leq l:f(x^i)=0}(s_1(f)-1)\right)$$

$$\leq\sum_{l<t\leq m}\sum_{x\in H^t,f(x)=1}\left(\sum_{k>m,f(x^k)=0}1+\sum_{i\leq l,f(x^i)=0}1\right)(s_1(f)-1)$$

$$\leq\sum_{l<t\leq m}\sum_{x\in H^t,f(x)=1}(s_1(f)-1)^2=(s_1(f)-1)^2\sum_{l<t\leq m}|H^t\cap f^{-1}(1)|.$$

In Step four, we substitute x^k with x for the second part. In Step six, for the first part, $\sum_{i\in S_k^1}1\leq s_1(f)-1$ because S_k^1 is a subset of N_k^1 of size $s_1(f)-1$. For the second part, $|\{k>m|i\in S_k^1\}|\leq|\{k>m|i\in N_k^1\}|=s_1(f)-1$.

5 Conclusion

In this work, we give a better upper bound of block sensitivity in terms of sensitivity. Our results are based on carefully exploiting the structure of the light cubes. However, our approach has an obvious limitation. In the extremal case, if there are no light cubes, then we can only get $bs(f)\leq C(f)\leq\frac{2}{3}s(f)2^{s(f)-1}$. Better understanding about the structure of heavy cubes is needed in order to conquer this limitation.

References

1. Ambainis, A., Bavarian, M., Gao, Y., Mao, J., Sun, X., Zuo, S.: Tighter relations between sensitivity and other complexity measures. In: Esparza, J., Fraigniaud, P., Husfeldt, T., Koutsoupias, E. (eds.) ICALP 2014. LNCS, vol. 8572, pp. 101–113. Springer, Heidelberg (2014). doi:10.1007/978-3-662-43948-7_9
2. Ambainis, A., Prūsis, K.: A tight lower bound on certificate complexity in terms of block sensitivity and sensitivity. In: Csuhaj-Varjú, E., Dietzfelbinger, M., Ésik, Z. (eds.) MFCS 2014. LNCS, vol. 8635, pp. 33–44. Springer, Heidelberg (2014). doi:10.1007/978-3-662-44465-8_4
3. Ambainis, A., Prūsis, K., Vihrovs, J.: Sensitivity versus certificate complexity of boolean functions. In: Kulikov, A.S., Woeginger, G.J. (eds.) CSR 2016. LNCS, vol. 9691, pp. 16–28. Springer, Cham (2016). doi:10.1007/978-3-319-34171-2_2
4. Ambainis, A., Sun, X.: New separation between $s(f)$ and $bs(f)$. CoRR, abs/1108.3494 (2011)
5. Ambainis, A., Vihrovs, J.: Size of sets with small sensitivity: a generalization of simon's lemma. In: Jain, R., Jain, S., Stephan, F. (eds.) TAMC 2015. LNCS, vol. 9076, pp. 122–133. Springer, Cham (2015). doi:10.1007/978-3-319-17142-5_12
6. Bafna, M., Lokam, S.V., Tavenas, S., Velingker, A.: On the sensitivity conjecture for read-k formulas. In: 41st International Symposium on Mathematical Foundations of Computer Science, MFCS, pp. 16:1–16:14 (2016)
7. Ben-David, S., Hatami, P., Tal, A.: Low-sensitivity functions from unambiguous certificates. In: 8th Innovations in Theoretical Computer Science, ITCS 9–11, 2017, Berkeley, USA, January 2017
8. Boppana, M.: Lattice variant of the sensitivity conjecture. In: Electronic Colloquium on Computational Complexity (ECCC), vol. 19, p. 89 (2012)
9. Buhrman, H., De Wolf, R.: Complexity measures and decision tree complexity: a survey. Theor. Comput. Sci. **288**(1), 21–43 (2002)
10. Chakraborty, S.: On the sensitivity of cyclically-invariant boolean functions. In: Proceedings of the 20th Annual IEEE Conference on Computational Complexity, CCC 2005, pp. 163–167 (2005)
11. Cook, S., Dwork, C.: Bounds on the time for parallel ram's to compute simple functions. In: Proceedings of the Fourteenth Annual ACM Symposium on Theory of Computing, STOC 1982, pp. 231–233 (1982)
12. Cook, S.A., Dwork, C., Reischuk, R.: Upper and lower time bounds for parallel random access machines without simultaneous writes. SIAM J. Comput. **15**(1), 87–97 (1986)
13. Gilmer, J., Koucký, M., Saks, M.E.: A new approach to the sensitivity conjecture. In: Proceedings of the 2015 Conference on Innovations in Theoretical Computer Science, ITCS 2015, Rehovot, Israel, January 11–13, 2015, pp. 247–254 (2015)
14. Gopalan, P., Nisan, N., Servedio, R.A., Talwar, K., Wigderson, A.: Smooth boolean functions are easy: efficient algorithms for low-sensitivity functions. In: Proceedings of the 2016 ACM Conference on Innovations in Theoretical Computer Science, Cambridge, MA, USA, January 14–16, 2016, pp. 59–70 (2016)
15. Gopalan, P., Servedio, R.A., Tal, A., Wigderson, A.: Degree and sensitivity: tails of two distributions. In: Electronic Colloquium on Computational Complexity (ECCC), vol. 23, p. 69 (2016)
16. Hatami, P., Kulkarni, R., Pankratov, D.: Variations on the sensitivity conjecture. Theor. Comput. Grad. Surv. **4**, 1–27 (2011)

17. Kenyon, C., Kutin, S.: Sensitivity, block sensitivity, and l-block sensitivity of boolean functions. Inf. Comput. **189**(1), 43–53 (2004)
18. Lin, C., Zhang, S.: Sensitivity conjecture and log-rank conjecture for functions with small alternating numbers. CoRR, abs/1602.06627 (2016)
19. Nisan, N.: Crew prams and decision trees. SIAM J. Comput. **20**(6), 999–1007 (1991)
20. Nisan, N., Szegedy, M.: On the degree of Boolean functions as real polynomials. Comput. Complex. **4**, 301–313 (1994)
21. Karthik, C.S., Tavenas, S.: On the sensitivity conjecture for disjunctive normal forms. CoRR, abs/1607.05189 (2016)
22. Zhan, J., Lovett, S., Tal, A.: Robust sensitivity. In: Electronic Colloquium on Computational Complexity (ECCC), vol. 23, p. 161 (2016)
23. Szegedy, M.: An $o(n^{0.4732})$ upper bound on the complexity of the GKS communication game. CoRR, abs/1506.06456 (2015)
24. Tal, A.: On the sensitivity conjecture. In: 43rd International Colloquium on Automata, Languages, Programming, ICALP 2016, July 11–15, 2016, Rome, Italy, pp. 38:1–38:13 (2016)
25. Turán, G.: The critical complexity of graph properties. Inf. Process. Lett. **18**(3), 151–153 (1984)
26. Turán, P., Paul, E.: On an extremal problem in graph theory. Matematikai és Fizikai Lapok (in Hungarian) **48**, 436–452 (1941)

Unfolding Some Classes of Orthogonal Polyhedra of Arbitrary Genus

Kuan-Yi Ho[1], Yi-Jun Chang[2], and Hsu-Chun Yen[1(✉)]

[1] Department of Electrical Engineering, National Taiwan University, Taipei, Taiwan
hcyen@ntu.edu.tw
[2] Department of EECS, University of Michigan, Ann Arbor, MI, USA

Abstract. Unfolding polyhedra beyond genus zero (i.e., with holes) is challenging, yet it has not been investigated until very recently. We show two types of orthogonal polyhedra of arbitrary genus, namely, well-separated orthographs and regular orthogonal polyhedra with x- and z-holes, to enjoy (2×1)-grid-unfoldings, generalizing some prior work in the literature by allowing holes (or more complicated holes) to exist. In addition to the development of new unfolding techniques, for the first time we identify classes of nontrivial orthogonal polyhedra of arbitrary genus to admit grid-unfoldings subject to a small amount of refinements.

1 Introduction

Unfolding a polyhedron refers to the process of cutting and flattening the surface of a polyhedron to yield a single connected planar piece. As documented in [9], constructing a polyhedron by folding from its unfolding can be traced back to several hundred years ago.

For the sake of simplicity, it is desirable to only cut along a subset of edges of a polyhedron when producing an unfolding. It is known, however, that even for orthogonal polyhedra (i.e., polyhedra with their faces orthogonal to x-, y- and z-axes), cutting along edges is not sufficient to guarantee an unfolding in general. *Grid-unfolding*, introduced in [11], provides additional flexibility by allowing new edges resulting from intersecting the polyhedron with all coordinate planes passing through vertices of the polyhedron to be cut. Even with such a relaxation, it remains a major open problem as to whether every orthogonal polyhedron admits a grid-unfolding [8,11].

To offer an even higher degree of flexibility, the notion of a *refinement* comes into play, by allowing each rectangular face of a polyhedron to be further refined to an $a \times b$ grid with all grid lines cuttable [3,7]. Such an unfolding style is called an $(a \times b)$-*grid-unfolding* (or grid-unfolding with $(a \times b)$-refinement). With $(O(n) \times O(n))$-refinement, general orthogonal polyhedra could always be grid-unfolded [2]. If we only allow constant refinements, orthostacks [11] and Manhattan towers [6] are unfoldable. On the other hand, only several special

H.-C Yen—Research supported in part by Ministry of Science and Technology, Taiwan, under grant MOST 103-2221-E-002-154-MY3.

classes of orthogonal polyhedra can be grid-unfolded without refinement, including well-separated orthotrees [5].

All the aforementioned results assume the underlying polyhedra to be of genus zero, i.e., without holes. As far as we know, only a scarcity of results are available for unfolding polyhedra of genus greater than zero. In [10], it was shown that several special classes of one-layer lattice polyhedra with cubic holes admit edge-unfoldings. For general orthogonal polyhedra, the only study [4] was for the case of genus at most 2, showing such polyhedra to have $(O(n) \times O(n))$-grid-unfoldings. It remains a challenging open problem whether orthogonal polyhedra of any genus could be grid-unfolded with refinements.

In this paper, we investigate the issue of unfolding the following two types of orthogonal polyhedra of arbitrary genus, namely, *well-separated orthographs* and *regular orthogonal polyhedra with x- and z-holes*.

An orthograph is a polyhedron made from gluing congruent faces (of the same shape and size) of rectangular boxes together. Figure 1(a) shows an orthograph with the so-called *junctions* colored in grey. An orthograph is well-separated if no two junctions are next to each other. A regular orthogonal polyhedron can be thought of as the outcome of pulling up an orthogonal polygon on the $x - z$ plane along the y direction to form a polyhedron. As illustrated in Fig. 1(b), holes are only allowed in the x and z directions. It turns out that both types of orthogonal polyhedra can be grid-unfolded with (2×1)-refinement.

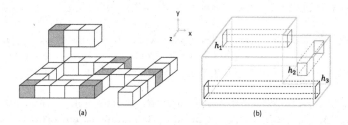

Fig. 1. (a) A well-separated orthograph in which junctions are colored in grey. (b) A regular orthogonal polyhedra with x- and z-holes.

Orthographs can be thought of as extensions of the so-called orthotrees investigated in [1,5], by allowing ends of an orthotree to be glued to the rest of the structure to form cycles (i.e., of genus greater than zero). In [5], well-separated orthotrees admit edge-unfoldings (assuming box edges are cuttable). Our result shows that allowing (2×1)-refinement is sufficient to grid-unfold well-separated orthographs - a much larger class in comparison with orthotrees.

Regular orthogonal polyhedra with x- and z-holes represent a natural extension/generalization of one-layer lattice polyhedra with cubic holes studied in [10], which were shown to enjoy edge-unfoldings. In particular, one-layer polyhedra with cubic holes with rectangular boundary (see Sect. 3.1 of [10]) is a subclass of regular orthogonal polyhedra with only x-holes. Again, allowing (2×1)-refinement enables orthogonal polyhedra with a more complicated structure of holes to be unfolded.

The significance of our results is two-fold. First, for the first time we identify classes of nontrivial orthogonal polyhedra of arbitrary genus that can be grid-unfolded with refinements. Second, we feel that the new techniques devised in this paper are applicable to unfolding generalized versions (by allowing holes to exist) of certain existing types of orthogonal polyhedra.

2   Preliminaries

An *orthograph* O is a polyhedron made from gluing congruent faces (of the same shape and size) of rectangular boxes together. With respect to an orthograph O, its dual graph $G_O = (V, E)$ is a graph in which each vertex of G_O corresponds to a box of O and an edge between two vertices of G_O exists if the corresponding two boxes are glued to each other. As there is a one-to-one correspondence between boxes of O and vertices of G_O, throughout this paper we use the name of a box to denote that of the corresponding vertex of G_O as well. A face of a box is said to be *open* if it is not glued to any other box; otherwise, it is called *closed*. A box $b_i \in O$ is called a

- *junction*: if there exist two boxes that are glued to b_i on the neighboring faces (i.e., faces sharing a common boundary edge),
- *connector*: if $\deg(b_i) = 2$ and the two boxes that are glued to b_i are on the opposite faces,
- *leaf*: if $\deg(b_i) = 1$.

Note that each box in O belongs to exactly one of the above three cases. An orthograph is *well-separated* if there is no junction that is the neighbor of another junction. As opposed to the orthotrees discussed in [5], the dual graph of an orthograph can have cycles.

Figure 1(a) displays an example of a well-separated orthograph with junctions colored in grey.

The boundary of an unfolding of a polyhedron is a polygon in the plane. We call a polygon Q *y-convex* if the intersection of any straight line parallel to the y-axis and the interior of Q is a single connected line segment.

For convenience of illustration, we adopt the notations used in [5] for labeling faces and edges of a box. The two faces perpendicular to the x-axis (y-axis and z-axis, respectively) are annotated by x^+ and x^- (y^+ and y^-, and z^+ and z^-, respectively). The edges of x^+ and x^- are labeled as u (up), d (down), f (front), and b (back). Similarly, edges of y^+ and y^- are labeled as f (front), b (back), l (left), and r (right), and z^+ and z^- are labeled as u, d, l, and r. See Fig. 2 for details. Note that an edge may carry different labels depending on the face it is referred to. For notational convenience, the w edge of a face F is denoted by $F_{(w)}$. For instance, the up edge of x^+ is denoted by $x^+_{(u)}$.

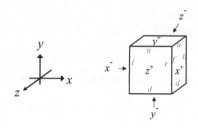

Fig. 2. Annotations of a single box.

3 Unfolding Paths on Modified Dual Graphs

A key ingredient in our analysis is the concept of an *unfolding path* on a modified version of the dual graph of an orthograph.

The intuition is that if there is an Euler path (meeting some mild requirements) on the dual graph, we can unfold a well-separated orthograph along this path. For any well-separated orthograph, however, there may not be an Euler path on its dual graph G. To overcome this difficulty, we allow our unfolding path to pass through each edge twice by simply doubling each edge in the dual graph of G.

To come up with an unfolding path which is simple enough to facilitate unfolding, we further modify the dual graph of a well-separated orthograph in the way explained below. In our modification, we will make use of the property that junctions could not touch each other in a well-separated orthograph.

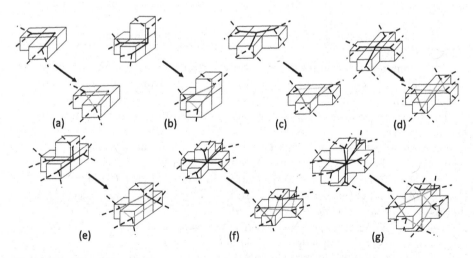

Fig. 3. Modifying the dual graph structure of a junction. Indirect augmented edges are drawn as thick edges.

Definition 1. *Given a well-separated orthograph O, a* modified dual graph *is a multigraph graph $G'_O = (V, E)$ obtained by applying the following operations to the dual graph G_O:*

(1) For each edge $e = (a, b)$ such that both a and b are not junctions, replace the edge (a, b) by two parallel edges $\{(a, b), (a, b)\}$.

(2) For each junction c, connect the neighborhood of c by a cycle as specified in Fig. 3, and then remove c from the multigraph.

Consider Fig. 3(b) for example. In the original dual graph, the junction (of degree 3) is glued to the three boxes in the x^-, y^+ and z^+ directions. Instead of having the edges linking the junction directly to its three neighbors, we link the three neighbors together in a cycle.

The remaining cases are explained in Fig. 3. Associated with a junction, each of the solid edges is called an *augmented edge* of the junction. At this point, the reader may ignore the meaning of edge colors in Fig. 3. It should be noted that just forming a cycle is not sufficient to guarantee an unfolding. For instance, in Fig. 3(g) if a cycle contains an edge connecting two opposite boxes, such a cycle does not yield any valid unfolding.

As we shall see later, augmented edges map to fragments of the "backbone" of the unfolding. Augmented edges are divided into two types, namely, *direct* and *indirect*. For an augmented edge e between two boxes B_1 and B_2, if e is direct, then e is realized by two connected squares $\boxed{Q_1 \,|\, Q_2}$ (where Q_1 and Q_2 are faces of B_1 and B_2, respectively) in the x^+ direction along the backbone. If e is indirect, it is realized by a sequence of squares connected in a line $\boxed{Q_1 |\cdots| Q_2}$ in the x^+ direction such that Q_1 and Q_2 are faces of B_1 and B_2, respectively, and the intermediate squares belong to faces of the junction.

Definition 2. *Given a well-separated orthograph O with its modified dual graph G'_O, a path P of G'_O is an* unfolding path *if it has the following properties.*

(1) P passes through all the edges of G'_O except one edge which links the two end vertices of P.

(2) P does not pass through two augmented edges consecutively.

(3) If there are vertices u and v such that uvu is a subpath of P, then v is a leaf.

The following result is easy to show:

Lemma 1. *For every well-separated orthograph, there is an unfolding path in the modified dual graph of the orthograph.*

4   Unfolding Well-Separated Orthographs

In this section, we show how to unfold a well-separated orthograph taking advantage of the unfolding path guaranteed by Lemma 1.

The notion of a *backbone* will serve for the purpose of a y-convex partial unfolding to which the remaining faces of a polyhedron could be attached. The reader should note the difference between an unfolding path and a backbone. An unfolding path only gives the sequence of boxes to be visited in the unfolding process. It does not tell the exact faces of the unfolding along the backbone of the unfolding. For ease of explanation, in our subsequent discussion we use (colored) line segments to represent fragments of the backbone in such a way that a line segment cuts through a face if the face is on the backbone.

Before proceeding further, we first give the intuitive idea behind our unfolding strategy. For every box B, each of its closed faces is associated with two edges called *ports*, and a set of *links* is to identify which pair of ports are connected together. A link between ports a and b is written as $a \sim b$. The idea is to associate each link (pairing different ports) with a y-convex partial unfolding in the way that any two such partial unfoldings are disjoint, and the union of all such partial unfoldings covers all the open faces of box B. Finally, the unfolding of the entire polyhedron is obtained by connecting all the y-convex partial unfoldings together piece by piece based on the unfolding path constructed in the beginning of the procedure. Consider Fig. 4 in which three boxes B_l (a leaf), B_c (a connector) and B_J (a junction) are given. Box B_l has link $\{k \sim m\}$, B_c has links $\{i \sim h, j \sim g\}$, and B_J has links $\{a \sim b, c \sim d, e \sim f\}$. The y-convex partial unfoldings realizing the links of a box are displayed at the top of the figure in Fig. 4. Note that for B_J, since $c = d$ and $a = b$ (i.e., c and d are identical edges, so are a and b), there need not be any partial unfolding associated with these two links. Such $a \sim b$ and $c \sim d$ are called *null links*. A null link could only appear on a junction box, serving as a turning point connecting two faces belonging to two neighbors of the junction. It is easy to see that if the unfolding starts at edge k, by joining m and i, and later h and a, a longer fragment of the backbone (highlighted by the sequence of line segments in yellow) is thus obtained, and the corresponding partial unfolding remains y-convex.

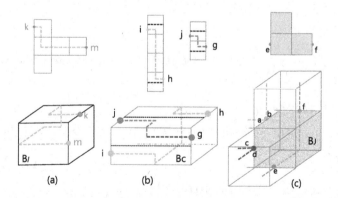

Fig. 4. Ports, links, and y-convex partial unfoldings of boxes. (Color figure online)

Suppose B is a box of degree d ($1 \leq d \leq 6$), and let $F_1, \cdots, F_d$ be the closed faces of B. Let E_i be the set of four edges of F_i. A *configuration* of B is a set of d links $L = \bigcup_{i=1}^{d} \{x_i \sim y_i\}$, where $| \bigcup_{i=1}^{d} \{x_i, y_i\} \cap E_j | = 2$, $\forall 1 \leq j \leq d$, i.e., each face F_i contributes exactly two edges (ports) to L, and unless $d = 1$ (i.e., the case of a leaf), x_i and y_i do not belong to the same face.

A configuration L of a box B is *realizable* if the following hold:

(1) a y-convex partial unfolding U_i (with possible refinements on faces) can be assigned to each non-null link $x_i \sim y_i$, and in the unfolding x_i and y_i are located at the left and right borders of U_i, and
(2) U_i and U_j are disjoint for $i \neq j$, and the union of all such U_i covers all the open faces of B.

Note that there are multiple configurations associated with a box of a polyhedron. We shall show in our subsequent discussion that regardless of the type of a junction, there always exists *one* configuration which is realizable. Furthermore, the configuration respects the structure of the modified dual graph described in Fig. 3, in the sense that

– if B is a junction, then for every link $u \sim v$ of B and suppose B_u (B_v, resp.) is the box glued to the face of B on which port u (v, resp.) resides, then in the modified dual graph (B_u, B_v) is a direct augmented edge if $u = v$ (i.e., a null link); otherwise, (B_u, B_v) is indirect.

For leaves and connectors, we are also able to show that *all* of their configurations are realizable. Since the polyhedron under consideration is well-separated, for two consecutive junctions B_1 and B_2 along the unfolding path, there must be a connector B_c in between, which has the ability to "rearrange" ports arbitrarily so that ports of B_1 and B_2 can match with each other in forming the backbone.

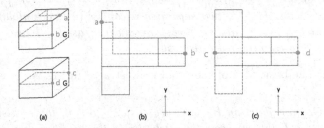

Fig. 5. Unfolding a single box.

Lemma 2. *Given a leaf B with closed face x^+ and for arbitrary ports p and q such that $p, q \in \{x_{(u)}^-, x_{(d)}^-, x_{(f)}^-, x_{(b)}^-\}$ and $p \neq q$, then it is always possible to unfold B (excluding the x^+ face) into a y-convex polygon such that p and q are the left and right borders of the polygon, respectively. Hence, all possible configurations of a leaf are realizable.*

Proof sketch. Figure 5(a) displays two cases of p and q (i.e., neighbored to each other, or on opposite sides). The remaining cases of p and q are symmetric. The corresponding two y-convex unfoldings are shown in Fig. 5(b) and (c).   □

Recall that in a well-separated orthograph, the unfolding path from one junction to another must pass through some connector, and each connector is visited twice along the unfolding path. Consider a connector B_c shown in Fig. 6(a). Without loss of generality we assume that the unfolding path first enters the connector from the left and at a later time reenters from the right. W.r.t. such an unfolding path, the backbone enters from c and exits at d, visits a portion of the orthograph attached to x^+ of B_c (i.e., the right face of B_c), and reenters B_c from b and leaves at a. That is, $c \sim d$ and $b \sim a$ serve as the two links of the connector B_c, while $a, b, c,$ and d are ports.

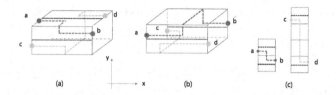

Fig. 6. (a) $c \sim d$ is unfolded in a clockwise fashion in the x^+ direction. (b) $c \sim d$ is unfolded in a counter-clockwise fashion in the x^+ direction. (c) Separated y-convex unfoldings of $c \sim d$ and $a \sim b$ in (a).

Lemma 3. *Let B_c be a connector and $\{c \sim d, a \sim b\}$ be the set of links of B_c, where c and a (b and d resp.) are ports of face x^- (x^+, resp.). For all possible combinations of $c, a \in \{x^-_{(u)}, x^-_{(d)}, x^-_{(f)}, x^-_{(b)}\}$, and $b, d \in \{x^+_{(u)}, x^+_{(d)}, x^+_{(f)}, x^+_{(b)}\}$ such that $a \neq c$ and $b \neq d$, two separated y-convex (2×1)-unfoldings of B_c (containing the $y^+, y^-, z^+,$ and z^- faces) are always feasible. Hence, all possible configurations of a connector are realizable.*

Proof sketch. The basic idea of the proof is illustrated in Fig. 6.   □

Lemma 4. *Given a junction B_J with degree d, $2 \leq d \leq 6$, there exists a realizable configuration L that respects the structure of the modified dual graph associated with B_J.*

Proof sketch. The proof is done by case analysis, as illustrated in Fig. 8. The reader should also consult Fig. 3. Note that there is a 1–1 correspondence between the seven structures in Figs. 3 and 8 (including the matching edge colors between corresponding structures).

To understand what the lemma says, consider Fig. 7 again. By letting $\{a, c\}$ ($\{b, f\}$ and $\{d, e\}$, resp.) be the two ports associated with face x^- (y^+ and z^+, resp.) and the set of links be $L = \{a \sim b, c \sim d, e \sim f\}$ (where $c = d$ and $a = b$),

$c \sim d$ admits a y-convex partial unfolding as Fig. 7(b) shows. Hence, such a configuration is realizable. In view of the colored line segments in Fig. 7(a), it is also easy to see that the configuration respects the structure of Fig. 3(b). From the colored lines in each of the cases of Fig. 8, it is not difficult to see that in each case a realizable configuration is available which also respects the respective structure displayed in Fig. 3. Note that direct and indirect augmented edges in each of the cases of Fig. 8 are easy to be identified.   □

Figure 7 illustrates how to piece y-convex partial unfoldings in the x^-, z^+ and y^+ directions to the y-convex partial unfolding of B_J.

We are in a position to prove the main result of this section.

Theorem 1. *Every well-separated orthograph can be* (2×1)-*grid-unfolded.*

Proof sketch. Given a well-separated orthograph, we first obtain the unfolding path P guaranteed by Lemma 1. Based on the unfolding path, we carry out the following:

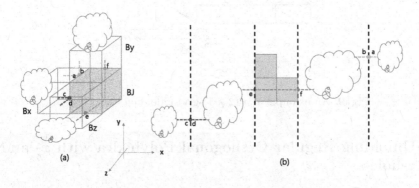

Fig. 7. Unfolding a junction of degree 3.

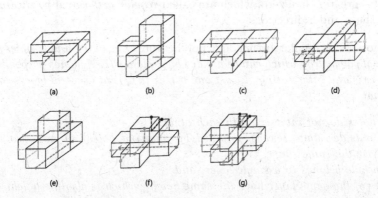

Fig. 8. Segments along the backbone associated with various cases in Fig. 3. (Color figure online)

(1) select ports/links for junctions in accordance to the modified dual graph G'_O,
(2) select configurations for connectors and leaves that match the configurations of junctions selected in Step (1), and
(3) for all boxes, compute unfoldings that realize their configurations (guaranteed by Lemmas 2, 3, and 4, see also Figs. 5, 6 and 7).

The final unfolding is formed by gluing these partial unfoldings together along the unfolding path. □

Figure 9 displays the backbone of an unfolding of a well-separated orthograph.

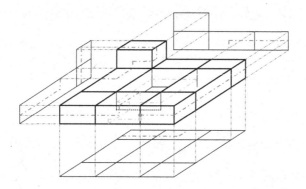

Fig. 9. An example of a (2×1)-grid-unfolding - 3D view.

5 Unfolding Regular Orthogonal Polyhedra with x- and z-holes

In this section, we turn our attention to another class of orthogonal polyhedra with genus greater than zero, which are called *regular orthogonal polyhedra* with holes in the x and z directions.

Definition 3. *A regular orthogonal polyhedron Q with x- and z-holes is an orthogonal polyhedron which can be divided by y-planes (i.e., planes perpendicular to the y-axis and intersecting at least one vertex in Q) into several layers in such a way that*

(1) the holes do not intersect with each other,
(2) the outside bands (bands obtained by ignoring holes) of any two layers enclose the same shape,
(3) each hole belongs to a single layer, and
(4) holes of the same layer have the same height, which is also the height of the layer.

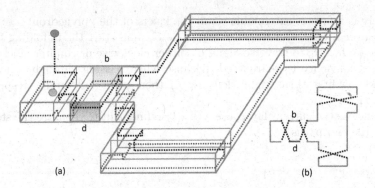

Fig. 10. An example of one-layer of a regular orthogonal polyhedron with holes and its unfolding cycle.

The reader may refer to Fig. 10(a) as an example of one layer of a regular orthogonal polyhedron with holes.

Note that in each layer, holes partition the layer into several *modules*; the band of each module may again be split into *segments*. See Fig. 10(a) for an example, in which b and d are two segments split by the two z-holes. In the way displayed in Fig. 10(b), it is straightforward to connect all segments split by holes into a cycle (called an *unfolding cycle*). Note that a segment always takes the neighboring hole to cross over to the next segment on the other side along an unfolding cycle. By doing so, the four faces of a hole can be unfolded in a y-convex fashion, as shown in Fig. 11. Analogous to the unfolding path in the previous section, the unfolding cycle plays a key role in unfolding regular orthogonal polyhedra with holes, as the cycle induces an unfolding backbone. We use similar notations as in Fig. 2 for a hole. We have the following result:

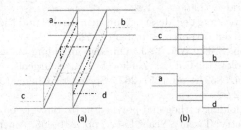

Fig. 11. The unfolding near a hole.

Theorem 2. *Every regular orthogonal polyhedron with x- and z-holes can be (2×1)-grid-unfolded.*

Proof sketch. The unfolding is carried out layer-by-layer in a top-down fashion. What follows are key steps in the unfolding. The reader is referred to Fig. 10.

(1) There is a face F extendable to all the layers of the polyhedron;
(2) For layer i, construct its unfolding cycle C_i, and start the unfolding from a node on face F to traverse along C_i either clockwise or counter-clockwise;
(3) The y-convexity is guaranteed, because when passing through a hole the backbone first turns right (left, resp.) then turns left (right, resp.). See Fig. 11;
(4) To enclose the current layer, use a (2×1)-refinement to separate the starting and the end paths.   □

References

1. Aloupis, G., et al.: Common unfoldings of polyominoes and polycubes. In: Akiyama, J., Bo, J., Kano, M., Tan, X. (eds.) CGGA 2010. LNCS, vol. 7033, pp. 44–54. Springer, Heidelberg (2011). doi:10.1007/978-3-642-24983-9_5

2. Chang, Y.-J., Yen, H.-C.: Unfolding orthogonal polyhedra with linear refinement. In: Elbassioni, K., Makino, K. (eds.) ISAAC 2015. LNCS, vol. 9472, pp. 415–425. Springer, Heidelberg (2015). doi:10.1007/978-3-662-48971-0_36

3. Damian, M., Demaine, E.D., Flatland, R.: Unfolding orthogonal polyhedra with quadratic refinement: the delta-unfolding algorithm. Graphs Comb. **30**(1), 125–140 (2014)

4. Damian, M., Demaine, E.D., Flatland, R., O'Rourke, J.: Unfolding genus-2 orthogonal polyhedra with linear refinement arXiv:1611.00106v1 [cs.CG] (2016)

5. Damian, M., Flatland, R., Meijer, H., O'Rourke, J.: Unfolding well-separated orthotrees. In: Proceedings of 15th Annual Fall Workshop on Computational Geometry, pp. 23–25 (2005)

6. Damian, M., Flatland, R., O'Rourke, J.: Unfolding Manhattan towers. Comput. Geom. **40**(2), 102–114 (2008)

7. Damian, M., Flatland, R., O'Rourke, J.: Epsilon-unfolding orthogonal polyhedra. Graphs Comb. **23**(1), 179–194 (2007)

8. Demaine, E.D., O'Rourke, J.: Geometric Folding Algorithms. Cambridge University Press, Cambridge (2007)

9. Dürer, A.: Underweysung der Messung, mit dem Zirckel und Richtscheyt, in Linien, Ebenen unnd gantzen corporen. Nürnberg (1525)

10. Liou, M.-H., Poon, S.-H., Wei, Y.-J.: On edge-unfolding one-layer lattice polyhedra with cubic holes. In: Cai, Z., Zelikovsky, A., Bourgeois, A. (eds.) COCOON 2014. LNCS, vol. 8591, pp. 251–262. Springer, Cham (2014). doi:10.1007/978-3-319-08783-2_22

11. O'Rourke, J.: Unfolding orthogonal polyhedra. In: Surveys on Discrete and Computational Geometry: Twenty Years Later, pp. 231–255. American Mathematical Society (2008)

Reconfiguration of Maximum-Weight
b-Matchings in a Graph

Takehiro Ito[1][(⊠)], Naonori Kakimura[2], Naoyuki Kamiyama[3],
Yusuke Kobayashi[4], and Yoshio Okamoto[5]

[1] Graduate School of Information Sciences, Tohoku University, Sendai, Japan
`takehiro@ecei.tohoku.ac.jp`
[2] Department of Mathematics, Keio University, Yokohama, Japan
`kakimura@math.keio.ac.jp`
[3] Institute of Mathematics for Industry, Kyushu University, Fukuoka, Japan
`kamiyama@imi.kyushu-u.ac.jp`
[4] Faculty of Engineering, Information and Systems,
University of Tsukuba, Tsukuba, Japan
`kobayashi@sk.tsukuba.ac.jp`
[5] Department of Communication Engineering and Informatics,
University of Electro-Communications, Chofu, Japan
`okamotoy@uec.ac.jp`

Abstract. Consider a graph such that each vertex has a nonnegative integer capacity and each edge has a positive integer weight. Then, a b-matching in the graph is a multi-set of edges (represented by an integer vector on edges) such that the total number of edges incident to each vertex is at most the capacity of the vertex. In this paper, we study a reconfiguration variant for maximum-weight b-matchings: For two given maximum-weight b-matchings in a graph, we are asked to determine whether there exists a sequence of maximum-weight b-matchings in the graph between them, with subsequent b-matchings obtained by removing one edge and adding another. We show that this reconfiguration problem is solvable in polynomial time for instances with no integrality gap. Such instances include bipartite graphs with any capacity function on vertices, and 2-matchings in general graphs. Thus, our result implies that the reconfiguration problem for maximum-weight matchings can be solved in polynomial time for bipartite graphs.

T. Ito – Supported by JST CREST Grant Number JPMJCR1402, Japan, and JSPS KAKENHI Grant Number JP16K00004.
N. Kakimura – Supported by JST ERATO Grant Number JPMJER1305, Japan, and by JSPS KAKENHI Grant Number JP17K00028.
N. Kamiyama – Supported by JST PRESTO Grant Number JPMJPR14E1, Japan.
Y. Kobayashi – Supported by JST ERATO Grant Number JPMJER1305, Japan, and by JSPS KAKENHI Grant Numbers JP16K16010 and JP16H03118.
Y. Okamoto – Supported by Kayamori Foundation of Informational Science Advancement, JST CREST Grant Number JPMJCR1402, Japan, and JSPS KAKENHI Grant Numbers JP24106005, JP24700008, JP24220003, JP15K00009.

Y. Cao and J. Chen (Eds.): COCOON 2017, LNCS 10392, pp. 287–296, 2017.
DOI: 10.1007/978-3-319-62389-4_24

1   Introduction

Recently, *reconfiguration problems* [11] have attracted much attention in the field of theoretical computer science. These problems arise when we wish to find a step-by-step transformation between two feasible solutions of a combinatorial problem such that all intermediate results are also feasible and each step conforms to a fixed reconfiguration rule (i.e., an adjacency relation defined on feasible solutions of the original combinatorial problem). For example, in the (cardinality) MATCHING RECONFIGURATION problem, feasible solutions are matchings in a graph having the same cardinality and one of the studied reconfiguration rules is to exchange an edge in the current matching with an edge which is not contained in the matching. This kind of reconfiguration problems has been studied extensively for several well-known combinatorial problems, including SATISFIABILITY [6,16,18], INDEPENDENT SET [4,5,14], VERTEX COVER [12,19], CLIQUE [13], DOMINATING SET [7,8], VERTEX-COLORING [1,3,9], and so on. (See also a survey [10]).

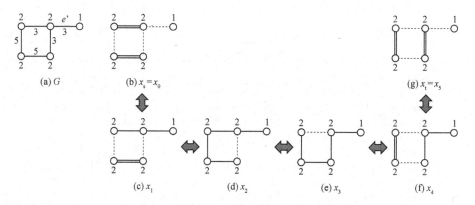

Fig. 1. (a) Graph G with vertex-capacities and edge-weights, and (b)–(g) a sequence $\langle x_s = x_0, x_1, \ldots, x_5 = x_t \rangle$ of maximum-weight b-matchings in G, where each $x_i(e)$, $0 \le i \le 5$, is represented as the number of parallel edges between the endpoints of the edge e.

1.1   Our Problem

In this paper, we generalize (cardinality) MATCHING RECONFIGURATION, and study a reconfiguration problem for MAXIMUM-WEIGHT b-MATCHING defined as follows.

For a graph G, we denote by $V(G)$ and $E(G)$ the vertex set and edge set of G, respectively. Let $b: V(G) \to \mathbb{Z}_{\ge 0}$ be a capacity function on vertices, where $\mathbb{Z}_{\ge 0}$ is the set of all nonnegative integers. Then, a vector $x \in \mathbb{Z}_{\ge 0}^{E(G)}$ is called a b-*matching* in G if $\sum_{e \in \delta(v)} x(e) \le b(v)$ holds for each vertex $v \in V(G)$, where $\delta(v)$ denotes the set of all edges incident to the vertex v. For example, Fig. 1(b)–(g) illustrate six b-matchings in the graph G of Fig. 1(a). Note that an ordinary matching in a graph G is a b-matching in G such that $b\colon V(G) \to \{1\}$. The *cardinality* of a b-matching x in G is defined as $\sum_{e \in E(G)} x(e)$. Let $w\colon E(G) \to \mathbb{Z}_+$ be a weight function on edges, where $\mathbb{Z}_+$ is the set of all positive integers. Then, the *weight* of a b-matching x in G is defined as $\sum_{e \in E(G)} w(e)x(e)$.

For two b-matchings x and x' in a graph G, we write $x \leftrightarrow x'$ if there exists a pair of edges e and f in G such that $x(e) - x'(e) = x'(f) - x(f) = 1$ and $x(g) = x(g)$ for all edges $g \in E(G) \setminus \{e, f\}$. Thus, both x and x' have the same cardinality. (See any two consecutive b-matchings in Fig. 1(b)–(g) as examples.) For two maximum-weight b-matchings x and x' in G, we write $x \overset{w}{\longleftrightarrow} x'$ if there exists a sequence $\langle x_0, x_1, \ldots, x_\ell \rangle$ of b-matchings in G such that

(i) $x_0 = x$ and $x_\ell = x'$;
(ii) all b-matchings $x_0, x_1, \ldots, x_\ell$ have the maximum weight in G; and
(iii) $x_{i-1} \leftrightarrow x_i$ holds for each $i \in \{1, 2, \ldots, \ell\}$.

Then, the MAXIMUM-WEIGHT b-MATCHING RECONFIGURATION problem is defined as follows:

> **Input:** A graph G, a capacity function $b\colon V(G) \to \mathbb{Z}_{\geq 0}$ on vertices, a weight function $w\colon E(G) \to \mathbb{Z}_+$ on edges, and two maximum-weight b-matchings x_s and x_t in G
> **Question:** Determine whether $x_\mathrm{s} \overset{w}{\longleftrightarrow} x_\mathrm{t}$ or not.

We denote by a 5-tuple $(G, b, w, x_\mathrm{s}, x_\mathrm{t})$ an instance of MAXIMUM-WEIGHT b-MATCHING RECONFIGURATION. Note that this is a decision problem and hence it does not ask for an actual sequence of maximum-weight b-matchings. For the particular instance of Fig. 1, it has a desired sequence $\langle x_\mathrm{s} = x_0, x_1, \ldots, x_5 = x_\mathrm{t} \rangle$ as illustrated in the figure, and hence the answer is yes.

1.2   Known and Related Results

Ito et al. [11, Proposition 2] studied (cardinality) MATCHING RECONFIGURA-TION, and gave a polynomial-time algorithm to solve the problem for any graph. Mühlenthaler [20] generalized MATCHING RECONFIGURATION to the reconfiguration problem for degree-constrained subgraphs in a graph G, where a degree-constrained subgraph is a subgraph of G such that the degree of each vertex satisfies both lower and upper bounds of the vertex. This generalized reconfiguration problem is also solvable in polynomial time for any graph [20].[1]

In the reconfiguration problem of Mühlenthaler [20], each edge can be chosen at most once in a degree-constrained subgraph. However, the algorithm of [20] can be easily extended so that it works correctly and runs in polynomial time even if multiplicities on edges are allowed. By setting the lower bound equal to zero and the upper bound equal to $b(v)$ for each vertex v in a graph

---

[1] Properly speaking, both Ito et al. [11] and Mühlenthaler [20] studied their reconfiguration problems under a more generalized reconfiguration rule, called the TAR (Token Addition and Removal) rule. Their results hold also under the reconfiguration rule of this paper, which is called the TJ (Token Jumping) rule.

G, b-matchings (and hence ordinary matchings) in G can be seen as degree-constrained subgraphs of G. Consider MAXIMUM-WEIGHT b-MATCHING RECONFIGURATION when restricted to identical edge-weight. Then, each maximum-weight b-matching in a graph G is simply a maximum-cardinality b-matching in G. Thus, the result by Mühlenthaler [20] implies the following proposition.

Proposition 1 [20]. MAXIMUM-WEIGHT b-MATCHING RECONFIGURATION *is solvable in polynomial time when restricted to identical edge-weight.*

As far as we know, reconfiguration problems have been studied mostly for unweighted instances. Note that SHORTEST PATH RECONFIGURATION [2] and STEINER TREE RECONFIGURATION [17] are defined on unweighted graphs, and hence they are cardinality variants. MATROID RECONFIGURATION [11, Proposition 1] is only the example in reconfiguration which admits a polynomial-time algorithm for weighted instances. However, matchings do not form matroid bases.

1.3 Our Contribution

In this paper, we show that MAXIMUM-WEIGHT b-MATCHING RECONFIGURATION is solvable in polynomial time for instances with no integrality gap. Such instances include bipartite graphs with any capacity function b on vertices, and general graphs G with the capacity function $b\colon V(G) \to \{2\}$. Thus, our result yields that the reconfiguration problem for maximum-weight matchings can be solved in polynomial time for bipartite graphs.

Our idea is to use the structure of maximum-weight b-matchings in a graph with no integrality gap. As an intuitive example, the edge e' in Fig. 1 would be "useless" if $w(e') \le 2$ because edges in two given maximum-weight b-matchings have weights at least three; indeed, it is a no-instance if $w(e') \le 2$. In Sect. 2, we formulate the problem of finding a maximum-weight b-matching in a graph as an integer program, and show that the complementary slackness condition gives a characterization of b-matchings that have the maximum weight (Lemma 1). Then, in Sect. 3, we will make use of Lemma 1, and reduce the problem of asking the existence of a desired sequence of maximum-*weight* b-matchings to the problem of asking that of maximum-*cardinality* b-matchings; recall that the cardinality variant is solvable in polynomial time (Proposition 1).

2 Maximum-Weight b-Matchings

In this section, we give a characterization of maximum-weight b-matchings which will play an important role in our algorithm in Sect. 3.

Let G be a graph, and let $b: V(G) \to \mathbb{Z}_{\geq 0}$ and $w: E(G) \to \mathbb{Z}_+$ be capacity and weight functions, respectively. We can formulate the problem of finding a maximum-weight b-matching in G as the following integer program **IP**:

$$
\begin{aligned}
\max. \quad & \sum_{e \in E(G)} w(e)x(e) \\
\text{s.t.} \quad & \sum_{e \in \delta(v)} x(e) \leq b(v) \quad (\forall v \in V(G)) \\
& x(e) \in \mathbb{Z}_{\geq 0} \quad (\forall e \in E(G)).
\end{aligned}
$$

We denote by a triple (G, b, w) an input to **IP**. Let **LP** be the following linear programming relaxation of **IP**:

$$
\begin{aligned}
\max. \quad & \sum_{e \in E(G)} w(e)x(e) \\
\text{s.t.} \quad & \sum_{e \in \delta(v)} x(e) \leq b(v) \quad (\forall v \in V(G)) \\
& x(e) \geq 0 \quad (\forall e \in E(G)).
\end{aligned}
$$

The dual program **DP** of **LP** can be described as follows:

$$
\begin{aligned}
\min. \quad & \sum_{v \in V(G)} b(v)y(v) \\
\text{s.t.} \quad & y(u) + y(v) \geq w(e) \quad (\forall e = \{u, v\} \in E(G)) \\
& y(v) \geq 0 \quad (\forall v \in V(G)).
\end{aligned}
$$

The complementary slackness condition (see, e.g., [15, Corollary 3.23]) implies the following theorem.

Theorem 1. *Suppose that x and y are feasible solutions of* **LP** *and* **DP***, respectively. Then, the following two statements (1) and (2) are equivalent.*

(1) *x and y are optimal solutions of* **LP** *and* **DP***, respectively.*
(2) *x and y satisfy the following (i) and (ii):*
  (i) *$y(u) + y(v) = w(e)$ for every edge $e = \{u, v\} \in E(G)$ with $x(e) > 0$; and*

  (ii) *$\displaystyle\sum_{e \in \delta(v)} x(e) = b(v)$ for every vertex $v \in V(G)$ with $y(v) > 0$.*

For each feasible solution y of **DP**, let $V_y = \{v \in V(G) \mid y(v) > 0\}$ and $E_y = \{e = \{u, v\} \in E(G) \mid y(u) + y(v) = w(e)\}$. Then, Theorem 1 implies the following corollary.

Corollary 1. *Assume that the optimal value of* **IP** *for (G, b, w) is equal to that of* **LP***. Let y be an optimal solution of* **DP***. Then, a b-matching $x \in \mathbb{Z}_{\geq 0}^{E(G)}$ in G has the maximum weight if and only if $\{e \in E(G) \mid x(e) > 0\} \subseteq E_y$ and $\sum_{e \in \delta(v)} x(e) = b(v)$ for every vertex $v \in V_y$.*

Proof. We first prove the only-if direction. Suppose that x is a maximum-weight b-matching in G. Then, because the optimal value of **LP** is assumed to be equal to that of **IP**, x is an optimal solution of **LP**. Since y is an optimal solution of **DP**, x and y satisfy Theorem 1(2). Therefore, $\{e \in E(G) \mid x(e) > 0\} \subseteq E_y$ holds. For every vertex $v \in V_y$, we have $y(v) > 0$ and hence Theorem 1(2)-(ii) yields $\sum_{e \in \delta(v)} x(e) = b(v)$.

We then prove the if direction. Suppose that a b-matching x in G satisfies $\{e \in E(G) \mid x(e) > 0\} \subseteq E_y$ and $\sum_{e \in \delta(v)} x(e) = b(v)$ for every vertex $v \in V_y$. Since each edge $e = \{u, v\} \in E(G)$ with $x(e) > 0$ is contained in E_y, we have $y(u) + y(v) = w(e)$. Therefore, Theorem 1(2)-(i) holds. We then claim that x and y satisfy Theorem 1(2)-(ii). Consider any vertex $v \in V(G)$ such that $y(v) > 0$. Then, we have $v \in V_y$, and hence $\sum_{e \in \delta(v)} x(e) = b(v)$ holds; thus, Theorem 1(2)-(ii) holds. In this way, x and y satisfy Theorem 1(2). Then, Theorem 1(1) yields that x is an optimal solution of **LP**. Since the optimal value of **IP** is assumed to be equal to that of **LP**, x is a maximum-weight b-matching in G.   □

We now rephrase Corollary 1 so that it can be easily applied to our algorithm in the next section. For a graph G and its edge subset $E' \subseteq E(G)$, we denote by $G[E']$ the subgraph of G induced by E', that is, the vertex set of $G[E']$ is $\{u, v \in V(G) \mid \{u, v\} \in E'\}$ and the edge set of $G[E']$ is E'. For a vertex subset $C \subseteq V(G)$, we say that a b-matching $x \in \mathbb{Z}_{\geq 0}^{E(G)}$ in G is *C-saturated* if $\sum_{e \in \delta(v)} x(e) = b(v)$ holds for every vertex $v \in C$. Then, Corollary 1 can be rephrased as the following lemma; recall that a *vertex cover* of a graph G is a vertex subset of G which contains at least one of the endpoints of every edge in G.

Lemma 1. *Assume that the optimal value of* **IP** *for* (G, b, w) *is equal to that of* **LP**. *Then, there exist a vertex subset* $C \subseteq V(G)$ *and an edge subset* $E' \subseteq E(G)$ *such that*

(a) C *is a vertex cover of* $G[E']$; *and*
(b) *a b-matching* $x \in \mathbb{Z}_{\geq 0}^{E(G)}$ *in G has the maximum weight if and only if* $\{e \in E(G) \mid x(e) > 0\} \subseteq E'$ *and x is C-saturated.*

Furthermore, such a pair of C and E' can be found in polynomial time.

Proof. Because an optimal solution y of **DP** can be computed in polynomial time [21], we can obtain V_y and E_y in polynomial time. Let $C = V_y$ and $E' = E_y$. Then, Condition (b) follows immediately from Corollary 1.

We now verify Condition (a). Consider any edge $e = \{u, v\} \in E' = E_y$. Then, $y(u) + y(v) = w(e)$ holds. Since $w(e) > 0$, we have $y(u) > 0$ or $y(v) > 0$. Therefore, $u \in V_y$ or $v \in V_y$, that is, at least one of the endpoints of e is contained in V_y. In this way, $C = V_y$ forms a vertex cover of $G[E']$.   □

Note that we use the assumption of (nonzero) positive edge-weights only in the proof of Lemma 1(a). Theorem 1 and Corollary 1 hold even for nonnegative edge-weights, that is, $w(e) \geq 0$ for all edges $e \in E(G)$.

3 Algorithm

In this section, we give the main result of the paper as the following theorem.

Theorem 2. MAXIMUM-WEIGHT b-MATCHING RECONFIGURATION *can be solved in polynomial time for any instance* (G, b, w, x_s, x_t) *such that the optimal value of* **IP** *for* (G, b, w) *is equal to that of* **LP**.

It is known that the optimal value of **IP** for (G, b, w) is equal to that of **LP** if G is bipartite [22, Theorem 21.2], or $b: V(G) \to \{2\}$ [22, Corollary 30.2a]. Then, we have the following corollary.

Corollary 2. MAXIMUM-WEIGHT b-MATCHING RECONFIGURATION *can be solved in polynomial time for bipartite graphs, or* $b: V(G) \to \{2\}$.

In the remainder of this section, we prove Theorem 2 by giving such an algorithm. As we mentioned in Introduction, we will reduce the problem of asking the existence of a desired sequence of maximum-*weight* b-matchings to the problem of asking that of maximum-*cardinality* b-matchings, by using the characterization of maximum-weight b-matchings (Lemma 1).

Let (G, b, w, x_s, x_t) be an instance of MAXIMUM-WEIGHT b-MATCHING RECONFIGURATION such that the optimal value of **IP** for (G, b, w) is equal to that of **LP**. Let $C \subseteq V(G)$ and $E' \subseteq E(G)$ be the pair obtained by Lemma 1. By Lemma 1(b), any maximum-weight b-matching $x \in \mathbb{Z}_{\geq 0}^{E(G)}$ in G satisfies $x(e) = 0$ for all edges $e \in E(G) \setminus E'$. Therefore, it suffices to consider only C-saturated b-matchings in the induced subgraph $G[E']$. Note that both x_s and x_t are C-saturated b-matchings in $G[E']$, because they are maximum-weight b-matchings in G.

For two C-saturated b-matchings $x, x' \in \mathbb{Z}_{\geq 0}^{E'}$ in $G[E']$, we write $x \overset{C, E'}{\longleftrightarrow} x'$ if there exists a sequence $\langle x_0, x_1, \ldots, x_\ell \rangle$ of b-matchings in $G[E']$ such that

(i) $x_0 = x$ and $x_\ell = x'$;
(ii) all b-matchings $x_0, x_1, \ldots, x_\ell$ are C-saturated; and
(iii) $x_{i-1} \leftrightarrow x_i$ holds for each $i \in \{1, 2, \ldots, \ell\}$.

By Lemma 1(b) we then have the following proposition.

Proposition 2. $x_s \overset{C, E'}{\longleftrightarrow} x_t$ *if and only if* $x_s \overset{w}{\longleftrightarrow} x_t$.

Therefore, for proving Theorem 2, it suffices to determine whether $x_s \overset{C, E'}{\longleftrightarrow} x_t$ or not, in polynomial time. Our algorithm can be outlined as Algorithm 1. By Lemma 1 and Proposition 1, Algorithm 1 runs in polynomial time. Thus, we will prove its correctness in the remainder of this section.

We first note that no edge in E_C can be touched by any transformation of C-saturated b-matchings, as in the following lemma.

Lemma 2. *Let* x *and* x' *be* C-*saturated* b-*matchings in* $G[E']$ *such that* $x \overset{C, E'}{\longleftrightarrow} x'$. *Then,* $x(e) = x'(e)$ *holds for each edge* $e \in E_C$.

Algorithm 1. Polynomial-time algorithm for MAXIMUM-WEIGHT b-MATCHING RECONFIGURATION

Input: An instance (G, b, w, x_s, x_t) of MAXIMUM-WEIGHT b-MATCHING RECONFIG-URATION such that the optimal value of **IP** for (G, b, w) is equal to that of **LP**

Output: yes/no

Step 1. Obtain a vertex subset $C \subseteq V(G)$ and an edge subset $E' \subseteq E(G)$ satisfying Conditions (a) and (b) of Lemma 1. Let E_C be the set of all edges e in $G[E']$ such that both endpoints of e belong to C.

Step 2. If there exists an edge $e \in E_C$ such that $x_s(e) \neq x_t(e)$, then return no.

Step 3. Let $b'(v) = b(v) - \sum_{e \in \delta(v) \cap E_C} x_s(e)$ for each vertex v in $G[E']$. Delete all edges in E_C from $G[E']$; let G' be the resulting graph. Let x'_s and x'_t be two b'-matchings in G' such that $x'_s(e) = x_s(e)$ and $x'_t(e) = x_t(e)$, respectively, for all edges e in G'.

Step 4. Apply Proposition 1 to the instance $(G', b', w_{\mathsf{id}}, x'_s, x'_t)$ where $w_{\mathsf{id}} \colon E(G') \to \{1\}$, and return its answer.

This lemma ensures the correctness of Step 2 of Algorithm 1.

We then show that the graph G' obtained by Step 3 of Algorithm 1 satisfies the following lemma.

Lemma 3. *The graph G' obtained by Algorithm 1 is a bipartite graph whose bipartition is C and $V(G') \setminus C$.*

Proof. Since all edges in E_C have been deleted, there is no edge joining two vertices in C. On the other hand, by Lemma 1(a) at least one of the endpoints of each edge in $G[E']$ is contained in C. Thus, there is no edge joining two vertices in $V(G') \setminus C = V(G[E']) \setminus C$. Therefore, G' is a bipartite graph with bipartition $(C, V(G') \setminus C)$. □

Finally, the correctness of Step 4 of Algorithm 1 can be verified by combining the following lemma with Lemma 2.

Lemma 4. *A b'-matching x in G' is C-saturated if and only if x has the maximum cardinality in G'.*

Proof. We first prove the only-if direction. Suppose that a b'-matching x in G' is C-saturated. Then, $\sum_{e \in E(G')} x(e) \geq \sum_{v \in C} b'(v)$ holds. Since G' is a bipartite graph whose one side of the bipartition is C, any b'-matching in G' is of cardinality at most $\sum_{v \in C} b'(v)$. Therefore, x is a maximum-cardinality b'-matching in G'.

We then prove the if direction. Suppose that a b'-matching x in G' has the maximum cardinality in G'. It suffices to prove

$$\sum_{e \in E(G')} x(e) \geq \sum_{e \in E(G')} x'_s(e) \geq \sum_{v \in C} b'(v);$$

then, all vertices in C must be saturated by x, because G' is a bipartite graph with bipartition $(C, V(G') \setminus C)$. The first inequality holds because x is a maximum-cardinality b'-matching in G'. We thus prove the second inequality, as follows. Since x_s is a maximum-weight b-matching in G, by Lemma 1(b) it satisfies $\{e \in E(G) \mid x_\mathrm{s}(e) > 0\} \subseteq E'$ and is C-saturated. Therefore, we have

$$\sum_{e \in E(G')} x_\mathrm{s}'(e) = \sum_{e \in E' \setminus E_C} x_\mathrm{s}(e) = \sum_{e \in E(G) \setminus E_C} x_\mathrm{s}(e)$$
$$\geq \sum_{v \in C} b(v) - \sum_{e \in E_C} x_\mathrm{s}(e) \geq \sum_{v \in C} b'(v),$$

as claimed.   □

In this way, Algorithm 1 correctly solves MAXIMUM-WEIGHT b-MATCHING RECONFIGURATION in polynomial time. This completes our proof of Theorem 2.

4   Conclusion

In this paper, we have shown that MAXIMUM-WEIGHT b-MATCHING RECONFIGURATION is solvable in polynomial time for instances with no integrality gap. We emphasize again that such instances include b-matchings (and hence ordinary matchings) in bipartite graphs and 2-matchings in general graphs.

As we have mentioned in Sect. 1.2, both Ito et al. [11] and Mühlenthaler [20] studied their reconfiguration problems under a more generalized reconfiguration rule, called the TAR rule. In the WEIGHTED b-MATCHING RECONFIGURATION problem under the TAR rule, we are given two b-matchings (which do not necessarily have the maximum weight) together with an integer threshold $k \in \mathbb{Z}_{\geq 0}$, and asked the existence of a sequence of b-matchings between them, obtained by either adding or deleting one edge at a time, with keeping weights at least k. It remains open to clarify the complexity status for WEIGHTED b-MATCHING RECONFIGURATION under the TAR rule; this open question was originally posed by Ito et al. [11] for WEIGHTED MATCHING RECONFIGURATION.

References

1. Bonamy, M., Johnson, M., Lignos, I., Patel, V., Paulusma, D.: Reconfiguration graphs for vertex colourings of chordal and chordal bipartite graphs. J. Comb. Optim. **27**, 132–143 (2014)
2. Bonsma, P.: Rerouting shortest paths in planar graphs. In: Proceedings of FSTTCS 2012, LIPIcs, vol. 18, pp. 337–349 (2012)
3. Bonsma, P., Cereceda, L.: Finding paths between graph colourings: PSPACE-completeness and superpolynomial distances. Theoret. Comput. Sci. **410**, 5215–5226 (2009)
4. Bonsma, P., Kamiński, M., Wrochna, M.: Reconfiguring independent sets in claw-free graphs. In: Ravi, R., Gørtz, I.L. (eds.) SWAT 2014. LNCS, vol. 8503, pp. 86–97. Springer, Cham (2014). doi:10.1007/978-3-319-08404-6_8

5. Demaine, E.D., Demaine, M.L., Fox-Epstein, E., Hoang, D.A., Ito, T., Ono, H., Otachi, Y., Uehara, R., Yamada, T.: Linear-time algorithm for sliding tokens on trees. Theoret. Comput. Sci. **600**, 132–142 (2015)
6. Gopalan, P., Kolaitis, P.G., Maneva, E.N., Papadimitriou, C.H.: The connectivity of Boolean satisfiability: computational and structural dichotomies. SIAM J. Comput. **38**, 2330–2355 (2009)
7. Haas, R., Seyffarth, K.: The k-dominating graph. Graphs Comb. **30**, 609–617 (2014)
8. Haddadan, A., Ito, T., Mouawad, A.E., Nishimura, N., Ono, H., Suzuki, A., Tebbal, Y.: The complexity of dominating set reconfiguration. Theoret. Comput. Sci. **651**, 37–49 (2016)
9. Hatanaka, T., Itô, T., Zhou, X.: The list coloring reconfiguration problem for bounded pathwidth graphs. IEICE Trans. Fundam. Electron. Commun. Comput. Sci. E98-A, 1168–1178 (2015)
10. van den Heuvel, J.: The complexity of change. In: Surveys in Combinatorics 2013. London Mathematical Society Lecture Notes Series, vol. 409 (2013)
11. Ito, T., Demaine, E.D., Harvey, N.J.A., Papadimitriou, C.H., Sideri, M., Uehara, R., Uno, Y.: On the complexity of reconfiguration problems. Theoret. Comput. Sci. **412**, 1054–1065 (2011)
12. Ito, T., Nooka, H., Zhou, X.: Reconfiguration of vertex covers in a graph. IEICE Trans. Inf. Syst. E99-D, 598–606 (2016)
13. Ito, T., Ono, H., Otachi, Y.: Reconfiguration of cliques in a graph. In: Jain, R., Jain, S., Stephan, F. (eds.) TAMC 2015. LNCS, vol. 9076, pp. 212–223. Springer, Cham (2015). doi:10.1007/978-3-319-17142-5_19
14. Kamiński, M., Medvedev, P., Milanič, M.: Complexity of independent set reconfigurability problems. Theoret. Comput. Sci. **439**, 9–15 (2012)
15. Korte, B., Vygen, J.: Combinatorial Optimization: Theory and Algorithms, 5th edn. Springer, Heidelberg (2012)
16. Makino, K., Tamaki, S., Yamamoto, M.: An exact algorithm for the Boolean connectivity problem for k-CNF. Theoret. Comput. Sci. **412**, 4613–4618 (2011)
17. Mizuta, H., Ito, T., Zhou, X.: Reconfiguration of Steiner trees in an unweighted graph. In: Mäkinen, V., Puglisi, S.J., Salmela, L. (eds.) IWOCA 2016. LNCS, vol. 9843, pp. 163–175. Springer, Cham (2016). doi:10.1007/978-3-319-44543-4_13
18. Mouawad, A.E., Nishimura, N., Pathak, V., Raman, V.: Shortest reconfiguration paths in the solution space of Boolean formulas. In: Halldórsson, M.M., Iwama, K., Kobayashi, N., Speckmann, B. (eds.) ICALP 2015. LNCS, vol. 9134, pp. 985–996. Springer, Heidelberg (2015). doi:10.1007/978-3-662-47672-7_80
19. Mouawad, A.E., Nishimura, N., Raman, V.: Vertex cover reconfiguration and beyond. In: Ahn, H.-K., Shin, C.-S. (eds.) ISAAC 2014. LNCS, vol. 8889, pp. 452–463. Springer, Cham (2014). doi:10.1007/978-3-319-13075-0_36
20. Mühlenthaler, M.: Degree-constrained subgraph reconfiguration is in P. In: Italiano, G.F., Pighizzini, G., Sannella, D.T. (eds.) MFCS 2015. LNCS, vol. 9235, pp. 505–516. Springer, Heidelberg (2015). doi:10.1007/978-3-662-48054-0_42
21. Schrijver, A.: Theory of Linear and Integer Programming. Wiley, Chichester (1986)
22. Schrijver, A.: Combinatorial Optimization: Polyhedra and Efficiency. Springer, Berlin (2003)

Constant Approximation for Stochastic Orienteering Problem with $(1 + \epsilon)$-Budget Relaxiation

Yiyao Jiang[✉]

Institute for Interdisciplinary Information Sciences,
Tsinghua University, Beijing, China
jiangyy13@mails.tsinghua.edu.cn

Abstract. In the Stochastic Orienteering Problem (SOP), we are given finite metric space (V, d) and a starting point $\rho \in V$. Each $v \in V$ has an associated reward $r_v \geq 0$ and random completion time S_v, where the distribution of each S_v is know (once a reward has been earned, it cannot be earned again); the time cost of traveling from $u \in V$ to $v \in V$ is $d(u, v)$. The goal is to sequentially visit vertices and complete tasks in order to maximize the total rewards within 1 unit of time (after normalization). In this paper, we present a nonadaptive $O(\epsilon^{-14})$-approximation for (the original, adaptive) SOP when we relax the unit time budget to $(1 + \epsilon)$, $0 < \epsilon < 1$.

1 Introduction

In the competitive practice of orienteering, participants are given a map with marked locations. Starting from a given point, they must try to visit as many locations as possible within a given time limit. Each point has a given reward, and the goal for competitors is to maximize the total reward within the time limit. This problem was first studied in [8] from an algorithmic perspective; see [13] for a more recent survey.

A similar problem to Orienteering Problem (OP) is the Prize Collecting Traveling Salesman Problem (PCTSP) [5]. In the latter, the salesman's costs are the time spent traveling between points, plus a point-dependent penalty that is incurred for each point that he fails to visit. PCTSP's goal is to minimize the traveling time plus the penalties of the points. Comparing with OP, PCTSP is not limited in time, whereas OP is strictly bounded with time limit.

1.1 The (Simple) Orienteering Problem

An instance of the Orienteering Problem (OP) consists of a metric space (V, d) with $|V| = n$ points and distances $d(u, v) \in \mathbb{R}^+$ for each $(u, v) \in V \times V$; moreover, the instance specifies a starting point $\rho \in V$, a total time budget B, and a reward $r_v \in \mathbb{R}^+$ for each $v \in V$. The object is to chart a path that maximizes

© Springer International Publishing AG 2017
Y. Cao and J. Chen (Eds.): COCOON 2017, LNCS 10392, pp. 297–308, 2017.
DOI: 10.1007/978-3-319-62389-4_25

the sum of rewards r_v while staying within the time budget B, where the distances $d(u, v)$ are interpreted as time costs. In this setting, and as all quantities are real numbers, we can assume without loss of generality that $B = 1$ after normalization.

In 1998 Arkin et al. [1] obtained a 2-approximation for OP with planar Euclidean distances. Furthermore, Chen and Har-Peled [7] obtained a polynomial-time approximation scheme (PTAS) for OP in fixed-dimensional Euclidean space.

For general metric spaces (V, d), Blum et al. [2] obtained the first constant-factor approximation (with factor 4) in 2007. The best result until now is a $(2 + \epsilon)$-approximation algorithm by Chekuri et al. [6].

The Orienteering Problem finds many natural applications. For example, a traveling salesman with limited time [12] (who must maximize his total profit in one day).

1.2   Stochastic Orienteering Problem

In the Stochastic Orienteering Problem (SOP) [9] a job is associated to each vertex $v \in V$, which must be completed at v in order to obtain the reward r_v. The time necessary to complete the job, moreover, is randomly distributed. For example, a salesman may need to wait for his customer to come downstairs, or a tourist need time to enjoy the scenery, and so on.

Generally, one is not allowed to give up a job until it is finished. In certain models one is allowed to give up a job and its reward after starting the job, while being forbidden from re-trying to complete the job later, once it has been given up.

The time required by job v is a random variable S_v, where S_v has a known discrete distribution π_v for each $v \in V$, where π_v modeled as a function from $\mathbb{R}^+$ into $[0, 1]$; the distributions $\{\pi_v : v \in V\}$ are thus part of an instance of SOP.

We note that SOP can be considered as a kind of combination of OP and of the Stochastic Knapsack Problem (SKP) [11]. The Stochastic Knapsack Problem, indeed, corresponds to an instance of SOP in which $d(u, v) = 0$ for all $(u, v) \in V \times V$. For SKP, moreover, a $(2 + \epsilon)$-approximation algorithm is known [4].

For a non-adaptive variant of SOP in which the path is chosen in advance and in which the distances $d(u, v)$ are restricted to integers, Gupta et al. [9] have shown a constant factor approximation algorithm. In that setting (and due to the integer restriction on distances) the time budget B is not normalized to 1, and remains a parameter of the problem. In fact, Gupta et al. show that their algorithm is an $O(\log \log B)$-approximation algorithm for SOP with arbitrary *adaptive* policies. In the same formal setting, moreover, Bansal and Nagarajan [3] have shown that a $\Omega((\log \log B)^{1/2})$ gap between adaptive and non-adaptive policies is unavoidable. This contrasts with the case of SKP, for which non-adaptive policies can approximate adaptive policies to within a constant factor [4] (In OP the distinction between adaptive and non-adaptive policies is moot.).

In this paper we present an $O(\epsilon^{-14})$-approximation for the (original, adaptive) SOP with time budget $(1+\epsilon)B = 1+\epsilon$, $0 < \epsilon < 1$. More precisely, we show a *non-adaptive* policy of time budget $1 + \epsilon$ such that the expected reward of the optimal *adaptive* policy at time budget 1 is at most $O(\epsilon^{-14})$ times the expected reward of our [non-adaptive, time $1 + \epsilon$] policy. Our algorithm therefore achieves an $O(1)$-approximation for any constant $0 < \epsilon < 1$.

Theorem 1. *There is a polynomial time, non-adaptive $O(\epsilon^{-14})$-approximation for SOP with respect to a relaxed time budget of $(1 + \epsilon)B$, $0 < \epsilon < 1$.*

2 Definitions and Notations

We write $\mathbb{R}^+$ for the nonnegative real numbers $[0, \infty)$. In this paper all quantities are real-valued.

2.1 Orienteering

As introduced in Sect. 1, an instance of the Orienteering Problem (OP) is defined on an underlying metric space (V, d) with $|V| = n$ points and distances $d(u, v) \in \mathbb{R}^+$ for each $(u, v) \in V \times V$. We are given a starting "root" point $\rho \in V$ and a total time budget B. An instance of OP thus corresponds to a tuple $I_O = (V, d, \{r_v\}_{v \in V}, B, \rho)$.

2.2 Stochastic Orienteering

In the Stochastic Orienteering Problem (SOP) each point $v \in V$ is related to a unique stochastic job. The job associated to v has a fixed reward $r_v \geq 0$, and a random variable of processing time size S_v with a known but arbitrary discrete probability distribution $\pi_v : U \to [0, 1]$, where $U \subseteq \mathbb{R}^+$ is a countable set, such that $\sum_{t \in U} \pi_v(t) = 1$ for all $v \in V$. (Time size is the time that we have to wait at the point before receiving the reward for job v. Note that we can take the same underlying time set U for all $v \in V$, wlog.) An instance of SOP thus corresponds to a tuple $I_{SO} = (\pi, V, d, \{r_v\}_{v \in V}, \{S_v\}_{v \in V}, B, \rho)$.

At each step we travel to a point v from a current point u (at time cost $d(u, v)$) and then process the job v, at time cost S_v, and once a job v is selected, one must wait for v to complete. If we are still within the time budget B when the job is completed we receive the reward r_v, and choose a new point as the next destination.

For generality, $\forall t \in U$ we assume $t \in [0, B]$. If one time size $t > B$, we can just truncated it to B, since we will always stop at time budget B before completing the job and never get the reward. Also, we can assume wlog that $d(u, \rho) \leq B$ for all $u \in V$, which also implies that $d(u, v) \in [0, 2B]$ for all $u, v \in V$.

2.3 Task Orienteering Problem

Our analysis refers to a simplified version of the Stochastic Orienteering Problem known as the "Task Orienteering Problem" (TOP); TOP corresponds to an instance of SOP in which each time size S_v is deterministic. An instance of TOP thus corresponds to a tuple $I_{TO} = (V, d, \{r_v\}_{v \in V}, \{s_v\}_{v \in V}, B, \rho)$ with $s_v \in \mathbb{R}^+$ for each $v \in V$.

3   Algorithm

We begin with an analysis of the Task Orienteering Problem described in Sect. 2.3, to be used as a component in our main analysis.

3.1   An Algorithm for TOP

We first provide a polynomial time algorithm AlgTO (Algorithm 1) and then prove that AlgTO gives an $O(1)$-approximation to TOP.

Definition 1 *(Valid OP Instance).* *Given an instance $I_{TO} = (V, d, \{r_v\}_{v \in V},$ $\{s_v\}_{v \in V}, B, \rho)$ of TOP, we define the following valid OP instance $I_O(I_{TO}) :=$ $(V', d', \{r_v\}_{v \in V'}, B, \rho)$, where*

1. $V' := \{v \in V : d(\rho, v) + s_v \leq B\}$;
2. $d'(u, v) := d(u, v) + s_v/2 + s_u/2$ *for all* $u, v \in V'$, $u \neq v$;
3. $d'(u, u) := 0$ *for all* $u \in V$;
4. $r'_v := r_v$ *for all* $v \in V'$.

Lemma 1. *The function $d' : V' \times V' \to \mathbb{R}^+$ constructed in Definition 1 is a metric.*

The proof is easy.

Algorithm 1. AlgTO for Task Orienteering on input $I_{TO} = (V, d, \{r_v\}_{v \in V},$ $\{s_v\}_{v \in V}, B, \rho)$

---

1. **let** $I_O(I_{TO}) := (V', d', \{r'_v\}_{v \in V'}, B, \rho)$;
2. **run** AlgOrient on the (Simple) Orienteering Problem I_O,
3.   **get** the path P, the reward $OPT'_O = \sum_{v \in P} r'_v$, and the ending point ρ';
4. **compare** $\sum_{v \in P \setminus \{\rho'\}} r'_v$ and $r'_{\rho'}$;
5. **output** the path

$$P' := \begin{cases} \rho \longrightarrow \rho', & \text{if } r'_{\rho'} \geq \sum_{v \in P \setminus \{\rho'\}} r'_v, \\ P \setminus \{\rho'\}, & \text{otherwise;} \end{cases}$$

6.   and the reward $OPT'_{TO} := \sum_{v \in P'} r'_v$.

Theorem 2. *Algorithm AlgTO (Algorithm 1) is a polynomial time $O(1)$-approximation for the Task Orienteering Problem.*

Proof. Assume the optimal result for Task Orienteering problem I_{TO} is OPT_{TO}. Now we prove that OPT'_{TO} returning by the polynomial time algorithm AlgTO (Algorithm 1) is an $\Omega(1)$-approximation to OPT_{TO}.

In Algorithm 1 line 1, as in Definition 1 item 1, we delete all points $v \in V$ which are obviously not in the optimal path of I_{TO}, because even we only reach to one point v from the starting point ρ, we still do not have enough time to reach the point v and complete the job with time $d(\rho, v) + s_v > B$.

In Definition 1 item 2, for each point $v \in V'$, we divide the fixed job time s_v by 2, and add $s_v/2$ to the lengths of all edges adjacent to point v. (Note that, although we add $s_v/2$ to the lengths of all edges adjacent to point v, in real algorithm path, we only count twice of $s_v/2$ as doing the job in time s_v: the edge into point v, and the edge out of point v). Define the new distance as $d'(u, v)$.

In Lemma 1, we prove that (V', d') is a metric, thus we create an Orienteering Problem instance $I_O = (V', d', \{r_v\}_{v \in V}, B, \rho)$ successfully in Algorithm 1 step 1.

In Algorithm 1 line 2, we use the polynomial time algorithm AlgOrient (by Blum et al. [2] mentioned in Sect. 2.1) on OP instance I_O. And in line 3, the path P of reward OPT'_O is a constant approximation of the optimal solution for I_O by algorithm AlgOrient. Note that any path in I_{TO} including the optimal path of I_{TO} is surely a feasible path in I_O (since the only change is the distances in I_{TO} are longer), so $OPT'_O = \Omega(OPT_{TO})$.

Note that in path P, we always get in and out of a point v except for the starting point ρ and ending point ρ'. So the $s_v/2$ are always counted twice, and in I_{TO} we do complete all the jobs on the path except for ρ and ρ'. Thus the only incidence we cannot complete path P in I_{TO} is the jobs on ρ and ρ' in path P.

At the starting point ρ in I_{TO}, we can always create a fake "starting point" $\rho^\star$ which is the same $d'(,)$ as ρ but $d'(\rho, \rho^\star) = d'(\rho^\star, \rho) = d'(\rho^\star, \rho^\star) = 0$, $s_{\rho^\star} = 0$ and $r_{\rho^\star} = 0$, and we set ρ as a normal point. Run AlgTO (Algorithm 1) on I_{TO} with $\rho^\star$. If ρ is still in path, we just start from ρ and use s_ρ time to get the reward r_ρ then go through path P; else, ρ is not in path, we also start from ρ, but give up the job and reward on ρ, then go through path P. All these changes do not change the OPT'_O, and do not effect on OPT_{TO}. Thus, we now only need to concern about ρ'.

In Algorithm 1 line 4, we compare the reward $r'_{\rho'}$ at the ending point ρ' (note that $d(\rho, \rho') + s_{\rho'} \leq B$ in Definition 1 indicates that path $\rho \longrightarrow \rho'$ is feasible in I_{TO}), and the rewards $\sum_{v \in P \setminus \{\rho'\}} r'_v$ of all the points in path $\widehat{P}$ except for ρ'. Since $OPT'_O = \sum_{v \in P} r'_v = r'_{\rho'} + \sum_{v \in P \setminus \{\rho'\}} r'_v$, $\max\{r'_{\rho'}, \sum_{v \in P \setminus \{\rho'\}} r'_v\} \geq OPT'_O/2 = \Omega(OPT_{TO})$.

In Algorithm 1 line 5 and 6, we thus choose the larger one between $r'_{\rho'}$ and $\sum_{v \in P \setminus \{\rho'\}} r'_v$ as OPT'_{TO}. If $r_{\rho'}$ is larger, then we just go to a single point ρ' and complete the job; if the other is larger, we just ignore ρ', go and do the jobs through the path P except for ρ'.

Thus, we have a polynomial time $O(1)$-approximation algorithm AlgTO for Task Orienteering problem.

3.2 The Algorithm for Stochastic Orienteering

Definition 2 (*Valid* TOP *Instance*). *Given an instance $I_{SO} = (\pi, V, d, \{r_v\}_{v \in V}, \{S_v\}_{v \in V}, B, \rho)$ of SOP, and a value $\epsilon > 0$, we define the following valid TOP* *instance with parameter ϵ to be $I_{TO}(\epsilon, I_{SO}) := (\widehat{V}, \widehat{d}, \{\widehat{r}_u\}_{\forall u \in \widehat{V}}, \{\widehat{s}_u\}_{\forall u \in \widehat{V}}, \widehat{B}, \widehat{\rho})$, where we*

1. *define $\widehat{V}$: let $\widehat{V} := V$,*
   for all $u \in V$, if $d(\rho, u) > B$, $\widehat{V} := \widehat{V} \backslash \{u\}$,
   create a virtual point $\widehat{\rho}$, $\widehat{V} := V \cup \{\widehat{\rho}\}$;
2. *for all $u, v \in \widehat{V}$, define distances $\widehat{d}(u, v)$: if $u, v \in V$, $\widehat{d}(u, v) := d(u, v)$,*
   for all $u \in V$, set $\widehat{d}(u, \widehat{\rho}) = \widehat{d}(\widehat{\rho}, u) = B$,
   set $d(\widehat{\rho}, \widehat{\rho}) = 0$;
3. *for all $u \in \widehat{V}$, define rewards $\widehat{r}_u$: if $u \in V$, $\widehat{r}_u = r_u$, set $\widehat{r}_{\widehat{\rho}} = 0$;*
4. *for all $u \in \widehat{V}$, define deterministic job times $\widehat{s}_u$: if $u \in V$, $\widehat{s}_u = E[S'_u]$,*
   where $S'_u = \min\{S_u, B\}$,
   set $\widehat{s}_{\widehat{\rho}} = 0$;
5. *define time bound $\widehat{B}$: let $\widehat{B} := (1 + \epsilon^{13})B$;*
6. *define the starting point: set $\widehat{\rho}$ to be the new starting point.*

Lemma 2. *The $(\widehat{V}, \widehat{d})$ of the TOP instance I_{TO} in Definition 2 is a metric space.*

The proof is easy.

Algorithm 2. Algorithm AlgSO for SOP on input $I_{SO} = (\pi, V, d, \{r_v\}_{v \in V}, \{S_v\}_{v \in V}, B, \rho)$ and parameter $0 < \epsilon < 1$

---
1. **for all** $v \in V$ **do**
2.   let $R_v := r_v \cdot \Pr_{S_v \sim \pi_v}[S_v \le (B - d(\rho, v))]$ be the expected reward of the single-node tour from ρ to v;
3. **w.p.** 1/2, just visit the point $\widehat{v}$ with the highest $R_{\widehat{v}}$ and **exit**.
4. **let** $I_{TO}(\epsilon, I_{SO}) := (\widehat{V}, \widehat{d}, \{\widehat{r}_u\}_{\forall u \in \widehat{V}}, \{\widehat{s}_u\}_{\forall u \in \widehat{V}}, \widehat{B}, \widehat{\rho})$;
5. **run** AlgTO (Algorithm 1) on the valid TOP instance $I_{TO}(\epsilon, I_{SO})$,
6.   **get** the path P, the reward $\widehat{OPT}_{TO}$;
7. replace the starting point $\widehat{\rho}$ in path P of $I_{TO}(\epsilon, I_{SO})$ by ρ and output path $\widehat{P}$;
8. go through path $\widehat{P}$ in I_{SO} with time budget $(1 + \epsilon)B$ and output the total reward $\widehat{OPT}_{SO}$.

We want to prove that Algorithm 2 is a polynomial time $O(\epsilon^{-14})$-approximation for Stochastic Orienteering problem with respect to a relaxed time budget of $(1 + \epsilon)B, 0 < \epsilon < 1$.

Define the expected reward on optimal adaptive policy on I_{SO} as OPT, and define the expected reward on optimal policy on $I_{TO}(\epsilon, I_{SO})$ as OPT_{TO}.

We design to have either a single-point tour $\rho \longrightarrow v$ of expected reward $R_v = \Omega(OPT)$ with 50 % chance, or the path $\widehat{P}$ on I_{SO} of reward $\widehat{OPT}_{SO} = \Omega(OPT)$ with 50 % chance. For this purpose, we want the following theorem.

Theorem 3. *Given an instance I_{SO} for which an optimal adaptive strategy has an expected reward of* OPT, *either there is a single-point tour with expected reward $\Omega(OPT)$, or the valid Task Orienteering instance $I_{TO}(\epsilon, I_{SO})$ has reward* $OPT_{TO} = \Omega(OPT)$, *with constant approximation parameter ϵ^{14}.*

We will prove Theorem 3 later in Sect. 4. For now, let us assume that Theorem 3 is correct, and use it to complete the proof of Theorem 1. The following proof is for Theorem 1.

3.3 Proof for Theorem 1

Suppose we enter Algorithm 2 line 4, and in line 6, AlgTO (Algorithm 1) finds a path $P = (\widehat{\rho}, v_1, v_2, \ldots, v_k)$ with reward $\widehat{OPT_{TO}}$. And then we get the path $\widehat{P} = (\rho, v_1, v_2, \ldots, v_k)$.

Lemma 3. *In $I_{SO} = (\pi, V, d, \{r_v\}_{v \in V}, \{S_v\}_{v \in V}, B, \rho)$, for any point $v_i \in \widehat{P}$, the probability of successfully reaching to v and finishing the job at v before violating the budget $(1 + \epsilon)B$ is at least $1 - \epsilon^{13}$.*

Proof. Note that now we relax the time budget on I_{SO} to $(1 + \epsilon)B$, but we still have time budget $(1 + \epsilon^{13})B$ on $I_{TO}(\epsilon, I_{SO})$ as in Definition 2.

In Definition 2, for $I_{TO}(\epsilon, I_{SO})$, $\sum_{i=1}^{k}(\widehat{s}_{v_i} + \widehat{d}(v_{i-1}, v_i)) \leq (1 + \epsilon^{13})B$, $S'_{v_i} = \min\{S_{v_i}, B\}$, and $\widehat{s}_{v_i} = E[S'_{v_i}]$. Since $\widehat{d}(\widehat{\rho}, v_1) = B$, $E[S'_{v_1}] + \sum_{i=2}^{k}(E[S'_{v_i}] + \widehat{d}(v_{i-1}, v_i)) \leq \epsilon^{13}B$.

Define $S^{\star}_{v_i} = \begin{cases} S'_{v_1} & \text{if } i = 1, \\ S'_{v_i} + \widehat{d}(v_{i-1}, v_i) & \text{if } 2 \leq i \leq k. \end{cases}$

Then $\sum_{i=1}^{k} E[S^{\star}_{v_i}] \leq \epsilon^{13}B$. By using Markov's inequality, for each $j \leq k$, we have

$$\Pr[\sum_{i=1}^{j} S^{\star}_{v_i} \leq \epsilon B] \geq 1 - \epsilon^{13}$$

The change from S_u to S'_u does not effect on optimal adaptive policy on I_{SO}, the distance $d(\rho, v_1) \leq B$, and $d(v_{i-1}, v_i) = \widehat{d}(v_{i-1}, v_i)$. So in path $\widehat{P}$ of I_{SO} with time budget $(1 + \epsilon)B$, for each point v_j $(1 \leq j \leq k)$, the probability we successfully reach to point v_j, and complete the job on point v_j is not less than $\Pr[\sum_{i=1}^{j} S^{\star}_{v_i} \leq \epsilon B]$. Thus the probability of successfully reaching to v and finishing the job at v before violating the budget $(1 + \epsilon)B$ is at least $1 - \epsilon^{13}$. And this complete the proof for Lemma 3.

Proof (The proof of Theorem 1). We prove that AlgSO (Algorithm 2) is a polynomial time $O(\epsilon^{-14})$-approximation for Stochastic Orienteering problem with respect to a relaxed time budget of $(1 + \epsilon)B$, $0 < \epsilon < 1$.

We assume that Theorem 3 is correct. Then either there is a single-point tour with expected reward $\Omega(OPT)$, or the valid Task Orienteering instance $I_{TO}(\epsilon, I_{SO})$ has reward $OPT_{TO} = \Omega(OPT)$, with a constant approximation parameter ϵ^{14}.

1. There is a single-point tour $\rho \longrightarrow v$ satisfies $R_v = \Omega(\epsilon^{14}\text{OPT})$.
   We have 50% chance to obtain the highest $R_{\widehat{v}}$. And $R_{\widehat{v}} \geq R_v = \Omega(\epsilon^{14}\text{OPT})$.
2. The valid Task Orienteering instance $I_{\text{TO}}(\epsilon, I_{\text{SO}})$ has reward $\text{OPT}_{\text{TO}} = \Omega(\text{OPT})$.
   We have 50% chance to get $\widehat{\text{OPT}_{\text{SO}}}$.
   In Algorithm 2 line 6, by applying Theorem 2, AlgTO (Algorithm 1) will find a path $P = (\widehat{\rho}, v_1, v_2, \ldots, v_k)$ with reward $\widehat{\text{OPT}_{\text{TO}}} = \Omega(\text{OPT}_{\text{TO}})$. By assuming Theorem 3 is correct, we have $\text{OPT}_{\text{TO}} = \Omega(\epsilon^{14}\text{OPT})$. Then $\widehat{\text{OPT}_{\text{TO}}} = \Omega(\text{OPT}_{\text{TO}}) = \Omega(\epsilon^{14}\text{OPT})$.
   Now applying Lemma 3 on path $\widehat{P}$ that we get in Algorithm 2 line 7, and that gives us an expected reward of at least $\widehat{\text{OPT}_{\text{SO}}} = \widehat{\text{OPT}_{\text{TO}}} \cdot (1 - \epsilon^{13}) = \Omega(\epsilon^{14}\text{OPT})$.

So we always have at least 50% probability to get $\Omega(\epsilon^{14}\text{OPT})$ by Algorithm 2, and thus we complete the proof of Theorem 1.

4   Optimal Policy for Stochastic Orienteering

The main method to deal with the huge optimal decision tree defined in Sect. 4.1 is Discretization and Block Policy [10]. Since all quantities are all real-valued, by scaling, we can assume $B = 1$ and the size of each item is distributed between 0 and B. The relaxed time budget $B + O(\varepsilon)$ should be less than $2B$.

4.1   Policy and Decision Tree

A policy σ for an SOP instance (π, V, d, B) can be represented as a decision tree $T_\sigma(\pi, V, d, B)$.

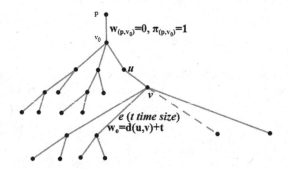

Fig. 1. Decision tree T_σ

·In Fig. 1, each node of T_σ is labeled by a vertex $v \in V$, with the root of T_σ being labeled by ρ. The vertices on a path from the root to a node are distinct, hence T_σ has depth at most $n = |V|$. A node of T_σ has countably many children, or one for each element of U, where $U = \{t \in \mathbb{R}^+ : \pi_v(t) > 0 \text{ for some } v \in V\}$ is the (countable) support set of the time size distributions. Let $z \in T_\sigma$ be a node of label $v \in V$, and let e be the t-th edge emanating from z, $t \in U$; then e has *probability* $\pi_e := \pi_v(t) = \Pr[S_v = t]$; moreover if z's parent node is node $y \in T_\sigma$ of label $u \in V$, then e has *weight* $w_e := d(u, v) + t$.

We will sometimes refer to nodes in T_σ by their label. (With little chance of confusion.) For a node $v \in T_\sigma$, we define $P(v)$ to be the path of edges of leading from the root to v. We define $W(v) := \sum_{e \in P(v)} w_e$ and $\Phi(v) := \prod_{e \in P(v)} \pi_e$. Thus $W(v)$ is the total time elapsed before reaching v and $\Phi(v)$ is the probability of reaching v.

We note that a given policy σ will be "compatible" with any tuple (π', V', d', B') such that $V' = V$ and such that π' has the same support U as π. Namely, the modified π' and d' will induce modified edge weights w'_e and modified edge probabilities p'_e. We emphasize this by including (π, V, d, B) as arguments to T_σ (though we will always use the same V for a given policy σ).

We let

$$R(\sigma, \pi, V, d, B) = \sum_{v \in T_\sigma, e=(v,u):W(v)+w_e \leq B} \pi_e \cdot r_v \Phi(v)$$

denote the expected reward that policy σ achieves with respect to the metric d, time size distributions π, and time budget B. We use **OPT** to denote the expected reward of the optimal adaptive policy.

For the following lemma, we also consider a modified node label $X_v := S_v + d(u, v)$, where u is the parent and v is the child. Thus, while $w_e = t + d(u, v)$ is a number, X_v is a random variable (Moreover, X_v "ignores" the fact that edge e is the t-th edge of v, i.e., that edge e is associated to a particular outcome of S_v.).

Lemma 4 (*part of Lemma 2.4 in* [4]). *For any policy σ on instance (π, V, d, B), there exists a policy σ' such that*

$$R(\sigma', \pi, V, d, B) = (1 - O(\epsilon))R(\sigma, \pi, V, d, B),$$

and for any realization path P in $T_{\sigma'}(\pi, V, d, B)$, $\sum_{v \in P} E[X_v] = O(B/\epsilon)$.

Here we mention that policy σ' is just cutting some branches on policy σ to ensure $\sum_{v \in P} E[X_v] = O(B/\epsilon)$. Note that policy σ is all designed by us, we surely may have a node v that $\sum_{v \in P} E[X_v] > O(B/\epsilon)$ though v has no contribution to reward.

4.2 Discretization

We present how to discretize the decision tree in Sect. 4.1 with given ϵ. We use the discretization method similar in paper [10] Sect. 2.1. We split all jobs into two parts: small size vertices and big size vertices, and discretize them separately.

Denote $\tilde{S}_v$ for value of vertex v after discretization, and the new distribution $\tilde{\pi}_v$. And for each node v in the tree, define $\tilde{X}_v := \tilde{S}_v + d(u, v)$ as a random variable which is similar in the definition of X_v in Sect. 4.1.

1. Small size region if $S_v \leq \epsilon^4$.
   There exists a value $0 \leq h \leq \epsilon^4$, such that $\Pr[S_v \geq h | S_v \leq \epsilon^4] \cdot \epsilon^4 = E[S_v | S_v \leq \epsilon^4]$. Then set: $\widetilde{S}_v = \begin{cases} 0, & 0 \leq S_v < h; \\ \epsilon^4, & h \leq S_v \leq \epsilon^4; \\ S_v, & S_v > \epsilon^4. \end{cases}$

2. Large size region if $S_v > \epsilon^4$.
   We simply discretize it as $\widetilde{S}_v = \lfloor \frac{S_v}{\epsilon^5} \rfloor \epsilon^5$.

We denote the set of the discretized size by $S = \{s_0, s_1, \cdots, s_{z-1}\}$ where $s_0 = 0$, $s_1 = \epsilon^5$, $s_2 = 2\epsilon^5, \ldots, s_{z-1}$. Note that $s_1 = \epsilon^5, s_2 = 2\epsilon^5, \ldots, s_{1/\epsilon-1} = \epsilon^4 - \epsilon^5$ are also in S though their probability is 0. The total number of values of time size is $|S| = z = O(B/\epsilon^5)$ which is a constant.

4.3   Canonical Policies

We need to use *canonical policies* introduced in [4]. A policy $\widetilde{\sigma}$ is a canonical policy if it makes decisions based on the discretized sized of vertices, but not their actual sizes. Under the canonical policy, we will keep trying new vertices if the total discretized size budget does not exceed, even the actual budget overflows. But no reward will get from these vertices which make actual budget exceeding.

Lemma 5 (*Lemma 4.2 in [10]*). *Let π be the distribution of vertice size and $\widetilde{\pi}$ be the discretized version of π. Then, the following statements hold:*

1. *For any policy σ, there exists a canonical policy $\widetilde{\sigma}$ such that*

$$R(\widetilde{\sigma}, \widetilde{\pi}, V, d, (1 + 4\epsilon)B) = (1 - O(\epsilon))R(\sigma, \pi, V, d, B);$$

2. *For any canonical policy $\widetilde{\sigma}$,*

$$R(\widetilde{\sigma}, \pi, V, d, (1 + 4\epsilon)B) = (1 - O(\epsilon))R(\widetilde{\sigma}, \widetilde{\pi}, V, d, B).$$

Proof sketch: for the first result, we prove that there is randomized canonical policy σ_r such that $R(\sigma_r, \widetilde{\pi}, V, \widetilde{d}, (1 + 4\epsilon)B) = (1 - O(\epsilon))R(\sigma, \pi, V, d, B)$. Thus such a deterministic policy $\widetilde{\sigma}$ exists. The randomized policy σ_r is derived from σ as follows. T_{σ_r} keeps the same tree structure as T_σ. If σ_r visits a vertex v and observes a discretized size $s \in S$, it chooses a random branch in σ among those sizes that are mapped to s. Let $t_1, t_2, \ldots, t_k$ are possible size realization of s as per size distribution π and let $\pi_v(t_1), \pi_v(t_2), \ldots, \pi_v(t_k)$ be the corresponding probabilities. Hence we choose one of the branches corresponding to size $t_1, t_2, \ldots, t_k$ w.p. $\pi_v(t_1), \pi_v(t_2), \ldots, \pi_v(t_k)$ (normalized) respectively. For more detail, see Lemma 4.2 in [10].

From Lemma 5, we conclude that the reward will not loss too much if we using canonical policies to applying the policies with $T_{\widetilde{\sigma}}$ described in Sect. 4.2, but not T.

4.4   Block Tree

Now we partition the decision tree $T_{\widetilde{\sigma}}$ created in Sect. 4.3 into blocks.

For any node v in the decision tree $T_{\widetilde{\sigma}}$, we define the *leftmost path of v* to be the realization path which starts at v, ends at a leaf, and consists of only edges corresponding to size zero. We define the *block* starting with node v (denote as seg(v)) in $T_{\widetilde{\sigma}}$ as the maximal prefix of the leftmost path of v such that: If $E[\widetilde{X}(v)] > \epsilon^{13}$, seg($v$) is the singleton node $\{v\}$. Otherwise, $E[\widetilde{X}(seg(v))] \leq \epsilon^{13}$.

We partition $T_{\widetilde{\sigma}}$ into blocks as follows. We say a node v is a *starting node* if v is a root or v corresponds to a non-zero size realization of its parent. For each starting node v, we greedily partition the leftmost path of v into blocks, i.e., delete seg(v) and recurse on the remaining part.

Lemma 6. *A canonical policy $\widetilde{\sigma}$ with expected reward $(1 - O(\epsilon))$OPT can be partitioned into several blocks that satisfy the following properties:*

1. *There are at most $O(\epsilon^{-14})$ blocks on any root-leaf path in the decision tree.*
2. *There are $|S| = O(B/\epsilon^5)$ children for each block.*
3. *Each block M with more than one node satisfies that $\sum_{b \in M} E[\widetilde{X}_b] \leq \epsilon^{13}$.*

Proof. 1.   Fix a particular root-to-leaf path R. Let us bound the number of blocks on R.

We have $\sum_{v \in R} E[\widetilde{X}(v)] = O(B/\epsilon) = O(1/\epsilon) = O(\epsilon^{-1})$ by Lemma 4. By definition of Block Policies, any single-point block v in $T_{\widetilde{\sigma}}$ satisfies $E[\widetilde{X}(v)] > \epsilon^{13}$. Then there are at most $O(\epsilon^{-1})/\epsilon^{13} = O(\epsilon^{-14})$ single-point in path R.

By definition of Discretization, any $\widetilde{X}(v)$ with non-zero size after discretization, is no less than ϵ^4. Then there are at most $O(1/\epsilon^4) = O(\epsilon^{-4})$ nodes w corresponding to a non-zero size realization.

This gives a bound $O(\epsilon^{-14} + \epsilon^{-4}) = O(\epsilon^{-14})$ blocks on any root-leaf path in the decision tree.
2. After Discretization in Sect. 4.2, we have only $O(B/\epsilon^5)$ number of values which means there are $|S| = O(B/\epsilon^5)$ children for each block.
3. This is exactly what we define our blocks.

And then we prove Theorem 3.

Proof (The proof of Theorem 3). By Properties 1,2 in Lemma 6 above, there are only constant number of blocks in the decision tree $T_{\widetilde{\sigma}}$ with block policies. Since there are at most $O(\epsilon^{-14})$ blocks in the optimal policy path in decision tree $T_{\widetilde{\sigma}}$ of reward $(1 - O(\epsilon))$OPT, there exists a block M with $\Omega(\epsilon^{14}$OPT$)$ reward.

We consider two situations in block M with different number of nodes.

Block M is a single-node block of node v.

And the node v has a huge reward. Then the single-node tour from the starting point ρ to v has an expected reward $\Omega(\epsilon^{14}$OPT$)$.

Otherwise, block M has at least two nodes.

And block M has an expected reward $\Omega(\epsilon^{14}$OPT$)$. By Property 3 in Lemma 6, we have $\sum_{b \in M} E[\widetilde{X}_b] \leq \epsilon^{13}$. Since the distance from the starting point $\widetilde{\rho}$ to the first node in M is at most 1, the path $P^\star$ which connects $\widetilde{\rho}$ and the block M in order has length at most $1 + \epsilon^{13}$. Thus $P^\star$ is a feasible path for $I_{TO}(\epsilon, I_{SO})$ as defined in Definition 2, and it has an expected reward $\Omega(\epsilon^{14}\text{OPT}) \leq \text{OPT}_{TO}$.

Thus, either there is a single-point tour with expected reward $\Omega(\text{OPT})$, or the valid Task Orienteering instance $I_{TO}(\epsilon, I_{SO})$ has reward $\text{OPT}_{TO} = \Omega(\text{OPT})$, with constant approximation parameter ϵ^{14}. And this completes the proof for Theorem 3.

5   Conclusion

We describe an $O(1)$-approximation algorithm with time budget $(1+\epsilon)B$. As the expected reward of the optimal policy at time budget B is at most $O(\epsilon^{-14})$ times the expected reward of our [time $(1+\epsilon)B$] policy, we get an $O(1)$-approximation for any constant $0 < \epsilon < 1$.

References

1. Arkin, E.M., Mitchell, J.S.B., Narasimhan, G.: Resource-constrained geometric network optimization. In: Proceedings of the Fourteenth Annual Symposium on Computational Geometry, pp. 307–316 (1998)
2. Avrim, B., Shuchi, C., Karger David, R., Terran, L., Adam, M., Maria, M.: Approximation algorithms for orienteering and discounted-reward TSP. SIAM J. Comput. **37**(2), 653 (2007)
3. Bansal, N., Nagarajan, V.: On the Adaptivity Gap of Stochastic Orienteering. Springer International Publishing, Cham (2014)
4. Bhalgat, A., Goel, A., Khanna, S.: Improved approximation results for stochastic knapsack problems. In: Proceedings of the Twenty Second Annual ACM SIAM Symposium on Discrete Algorithms, vol. 27(2), pp. 1647–1665 (2011)
5. Bienstock, D., Goemans, M.X., Simchi-Levi, D., Williamson, D.: A note on the prize collecting traveling salesman problem. Math. Program. **59**(3), 413–420 (1993)
6. Chekuri, C., Korula, N., Pal, M.: Improved algorithms for orienteering and related problems. ACM Trans. Algorithms **8**(3), 661–670 (2008)
7. Chen, K., Har-Peled, S.: The orienteering problem in the plane revisited. In: Proceedings of Symposium on Computational Geometry, pp. 247–254 (2006)
8. Golden, B.L., Levy, L., Vohra, R.: The orienteering problem. Nav. Res. Logist. **34**(34), 307–318 (1987)
9. Gupta, A., Krishnaswamy, R., Nagarajan, V., Ravi, R.: Approximation algorithms for stochastic orienteering. In: Proceedings of the Twenty-Third Annual ACM-SIAM Symposium on Discrete Algorithms, pp. 1522–1538 (2012)
10. Li, J., Yuan, W.: Stochastic combinatorial optimization via poisson approximation. In: Proceedings of the Annual ACM Symposium on Theory of Computing, pp. 971–980 (2012)
11. Ross, K.W., Tsang, D.H.K.: Stochastic knapsack problem. IEEE Trans. Commun. **37**(7), 740–747 (1989)
12. Tsiligirides, T.: Heuristic methods applied to orienteering. J. Oper. Res. Soc. **35**(9), 797–809 (1984)
13. Vansteenwegen, P., Souffriau, W., Van Oudheusden, D.: The orienteering problem: a survey. Mitteilungen Klosterneuburg Rebe Und Wein Obstbau Und Fruchteverwertung **209**(209), 1–10 (2011)

Quantum Query Complexity of Unitary Operator Discrimination

Akinori Kawachi[1], Kenichi Kawano[1]($\boxtimes$), François Le Gall[2],
and Suguru Tamaki[2]

[1] Tokushima University, 2-1 Minamijyousanjima, Tokushima 770-8506, Japan
c501637039@tokushima-u.ac.jp
[2] Kyoto University, Yoshida Honmachi, Sakyo-ku, Kyoto 606-8501, Japan

Abstract. Unitary operator discrimination is a fundamental problem in quantum information theory. The basic version of this problem can be described as follows: given a black box implementing a quantum operator U, and the promise that the black box implements either the unitary operator U_1 or the unitary operator U_2, the goal is to decide whether $U = U_1$ or $U = U_2$. In this paper, we consider the query complexity of this problem. We show that there exists a quantum algorithm that solves this problem with bounded-error probability using $\left\lceil \frac{\pi}{3\theta_{\mathrm{cover}}} \right\rceil$ queries to the black-box, where θ_{cover} represents the "closeness" between U_1 and U_2 (this parameter is determined by the eigenvalues of the matrix $U_1^\dagger U_2$). We also show that this upper bound is essentially tight: we prove that there exist operators U_1 and U_2 such that any quantum algorithm solving this problem with bounded-error probability requires at least $\left\lceil \frac{2}{3\theta_{\mathrm{cover}}} \right\rceil$ queries.

Keywords: Quantum algorithms · Quantum information theory · Query complexity

1 Introduction

Background. Quantum state discrimination is one of the most fundamental problems in quantum information theory [3,4,10,14]. In a typical setting, the goal of this problem is to discriminate two quantum states. The success probability of this problem is known to be characterized by the orthogonality of the two states. In particular, non-orthogonal states cannot be discriminated with probability 1, no matter how many independent copies of the input state are given.

A problem closely related to quantum state discrimination is the quantum operator discrimination problem [1,5–9,12,15,16]. Similarly to quantum state discrimination, in a typical setting the goal of quantum operator discrimination is to discriminate two operators: given a black box implementing a quantum operator O, and the promise that the black box implements either operator O_1 or operator O_2, the goal is to decide whether $O = O_1$ or $O = O_2$. A specificity of the quantum operator discrimination problem, which is not present in the quantum state discrimination problem, is that we can choose an arbitrary

© Springer International Publishing AG 2017
Y. Cao and J. Chen (Eds.): COCOON 2017, LNCS 10392, pp. 309–320, 2017.
DOI: 10.1007/978-3-319-62389-4_26

input state on which the operator O is applied. Additionally, the operator O can be applied more than once in various combinations (parallel, sequential, or in any other scheme physically allowed). In contrast to quantum state discrimination, it is known that for several classes of operators discrimination is possible without error [1, 6–9, 12]. For example, any two unitary operators can be discriminated without error by applying the operator in parallel one some well-chosen entangled quantum state [1, 7], or by applying the operator sequentially on a non-entangled state [8]. Besides unitary operators, it is also known that projective measurements can be discriminated without error [12]. Necessary and sufficient conditions for discriminating trace preserving completely positive (TPCP) maps without errors are given in [9] as well.

Many computational problems solved by quantum algorithms can be recast as discrimination problems. Here the quantum operator given as input typically implements a classical operation on the basis of the corresponding Hilbert space and the goal is to (possibly partially) identify which classical operation it implements. Such problems generalize Grover's original quantum search problem [11] and have been studied under the name of the oracle identification problem [2, 13]. For instance, Grover's algorithm for search solves the following problem: given an oracle U_x corresponding to an unknown string $x \in \{0, 1\}^n$ that maps any quantum basis state $|i\rangle |b\rangle$, with $i \in \{1, \ldots, n\}$ and $b \in \{0, 1\}$, to the quantum state $|i\rangle |b \oplus x_i\rangle$, determine if x contains at least one non-zero coordinate. These problems have been mainly considered in the query complexity setting, where the complexity is defined as the number of calls of the operator U_x used by the algorithm. Upper and lower bounds on the query complexity of several oracle identification problems have been obtained [2, 13].

Quantum operator discrimination problems relevant to quantum information theory, on the other hand, have not yet been the subject of much study in the framework of quantum query complexity.

Our Results. In this paper we investigate the query complexity of quantum operator discrimination problem for general quantum unitary operators (i.e., quantum unitary operators not necessarily corresponding to classical operations) in bounded-error settings. More precisely, we consider the following problem: given an unknown unitary operator $U \in \{U_1, U_2\}$, where U is given as a (quantum) black-box and both U_1 and U_2 are known unitary operators, determine whether $U = U_1$ or $U = U_2$ correctly with bounded-error probability.

Our main contribution is a characterization of the query complexity of this problem (i.e., the number of times the black-box U has to be applied to solve the problem) in terms of a parameter θ_{cover}, which is defined formally later (Definition 8 in Sect. 3), representing the "closeness" between U_1 and U_2 by showing a tradeoff between the number of queries and success probability. We show the following upper and lower bounds:

- there exists a quantum algorithm that makes $\left\lceil \frac{\pi}{3\theta_{\mathrm{cover}}} \right\rceil$ queries and can correctly discriminates U_1 from U_2, for any unitary operators U_1 and U_2, with probability $2/3$ (Theorem 4);

– there exist unitary operators U_1 and U_2 such that any quantum algorithm requires at least $\left\lceil \frac{2}{3\theta_{\text{cover}}} \right\rceil$ queries to discriminate U_1 from U_2 with probability 2/3 (Theorem 5).

We thus obtain a tight (up to possible constant factors) characterization of the query complexity of unitary operator discrimination. Our upper bound is actually achieved by a quantum algorithm that makes only non-adaptive queries, i.e., a quantum algorithm in which all queries to the black-box can be made in parallel. On the other hand, our lower bound holds even for adaptive quantum algorithms. Our results thus show that for the quantum unitary operator discrimination problem making adaptive query cannot (significantly) reduce the query complexity of the algorithm. This contrasts with oracle identification problems such as quantum search, where adaptive queries are necessary for achieving the speed-up exhibited by Grover's algorithm.

Relation with Other Works. Note that Acín actually showed the same upper bound on the query complexity of the problem with non-adaptive queries [1]. Acín provided a geometric interpretation for the statistical distinguishability of unitary operators based on sophisticated metrics of Fubini-Study and Bures. D'Ariano et al. briefly noted a simpler geometric interpretation based on the Euclidean metric on the complex plane for the same result [7]. Duan et al. also showed the same upper bound with adaptive queries and a non-entangled input state [8]. Our algorithm of the upper bound uses an entangled input state with non-adaptive queries to analyze a tradeoff between the number of queries and success probability for the upper bound from a geometric interpretation similar to D'Ariano et al.'s. Duan et al. also showed the optimality of their lower bounds to perfectly discrimate two unitary operators [8]. Our proof of the lower bound additionally analyzes the necessary number of queries for imperfect discrimination in details by showing a tradeoff between the number of queries and success probability.

Organization of the Paper. The organization of this paper is as follows. In Sect. 2, we define formally the quantum state discrimination problem, the unitary operator discrimination problem, the error probability of a discriminator, and the query complexity. In Sect. 3, we construct a discriminator for arbitrary unitary operators U_1 and U_2, and give our upper bound on the query complexity by estimating the error probability of this discriminator. Finally, in Sect. 4, we show the lower bound on the query complexity of any quantum algorithm solving the quantum unitary operator discrimination problem.

2   Preliminaries

First, we define the quantum state discrimination problem since the unitary operator discrimination problem can be reduced to the quantum state discrimination problem.

Definition 1 (Quantum state discrimination problem). *The quantum state discrimination problem is defined as the problem of determining whether an unknown state $|\phi\rangle$ is $|\phi_1\rangle$ or $|\phi_2\rangle$, where $|\phi\rangle$ is given from a candidate set $S = \{|\phi_1\rangle, |\phi_2\rangle\}$ of arbitrary two quantum states. Below, we denote the quantum state discrimination problem of $S = \{|\phi_1\rangle, |\phi_2\rangle\}$ by $\mathrm{QSDP}\,(\{|\phi_1\rangle, |\phi_2\rangle\})$. Unless otherwise noted, $|\phi_1\rangle$ and $|\phi_2\rangle$ are given with probability $1/2$.*

Definition 2 (Error probability of $\mathrm{QSDP}\,(\{|\phi_1\rangle, |\phi_2\rangle\})$). *For a quantum state discriminator $\mathcal{A}$, when $|\phi_1\rangle$ and $|\phi_2\rangle$ are given with probability $1/2$, the error probability p_{error} of $\mathcal{A}$ is defined as follows:*

$$p_{\mathrm{error}} = \frac{1}{2}\Pr\left[\mathcal{A}\left(|\phi_1\rangle\right) = \text{``}2\text{''}\right] + \frac{1}{2}\Pr\left[\mathcal{A}\left(|\phi_2\rangle\right) = \text{``}1\text{''}\right].$$

Namely, the Error probability p_{error} is the expected value of the probability that $\mathcal{A}$ mistakes $|\phi_1\rangle$ for $|\phi_2\rangle$ and $|\phi_2\rangle$ for $|\phi_1\rangle$.

The error probability is characterized by the closeness between two quantum states such as the trace distance and the fidelity.

Definition 3 (Trace distance). *Let ρ and σ be pure quantum states. The trace distance $d\,(\rho, \sigma)$ is defined as $d\,(\rho, \sigma) := \frac{1}{2}\,\|\rho - \sigma\|_{\mathrm{tr}}$, where $\|X\|_{\mathrm{tr}} := \mathrm{Tr}\,|X| = \mathrm{Tr}\sqrt{X^\dagger X}$.*

Definition 4 (Fidelity). *Let $\rho = |\psi\rangle\langle\psi|$ and $\sigma = |\phi\rangle\langle\phi|$ be pure quantum states. The fidelity $F\,(\rho, \sigma)$ is defined as $F\,(\rho, \sigma) = |\langle\psi|\phi\rangle|$. Since it is an absolute value of inner product, it holds that $0 \leq F\,(\rho, \sigma) \leq 1$, $F\,(\rho, \sigma) = 1 \Leftrightarrow |\psi\rangle = |\phi\rangle$, and $F\,(\rho, \sigma) = 0 \Leftrightarrow |\psi\rangle \perp |\phi\rangle$.*

Also, the trace distance and the fidelity satisfy the following relation.

Lemma 1. *Let ρ and σ be pure quantum states. We then have $d\,(\rho, \sigma) = \sqrt{1 - F\,(\rho, \sigma)^2}$.*

Using Definitions 3, 4 and Lemma 1, the error probability p_{error} can be characterized as follows.

Theorem 1 ([16]). *Suppose the quantum states $\rho = |\psi\rangle\langle\psi|$ and $\sigma = |\phi\rangle\langle\phi|$ are given from the candidate set $S = \{\rho, \sigma\}$ of pure quantum states with probability $1/2$. Then there exists a discriminator $\mathcal{A}$ such that its error probability p_{error} is given as follows:*

$$p_{\mathrm{error}} = \frac{1}{2}\left(1 - \left\|\frac{1}{2}\rho - \frac{1}{2}\sigma\right\|_{\mathrm{tr}}\right) = \frac{1}{2}\left(1 - d\,(\rho, \sigma)\right) = \frac{1}{2}\left(1 - \sqrt{1 - F\,(\rho, \sigma)^2}\right).$$

The error probability is zero if and only if two quantum states are orthogonal.

Next, we define the unitary operator discrimination problem.

Definition 5 (Unitary operator discrimination problem). *The unitary operator discrimination problem is defined as the problem of determining whether an unknown operator U is U_1 or U_2, where U is given from a candidate set $S = \{U_1, U_2\}$ of arbitrary two unitary operators. Below, we denote the unitary operator discrimination problem of $S = \{U_1, U_2\}$ by $\mathrm{UODP}\,(\{U_1, U_2\})$. Unless otherwise noted, U_1 and U_2 are given with probability $1/2$.*

Definition 6 (Error probability of $\mathrm{UODP}\,(\{U_1, U_2\})$). *For a unitary operator discrimination $\mathcal{A}$, when U_1 and U_2 are given with probability $1/2$, the error probability p_{error}^U of $\mathcal{A}$ is defined as follows:*

$$p_{\mathrm{error}}^U = \frac{1}{2} \Pr\left[\mathcal{A}^{U_1} = \text{``2''}\right] + \frac{1}{2} \Pr\left[\mathcal{A}^{U_2} = \text{``1''}\right].$$

Namely, the error probability p_{error}^U is the expected value of the probability that $\mathcal{A}$ mistakes U_1 for U_2 and U_2 for U_1.

Generally, the unitary operator discrimination problem can be reduced to the quantum state discrimination problem by applying an unknown unitary operator U to an arbitrary $|\phi\rangle$. An application of U to $|\phi\rangle$ is called a *query*. Then, we denote by $\left|\phi_q^U\right\rangle$ the quantum state generated by q queries to U and unitary operators $\{V_i\}_{i=1}^{q}$, which are independent of U (however, possibly depend on $\{U_1, U_2\}$), specified by $\mathcal{A}$. It is given as follows:

$$\left|\phi_q^U\right\rangle = V_q \left(U \otimes \mathbb{I}\right) V_{q-1} \left(U \otimes \mathbb{I}\right) \cdots V_1 \left(U \otimes \mathbb{I}\right) |\phi\rangle.$$

Here, $|\phi\rangle$ and $\left|\phi_q^U\right\rangle$ are quantum states of $m + n$ qubits, V_i is a unitary operator over $m + n$ qubits for each i, and U and $\mathbb{I}$ are unitary operators over n qubits and m qubits, respectively. Let $\hat{U}$ is a sequence of unitary operators of $\mathcal{A}$. Then we have $\left|\phi_q^U\right\rangle = \hat{U} |\phi\rangle$. Also, we define $\hat{U}_1$ and $\hat{U}_2$ as unitary operators satisfying $\left|\phi_q^{U_1}\right\rangle = \hat{U}_1 |\phi\rangle$ and $\left|\phi_q^{U_2}\right\rangle = \hat{U}_2 |\phi\rangle$, respectively.

The unitary operator discriminator that makes all the queries at once (regardless of the answers to other queries) is called a *non-adaptive* unitary operator discriminator. Otherwise it is called an *adaptive* unitary operator discriminator. The error probability of a unitary operator discriminator is given as follows.

Theorem 2. *For unitary operators $\hat{U}_1$ and $\hat{U}_2$, let $U' = V^\dagger \hat{U}_1^\dagger \hat{U}_2 V$, where U' is a diagonal matrix, and let $|\phi'\rangle = V^\dagger |\phi\rangle$. When the probability given U_1 and U_2 from the candidate set $S = \{U_1, U_2\}$ is each probability $1/2$, the error probability of U_1 and U_2 is given as follows:*

$$p_{\mathrm{error}}^U = \frac{1}{2}\left(1 - \sqrt{1 - \min_{|\phi\rangle}\left|\langle\phi|\,\hat{U}_1^\dagger \hat{U}_2\,|\phi\rangle\right|^2}\right) = \frac{1}{2}\left(1 - \sqrt{1 - \min_{|\phi'\rangle}|\langle\phi'|\,U'\,|\phi'\rangle|^2}\right).$$

This is shown immediately from Theorem 1.

Definition 7. *We say a discriminator $\mathcal{A}$ makes q queries if $\mathcal{A}$ applies a unitary operator U q times to the initial state $|\phi\rangle$ in total. We say $\mathcal{A}$ solves*

UODP $(\{U_1, U_2\})$ *with* q *queries if* $\mathcal{A}$ *correctly discriminates unitary opera-tors* U_1 *and* U_2 *with probability at least* $2/3$ *with* q *queries. The query com-plexity of* UODP $(\{U_1, U_2\})$ *is the minimum number of queries where* $\mathcal{A}$ *solves* UODP $(\{U_1, U_2\})$.

3   Upper Bounds on Query Complexity

Let U_1 and U_2 be unitary operators acting on n qubits. Below we give a construc-tion of a non-adaptive q-query discriminator for UODP $(\{U_1, U_2\})$; see Fig. 1 for its schematic description.

Construction 3 (Non-adaptive q-query discriminator $\mathcal{A}$ for UODP $(\{U_1, U_2\})$)

1. Generate an initial nq-bit quantum state $|\phi\rangle$ that is determined by U_1, U_2 and q as in the proof of Lemma 2.
2. Apply $U^{\otimes q}$ to $|\phi\rangle$, where $U \in \{U_1, U_2\}$ is the unknown unitary operator.
3. Apply the quantum state discriminator for QSDP $(\{U_1^{\otimes q} |\phi\rangle, U_2^{\otimes q} |\phi\rangle\})$, due to Theorem 1, to the quantum state $U^{\otimes q} |\phi\rangle$.
4. Output the result of the quantum state discriminator.

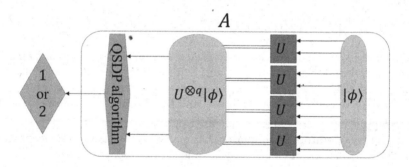

Fig. 1. Non-adaptive unitary operator discriminator $\mathcal{A}$

To analyze the error probability of the non-adaptive q-query discriminator $\mathcal{A}$, let us introduce the notion of covering angle.

Definition 8 (Covering angle θ_{cover}). *For real numbers* α *and* $0 \le \beta < 2\pi$, *let* $D_{\alpha,\beta} := \{z \in \mathbb{C} : \alpha \le \arg z \le \alpha + \beta\}$. *Let* $e^{i\theta_1}, \ldots, e^{i\theta_n}$ *be complex numbers of absolute value* 1, *where* $0 \le \theta_i < 2\pi, i = 1, \ldots n$. *Then, the* covering angle *of the set* $\{e^{i\theta_1}, \ldots, e^{i\theta_n}\}$, *denoted by* θ_{cover}, *is defined as*

$$\theta_{\text{cover}} := \min\{\beta : \exists \alpha \text{ s.t. } e^{i\theta_1}, \ldots, e^{i\theta_n} \in D_{\alpha,\beta}\}.$$

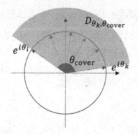

Fig. 2. Covering angle θ_{cover}

The covering angle θ_{cover} is *formed* by $e^{i\theta_k}$ and $e^{i\theta_l}$ if $\theta_l = \theta_k + \theta_{\text{cover}}$ and $e^{i\theta_1}, \ldots, e^{i\theta_n} \in D_{\theta_k, \theta_{\text{cover}}}$ hold. See Fig. 2 for an illustrated example of a covering angle.

In the rest of this section, we assume that the set of eigenvalues of the unitary operator $U_1^{\dagger} U_2$ is $\{e^{i\theta_1}, \ldots, e^{i\theta_n}\}$ and the covering angle of it is θ_{cover}. The following lemma is the main technical result of this section.

Lemma 2. *The error probability p_{error}^U of the non-adaptive q-query discriminator $\mathcal{A}$ is given as follows:*

$$
p_{\text{error}}^U = \begin{cases} \dfrac{1}{2}\left(1 - \sin\dfrac{q\theta_{\text{cover}}}{2}\right) & (0 \le q\theta_{\text{cover}} < \pi), \\ 0 & (\pi \le q\theta_{\text{cover}}). \end{cases}
$$

The lemma immediately implies the following:

Theorem 4. *The non-adaptive q-query discriminator $\mathcal{A}$ solves* UODP $(\{U_1, U_2\})$ *if $q \ge \left\lceil \frac{\pi}{3\theta_{\text{cover}}} \right\rceil$. Furthermore, the error probability of $\mathcal{A}$ is zero if $q \ge \left\lceil \frac{\pi}{\theta_{\text{cover}}} \right\rceil$.*

Proof (of Theorem 4). If $q \ge \left\lceil \frac{\pi}{\theta_{\text{cover}}} \right\rceil$, then $\pi \le q\theta_{\text{cover}}$ holds and the error probability of $\mathcal{A}$ is zero by Lemma 2. For $0 \le q\theta_{\text{cover}} < \pi$, again by Lemma 2, it suffices to find q such that

$$
p_{\text{error}}^U = \frac{1}{2}\left(1 - \sin\frac{q\theta_{\text{cover}}}{2}\right) \le \frac{1}{3} \tag{1}
$$

holds according to Definition 7. By calculation, we have that the inequality in (1) holds if $q \ge \left\lceil \frac{\pi}{3\theta_{\text{cover}}} \right\rceil$. $\qquad\square$

It remains to prove Lemma 2. It is instructive to first analyze a special case of $q = 1$ as it captures the essence of general cases.

Lemma 3. *The error probability of the non-adaptive 1-query discriminator $\mathcal{A}$ is given as follows:*

$$
p_{\text{error}}^U = \begin{cases} \dfrac{1}{2}\left(1 - \sin\dfrac{\theta_{\text{cover}}}{2}\right) & (0 \le \theta_{\text{cover}} < \pi), \\ 0 & (\pi \le \theta_{\text{cover}} \le 2\pi). \end{cases}
$$

Proof (of Lemma 3). We consider two cases according to the value of θ_{cover}.

Case (i) $0 \leq \theta_{\mathrm{cover}} < \pi$.
Let $U' = \mathrm{diag}\left(e^{i\theta_1}, \ldots, e^{i\theta_n}\right)$ and $|\phi'\rangle = (\alpha_1, \ldots, \alpha_n)$. By Theorem 2, the minimum value of the square of the fidelity of $U_1 |\phi\rangle$ and $U_2 |\phi\rangle$ is represented as follows:

$$\min_{|\phi\rangle} \left| \langle \phi | U_1^\dagger U_2 |\phi\rangle \right|^2 = \min_{|\phi'\rangle} |\langle \phi' | U' |\phi'\rangle|^2 = \min_{\sum_{j=1}^n |\alpha_j|^2 = 1} \left| \sum_{j=1}^n |\alpha_j|^2 e^{i\theta_j} \right|^2.$$

The last term above is equal to the square of the shortest distance from the origin of the complex plane to the convex set $C := \left\{ \sum_{j=1}^n |\alpha_j|^2 e^{i\theta_j} : \sum_{j=1}^n |\alpha_j|^2 = 1 \right\}$. The shortest distance from the origin of the complex plane to C is equal to the shortest distance from the origin of the complex plane to the line segment $C' := \left\{ |\alpha_k|^2 e^{i\theta_k} + |\alpha_l|^2 e^{i\theta_l} : |\alpha_k|^2 + |\alpha_l|^2 = 1 \right\}$, where $e^{i\theta_k}$ and $e^{i\theta_l}$ form the covering angle θ_{cover}, see Fig. 3 for illustration. Thus, we have

$$\min_{\sum_{j=1}^n |\alpha_j|^2 = 1} \left| \sum_{j=1}^n |\alpha_j|^2 e^{i\theta_j} \right|^2 = \min_{|\alpha_k|^2 + |\alpha_l|^2 = 1} \left| |\alpha_k|^2 e^{i\theta_k} + |\alpha_l|^2 e^{i\theta_l} \right|^2.$$

The minimum of the right hand side above is achieved by setting $|\alpha_k|^2 = \frac{1}{2}$, $|\alpha_l|^2 = \frac{1}{2}$. Hence

$$\min_{|\alpha_k|^2 + |\alpha_l|^2 = 1} \left| |\alpha_k|^2 e^{i\theta_k} + |\alpha_l|^2 e^{i\theta_l} \right|^2 = \left(\frac{1}{2} \left| e^{i\theta_k} + e^{i\theta_l} \right| \right)^2 = \cos^2 \frac{\theta_{\mathrm{cover}}}{2}.$$

Thus, p_{error}^U of $\mathcal{A}$ is represented as follows:

$$p_{\mathrm{error}}^U = \frac{1}{2} \left(1 - \sqrt{1 - \cos^2 \frac{\theta_{\mathrm{cover}}}{2}} \right) = \frac{1}{2} \left(1 - \sin \frac{\theta_{\mathrm{cover}}}{2} \right).$$

Case (ii) $\pi \leq \theta_{\mathrm{cover}} \leq 2\pi$.
The error probability can be calculated in the same way as Case (i). In this case, the convex set C contains the origin of the complex plane, i.e., the shortest distance from the origin to C is zero, hence we have $p_{\mathrm{error}}^U = 0$.   □

We are prepared to prove Lemma 2.

Proof (of Lemma 2)

Case (i) $0 \leq q\theta_{\mathrm{cover}} \leq \pi$.
Let $U' = \mathrm{diag}\left(e^{i\theta_1}, \ldots, e^{i\theta_n}\right)$. Then the set of eigenvalues of $U'^{\otimes q}$ is

$$\Lambda = \left\{ \prod_{j=1}^q e^{i\theta_{k_j}} : 1 \leq k_1, \ldots, k_q \leq n \right\}.$$

Note that if the covering angle θ_{cover} of $\{e^{i\theta_1}, \ldots, e^{i\theta_n}\}$ is formed by $e^{i\theta_k}$ and $e^{i\theta_l}$, then the covering angle of Λ is $q\theta_{\text{cover}}$ and formed by $e^{iq\theta_k}$ and $e^{iq\theta_l}$, as long as q is not too large.

As the proof of Lemma 3, the minimum value of the square of the fidelity of $U_1^{\otimes q} |\phi\rangle$ and $U_2^{\otimes q} |\phi\rangle$ is

$$\min_{|\phi\rangle} \left| \langle \phi | U_1^{\otimes q\dagger} U_2^{\otimes q} |\phi\rangle \right|^2 = \min_{|\phi'\rangle} \left| \langle \phi' | U'^{\otimes q} |\phi'\rangle \right|^2$$

$$= \min_{\sum_{j=1}^{qn} |\alpha_j|^2 = 1} \left| |\alpha_1|^2 e^{iq\theta_1} + |\alpha_2|^2 e^{i((q-1)\theta_1 + \theta_2)} + \cdots + |\alpha_{qn}|^2 e^{iq\theta_n} \right|^2.$$

This is the square of the shortest distance from the origin of the complex plane to the convex set

$$C_q := \left\{ |\alpha_1|^2 e^{iq\theta_1} + |\alpha_2|^2 e^{i((q-1)\theta_1 + \theta_2)} + \cdots + |\alpha_{qn}|^2 e^{iq\theta_{\tilde{n}}} : \sum_{j=1}^{qn} |\alpha_j| = 1 \right\}.$$

The shortest distance from the origin of the complex plane to C_q is equal to the line segment

$$C_q' := \left\{ |\alpha_x|^2 e^{iq\theta_k} + |\alpha_y|^2 e^{iq\theta_l} : |\alpha_x|^2 + |\alpha_y|^2 = 1 \right\},$$

see Fig. 4 for illustration. Hence, in the same way as the proof of Lemma 3, we have

$$p_{\text{error}}^U = \frac{1}{2} \left(1 - \sin \frac{q\theta_{\text{cover}}}{2} \right) \quad (0 \le q\theta_{\text{cover}} \le \pi).$$

Case (ii) $\pi \le q\theta_{\text{cover}}$.
Since the convex set C_q contains the origin of the complex plane, the error probability of $\mathcal{A}$ can be made zero as the case of $q = 1$. $\square$

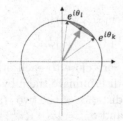

Fig. 3. The shortest distance to the convex set C

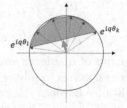

Fig. 4. The shortest distance to the convex set C_q

4   Lower Bounds of Query Complexity

Every unitary operator discriminator can be represented as an adaptive discriminator given in Fig. 5. We now evaluate the necessary number q of queries to solve UODP $(\{U_1, U_2\})$ for some U_1, U_2, where $\hat{U}$ in Fig. 5 is constructed as

$$\hat{U} := V_q \left(U \otimes \mathbb{I}\right) V_{q-1} \left(U \otimes \mathbb{I}\right) \cdots V_1 \left(U \otimes \mathbb{I}\right)$$

from a given $U \in \{U_1, U_2\}$ and fixed unitary operators V_i $(i = 1, 2, \ldots, q)$.

The following theorem shows the necessary number of queries for some specific U_1, U_2.

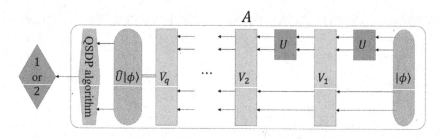

Fig. 5. The arbitrary adaptive discriminator

Theorem 5. *For every θ $(0 < \theta < \pi)$, we define the unitary operators U_1 and U_2 as follows:*

$$U_1 = \begin{pmatrix} 1 & & & 0 \\ & 1 & & \\ & & \ddots & \\ 0 & & & 1 \end{pmatrix}, \quad U_2 = \begin{pmatrix} e^{i\theta} & & & 0 \\ & 1 & & \\ & & \ddots & \\ 0 & & & 1 \end{pmatrix}.$$

If $\mathcal{A}$ solves UODP $(\{U_1, U_2\})$ with adaptive q queries, $q \geq \left\lceil \dfrac{2}{3\theta} \right\rceil$ holds.

Proof. Let $|\phi\rangle$ be any initial state. Applying $\hat{U}$ to the initial state, we obtain the following states for U_1 and U_2 respectively:

$$\left|\phi_q^{U_1}\right\rangle = V_q \left(U_1 \otimes \mathbb{I}\right) V_{q-1} \left(U_1 \otimes \mathbb{I}\right) \cdots V_1 \left(U_1 \otimes \mathbb{I}\right) |\phi\rangle$$
$$\left|\phi_q^{U_2}\right\rangle = V_q \left(U_2 \otimes \mathbb{I}\right) V_{q-1} \left(U_2 \otimes \mathbb{I}\right) \cdots V_1 \left(U_2 \otimes \mathbb{I}\right) |\phi\rangle.$$

If the number of queries is small, $\left|\phi_q^{U_1}\right\rangle$ and $\left|\phi_q^{U_2}\right\rangle$ are close to each other. We estimate the necessary number q of queries of $\mathcal{A}$ to discriminate them by analyzing the distance between the two states. The trace distance of the two states can be given as $\|\rho_q - \sigma_q\|_{\mathrm{tr}}$, where $\rho_q := \left|\phi_q^{U_2}\right\rangle\left\langle\phi_q^{U_2}\right|$, and $\sigma_q := \left|\phi_q^{U_1}\right\rangle\left\langle\phi_q^{U_1}\right|$. Then, we have

$$\|\rho_q - \sigma_q\|_{\mathrm{tr}} = \left\| V_q \left(U_2 \otimes \mathbb{I}\right) \rho_{q-1} \left(U_2 \otimes \mathbb{I}\right)^\dagger V_q^\dagger - V_q \left(U_1 \otimes \mathbb{I}\right) \sigma_{q-1} \left(U_1 \otimes \mathbb{I}\right)^\dagger V_q^\dagger \right\|_{\mathrm{tr}}$$
$$= \left\| V_q \left(U_2 \otimes \mathbb{I}\right) \rho_{q-1} \left(U_2 \otimes \mathbb{I}\right)^\dagger V_q^\dagger - V_q \sigma_{q-1} V_q^\dagger \right\|_{\mathrm{tr}}.$$

Since unitary operators do not change the trace distance, we have

$$\left\| V_q \left(U_2 \otimes \mathbb{I} \right) \rho_{q-1} \left(U_2 \otimes \mathbb{I} \right)^\dagger V_q^\dagger - V_q \sigma_{q-1} V_q^\dagger \right\|_{\mathrm{tr}} = \left\| \left(U_2 \otimes \mathbb{I} \right) \rho_{q-1} \left(U_2 \otimes \mathbb{I} \right)^\dagger - \sigma_{q-1} \right\|_{\mathrm{tr}}.$$

By the triangle inequality,

$$\left\| \left(U_2 \otimes \mathbb{I} \right) \rho_{q-1} \left(U_2 \otimes \mathbb{I} \right)^\dagger - \sigma_{q-1} \right\|_{\mathrm{tr}}$$
$$= \left\| \left(U_2 \otimes \mathbb{I} \right) \rho_{q-1} \left(U_2 \otimes \mathbb{I} \right)^\dagger - \left(U_2 \otimes \mathbb{I} \right) \sigma_{q-1} \left(U_2 \otimes \mathbb{I} \right)^\dagger + \left(U_2 \otimes \mathbb{I} \right) \sigma_{q-1} \left(U_2 \otimes \mathbb{I} \right)^\dagger - \sigma_{q-1} \right\|_{\mathrm{tr}}$$
$$\leq \left\| \left(U_2 \otimes \mathbb{I} \right) \rho_{q-1} \left(U_2 \otimes \mathbb{I} \right)^\dagger - \left(U_2 \otimes \mathbb{I} \right) \sigma_{q-1} \left(U_2 \otimes \mathbb{I} \right)^\dagger \right\|_{\mathrm{tr}}$$
$$+ \left\| \left(U_2 \otimes \mathbb{I} \right) \sigma_{q-1} \left(U_2 \otimes \mathbb{I} \right)^\dagger - \sigma_{q-1} \right\|_{\mathrm{tr}}$$
$$= \left\| \rho_{q-1} - \sigma_{q-1} \right\|_{\mathrm{tr}} + \left\| \left(U_2 \otimes \mathbb{I} \right) \sigma_{q-1} \left(U_2 \otimes \mathbb{I} \right)^\dagger - \sigma_{q-1} \right\|_{\mathrm{tr}}.$$

By Definitions 3, 4 and Lemma 1, $\left\| \left(U_2 \otimes \mathbb{I} \right) \sigma_{q-1} \left(U_2 \otimes \mathbb{I} \right)^\dagger - \sigma_{q-1} \right\|_{\mathrm{tr}}$ is:

$$\left\| \left(U_2 \otimes \mathbb{I} \right) \sigma_{q-1} \left(U_2 \otimes \mathbb{I} \right)^\dagger - \sigma_{q-1} \right\|_{\mathrm{tr}} = 2d \left(\left(U_2 \otimes \mathbb{I} \right) \sigma_{q-1} \left(U_2 \otimes \mathbb{I} \right)^\dagger, \sigma_{q-1} \right)$$
$$= 2\sqrt{ 1 - F \left(\left(U_2 \otimes \mathbb{I} \right) \sigma_{q-1} \left(U_2 \otimes \mathbb{I} \right)^\dagger, \sigma_{q-1} \right)^2 }$$
$$= 2\sqrt{ 1 - \left| \left\langle \phi_{q-1}^{U_1} \right| \left(U_2 \otimes \mathbb{I} \right) \left| \phi_{q-1}^{U_1} \right\rangle \right|^2 }$$
$$\leq 2\sqrt{ 1 - \min_{\left| \phi_{q-1}^{U_1} \right\rangle} \left| \left\langle \phi_{q-1}^{U_1} \right| \left(U_2 \otimes \mathbb{I} \right) \left| \phi_{q-1}^{U_1} \right\rangle \right|^2 }$$
$$= 2\sqrt{ 1 - \cos^2 \frac{\theta}{2} } = 2 \sin \frac{\theta}{2}.$$

Since $\sin \theta \leq \theta$, we have $2 \sin \frac{\theta}{2} \leq \theta$ from the above equations. Therefore,

$$\left\| \rho_q - \sigma_q \right\|_{\mathrm{tr}} \leq \left\| \rho_{q-1} - \sigma_{q-1} \right\|_{\mathrm{tr}} + \theta.$$

If the number of queries is zero, we have $\left\| \rho_0 - \sigma_0 \right\|_{\mathrm{tr}} = 0$. After q queries, the distance between the two states is at most $q\theta$. If the number of queries is q, the error probability of $\mathcal{A}$ is at least $\frac{1}{2} \left(1 - \frac{q\theta}{2} \right)$ from Theorem 1. Thus, from $\frac{1}{2} \left(1 - \frac{q\theta}{2} \right) < \frac{1}{3}$, we have $q > \frac{2}{3\theta}$. Since the number of queries is an integer, $q \geq \left\lceil \frac{2}{3\theta} \right\rceil$ holds. $\qquad\square$

Acknowledgments. AK was partially supported by MEXT KAKENHI (24106009) and JSPS KAKENHI (16H01705, 17K12640). ST was supported in part by MEXT KAKENHI (24106003) and JSPS KAKENHI (26330011, 16H02782). FLG was partially supported by MEXT KAKENHI (24106009) and JSPS KAKENHI (16H01705, 16H05853). The authors are grateful to Akihito Soeda for helpful discussions.

References

1. Acín, A.: Statistical distinguishability between unitary operations. Phys. Rev. Lett. **87**(17), 177901 (2001)
2. Ambainis, A., Iwama, K., Kawachi, A., Raymond, R., Yamashita, S.: Improved algorithms for quantum identification of boolean oracles. Theoret. Comput. Sci. **378**(1), 41–53 (2007)
3. Audenaert, K.M.R., Calsamiglia, J., Masanes, L., Muñoz-Tapia, R., Acín, A., Bagan, E., Verstraete, F.: The quantum Chernoff bound. Phys. Rev. Lett. **98**(16), 160501 (2007)
4. Chefles, A.: Unambiguous discrimination between linearly independent quantum states. Phys. Lett. A **239**(6), 339–347 (1998)
5. Chefles, A., Kitagawa, A., Takeoka, M., Sasaki, M., Twamley, J.: Unambiguous discrimination among oracle operators. J. Phys. A: Math. Theor. **40**(33), 10183 (2007)
6. Childs, A.M., Preskill, J., Renes, J.: Quantum information and precision measurement. J. Mod. Opt. **47**, 155–176 (2000)
7. D'Ariano, G.M., Lo Presti, P., Paris, M.G.A.: Using entanglement improves the precision of quantum measurements. Phys. Rev. Lett. **87**(27), 270404 (2001)
8. Duan, R., Feng, Y., Ying, M.: Entanglement is not necessary for perfect discrimination between unitary operations. Phys. Rev. Lett. **98**, 100503 (2007)
9. Duan, R., Feng, Y., Ying, M.: The perfect distinguishability of quantum operations. Phys. Rev. Lett. **103**, 210501 (2009)
10. Feng, Y., Duan, R., Ying, M.: Unambiguous discrimination between quantum mixed states. Phys. Rev. A **70**, 012308 (2004)
11. Grover, L.K.: Quantum mechanics helps in searching for a needle in a haystack. Phys. Rev. Lett. **79**, 325 (1997)
12. Ji, Z., Feng, Y., Duan, R., Ying, M.: Identification and distance measures of measurement apparatus. Phys. Rev. Lett. **96**, 200401 (2006)
13. Kothari, R.: An optimal quantum algorithm for the oracle identification problem. Proceedings of the 31st International Symposium on Theoretical Aspects of Computer Science (STACS 2014), Leibniz International Proceedings in Informatics, vol. 25, pp. 482–493 (2014)
14. Mochon, C.: Family of generalized "pretty good" measurements and the minimal-error pure-state discrimination problems for which they are optimal. Phys. Rev. A **73**, 012308 (2006)
15. Sacchi, M.F.: Optimal discrimination of quantum operations. Phys. Rev. A **71**, 062340 (2005)
16. Ziman, M., Sedlák, M.: Single-shot discrimination of quantum unitary processes. J. Mod. Opt. **57**(3), 253–259 (2010)

Randomized Incremental Construction for the Hausdorff Voronoi Diagram Revisited and Extended

Elena Khramtcova[1] and Evanthia Papadopoulou[2]($\boxtimes$)

[1] Computer Science Department,
Université Libre de Bruxelles (ULB), Brussels, Belgium
[2] Faculty of Informatics,
Università Della Svizzera Italiana (USI), Lugano, Switzerland
`evanthia.papadopoulou@usi.ch`

Abstract. The Hausdorff Voronoi diagram of clusters of points in the plane is a generalization of Voronoi diagrams based on the Hausdorff distance function. Its combinatorial complexity is $O(n + m)$, where n is the total number of points and m is the number of *crossings* between the input clusters ($m = O(n^2)$); the number of clusters is k. We present efficient algorithms to construct this diagram via the randomized incremental construction (RIC) framework [Clarkson et al. 89,93]. For *non-crossing* clusters ($m = 0$), our algorithm runs in expected $O(n \log n + k \log n \log k)$ time and deterministic $O(n)$ space. For arbitrary clusters the algorithm runs in expected $O((m + n \log k) \log n)$ time and $O(m + n \log k)$ space. The two algorithms can be combined in a crossing-oblivious scheme within the same bounds. We show how to apply the RIC framework efficiently to handle non-standard characteristics of generalized Voronoi diagrams, including sites (and bisectors) of non-constant complexity, sites that are not enclosed in their Voronoi regions, empty Voronoi regions, and finally, disconnected bisectors and Voronoi regions. The diagram finds direct applications in VLSI CAD.

1 Introduction

The Voronoi diagram is a powerful geometric partitioning structure [2] with diverse applications in science and engineering. We consider the *Hausdorff Voronoi diagram* of *clusters of points* in the plane, a generalization of Voronoi diagrams based on the Hausdorff distance function, which finds applications in predicting (and evaluating) faults in VLSI design and other geometric networks.

Research supported in part by the Swiss National Science Foundation, projects SNF 20GG21-134355 (ESF EUROCORES EuroGIGA/VORONOI) and SNF 200021E-154387. E. K. was also supported partially by F.R.S.-FNRS and the SNF P2TIP2-168563 under the Early PostDoc Mobility program.
Research performed mainly while E. K. was at the Università della Svizzera italiana (USI).

© Springer International Publishing AG 2017
Y. Cao and J. Chen (Eds.): COCOON 2017, LNCS 10392, pp. 321–332, 2017.
DOI: 10.1007/978-3-319-62389-4_27

Given a family F of clusters of points in the plane, the *Hausdorff Voronoi diagram* of F is a subdivision of the plane into maximal regions such that all points within one region have the same *nearest cluster* (see Fig. 1a). Distance between a point $t \in \mathbb{R}^2$ and a cluster $P \in F$ is measured as their Hausdorff distance, which equals the *farthest distance* between t and P, $\mathsf{d_f}(t, P) = \max_{p \in P} d(t, p)$, where $d(\cdot, \cdot)$ denotes the Euclidean distance[1] between two points in the plane. The total number of points is n and the number of clusters is k. No two clusters in F share a common point.

Informally, the Hausdorff Voronoi diagram is a *min-max* type of diagram. The opposite *max-min* type, where distance is minimum and the diagram is farthest, is also of interest, see e.g., [1,7,13]. Recently both types of diagrams were combined to determine *stabbing circles* for sets of segments in the plane [10]. We remark that the Hausdorff diagram is different in nature from the *clustering induced Voronoi diagram* by Chen et al. [6], where sites can be all subsets of points and the *influence function* reflects a collective effect of all points in a site.

The Hausdorff Voronoi diagram finds direct applications in Very Large Scale Integration (VLSI) circuit design. It can be used to model the location of defects falling over parts of a network that have been embedded in the plane, destroying its connectivity. It has been extensively used by the semiconductor industry to estimate the *critical area* of a VLSI layout for various types of *open faults*, see e.g., [4,16]. Critical area is a measure reflecting the sensitivity of a design to random manufacturing defects. The diagram can find applications in geometric networks embedded in the plane, such as transportation networks, where critical area may need to be extracted for the purpose of flow control and disaster avoidance.

Previous work. The Hausdorff Voronoi diagram was first considered by Edelsbrunner et al. [12] under the name *cluster Voronoi diagram*. The authors showed that its combinatorial complexity is $O(n^2\alpha(n))$, and gave a divide and conquer construction algorithm of the same time complexity, where $\alpha(n)$ is the inverse Ackermann function. These bounds were later improved to $O(n^2)$ by Papadopoulou and Lee [18]. When the convex hulls of the clusters are disjoint [12] or *non-crossing* [18] (see Definition 1), the combinatorial complexity of the diagram is $O(n)$. The $O(n^2)$-time algorithm of Edelsbrunner et al. exploits the equivalence of the Hausdorff diagram of k clusters to the upper envelope of a family of k lower envelopes in an arrangement of planes in $\mathbb{R}^3$, and it is optimal in the worst case. However, it remains quadratic even if the diagram has complexity $O(n)$.

Definition 1. *Two clusters P and Q are called* non-crossing, *if the convex hull of $P \cup Q$ admits at most two supporting segments with one endpoint in P and one endpoint in Q. In the convex hull of $P \cup Q$ admits more than two such supporting segments, then P and Q are called* crossing *(see Fig. 1b).*

---

[1] Other metrics, such as the L_p metric, are also possible.

The combinatorial complexity of the Hausdorff Voronoi diagram is $O(n+m)$, where m is the number of *crossings* between pairs of crossing clusters (see Definition 2), and this is tight [17]. The number of crossings m is upper-bounded by the number of supporting segments between pairs of crossing clusters. In the worst case, m is $O(n^2)$. Computing the Hausdorff Voronoi diagram in sub-quadratic time when m is $O(n)$ (even if $m = 0$) has not been an easy task. For non-crossing clusters $(m = 0)$, the Hausdorff Voronoi diagram is an instance of *abstract Voronoi diagrams* [14]. But a bisector can have complexity $\Theta(n)$, thus, if we directly apply the randomized incremental construction for abstract Voronoi diagrams [15] we get an $O(n^2 \log n)$-time algorithm, and this is not easy to overcome (see [5]). When clusters are crossing, their bisectors are disconnected curves [18], and thus, they do not satisfy the basic axioms of abstract diagrams.

For non-crossing clusters, Dehne et al. [11] gave the first subquadratic-time algorithm to compute the Hausdorff diagram, in time $O(n \log^4 n)$ and space $O(n \log^2 n)$. Recently, we gave a randomized incremental algorithm as based on *point location* in a *hierarchical dynamic data structure* [5]. The expected running time of this algorithm is $O(n \log n \log k)$ and the expected space complexity is $O(n)$ [5]. However, it does not easily generalize to crossing clusters.

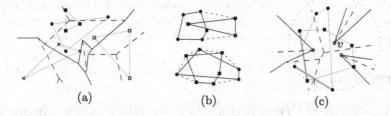

(a)   (b)   (c)

Fig. 1. (a) The Hausdorff Voronoi diagram of a family of four clusters, where each cluster has three points. (b) A pair of clusters. Above: clusters are non-crossing; below: clusters are crossing. (c) Two crossing clusters P and Q (filled disks and squares, resp.), and their Hausdorff Voronoi diagram (black lines). The region of P is disconnected into three faces. A crossing mixed vertex v, the circle passing through the three points that induce v (dotted lines), and two diagonals of P and Q related to v (bold, grey).

Our Contribution. In this paper we revisit the randomized incremental construction for the Hausdorff diagram obtaining three new results that complete our investigation on efficient construction algorithms for this diagram, especially in the setting driven by our application, where crossings may be present but their number is expected to be small (typically, $m = O(n)$). We follow the randomized incremental construction (RIC) framework introduced by Clarkson et al. [8,9]. We show how to efficiently apply this framework to construct a generalized Voronoi diagram in the presence of several non-standard features: (1) bisectors can each have complexity $\Theta(n)$; (2) sites need not be enclosed in their Voronoi regions; (3) Voronoi regions can be empty; and (4) bisector curves may be disconnected. Note that a direct application of the framework would yield an $O(n^2 \log n)$ (or $O(kn \log n)$) algorithm, even for a diagram of complexity $O(n)$.

First, we consider non-crossing clusters, when the complexity of the diagram is $O(n)$. Our algorithm runs in expected $O(n \log n + k \log n \log k)$ time and deterministic $O(n)$ space. In comparison to [5], the construction is considerably simpler and it slightly improves its time complexity; more importantly, it extends to arbitrary clusters, unlike [5]. We give the construction for both a *conflict* and a *history graph*, where the latter is an on-line variation of the algorithm.

Then, we consider arbitrary clusters of points. Allowing clusters to cross adds an entire new challenge to the construction algorithm: a bisector between two clusters consists of multiple components, and one Voronoi region may disconnect in several faces. We show how to overcome this challenge on a conflict graph and derive an algorithm whose expected time and space requirements are respectively $O(m \log n + n \log k \log n)$ and $O(m + n \log k)$. To the best of our knowledge, this is the first time the RIC framework is applied to the construction of a Voronoi diagram with disconnected bisectors and disconnected regions.

Finally, we address the question of deciding which algorithm to use on a given input, without a prior knowledge on the existence of crossings. Deciding whether the input clusters are crossing or not, may require quadratic time by itself, because the convex hulls of the input clusters may have quadratic intersections, even if the clusters are non-crossing. In Sect. 5, we show how to only detect crossings which are relevant to the construction of the Hausdorff Voronoi diagram, and thus, provide a crossing-oblivious algorithm that combines our two previous algorithms, while keeping intact the time complexity bounds.

2   Preliminaries

Let F be a family of k clusters of points in the plane, and let n be the total number of points in F. No two clusters share a point. We assume that each cluster equals the vertices on its convex hull, as only points on a convex hull may have non-empty regions in the Hausdorff diagram. For simplicity of presentation, we follow a general position assumption that no four points lie on the same circle.

The *farthest Voronoi diagram* of a cluster C, for brevity $\mathsf{FVD}(C)$, is a partitioning of the plane into regions, where the *farthest Voronoi region* of a point $c \in C$ is $\mathsf{freg}_C(c) = \{t \mid \forall c' \in C \setminus \{c\} \colon d(t, c) > d(t, c')\}$. Let $\mathcal{T}(C)$ denote the graph structure of $\mathsf{FVD}(C)$, $\mathcal{T}(C) = \mathbb{R}^2 \setminus \bigcup_{c \in C} \mathsf{freg}_C(c)$. $\mathcal{T}(C)$ is well known to be a tree. If $C = \{c\}$, let $\mathcal{T}(C) = c$. If $|C| > 1$, we assume that $\mathcal{T}(C)$ is rooted at a point at infinity on an unbounded Voronoi edge.

The *Hausdorff Voronoi diagram*, for brevity $\mathsf{HVD}(F)$, is a partitioning of the plane into regions, where the *Hausdorff Voronoi region* of a cluster $C \in F$ is $\mathsf{hreg}_F(C) = \{p \mid \forall C' \in F \setminus \{C\} \colon \mathsf{d_f}(p, C) < \mathsf{d_f}(p, C')\}$. Region $\mathsf{hreg}_F(C)$ is further subdivided into subregions by $\mathsf{FVD}(C)$. In particular, the *Hausdorff Voronoi region* of a point $c \in C$ is $\mathsf{hreg}_F(c) = \mathsf{hreg}_F(C) \cap \mathsf{freg}_C(c)$. Fig. 1a illustrates the Hausdorff Voronoi diagram of a family of four clusters. The convex hulls of the clusters are shown in grey lines. Solid black lines indicate the *Hausdorff Voronoi edges* bounding the Voronoi regions of individual clusters; the dashed lines indicate the finer subdivision by the farthest Voronoi diagram of each cluster.

The Hausdorff Voronoi edges are portions of *Hausdorff bisectors* between pairs of clusters; the bisector of two clusters $P, Q \in F$ is $b_h(P, Q) = \{y \mid d_f(y, P) = d_f(y, Q)\}$. See the solid black lines in Fig. 1c. The Hausdorff bisector is a subgraph of $T(P \cup Q)$; it consists of one (if P, Q are non-crossing) or more (if P, Q are crossing) unbounded polygonal chains [18]. In Fig. 1c the Hausdorff bisector has three such chains. Each vertex of $b_h(P, Q)$ is the center of a circle passing through two points of one cluster and one point of another that entirely encloses P and Q, see e.g. the dotted circle around vertex v in Fig. 1c.

Definition 2. *A vertex on the bisector $b_h(C, P)$, induced by two points $c_i, c_j \in C$ and a point $p_l \in P$, is called* crossing, *if there is a diagonal $p_l p_r$ of P that crosses the diagonal $c_i c_j$ of C, and all points c_i, c_j, p_l, p_r are on the convex hull of $C \cup P$. (See vertex v in Fig. 1c.) The total number of crossing vertices along the bisectors of all pairs of clusters is the* number of crossings *and is denoted by m.*

The Hausdorff Voronoi diagram contains three types of vertices [17] (see Figs. 1a and c): (1) *pure* Voronoi vertices, equidistant to three clusters; (2) *mixed* Voronoi vertices, equidistant to three points of two clusters; and (3) *farthest* Voronoi vertices, equidistant to three points of one cluster. The mixed vertices, which are induced by two points of cluster C (and one point of another cluster), are called *C-mixed* vertices; they are incident to edges of $\mathsf{FVD}(C)$. The Hausdorff Voronoi edges are polygonal lines (portions of Hausdorff bisectors) that connect pure Voronoi vertices. Mixed Voronoi vertices are vertices of Hausdorff bisectors. They are characterized as crossing or non-crossing according to Definition 2. The following property is crucial for our algorithms.

Lemma 1 [17]. *Each face of a (non-empty) region $\mathsf{hreg}_F(C)$ intersects $T(C)$ in one non-empty connected component. The intersection points delimiting this component are C-mixed vertices.*

Unless stated otherwise, we use a refinement of the Hausdorff Voronoi diagram as derived by the *visibility decomposition* [18] of each region $\mathsf{hreg}_F(p)$ (see Fig. 2a): for each vertex v on the boundary of $\mathsf{hreg}_F(p)$ draw the line segment pv, as restricted within $\mathsf{hreg}_F(p)$. In Fig. 2a, the edges of the visibility decomposition in $\mathsf{hreg}_F(p)$ are shown in bold. Each face f within $\mathsf{hreg}_F(p)$ is convex; point p is called the *owner* of f.

Observation 1. *A face f of $\mathsf{hreg}_F(p)$, $p \in P$, borders the regions of $O(1)$ (at most 3) other clusters in $F \setminus \{P\}$.*

Overview of the RIC framework [8,9]. The randomized incremental construction framework to compute a Voronoi diagram, inserts sites (or objects) one by one in random order, each time recomputing the target diagram. The diagram is viewed as a collection of *ranges*, *defined* and without *conflicts* with respect to the set of sites inserted so far. To update the diagram efficiently, a *conflict* or *history* graph is maintained. An important prerequisite for using the framework is: *each range must be defined by a constant number of objects.*

The conflict graph is a bipartite graph where one group of nodes correspond to the ranges of the diagram of sites inserted so far, and the other group correspond to sites that are not yet inserted. In the conflict graph, a range and a site are connected by an arc if and only if they are in conflict. A RIC algorithm that uses a conflict graph is efficient if the following *update condition* is satisfied at each incremental step: (1) Updating the set of ranges defined and without conflicts over the current subset of objects requires time proportional to the number of ranges killed or created during this step; and (2) Updating the conflict graph requires time proportional to the number of arcs of the conflict graph added or removed during this step.

The expected time and space complexity for a RIC algorithm is as stated in [3, Theorem 5.2.3]: Let $f_0(r)$ be the expected number of ranges in the target diagram of a random sample of r objects, and let k be the number of insertion steps of the algorithm. Then (1) Expected number of ranges created during the algorithm is $O\left(\sum_{r=1}^{k}(f_0(r)/r)\right)$. (2) If the update condition holds, the expected time and space of the algorithm is $O\left(k\sum_{r=1}^{k}(f_0(r)/r^2)\right)$.

3   Constructing HVD(F) for Non-crossing Clusters

Let the clusters in F be pairwise non-crossing. Then the Voronoi regions are connected and the combinatorial complexity of the Hausdorff diagram is $O(n)$.

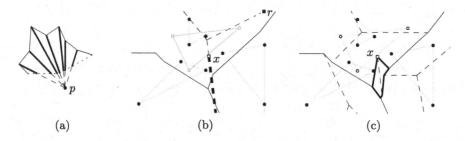

$$(a) \qquad\qquad\qquad (b) \qquad\qquad\qquad (c)$$

Fig. 2. Left: (a) Visibility decomposition of the diagram. Right: Insertion of a cluster C (unfilled disks): (b) $\mathcal{T}(C)$ (dashed) rooted at r, its active subtree (bold) rooted at x. (c) After the insertion: x is a C-mixed vertex of the HVD.

Let $S \subset F$ and $C \in F \setminus S$. Suppose that HVD(S) has been computed; our goal is to insert C and obtain HVD($S \cup \{C\}$). We introduce the following definition. Consider $\mathrm{hreg}_{S \cup \{C\}}(C)$. As we traverse $\mathcal{T}(C)$ starting at its root, let x be the first point we encounter in the closure of $\mathrm{hreg}_{S \cup \{C\}}(C)$. We refer to the subtree of $\mathcal{T}(C)$ rooted at x as the *active subtree* of $\mathcal{T}(C)$, and denote it by $\mathcal{T}_a(C, S)$. In Fig. 2b, $\mathcal{T}_a(C, S)$ is shown in bold dashed lines superimposed on HVD(S). Figure 2c illustrates HVD($S \cup \{C\}$).

By Lemma 1, since Voronoi regions of non-crossing clusters are connected, $\mathsf{hreg}_{S \cup \{C\}}(C) \neq \emptyset$ if and only if $\mathcal{T}_a(C, S) \neq \emptyset$. Further, the root of $\mathcal{T}_a(C, S)$ is a C-mixed vertex of $\mathsf{HVD}(S \cup \{C\})$, unless $\mathcal{T}_a(C, S) = \mathcal{T}(C)$.

3.1   Objects, Ranges and Conflicts

We formulate the problem of computing $\mathsf{HVD}(F)$ in terms of *objects*, *ranges* and *conflicts*. Objects are the clusters in F. Ranges are the faces of $\mathsf{HVD}(S)$ as refined by the visibility decomposition (see Sect. 2 and Fig. 1c). A range corresponding to face f, $f \subset \mathsf{hreg}_S(p)$, where $p \in P$, is said to be defined by the cluster P and by the remaining clusters in S whose Voronoi regions border f. By Observation 1, there are at most three such clusters, and thus, we can apply the RIC framework. Point p is called the *owner* of range f, $f \subset \mathsf{hreg}_S(p)$.

Definition 3 (Conflict for non-crossing clusters). *A range f is in conflict with a cluster $C \in F \setminus S$ if $\mathcal{T}_a(C, S)$ is not empty and its root x lies in f. A conflict is a triple (f, x, C). The list of conflicts of range f is denoted as $\mathcal{L}(f)$.*

The following is implied by Lemma 1 and it is the basis of our algorithm.

Lemma 2. *Each cluster $C \in F \setminus S$ has at most one conflict. If a cluster C has no conflicts, then $\mathsf{hreg}_{S \cup \{C\}}(C) = \emptyset$, thus, $\mathsf{hreg}_F(C) = \emptyset$.*

Due to space constraints, we only present here the variant of the algorithm that uses the conflict graph. It is not hard to adapt it to use a history graph using the same definitions of ranges and conflicts.

3.2   Insertion of a Cluster

Suppose that $\mathsf{HVD}(S)$, $S \subset F$, and the conflict graph have been constructed. Let $C \in F \setminus S$. Using the conflict of C, we can easily compute $\mathsf{hreg}_{S \cup \{C\}}(C)$. Starting at the root of $\mathcal{T}_a(C, S)$ (stored with the conflict) trace the boundary of the region as in [18]. The remaining problem is to identify conflicts for the new ranges of $\mathsf{hreg}_{S \cup \{C\}}(C)$ and to update the conflict graph. The algorithm is given in Fig. 3 as pseudocode.

To identify new conflicts we use the information of conflicts stored with the ranges that get deleted. Let f be a deleted range of owner p ($f \subset \mathsf{hreg}_S(p)$). For each conflict (f, y, Q), we compute the new root of $\mathcal{T}_a(Q, S \cup \{C\})$, if different from $\mathcal{T}_a(Q, S)$, and identify the new range that contains it. To compute the root of $\mathcal{T}_a(Q, S \cup \{C\})$, it is enough to traverse $\mathcal{T}_a(Q, S)$, searching for an edge uv that contains a point equidistant from Q and C (see Line 12). If such an edge uv exists, we compute the point equidistant from C and Q on uv (a Q-mixed vertex) by performing a *segment query* [5] for uv in $\mathsf{FVD}(C)$. If no such edge exists then $\mathcal{T}_a(Q, S \cup \{C\}) = \emptyset$, and no conflicts for Q should be created.

Given a segment uv of $\mathcal{T}(C)$, such that u is closer to C then to Q, and v is closer to Q than to C, the *segment query* finds the unique point on uv, which is equidistant to Q and C. This query can be answered in $O(\log |P|)$ time using the *centroid decomposition* of $\mathsf{FVD}(P)$, as detailed in [5].

The remaining algorithm is straightforward (see Fig. 3).

Algorithm *Insert-NonCrossing*
(∗ Inserts a cluster C in HVD(S); updates the conflict graph ∗)
1. Let (g, x, C) be the conflict of C.
2. Compute $\mathsf{hreg}_{S \cup \{C\}}(C)$ by tracing its boundary starting at x.
3. Update HVD(S) to HVD$(S \cup \{C\})$.
4. **for** $f \in$ HVD$(S) \setminus$ HVD$(S \cup \{C\})$ **do**
5. $\quad$ Let p be the owner of f ($f \subset \mathsf{hreg}_S(p)$).
6. $\quad$ **for** each conflict $(f, y, Q) \in \mathcal{L}(f)$ **do**
7. $\quad\quad$ Discard the conflict (f, y, Q).
8. $\quad\quad$ **if** $d(y, p) < \mathsf{d_f}(y, C)$ **then**
9. $\quad\quad\quad$ Locate the range $f' \subset \mathsf{hreg}_{S \cup \{C\}}(p)$ that contains y by binary search in the visibility decomposition of $\mathsf{hreg}_{S \cup \{C\}}(p)$.
10. $\quad\quad\quad$ Create conflict (f', y, Q).
11. $\quad\quad$ **else**
12. $\quad\quad\quad$ Search for an edge uv in $\mathcal{T}_a(Q, S)$ such that $\mathsf{d_f}(u, Q) > \mathsf{d_f}(u, C)$ and $\mathsf{d_f}(v, Q) < \mathsf{d_f}(v, C)$.
13. $\quad\quad\quad$ **if** uv is found **then**
14. $\quad\quad\quad\quad$ Find root z of $\mathcal{T}_a(Q, S \cup \{C\})$ by a segment query for uv in FVD(C).
15. $\quad\quad\quad\quad$ Find point $c \in C$ such that $z \in \mathsf{hreg}_{S \cup \{C\}}(c)$ by a point location in FVD(C).
16. $\quad\quad\quad\quad$ Locate the range $h \subset \mathsf{hreg}_{S \cup \{C\}}(c)$ that contains z by a binary seacrh in the visibility decomposition of $\mathsf{hreg}_{S \cup \{C\}}(c)$.
17. $\quad\quad\quad\quad$ Create conflict (h, z, Q).

Fig. 3. Algorithm to insert cluster C; case of non-crossing clusters.

Theorem 1. *The Hausdorff Voronoi diagram of k non-crossing clusters of total complexity n can be computed in expected $O(n \log n + k \log n \log k)$ time and $O(n)$ (deterministic) space.*

Proof (Sketch). Let $\{C_1, \ldots, C_k\}$ be a random permutation of F, $\{F_0, \ldots, F_k\}$ be such that $F_0 = \emptyset$, $F_i = F_{i-1} \cup \{C_i\}$ for $1 \le i \le k - 1$, and $F_k = F$. At step i we insert cluster C_i by the algorithm in Fig. 3.

By Lemma 2, at any step i the conflict graph has $O(k)$ arcs; the complexity of HVD(F_i) is $O(n)$. Thus the space required by the algorithm is $O(n)$.

The expected total number of ranges created and deleted during the algorithm is $O(n)$ [5]. Updating the diagram can be done in time proportional to this number times $O(\log n)$ [18]. Updating the conflict graph at step i requires $O(\log n(R_i + N_i))$ time, where R_i is the number of conflicts deleted at step i, and N_i is the number of edges dropped out of the active subtrees at step i. Since an edge is dropped out at most once, and there is $O(n)$ edges in total, the claimed time bound follows by applying the Clarkson-Shor analysis (see Sect. 2). □

4   Computing HVD(F) for Arbitrary Clusters of Points

In this section we drop the assumption that clusters are pairwise non-crossing. This raises a major difficulty that Voronoi regions may be disconnected and

Hausdorff bisectors may consist of more than one polygonal curve. The definition of conflict from Sect. 3.1 no longer guarantees a correct diagram. For a region $r \subset \mathbb{R}^2$, its boundary and its closure are denoted, respectively, ∂r and $\bar{r}$.

Let $C_1, \ldots, C_k$ be a random permutation of clusters in F. We incrementally compute $\mathsf{HVD}(F_i)$, $i = 1, \ldots, k$, where $F_i = \{C_1, C_2, \ldots, C_i\}$. At each step i, cluster C_i is inserted in $\mathsf{HVD}(F_{i-1})$. We maintain the conflict graph (see Sect. 2) between the ranges of $\mathsf{HVD}(F_i)$ and the clusters in $F \setminus F_i$. Like in Sect. 3, ranges correspond to faces of $\mathsf{HVD}(F_i)$ as partitioned by the visibility decomposition.

To correctly insert cluster C_i in $\mathsf{HVD}(F_{i-1})$, one must know at least one point in each face of the (new) region of C_i in $\mathsf{HVD}(F_i) = \mathsf{HVD}(F_{i-1} \cup \{C_i\})$. By Lemma 1, it is sufficient to maintain information on the Q-mixed vertices of $\mathsf{HVD}(F_i \cup \{Q\})$, for every $Q \in F \setminus F_i$. But to apply the Clarkson-Shor analysis we need the ability to determine the new conflicts from the conflicts of ranges that get deleted. Due to this requirement, it is essential that we maintain not only conflicts between Q and the ranges of $\mathsf{HVD}(F_i)$ that contain Q-mixed vertices in $\mathsf{HVD}(F_i \cup \{Q\})$, as Lemma 1 suggests, but also conflicts between Q and every range that intersects the boundary of $\mathsf{hreg}_{F_i \cup \{Q\}}(Q)$.

Let f be a range of $\mathsf{HVD}(F_i)$, $f \subset \mathsf{hreg}_{F_i}(p)$, where $p \in P$ and $P \in F_i$. Let Q be a cluster in $F \setminus F_{i-1}$, We define a conflict between f and Q:

Definition 4 (Conflict for arbitrary clusters). *Range f is in conflict with cluster Q, if f intersects the boundary of $\mathsf{hreg}_{F_i \cup \{Q\}}(Q)$. The vertex list of a conflict $V(f, Q)$ is the list of all vertices and all endpoints of $\mathsf{b_h}(P, Q) \cap \bar{f}$, ordered in clockwise angular order around p.*

Observation 2 $\mathsf{b_h}(P, Q) \cap \bar{f} = \mathsf{b_h}(p, Q) \cap \bar{f}$. *The order of vertices in $V(f, Q)$ coincides with the natural order of vertices along $\mathsf{b_h}(p, Q)$.*

Note $\mathsf{b_h}(p, Q)$ is a single convex chain [18], unlike $\mathsf{b_h}(P, Q)$. Figure 4 shows bisector $\mathsf{b_h}(P, Q)$ and a range f intersected by it. ∂f consists of four parts: the top side is a portion of a pure edge of $\mathsf{HVD}(F_i)$; the bottom chain, shown in bold, is a portion of $\mathcal{T}(P)$; the two sides are edges of the visibility decomposition.

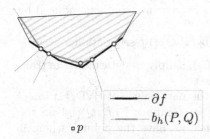

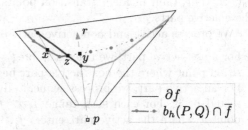

Fig. 4. A range $f \subset \mathsf{hreg}_{F_i}(p)$ (shaded); Bisector $\mathsf{b_h}(P, Q)$, $Q \in F \setminus F_i$; vertices of $\mathsf{b_h}(P, Q)$ (unfilled circle marks).

Fig. 5. An old range $f \subset \mathsf{hreg}_{F_{i-1}}(p)$; New ranges of $\mathsf{hreg}_{F_i}(p)$ derived from f (bounded by light solid lines). $\mathsf{b_h}(P, Q) \cap \bar{f}$ where Q is a cluster in conflict with f.

4.1   Inserting a Cluster

Insert C_i into $\mathsf{HVD}(F_{i-1})$. We compute all faces of $\mathsf{hreg}_{F_i}(C_i)$ by tracing their boundary, starting at the vertices in the vertex lists of the conflicts of C_i. The insertion of $\mathsf{hreg}_{F_i}(C_i)$ results in deleting some *old* ranges and inserting some *new* ones. We have two types of new ranges: *type (1)*: ranges in $\mathsf{hreg}_{F_i}(C_i)$; and *type (2)*: updated ranges of clusters in F_{i-1}.

Update the conflict graph. For each cluster $Q \in F \setminus F_i$ in conflict with a deleted range, compute the conflicts of Q with the new ranges of type (1) and (2).

Ranges of type (1). Consider the ranges in $\mathsf{hreg}_{F_i}(C_i)$. We follow the bisector $\mathsf{b_h}(Q, C_i)$ within $\mathsf{hreg}_{F_i}(C_i)$, while computing it on the fly. For each face $f_i \subset \mathsf{hreg}_{F_i}(C_i)$ that we encounter as we walk on $\mathsf{b_h}(Q, C_i)$, we update the vertex list $V(f_i, Q)$. To update $V(f_i, Q)$ we insert the vertices on the branch of $\mathsf{b_h}(Q, C_i) \cap f_i$ that was just encountered, including its endpoints on ∂f_i. The insertion point in $V(f_i, Q)$ can be determined by binary search. Note that $\mathsf{b_h}(Q, C_i) \cap f_i$ may consist of several components, thus, f_i can be encountered a number of times; each time, we augment $V(f_i, Q)$ independently.

Ranges of type (2). Let $f \subset \mathsf{hreg}_{F_{i-1}}(p)$, $p \in P$, be a deleted range that was in conflict with Q. The vertex list $V(f, Q)$ reveals $\mathsf{b_h}(P, Q) \cap \overline{f}$. For each new range f' of type (2) such that $f' \subset f$ and f' is in conflict with Q, we need to compute $V(f', Q)$. Observe, that the union of $\mathsf{b_h}(P, Q) \cap \overline{f'}$ for all such ranges f' is $(\mathsf{b_h}(P, Q) \cap \overline{f}) \setminus \mathsf{hreg}_{F_i}(C_i)$. We call the maximal contiguous portions of $\mathsf{b_h}(P, Q) \cap \overline{f}$, outside $\mathsf{hreg}_{F_i}(C_i)$, the *active parts* of $\mathsf{b_h}(P, Q) \cap \overline{f}$. The *non-active parts* of $\mathsf{b_h}(P, Q) \cap \overline{f}$ are its maximal contiguous portions inside $\mathsf{hreg}_{F_i}(C_i)$. Figure 5 shows the active and the non-active parts of $\mathsf{b_h}(P, Q) \cap \overline{f}$ by (red) bold and dotted lines respectively. Note that one active (resp., non-active) part may consist of multiple polygonal curves. A point incident to one active and one non-active part is called a *transition point*, see e.g. point z in Fig. 5. Transition points lie in $\mathsf{b_h}(Q, C_i) \cap \partial \mathsf{hreg}_{F_i}(C_i)$; they are used as starting points to compute the conflicts of ranges of type (1). Our task is to determine all active parts of $\mathsf{b_h}(P, Q) \cap \overline{f}$, their incident transition points, and to create conflicts induced by these active parts.

We process active and non-active parts of $\mathsf{b_h}(P, Q) \cap \overline{f}$ sequentially.

- For a non-active part, we trace it in $\mathsf{hreg}_{F_i}(C_i)$, simply to determine the transition point where the next active part begins.
- For an active part, we process sequentially the new ranges of $\mathsf{HVD}(F_i)$ intersected by it. For each such range $f_j \subset f$, we compute $V(f_j, Q)$, given the point x where the active part enters f_j. Once we have the point z where it exits f_j, list $V(f_j, Q)$ is easily derived from the portion of $\mathsf{b_h}(P, Q)$ between x and z. To find z, consider the rightmost ray r originating at p and passing through ∂f_j. If $\mathsf{b_h}(P, Q) \cap \overline{f}$ intersects r (see Fig. 5), let y be the point of this intersection. Otherwise, we let t be the rightmost endpoint of $\mathsf{b_h}(P, Q) \cap \overline{f}$ to the left of r. If $t \in \mathcal{T}(P)$, set $z = t$, otherwise set $y = t$.

- If $y \in \partial f_j$, then we set $z = y$. In this case the active part of $\mathsf{b}_\mathsf{h}(P, Q) \cap \overline{f}$ enters the next new range f_k at point y. In Fig. 5, this case is illustrated by point x that plays the role of $z = y$.
- If $y \notin \overline{f_j}$, we determine point z as the unique point on $\mathsf{b}_\mathsf{h}(P, Q)$, such that z is between x and y, and $z \in \partial f_j$. See Fig. 5. Point z is the endpoint of the active part that we were processing, thus z is a transition point.

The following theorem states the complexity of our algorithm. The key ingredient of the proof is charging the time required to update the conflict graph at step i to the number of conflicts created and deleted at that step, the total size of vertex lists of conflicts of the new ranges of type (1), and the vertices of the vertex lists of deleted conflicts that do not appear in vertex lists of new conflicts.

Theorem 2. *The Hausdorff Voronoi diagram of a family F of k clusters of total complexity n can be computed in $O((m + n \log k) \log n)$ expected time and $O(m + n \log k)$ expected space.*

5   A Crossing-Oblivious Algorithm

Below we discuss how to compute $\mathsf{HVD}(F)$ if it is not known whether clusters in F have crossings. Deciding fast whether F has crossings is not easy, as the convex hulls of the clusters may have a quadratic total number of intersections, even if the clusters are actually non-crossing.

Our algorithm combines the light algorithm for non-crossing clusters (Sect. 3) with the heavy algorithm for arbitrary clusters (Sect. 4), and overcomes the above issue, staying within best complexity bound (see Theorem 3). We start with the algorithm of Sect. 3 and run it until we realize that the diagram cannot be updated correctly. If this happens, we terminate the algorithm of Sect. 3, and run the one of Sect. 4. In particular, after each insertion of a cluster C_i we perform a check. A positive answer to this check guarantees that the region of C_i in $\mathsf{HVD}(F_i)$ is connected, and thus, it is computed correctly. A negative answer indicates that C_i has a crossing with some cluster which has already been inserted in the diagram and its region may be disconnected. At the first negative check, we restart the computation of the diagram from scratch using the algorithm of Sect. 4. We can afford running the latter algorithm, since it is certain that the input set of clusters has crossings.

The procedure of the check is based on properties established in [17], which guarantee that if $\mathsf{hreg}_{F_i}(C_i)$ is disconnected then each of its connected components is incident to a crossing C_i-mixed vertex of $\mathsf{HVD}(F_i)$. It is possible in $O(|C_i| \log n)$ to check whether all C_i-mixed vertices adjacent to a connected component of $\mathsf{hreg}_{F_i}(C_i)$ are non-crossing. We thus obtain the following.

Theorem 3. *Let F be a family of k clusters of total complexity n. There is an algorithm that computes $\mathsf{HVD}(F)$. If the clusters in F are non-crossing, the algorithm requires $O(n)$ space, and expected $O(n \log n + k \log n \log k)$ time. If the clusters in F are crossing, the algorithm requires $O((m + n \log k) \log n)$ expected time and $O(m + n \log k)$ expected space, where m is the total number of crossings between pairs of clusters in F.*

References

1. Abellanas, M., Hurtado, F., Icking, C., Klein, R., Langetepe, E., Ma, L., Palop, B., Sacristán, V.: The farthest color Voronoi diagram and related problems. In: 17th European Workshop on Computational Geometry (EWCG), pp. 113–116, full version: Technical Report 002 2006, Universität Bonn (2006)
2. Aurenhammer, F., Klein, R., Lee, D.T.: Voronoi Diagrams and Delaunay Triangulations. World Scientific, Singapore (2013)
3. Boissonnat, J.D., Yvinec, M.: Algorithmic Geometry. Cambridge University Press, New York (1998)
4. Voronoi, C.A.A.: Voronoi Critical Area Analysis. IBM VLSI CAD Tool, IBM Microelectronics Division, Burlington, VT, distributed by Cadence. Patents: US6178539, US6317859, US7240306, US7752589, US7752580, US7143371, US20090125852
5. Cheilaris, P., Khramtcova, E., Langerman, S., Papadopoulou, E.: A randomized incremental algorithm for the Hausdorff Voronoi diagram of non-crossing clusters. Algorithmica **76**(4), 935–960 (2016)
6. Chen, D.Z., Huang, Z., Liu, Y., Xu, J.: On clustering induced Voronoi diagrams. In: 2013 IEEE 54th Annual Symposium on Foundations of Computer Science (FOCS), pp. 390–399. IEEE (2013)
7. Cheong, O., Everett, H., Glisse, M., Gudmundsson, J., Hornus, S., Lazard, S., Lee, M., Na, H.S.: Farthest-polygon Voronoi diagrams. Comput. Geom. **44**(4), 234–247 (2011)
8. Clarkson, K., Shor, P.: Applications of random sampling in computational geometry, II. Discrete Comput. Geom. **4**, 387–421 (1989)
9. Clarkson, K.L., Mehlhorn, K., Seidel, R.: Four results on randomized incremental constructions. Comput. Geom. Theory Appl. **3**(4), 185–212 (1993)
10. Claverol, M., Khramtcova, E., Papadopoulou, E., Saumell, M., Seara, C.: Stabbing circles for sets of segments in the plane. Algorithmica (2017). doi:10.1007/s00453-017-0299-z
11. Dehne, F., Maheshwari, A., Taylor, R.: A coarse grained parallel algorithm for Hausdorff Voronoi diagrams. In 35th ICPP, pp. 497–504 (2006)
12. Edelsbrunner, H., Guibas, L., Sharir, M.: The upper envelope of piecewise linear functions: algorithms and applications. Discrete Comput. Geom. **4**, 311–336 (1989)
13. Huttenlocher, D.P., Kedem, K., Sharir, M.: The upper envelope of Voronoi surfaces and its applications. Discrete Comput. Geom. **9**, 267–291 (1993)
14. Klein, R.: Concrete and Abstract Voronoi diagrams. LNCS, vol. 400. Springer, Heidelberg (1989)
15. Klein, R., Mehlhorn, K., Meiser, S.: Randomized incremental construction of abstract Voronoi diagrams. Comput. Geom. **3**(3), 157–184 (1993)
16. Papadopoulou, E.: Net-aware critical area extraction for opens in VLSI circuits via higher-order Voronoi diagrams. IEEE Trans. CAD Integrated Circuits and Systems **30**(5), 704–717 (2011)
17. Papadopoulou, E.: The Hausdorff Voronoi diagram of point clusters in the plane. Algorithmica **40**(2), 63–82 (2004)
18. Papadopoulou, E., Lee, D.T.: The Hausdorff Voronoi diagram of polygonal objects: a divide and conquer approach. Int. J. Comput. Geom. Ap. **14**(6), 421–452 (2004)

NP-completeness Results for Partitioning a Graph into Total Dominating Sets

Mikko Koivisto[1], Petteri Laakkonen[2], and Juho Lauri[2,3P (✉)]

[1] University of Helsinki, Helsinki, Finland
mikko.koivisto@helsinki.fi
[2] Tampere University of Technology, Tampere, Finland
{petteri.laakkonen,juho.lauri}@tut.fi
[3] Bell Labs, Dublin, Ireland
juho.lauri@nokia.com

Abstract. A *total domatic k-partition* of a graph is a partition of its vertex set into k subsets such that each intersects the open neighborhood of each vertex. The maximum k for which a total domatic k-partition exists is known as the *total domatic number* of a graph G, denoted by $d_t(G)$. We extend considerably the known hardness results by showing it is NP-complete to decide whether $d_t(G) \geq 3$ where G is a bipartite planar graph of bounded maximum degree. Similarly, for every $k \geq 3$, it is NP-complete to decide whether $d_t(G) \geq k$, where G is a split graph or k-regular. In particular, these results complement recent combinatorial results regarding $d_t(G)$ on some of these graph classes by showing that the known results are, in a sense, best possible. Finally, for general n-vertex graphs, we show the problem is solvable in $2^n n^{O(1)}$ time, and derive even faster algorithms for special graph classes.

1 Introduction

Domination is undoubtedly one of the most intensively studied concepts in graph theory. Besides being a fundamental graph property, domination also routinely appears in real-world applications related to data transfer (see e.g., [9,16,31]).

Let $G = (V, E)$ be a graph. A *dominating set* of G is a set of vertices $S \subseteq V$ such that every vertex in V either is in S or is adjacent to a vertex in S. The *domination number* of a graph G, denoted by $\gamma(G)$, is the size of a minimum dominating set in G. A classical variant of domination is the concept of *total domination*, introduced by Cockayne et al. [7] in 1980. Here, a set of vertices $D \subseteq V$ is a *total dominating set* of G if every vertex in V is adjacent to some vertex in D. The *total domination number* of G, denoted by $\gamma_t(G)$, is the size of a minimum total dominating set in G. Note that every total dominating set is also a dominating set, but the converse is not true in general. Given the centrality of the topic, much is known about total domination in graphs. For instance, Bollobás and Cockayne [5] proved that $\gamma_t(G) \leq 2\gamma(G)$. For other combinatorial results, we refer the reader to the books [17,18] and the recent monograph [22]. From the viewpoint of complexity, the problem of deciding whether a graph G

© Springer International Publishing AG 2017
Y. Cao and J. Chen (Eds.): COCOON 2017, LNCS 10392, pp. 333–345, 2017.
DOI: 10.1007/978-3-319-62389-4_28

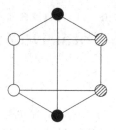

Fig. 1. The graph $G = \overline{C}_6$ has $d_t(G) = 3$.

has a total dominating set of size at most k is well-known to be NP-complete, even for bipartite graphs [27]. The reader interested in computational results is referred to the survey [20].

In this work, we focus on the problem of partitioning the vertex set of a graph into a maximum number of total dominating sets, a concept also introduced by Cockayne et al. [7]. For ease of presentation, we follow Zelinka [32] and use a definition rooted in graph colorings. A *total domatic k-coloring* is a partition of V into k color classes such that each vertex in V is adjacent to a member from each class. The maximum k for which a graph G has a total domatic k-coloring is known as the *total domatic number* of G, and is denoted by $d_t(G)$. This is equivalent to the maximum number of total dominating sets the vertex set of G can be partitioned into. For convenience, we can say a graph G is *total domatic k-colorable* if $d_t(G) \geq k$. It is immediate from the definition that for any nontrivial graph G, we have that $1 \leq d_t(G) \leq \delta(G) \leq \Delta(G)$, where $\delta(G)$ and $\Delta(G)$ denote the minimum and maximum degree, respectively. For an illustration of the concept, consider the graph $G = \overline{C}_6$, i.e., the complement of a 6-cycle (see Fig. 1). As G is 3-regular, we have that $d_t(G) \leq 3$. On the other hand, a total domatic 3-coloring of G is straightforward to construct proving that $d_t(G) = 3$.

Zelinka [33] proved that no minimum degree condition suffices to guarantee $d_t(G) = 2$. Later on, Heggernes and Telle [19] showed that deciding whether $d_t(G) \geq 2$ is NP-complete for bipartite graphs. The study of total domatic number has regained attention lately. Recently, motivated by communication problems in large multi-robot networks, Chen et al. [6] reintroduced the concept under the name coupon coloring. In particular, they called the total domatic number the *coupon coloring number*. Subsequently, and seemingly unaware of the earlier work by Zelinka [32–34], the coupon coloring number was studied by Shi et al. [30]. In particular, Zelinka [32] determined the total domatic number of cactus graphs, majorizing Shi's et al. [30] result for cycles.

As further motivation, we mention a slight variation of an application in network science described by Abbas et al. [1]. Imagine a network of agents where each agent holds an instrument (e.g., thermometer, hygrometer, or so on). Moreover, interaction between agents is limited such that an agent can only communicate with its neighbors. As an agent requires access to each instrument, it needs to be copied across the agents to form a particular kind of dominating set. Indeed, to accommodate e.g., power limitations or system failures, we do not rely on the instrument an agent itself holds. Now, each instrument forms a total dominating set, and the maximum number of instruments that can be made available is precisely the total domatic number of the network. Abbas et al. [1] also stress that "... the underlying network topology of multirobot networks plays a crucial role in achieving the system level objectives within the network in a distributed manner." This motivates the study of the complexity of computing the total domatic number on restricted topologies. Furthermore, we also remark that Henning [20, Problem 12] calls for a deeper investigation of total domination on planar graphs.

Our results. We considerably extend the known hardness results for computing the total domatic number of a graph.

- For Sect. 3, our main result is that it is NP-complete to decide whether a bipartite planar graph G of bounded maximum degree has $d_t(G) \geq 3$. This complements the recent combinatorial results of Goddard and Henning [14] who showed that no planar graph G' has $d_t(G') = 5$. In other words, our result shows that it is unlikely one could have a polynomial-time characterization of planar graphs with described total domatic numbers.
- In Sect. 4, we prove that for every $k \geq 3$, it is NP-complete to decide whether a k-regular graph G has $d_t(G) \geq 3$. In contrast, Akbari et al. [2] characterized the 3-regular graphs H with $d_t(H) \geq 2$. This is best possible in the sense that our hardness result gives strong evidence for the non-existence of a polynomial-time characterization of k-regular graphs with $d_t(G) \geq k$.
- In Sect. 5 we focus on chordal graphs. We begin by proving that it is NP-complete to decide whether $d_t(G) \geq 2$ where G is a split graph. On a positive side, we show that the total domatic number can be computed in polynomial-time for threshold graphs.
- Finally, in Sect. 6, we give fast exact exponential-time algorithms for the problem. In particular, we show how the problem can be solved in $2^n n^{O(1)}$ time for general graphs, and derive even faster algorithms for special graph classes.

Statements whose proofs are located in the appendix are marked with $\star$.

2 Preliminaries

For a positive integer n, we write $[n] = \{1, 2, \ldots, n\}$.

In what follows, we define the graph-theoretic concepts most central to our work. For graph-theoretic notation not defined here, we refer the reader to [10]. We also briefly introduce decision problems our hardness results depend on.

Graph parameters and classes. All graphs we consider are undirected and simple. For a graph G, we denote by $V(G)$ and $E(G)$ its vertex set and edge set, respectively. To reduce clutter, an edge $\{u, v\}$ is often denoted as uv. Two vertices x and y are *adjacent* (or *neighbors*) if xy is an edge of G. The *neighborhood* of a vertex v, denoted by $N(v)$, is the set of all vertices adjacent to v. Let $G = (V, E)$. When we *identify* two distinct vertices $v, w \in V$, we obtain the graph with vertex set $V \setminus \{v, w\} \cup \{u\}$ and edge set $E \setminus \{\{v', v''\} \mid v' \in V, v'' \in \{v, w\}\} \cup \{\{v', u\} \mid v \notin \{v, w\} \text{ and } \{v', v\} \in E \text{ or } \{v', w\} \in E\}$.

A *vertex-coloring* is a function $c : V \to [k]$ assigning a color from $[k]$ to each vertex of a graph $G = (V, E)$. The coloring is said to be *proper* if $c(u) \neq c(v)$ for every $uv \in E$. A graph G is said to be k-*colorable* if there exists a proper vertex-coloring using k colors for it. The minimum k for which a graph G is k-colorable is known as its *chromatic number*, denoted by $\chi(G)$. In particular, a 2-colorable graph is *bipartite*. Similarly, an *edge-coloring* is a function $f : E \to [k]$ assigning a color from $[k]$ to each edge. We say that f is *proper* if two adjacent edges receive a distinct color under f. A graph G is said to be k-*edge-colorable* if there exists a proper edge-coloring using k colors for it. The minimum k for which G is k-edge-colorable is known as its *chromatic index*, denoted by $\chi'(G)$.

A *chord* is an edge joining two non-consecutive vertices in a cycle. A graph is *chordal* if every cycle of length 4 or more has a chord. Equivalently, a graph is chordal if it contains no induced cycle of length 4 or more. A graph is a *split graph* if its vertex set can be partitioned into a clique and an independent set. It is known that all split graphs are chordal. A well-known subclass of split graphs is formed by *threshold graphs*, which are the graphs that can be formed from the empty graph by repeatedly adding either an isolated vertex or a dominating vertex (see [25, Theorem 1.2.4]).

Finally, we mention the following well-known structural measure for "tree-likeness" of graphs. A *tree decomposition* of G is a pair $(T, \{X_i : i \in I\})$ where $X_i \subseteq V$, $i \in I$, and T is a tree with elements of I as nodes such that:

1. for each edge $uv \in E$, there is an $i \in I$ such that $\{u, v\} \subseteq X_i$, and
2. for each vertex $v \in V$, $T[\{i \in I \mid v \in X_i\}]$ is a tree with at least one node.

The *width* of a tree decomposition is $\max_{i \in I} |X_i| - 1$. The *treewidth* of G, denoted by $\mathrm{tw}(G)$, is the minimum width taken over all tree decompositions of G.

Decision problems. Our main focus is the computational complexity of the following problem.

> TOTAL DOMATIC k-PARTITION
> **Instance:** A graph $G = (V, E)$.
> **Question:** Can V be partitioned into k total dominating sets, i.e., is $d_t(G) \geq k$?

Our NP-completeness results are established by polynomial-time reductions from well-known coloring problems. We introduce them here for completeness.

In both k-COLORING and EDGE k-COLORING, the input is a graph $G = (V, E)$. In the former, we must decide whether $\chi(G) \leq k$, while in the latter, we are asked whether $\chi'(G) \leq k$. Both problems are NP-complete for every $k \geq 3$ (see e.g., [12,24]). Finally, in the SET SPLITTING problem, we are given a universe $U = [n]$ and a set system $\mathcal{F} \subseteq 2^U$. The task is to decide whether the elements of U can be colored in 2 colors such that each member of $\mathcal{F}$ contains both colors. This problem is well-known to be NP-complete as well.

3   Total Domatic Partitioning of Planar Graphs

It is well-known that every planar graph G has a vertex with degree at most five, so we have that $1 \leq d_t(G) \leq 5$. However, very recently, Goddard and Henning [14] proved that no planar graph G has $d_t(G) = 5$ by establishing the following tight bounds.

Theorem 1 (Goddard and Henning [14]). *Any planar graph G has $1 \leq d_t(G) \leq 4$. Moreover, these bounds are tight.*

In the following, we prove that there is likely no straightforward (i.e., polynomial-time) characterization of planar graphs with specified total domatic numbers. More precisely, we show that it is NP-complete to decide whether a planar graph G of bounded maximum degree has $d_t(G) \geq 3$. We first make the idea of the reduction clear, and then introduce slightly more complex gadgets that establish, using the same correctness argument, the same result for bipartite planar graphs of bounded maximum degree.

Lemma 2. 3-COLORING *reduces in polynomial time to* TOTAL DOMATIC 3-PARTITION.

Proof. Let G be an instance of 3-COLORING. In polynomial time, we will create the following instance G' of TOTAL DOMATIC 3-PARTITION, such that G is 3-colorable if and only if G' is total domatic 3-colorable.

The graph $G' = (V', E')$ is obtained from G by replacing each edge with a diamond (see Fig. 2), and by attaching to each vertex of G a copy of $\overline{C}_6$ (see Fig. 1). Formally, we let $V' = V \cup \{x_{ij}, x'_{ij} \mid ij \in E\} \cup \{w_{vi} \mid v \in V, i \in [5]\}$. Set $E' = D \cup C$, where

$$D = \{ix_{ij}, ix'_{ij}, x_{ij}x'_{ij}, jx_{ij}, jx'_{ij} \mid ij \in E\},$$
$$C = \{w_{vi}w_{vi+1}, w_{v1}w_{v5}, w_{v2}w_{v4}, vw_{v1}, vw_{v3}, vw_{v5} \mid 1 \leq i < 4, v \in V\}.$$

This finishes the construction of G'.

Let $c : V \to [3]$ be a proper vertex-coloring of G. We will construct a total domatic 3-coloring $c' : V' \to [3]$ as follows. We retain the coloring of the vertices in V, that is, $c'(v) = c(v)$ for every $v \in V$. Then, as the degree of x_{ij} is 3, it holds that in any valid total domatic 3-coloring of G', the colors from [3] must be bijectively mapped to the neighborhood of x_{ij}; by symmetry, the same holds

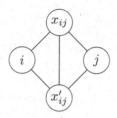

Fig. 2. The edge ij replaced by a diamond in the construction of Lemma 2.

for x'_{ij}. Then consider two adjacent vertices i and j in G. We set $c'(x_{ij}) = c'(x'_{ij}) = f$, where f is the unique color in [3] neither $c(i)$ nor $c(j)$. Clearly, the neighborhood of both x_{ij} and x'_{ij} contains every color from [3]. Finally, consider an arbitrary vertex $v \in V$. Without loss of generality, suppose $c(v) = 1$. We will then finish the vertex-coloring c' as follows (see also Fig. 1):

$$c'(w_{v5}) = 1\,, \ c'(w_{v1}) = 2\,, \ c'(w_{v3}) = 2\,, \ c'(w_{v2}) = 3\,, \ c'(w_{v4}) = 3\,.$$

It is straightforward to verify that c' is indeed a total domatic 3-coloring of G'.

For the other direction, suppose that there is a total domatic 3-coloring c' of G'. Again, it holds that every color from [3] is bijectively mapped to the neighborhood of x_{ij}, for every i and j. Moreover, it holds that $c'(x_{ij}) = c'(x'_{ij})$, implying that $c'(i) \neq c'(j)$. Thus, c' restricted to G gives a proper 3-coloring for G. This concludes the proof. □

It is known that deciding whether a planar graph of maximum degree 4 can be properly 3-colored is NP-complete [12]. This combined with the previous lemma establishes the following.

Theorem 3. *It is* NP-*complete to decide whether a planar graph G of maximum degree 11 has $d_t(G) \geq 3$.*

Corollary 4. *It is* NP-*complete to decide whether a bipartite planar graph G of maximum degree 19 has $d_t(G) \geq 3$.*

Proof. To build G', we proceed with a construction similar to Lemma 2. However, for each edge $ij \in E(G)$, instead of a diamond, we construct the gadget shown in Fig. 3 (a). For each vertex $v \in V$, instead of $\overline{C}_6$, we identify v with the gadget shown in Fig. 3 (b).

It is straightforward to verify that both gadgets are planar and bipartite. Clearly, G' is planar. Moreover, G' is bipartite as an odd cycle of G has even length in G. Correctness follows by the same argument as in Lemma 2. □

We remark that Lemma 2 has consequences for the complexity of total domatic 3-coloring graphs of bounded treewidth, and further consequences for planar graphs as well. Before proceeding, we briefly recall that a *parameterized problem I* is a pair (x, k), where x is drawn from a fixed, finite alphabet and k

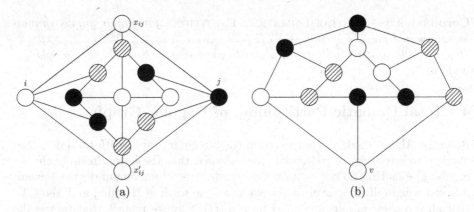

Fig. 3. (a) A replacment gadget for edge ij, and (b) a gadget each vertex v is identified with. Both gadgets are planar and bipartite.

is an integer called the *parameter*. Then, a *kernel* for (x, k) is a polynomial-time algorithm that returns an instance (x', k') of I such that (x, k) is a YES-instance if and only if (x', k') is a YES-instance, and $|x'| \leq g(k)$, for some computable function $g : \mathbb{N} \to \mathbb{N}$. If $g(k)$ is a polynomial (exponential) function of k, we say that I admits a polynomial (exponential) kernel (for more, see Cygan et al. [8]). We then recall the following earlier result.

Theorem 5 (van Rooij et al. [29]). *For every $k \geq 1$,* TOTAL DOMATIC *k-PARTITION parameterized by treewidth admits an exponential kernel.*

In the following, we show that this is the best possible, i.e., there is no polynomial kernel under reasonable complexity-theoretic assumptions. We make the following observations regarding the gadget construction in Lemma 2.

Observation 6. *It holds that* $\mathrm{tw}(\overline{C}_6) = 3$.

By identifying a $\overline{C}_6$ with a vertex of a bounded treewidth graph G, we do not considerably increase the treewidth of G.

Observation 7. *Let G be a graph of treewidth k, let G' be a graph of treewidth k', and let H be the graph obtained through the identification of two vertices $v \in V(G)$ and $v' \in V(G')$. Then* $\mathrm{tw}(H) \leq \max\{k, k'\}$.

Finally, Bodlaender et al. [4] proved that 3-COLORING does not admit a polynomial kernel parameterized by treewidth unless NP $\subseteq$ coNP/poly. As the proof of Lemma 2 gives a parameter-preserving transformation guaranteeing $\mathrm{tw}(G') \leq \mathrm{tw}(G) + 3$, we have the following.

Theorem 8. TOTAL DOMATIC 3-PARTITION *parameterized by treewidth does not admit a polynomial kernel unless* NP $\subseteq$ coNP/poly.

Another consequence of Lemma 2 is captured by the following observation. For its statement, we recall the well-known *exponential time hypothesis* (ETH), which is a conjecture stating that there is a constant $c > 0$ such that 3-SAT cannot be solved in time $O(2^{cn})$, where n is the number of variables.

Corollary 9 ($\star$). TOTAL DOMATIC 3-PARTITION *for planar graphs cannot be solved in time $2^{o(\sqrt{n})}$ unless ETH fails, where n is the number of vertices. However, the problem admits an algorithm running in time $2^{O(\sqrt{n})}$ for planar graphs.*

4   Total Domatic Partitioning of Regular Graphs

Recently, Akbari et al. [2] characterized the 3-regular graphs with total domatic number at least two. In particular, they showed that these are precisely the 3-regular graphs that do not contain a particular tree of maximum degree 3 as an induced subgraph. Moreover, it follows from the work of Henning and Yeo [21] that all k-regular graphs for $k \geq 4$ have $d_t(G) \geq 2$. We remark that the result of Akbari et al. [2] is the best possible in the sense that it is NP-complete to decide whether a k-regular graph G has $d_t(G) \geq k$, for every $k \geq 3$. We establish this by reducing from EDGE k-COLORING on k-regular graphs, where $k \geq 3$, a problem shown to be NP-complete by Leven and Galil [24].

Essentially, our construction to follow is used already by Heggernes and Telle [19, Theorem 5], but we describe it in the appendix for completeness.

Theorem 10 ($\star$). *For every $k \geq 3$, it is NP-complete to decide whether a k-regular graph G has $d_t(G) \geq k$.*

5   Total Domatic Partitioning of Chordal Graphs

In the spirit of the previous sections, we begin by proving that total domatic coloring of chordal graphs is computationally difficult. More precisely, we show that for every $k \geq 2$, deciding whether $d_t(G) \geq k$ is NP-complete already for split graphs. The result is obtained by showing polynomial-time equivalence between this problem and HYPERGRAPH RAINBOW k-COLORING, studied by Guruswami and Lee [15] among others. In the latter, we are given a universe $U = [n]$ and a set family $\mathcal{F} \subseteq 2^U$. The goal is to decide whether the elements of U can be colored in k colors such that each member of $\mathcal{F}$ contains each of the k colors. The problem is equivalent to SET SPLITTING for $k = 2$.

Theorem 11 ($\star$). *For every $k \geq 1$, TOTAL DOMATIC k-PARTITION for split graphs is equivalent to HYPERGRAPH RAINBOW k-COLORING.*

Theorem 12 ($\star$). *For every $k \geq 2$, it is NP-complete to decide whether a split graph G has $d_t(G) \geq k$.*

Given this negative result, it is interesting to consider split graphs with further restrictions on their structure. For instance, every complete graph is a split graph, and the total domatic number of complete graphs is known.

Proposition 13 (Shi et al. [30]). *Let G be the complete graph with n vertices. Then $d_t(G) = \lfloor n/2 \rfloor$.*

However, complete graphs have a very special structure: can we impose a weaker structural requirement on split graphs to obtain a graph class G for which $d_t(G)$ can be computed in polynomial time? Before giving a positive answer to this question, we show the following.

Lemma 14. *Let $G = (V, E)$ be a graph, and let S be a subset of V such that every $s \in S$ is a dominating vertex. Then $d_t(G) \geq \min\{\lfloor n/2 \rfloor, |S|\}$.*

Proof. Denote $\ell = |S|$ and $h = |V \setminus S|$. Suppose first that $h \geq \ell$. We will prove that $d_t(G) \geq \ell$. Construct a total domatic ℓ-coloring $c : V \to [\ell]$ as follows. Map $[\ell]$ bijectively to S and surjectively to $V \setminus S$. Then, each $v \notin S$ is dominated by each $s \in S$, and thus each color from $[\ell]$ appears in the neighborhood of v. Similarly, by construction, $[\ell]$ is mapped surjectively to $V \setminus S$ and $s \in S$ dominates each vertex not in S. Therefore, when $h \geq \ell$, we conclude that $d_t(G) \geq \ell$.

Finally, suppose that $h < \ell$. Let $q = \lfloor (\ell + h)/2 \rfloor = \lfloor n/2 \rfloor$, and choose $A \subseteq S$ such that $|A| = q$. Map $[\ell]$ bijectively to the vertices in A and surjectively to $V \setminus A$. By the above argument, this is a valid total domatic q-coloring. Therefore, we have that $d_t(G) \geq q$, concluding the proof. □

Theorem 15. *Let $G = (V, E)$ be a connected threshold graph with n vertices, and let S be a subset of V such that every $s \in S$ is a dominating vertex. Then $d_t(G) = \min\{\lfloor n/2 \rfloor, |S|\}$.*

Proof. It is well-known that every threshold graph G can be represented as a string $s(G)$ of characters u and j, where u denotes the addition of an isolated vertex, and j the addition of a dominating vertex (see [25, Theorem 1.2.4]). Observe that if the last symbol of $s(G)$ is u, then G is not connected and consequently $d_t(G) = 0$. Thus, it holds that the last symbol of $s(G)$ is j.

Denote $\ell = |S|$ and $h = |V \setminus S|$. Suppose $h \geq \ell$, and consider the first occurrence of symbol u in $s(G)$. The corresponding vertex has degree ℓ, and no other vertex has degree less than ℓ. Thus, $d_t(G) \leq \ell$. By Lemma 14, it follows that in this case, $d_t(G) = \ell$. Finally, suppose $h < \ell$. In the extremal case, $\ell = n$, i.e., each vertex of G is a dominating vertex. By Proposition 13, we have that $d_t(G) \leq (\lfloor \ell + h \rfloor)/2 = \lfloor n/2 \rfloor$. On the other hand, Lemma 14 gives us a matching lower bound. This concludes the proof. □

By the previous theorem, the total domatic number can be computed efficiently for threshold graphs.

Finally, despite Theorem 12, we observe that for some graphs G we can always guarantee $d_t(G) \geq 2$.

Proposition 16. *Let G be an n-vertex Hamiltonian graph where n is a multiple of four. Then $d_t(G) \geq 2$.*

Proof. Let $v_0 v_1 \cdots v_n$ be a Hamiltonian cycle in G such that $n \bmod 4 = 0$. Consider a vertex-coloring $c : V \to [2]$ such that $c(v_i) = c(v_{i+1}) = 1$ and $c(v_{i+2}) = c(v_{i+3}) = 2$ for $i = 0, 4, \ldots, n - 4$. By construction, each neighborhood contains both colors 1 and 2, so we are done. □

6   On Exact Algorithms for Total Domatic Partitioning

When a problem of interest is shown to be NP-complete, it motivates the consideration of alternative algorithmic approaches and easier special cases. Our results show that TOTAL DOMATIC k-PARTITION remains hard for several special cases. In addition, parameterization (see, e.g., [8]) by the number of colors k seems uninteresting since the problem remains NP-complete for constant values of k. These observations further motivate the study of exact (exponential-time) algorithms. A brute-force algorithm tries every possible k-coloring and outputs YES if and only if one of the k-colorings is a total domatic k-coloring. Such an algorithm runs in time $k^n n^{O(1)}$, where n is the number of vertices. Can we do considerably better? In what follows, we show this to be the case, and give even faster algorithms for special graph classes.

To obtain a moderately exponential algorithm, we apply a result of Björklund et al. [3] for the SET PARTITION problem. In this problem we are given a universe $U = [n]$, a set family $\mathcal{F} \subseteq 2^U$, and an integer k. The task is to decide whether U admits a partition into k members of $\mathcal{F}$. Using algebraic methods, Björklund et al. showed the following.

Theorem 17 (Björklund et al. [3, Theorems 2 and 5]). SET PARTITION *can be solved in* $2^n n^{O(1)}$ *time. If membership in* $\mathcal{F}$ *can be decided in* $n^{O(1)}$ *time, then* SET PARTITION *can be solved in* $3^n n^{O(1)}$ *time and* $n^{O(1)}$ *space.*

We apply this result to the set family consisting of all total dominating sets of a given graph. Since we can decide in polynomial time whether a given set of vertices is a total dominating set, we get the following.

Corollary 18. TOTAL DOMATIC k-PARTITION *can be solved in* $3^n n^{O(1)}$ *time and polynomial space. In exponential space, the time can be improved to* $2^n n^{O(1)}$.

We note that Björklund et al. give a similar application to domatic number. Relying on sophisticated algorithms for enumerating minimal dominating sets due to Fomin et al. [11], they further improve the polynomial-space results by lowering the constant of the exponential from 3 to 2.8718. Currently, the lowest constant is 2.7139, due to Nederlof et al. [26].

For total domatic number we discover another way to improve the polynomial-space algorithm, however, restricting ourselves to regular graphs. We get the following result by a simple reduction to graph coloring, for which the constant was recently improved to 2.2355 by Gaspers and Lee [13].

Theorem 19. *One can decide in* $O(2.2355^n)$ *time and polynomial space whether a given* k-regular graph G with n vertices has $d_t(G) \geq k$.

Proof. Define a graph $G' = (V, E')$, where we put an edge between two vertices $u, v \in V$ exactly when they occur in the same neighborhood in G. It holds that $d_t(G) = k$ if and only if $\chi(G') = k$. $\square$

As noted by Chen et al. [6], TOTAL DOMATIC k-PARTITION for $k = 2$ corresponds to the well-known problem of hypergraph 2-coloring, also known as

SET SPLITTING. To see this, we construct an instance of SET SPLITTING with the universe corresponding to the vertex set of the graph, and the set family to the neighborhoods. We obtain the following bound exploiting the algorithms of Nederlof et al. [26] for SET SPLITTING.

Theorem 20. *One can decide in $O(1.8213^n)$ time and polynomial space whether a given graph G with n vertices has $d_t(G) \geq 2$. In exponential-space, the time can be improved to $O(1.7171^n)$.*

7   Concluding Remarks

Our hardness results mirror those known for domatic number. Indeed, the computation of the domatic number was shown to be NP-complete for split graphs by Kaplan and Shamir [23], while hardness for bipartite planar graphs was proved by Poon et al. [28]. For positive results on special graph classes, results appear more scattered. For instance, it seems unknown whether domatic number can be solved in polynomial time for threshold graphs.

Concerning exact algorithms for total domatic number, an intriguing question is whether one can beat the 3^n time bound in polynomial space in general graphs. For domatic number the known algorithms achieve that via inclusion–exclusion and a branching algorithm that either lists minimal dominating sets or counts dominating sets. Currently we do not know whether these branching algorithms can be effectively adapted to total dominating sets.

Acknowledgments. This work was supported in part by the Academy of Finland, under Grant 276864 (M.K.), and by the Emil Aaltonen Foundation, under Grant 160138 N (J.L.).

References

1. Abbas, W., Egerstedt, M., Liu, C.H., Thomas, R., Whalen, P.: Deploying robots with two sensors in $K_{1,6}$-free graphs. J. Graph Theor. **82**(3), 236–252 (2016)
2. Akbari, S., Motiei, M., Mozaffari, S., Yazdanbod, S.: Cubic graphs with total domatic number at least two. arXiv preprint arXiv:1512.04748 (2015)
3. Björklund, A., Husfeldt, T., Koivisto, M.: Set partitioning via inclusion-exclusion. SIAM J. Comput. **39**(2), 546–563 (2009)
4. Bodlaender, H.L., Downey, R.G., Fellows, M.R., Hermelin, D.: On problems without polynomial kernels. J. Comput. Syst. Sci. **75**(8), 423–434 (2009)
5. Bollobás, B., Cockayne, E.J.: Graph-theoretic parameters concerning domination, independence, and irredundance. J. Graph Theor. **3**(3), 241–249 (1979)
6. Chen, B., Kim, J.H., Tait, M., Verstraete, J.: On coupon colorings of graphs. Discr. Appl. Math. **193**, 94–101 (2015)
7. Cockayne, E.J., Dawes, R.M., Hedetniemi, S.T.: Total domination in graphs. Networks **10**(3), 211–219 (1980)
8. Cygan, M., Fomin, F.V., Kowalik, Ł., Lokshtanov, D., Marx, D., Pilipczuk, M., Pilipczuk, M., Saurabh, S.: Parameter. Algorithms. Springer, Heidelberg (2015)

9. Dai, F., Wu, J.: An extended localized algorithm for connected dominating set formation in ad hoc wireless networks. IEEE Trans. Parallel Distrib. Syst. **15**(10), 908–920 (2004)

10. Diestel, R.: Graph Theory. Springer, Heidelberg (2010)

11. Fomin, F.V., Grandoni, F., Pyatkin, A.V., Stepanov, A.A.: Combinatorial bounds via measure and conquer: bounding minimal dominating sets and applications. ACM Trans. Algorithms **5**(1), 9:1–9:17 (2008)

12. Garey, M., Johnson, D., Stockmeyer, L.: Some simplified NP-complete graph problems. Theor. Comput. Sci. **1**(3), 237–267 (1976)

13. Gaspers, S., Lee, E.: Faster graph coloring in polynomial space. ArXiv e-prints arXiv:1607.06201 (2016)

14. Goddard, W., Henning, M.A.: Thoroughly distributed colorings. arXiv preprint arXiv:1609.09684 (2016)

15. Guruswami, V., Lee, E.: Strong inapproximability results on balanced rainbow-colorable hypergraphs. In: Proceedings of the 26th Annual ACM-SIAM Symposium on Discrete Algorithms (SODA), pp. 822–836. SIAM (2015)

16. Han, B., Jia, W.: Clustering wireless ad hoc networks with weakly connected dominating set. J. Parallel Distrib. Comput. **67**(6), 727–737 (2007)

17. Haynes, T.W., Hedetniemi, S.T., Slater, P.J.: Domination in Graphs: Advanced Topics. Marcel Dekker Inc., New York (1998)

18. Haynes, T.W., Hedetniemi, S.T., Slater, P.J.: Fundamentals of Domination in Graphs. CRC Press, Boca Raton (1998)

19. Heggernes, P., Telle, J.A.: Partitioning graphs into generalized dominating sets. Nordic J. Comput. **5**(2), 128–142 (1998)

20. Henning, M.A.: A survey of selected recent results on total domination in graphs. Discr. Math. **309**(1), 32–63 (2009)

21. Henning, M.A., Yeo, A.: 2-colorings in k-regular k-uniform hypergraphs. Eur. J. Comb. **34**(7), 1192–1202 (2013)

22. Henning, M.A., Yeo, A.: Total Domination in Graphs. Springer, New York (2013)

23. Kaplan, H., Shamir, R.: The domatic number problem on some perfect graph families. Inf. Process. Lett. **49**(1), 51–56 (1994)

24. Leven, D., Galil, Z.: NP-completeness of finding the chromatic index of regular graphs. J. Algorithms **4**(1), 35–44 (1983)

25. Mahadev, N.V.R., Peled, U.N.: Threshold Graphs and Related Topics, vol. 56. Elsevier, Amsterdam (1995)

26. Nederlof, J., van Rooij, J.M.M., van Dijk, T.C.: Inclusion/exclusion meets measure and conquer. Algorithmica **69**(3), 685–740 (2014)

27. Pfaff, J., Laskar, R., Hedetniemi, S.T.: NP-completeness of total and connected domination and irredundance for bipartite graphs. Technical report, Clemson University, Department of Mathematical Sciences 428 (1983)

28. Poon, S.-H., Yen, W.C.-K., Ung, C.-T.: Domatic partition on several classes of graphs. In: Lin, G. (ed.) COCOA 2012. LNCS, vol. 7402, pp. 245–256. Springer, Heidelberg (2012). doi:10.1007/978-3-642-31770-5_22

29. Rooij, J.M.M., Bodlaender, H.L., Rossmanith, P.: Dynamic programming on tree decompositions using generalised fast subset convolution. In: Fiat, A., Sanders, P. (eds.) ESA 2009. LNCS, vol. 5757, pp. 566–577. Springer, Heidelberg (2009). doi:10.1007/978-3-642-04128-0_51

30. Shi, Y., Wei, M., Yue, J., Zhao, Y.: Coupon coloring of some special graphs. J. Comb. Optim. **33**(1), 156–164 (2017)

31. Stojmenovic, I., Seddigh, M., Zunic, J.: Dominating sets and neighbor elimination-based broadcasting algorithms in wireless networks. IEEE Trans. Parallel Distrib. Syst. **13**(1), 14–25 (2002)
32. Zelinka, B.: Total domatic number of cacti. Math. Slovaca **38**(3), 207–214 (1988)
33. Zelinka, B.: Total domatic number and degrees of vertices of a graph. Math. Slovaca **39**(1), 7–11 (1989)
34. Zelinka, B.: Domination in generalized Petersen graphs. Czech. Math. J. **52**(1), 11–16 (2002)

Strong Triadic Closure in Cographs and Graphs of Low Maximum Degree

Athanasios L. Konstantinidis[1], Stavros D. Nikolopoulos[2],
and Charis Papadopoulos[1($\boxtimes$)]

[1] Department of Mathematics, University of Ioannina, Ioannina, Greece
konsakis@yahoo.com, charis@cs.uoi.gr
[2] Department of Computer Science and Engineering,
University of Ioannina, Ioannina, Greece
stavros@cs.uoi.gr

Abstract. The MAXSTC problem is an assignment of the edges with strong or weak labels having the maximum number of strong edges such that any two vertices that have a common neighbor with a strong edge are adjacent. The CLUSTER DELETION problem seeks for the minimum number of edge removals of a given graph such that the remaining graph is a disjoint union of cliques. Both problems are known to be NP-hard and an optimal solution for the CLUSTER DELETION problem provides a solution for the MAXSTC problem, however not necessarily an optimal one. In this work we give the first systematic study that reveals graph families for which the optimal solutions for MAXSTC and CLUSTER DELETION coincide. We first show that MAXSTC coincides with CLUSTER DELETION on cographs and, thus, MAXSTC is solvable in quadratic time on cographs. As a side result, we give an interesting computational characterization of the maximum independent set on the cartesian product of two cographs. Furthermore we study how low degree bounds influence the complexity of the MAXSTC problem. We show that this problem is polynomial-time solvable on graphs of maximum degree three, whereas MAXSTC becomes NP-complete on graphs of maximum degree four. The latter implies that there is no subexponential-time algorithm for MAXSTC unless the Exponential-Time Hypothesis fails.

1 Introduction

The principle of strong triadic closure is an important concept in social networks [8]. It states that it is not possible for two individuals to have a strong relationship with a common friend and not know each other [11]. Towards the prediction of the behavior of a network, such a principle has been recently proposed as a maximization problem, called MAXSTC, in which the goal is to maximize the number of strong edges of the underlying graph that satisfy the strong triadic closure [20]. Closely related to the MAXSTC problem is the CLUSTER DELETION problem which finds important applications in areas involving information clustering [1]. In the second problem the goal is to remove the minimum number of edges such that the resulting graph consists of vertex-disjoint union of cliques.

© Springer International Publishing AG 2017
Y. Cao and J. Chen (Eds.): COCOON 2017, LNCS 10392, pp. 346–358, 2017.
DOI: 10.1007/978-3-319-62389-4_29

The connection between MaxSTC and Cluster Deletion arises from the fact that the edges inside the cliques in the resulting graph for Cluster Deletion can be seen as strong edges for MaxSTC which satisfy the strong triadic closure. Thus the number of edges in an optimal solution for Cluster Deletion consists a lower bound for the number of strong edges in an optimal solution for MaxSTC. However there are examples of graphs showing that an optimal solution for MaxSTC contains larger number of edges than an optimal solution for Cluster Deletion [16]. Interestingly there are also families of graphs in which their optimal value for MaxSTC matches such a lower bound. For instance any solution on graphs that do not contain triangles coincide for both problems. Here we initiate a systematic study on other non-trivial classes of graphs for which the optimal solutions for both problems have exactly the same value.

Our main motivation is to further explore the complexity of the MaxSTC problem when restricted to graph classes. As MaxSTC has been recently introduced, there are few results concerning its complexity. The problem has been shown to be NP-complete for general graphs [20] and split graphs [16] whereas it becomes polynomial-time tractable on proper interval graphs and trivially perfect graphs [16]. The NP-completeness on split graphs shows an interesting algorithmic difference between the two problems, since Cluster Deletion on such graphs can be solved in polynomial time [2]. Cluster Deletion is known to be NP-complete on general graphs [19]. It remains NP-complete on chordal graphs and also on graphs of maximum degree four [2,15]. On the positive side Cluster Deletion admits polynomial-time algorithms on proper interval graphs [2], graphs of maximum degree three [15], and cographs [10]. In fact for cographs a greedy algorithm that finds iteratively maximum cliques gives an optimal solution, although no running time was explicitly given in [10].

Such a greedily approach is also proposed for computing a maximal independent set of the cartesian product of general graphs. Summing the partial products between iteratively maximum independent sets consists a lower bound for the cardinality of the maximum independent set of the cartesian product [13,14]. Here we prove that a maximum independent set of the cartesian product of two cographs matches such a lower bound. We would like to note that a polynomial-time algorithm for computing such a maximum independent set is already claimed [12]. However no characterization is given, nor an explicit running time of the algorithm is reported.

Our results. In this work we further explore the complexity of the MaxSTC problem. We consider two unrelated families of graphs, cographs and graphs of bounded degree. Cographs are characterized by the absence of a chordless path on four vertices. For such graphs we prove that the optimal value for MaxSTC matches the optimal value for Cluster Deletion. For doing so, we reveal an interesting vertex partitioning with respect to their maximum clique and maximum independent set. This result enables us to give an $O(n^2)$-time algorithm for MaxSTC on cographs. As a byproduct we characterize a maximum independent set of the cartesian product of two cographs which implies a polynomial-time algorithm for computing such a maximum independent set. Moreover we study the influence of low maximum degree for the MaxSTC problem. We show an interesting complexity dichotomy result: for graphs of maximum degree four MaxSTC remains NP-complete, whereas for graphs of maximum degree three the problem is solved in polynomial time. Our reduction for the NP-completeness on graphs of maximum degree four implies that, under the Exponential-Time Hypothesis, there is no subexponential time algorithm for MaxSTC.

2 Preliminaries

We refer to [3] for our standard graph terminology. For $R \subseteq E(G)$, $G \setminus R$ denotes the graph $(V(G), E(G) \setminus R)$, that is a subgraph of G and for $S \subseteq V(G)$, $G - S$ denotes the graph $G[V(G) - S]$, that is an induced subgraph of G. Given a graph $G = (V, E)$, a *strong-weak labeling* on the edges of G is a bijection $\lambda : E(G) \to \{strong, weak\}$; i.e., λ assigns to each edge of $E(G)$ one of the labels *strong* or *weak*. The *strong triadic closure* of a graph G is a strong-weak labeling λ such that for any two strong edges $\{u, v\}$ and $\{v, w\}$ there is a (weak or strong) edge $\{u, w\}$. The problem of computing the maximum strong triadic closure, denoted by MaxSTC, is to find a strong-weak labeling on the edges of $E(G)$ that satisfies the strong triadic closure and has the maximum number of strong edges. We denote by (E_S, E_W) the partition of $E(G)$ into strong edges E_S and weak edges E_W. The graph spanned by E_S is the graph $G \setminus E_W$. For a strong edge $\{u, v\}$, we say that u (resp., v) is a strong neighbor of v (resp., u). We denote by $N_S(v) \subseteq N(v)$ the strong neighbors of v. Given an optimal solution for MaxSTC that consists of the strong edges E_S, the graph spanned by the edges of E_S is denoted by $E_S(G)$. Whenever we write $|E_S(G)|$ we refer to its number of edges, that is $|E_S(G)| = |E_S|$.

In the CLUSTER DELETION problem the goal is to partition the vertices of a given graph G into vertex-disjoint cliques with the minimum number of edges outside the cliques, or, equivalently, with the maximum number of edges inside the cliques. A *cluster graph* is a graph in which every connected component is a clique. Cluster graphs are characterized as exactly the graphs that do not contain a P_3 as an induced subgraph. Given an optimal solution for CLUSTER DELETION, the cluster graph spanned by the edges that are inside the cliques is denoted by $E_C(G)$. We write $|E_C(G)|$ to denote the number of edges in the cluster graph. Notice that labeling strong all the edges of a cluster graph satisfies the strong triadic closure, so that $|E_C(G)| \leq |E_S(G)|$ holds for any graph G.

3 Computing MaxSTC on Cographs

Let $G = (V, E)$ and $H = (W, F)$ be two undirected graphs with $V \cap W = \emptyset$. The *disjoint union* of G and H is the graph obtained from the union of G and H, denoted by $G \oplus H = (V \cup W, E \cup F)$. The *complete join* of G and H is the graph obtained from the union of G and H and adding edges between every pair of vertices that belong to different graphs, denoted by $G \otimes H = (V \cup W, E \cup F \cup \{vw \mid v \in V, w \in W\})$. A graph is a *cograph* if it can be generated from single-vertex graphs and recursively applying the disjoint union and complete join operations. The complement of a cograph is also a cograph. Cographs are exactly the graphs that do not contain any chordless path on four vertices [5], and they can be recognized in linear time [6].

Let G be the given cograph. Our main goal is to show that there is an optimal solution for MaxSTC on G that coincides with an optimal solution for CLUSTER DELETION on G. The strong edges that belong to an optimal solution for MaxSTC span the graph $E_S(G)$. An optimal solution for CLUSTER DELETION consists of a cluster graph $E_C(G)$ by removing a minimum number of edges of G. Labeling all edges of a cluster graph as strong, results in a strong-weak labeled graph that satisfy the strong triadic closure. Thus our goal is to show that there is an optimal solution $E_S(G)$ for MaxSTC that is a cluster graph.

A clique (resp. independent set) of G having the maximum number of vertices is denoted by $C_{\max}(G)$ (resp., $I_{\max}(G)$). A *greedy clique partition* of G, denoted by $\mathcal{C}$, is the ordering of cliques $(C_1, C_2, \ldots, C_p)$ in G such that $C_1 = C_{\max}(G)$ and $C_i = C_{\max}\left(G - \bigcup_{j=1}^{i-1} C_j\right)$ for $i = 2, 3, \ldots, p$. Similarly, a *greedy independent set partition* of G, denoted by $\mathcal{I}$, is the ordering of independent sets $(I_1, I_2, \ldots, I_q)$ in G such that $I_1 = I_{\max}(G)$ and $I_i = I_{\max}\left(G - \bigcup_{j=1}^{i-1} I_j\right)$ for $i = 2, 3, \ldots, q$. Observe that the subgraph spanned by the edges of $\mathcal{C}$ does not contain any P_3 and, thus, forms a solution for CLUSTER DELETION. Although in general a greedy clique partition does not necessarily imply an optimal solution for CLUSTER DELETION, when restricted to cographs the optimal solution is characterized by the greedy clique partition.

Lemma 1 ([10]). *Let G be a cograph with a greedy clique partition $\mathcal{C}$. Then the edges of $\mathcal{C}$ span an optimal solution $E_C(G)$ for CLUSTER DELETION.*

We will use such a characterization of CLUSTER DELETION in order to give its equivalence with the MaxSTC problem. Notice, however, that due to the freedom of the adjacencies between the cliques of a greedy clique partition, it is not sufficient to consider such a partition of the vertices. For doing so, we will further decompose the cliques of a greedy clique partition. It is known that a graph G is a cograph if and only if for any maximal clique C and any maximal independent set I of every induced subgraph of G, $|C \cap I| = 1$ holds (also known as the *clique-kernel intersection* property) [5]. Thus we state the following lemma.

Lemma 2 ([5]). *Let G be a cograph. Then $C_{\max}(G) \cap I_{\max}(G) = \{v\}$ for some vertex v.*

We recursively apply Lemma 2, to obtain the following result.

Lemma 3. *Let G be a cograph with a greedy clique partition $\mathcal{C} = (C_1, \ldots, C_p)$ and a greedy independent set partition $\mathcal{I} = (I_1, \ldots, I_q)$. For every i, j with $1 \leq i \leq p$ and $1 \leq j \leq q$, if $|C_i| \geq j$ or $|I_j| \geq i$ then $C_i \cap I_j \neq \emptyset$.*

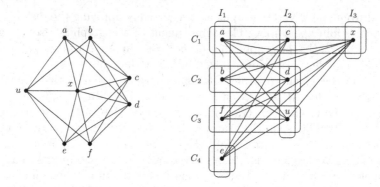

Fig. 1. A cograph and its greedy canonical partition $(\mathcal{C}, \mathcal{I})$ where $\mathcal{C} = (C_1, C_2, C_3, C_4)$ and $\mathcal{I} = (I_1, I_2, I_3)$.

Proof. We prove the statement by induction on p and q. If $p = 1$ (resp., $q = 1$) then $|C_1| \geq 1$ (resp., $|I_1| \geq 1$) and $C_1 \cap I_1 \neq \emptyset$ trivially holds by Lemma 2. Consider the sets C_i and I_j. Let $G_{i,j}$ be the graph obtained from G by removing the sets of vertices $C_1, \ldots, C_{i-1}$ and $I_1, \ldots, I_{j-1}$. By the induction hypothesis we know that $C_i \cap I_1 = \{u_1\}, \ldots, C_i \cap I_{j-1} = \{u_{j-1}\}$ and $C_1 \cap I_j = \{v_1\}, \ldots, C_{i-1} \cap I_j = \{v_{i-1}\}$. We will prove that $C_{\max}(G_{i,j}) = C_i \setminus \{u_1, \ldots, u_{j-1}\}$ and $I_{\max}(G_{i,j}) = I_j \setminus \{v_1, \ldots, v_{i-1}\}$.

Observe that $|C_i \setminus \{u_1, \ldots, u_{j-1}\}| = |C_i| - j + 1$. Assume for contradiction that there is clique C' in $G_{i,j}$ such that $|C'| > |C_i| - j + 1$. Then $C' \cap I_{j-1} = \emptyset$. We consider the subgraph G' of G induced by the vertices of I_{j-1} and the vertices of $G_{i,j}$. Since I_{j-1} is an independent set, at most one vertex of I_{j-1} is adjacent to all the vertices of a maximum clique of $G_{i,j}$. Thus a maximum clique of G' has size at most $|C_i| - j + 2$, so that C' is a maximum clique of G'. Now observe that I_{j-1} is a maximum independent set of G' by the greedy choice of I_{j-1}. Then, however, we reach a contradiction, since by Lemma 2 we know that $C' \cap I_{j-1} \neq \emptyset$. Therefore $C_i \setminus \{u_1, \ldots, u_{j-1}\}$ is a maximum clique of $G_{i,j}$.

Symmetric arguments show that $I_{\max}(G_{i,j}) = I_j \setminus \{v_1, \ldots, v_{i-1}\}$. By the choices of i and j notice that $\{u_1, \ldots, u_{j-1}\} \cap \{v_1, \ldots, v_{i-1}\} = \emptyset$. Then Lemma 2 applies for $G_{i,j}$ which shows that $(C_i \setminus \{u_1, \ldots, u_{j-1}\}) \cap (I_j \setminus \{v_1, \ldots, v_{i-1}\}) \neq \emptyset$. Therefore $C_i \cap I_j \neq \emptyset$, showing the desired statement. □

Lemma 3 suggests a partition of the vertices of G with respect to $\mathcal{C}$ and $\mathcal{I}$ as follows. We call *greedy canonical partition* a pair $(\mathcal{C}, \mathcal{I})$ with elements $\langle v_{i,j} \rangle$ where $1 \leq i \leq p$ and $1 \leq j \leq |C_i|$, such that $V(G) = \{v_{1,1}, \ldots, v_{p,|C_p|}\}$ and $v_{i,j} \in C_i \cap I_j$. Figure 1 shows such a greedy canonical partition of a given cograph. Observe that such a partition corresponds to a 2-dimensional representation of G. By Lemma 3 it follows that a cograph admits a greedy canonical partition.

Let us turn our attention back to the initial MaxSTC problem. We first consider the disjoint union of cographs.

Lemma 4. *Let G and H be vertex-disjoint cographs. Then $E_S(G \oplus H) = E_S(G) \oplus E_S(H)$ and $E_C(G \oplus H) = E_C(G) \oplus E_C(H)$.*

Proof. There are no edges between G and H so that a strong edge of G and a strong edge of H have no common endpoint. Thus the union of the solutions for G and H satisfy the strong triadic closure. By Lemma 1, $E_C(G \oplus H)$ contains the edges of a greedy clique partition which is obtained from the corresponding cliques of G and H.   □

We next consider the complete join of cographs. Given two vertex-disjoint cographs G and H with greedy clique partitions $\mathcal{C} = (C_1, \ldots, C_p)$ and $\mathcal{C}' = (C_1', \ldots, C_{p'}')$, respectively, we denote by $C_i(G, H)$ the edges that have one endpoint in C_i and the other endpoint in C_i', for every $1 \le i \le \min\{p, p'\}$.

Lemma 5. *Let G and H be vertex-disjoint cographs with greedy clique partitions $\mathcal{C} = (C_1, \ldots, C_p)$ and $\mathcal{C}' = (C_1', \ldots, C_{p'}')$, respectively. Then,*

- *$E_S(G \otimes H) = (E_S(G) \oplus E_S(H)) \cup E(G, H)$ and*
- *$E_C(G \otimes H) = (E_C(G) \oplus E_C(H)) \cup E(G, H)$,*

where $E(G, H) = C_1(G, H) \cup \cdots \cup C_k(G, H)$ and $k = \min\{p, p'\}$.

Proof. For the edges of $E_C(G \otimes H)$ we know that a greedy clique partition of $G \otimes H$ forms an optimal solution by Lemma 1. A greedy clique partition of $G \otimes H$ is obtained from the cliques $C_i \cup C_i'$, for every $1 \le i \le k$, since all the vertices of G are adjacent to all the vertices of H. The edges of $C_i \cup C_i'$ can be partitioned into the sets $E(C_i), E(C_i'), C_i(G, H)$ giving the desired formulation for $E_C(G \otimes H)$.

We consider the optimal solution for MaxSTC described by the edges of $E_S(G \otimes H)$. Let us show that any solution on the edges of G satisfy the strong triadic closure in the graph $G \otimes H$. Consider a strong edge $\{x, y\}$ of G. If the resulting labeling does not satisfy the strong triadic closure then there is a strong edge $\{x, w\}$ such that y and w are non-adjacent. As G and H are vertex-disjoint graphs, $w \in V(G)$ or $w \in V(H)$. If $w \in V(G)$ then we already know that the labeling of $E_S(G)$ satisfies the strong triadic closure so that y and w are adjacent. If $w \in V(H)$ then by the complete join operation w is adjacent to y. Thus maximizing the number of strong edges that belong in G and H results in an optimal solution for $G \otimes H$.

We next consider the edges that have one endpoint in G and the other in H, denoted by $E(G, H)$. Our goal is to show that edges of $C_1(G, H) \cup \cdots \cup C_k(G, H)$ belong to an optimal solution. Let $(\mathcal{C}, \mathcal{I})$ and $(\mathcal{C}', \mathcal{I}')$ be the greedy canonical partitions of G and H, respectively, where $\mathcal{C} = (C_1, \ldots, C_p)$, $\mathcal{I} = (I_1, \ldots, I_q)$, and $\mathcal{C}' = (C_1', \ldots, C_{p'}')$, $\mathcal{I}' = (I_1', \ldots, I_{q'}')$. In the forthcoming arguments we prove that $|E(G, H)| \le \sum |C_i||C_i'|$.

We consider the vertices of I_j, $1 \le j \le q$, and count the number of strong edges that have one endpoint in I_j and the other endpoint on a vertex of H. Without loss of generality assume that $|I_1| \le |I_1'|$. Then notice that $k = |I_1|$, since $p = |I_1|$ and $p' = |I_1'|$ by Lemma 3. For a subset W of vertices of G, we denote by $s(W)$ the number of strong edges of $E(G,H)$ that are incident to the vertices of W. By the strong triadic closure, any vertex of H has at most one strong neighbor in I_j and any vertex of G has at most one strong neighbor in $I_{j'}'$, $1 \leq j' \leq q'$. Thus for any $I_{j'}'$ of H there are at most $\min\{|I_j|, |I_{j'}'|\}$ strong edges between the vertices of I_j and $I_{j'}'$. Let r_j be the largest index of $\{1, \ldots, q'\}$ for which $|I_{r_j}'| \geq |I_j|$; notice that r_j exists since $|I_j| \leq |I_1| \leq |I_1'|$. Then, since $|I_1'| \geq \cdots \geq |I_{q'}'|$, it is clear that $|I_j|$ is smaller or equal than any of $|I_1'|, \ldots, |I_{r_j}'|$ and greater than any of $|I_{r_j+1}'|, \ldots, |I_{q'}'|$. Moreover by Lemma 3 we know that for every $1 \leq i \leq |I_j|$, C_i' contains exactly one vertex from each of $I_1', \ldots, I_{r_j}', I_{r_j+1}', \ldots, I_{|C_i'|}'$. Thus we get the following inequality.

$$s(I_j) \leq \sum_{j'=1}^{q'} \min\{|I_j|, |I_{j'}'|\} = \sum_{j'=1}^{r_j} |I_j| + \sum_{j'=r_j+1}^{q'} |I_{j'}'| = \sum_{i=1}^{|I_j|} |C_i'|.$$

Summing up each of $s(I_j)$ for every I_j, $1 \leq j \leq q$, we obtain:

$$|E(G,H)| = \sum_{j=1}^{q} s(I_j) \leq \sum_{j=1}^{q} \sum_{i=1}^{|I_j|} |C_i'|$$

Observe that in the described sum each $|C_i'|$ is counted for all $1 \leq j \leq q$ such that $|I_j| \geq i$. Thus by Lemma 3 the number that $|C_i'|$ appears in the formula is exactly $|C_i|$, since for such $|I_j|$ and i, we have $C_i \cap I_j \neq \emptyset$. Therefore we get the desired upper bound for the number of strong edges in $E(G,H)$:

$$|E(G,H)| \leq \sum_{j=1}^{q} \sum_{i=1}^{|I_j|} |C_i'| = \sum_{i=1}^{|I_1|} |C_i||C_i'|.$$

Notice that the edges of $C_1(G,H) \cup \cdots \cup C_k(G,H)$ satisfy the strong triadic closure, since every two strong edges incident to a vertex of G belong to $C_i(G,H)$ which implies that the endpoints of H belong to a clique C_i' and, thus, are adjacent in $G \otimes H$. Also observe that the number of edges of $C_1(G,H) \cup \cdots \cup C_k(G,H)$ matches the given upper bound. Therefore the claimed formula holds for the strong edges of $E_S(G \otimes H)$ and this concludes the proof. $\square$

Now we are ready to state our claimed result, namely that the solutions for MAXSTC and CLUSTER DELETION coincide for the class of cographs.

Theorem 1. *Let G be a cograph. There is an optimal solution for MAXSTC on G that is a cluster graph. Moreover MAXSTC can be solved in $O(n^2)$ time.*

Proof. An optimal solution for MAXSTC coincides with an optimal solution for CLUSTER DELETION trivially for graphs that consist of a single vertex. If G is a non-trivial cograph then it is constructed by the disjoint union or the complete join operation. In the former case Lemma 4 applies, whereas in the later Lemma 5 applies showing that in all cases $E_S(G) = E_C(G)$.

Regarding the running time notice that given a suitable data structure, called a *cotree*, a maximum clique C_1 of G can be found in $O(n)$ time [5]. At the beginning we construct the cotree of G which takes time $O(n+m)$ [6]. Removing a vertex v from a cograph G and updating the cotree takes $O(d(v))$ time, where $d(v)$ is the degree of v in G [18]. Thus after removing all vertices from G we can maintain the cotree in an overall $O(n+m)$ time. In every intermediate step we first remove the set of vertices C_i in $O(d(C_i))$ time where $d(C_i)$ is the sum of the degree of the vertices of C_i, and then spend $O(n)$ time to compute a maximum clique by using the resulting cotree. Therefore a greedy clique partition of G can be found in total $O(n^2)$ time, since there are at most n such cliques in $\mathcal{C}$.   □

3.1   Maximum Independent Set of the Cartesian Product of Cographs

In this section we apply the characterization of Theorem 1 in order to show an interesting computational characterization of the cartesian product of cographs. Towards such a characterization we take advantage of an equivalent transformation of an optimal solution for MAXSTC in terms of a maximum independent set of an auxiliary graph that is called the *line-incompatibility* graph. The line-incompatibility graph (also known under the term *Gallai graph* [4,17]), denoted by $\Gamma(G)$, has a node uv in $\Gamma(G)$ for every edge $\{u,v\}$ of G, and two nodes uv, vw of $\Gamma(G)$ are adjacent if and only if the vertices u, v, w induce a P_3 in G. The connection between $\Gamma(G)$ and MAXSTC is given in the following result.

Proposition 1 ([16]). *For any graph G, a subset E_S of edges span $E_S(G)$ if and only if the nodes corresponding to E_S form $I_{\max}(\Gamma(G))$.*

Let G and H be two vertex-disjoint graphs. The *cartesian product* of G and H, denoted by $G \times H$, is the graph with the vertex set $V(G) \times V(H)$ and any two vertices (u,u') and (v,v') are adjacent in $G \times H$ if and only if either $u = v$ and u' is adjacent to v' in H, or $u' = v'$ and u is adjacent to v in G. We are interested in computing a maximum independent set of $G \times H$ whenever G and H are cographs. We first characterize the graph $\Gamma(G \otimes H)$ in terms of $G \times H$.

Lemma 6. *Let G and H be two vertex-disjoint cographs. Then, $\Gamma(G \otimes H) = \Gamma(G) \oplus \Gamma(H) \oplus (\overline{G} \times \overline{H})$.*

Proof. Notice that $G \otimes H$ is a connected cograph, as every vertex of G is adjacent to every vertex of H. The edges of $G \otimes H$ can be partitioned into the following sets of edges: $E(G)$, $E(H)$, and $E(G,H)$ where $E(G,H)$ is the set of edges between G and H in $G \otimes H$. By definition the nodes of $\Gamma(G)$ and $\Gamma(H)$ correspond to the sets $E(G)$ and $E(H)$. Moreover since G and H are vertex-disjoint graphs, $\Gamma(G)$ and $\Gamma(H)$ are also node-disjoint. This means that there are no common endpoints in the edges inside G and H. Hence every node of $\Gamma(G)$ is non-adjacent to all nodes of $\Gamma(H)$.

Next we show that every node of $\Gamma(G \otimes H)$ that corresponds to an edge of $E(G,H)$ is non-adjacent to the nodes of $\Gamma(G)$ and $\Gamma(H)$. If a node xy of $\Gamma(G)$

is adjacent to a node xa of $E(G, H)$ then a is a vertex of H and $\{y, a\}$ is not an edge of $G \otimes H$ contradicting the adjacency between the vertices of G and H. Symmetric arguments show that any node of $\Gamma(H)$ is non-adjacent to any node of $E(G, H)$. Thus no node that corresponds to an edge of $E(G, H)$ is adjacent to any node of $\Gamma(G) \oplus \Gamma(H)$.

To complete the proof we need to show that graph of $\Gamma(G \otimes H)$ induced by the nodes of $E(G, H)$ is exactly the graph $\overline{G} \times \overline{H}$. Let x, y be two vertices of G and let w, z be two vertices of H. By definition of $\Gamma(G \otimes H)$ two nodes xw, yz are adjacent if and only if either $x = y$ and w is non-adjacent to z in H (so that w is adjacent to z in $\overline{H}$), or $w = z$ and x is non-adjacent to y in G (so that x is adjacent to y in $\overline{G}$). Such an adjacency corresponds exactly to the definition of the cartesian product of $\overline{G}$ and $\overline{H}$. Therefore the graph of $\Gamma(G \otimes H)$ induced by the nodes of $E(G, H)$ is exactly the graph $\overline{G} \times \overline{H}$.   □

Now we are ready to give the characterization of a maximum independent set of the cartesian product of cographs, in terms of their greedy independent set partition. Although a polynomial-time algorithm for computing such a maximum independent set has already been claimed earlier [12], no characterization is proposed nor an explicit bound on the running time is reported.

Theorem 2. *Let G and H be two vertex-disjoint cographs with greedy independent set partitions $\mathcal{I} = (I_1, \ldots, I_q)$ and $\mathcal{I}' = (I'_1, \ldots, I'_{q'})$, respectively. Then the vertices of $(I_1 \times I'_1) \oplus \cdots \oplus (I_\ell \times I'_\ell)$ form a maximum independent set of $G \times H$, where $\ell = \min\{q, q'\}$. Moreover $I_{\max}(G \times H)$ can be computed in $O(n^2)$ time, where $n = \max\{|V(G)|, |V(H)|\}$.*

Proof. Let $(C_1, \ldots, C_p)$ and $(C'_1, \ldots, C'_{p'})$ be greedy clique partitions of G and H, respectively. By Lemma 5 we know that $E_S(G \otimes H) = E_S(G) \oplus E_S(H) \cup E(G, H)$, where $E(G, H) = C_1(G, H) \cup \cdots \cup C_k(G, H)$ and $k = \min\{p, p'\}$. Notice that if $(C_1, \ldots, C_p)$ is a greedy clique partition for G then $(C_1, \ldots, C_p)$ is a greedy independent set partition for $\overline{G}$. Moreover by Proposition 1 we know that the edges of $E_S(G \otimes H)$ correspond to the nodes of $I_{\max}(\Gamma(G \otimes H))$. Since $\Gamma(G \otimes H) = \Gamma(G) \oplus \Gamma(H) \oplus (\overline{G} \times \overline{H})$ by Lemma 6, we get $E(G, H) = I_{\max}(\overline{G} \times \overline{H})$. Therefore the vertices of $(I_1 \times I'_1) \oplus \cdots \oplus (I_\ell \times I'_\ell)$ consist a $I_{\max}(\Gamma(G \otimes H))$.

For the running time we need to compute two greedy independent set partitions $(I_1, \ldots, I_q)$ and $(I'_1, \ldots, I'_{q'})$ for G and H, respectively, and then combine each of I_j with I'_j, for $1 \le j \le \ell$. Computing a greedy independent set partition for a cograph G can be done in $O(n^2)$ time by applying the algorithm on $\overline{G}$ given in the proof of Theorem 1. Therefore the total running time is bounded by $O(|V(G)|^2 + |V(H)|^2)$.   □

4   Graphs of Low Maximum Degree

Here we study the influence of the bounded degree in a graph for the MAXSTC problem. We show an interesting complexity dichotomy result: for graphs of maximum degree four MaxSTC remains NP-complete, whereas for graphs of maximum degree three the problem is solved in polynomial time.

We prove the hardness result even on a proper subclass of graphs with maximum degree four. A graph G is a *4-regular K_4-free graph*, if every vertex of G has degree four and there is no K_4 in G. The decision version of MaxSTC takes as input a graph G and an integer k and asks whether there is a strong-weak labeling of G that satisfies the strong triadic closure with at least k strong edges. Similarly the decision version of CLUSTER DELETION takes as input a graph G and an integer k and asks whether G has a spanning cluster subgraph by removing at most k edges. It is known that the decision version of CLUSTER DELETION on connected 4-regular K_4-free graphs is NP-complete [15].

Theorem 3. *The decision version of MaxSTC is NP-complete on connected 4-regular K_4-free graphs.*

Proof. We give a polynomial-time reduction to MaxSTC from the CLUSTER DELETION problem on connected 4-regular K_4-free graphs which is already known to be NP-complete [15]. Let $G = (V, E)$ be a connected 4-regular K_4-free graph with $n = 3q$ and $2n$ edges. Let $E_C(G)$ be a solution for the CLUSTER DELETION with $k = n$ edges. It is not difficult to see that every connected component of $E_C(G)$ is a triangle, since the graph is 4-regular and K_4 is a forbidden graph [15]. Then $E_C(G)$ is a solution for MaxSTC with at least n strong edges.

For the opposite direction, assume that $E_S(G)$ is a solution for MaxSTC with at least n strong edges. We show that the graph spanned by the strong edges of $E_S(G)$ is a two-regular graph. That is, every vertex of G has exactly two strong neighbors. Assume that there is a vertex v that has at least three strong neighbors. By the strong triadic closure all its strong neighbors must induce a clique in G. Then $N[v]$ induces a K_4 which is a forbidden subgraph. Thus every vertex has at most two strong neighbors. Furthermore if there is a vertex having only one strong neighbor then $|E_S(G)| < n$ which contradicts the assumption of n strong edges. Hence every vertex has exactly two strong neighbors in $E_S(G)$.

Since $E_S(G)$ is a 2-regular graph we know that the graph spanned by the strong edges is the disjoint union of triangles or chordless cycles C_p, with $4 \le p \le n$. Let us also rule out that a connected component of $E_S(G)$ is a chordless cycle on four vertices C_4. To see this, observe that if there is a C_4 in $E_S(G)$ then the four vertices of the C_4 induce a K_4 in G. Now assume that there is a connected component of $E_S(G)$ that is a chordless cycle C_p with $4 < p < n$. In such a connected component, every vertex belongs to two distinct P_3's as an endpoint. More precisely, let $v_1, \ldots, v_p$ be the vertices of C_p such that $\{v_i, v_{i+1}\}$ and $\{v_p, v_1\}$ are strong edges with $1 \le i < p$. Then for every vertex v_i of C_p there two P_3's v_{i-2}, v_{i-1}, v_i and v_i, v_{i+1}, v_{i+2} such that $v_{i-2} \ne v_{i+2}$. By the strong triadic closure we know that v_i is adjacent to both v_{i-2} and v_{i+2} in G. Since G is a 4-regular graph, there are no more edges incident to any vertex of C_p. Thus every vertex of C_p is non-adjacent to any other vertex of $G-C_p$ which contradicts the original connectivity of G. Therefore either every connected component of $E_S(G)$ is a triangle, or $E_S(G)$ is connected and $E_S(G) = C_n$.

If every connected component of $E_S(G)$ is a triangle then clearly $E_S(G)$ spans a cluster graph. Suppose that $E_S(G) = C_n$. Since $n = 3q$, we can partition the vertices of C_n into q triangles with the same number of strong edges as follows. For every triplet of vertices v_i, v_{i+1}, v_{i+2}, $1 \leq i \leq n - 2$, we further label the edge $\{v_i, v_{i+2}\}$ strong and the edges $\{v_{i+2}, v_{i+3}\}$ and $\{v_n, v_1\}$ are labeled weak. Observe that $\{v_i, v_{i+2}\}$ is an edge of G, since both $\{v_i, v_{i+1}\}, \{v_{i+1}, v_{i+2}\}$ are strong edges. Such a labeling satisfies the strong triadic closure property and maintain the same number of strong edges. Therefore in every case a solution for MaxSTC with n edges can be equivalently transformed into a solution for CLUSTER DELETION with n edges.   □

We can also obtain lower bounds for the running time of MaxSTC with respect to the integer k (size of the solution) or the number of vertices n. A subexponential-time algorithm for MaxSTC would imply an algorithm for solving CLUSTER DELETION that has running time subexponential in the size of the solution k or the number of vertices n. However CLUSTER DELETION does not admit such subexponential-time algorithms even if we restrict to graphs of maximum degree four [15]. Since we can reduce CLUSTER DELETION to MaxSTC instances on the same graph with $k = n$, we arrive at the following.

Corollary 1. MaxSTC *cannot be solved in* $2^{o(k)} \cdot poly(n)$ *time or in* $O(2^{o(n)})$ *time unless the exponential-time hypothesis fails.*

We stress that due to Proposition 1, MaxSTC reduces to finding a minimum vertex cover of $\Gamma(G)$ corresponding to the weak edges in an optimal solution. Thus MaxSTC admits algorithms with running times $2^{\Omega(k)}$ poly(n) or $O^*(c^n)^1$ where k is the minimum number of weak edges and $c < 2$ is a constant [7,9].

Now let us show that if we restrict to graphs of maximum degree three then MaxSTC becomes polynomial-time solvable. Our goal is to show that there is an optimal solution for MaxSTC that is a cluster graph, since CLUSTER DELETION is solved in polynomial time on such graphs [15].

Theorem 4. *Let G be a graph with maximum degree three. Then there is an optimal solution for MaxSTC on G that is a cluster graph.*

Since CLUSTER DELETION can be solved in $O(n^{1.5} \cdot \log^2 n)$ on graphs with maximum degree three [15], we get the following result.

Corollary 2. MaxSTC *can be solved in* $O(n^{1.5} \cdot \log^2 n)$ *time when the input graph has maximum degree three.*

5   Concluding Remarks

We have performed a systematic study on families of graphs for which the optimal solutions for MaxSTC and CLUSTER DELETION problems coincide. As an

---

[1] The O^* notation suppresses polynomial factors of n.

important outcome, we have complemented previous results regarding the complexity of MaxSTC when restricted to cographs or graphs of bounded degree. Some open questions arise from our work. It is interesting to completely characterize graphs by forbidden subgraphs for which MaxSTC and Cluster Deletion solutions coincide. Towards such an approach, Proposition 1 seems a useful tool. Moreover, despite the fact that the optimal solutions for MaxSTC and Cluster Deletion do not coincide in a superclass of cographs, namely to that of permutation graphs, both problems restricted to such graphs have unresolved complexity status which is interesting to settle.

A more general and realistic scenario for both problems is to restrict the choice of the considered edges. Assume that a subset F of edges is required to be included in the same clusters for Cluster Deletion or those edges are required to be strong for MaxSTC. Then it is natural to ask for a suitable set of edges $E' \subseteq E \setminus F$ with $|E'|$ as large as possible such that the edges of $E' \cup F$ span a cluster graph or satisfy the strong triadic closure. Clarifying the complexity of such generalized problems is interesting on graphs for which Cluster Deletion or MaxSTC are solved in polynomial time.

References

1. Bansal, N., Blum, A., Chawla, S.: Correlation clustering. Mach. Learn. **56**, 89–113 (2004)
2. Bonomo, F., Durán, G., Valencia-Pabon, M.: Complexity of the cluster deletion problem on subclasses of chordal graphs. Theoret. Comput. Sci. **600**, 59–69 (2015)
3. Brandstädt, A., Le, V.B., Spinrad, J.: Graph Classes: A Survey. Society for Industrial and Applied Mathematics (1999)
4. Cochefert, M., Couturier, J.-F., Golovach, P.A., Kratsch, D., Paulusma, D.: Parameterized algorithms for finding square roots. Algorithmica **74**, 602–629 (2016)
5. Corneil, D.G., Lerchs, H., Stewart, L.K.: Complement reducible graphs. Discret. Appl. Math. **3**, 163–174 (1981)
6. Corneil, D.G., Perl, Y., Stewart, L.K.: A linear recognition algorithm for cographs. SIAM J. Comput. **14**, 926–934 (1985)
7. Cygan, M., Fomin, F.V., Kowalik, L., Lokshtanov, D., Marx, D., Pilipczuk, M., Pilipczuk, M., Saurabh, S.: Parameterized Algorithms. Springer, Cham (2015)
8. Easley, D., Kleinberg, J.: Networks, Crowds, and Markets: Reasoning About a Highly Connected World. Cambridge University Press, New York (2010)
9. Fomin, F.V., Kratsch, D.: Exact Exponential Algorithms. Springer, Heidelberg (2010)
10. Gao, Y., Hare, D.R., Nastos, J.: The cluster deletion problem for cographs. Discret. Math. **313**, 2763–2771 (2013)
11. Granovetter, M.: The strength of weak ties. Am. J. Sociol. **78**, 1360–1380 (1973)
12. Hon, W.-K., Kloks, T., Liu, H.-H., Poon, S.-H., Wang, Y.-L.: On independence domination. In: Gąsieniec, L., Wolter, F. (eds.) FCT 2013. LNCS, vol. 8070, pp. 183–194. Springer, Heidelberg (2013). doi:10.1007/978-3-642-40164-0_19
13. Imrich, W., Klavzar, S., Rall, D.F.: Topics in Graph Theory: Graphs and Their Cartesian Product. AK Peters Ltd., Wellesley (2008)
14. Jha, P.K., Slutzki, G.: Independence numbers of product graphs. Appl. Math. Lett. **7**, 91–94 (1994)

15. Komusiewicz, C., Uhlmann, J.: Cluster editing with locally bounded modifications. Discret. Appl. Math. **160**, 2259–2270 (2012)
16. Konstantinids, A., Papadopoulos, C.: Maximizing the strong triadic closure in split graphs and proper interval graphs. CoRR, abs/1609.09433 (2016)
17. Le, V.B.: Gallai graphs and anti-gallai graphs. SIAM J. Discret. Math. **159**, 179–189 (1996)
18. Shamir, R., Sharan, R.: A fully dynamic algorithm for modular decomposition and recognition of cographs. Discret. Appl. Math. **136**, 329–340 (2004)
19. Shamir, R., Sharan, R., Tsur, D.: Cluster graph modification problems. Discret. Appl. Math. **144**, 173–182 (2004)
20. Sintos, S., Tsaparas, P.: Using strong triadic closure to characterize ties in social networks. In: Proceedings of KDD 2014, pp. 1466–1475 (2014)

Hardness and Structural Results for Half-Squares of Restricted Tree Convex Bipartite Graphs

Hoang-Oanh Le[1] and Van Bang Le[2]([⊠])

[1] Berlin, Germany
lehoangoanh@web.de
[2] Institut Für Informatik, Universität Rostock, Rostock, Germany
van-bang.le@uni-rostock.de

Abstract. Let $B = (X, Y, E)$ be a bipartite graph. A half-square of B has one color class of B as vertex set, say X; two vertices are adjacent whenever they have a common neighbor in Y. Every planar graph is half-square of a planar bipartite graph, namely of its subdivision. Until recently, only half-squares of planar bipartite graphs (the map graphs) have been investigated, and the most discussed problem is whether it is possible to recognize these graphs faster and simpler than Thorup's $O(n^{120})$ time algorithm.

In this paper, we identify the first hardness case, namely that deciding if a graph is a half-square of a balanced bisplit graph is NP-complete. (Balanced bisplit graphs form a proper subclass of star convex bipartite graphs.) For classical subclasses of tree convex bipartite graphs such as biconvex, convex, and chordal bipartite graphs, we give good structural characterizations of their half-squares that imply efficient recognition algorithms. As a by-product, we obtain new characterizations of unit interval graphs, interval graphs, and of strongly chordal graphs in terms of half-squares of biconvex bipartite, convex bipartite, and chordal bipartite graphs, respectively.

1 Introduction

The square of a graph H, denoted H^2, is obtained from H by adding new edges between two distinct vertices whenever their distance is two. Then, H is called a square root of $G = H^2$. Given a graph G, it is NP-complete to decide if G is the square of some graph H [23], even for a split graph H [16].

Given a bipartite graph $B = (X, Y, E_B)$, the subgraphs of the square B^2 induced by the color classes X and Y, $B^2[X]$ and $B^2[Y]$, are called the two *half-squares* of B [4].

While not every graph is the square of a graph and deciding if a graph is the square of a graph is hard, *every* graph $G = (V, E_G)$ is half-square of a bipartite graph: if $B = (V, E_G, E_B)$ is the bipartite graph with $E_B = \{ve \mid v \in V, e \in E_G, v \in e\}$, then clearly $G = B^2[V]$. So one is interested in half-squares of special

© Springer International Publishing AG 2017
Y. Cao and J. Chen (Eds.): COCOON 2017, LNCS 10392, pp. 359–370, 2017.
DOI: 10.1007/978-3-319-62389-4_30

bipartite graphs. Note that B is the subdivision of G, hence every planar graph is half-square of a planar bipartite graph.

Let $\mathcal{B}$ be a class of bipartite graphs. A graph $G = (V, E_G)$ is called *half-square* of $\mathcal{B}$ if there exists a bipartite graph $B = (V, W, E_B)$ in $\mathcal{B}$ such that $G = B^2[V]$. Then, B is called a $\mathcal{B}$-*half root* of G. With this notion, the following decision problem arises.

HALF-SQUARE OF $\mathcal{B}$

Instance: A graph $G = (V, E_G)$

Question: Is G half-square of a bipartite graph in $\mathcal{B}$, i.e., does there exist a bipartite graph $B = (V, W, E_B)$ in $\mathcal{B}$ such that $G = B^2[V]$?

In this paper, we discuss HALF-SQUARE OF $\mathcal{B}$ for several restricted bipartite graph classes $\mathcal{B}$.

Previous results and related work. Half-squares of bipartite graphs have been introduced in [4] in order to give a graph-theoretical characterization of the so-called *map graphs*. It turns out that map graphs are exactly half-squares of planar bipartite graphs. As we have seen at the beginning, every planar graph is a map graph. The main problem concerning map graphs is to recognize if a given graph is a map graph. In [27], Thorup shows that HALF-SQUARE OF PLANAR, that is, deciding if a graph is a half-square of a planar bipartite graph, can be solved in polynomial time[1]. Very recently, in [22], it is shown that HALF-SQUARES OF OUTERPLANAR and HALF-SQUARE OF TREE are solvable in linear time. Other papers deal with solving hard problems in map graphs include [3,5, 6,8]. Some applications of map graphs have been addressed in [1].

Our results. We identify the first class $\mathcal{B}$ of bipartite graphs for which HALF-SQUARE OF $\mathcal{B}$ is NP-hard. Our class $\mathcal{B}$ is a subclass of the class of the bisplit bipartite graphs and of star convex bipartite graphs (all terms are given later). For some other subclasses of bipartite graphs, such as biconvex, convex, and chordal bipartite graphs, we give structural descriptions for their half-squares, that imply polynomial-time recognition algorithms:

- Recognizing half-squares of balanced bisplit graphs (a proper subclass of star convex bipartite graphs) is hard, even when restricted to co-bipartite graphs;
- Half-squares of biconvex bipartite graphs are precisely the unit interval graphs;
- Half-squares of convex bipartite graphs are precisely the interval graphs;
- Half-squares of chordal bipartite graphs are precisely the strongly chordal graphs.

2   Preliminaries

Let $G = (V, E_G)$ be a graph with vertex set $V(G) = V$ and edge set $E(G) = E_G$. A *stable set* (a *clique*) in G is a set of pairwise non-adjacent (adjacent) vertices.

---

[1] Thorup did not give the running time explicitly, but it is estimated to be roughly $O(n^{120})$ with n being the vertex number of the input graph.

The complete graph on n vertices, the complete bipartite graph with s vertices in one color class and t vertices in the other color class, the cycle with n vertices are denoted $K_n, K_{s,t}$, and C_n, respectively. A K_3 is also called a *triangle*, a complete bipartite graph is also called a *biclique*, a complete bipartite graph $K_{1,n}$ is also called a *star*.

The neighborhood of a vertex v in G, denoted by $N_G(v)$, is the set of all vertices in G adjacent to v; if the context is clear, we simply write $N(v)$. A *universal vertex* v in G is one with $N(v) = V \setminus \{v\}$, i.e., v is adjacent to all other vertices in G.

For a subset $W \subseteq V$, $G[W]$ is the subgraph of G induced by W, and $G - W$ stands for $G[V \setminus W]$. We write $B = (X, Y, E_B)$ for bipartite graphs with a bipartition into stable sets X and Y. For subsets $S \subseteq X$, $T \subseteq Y$ we denote $B[S, T]$ for the bipartite subgraph of B induced by $S \cup T$.

We will consider half-squares of the following well-known subclasses of bipartite graphs: Let $B = (X, Y, E_B)$ be a bipartite graph.

- B is *X-convex* if there is a linear ordering on X such that, for each $y \in Y$, $N(y)$ is an interval in X. Being *Y-convex* is defined similarly. B is *convex* if it is X-convex or Y-convex. B is *biconvex* if it is both X-convex and Y-convex. We write CONVEX and BICONVEX to denote the class of convex bipartite graphs, respectively, the class of biconvex bipartite graphs.
- B is *chordal bipartite* if B has no induced cycle of length at least six. CHORDAL BIPARTITE stands for the class of chordal bipartite graphs.
- B is *tree X-convex* if there exists a tree $T = (X, E_T)$ such that, for each $y \in Y$, $N(y)$ induces a subtree in T. Being *tree Y-convex* is defined similarly. B is *tree convex* if it is tree X-convex or tree Y-convex. B is *tree biconvex* if it is both tree X-convex and tree Y-convex. When T is a star, we also speak of *star convex* and *star biconvex* bipartite graphs.
  TREE CONVEX and TREE BICONVEX are the class of all tree convex and all tree biconvex bipartite graphs, respectively, and STAR CONVEX and STAR BICONVEX are the class of all star convex and all star biconvex bipartite, respectively.

It is known that BICONVEX $\subset$ CONVEX $\subset$ CHORDAL BIPARTITE $\subset$ TREE BICONVEX $\subset$ TREE CONVEX. All inclusions are proper; see [2, 19, 26] for more information on these graph classes.

Given a graph G, we often use the following two kinds of bipartite graphs associated to G:

Definition 1. *Let $G = (V, E_G)$ be an arbitrary graph.*

- *The bipartite graph $B = (V, E_G, E_B)$ with $E_B = \{ve \mid v \in V, e \in E_G, v \in e\}$ is the* subdivision *of G.*
- *Let $\mathcal{C}(G)$ denote the set of all maximal cliques of G. The bipartite graph $B = (V, \mathcal{C}(G), E_B)$ with $E_B = \{vQ \mid v \in V, Q \in \mathcal{C}(G), v \in Q\}$ is the* vertex-clique incidence bipartite graph *of G.*

Note that the subdivision of a planar graph is planar, and subdivisions and vertex-clique incidence graphs of triangle-free graphs coincide.

Proposition 1. *Every graph is half-square of its vertex-clique incidence bipartite graph. More precisely, if $B = (V, \mathcal{C}(G), E_B)$ is the vertex-clique incidence bipartite graph of $G = (V, E_G)$, then $G = B^2[V]$. Similar statement holds for subdivisions.*

Proof. For distinct vertices $u, v \in V$, $uv \in E_G$ if and only if $u, v \in Q$ for some $Q \in \mathcal{C}(G)$, if and only if u and v are adjacent in $B^2[V]$. That is, $G = B^2[V]$. □

3   Recognizing Half-Squares of Balanced Bisplit Graphs Is Hard

Recall that a biclique is a complete bipartite graph. Following the concept of split graphs, we call a bipartite graph *bisplit* if it can be partitioned into a biclique and a stable set. In this section, we show that HALF-SQUARE OF BALANCED BISPLIT is NP-hard. Balanced bisplit graphs form a proper subclass of bisplit graphs, and are defined as follows.

Definition 2. *A bipartite graph $B = (X, Y, E_B)$ is called* balanced bisplit *if it satisfies the following properties:*

(i) $|X| = |Y|$;
(ii) *there is partition $X = X_1 \,\dot{\cup}\, X_2$ such that $B[X_1, Y]$ is a biclique;*
(iii) *there is partition $Y = Y_1 \,\dot{\cup}\, Y_2$ such that the edge set of $B[X_2, Y_2]$ is a perfect matching.*

Note that by (i) and (iii), $|X_1| = |Y_1|$, and by (ii) and (iii), every vertex in X_1 is universal in $B^2[X]$.

In order to prove the NP-hardness of HALF-SQUARE OF BALANCED BISPLIT, we will reduce the following well-known NP-complete problem EDGE CLIQUE COVER to it.

EDGE CLIQUE COVER

Instance: A graph $G = (V, E_G)$ and a positive integer k.
Question: Do there exist k cliques in G such that each edge of G is contained in some of these cliques?

EDGE CLIQUE COVER is NP-complete [12,14,24], even when restricted to co-bipartite graphs [17]. (A co-bipartite graph is the complement of a bipartite graph.)

Theorem 1. HALF-SQUARE OF BALANCED BISPLIT *is NP-complete, even when restricted to co-bipartite graphs.*

Proof. It is clear that HALF-SQUARE OF BALANCED BISPLIT is in NP, since guessing a bipartite-half root $B = (V, W, E_B)$ with $|W| = |V|$, verifying that B is balanced bisplit, and $G = B^2[V]$ can obviously be done in polynomial time. Thus, by reducing EDGE CLIQUE COVER to HALF-SQUARE OF BALANCED BISPLIT, we will conclude that HALF-SQUARE OF BALANCED BISPLIT is NP-complete.

Let $(G = (V, E_G), k)$ be an instance of EDGE CLIQUE COVER. Note that we may assume that $k \leq |E_G|$, and that G is connected and has no universal vertices. We construct an instance $G' = (V', E_{G'})$ as follows: G' is obtained from G by adding a set U of k new vertices, $U = \{u_1, \ldots, u_k\}$, and all edges between vertices in U and all edges uv with $u \in U$, $v \in V$. Thus, $V' = V \cup U$, $G'[V] = G$ and the k new vertices in U are exactly the universal vertices of G'. Clearly, G' can be constructed in polynomial time $O(k|V|) = O(|E_G| \cdot |V|)$, and in addition, if G is co-bipartite, then G' is co-bipartite, too. We now show that $(G, k) \in$ EDGE CLIQUE COVER if and only if $G' \in$ HALF-SQUARE OF BALANCED BISPLIT.

First, suppose that the edges of $G = (V, E_G)$ can be covered by k cliques $Q_1, \ldots, Q_k$ in G. We are going to show that G' is half-square of some balanced bisplit bipartite graph. Consider the bipartite graph $B = (V', W, E_B)$ (see also Fig. 1) with

$$W = W_1 \cup W_2, \text{ where } W_1 = \{w_1, \ldots, w_k\}, \text{ and } W_2 = \{w_v \mid v \in V\}.$$

In particular, $|V'| = |W| = k + |V|$. The edge set E_B is as follows:

- B has all edges between U and W, i.e., $B[U, W]$ is a biclique,
- B has edges vw_v, $v \in V$. Thus, the edge set of $B[V, W_2]$ forms a perfect matching, and
- B has edges vw_i, $v \in V$, $1 \leq i \leq k$, whenever $v \in V$ is contained in clique Q_i, $1 \leq i \leq k$.

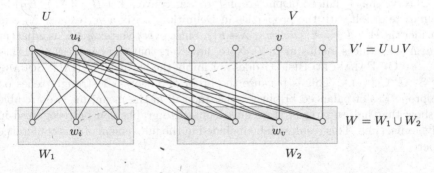

Fig. 1. The balanced bisplit graph $B = (V', W, E_B)$ proving $G' = B^2[V']$; $v \in V$ is adjacent to $w_i \in W_1$ if and only if $v \in Q_i$.

Thus, B is a balanced bisplit graph. Moreover, by the construction of B, we have in $B^2[V']$:

- $U = \{u_1, \ldots, u_k\}$ is a clique (as $B[U, W]$ is a biclique);
- every vertex $u \in U$ is adjacent to all vertices $v \in V$ (recall that G is connected, so every $v \in V$ is in some Q_i, and $w_i \in W_1$ is a common neighbor of u and v), and
- no two distinct vertices $v, z \in V$ have common neighbor in W_2. So u and z are adjacent in $B^2[V']$ if and only if v and z have a common neighbor w_i in W_1, if and only if v and z belong to clique Q_i in G, if and only if u and z are adjacent in G.

That is, $G' = B^2[V']$.

Conversely, suppose $G' = H^2[V']$ for some balanced bisplit graph $H = (V', Y, E_H)$ with $|V'| = |Y|$ and partitions $V' = X_1 \,\dot\cup\, X_2$ and $Y = Y_1 \,\dot\cup\, Y_2$ as in Definition 2. We are going to show that the edges of G can be covered by k cliques. As $H[X_1, Y]$ is a biclique, all vertices in X_1 are universal in $H^2[V'] = G'$. Hence

$$X_1 = U$$

because no vertex in $V = V' \setminus U$ is universal in G' (recall that G has no universal vertex). Therefore

$$X_2 = V \text{ and } G = H^2[V].$$

Note that, as H is a balanced bisplit graph, $|Y_1| = |U| = k$. Write $Y_1 = \{q_1, \ldots, q_k\}$ and observe that no two vertices in V have a common neighbor in Y_2. Thus, for each edge vz in $G = H^2[V]$, v and z have a common neighbor q_i in Y_1. Therefore, the k cliques Q_i in $H^2[V]$, $1 \le i \le k$, induced by the neighbors of q_i in V, cover the edges of G.   □

Theorem 1 indicates that recognizing half-squares of restricted bipartite graphs is algorithmically much more complex than recognizing squares of bipartite graphs; the latter can be done in polynomial time [15].

Observe that balanced bisplit graphs are star convex: Let $B = (X, Y, E_B)$ be a bipartite graph with the properties in Definition 2. Fix a vertex $u \in X_1$ and consider the star $T = (X, \{uv \mid v \in X - u\})$. Since every vertex $y \in Y$ is adjacent to u, $N(y)$ induces a substar in T. Note, however, that the hardness of HALF-SQUARE OF BALANCED BISPLIT does not imply the hardness of HALF-SQUARE OF STAR CONVEX. This is because the proof of Theorem 1 strongly relies on the properties of balanced bisplit graphs. Indeed, in the meantime, we are able to show that half-squares of star-convex bipartite graphs can be recognized in polynomial time. This result will be included in the full version of this conference paper.

4   Half-Squares of Biconvex and Convex Bipartite Graphs

In this section, we show that half-squares of convex bipartite graphs are precisely the interval graphs and half-squares of biconvex bipartite graphs are precisely the unit interval graphs.

Recall that $G = (V, E_G)$ is an interval graph if it admits an interval representation $I(v), v \in V$, such that two vertices in G are adjacent if and only if the corresponding intervals intersect. Let G be an interval graph. It is well-known [9,10] that there is a linear ordering of the maximal cliques of G, say $\mathcal{C}(G) = (Q_1, \ldots, Q_q)$, such that every vertex of G belongs to maximal cliques that are consecutive in that ordering, that is, for every vertex u of G, there are indices $\ell(u)$ and $r(u)$ with

$$\{i \mid 1 \leq i \leq q \text{ and } u \in Q_i\} = \{i \mid \ell(u) \leq i \leq r(u)\}.$$

If C and D are distinct maximal cliques of G, then $C \setminus D$ and $D \setminus C$ are both not empty, that is, for every $j \in \{1, \ldots, q\}$, there are vertices u and v such that $r(u) = \ell(v) = j$.

Recall that unit interval graphs are those interval graphs admitting an interval representation in which all intervals have the same length. It is well known [25] that a graph is a unit interval graphs if and only if it has an interval representation in which no interval is properly contained in another interval (a proper interval graph), if and only if it is a $K_{1,3}$-free interval graph.

Lemma 1. *The half-squares of a biconvex bipartite graph are $K_{1,3}$-free.*

Proof. Let $B = (X, Y, E_B)$ be a biconvex bipartite graph. By symmetry we need only to show that $B^2[X]$ is $K_{1,3}$-free. Suppose, by contradiction, that x_1, x_2, x_3, x_4 induce a $K_{1,3}$ in $B^2[X]$ with edges $x_1 x_2, x_1 x_3$ and $x_1 x_4$. Let y_i be a common neighbor of x_1 and x_2, y_j be a common neighbor of x_1 and x_3, and y_k be a common neighbor of x_1 and x_4. Then, y_i, y_j, y_k are pairwise distinct and induce, in B, a subdivision of $K_{1,3}$. This is a contradiction because biconvex bipartite graphs do not contain an induced subdivision of the $K_{1,3}$. Thus, the half-squares of a biconvex bipartite graph are $K_{1,3}$-free.   □

Lemma 2

(i) *Every interval graph is half-square of a convex bipartite graph. More precisely, if $G = (V, E_G)$ is an interval graph and $B = (V, \mathcal{C}(G), E_B)$ is the vertex-clique incidence bipartite graph of G, then $G = B^2[V]$ and B is $\mathcal{C}(G)$-convex.*

(ii) *If $B = (X, Y, E_B)$ is X-convex, then $B^2[Y]$ is an interval graph.*

Proof. (i) Let $G = (V, E_G)$ be an interval graph, and let $B = (V, \mathcal{C}(G), E_B)$ be the vertex-clique incidence bipartite graph of G. Since each $v \in V$ appears in the interval $\{Q_i \mid \ell(v) \leq i \leq r(v)\}$ in $\mathcal{C}(G) = (Q_1, \ldots, Q_q)$, B is $\mathcal{C}(G)$-convex. Moreover, by Proposition 1, $G = B^2[V]$.

(ii) This is because X admits a linear ordering such that, for each $y \in Y$, $N(y)$ is an interval in X. This collection is an interval representation of $B^2[Y]$ because y and y' are adjacent in $B^2[Y]$ if and only if $N(y) \cap N(y') \neq \emptyset$.   □

Theorem 2. *A graph is half-square of a biconvex bipartite graph if and only if it is a unit interval graph.*

Proof. First, by Lemma 2 (ii), half-squares of biconvex bipartite graphs are interval graphs, and then by Lemma 1, half-squares of biconvex bipartite graphs are unit interval graphs.

Next we show that every unit interval graph is half-square of some biconvex bipartite graph. Let $G = (V, E_G)$ be a unit interval graph. Let $B = (V, \mathcal{C}(G), E_B)$ be the vertex-clique incidence bipartite graph of G. By Lemma 2 (i), $G = B^2[V]$ and B is $\mathcal{C}(G)$-convex. We now are going to show that B is V-convex, too.

Consider a linear order in $\mathcal{C}(G)$, $\mathcal{C}(G) = (Q_1, \ldots, Q_q)$, such that each $v \in V$ is contained in exactly the cliques Q_i, $\ell(v) \le i \le r(v)$. Let $v \in V$ be lexicographically sorted according $(\ell(v), r(v))$. We claim that B is V-convex with respect to this ordering. Assume, by a contradiction, that some Q_i has neighbors v, u and non-neighbor x with $v < x < u$ in the sorted list, say. In particular, v, u belong to Q_i, but x not; see also Fig. 2.

$Q_{\ell(v)}$		$Q_{\ell(x)-1}$	$Q_{\ell(x)}$		$Q_{r(x)}$	$Q_{r(x)+1}$		Q_i
	...			...			...	
v		v	v		v	v		v
$\star$		y	x		x	z		u
$\vdots$		$\vdots$	$\vdots$		$\vdots$	$\vdots$		$\vdots$

Fig. 2. Assuming $v < x < u$, and $v, u \in Q_i$, but $x \notin Q_i$.

Since $x < u$ and $\ell(u) \le i$, we have $\ell(x) < i$. Since u is not in Q_i, we therefore have
$$\ell(x) \le r(x) < i.$$
In particular, $r(x) + 1 \le i$. Since $v < x$ and $r(v) \ge i$, we have
$$\ell(v) < \ell(x).$$
In particular, $\ell(x) - 1 \ge 1$. Now, by the maximality of the cliques, there exists $y \in Q_{\ell(x)-1}$ with $r(y) = \ell(x) - 1$ (hence y is non-adjacent to x), and there exists $z \in Q_{r(x)+1}$ with $\ell(z) = r(x) + 1$ (hence z is non-adjacent to x and y; note that $r(x) + 1 = i$ and $z = u$ are possible). But then v, x, y, and z induce a $K_{1,3}$ in G, a contradiction.

Thus, we have seen that every unit interval graph is half-square of a biconvex bipartite graph.   □

We next characterize half-squares of convex bipartite graphs as interval graphs. This is somehow surprising because the definition of being convex bipartite is asymmetric with respect to the two half-squares.

Theorem 3. *A graph is a half-square of a convex bipartite graph if and only if it is an interval graph.*

Proof. By Lemma 2 (i), interval graphs are half-squares of convex bipartite graphs. It remains to show that half-squares of convex bipartite graphs are interval graphs. Let $B = (X, Y, E_B)$ be an X-convex bipartite graph. By Lemma 2 (ii), $B^2[Y]$ is an interval graph. We now are going to show that $B^2[X]$ is an interval graph, too.

Let $B' = (X, Y', E_{B'})$ be obtained from B by removing all vertices $y \in Y$ such that $N_B(y)$ is properly contained in $N_B(y')$ for some $y' \in Y$. Clearly, $B^2[X] = B'^2[X]$. We show that B' is Y'-convex, hence, by Lemma 2 (ii), $B^2[X] = B'^2[X]$ is an interval graph, as claimed. To this end, let $X = \{x_1, \ldots, x_n\}$ such that, for every $y \in Y'$, $N_{B'}(y)$ is an interval in X. (Recall that B, hence B' is X-convex.) For $y \in Y'$ let $\text{left}(y) = \min\{i \mid x_i \in N_{B'}(y)\}$, and sort $y \in Y'$ increasing according $\text{left}(y)$. Then, for each $x \in X$, $N_{B'}(x)$ is an interval in Y': Assume, by contradiction, that there is some $x \in X$ such that $N_{B'}(x)$ is not an interval in Y'. Let y be a leftmost and y' be a rightmost vertex in $N_{B'}(x)$. By the assumption, there is some $y'' \in Y' \setminus N_{B'}(x)$ with $\text{left}(y) < \text{left}(y'') < \text{left}(y')$. Then, as $N_{B'}(y), N_{B'}(y'')$ and $N_{B'}(y')$ are intervals in X, $N_{B'}(y'')$ must be a subset of $N_{B'}(y)$; see also Fig. 3. Since $x \in N_{B'}(y)$ but $x \notin N_{B'}(y'')$, $N_{B'}(y'')$ must be a proper subset of $N_{B'}(y)$, contradicting to the fact that in B', no such pair of vertices exists in Y'. Thus, B' is Y'-convex.

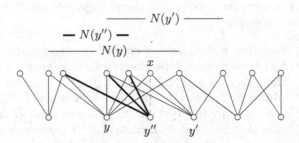

Fig. 3. Assuming $\text{left}(y) < \text{left}(y'') < \text{left}(y')$, and x is adjacent to y and y', but non-adjacent to y''.

Note that B' is indeed biconvex, hence, by Theorem 2, $B^2[X] = B'^2[X]$ is even a unit interval graph.   □

Since (unit) interval graph with n vertices and m edges can be recognized in linear time $O(n + m)$ and all maximal cliques of an (unit) interval graph can be listed in the same time complexity (cf. [11]), Theorems 3 and 2 imply:

Corollary 1. HALF-SQUARE OF CONVEX *and* HALF-SQUARE OF BICONVEX *can be solved in linear time. A (bi)convex bipartite half-root, if any, can be constructed in linear time.*

5   Half-Squares of Chordal Bipartite Graphs

In this section, we show that half-squares of chordal bipartite graphs are precisely the strongly chordal graphs. Recall that a graph is chordal if it has no induced cycle of length at least four. It is well-known (see, e.g., [11,21,26]) that a graph $G = (V, E_G)$ is chordal if and only if it admits a tree representation, that is, there exists a tree T such that, for each vertex $v \in V$, T_v is a subtree of T and two vertices in G are adjacent if and only if the corresponding subtrees in T intersect. Moreover, the vertices of T can be taken as the maximal cliques of the chordal graph (a clique tree). Recall also that a graph is strongly chordal if it is chordal and has no induced k-sun, $k \geq 3$. Here a k-sun consists of a stable set $\{s_1, \ldots, s_k\}$ and a clique $\{t_1, \ldots, t_k\}$ and edges $s_i t_i$, $s_i t_{i+1}$, $1 \leq i \leq k$. (Indices are taken modulo k.)

We first begin with the following fact.

Lemma 3. Let $B = (V, W, E_B)$ be bipartite graph without induced C_6 and let $k \geq 3$. If $B^2[V]$ contains an induced k-sun, then B contains an induced cycle of length $2k$.

The proof of Lemma 3 will be given in the full version of this paper.

Theorem 4. A graph is half-square of a chordal bipartite graph if and only if it is a strongly chordal graph.

Proof. We first show that half-squares of chordal bipartite graphs are chordal. Let $B = (X, Y, E_B)$ be a chordal bipartite graph. It is known that B is tree convex [13,18]. Thus, there is a tree $T = (X, E_T)$ such that, for each $y \in Y$, $N(y)$ induces a subtree in T. Then, for distinct vertices $y, y' \in Y$, y and y' are adjacent in $B^2[Y]$ if and only if $N(y) \cap N(y') \neq \emptyset$, and thus, $B^2[Y]$ has a tree representation, hence chordal. Now, by Lemma 3, $B^2[Y]$ cannot contain any sun k-sun, $k \geq 3$, showing that it is a strongly chordal graph. By symmetry, $B^2[X]$ is also strongly chordal. We have seen that half-squares of chordal bipartite graphs are strongly chordal graphs.

Next, let $G = (V, E_G)$ be a strongly chordal graph, and let $B = (V, \mathcal{C}(G), E_B)$ be the vertex-clique incidence bipartite graph of G. By Proposition 1, $G = B^2[V]$. Moreover, it is well-known [7] that B is chordal bipartite. Thus, every strongly chordal graph is a half-square of some chordal bipartite graph, namely of its vertex-clique incidence bipartite graph. $\square$

Testing if G is strongly chordal can be done in $O(\min\{n^2, m \log n\})$ time [7,20,26]. Assuming G is strongly chordal, all maximal cliques $Q_1, \ldots, Q_q$ of G can be listed in linear time (cf. [11,26]); note that $q \leq n$. So, Theorem 4 implies:

Corollary 2. HALF-SQUARE OF CHORDAL BIPARTITE can be solved in time $O(\min\{n^2, m \log n\})$, where n and m are the vertex and edge number of the input graph, respectively. A chordal bipartite half-root, if any, can be constructed in the same time complexity.

Theorem 4 (and its proof) gives another proof for a characterization of half-squares of trees found in [22]. A *block graph* is one in which every maximal 2-connected subgraph (a block) is a complete graph; equivalently, a block graph is a chordal graph without induced $K_4 - e$, a K_4 minus an edge.

Theorem 5 ([22]). *Half-squares of trees are exactly the block graphs.*

6   Conclusions

Until recently, only half-squares of planar bipartite graphs (the map graphs) have been investigated, and the most considered problem is if it is possible to recognize these graphs faster and simpler than Thorup's $O(n^{120})$ time algorithm.

In this paper, we have shown the first NP-hardness result, namely that recognizing if a graph is half-square of a balanced bisplit graph is NP-complete. For classical subclasses of tree convex bipartite graphs such as biconvex, convex, and chordal bipartite graphs, we have given good structure characterizations for their half-squares. These structural results imply that half-squares of these restricted classes of bipartite graphs can be recognized efficiently.

Recall that chordal bipartite graphs form a subclass of tree biconvex bipartite graphs [13,18], and that half-squares of chordal bipartite graphs can be recognized in polynomial time, while the complexity of recognizing half-squares of tree (bi)convex bipartite graphs is unknown. So, an obvious question is: what is the computational complexity of HALF-SQUARE OF TREE (BI)CONVEX?

Acknowledgment. We thank Hannes Steffenhagen for his careful reading and very helpful remarks.

References

1. Brandenburg, F.J.: On 4-map graphs and 1-planar graphs and their recognition problem. ArXiv (2015). http://arxiv.org/abs/1509.03447
2. Brandstädt, A., Le, V.B., Spinrad, J.P.: Graph Classes: A Survey. Society for Industrial and Applied Mathematics, Philadelphia (1999)
3. Chen, Z.-Z.: Approximation algorithms for independent sets in map graphs. J. Algorithms **41**, 20–40 (2001). doi:10.1006/jagm.2001.1178
4. Chen, Z.-Z., Grigni, M., Papadimitriou, C.H.: Map graphs. J. ACM **49**(2), 127–138 (2002). doi:10.1145/506147.506148
5. Demaine, E.D., Fomin, F.V., Hajiaghayi, M.T., Thilikos, D.M.: Fixed-parameter algorithms for (k, r)-center in planar graphs and map graphs. ACM Trans. Algorithms **1**(1), 33–47 (2005). doi:10.1145/1077464.1077468
6. Demaine, E.D., Hajiaghayi, M.T.: The bidimensionality theory and its algorithmic applications. Comput. J. **51**(3), 292–302 (2007). doi:10.1093/comjnl/bxm033
7. Farber, M.: Characterizations of strongly chordal graphs. Discrete Math. **43**, 173–189 (1983). doi:10.1016/0012-365X(83)90154-1
8. Fomin, F.V., Lokshtanov, D., Saurabh, S.: Bidimensionality and geometric graphs. In: Proceedings of the Twenty-Third Annual ACM-SIAM Symposium on Discrete Algorithms 2012 (SODA 2012), pp. 1563–1575 (2012)

9. Fulkerson, D.R., Gross, O.A.: Incidence matrices and interval graphs. Pacific J. Math. **15**, 835–855 (1965)

10. Gilmore, P.C., Hoffman, A.J.: A characterization of comparability graphs and of interval graphs. Canad. J. Math. **16**, 539–548 (1964). doi:10.4153/CJM-1964-055-5

11. Golumbic, M.C.: Algorithmic Graph Theory and Perfect Graphs. Academic Press, New York (1980)

12. Holyer, I.: The NP-completeness of some edge-partition problems. SIAM J. Comput. **4**, 713–717 (1981). doi:10.1137/0210054

13. Jiang, W., Liu, T., Wang, C., Ke, X.: Feedback vertex sets on restricted bipartite graphs. Theoret. Comput. Sci. **507**, 41–51 (2013). doi:10.1016/j.tcs.2012.12.021

14. Kou, L.T., Stockmeyer, L.J., Wong, C.-K.: Covering edges by cliques with regard to keyword conflicts and intersection graphs. Comm. ACM **21**, 135–139 (1978). doi:10.1145/359340.359346

15. Lau, L.C.: Bipartite roots of graphs. ACM Trans. Algorithms **2**(2), 178–208 (2006). doi:10.1145/1150334.1150337. Proceedings of the 15th Annual ACM-SIAM Symposium on Discrete Algorithms (SODA 2004), pp. 952–961

16. Lau, L.C., Corneil, D.G.: Recognizing powers of proper interval, split, and chordal graphs. SIAM J. Discrete Math. **18**(1), 83–102 (2004). doi:10.1137/S0895480103425930

17. Le, V.B., Peng, S.-L.: On the complete width and edge clique cover problems. In: Xu, D., Du, D., Du, D. (eds.) COCOON 2015. LNCS, vol. 9198, pp. 537–547. Springer, Cham (2015). doi:10.1007/978-3-319-21398-9_42

18. Lehel, J.: A characterization of totally balanced hypergraphs. Discrete Math. **57**, 59–65 (1985). doi:10.1016/0012-365X(85)90156-6

19. Liu, T.: Restricted bipartite graphs: comparison and hardness results. In: Gu, Q., Hell, P., Yang, B. (eds.) AAIM 2014. LNCS, vol. 8546, pp. 241–252. Springer, Cham (2014). doi:10.1007/978-3-319-07956-1_22

20. Lubiw, A.: Doubly lexical orderings of matrices. SIAM J. Comput. **16**, 854–879 (1987). doi:10.1137/0216057

21. McKee, T.A., McMorris, F.R.: Topics in Intersection Graph Theory. Society for Industrial and Applied Mathematics, Philadelphia (1999)

22. Mnich, M., Rutter, I., Schmidt, J.M.: Linear-time recognition of map graphs with outerplanar witness. In: Proceedings of the 15th Scandinavian Symposium and Workshops on Algorithm Theory 2016 (SWAT 2016), article no. 5, pp. 5:1–5:14 (2016). http://www.dagstuhl.de/dagpub/978-3-95977-011-8

23. Motwani, R., Sudan, M.: Computing roots of graphs is hard. Discrete Appl. Math. **54**(1), 81–88 (1994). doi:10.1016/0166-218X(94)00023-9

24. Orlin, J.: Contentment in graph theory: covering graphs with cliques. Indagationes Math. **80**(5), 406–424 (1977). doi:10.1016/1385-7258(77)90055-5

25. Roberts, F.S.: Indifference graphs. In: Harary, F. (ed.) Proof Techniques in Graph Theory, pp. 139–146. Academic Press, New York (1969)

26. Spinrad, J.P.: Efficient Graph Representations. Fields Institute Monographs, Toronto (2003)

27. Thorup, M.: Map graphs in polynomial time. In: Proceedings of the 39th IEEE Symposium on Foundations of Computer Science 1998 (FOCS 1998), pp. 396–405 (1998). doi:10.1109/SFCS.1998.743490

Faster Graph Coloring in Polynomial Space

Serge Gaspers[1,2] and Edward J. Lee[1,2($\boxtimes$)]

[1] UNSW, Sydney, Australia
`sergeg@cse.unsw.edu.au`, `e.lee@unsw.edu.au`
[2] Data61, CSIRO, Sydney, Australia

Abstract. We present a polynomial-space algorithm that computes the number of independent sets of any input graph in time $O(1.1389^n)$ for graphs with maximum degree 3 and in time $O(1.2356^n)$ for general graphs, where n is the number of vertices. Together with the inclusion-exclusion approach of Björklund, Husfeldt, and Koivisto [SIAM J. Comput. 2009], this leads to a faster polynomial-space algorithm for the graph coloring problem with running time $O(2.2356^n)$. As a byproduct, we also obtain an exponential-space $O(1.2330^n)$ time algorithm for counting independent sets.

Our main algorithm counts independent sets in graphs with maximum degree 3 and no vertex with three neighbors of degree 3. This polynomial-space algorithm is analyzed using the recently introduced Separate, Measure and Conquer approach [Gaspers & Sorkin, ICALP 2015]. Using Wahlström's compound measure approach, this improvement in running time for small degree graphs is then bootstrapped to larger degrees, giving the improvement for general graphs. Combining both approaches leads to some inflexibility in choosing vertices to branch on for the small-degree cases, which we counter by structural graph properties. The main complication is to upper bound the number of times the algorithm has to branch on vertices all of whose neighbors have degree 2, while still decreasing the size of the separator each time the algorithm branches.

1 Introduction

Graph coloring is a central problem in discrete mathematics and computer science. In exponential time algorithmics [16], graph coloring is among the most well studied problems, and it is an archetypical partitioning problem. Given a graph G and an integer k, the problem is to determine whether the vertex set of G can be partitioned into k independent sets. Already in 1976, Lawler [26] designed a dynamic programming algorithm for graph coloring and upper bounded its running time by $O(2.4423^n)$, where n is the number of vertices of the input graph. This was the best running time for graph coloring for 25 years, when Eppstein [10,11] improved the running time to $O(2.4150^n)$ by using better bounds on the number of small maximal independent sets in a graph. Based on bounds on the number of maximal induced bipartite subgraphs and refined bounds on the number of size-constrained maximal independent sets, Byskov [7] improved

© Springer International Publishing AG 2017
Y. Cao and J. Chen (Eds.): COCOON 2017, LNCS 10392, pp. 371–383, 2017.
DOI: 10.1007/978-3-319-62389-4_31

the running time to $O(2.4023^n)$. An algorithm based on fast matrix multiplication by Björklund and Husfeldt [3] improved the running time to $O(2.3236^n)$. The current fastest algorithm for graph coloring, by Björklund et al. [2,4,25], is based on the principle of inclusion–exclusion and Yates' algorithm for the fast zeta transform. This breakthrough algorithm solves graph coloring in $O^*(2^n)$ time, where the O^*-notation is similar to the O-notation but ignores polynomial factors.

A significant drawback of the aforementioned algorithms is that they use exponential space. Often, the space bound is the same as the time bound, up to polynomial factors. This is undesirable [31], certainly for modern computing devices. Polynomial-space algorithms for graph coloring have been studied extensively as well with successive running times $O^*(n!)$ [8], $O((k/e)^n)$ (randomized) [12], $O((2 + \log k)^n)$ [1], $O(5.283^n)$ [6], $O(2.4423^n)$ [3], and $O(2.2461^n)$ [4]. The latter algorithm is an inclusion-exclusion algorithm relying on a $O(1.2461^n)$ time algorithm [17] for computing the number of independent sets in a graph. Their method transforms any polynomial-space $O(c^n)$ time algorithm for counting independent sets into a polynomial-space $O((1 + c)^n)$ time algorithm for graph coloring. The running time bound for counting independent sets was subsequently improved by Fomin et al. [13] to $O(1.2431^n)$ and by Wahlström [30] to $O(1.2377^n)$. Wahlström's algorithm is the current fastest published algorithm for counting independent sets of a graph, it uses polynomial space, and it works for the more general problem of computing the number of maximum-weight satisfying assignments of a 2-CNF formula. For a reduction from counting independent sets to counting maximum-weight satisfying assignments of a 2-CNF formula where the number of variables equals the number of vertices, see [9].

We note that Junosza-Szaniawski and Tuczynski [24] present an algorithm for counting independent sets with running time $O(1.2369^n)$ in a technical report that also strives to disconnect low-degree graphs. For graphs with maximum degree 3 that have no degree-3 vertex with all neighbors of degree 3, they present a new algorithm with running time $2^{n_3/5+o(n)}$, where n_3 is the number of degree-3 vertices, and the overall running time improvement comes from plugging this result into Wahlström's [30] previously fastest algorithm for the problem. However, we note that the $2^{n_3/5+o(n)}$ running time for counting independent sets can easily be obtained from previous results. Namely, the problem of counting independent sets is a polynomial PCSP with domain size 2, as shown in [28], and the algorithm of [21] for polynomial PCSPs preprocesses all degree-2 vertices, leaving a cubic graph on n_3 vertices that is solved in $2^{n_3/5+o(n)}$ time. Improving on this bound is challenging, and degree-3 vertices with all neighbors of degree 2 need special attention since branching on them affects the degree-3 vertices of the graph exactly the same way as for the much more general polynomial PCSP problem, whereas for other degree-3 vertices one can take advantage of the asymmetric nature of the typical independent set branching (i.e., we can delete the neighbors when counting the independent sets containing the vertex we branch on).

Our Results. We present a polynomial-space algorithm computing the number of independent sets of any input graph G in time $O(1.2356^n)$, where n is the number of vertices of G. Our algorithm is a branching algorithm that works initially similarly as Wahlström's algorithm, where we slightly improve the analysis using potentials (as, e.g., in [20,23,29]) to amortize some of the worst branching cases with better ones. This algorithm uses a branching strategy that basically ensures that both the maximum degree and the average degree of the graph do not increase. This makes it possible to divide the analysis of the algorithm into sections depending on what local structures can still occur in the graph, use a separate measure for the analysis of each section, and combine these measures giving a compound (piecewise linear) measure for the analysis of the overall algorithm.

For instances where the maximum degree is 3 and no vertex has three neighbors with degree 3, we substitute a subroutine that is designed and analyzed using the recently introduced *Separate, Measure and Conquer* technique [21]. It computes a small balanced separator of the graph and prefers to branch on vertices in the separator, adjusting the separator as needed by the analysis, and reaping a huge benefit when the separator is exhausted and the resulting connected components can be handled independently. The Separate, Measure and Conquer technique helps to amortize this sudden gain with the analysis of the previous branchings, for an overall running time improvement.

Since using a separator restricts our choice in the vertices to branch on, we use the structure of the graph and its separation to upper bound the number of unfavorable branching situations and adapt our measure accordingly. Namely, the algorithm avoids branching on degree-3 vertices in the separator with all neighbors of degree 2 as long as possible, often rearranging the separator to avoid this case. In our analysis we can then upper bound the number of unfavorable branchings and give the central vertex involved in such a branching a special weight and role in the analysis. We call these vertices *spider vertices*. Our meticulous analysis of this subroutine upper bounds its running time by $O(1.0963^n)$. For graphs with maximum degree at most 3, we obtain a running time of $O(1.1389^n)$. This improvement for small degree graphs is bootstrapped, using Wahlström's compound measure analysis, to larger degrees, and gives a running time improvement to $O(1.2356^n)$ for counting independent sets of arbitrary graphs and to $O(2.2356^n)$ for graph coloring. Bootstrapping an exponential-space pathwidth-based $O(1.1225^n)$ time algorithm [15] for cubic graphs instead, we obtain an exponential-space algorithm for counting independent sets with running time $O(1.2330^n)$. We refer to [19] for a full version of the paper with all details.

2   Methods

Measure and Conquer. The analysis of our algorithm is based on the Measure and Conquer method [14]. A *measure* for a problem (or its instances) is a function from the set of all instances of the problem to the set of non-negative reals. Modern branching analyses often use a potential function as measure that gives a more fine-grained way of tracking the progress of a branching algorithm than a measure that is merely the number of vertices or edges of the graph. The following lemma is at the heart of our analysis. It generalizes a similar lemma from [20] to the treatment of subroutines.

Lemma 1 ([18]). *Let A be an algorithm for a problem P, B be an algorithm for a class C of instances of P, $c \geq 0$ and $r > 1$ be constants, and $\mu(\cdot), \mu_B(\cdot), \eta(\cdot)$ be measures for P, such that for any input instance I from C, $\mu_B(I) \leq \mu(I)$, and for any input instance I, A either solves P on $I \in C$ by invoking B with running time $O(\eta(I)^{c+1} r^{\mu_B(I)})$, or reduces I to k instances $I_1, \ldots, I_k$, solves these recursively, and combines their solutions to solve I, using time $O(\eta(I)^c)$ for the reduction and combination steps (but not the recursive solves),*

$$(\forall i) \quad \eta(I_i) \leq \eta(I) - 1, \quad and \quad \sum_{i=1}^{k} r^{\mu(I_i)} \leq r^{\mu(I)}. \tag{1}$$

Then A solves any instance I in time $O(\eta(I)^{c+1} r^{\mu(I)})$.

When Algorithm A does not invoke Algorithm B, we have the usual Measure and Conquer analysis. Here, μ is used to upper bound the number of leaves of the search tree and deserves the most attention, while η is usually a polynomial measure to upper bound the depth of the search tree. For handling subroutines, it is crucial that the measure does not increase when Algorithm A hands over the instance to Algorithm B and we constrain that $\mu_B(I) \leq \mu(I)$.

Compound Analysis. We can view Wahlström's compound analysis [30] as a repeated application of Lemma 1. For example, there is a subroutine A_3 for when the maximum degree of the graph is 3. The algorithm prefers then to branch on a degree-3 vertex with all neighbors of degree 3. After all such vertices have been exhausted, the algorithm calls a new subroutine $A_{8/3}$ that takes as input a graph with maximum degree 3 where no degree-3 vertex has only degree 3 neighbors. In this case the average degree of the graph is at most 8/3, and the algorithm prefers to branch on vertices of degree 3 that have 2 neighbors of degree 3, etc. The analysis constrains that the measure for the analysis of $A_{8/3}$ is at most the measure for A_3 for the instance that is handed to $A_{8/3}$. In an optimal analysis, we expect the measure for such an instance to be equal in the analysis of A_3 and $A_{8/3}$, and Wahlström actually imposes equality at the *pivot point* 8/3.

Separate, Measure and Conquer. In our case, the $A_{8/3}$ algorithm is based on *Separate, Measure and Conquer.* For small-degree graphs, we can compute small balanced separators in polynomial time. The algorithm then prefers to branch on vertices in the separator. The Separate, Measure and Conquer technique allows to distribute the large gain obtained by disconnecting the instance onto the previous branching vectors. While, often, the measure is made up of weights that are assigned to each vertex, this method assigns these weights only to the larger part of the graph that is separated from the rest by the separator, and somewhat larger weights to the vertices in the separator. See (3) on page 8. Thus, after exhausting the separator, the measure accurately reflects the "amount of work" left to do. We artificially increase the measure of very balanced instances by small penalty weights – this is so because in this case it is more difficult for the branching strategy to make most of its progress on what ends uo being the large side. Since we may exhaust the separators a logarithmic number of times, and computing a new separator might introduce a penalty term each time, the measure also includes a logarithmic term that counteracts these artificial increases in measure, and will in the end only contribute a polynomial factor to the running time. For an in-depth treatment of the method we refer to [21]. Since we use the *Separate, Measure and Conquer* method when the average degree drops to at most $8/3$, we slightly generalize the separation computation from [21], where the bound on the separator size depended only on the maximum degree. A separation (L, S, R) of a graph G is a partition of the vertex set into (possibly empty) sets L, S, R such that every path from a vertex in L to a vertex in R contains a vertex from S.

Lemma 2. *Let $B \in \mathbb{R}$. Let μ be a measure for graph problems such that for every graph $G = (V, E)$, every $R \subseteq V$, and every $v \in V$, we have that $|\mu(R \cup \{v\}) - \mu(R)| \leq B$. Assume that $\mu(R)$, the restriction of μ to R, can be computed in polynomial time. If there is an algorithm computing a path decomposition of width at most k of a graph G in polynomial time, then there is a polynomial time algorithm computing a separation (L, S, R) of G with $|S| \leq k$ and $|\mu(L) - \mu(R)| \leq B$.*

Proof. The proof is basically the same as for the separation computation from [21], but we repeat it here for completeness. First, compute a path decomposition of width k in polynomial time. We view a path decomposition as a sequence of bags $(B_1, \ldots, B_b)$ which are subsets of vertices such that for each edge of G, there is a bag containing both endpoints, and for each vertex of G, the bags containing this vertex form a non-empty consecutive subsequence. The width of a path decomposition is the maximum bag size minus one. We may assume that every two consecutive bags B_i, B_{i+1} differ by exactly one vertex, otherwise we insert between B_i and B_{i+1} a sequence of bags where the vertices from $B_i \setminus B_{i+1}$ are removed one by one followed by a sequence of bags where the vertices of $B_{i+1} \setminus B_i$ are added one by one; this is the standard way to transform a path decomposition into a *nice* path decomposition of the same width where the number of bags is polynomial in the number of vertices [5]. Note that each bag is a separator and a bag B_i defines the separation (L_i, B_i, R_i) with $L_i = (\bigcup_{j=1}^{i-1} B_j) \setminus B_i$ and $R_i = V \setminus (L_i \cup B_i)$. Since the first of these separations has $L_1 = \emptyset$ and the last one has $R_b = \emptyset$, at least one of these separations has $|\mu_r(L_i) - \mu_r(R_i)| \leq B$. Finding such a bag can clearly be done in polynomial time.   □

We will use the lemma for graphs with maximum degree 3 and graphs with maximum degree 3 and average degree at most $8/3$, for which path decompositions of width at most $n/6 + o(n)$ and $n/9 + o(n)$ can be computed in polynomial time, [13, 15].

One disadvantage of using the Separate, Measure and Conquer method for $A_{8/3}$ is that the algorithm needs to choose vertices for branching so that the size of the separator decreases in each branch. However, Wahlström's algorithm defers to branch on degree-3 vertices with all neighbors of degree 2 until this is no longer possible, since this case leads to the largest branching factor for degree 3. For our approach, we instead rearrange the separator in some cases until we are only left with spider vertices, a structure where our algorithm cannot avoid branching on a degree-3 vertex with all neighbors of degree 2, we give a special weight to these spider vertices and upper bound their number.

Potentials. To optimize the running time further, we use potentials; see [20, 23, 29]. These constant weights are added to the measure if certain global properties of the instance hold. We may use them to slightly increase the measure when an unfavorable branching strategy needs to be used. The constraint (1) for this unfavorable case then becomes less constraining, while all branchings that can lead to this unfavorable case get tighter constraints. This allows to amortize unfavorable cases with favorable ones.

3   Algorithm

We first introduce notation necessary to present the algorithm. Let $V(G)$ and $E(G)$ denote the vertex set and the edge set of the input graph G. For a vertex $v \in V(G)$, its neighborhood, $N_G(v)$, is the set of vertices adjacent to v. The *closed neighborhood* of a vertex v is $N_G[v] = N_G(v) \cup \{v\}$. If G is clear from context, we use $N(v)$ and $N[v]$.

The degree of v is denoted $d(v) = |N_G(v)|$. A (d_1, d_2, d_3)-vertex is a degree 3 vertex that has neighbors of degree d_1, d_2, d_3. An edge $uv \in E(G)$ is adjacent to vertex $u \in V(G)$ and $v \in V(G)$. For two vertices u and v connected by a path, let $P \subset V(G)$ with $u, v \notin P$ be the intermediate vertices between u and v on the path. If P consists only of degree-2 vertices then we call P a *2-path* of u and v.

The maximum degree of G is denoted $\Delta(G)$ and $d(G) = 2|E(G)|/|V(G)|$ is its *average degree*. A *cubic* graph consists only of degree-3 vertices. A *subcubic* graph has maximum degree at most 3. A $(k_1, k_2, ..., k_d)$ *vertex* is a degree-d vertex with neighbors of degree $k_1, k_2, ..., k_d$. A separation (L, S, R) of G is a partition of its vertex set into the three sets L, S, R such that no vertex in L is adjacent to any vertex in R. The sets L, S, R are also known as the *left set*, *separator*, and *right set*. Using a similar notion to [21], a separation (L, S, R) of G is *balanced* with respect to some measure μ, and a branching constant B if $|\mu(R) - \mu(L)| \leq 2B$ and *imbalanced* if $|\mu(R) - \mu(L)| > 2B$. By convention, $\mu(R) \geq \mu(L)$ otherwise, we swap L and R. We use the measure μ_r defined on page 8 to compute the separation in our algorithm. We will now describe the algorithm #IS which takes as input a graph G, a separation (L, S, R), and a cardinality function $\mathbf{c} : \{0, 1\} \times V(G) \to \mathbb{N}$, and computes the number of independent sets of G weighted by the cardinality function $\mathbf{c}$. For clarity, let $\mathbf{c}_{out}(v) = \mathbf{c}(0, v)$ and $\mathbf{c}_{in}(v) = \mathbf{c}(1, v)$. More precisely, it computes

$$ind(G, \mathbf{c}) = \sum_{X \subseteq V(G)} \mathbb{1}(X \text{ is an independent set in } G) \cdot \prod_{v \in X} \mathbf{c}_{in}(v) \cdot \prod_{v \in V \setminus X} \mathbf{c}_{out}(v)$$

where $\mathbb{1}(\cdot)$ is an indicator function which returns 1 if its argument is true and 0 otherwise. Note that for a cardinality function $\mathbf{c}$ initialized to $\mathbf{c}(0, v) = \mathbf{c}(1, v) = 1$ for every vertex $v \in V(G)$, we have that $ind(G, \mathbf{c})$ is the number of independent sets of G. Cardinality functions are used for bookkeeping during the branching process and have been used in this line of work before. The separation (L, S, R) is initialized to $(\emptyset, \emptyset, V(G))$ and will only come into play when G is subcubic and has no (3,3,3)-vertex. In this case, the algorithm calls a subroutine #3IS, which constitutes the main contribution of this paper. #3IS computes a balanced separation of G, preferring to branch on vertices in the separator, readjusting the separator as needed, and is analyzed using the Separate, Measure and Conquer method. The full description of the algorithms #IS, #3IS and associated helper algorithms can be found in [19].

Skeleton Graph. The skeleton graph $\Gamma(G)$, or just Γ, of a subcubic graph G is a graph where the degree-3 vertices of G are in bijection with the vertices in Γ. Two vertices in Γ are adjacent if the corresponding vertices are adjacent in G, or there exists a 2-path between the corresponding vertices in G. If G has a separation (L, S, R) then denote $(L_\Gamma, S_\Gamma, R_\Gamma)$ to be the same separation of G in Γ consisting of only degree-3 vertices. *Dragging* refers to moving vertices or a set of vertices of G from one component of (L, S, R) to another, creating a new separation (L', S', R') such that S' is still a separator of G.

Spider Vertices. As Wahlström's [30] analysis showed, an unfavorable branching case occurs on $(2, 2, 2)$ vertices. Due to our algorithm's handling of these vertices we narrowed down the undesirable vertices called *spider vertices* to a specific list of properties. If s is a spider vertex then:

- $s \in S$
- s has neighbors of degree (2,2,2)
- Either:
    - $|N_\Gamma(s) \cap L| = 2$ and $N_\Gamma(s) \cap R = \{r\}$ with r having neighbors of degree (2,2,2). In this case we call s a *left spider vertex*
    - $|N_\Gamma(s) \cap R| = 2$ and $N_\Gamma(s) \cap L = \{l\}$ with l having neighbors of degree (2,2,2). In this case we call s a *right spider vertex*
    - $|N_\Gamma(s) \cap L| = 1$, $|N_\Gamma(s) \cap R| = 1$, $N_\Gamma(s) \cap S = \{s'\}$ and s' has neighbors of degree (2,2,2). In this case we call both s and s' a *center spider vertex*, which occur in pairs.

A left spider vertex $s \in S$ can be dragged to the left along with the 2-path from s to r. If this were ever to occur, then r would be a right spider vertex, and vice versa (Fig. 1).

Multiplier Reduction. We use a reduction called multiplier reduction to simplify graphs that have a cut vertex efficiently. Suppose G has a separation $(V_1, \{x\}, V_2)$ and $G_1 = G[V_1 \cup \{x\}]$ has measure at most a constant B. The *multiplier reduction* can be applied to compute #IS$(G, (L, S, R), \mathbf{c})$ as follows.

Fig. 1. A left spider vertex s.

1. Let:
   - $G_{out} = G_1 \setminus \{x\}$
   - $G_{in} = G_1 \setminus N_{G_1}[x]$
   - $c_{out} = \#IS(G_{out},(L[G_{out}], S[G_{out}], R[G_{out}]),\mathbf{c})$
   - $c_{in} = \#IS(G_{in}, (L[G_{in}], S[G_{in}], R[G_{in}]), \mathbf{c})$
2. Modify $\mathbf{c}$ such that $\mathbf{c}_{in}(x) = \mathbf{c}_{in}(x) \cdot c_{in}$ and $\mathbf{c}_{out}(x) = \mathbf{c}_{out}(x) \cdot c_{out}$
3. Return $\#IS(G[V_2 \cup \{x\}], (L, S, R), \mathbf{c})$

Since G_1 has a measure of constant size, both steps 1 and 2 take polynomial time.

Lazy 2-separator. Suppose there is a vertex x initially chosen to branch on as well as two vertices $\{y, z\} \subset V(G)$ with $d(y) \geq 3$ and $d(z) \geq 3$ such that $\{y, z\}$ is a separator which separates x from G in a constant measure subgraph. We call such vertices *lazy 2-separators*, for a vertex x. Similar to Wahlström's elimination of separators of size 2 in [29], in line 15 of #IS ([19]) instead of branching on x, if there exists a lazy 2-separator $\{y, z\}$ for x we branch on y. A multiplier reduction will be performed on z in the recursive calls. Prioritizing *lazy 2-separators* allows to exclude some unfavorable cases when branching on x.

Associated Average Degree. Similar to [30], we define the *associated average degree* of a vertex $x \in V(G)$ as $\alpha(x)/\beta(x)$, in G with average degree $d(G) = k$ where

$$\alpha(x) = d(x) + |\{y \in N(x) : d(y) < k\}|, \text{ and } \beta(x) = 1 + \sum_{\{y \in N(x) | d(y) < k\}} 1/d(y) . \quad (2)$$

By selecting vertices with high associated average degree, our algorithm prioritizes branching on vertices with larger decreases in measure.

Branching. We now outline the branching routine used to recursively solve smaller instances of the problem. Suppose we have a graph G, a separation (L, S, R), and a cardinality function $\mathbf{c}$. For a vertex x we denote the following steps as *branching on x*.

1. Let:
   - $G_{out} = G \setminus \{x\}$
   - $G_{in} = G \setminus (N(x) \cup \{x\})$
   - $c_{out} = \#IS(G_{out}, (L[G_{out}], S[G_{out}], R[G_{out}]), \mathbf{c})$
   - $c_{in} = \#IS(G_{in}, (L[G_{in}], S[G_{in}], R[G_{in}]), \mathbf{c})$
   - $c'_{out} = \mathbf{c}_{out}(x)$
   - $c'_{in} = \mathbf{c}_{in}(x) \cdot \prod_{v \in N(x)} \mathbf{c}_{out}(v)$
2. Return $c'_{out} \cdot c_{out} + c'_{in} \cdot c_{in}$

4   Running Time Analysis

This section describes the running time analysis for #IS and #3IS, conducted via compound measures. Constraints are presented as branching vectors (δ_1, δ_2) which equates to the constraints $2^{-\delta_1} + 2^{-\delta_2} \leq 1$. We first describe some special vertex weights.

4.1   Measures

Measure with no (3,3,3) vertex. When using the Separate, Measure and Conquer technique from [21] the measure of a cubic graph instance G with no (3,3,3) vertices consists of additive components μ_s and μ_r, the measure of vertices in the separator, and those in either L or R, respectively. Let $S' \subseteq S$ be the set of all spider vertices, s_i and r_i refer to the weight attributed to a separator vertex and a right vertex, in R or L, respectively, of degree i. Left and right spider vertices have weight s'_3. In a center spider vertex pair s and s', one of them has weight s'_3 while the other takes on an ordinary weight of s_3. These structurally applied weights allow amortization of the spider vertex cases against non-spider vertices. Define the measure $\mu_{8/3}$ as

$$\mu_{8/3} = \mu_s(S) + \mu_r(R) + \mu_o(L, S, R), \tag{3}$$

where $\mu_s(S) = |S'| \cdot s'_3 + \sum_{v \in S \setminus S'} s_{d(v)}$, $\mu_r(R) = \sum_{v \in R} r_{d(v)}$, $B = 6s_3$ and

$$\mu_o(L, S, R) = \max \left\{ 0, B - \frac{\mu_r(R) - \mu_r(L)}{2} \right\} + (1 + B) \cdot \log_{1+\epsilon}(\mu_r(R) + \mu_s(S)).$$

We also require that $s_i \geq s_{i-1}$ and $r_i \geq r_{i-1}$ for $i \in \{1, 2, 3\}$. The constant B is larger than the maximum change in imbalance in each transformation in the analysis, except the separation transformation.

Lemma 3. *For a balanced separation (L, S, R) of a graph G with average degree $d = d(G)$, an upper bound for the measure $\mu_{8/3}$ is:*

$$\mu_{8/3}(d) \leq \begin{cases} \frac{n}{6}(d - 2)s'_3 \\ + \frac{1}{2}\left(\frac{5n}{6}(d-2)r_3 + n(3-d)r_2\right) \\ + \mu_o(L, S, R) + o(n) & \text{if } 2 \leq d \leq \frac{28}{11} \\ \frac{n}{4}(8 - 3d)s'_3 + \frac{n}{12}(11d - 28)s_3 \\ + \frac{1}{2}\left(\frac{5n}{6}(d-2)r_3 + n(3-d)r_2\right) \\ + \mu_o(L, S, R) + o(n) & \text{if } \frac{28}{11} < d \leq \frac{8}{3} \end{cases}$$

which is maximised when $d = \frac{8}{3}$ with the value

$$\mu_{8/3} \leq \frac{n}{9}s_3 + \frac{1}{2}\left(\frac{5n}{9}r_3 + \frac{n}{3}r_2\right) + \mu_o(L, S, R) + o(n)$$

if constraints $\frac{r_2}{2} \leq \frac{s'_3}{11} + \frac{5r_3}{22} + \frac{5r_2}{22} \leq \frac{s_3}{9} + \frac{5r_3}{18} + \frac{r_2}{3}$ are satisfied.

General Measure. In order to analyze higher degree cases, we use a measure of the form

$$\mu_i(G) = \sum_{v \in G} r_{d(v)} + \mu_o(L, S, R) \quad \text{where } \Delta(G) = i$$

for each part of the compound measure. The term $\mu_o(L, S, R)$ is the same sublinear term from the Separate, Measure and Conquer analysis on cubic graphs which needs to be propagated into the higher degree analyses.

4.2   Degree 3 Analysis

#IS can be solved in polynomial time when $\Delta(G) \leq 2$ [27]. However, stepping up to cubic graphs is a much harder problem. Greenhill [22] proves that counting independent sets is actually a #P-hard problem even for graphs with maximum degree 3.

Lemma 4. *Algorithm #IS applied to a graph G with $\Delta(G) \leq 3$ and no $(3,3,3)$ vertex has running time $O(1.0963^n)$.*

Proof (sketch). We present a sketch of the proof, emphasizing the tight constraints generated from algorithms #3IS, simplify and spider described in detail in [19], where full proofs can be found. As suggested in [21], each case will provide constraints that the weights described above will need to satisfy.

Some trivial constraints we must satisfy are $r_0 = r_1 = s_0 = s_1 = 0$ since these vertices can easily be eliminated and require no branching rules. Our algorithm considers skeleton graph vertices, and several rules drag entire 2-paths from one separation to another, requiring $r_2 = 0$. In simplify, line 8 implies the constraint $s_2 + s_3' + \frac{1}{2}(r_2 - r_3) \leq 0$, enabling us to move a degree-3 vertex into the separator by dragging out a degree-2 vertex.

From #3IS, line 2 imposes the constraint $\frac{1}{6}s_3' + \frac{5}{12}r_3 \leq r_3$. If a $(2,2,3)$ vertex s is chosen to branch on in line 19 as shown in Fig. 2(a) ([19]), then we get the constraint $\left(s_3 + \Delta s_3 + \frac{1}{2}(2\Delta r_3) - 3\delta, s_3 + 2\Delta s_3 + \frac{1}{2}(r_3 + 2\Delta r_3) - 4\delta\right)$. The last tight constraint is from spider line 17, displayed in Fig. 3(a) ([19]), giving the constraint $\left(s_3' + \frac{3}{2}\Delta r_3, s_3' + \frac{3}{2}\Delta r_3\right)$.

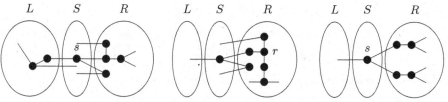

(a) Balanced branching   (b) Imbalanced branching on r (c) Imbalanced branching on s

Fig. 2. Worst case configurations for non-spider vertex branching in #3IS

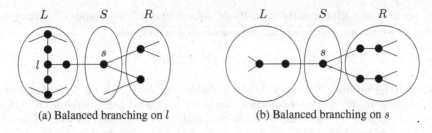

(a) Balanced branching on l (b) Balanced branching on s

Fig. 3. Worst case configurations for spider vertex branching in `spider`

While the cases in Figs. 2(b) and 3(b) ([19]) are not tight, they are of interest since these cases branch on vertices located outside the separator and it is guaranteed that s is removed from the separator after branching.   □

Lemma 5. *Algorithm #IS applied to a graph G with $d(G) \leq 3$ has running time $O(1.1389^n)$ and uses polynomial space.*

The algorithm `#IS` uses subroutine `#3IS`, which we analyze the measure and the weights for. We equate the Separate, Measure and Conquer weights with weights of the measure μ_3, based on the compound analysis from Wahlström [30]. As Wahlström's analysis only contains weights w'_3 and w'_2, for vertices of degree 3 and degree 2 respectively, the measure is $\mu_3(G) = ((d-2)w'_3 + (3-d)w'_2)\,n + \mu_o(L,S,R)$ where $d = d(G)$ is the average degree of a cubic graph, and $\mu_o(L,S,R)$ is the sub-linear term left over from the average degree $8/3$ analysis.

In the case of a graph G with no (3,3,3) vertex, in order to use Lemma 1, the values of w_1 and w_2 must satisfy inequalities $\frac{r_2}{2} \leq w_2$, $\frac{s'_3}{11} + \frac{5r_3}{22} + \frac{5r_2}{22} \leq \frac{6w_3}{11} + \frac{5w_2}{11}$, $\frac{s_3}{9} + \frac{5r_3}{18} + \frac{r_2}{3} \leq \frac{2w_3}{3} + \frac{w_2}{3}$, induced when $d = 2$, $\frac{28}{11}$, and $\frac{8}{3}$ for $\mu_{8/3}$ respectively. This results in the weights $w_3 = 0.1973$ and $w_2 = 0.0033$ when G has no (3,3,3) vertex. We also let $w'_3 \geq 0$ and $w'_2 \geq 0$ be the weights associated with vertices of degree 3 and degree 2 respectively, for a subcubic graph G. Using the analysis by compound measures with $\mu_3(G) = \sum_{i \in \{2,3\}} w'_i \cdot n_i$, the following constraint $\mu_{8/3}(G) = \mu_3(G)$ when $d(G) = 8/3$ is required for a valid compound measure. This can be rewritten as $2w_3 + w_2 = 2w'_3 + w'_2$. Branching on a (3,3,3) vertex, the only type of degree-3 that `#IS` branches on, gives a branching vector of $(4w'_3 - 3w'_2, 8w'_3 - 4w'_2)$. Setting the weights $w'_3 = 0.1876$ and $w'_2 = 0.0228$ satisfies the system of constraints described above and by using the measure $\mu_3(G)$, results in a running time of $O^*(1.1389^n)$.

Lemma 6. *#IS can be solved in time $O^*(1.2070^n)$ for graphs with maximum degree 4.*

Theorem 1. *#IS can be solved in time $O^*(1.2356^n)$ and polynomial space.*

If we plug in a simple pathwidth-based subroutine [15] for graphs of maximum degree 3, we obtain the following exponential-space result.

Theorem 2. *#IS can be solved in time* $O^*(1.2330^n)$.

Acknowledgements. We thank Magnus Wahlström for clarifying an issue of the case analysis in [30] and an anonymous reviewer for useful comments on an earlier version of the paper. Serge Gaspers is the recipient of an Australian Research Council (ARC) Future Fellowship (FT140100048) and acknowledges support under the ARC's Discovery Projects funding scheme (DP150101134).

References

1. Angelsmark, O., Thapper, J.: Partitioning based algorithms for some colouring problems. In: Hnich, B., Carlsson, M., Fages, F., Rossi, F. (eds.) CSCLP 2005. LNCS (LNAI), vol. 3978, pp. 44–58. Springer, Heidelberg (2006). doi:10.1007/11754602_4

2. Björklund, A., Husfeldt, T.: Inclusion-exclusion algorithms for counting set partitions. In: Proceedings of the 47th Annual IEEE Symposium on Foundations of Computer Science (FOCS 2006), pp. 575–582. IEEE Computer Society (2006)

3. Björklund, A., Husfeldt, T.: Exact algorithms for exact satisfiability and number of perfect matchings. Algorithmica **52**(2), 226–249 (2008)

4. Björklund, A., Husfeldt, T., Koivisto, M.: Set partitioning via inclusion-exclusion. SIAM J. Comput. **39**(2), 546–563 (2009)

5. Bodlaender, H.L., Kloks, T.: Efficient and constructive algorithms for the pathwidth and treewidth of graphs. J. Algorithm. **21**(2), 358–402 (1996)

6. Bodlaender, H.L., Kratsch, D.: An exact algorithm for graph coloring with polynomial memory. Technical report UU-CS-2006-015, Department of Information and Computing Sciences, Utrecht University (2006)

7. Byskov, J.M.: Enumerating maximal independent sets with applications to graph colouring. Oper. Res. Lett. **32**(6), 547–556 (2004)

8. Christofides, N.: An algorithm for the chromatic number of a graph. Comput. J. **14**(1), 38–39 (1971)

9. Dahllöf, V., Jonsson, P., Wahlström, M.: Counting models for 2SAT and 3SAT formulae. Theor. Comput. Sci. **332**(1–3), 265–291 (2005)

10. Eppstein, D.: Small maximal independent sets and faster exact graph coloring. In: Dehne, F., Sack, J.-R., Tamassia, R. (eds.) WADS 2001. LNCS, vol. 2125, pp. 462–470. Springer, Heidelberg (2001). doi:10.1007/3-540-44634-6_42

11. Eppstein, D.: Small maximal independent sets and faster exact graph coloring. J. Graph Algorithms Appl. **7**(2), 131–140 (2003)

12. Feder, T., Motwani, R.: Worst-case time bounds for coloring and satisfiability problems. J. Algorithm. **45**(2), 192–201 (2002)

13. Fomin, F.V., Gaspers, S., Saurabh, S., Stepanov, A.A.: On two techniques of combining branching and treewidth. Algorithmica **54**(2), 181–207 (2009)

14. Fomin, F.V., Grandoni, F., Kratsch, D.: A measure & conquer approach for the analysis of exact algorithms. J. ACM **56**(5), 25 (2009)

15. Fomin, F.V., Høie, K.: Pathwidth of cubic graphs and exact algorithms. Inform. Process. Lett. **97**(5), 191–196 (2006)

16. Fomin, F.V., Kratsch, D.: Exact Exponential Algorithms. Springer Science & Business Media, New York (2010)
17. Fürer, M., Kasiviswanathan, S.P.: Algorithms for counting 2-SAT solutions and colorings with applications. In: Kao, M.-Y., Li, X.-Y. (eds.) AAIM 2007. LNCS, vol. 4508, pp. 47–57. Springer, Heidelberg (2007). doi:10.1007/978-3-540-72870-2_5
18. Gaspers, S.: Exponential Time Algorithms. VDM Verlag, Saarbrücken (2010)
19. Gaspers, S., Lee, E.: Faster graph coloring in polynomial space. CoRR, abs/1607.06201 (2016)
20. Gaspers, S., Sorkin, G.B.: A universally fastest algorithm for Max 2-Sat, Max 2-CSP, and everything in between. J. Comput. Syst. Sci. **78**(1), 305–335 (2012)
21. Gaspers, S., Sorkin, G.B.: Separate, measure and conquer: faster polynomial-space algorithms for Max 2-CSP and counting dominating sets. In: Halldórsson, M.M., Iwama, K., Kobayashi, N., Speckmann, B. (eds.) ICALP 2015. LNCS, vol. 9134, pp. 567–579. Springer, Heidelberg (2015). doi:10.1007/978-3-662-47672-7_46
22. Greenhill, C.: The complexity of counting colourings and independent sets in sparse graphs and hypergraphs. Comput. Complex. **9**(1), 52–72 (2000)
23. Iwata, Y.: A faster algorithm for dominating set analyzed by the potential method. In: Marx, D., Rossmanith, P. (eds.) IPEC 2011. LNCS, vol. 7112, pp. 41–54. Springer, Heidelberg (2012). doi:10.1007/978-3-642-28050-4_4
24. Junosza-Szaniawski, K., Tuczynski, M.: Counting independent sets via divide measure and conquer method. Technical report abs/1503.08323, arXiv CoRR (2015)
25. Koivisto, M.: An $O^*(2^n)$ algorithm for graph coloring and other partitioning problems via inclusion-exclusion. In: Proceedings of the 47th Annual IEEE Symposium on Foundations of Computer Science (FOCS 2006), pp. 583–590. IEEE Computer Society (2006)
26. Lawler, E.L.: A note on the complexity of the chromatic number problem. Inform. Process. Lett. **5**(3), 66–67 (1976)
27. Roth, D.: On the hardness of approximate reasoning. Artif. Intell. **82**(1), 273–302 (1996)
28. Scott, A.D., Sorkin, G.B.: Polynomial constraint satisfaction problems, graph bisection, and the ising partition function. ACM T. Algorithms **5**(4), 45 (2009)
29. Wahlström, M.: Exact algorithms for finding minimum transversals in rank-3 hypergraphs. J. Algorithm **51**(2), 107–121 (2004)
30. Wahlström, M.: A tighter bound for counting max-weight solutions to 2SAT instances. In: Grohe, M., Niedermeier, R. (eds.) IWPEC 2008. LNCS, vol. 5018, pp. 202–213. Springer, Heidelberg (2008). doi:10.1007/978-3-540-79723-4_19
31. Woeginger, G.J.: Open problems around exact algorithms. Discret. Appl. Math. **156**(3), 397–405 (2008)

On the Modulo Degree Complexity
of Boolean Functions

Qian Li[1,2]([✉]) and Xiaoming Sun[1,2]

[1] CAS Key Lab of Network Data Science and Technology, Institute of Computing
Technology, Chinese Academy of Sciences, Beijing 100190, China
{liqian,sunxiaoming}@ict.ac.cn
[2] University of Chinese Academy of Sciences, Beijing 100049, China

Abstract. For each integer $m \geq 2$, every Boolean function f can be
expressed as a unique multilinear polynomial modulo m, and the degree
of this multilinear polynomial is called its *modulo m degree*. In this paper
we investigate the modulo degree complexity of total Boolean functions
initiated by Parikshit Gopalan et al. [8], in which they asked the following question: whether the degree complexity of a Boolean function
is polynomially related with its modulo m degree. For m be a power of
primes, it is already known that the module m degree can be arbitrarily
smaller compare to the degree complexity (see Sect. 2 for details). When
m has at least two distinct prime factors, the question remains open.
Towards this question, our results include: (1) we obtain some nontrivial
equivalent forms of this question; (2) we affirm this question for some
special classes of functions; (3) we prove a no-go theorem, explaining
why this problem is difficult to attack from the computational complexity point of view; (4) we show a super-linear separation between the
degree complexity and the modulo m degree.

1 Introduction

The polynomial representation of Boolean functions in different characteristics is
a powerful tool in extensive areas of computer science, such as machine learning
[12,13,16,18], computational complexity [1–3,19–21,23,25], explicit combinatorial constructions [5,7,9,10]. In this paper, we investigate the polynomial degree
of a function.

The *modulo m degree* of a Boolean function f, denoted by $\deg_m(f)$, is the
degree of the unique multilinear polynomial representing f over $\mathbb{Z}/m\mathbb{Z}$. In addition, we denote $\deg_0(f)$ (where the underlie ring is $\mathbb{Z}$) simply by $\deg(f)$. A central
topic here is to investigate the relationship between module m degrees for different m. From the definition it is clear that for any f, $\deg_m(f) \geq \deg_{m'}(f)$
if m' is a factor of m, particularly, $\deg(f) \geq \deg_m(f)$. This is because the

This work was supported in part by the National Natural Science Foundation of
China Grant 61433014, 61502449, 61602440, the 973 Program of China Grants No.
2016YFB1000201 and the China National Program for support of Top-notch Young
Professionals.

© Springer International Publishing AG 2017
Y. Cao and J. Chen (Eds.): COCOON 2017, LNCS 10392, pp. 384–395, 2017.
DOI: 10.1007/978-3-319-62389-4_32

polynomial representing f over $\mathbb{Z}/m\mathbb{Z}$ can be obtained from the representation over $\mathbb{Z}$ by taking each coefficient modulo m. The gap between $\deg(f)$ and $\deg_m(f)$ can be arbitrarily large when m is a prime: consider the function $f(x) = (x_1 + \cdots + x_n)^{m-1} \mod m$, it is easy to see f is Boolean due to Fermat's little theorem, $\deg_m(f) \leq m - 1$, and $\deg(f) = \Omega(n)$. Actually, the gap can be arbitrarily large even when m is a prime power [6].

In the seminal paper, Gopalan et al. [8] showed a general principle: low degree polynomials modulo p are hard to compute by polynomials in other characteristics. More precisely, let f be a Boolean function which depends on n variables, p and q be distinct primes, then

$$\deg_q(f) \geq \frac{n}{\lceil \log_2 p \rceil \deg_p(f) p^{2 \deg_p(f)}}.$$

Moreover, they also showed that it's still hard even to approximate, which implies most known lower bounds for $AC_0[q]$ circuits.

In this work, we focus on the relation between $\deg(f)$ and $\deg_m(f)$. As mentioned above, $\deg(f) \geq \deg_m(f)$, and the equality can be achieved by AND function. For the other direction, the gap can be arbitrarily large for prime powers [6]. The situation becomes different when m has at least two distinct prime factors p and q: according to the result in [8] as mentioned above, we have $\deg_m(f) \geq \max\{\deg_p(f), \deg_q(f)\} = \Omega(\log n) = \Omega(\log \deg(f))$. Gopalan et al. [8] asked what is the largest possible separation between $\deg(f)$ and $\deg_m(f)$. Here we conjecture these quantities are polynomially related:

Conjecture 1. Let f be a boolean function and m be an integer which has at least two distinct prime factors, then

$$\deg(f) \leq \mathrm{poly}(\deg_m(f)).$$

Our Results. Towards Conjecture 1, we first give some equivalent conjectures that might easier to solve. More precisely, we can replace the degree complexity on the left side by some other complexity measures that could be exponentially smaller than $\deg(f)$, such as the minimum certificate complexity etc.

We also confirm the conjecture for some special classes of functions, such as k-uniform hypergraph properties and functions with small alternating numbers.

Theorem 1. *For any non-trivial k-uniform hypergraph property f on n vertices and any integer m with at least two distinct prime factors, we have*

$$\deg(f) = O(\deg_m(f)^k).$$

Theorem 2. *Let f be a boolean function, then for any $m \geq 2$,*

$$\deg(f) = O(\mathrm{alt}(f) \cdot \deg_m(f)^2),$$

where $\mathrm{alt}(f)$ is the alternating number of f.

Note that $\deg_6(f) = \max\{\deg_2(f), \deg_3(f)\}$ according to the Chinese Remainder Theorem, thus Conjecture 1 for the case $m = 6$ is equivalent to conjecture that the module 3 degree of any polynomial P_2 over $\mathbb{F}_2$ with low degree must be large if the degree of the function represented by P_2 is large. The following no-go theorem somehow explains why this problem is hard to solve even for this simplest case from the computational complexity point of view.

Theorem 3. *Given a polynomial $P_2(x_1, x_2, \ldots, x_n)$ over $\mathbb{F}_2$ with $\text{poly}(n)$ monomials, it's impossible to decide whether $\deg_3(P_2) = n$ or not in polynomial time, unless $NP = RP$.*

Finally, in the direction to disprove this conjecture, we provide a quadratic separation. As we will see in Sect. 2, Conjecture 1 doesn't lose generality only focusing on the case $m = p_1p_2$, where p_1 and p_2 are two distinct primes.

Theorem 4. *For any two distinct primes p_1 and p_2, there exists a sequence of boolean functions f, s.t:*

$$\deg_{p_1p_2}(f) = O(\deg(f)^{1/2}).$$

We wonder whether this is the largest separation between $\deg_{p_1p_2}(f)$ and $\deg(f)$.

Organization. We present some preliminaries in Sect. 2, and give other equivalent conjectures in Sect. 3. We confirm this conjecture for k-uniform hypergraph properties and functions with small alternating number in Sect. 4 and present a no-go theorem and a super-linear separation in Sect. 5. Finally, we conclude this paper in Sect. 6.

2   Preliminaries

Let $f : \{0,1\}^n \rightarrow \{0,1\}$ be a Boolean function, and R be a commutative ring containing $\{0,1\}$ with characteristic m, we say a multilinear polynomial $P(x_1, \cdots, x_n) \in R[x_1, \cdots, x_n]$ represents f if $P(x) = f(x)$ for any $x \in \{0,1\}^n$. From the Mobius inversion formula, such a polynomial always exists and is unique. Moreover, the degree of P only depends on the characteristic of R[6], thus we can denote the degree of P by $\deg_m(f)$. In the paper we will only consider the case where $R = \mathbb{Z}/m\mathbb{Z}$ and denote such polynomials by $P_m(x)$.

We list some basic facts in the following, proofs of which can be found in [6].

Fact 1. *Suppose the polynomial representation of f is $\sum_{S \subseteq [n]} C_S \prod_{i \in S} x_i$, then the representation over $\mathbb{Z}/m\mathbb{Z}$ should be $\sum_{S \subseteq [n]} (C_S \bmod m) \prod_{i \in S} x_i$.*

For example, let f be the parity function, i.e., $x_1 \oplus \cdots \oplus x_n$. The polynomial representing f over $\mathbb{Z}$ is $\sum_{\emptyset \neq S \subseteq [n]} (-2)^{|S|} \prod_{i \in S} x_i$ with $\deg(f) = n$ from the Mobius inverse formula, the representation over $\mathbb{F}_2$ is $\sum_i x_i$ with $\deg_2(f) = 1$ by taking each coefficient modulo 2, and similarly the representation over $\mathbb{F}_3$ is $\sum_{\emptyset \neq S \subseteq [n]} \prod_{i \in S} x_i$ with $\deg_3(f) = n$. Indeed, it is not hard to see that $\deg_p(f) = n$ for every prime $p \neq 2$.

Fact 2. *For any Boolean function f, we have $\deg(f) \geq \deg_m(f)$ for all m. Similarly $\deg_m(f) \geq \deg_{m'}(f)$ if $m'|m$.*

The above fact implies $\deg_m(f) \leq \deg_{m^k}(f)$. The following fact shows that they are always within a factor $2k - 1$ of each other.

Fact 3. *For any Boolean function f, and any integers $m \geq 2$, $k \geq 1$, we have*

$$\deg_m(f) \leq \deg_{m^k}(f) \leq (2k - 1)\deg_m(f).$$

Now recall the function $f(x) = (x_1 + \cdots + x_n)^{m-1} \mod m$ with $\deg_m(f) \leq m-1$ and $\deg(f) = \Omega(n)$ for prime m, as mentioned in the introduction. Indeed, such functions also exist for power of primes.

Fact 4. *For any prime power m, there exists a sequence of functions f such that $\deg_m(f) = O(1)$ and $\deg(f) = \Omega(n)$.*

The following fact is a consequence of the Chinese Remainder Theorem,

Fact 5. *For any Boolean function f and any m and m' with $\gcd(m, m') = 1$, we have*

$$\deg_{m'm}(f) = \max\{\deg_{m'}(f), \deg_m(f)\}.$$

Due to Facts 2 and 5, we get an equivalent form of Conjecture 1 straightforwardly:

Conjecture 2. Let f be a boolean function, p and q be two distinct primes, then

$$\deg(f) \leq \text{poly}(\deg_p(f), \deg_q(f)).$$

Next, we give the definitions of some other complexity measures which will be used in this paper. For an input $x \in \{0,1\}^n$ and a subset B, x^B denotes the input obtained by flipping all the bit x_j such that $j \in B$.

Definition 1. *The sensitivity complexity of f on input x is defined as $s(f,x) := |\{i : f(x) \neq f(x^i)\}|$. The sensitivity complexity of the function f is defined as $s(f) := max_x s(f,x)$.*

It has been shown that $s(f) = O(\deg(f)^2)$[19], but whether $\deg(f)$ can be polynomially bounded in terms of $s(f)$ is still open today, actually it is what the famous *sensitivity conjecture* asks [11].

Definition 2. *The block sensitivity $bs(f,x)$ of f on input x is the maximum number of disjoint subsets $B_1, B_2, \cdots, B_r$ of $[n]$ such that for all j, $f(x) \neq f(x^{B_j})$. The block sensitivity of f is defined as $bs(f) = max_x bs(f,x)$, and the minimum block sensitivity of f is defined as $bs_{min}(f) = min_x bs(f)$.*

Definition 3. *Let C be an assignment $C : S \to \{0,1\}$ of values to some subsets $S \subseteq [n]$. We say C is consistent with $x \in \{0,1\}^n$ if $x_i = C(i)$ for all $i \in S$.*

For $b \in \{0,1\}$, a $b-$certificate for f is an assignment C such that $f(x) = b$ whenever x is consistent with C. The size of C is $|S|$.

The certificate complexity $C(f,x)$ of f on input x is the size of a smallest $f(x)$-certificate that is consistent with x. The certificate complexity of f is $C(f) = \max_x C(f,x)$. The minimum certificate complexity of f is $C_{min}(f) = \min_x C(f,x)$.

Definition 4. *Let $m \geq 2$ be an integer, the mod-m rank of a boolean function f, denoted by $\mathrm{rank}_m(f)$, is the minimum integer r s.t. f can be expressed as*

$$f = x_{i_1} f_1 + \cdots + x_{i_r} f_r + f_0 \qquad (\mathrm{mod}\ m),$$

where $\deg_m(f_i) < \deg_m(f)$ for all $0 \leq i \leq r$. Equivalently, $\mathrm{rank}_m(f)$ is the minimum number of variables to hit all largest monomials in $P_m(x)$. Here we say a monomial is largest if it has maximal degree.

Since we have to fix at least $\mathrm{rank}_m(f)$ variables to make all the largest monomials in $P_m(x)$ vanish, thus $\mathrm{rank}_m(f) \leq C_{min}(f)$ for any m. $C_{min}(f)$, $bs_{min}(f)$ and $\mathrm{rank}_m(f)$ are all polynomially bounded by $\deg(f)$, since $\{bs_{min}(f), \mathrm{rank}_m(f)\} \leq C_{min}(f) \leq C(f) = O(\deg(f)^3)[17]$, and sometimes they can be very small: $\mathrm{rank}_m(AND_n) = bs_{min}(AND_n) = C_{min}(AND_n) = 1 \ll n \doteq \deg(AND_n)$.

Definition 5. *For a Boolean function $f : \{0,1\}^n \to \{0,1\}$, we define the alternating number $\mathrm{alt}(f)$ of f to be the largest k such that there exist a list $\{x^{(1)}, x^{(2)}, \cdots, x^{(k+1)}\}$ with $x^{(i)} \preceq x^{(i+1)}$ and $f(x^{(i)}) \neq f(x^{(i+1)})$ for any $i \in [k]$. Here we say $x \preceq y$ if $x_i \leq y_i$ for all i.*

Definition 6. *A Boolean function f is symmetric if for every input $x = x_1, \cdots, x_n$ and every permutation $\sigma \in S_n$,*

$$f(x_1, \cdots, x_n) = f(x_{\sigma(1)}, \cdots, x_{\sigma(n)}).$$

A Boolean string can represent a graph in the following manner: $x_{(i,j)} = 1$ means there is an edge connecting vertex i and vertex j, and $x_{(i,j)} = 0$ means there is no such edge. Graph properties are functions which are independent with the labeling of vertices, i.e. two isomorphic graphs have the same function value.

Definition 7. *A Boolean function $f : \{0,1\}^{\binom{n}{2}} \to \{0,1\}$ is called a graph property if for every input $x = (x_{(1,2)}, \cdots, x_{(n-1,n)})$ and every permutation $\sigma \in S_n$,*

$$f(x_{(1,2)}, \cdots, x_{(n-1,n)}) = f(x_{(\sigma(1),\sigma(2))}, \cdots, x_{(\sigma(n-1),\sigma(n))}).$$

Similarly, we define k-uniform hypergraph properties.

Definition 8. *A Boolean function $f : \{0,1\}^{\binom{n}{k}} \to \{0,1\}$ is called a k-uniform hypergraph property if for every input $x = (x_{(1,2,\ldots,k)}, \cdots, x_{(n-k+1,\ldots,n-1,n)})$ and every permutation $\sigma \in S_n$,*

$$f(x_{(1,2,\ldots,k)}, \cdots, x_{(n-k+1,\ldots,n-1,n)}) = f(x_{(\sigma(1),\sigma(2),\ldots,\sigma(k))}, \cdots, x_{(\sigma(n-k+1),\ldots,\sigma(n-1),\sigma(n))}).$$

It is easy to see graph property is 2-uniform hypergraph property.

3   Equivalent Conjectures

Observe that the $\deg(f)$ on the left side in Conjectures 1 and 2 can be replaced by any other complexity measures which are polynomially related with $\deg(f)$, such as $D(f)$, $bs(f)$ etc. [4], to get equivalent conjectures. Surprisingly, we find that we can also replace it with some smaller complexity measures, such as $\text{rank}_p(f)$, $C_{min}(f)$, $bs_{min}(f)$ and $s(f)$. In the following, we prove them one by one.

Conjecture 3. Let f be a boolean function, p and q be two distinct primes, then

$$\text{rank}_p(f) \leq \text{poly}(\deg_p(f), \deg_q(f)).$$

Theorem 5. *Conjecture 3 $\Longleftrightarrow$ Conjecture 2.*

Proof. $\Longleftarrow$: Trivial, since $\text{rank}_p(f) = O(\deg(f)^3)$, as mentioned above.

$\Longrightarrow$: We design an algorithm to query f, which contains at most $\deg_p(f)$ rounds and each round reduces $\deg_p$ by at least one. Denote the function at round t by $f^{(t)}$. Note that $f^{(t)}$ is a subfunction of f, hence $\deg_p(f^{(t)}) \leq \deg_p(f)$ and $\deg_q(f^{(t)}) \leq \deg_q(f)$. For each round, we can query $\text{rank}_p(f^{(t)})$ variables to make the largest monomials in $P_p(x)$ vanish, which means $\deg_p(f^{(t)})$ is reduced by at least one. Therefore assuming Conjecture 3, we have $\text{rank}_p(f^{(t)}) \leq \text{poly}(\deg_p(f^{(t)}), \deg_q(f^{(t)})) \leq \text{poly}(\deg_p(f), \deg_q(f))$, which implies $\deg(f) \leq D(f) \leq \text{poly}(\deg_p(f), \deg_q(f))$.

Recall that $\text{rank}_p(f) \leq C_{min}(f) = O(\deg(f)^3)$, we get another equivalent conjecture.

Conjecture 4. Let f be a boolean function, p and q be two distinct primes, then

$$C_{min}(f) \leq \text{poly}(\deg_p(f), \deg_q(f)).$$

Now, we show $\deg(f)$ in Conjecture 2 can be replaced with $bs_{min}(f)$:

Conjecture 5. Let f be a boolean function, p and q be two distinct primes, then

$$bs_{min}(f) \leq \text{poly}(\deg_p(f), \deg_q(f)).$$

Theorem 6. *Conjecture 5 $\Longleftrightarrow$ Conjecture 2.*

Proof. $\Longleftarrow$: Directly follows from $bs_{min}(f) \leq bs(f) = O(\deg(f)^2)$ [19].

$\Longrightarrow$: We call monomial M maximal in $P_p(x)$ if no other monomials contains it. Observe that for any input x and any maximal monomial M, there exist a block $B \subseteq \text{supp}(M)$ such that $f(x) \neq f(x^B)$, because for any restriction S: $[n]\backslash M \to \{0,1\}$ monomial M can't be cancelled, which implies $f|_S$ is a non-constant function. In addition, according to the definition of $\text{rank}_p(f)$, there exists at least $\text{rank}_p(f)/\deg_p(f)$ disjoint largest monomials in $P_p(x)$. Therefore we get $bs_{min}(f) \geq \text{rank}_p(f)/\deg_p(f)$, which implies Conjecture 3 assuming Conjecture 5.

Finally, we show $\deg(f)$ in Conjecture 2 can be replaced with $s(f)$. The key technique is called *"replacing"*: just replace the occurrences of x_i with x_j, i.e., the new function is $f(\cdots, x_i, \cdots, x_i, \cdots)$. Note that x_i in the corresponding $P_m(x)$ are also replaced with x_j, thus $\deg_m(f)$ cannot increase, and the new function is still boolean.

For example, let $P_2(x) = x_1x_2 + x_1x_3 + x_2x_3$ and the corresponding $P_3(x)$ is $x_1x_2 + x_1x_3 + x_2x_3 + x_1x_2x_3$. If we replace x_2 with x_1, the new $P_2(x)$ is $x_1x_1 + x_1x_3 + x_1x_3 = x_1$ and the new $P_3(x)$ is $x_1x_1 + x_1x_3 + x_1x_3 + x_1x_1x_3 = x_1$.

Conjecture 6. Let f be a boolean function, p and q be two distinct primes, then

$$s(f) \le \operatorname{poly}(\deg_p(f), \deg_q(f)).$$

Theorem 7. *Conjecture 2 $\Longleftrightarrow$ Conjecture 6.*

The following simplified proof is observed by Shachar Lovett.

Proof. $\Longrightarrow$: Recall $s(f) = O(\deg(f)^2)$ [19], thus Conjecture 3 $\Rightarrow$ Conjecture 2 $\Rightarrow$ Conjecture 6.

$\Longleftarrow$: W.L.O.G, assume $bs(f, \mathbf{0}) = bs(f) = r$, thus there exist r disjoint blocks $B_1, \cdots, B_r \subseteq [n]$ such that for all i, $f(\mathbf{0}) \ne f(\mathbf{0}^{B_i})$. Further, we assume that $i \in B_i$. Now, we "replace" all variables in B_i with x_i to get a new function f'. It is easy to see that $f'(\mathbf{0}) = f(\mathbf{0}) \ne f(\mathbf{0}^{B_i}) = f'(\mathbf{0}^i)$, thus

$$bs(f) = s(f') \le \operatorname{poly}(\deg_p(f'), \deg_q(f')) \le \operatorname{poly}(\deg_p(f), \deg_q(f)),$$

Now we get the conclusion immediately by noting that $bs(f)$ and $\deg(f)$ are polynomially related [4].

4   Special Classes of Functions

In this section, we confirm Conjecture 1 for some special classes of functions.

4.1   Symmetric Functions

Chia-Jung Lee et al. [14] already confirmed the case of symmetric functions by showing that $2\deg_{p_1}(f)\deg_{p_2}(f) > n$. Here, we give another proof with better parameters.

Theorem 8. *Let $f : \{0,1\}^n \to \{0,1\}$ be symmetric and nonconstant, and p_1, p_2 are two distinct primes. Then*

$$\deg(f) \le n < p_1 \deg_{p_1}(f) + p_2 \deg_{p_2}(f).$$

Proof. For the sake of the presentation, let $d_i = \deg_{p_i}(f)$ and $L_i = p_i^{1+\lfloor \log_{p_i} d_i \rfloor}$.
Since f is symmetric, each $P_{p_i}(x)$ can be written as $\sum_{k=0}^{d_i} c_{i,k} \binom{|x|}{k}$. Then according to Lucas formula, for any nonnegative integers s, j and $k \leq d_i$, we have

$$\binom{sL_i + j}{k} \equiv_{p_i} \binom{j}{k}.$$

Define $g(|x|) = f(x)$, the above equality says $g(k + L_i) = g(k)$. Next, we want to show $n < L_1 + L_2$, which implies $n < p_1 d_1 + p_2 d_2$. Note that $L_1 \neq L_2$, w.l.o.g., assume $L_1 < L_2$.

Suppose $n \geq L_1 + L_2$, we claim that $\forall k \leq L_2$, $g(k) = g(k + L_1 \mod L_2)$, this is because if $k + L_1 \leq L_2$, it's trivial, otherwise, $g(k) = g(k + L_1) = g(k + L_1 - L_2) = g(k + L_1 \mod L_2)$. Moreover, $\gcd(L_1, L_2) = 1$, hence $\forall l \leq L_2$, there exists a integer t such that $l - k \equiv_{L_2} tL_1$, i.e. $g(k) = g(k + tL_1 \mod L_2) = g(l)$, which means f is constant, a contradiction.

Corollary 1. *Let* $f : \{0,1\}^n \to \{0,1\}$ *be symmetric and nonconstant,* $p_1 < p_2 < \cdots p_r$ *be distinct primes, and* r *and* e_i *'s be positive integers. Let* $m = \Pi_{i=1}^{r} p_i^{e_i}$. *Then*

$$\deg(f) \leq n < \deg_m(f)(p_1 + p_2).$$

Proof. First we have $\deg_m(f) \geq \deg_{p_1 p_2}(f) = \max\{\deg_{p_1}(f), \deg_{p_2}(f)\}$. Then according to the above theorem, $(p_1 + p_2)\deg_{p_1 p_2}(f) \geq p_1 \deg_{p_1}(f) + p_2 \deg_{p_2}(f) > n$, as expected.

4.2 Uniform Hypergraph Properties

Using Theorem 8, we can confirm Conjecture 2 for all k-uniform hypergraph properties, where k is a constant. For the reader's convenience, we restate Theorem 1 here.

Theorem 1. *For any non-trivial k-uniform hypergraph property f on n vertices and any integer m with distinct prime factors p_1 and p_2, we have*

$$\frac{1}{p_1 + p_2 + k} n \leq \deg_m(f),$$

which implies

$$\deg(f) \leq \binom{n}{k} = O(\deg_m(f)^k).$$

Proof. (The proof is similar with Lemma 8 in [22].) W.l.o.g., we assume that for the empty graph $\overline{K_n}$, $f(\overline{K_n}) = 0$. Since f is non-trivial, there must exist a graph G such that $f(G) = 1$. Let's consider graphs in $f^{-1}(1) = \{G : f(G) = 1\}$ with the minimum number of edges. Define $m = \min\{|E(G)| : f(G) = 1\}$.

We claim that if $m \geq \frac{1}{p_1 + p_2 + k} n$, then $\deg_m(f) \geq \frac{1}{p_1 + p_2 + k} n$. Let G be a graph in $f^{-1}(1)$ and $|E(G)| = m$. Consider the subfunction f' where $\forall e \notin E(G)$, x_e

is restricted to 0, since G has the the minimum number of edges, deleting any edges from G will change the values of $f(G)$, therefore, f' is a AND function. Thus, $\deg_m(f) \geq \deg_m(f') = m \geq \frac{1}{p_1+p_2+k}n$.

In the following we assume $m < \frac{1}{p_1+p_2+k}n$. Again let G be a graph in $f^{-1}(1)$ with $|E(G)| = m$. Let us consider the isolated vertices set I, as

$$\sum_{v \in V} deg(v) = k|E(G)| < \frac{k}{p_1 + p_2 + k}n.$$

We have

$$|I| \geq n - \sum_{v \in V} deg(v) > \frac{p_1 + p_2}{p_1 + p_2 + k}n.$$

Suppose $\deg_m(f) < \frac{1}{p_1+p_2+k}n$, we will deduce that there exists another graph with fewer edges and the same value, against the assumption that G has the minimum number of edges in $f^{-1}(1)$, which ends the whole proof.

Pick a vertex u with $deg(u) = d > 0$. Suppose in the graph G vertex u is adjacent to $(k-1)$-edges $\{e_1^{(k-1)}, e_2^{(k-1)}, \cdots, e_d^{(k-1)}\}$ and $I = \{u_1, u_2, \cdots, u_t\}$, where $t = |I|$.

Consider the t-variable Boolean function $g_1 : \{0,1\}^t \to \{0,1\}$, where

$$g_1(x_1, \cdots, x_t) = f(G + x_1(e_1^{(k-1)}, u_1) + \cdots + x_t(e_1^{(k-1)}, u_t)).$$

It is easy to see that g_1 is a symmetric function. We claim that g_1 is a constant function: if not, we have $\deg_m(g_1) \geq \frac{1}{p_1+p_2}t$ according to Corollary 1, which implies $\deg_m(f) \geq \frac{1}{p_1+p_2+k}n$ since g_1 is a restriction of f. In particular, $g_1(1, \cdots, 1) = g_1(0, \cdots, 0)$, i.e. $f(G_1) = f(G)$, where $G_1 = G + \sum_{i=1}^{t}(e_1^{(k-1)}, u_i)$.

Define $G_i = G_{i-1} + \sum_{j=1}^{t}(e_i^{(k-1)}, u_j)$ $(i = 2, \cdots, d)$. Similarly, we can show that

$$f(G) = f(G_1) = \cdots = f(G_d).$$

Next   we   will   delete all the edges between $\{u, u_1, \cdots, u_t\}$ and $\{e_1^{(k-1)}, e_2^{(k-1)}, \cdots, e_d^{(k-1)}\}$ from G_d by reversing the adding edge procedure of $G \to G_1 \to \cdots \to G_d$. More precisely, define $H_1 = G_d$; for $i = 2, \cdots, d$, define

$$H_i = H_{i-1} - (e_i^{(k-1)}, u) - (e_i^{(k-1)}, u_1) - \cdots - (e_i^{(k-1)}, u_t),$$

and

$$h_i(y_0, y_1, \cdots, y_t) = f(H_i + y_0(e_i^{(k-1)}, u) + y_1(e_i^{(k-1)}, u_1) + \cdots + y_t(e_i^{(k-1)}, u_t)).$$

Similarly, by the fact $\deg_m(f) < \frac{1}{p_1+p_2+k}n$ we can show that all the functions $h_2, \cdots, h_d$ are constant, which implies $f(H_1) = f(H_2) = \cdots = f(H_d)$. So we find another graph H_d with fewer edges than G and $f(H_d) = 1$.

4.3   Functions with Small Alternating Numbers

We can also confirm the functions with small alternating numbers.

Theorem 2. *Let f be a boolean function, then for any $m \geq 2$,*

$$D(f) = O(\mathrm{alt}(f) \cdot \deg_m(f)^2),$$

which implies

$$\deg(f) = O(\mathrm{alt}(f) \cdot \deg_m(f)^2).$$

Recall that $\deg_m(f) = \Omega(\log n)$ when m has two distinct prime factors [8], thus the above theorem confirms Conjecture 1 for non-degenerate functions with $\mathrm{alt}(f) = \mathrm{poly}\log(n)$.

Lin and Zhang [15] have shown the case $m = 2$, and their argument applies to general m as well. We omit the proof here.

5   A No-Go Theorem and a Super-Linear Separation

The following theorem somehow explains why it's hard to solve Conjecture 2, even for the simplest case where $p = 2$ and $q = 3$.

Theorem 3. *Given a polynomial $P_2(x_1, x_2, \ldots, x_n)$ over $\mathbb{F}_2$ with $\mathrm{poly}(n)$ monomials, it's impossible to decide whether $\deg_3(P_2) = n$ or not in polynomial time, unless $NP = RP$.*

Proof. It's sufficient to give a reduction to Unique-3CNF, since Unique-3CNF can't be solved in polynomial time unless $NP = RP$ [24].

Given a Unique-3CNF formula $\phi(x_1, \cdots, x_n)$ with m clauses, we first remove negated literals to make the formula monotone: for any variable x_i replace the occurrences of its negation by a new variable $x_i^\star$. Also introduce new variables x_i' and x_i'' and conjoin ϕ with the clauses $(x_i \vee x_i^\star) \wedge (x_i \vee x_i') \wedge (x_i^\star \vee x_i'') \wedge (x_i' \vee x_i'')$. Denote the new formula by ϕ' with n' variables and m' clauses. It is easy to see that $\sharp\phi \equiv -\sharp\phi' \mod 3$. Here $\sharp\phi$ is the number of solutions of ϕ, i.e., $\sharp\phi = \sharp\{x : \phi(x) = 1\}$.

Then we construct a polynomial P_2 over $\mathbb{F}_2$ from ϕ': There are m' variables $y_1, y_2, \cdots, y_{m'}$ and n' monomials $t_1, t_2, \cdots, t_{n'}$ in P_2, and t_i contains y_j if the jth clause contains x_i in ϕ'.

Note that the corresponding polynomial over $\mathbb{F}_3$ is

$$P_3 = \frac{1}{2}\left[1 - \prod_{i=1}^{m'}(1 - 2t_i)\right] = \prod_{i=1}^{m'}(1 + t_i) - 1,$$

According to the fact that $y_i^l = y_i$ for any integer $l \geq 1$ and any $y_i \in \{0, 1\}$, it is not hard to see the coefficient of $\prod_{i=1}^{m'} y_i$ is $\sharp\phi' \mod 3$. Note that ϕ has at most one solution, then ϕ is satisfiable if and only if $\sharp\phi \equiv -\sharp\phi' \equiv 1 \mod 3$, which means $\deg_3(P_2) = n$.

In the direction to disprove Conjecture 1, we give a quadratic separation.

Theorem 4. *For any two distinct prime p_1 and p_2, there exists a sequence of boolean functions f, s.t:*

$$\deg_{p_1 p_2}(f) = O(\deg(f)^{1/2}).$$

Proof. Let $f = \text{Mod}_3(\text{Mod}_2(x_1, \cdots, x_{\sqrt{n}}), \cdots, \text{Mod}_2(x_{n-\sqrt{n}+1}, \cdots, x_n))$.   Here, $\text{Mod}_{p_i}(\cdot) = 0$, if the sum of inputs can be divided by p_i, otherwise $\text{Mod}_{p_i}(\cdot) = 1$. On one hand, since $\text{Mod}_{p_i}(\cdot)$ is symmetric, it is easy to see $\deg(f) = \Omega(n)$. On the other hand, it is also not hard to see that $\deg_{p_i}(f) = O(\sqrt{n})$ for each i, which implies $\deg_{p_1 p_2}(f) = O(\sqrt{n})$.

6   Conclusion

In this work, we investigate the relationship between $\deg(f)$ and $\deg_m(f)$, more specifically, we focus on an open problem proposed by Gopalan et al. in [8], which asks whether $\deg(f)$ and $\deg_m(f)$ are polynomially related, when m has at least two distinct prime factors. First we present some nontrivial equivalent forms of this problem, then we affirm it for some special classes of functions. Finally we show a no-go theorem by which try to explain why this problem is hard, as well as a super-linear separation. Most of the problems remain open, here we list some of them:

1. Can we prove Conjecture 1 for cyclically invariant functions first?
2. Given a polynomial P_2 over $\mathbb{F}_2$ with $\text{poly}(n)$ size, we have shown that there's no efficient algorithms to compute its modulo 3 degree exactly unless $NP = RP$. Is it still hard to approximate that?

Acknowledgments. We thank the anonymous reviewer for pointing out the better construction in Theorem 4, and Shachar Lovett for providing us the simple proof of Theorem 7.

References

1. Aspnes, J., Beigel, R., Furst, M.L., Rudich, S.: The expressive power of voting polynomials. Combinatorica **14**(2), 135–148 (1994)
2. Beigel, R., Reingold, N., Spielman, D.A.: The perceptron strikes back. In: Proceedings of the Sixth Annual Structure in Complexity Theory Conference, pp. 286–291 (1991)
3. Bhatnagar, N., Gopalan, P., Lipton, R.J.: Symmetric polynomials over z_m and simultaneous communication protocols. J. Comput. Syst. Sci. **72**(2), 252–285 (2006)
4. Buhrman, H., De Wolf, R.: Complexity measures and decision tree complexity: a survey. Theor. Comput. Sci. **288**(1), 21–43 (2002)
5. Efremenko, K.: 3-query locally decodable codes of subexponential length. SIAM J. Comput. **41**(6), 1694–1703 (2012)

6. Gopalan, P.: Computing with Polynomials over Composites. Ph.D. thesis (2006)
7. Gopalan, P.: Constructing ramsey graphs from boolean function representations. Combinatorica **34**(2), 173–206 (2014)
8. Gopalan, P., Shpilka, A., Lovett, S.: The complexity of boolean functions in different characteristics. Comput. Complex. **19**(2), 235–263 (2010)
9. Grolmusz, V.: Superpolynomial size set-systems with restricted intersections mod 6 and explicit ramsey graphs. Combinatorica **20**(1), 71–86 (2000)
10. Grolmusz, V.: Constructing set systems with prescribed intersection sizes. J. Algorithms **44**(2), 321–337 (2002)
11. Hatami, P., Kulkarni, R., Pankratov, D.: Variations on the sensitivity conjecture. Theor. Comput. Grad. Surv. **4**, 1–27 (2011)
12. Klivans, A.R., Servedio, R.A.: Learning DNF in time $2^{\tilde{o}(n^{1/3})}$. J. Comput. Syst. Sci. **68**(2), 303–318 (2004)
13. Kushilevitz, E., Mansour, Y.: Learning decision trees using the fourier spectrum. SIAM J. Comput. **22**(6), 1331–1348 (1993)
14. Lee, C.J., Lokam, S.V., Tsai, S.C., Yang, M.C.: Restrictions of nondegenerate boolean functions and degree lower bounds over different rings. In: Proceedings of IEEE International Symposium on Information Theory, pp. 501–505 (2015)
15. Lin, C., Zhang, S.: Sensitivity conjecture and log-rank conjecture for functions with small alternating numbers. CoRR, abs/1602.06627 (2016)
16. Linial, N., Mansour, Y., Nisan, N.: Constant depth circuits, fourier transform, and learnability. J. ACM **40**(3), 607–620 (1993)
17. Midrijanis, G.: Exact quantum query complexity for total boolean functions. arXiv preprint quant-ph/0403168 (2004)
18. Mossel, E., O'Donnell, R., Servedio, R.A.: Learning juntas. In: Proceedings of the 35th Annual ACM Symposium on Theory of Computing, pp. 206–212 (2003)
19. Nisan, N., Szegedy, M.: On the degree of boolean functions as real polynomials. Comput. Complex. **4**, 301–313 (1994)
20. Razborov, A.A.: Lower bounds for the size of circuits of bounded depth with basis $\{\wedge, \oplus\}$. Math. Notes Acad. Sci. USSR **41**, 333–338 (1987)
21. Smolensky, R.: Algebraic methods in the theory of lower bounds for boolean circuit complexity. In: Proceedings of the 19th Annual ACM Symposium on Theory of Computing, pp. 77–82 (1987)
22. Sun, X.: An improved lower bound on the sensitivity complexity of graph properties. Theor. Comput. Sci. **412**(29), 3524–3529 (2011)
23. Tsang, H.Y., Wong, C.H., Xie, N., Zhang, S.: Fourier sparsity, spectral norm, and the log-rank conjecture. In: Proceedings of 54th Annual IEEE Symposium on Foundations of Computer Science, pp. 658–667 (2013)
24. Valiant, L.G., Vazirani, V.V.: NP is as easy as detecting unique solutions. Theor. Comput. Sci. **47**, 85–93 (1986)
25. Viola, E.: The sum of D small-bias generators fools polynomials of degree D. Comput. Complex. **18**(2), 209–217 (2009)

Approximating Weighted Duo-Preservation in Comparative Genomics

Saeed Mehrabi[(✉)]

Cheriton School of Computer Science, University of Waterloo, Waterloo, Canada
smehrabi@uwaterloo.ca

Abstract. Motivated by comparative genomics, Chen et al. [9] introduced the Maximum Duo-preservation String Mapping (MDSM) problem in which we are given two strings s_1 and s_2 from the same alphabet and the goal is to find a mapping π between them so as to maximize the number of duos preserved. A *duo* is any two consecutive characters in a string and it is *preserved* in the mapping if its two consecutive characters in s_1 are mapped to same two consecutive characters in s_2. The MDSM problem is known to be NP-hard and there are approximation algorithms for this problem [3,5], all of which consider only the "unweighted" version of the problem in the sense that a duo from s_1 is preserved by mapping to any same duo in s_2 regardless of their positions in the respective strings. However, it is well-desired in comparative genomics to find mappings that consider preserving duos that are "closer" to each other under some distance measure [18].

In this paper, we introduce a generalized version of the problem, called the Maximum-Weight Duo-preservation String Mapping (MWDSM) problem, capturing both duos-preservation and duos-distance measures in the sense that mapping a duo from s_1 to each preserved duo in s_2 has a weight, indicating the "closeness" of the two duos. The objective of the MWDSM problem is to find a mapping so as to maximize the total weight of preserved duos. We give a polynomial-time 6-approximation algorithm for this problem.

1 Introduction

Strings comparison is one of the central problems in the field of stringology with many applications such as in Data Compression and Bioinformatics. One of the most common goals of strings comparison is to measure the similarity between them, and one of the many ways in doing so is to compute the *edit distance* between them. The edit distance between two strings is defined as the minimum number of edit operations to transform one string into the other. In biology, during the process of DNA sequencing for instance, computing the edit distance between the DNA molecules of different species can provide insight about the level of "synteny" between them; here, each edit operation is considered as a single mutation.

In the simplest form, the only edit operation that is allowed in computing the edit distance is to shift a block of characters; that is, to change the order of

© Springer International Publishing AG 2017
Y. Cao and J. Chen (Eds.): COCOON 2017, LNCS 10392, pp. 396–406, 2017.
DOI: 10.1007/978-3-319-62389-4_33

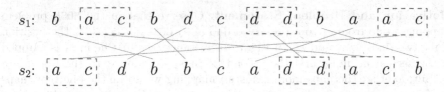

Fig. 1. An instance of the MDSM problem in which the mapping π preserves three duos.

the characters in the string. Computing the edit distance under this operation reduces to the Minimum Common String Partition (MCSP) problem, which was introduced by Goldstein et al. [14] (see also [20]) and is defined as follows. For a string s, let $P(s)$ denote a partition of s. Given two strings s_1 and s_2 each of length n, where s_2 is a permutation of s_1, the objective of the MCSP problem is to find a partition $P(s_1)$ of s_1 and $P(s_2)$ of s_2 of minimum cardinality such that $P(s_2)$ is a permutation of $P(s_1)$. The problem is known to be NP-hard and even APX-hard [14].

Recently, Chen et al. [9] introduced a maximization version of the MCSP problem, called the *Maximum Duo-preservation String Mapping* (MDSM) problem. A *duo* in a string s is a pair of consecutive characters in s. For two strings s_1 and s_2, where s_2 is a permutation of s_1 under a mapping π, we say that a *duo is preserved* in the mapping π, if its two consecutive characters are mapped to same two consecutive characters in s_2. Notice that if partitions $P(s_1)$ and $P(s_2)$ are a solution of size r for an instance of the MCSP problem, then this solution can be used to obtain a mapping π from s_1 to s_2 that preserve $n - r$ duos. As such, given two strings s_1 and s_2, the objective of the MDSM problem is to compute a mapping π from s_1 to s_2 that preserves a maximum number of duos. See Fig. 1 for an example.

Related Work. Since the MCSP problem is NP-hard [14], there has been many works on designing polynomial-time approximation algorithms for this problem [10–12,14,17]. The best approximation results thus far are an $O(\log n \log^* n)$-approximation algorithm for the general version of the problem [12], and an $O(k)$-approximation for the k-MCSP problem [12] (the k-MCSP is a variant of the problem in which each character can appear at most k times in each string). In terms of the parameterized complexity, the problem is known to be fixed-parameter tractable with respect to k and the size of an optimal partition [6,16], as well as the size of an optimal solution only [7]. For the MDSM problem, we observe that since the MCSP problem is NP-hard [14], the MDSM problem (i.e., its maximization version) is also NP-hard, and in fact even APX-hard [4]. Moreover, the problem is also shown to be fixed-parameter tractable with respect to the number of duos preserved [2]. Boria et al. [4] gave a 4-approximation algorithm for the MDSM problem, which was subsequently improved to algorithms with approximation factors of $7/2$ [3], 3.25 [5] and (recently) $(2 + \epsilon)$ for any $\epsilon > 0$ [13].

Motivation and Problem Statement. Observe that in the MDSM problem mapping a duo from s_1 to a preserved duo in s_2 does not consider the position of the two duos in s_1 and s_2. In Fig. 1, for instance, the first ac in s_1 is mapped to the second ac in s_2 and the second ac in s_1 is mapped to the first ac in s_2. But, another (perhaps more realistic) mapping would be the one that maps the first ac in s_1 to the first ac in s_2 and the second one in s_1 to the second one in s_2. The latter mapping would become more desirable when solving the problem on strings of extremely long length. In fact, considering the applications of the MDSM problem in comparative genomics, it is much more desirable to find mappings that take into account the position of the preserved features in the two sequences [15,18]. One reason behind this is the fact that focusing on giving priority to preserving features that are "closer" to each other (distance-wise under some distance measure) provides better information about the "synteny" of the corresponding species [15,18].

In this paper, we introduce a more general variant of the MDSM problem, called the Maximum-Weight Duo-preservation String Mapping (MWDSM) problem. In this problem, in addition to s_1 and s_2, we are also given a *weight function* defined on pairs of duos that considers the position of the duos in s_1 and s_2, and so better captures the concept of "synteny" in comparative genomics. Now, the objective becomes maximizing the total weight of the preserved duo (instead of maximizing the number of them). Let us define the problem more formally. For a string s, we denote by $D(s)$ the set of all duos in s ordered from left to right. For example, if $s = acbbda$, then $D(s) = \{ac, cb, bb, bd, da\}$.

Definition 1 (The MWDSM Problem). *Let s_1 and s_2 be two strings of length n. Moreover, let $\mathbf{w} : D(s_1) \times D(s_2) \to \mathbb{R}^+$ denote a weight function. Then, the MWDSM problem asks for a mapping π from s_1 to s_2 that preserve a set S of duos such that*

$$\sum_{d \in S} w(d, \pi(d))$$

is maximized over all such sets S, where $\pi(d)$ denotes the duo in s_2 to which $d \in s_1$ is mapped.

We note that the weight function is very flexible in the sense that it can capture any combination of duos-preservation and duos-distance measures. To our knowledge, this is the first formal study of a "weighted version" of the MDSM problem.

Our Result. Notice that the MWDSM problem is NP-hard as its unweighted variant (i.e., the MDSM problem) is known to be NP-hard [14]. We note that the previous approximation algorithms for the MDSM problem do not apply to the MWDSM problem. In particular, both 7/2-approximation algorithm of Boria et al. [3] and $(2 + \epsilon)$-approximation algorithm of Dudek et al. [13] are based on the local search technique, which is known to fail for weighted problems [8,19]. Moreover, the 3.25-approximation algorithm of Brubach [5] relies on a triplet matching approach, which involves finding a weighted matching (with specialized weights) on a particular graph, but it is not clear if the approach could handle the MWDSM problem with any arbitrary weight function w. Finally, while the linear programming algorithm of Chen et al. [9] might apply to the MWDSM problem, the approximation factor will likely stay the same, which is k^2, where k is the maximum number of times each character appears in s_1 and s_2.

In this paper, we give a polynomial-time 6-approximation algorithm for the MWDSM problem for any arbitrary weight function w. To this end, we construct a vertex-weighted graph corresponding to the MWDSM problem and show that the problem reduces to the *Maximum-Weight Independent Set (MWIS)* problem on this graph. Then, we apply the *local ratio technique* to approximate the MWIS problem on this graph. The local ratio technique was introduced by Bar-Yehuda and Even [1], and is used for designing approximation algorithms for mainly weighted optimization problems (see Sect. 2 for a formal description of this technique). While the approximation factor of our algorithm is slightly large in compare to that of algorithms for the unweighted version of the problem [3, 5], as we now have weights, our algorithm is much simpler as it benefits from the simplicity of the local ratio technique. To our knowledge, this is the first application of the local ratio technique to problems in stringology.

Organization. The paper is organized as follows. We first give some definitions and preliminary results in Sect. 2. Then, we present our 6-approximation algorithm in Sect. 3, and will conclude the paper with a discussion on future work in Sect. 4.

2   Preliminaries

In this section, we give some definitions and preliminaries. For a graph G, we denote the set of vertices and edges of G by $V(G)$ and $E(G)$, respectively. For a vertex $u \in V(G)$, we denote the set of neighbours of u by $N[u]$; note that $u \in N[u]$.

Let $\mathbf{w} \in \mathbb{R}^n$ be a weight vector, and let F be a set of feasibility constraints on vectors $\mathbf{x} \in \mathbb{R}^n$. A vector $\mathbf{x} \in \mathbb{R}^n$ is a feasible solution to a given problem $(F, \mathbf{p})$ if it satisfies all of the constraints in F. The value of a feasible solution $\mathbf{x}$ is the inner product $\mathbf{w} \cdot \mathbf{x}$. A feasible solution is *optimal* for a maximization (resp., minimization) problem if its value is maximal (resp., minimal) among all feasible solutions. A feasible solution $\mathbf{x}$ is an α-approximation solution, or simply an α-approximation, for a maximization problem if $\mathbf{w} \cdot \mathbf{x} \geq \alpha \cdot \mathbf{w} \cdot \mathbf{x}^*$, where $\mathbf{x}^*$ is an optimal solution. An algorithm is said to have an *approximation factor* of α (or, it is called an α-*approximation algorithm*), if it always computes α-approximation solutions.

Local Ratio. Our approximation algorithm uses the *local ratio technique*. This technique was first developed by Bar-Yehuda and Even [1]. Let us formally state the local ratio theorem.

Theorem 1. [1] *Let F be a set of constraints, and let $\mathbf{w}, \mathbf{w}_1$ and $\mathbf{w}_2$ be weight vectors where $\mathbf{w} = \mathbf{w}_1 + \mathbf{w}_2$. If $\mathbf{x}$ is an α-approximation solution with respect to $(F, \mathbf{w}_1)$ and with respect to $(F, \mathbf{w}_2)$, then $\mathbf{x}$ is an α-approximation solution with respect to $(F, \mathbf{w})$.*

We now describe how the local ratio technique is usually used for solving a problem. First, the solution set is empty. The idea is to find a decomposition of the weight vector $\mathbf{w}$ into $\mathbf{w}_1$ and $\mathbf{w}_2$ such that $\mathbf{w}_1$ is an "easy" weight function in some sense (we will discuss this in more details later). The local ratio algorithm continues recursively on the instance $(F, \mathbf{w}_2)$. We assume inductively that the solution returned recursively for the instance $(F, \mathbf{w}_2)$ is a good approximation and need to prove that it is also a good approximation for $(F, \mathbf{w})$. This requires proving that the solution returned recursively for the instance $(F, \mathbf{w}_2)$ is also a good approximation for the instance $(F, \mathbf{w}_1)$. This step is usually the main part of the proof of the approximation factor.

Graph G_I. Given an instance of the MWDSM problem, we first construct a bipartite graph $G_I = (A \cup B, E)$ as follows. The vertices in the left-side set A are the duos in $D(s_1)$ in order from top to bottom and the vertices in the right-side set B are the duos in $D(s_2)$ in order from top to bottom. There exists an edge between two vertices if and only if they represent the same duo. See Fig. 2 for an example. Boria et al. [4] showed that the MDSM problem on s_1 and s_2 reduces to the Maximum Constrained Matching (MCM) problem on G_I, which is defined as follows. Let $A = a_1, \ldots, a_n$ and $B = b_1, \ldots, b_n$. Then, we are interested in computing a maximum-cardinality matching M such that if $(a_i, b_j) \in M$, then a_{i+1} can be only matched to b_{j+1} and b_{j+1} can be only matched to a_{i+1}. In the following, we first assign weights to the edges of G_I and will then show that a similar reduction holds between the MWDSM problem on s_1 and s_2, and a weighted version of the MCM problem on G_I.

To weigh the edges of G_I, we simply assign $w(a_l, b_r)$ as the weight of e, for all $e \in E(G_I)$, where $a_l \in A$ and $b_r \in B$ are the endpoints of e and $w(a_l, b_r)$ is given by Definition 1. Now, we define the Maximum-Weight Constrained Matching (MWCM) problem as the problem of computing a maximum-weight matching M in G_I such that if $(a_i, b_j) \in M$, then a_{i+1} can be only matched to b_{j+1} and b_{j+1} can be only matched to a_{i+1}. To see the equivalence between the MWDSM problem on s_1 and s_2 and the MWCM problem on G_I, let S be a feasible solution to the MWDSM problem with total weight $w(S)$ determined by a mapping π. Then, we can obtain a feasible solution S' for the MWCM problem on G_I by selecting the edges in G_I that correspond to the preserved duos in S determined by π such that $w(S') = w(S)$. Moreover, it is not too hard to see that any feasible solution for the MWCM problem on G_I gives a feasible solution for the MWDSM problem with the same weight. This gives us the following lemma.

Lemma 1. *The MWDSM problem on s_1 and s_2 reduces to the MWCM problem on G_I.*

By Lemma 1, any feasible solution M with total weight $w(M)$ for the MWCM problem on G_I gives a mapping π between the strings s_1 and s_2 that preserves a set of duos with total weight $w(M)$. As such, for the rest of this paper, we focus on the MWCM problem on G_I and give a polynomial-time 6-approximation algorithm for this problem on G_I, which by Lemma 1, results in an approximation algorithm with the same approximation factor for the MWDSM problem on s_1 and s_2.

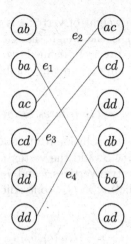

Fig. 2. The graph G_I corresponding to $s_1 = abacddd$ and $s_2 = acddbad$.

3   Approximation Algorithm

In this section, we give a 6-approximation algorithm for the MWCM problem. We were unable to apply the local ratio directly to the MWCM problem on G_I due to the constraint involved in the definition of the problem. Instead, we construct a vertex-weighted graph G_C, called the *conflict graph*, and show that the MWCM problem on G_I reduces to the Maximum-Weight Independent Set (MWIS) problem on G_C. We then apply the local ratio to approximate the MWIS problem on G_C, which results in an approximation algorithm for the MWCM problem on G_I. Consequently, this gives us an approximation algorithm for the MWDSM problem on s_1 and s_2 by Lemma 1.

Graph G_C. We now describe the concept of *conflict*. We say that two edges in $E(G_I)$ are *conflicting* if they both cannot be in a feasible solution for the MWCM problem at the same time, either because they share an endpoint or their endpoints are consecutive on one side of the graph, but not on the other side. The following observation is immediate.

Observation 1. *Let $e_1 = (a_i, b_j)$ and $e_2 = (a_k, b_l)$ be two conflicting edges in $E(G_I)$. Then, either $k \in \{i-1, i, i+1\}$ or $l \in \{j-1, j, j+1\}$.*

We define the *conflict graph* G_C as follows. Let $V(G_C)$ be $E(G_I)$; that is, $V(G_C)$ is the set of all edges in G_I. For a vertex $i \in V(G_C)$, we denote the edge in $E(G_I)$ corresponding to i by e_i. Two vertices i and j are adjacent in G_C if and only if e_i and e_j are conflicting in G_I. The conflict graph G_C corresponding to the graph G_I in Fig. 2 is shown on the right.

To assign weights to the vertices of G_C, let i be a vertex of G_C. Notice that i corresponds to the edge $e_i = (a_l, b_r)$ in $E(G_I)$, where $a_l \in A$ and $b_r \in B$

are preserved duos. Then, the weight of vertex i is defined as $w(i) := w(a_l, b_r)$ in which recall that $w(a_l, b_r)$ is the weight assigned to these preserved duos by Definition 1. Although not precisely defined, we again note that the weight function is very flexible and it can capture any combination of duos-preservation and duos-distance measures.

Lemma 2. *The MWCM problem on G_I reduces to the MWIS problem on G_C.*

Proof. Suppose that S is a feasible solution to the MWCM problem on G_I with total weight $w(S)$. For each edge $e_i \in S$, add the vertex $i \in V(G_C)$ to S'. Clearly, S' is an independent set because two vertices i and j in S' being adjacent would imply that e_i and e_j are conflicting in G_I, contradicting the feasibility of S. To see $w(S')$, notice that

$$w(S') = \sum_{i \in S'} w(i) = \sum_{e_i = (a_l, b_r) \in S} w(a_l, b_r) = \sum_{e_i \in S} w(e_i) = w(S).$$

Now, suppose that S' is an independent set in G_C with total weight $w(S')$. For each $u \in S'$, add $e_u \in E(G_I)$ to S. First, S is a feasible solution for the MWCM problem on G_I because the vertices of G_C corresponding to any two conflicting edges in S would be adjacent in G_C, contradicting the fact that S' is an independent set. Moreover,

$$w(S) = \sum_{e_i \in S} w(e_i) = \sum_{e_i = (a_l, b_r) \in S} w(a_l, b_r) = \sum_{i \in S'} w(i) = w(S').$$

This completes the proof of the lemma.   □

By Lemma 2, any approximation algorithm for the MWIS problem on G_C results in an approximation algorithm with the same factor for the MWCM problem on G_I. As such, for the rest of this section, we focus on the MWIS problem on G_C and show how to apply the local ratio technique to compute a 6-approximation algorithm for this problem on G_C.

Approximating the MWIS Problem on G_C. We first formulate the MWIS problem on G_C as a linear program. We define a variable $x(u)$ for each vertex $u \in V(G_C)$; if $x(u) = 1$, then vertex u belongs to the independent set. The integer program assigns the binary values to the vertices with the constraint that for each clique Q, the sum of the values assigned to all vertices in Q is at most 1.

$$\text{maximize} \quad \sum_{u \in V(G_C)} w(u) \cdot x(u) \tag{1}$$

$$\text{subject to} \quad \sum_{v \in Q} x(v) \le 1 \qquad \forall \text{ cliques } Q \in G_C,$$

$$x(u) \in \{0, 1\} \qquad \forall u \in V(G_C)$$

Note that the number of constraints in (1) can be exponential in general, as the number of cliques in G_C could be exponential. However, for the MWIS problem on G_C, we can consider only a polynomial number of cliques in G_C. To this end, let $u = (a_i, b_j)$ be a vertex in G_C. By Observation 1, if $v = (a_k, b_l)$ is in conflict with $u = (a_i, b_j)$, then either $k \in \{i - 1, i, i + 1\}$ or $l \in \{j - 1, j, j + 1\}$. Let S_u^{i-1} denote the set of all neighbours v of u in G_C such that $v = (a_{i-1}, b_s)$ for some $b_s \in B$ (recall the bipartite graph $G_I = (A \cup B, E)$). Define S_u^i and S_u^{i+1} analogously. Similarly, let S_u^{j-1} be the set of all neighbours v of u such that $v = (a_s, b_{j-1})$ for some $a_s \in A$, and define S_u^j and S_u^{j+1} analogously. Let $M := \{i - 1, i, i + 1, j - 1, j, j + 1\}$. Then, by relaxing the integer constraint of the above integer program, we can formulate the MWIS problem on G_C as the following linear program.

$$\text{maximize} \quad \sum_{u \in V(G_C)} w(u) \cdot x(u) \tag{2}$$

$$\text{subject to} \quad \sum_{v \in S_u^r} x(v) \leq 1 \qquad \forall u \in V(G_C), \forall r \in M,$$

$$x(u) \geq 0 \qquad \forall u \in V(G_C)$$

Notice that the linear program (2) has a polynomial number of constraints. These constraints suffice for the MWIS problem on G_C because, by Observation 1, the vertices $u = (a_i, b_j)$ and v of G_C corresponding to the two conflicting edges e_u and e_v in G_I belong to S_u^r, for some $r \in M$. Moreover, we observe that any independent set in G_C gives a feasible integral solution to the linear program. Therefore, the value of an optimal (not necessarily integer) solution to the linear program is an upper bound on the value of an optimal integral solution.

We are now ready to describe the algorithm. We first compute an optimal solution $\mathbf{x}$ for the above linear program. Then, the rounding algorithm applies a local ratio decomposition of the weight vector $\mathbf{w}$ with respect to $\mathbf{x}$. See Algorithm 1. The key to our rounding algorithm is the following lemma.

Lemma 3. *Let $\mathbf{x}$ be a feasible solution to (2). Then, there exists a vertex $u \in V(G_C)$ such that*

$$\sum_{v \in N[u]} x(v) \leq 6.$$

Proof. Let $u \in V(G_C)$. Notice that u corresponds to an edge in G_I; that is, $u = (a_i, b_j)$, where a_i and b_j are a pair of preserved duos in the mapping π from s_1 to s_2. Observe that $v \in N[u]$ for some $v = (a_k, b_l) \in V(G_C)$ if and only if (a_k, b_l) conflicts with (i, j) in G_I. Let $M := \{i - 1, i, i + 1, j - 1, j, j + 1\}$ and define the set S_u^r as above, for all $r \in M$. Note that the vertices in S_u^r form a clique in G_C, for all $r \in M$, because the set of edges corresponding to the vertices of S_u^r in G_I all share one endpoint (in particular, this endpoint is in A if $r \in \{i - 1, i, i + 1\}$ or it is in B if $r \in \{j - 1, j, j + 1\}$). See Fig. 3 for an illustration. This means by the first constraint of the linear program (2) that

$$\sum_{v \in S_u^r} x(v) \leq 1,$$

Algorithm 1. APPROXIMATEMWIS(G_C)

1: Delete all vertices with non-positive weight. If no vertex remains, then return the empty set;

2: Let $u \in V(G_C)$ be a vertex satisfying

$$\sum_{v \in N[u]} x(v) \leq 6.$$

Then, decompose $\mathbf{w}$ into $\mathbf{w} := \mathbf{w_1} + \mathbf{w_2}$ as follows:

$$w_1(v) := \begin{cases} w(u) & \text{if } v \in N[u], \\ 0 & \text{otherwise.} \end{cases}$$

3: Solve the problem recursively using $\mathbf{w_2}$ as the weight vector. Let S' be the independent set returned;

4: If u is not adjacent with some vertex in S', then return $S := S' \cup \{u\}$; otherwise, return $S := S'$.

---

for all $r \in M$. Therefore, we have

$$\sum_{v \in N[u]} x(v) \leq \sum_{r \in M} \sum_{v \in S_u^r} x(v) = |M| = 6.$$

This completes the proof of the lemma.   □

We now analyze the algorithm. First, the set S returned by the algorithm is clearly an independent set. The following lemma establishes the approximation factor of the algorithm.

Lemma 4. *Let $\mathbf{x}$ be a feasible solution to* (2). *Then,* $w(S) \geq \frac{1}{6}(\mathbf{w} \cdot \mathbf{x})$.

Proof. We prove the lemma by induction on the number of recursive calls. In the base case, the set returned by the algorithm satisfies the lemma because no vertices have remained. Moreover, the first step that removes all vertices with non-positive weight cannot decrease the right-hand side of the above inequality.

We now prove the induction step. Suppose that $\mathbf{z}$ and $\mathbf{z}'$ correspond to the indicator vectors for S and S', respectively. By induction, $\mathbf{w_2} \cdot \mathbf{z}' \geq \frac{1}{6}(\mathbf{w_2} \cdot \mathbf{x})$. Since $w_2(u) = 0$, we have $\mathbf{w_2} \cdot \mathbf{z} \geq \frac{1}{6}(\mathbf{w_2} \cdot \mathbf{x})$. From the last step of the algorithm, we know that at least one vertex from $N[u]$ is in S and so we have

$$\mathbf{w_1} \cdot \mathbf{z} = w(u) \sum_{v \in N[u]} z(v) \geq w(u).$$

Moreover, by Lemma 3,

$$\mathbf{w_1} \cdot \mathbf{x} = w(u) \sum_{v \in N[u]} x(v) \leq 6w(u).$$

Hence, $\mathbf{w_1} \cdot \mathbf{x} \leq 6w(u) \leq 6(\mathbf{w_1} \cdot \mathbf{z})$, which gives $\mathbf{w_1} \cdot \mathbf{z} \geq \frac{1}{6}(\mathbf{w_1} \cdot \mathbf{x})$. Therefore, we conclude that $(\mathbf{w_1} + \mathbf{w_2}) \cdot \mathbf{z} \geq \frac{1}{6}(\mathbf{w_1} + \mathbf{w_2}) \cdot \mathbf{x}$ and so $w(S) \geq \frac{1}{6}\mathbf{w} \cdot \mathbf{x}$. This completes the proof of the lemma.   □

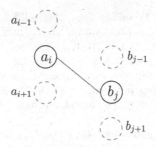

Fig. 3. Graph G_I with edge $u = (a_i, b_j)$. The edge corresponding to any vertex $v \in N[u]$ in G_C is incident to at least one of the six vertices in $\{i-1, i, i+1, j-1, j, j+1\}$.

Since there exists at least one vertex u for which $w_2(u) = 0$ in each recursive step, Algorithm 1 terminates in polynomial time. Therefore, by Lemmas 1, 2 and 4, we have the main result of this paper.

Theorem 2. *There exists a polynomial-time 6-approximation algorithm for the MWDSM problem on s_1 and s_2.*

4   Conclusion

In this paper, we studied a weighted version of the MDSM problem [9] that considers the position of the preserved duos in the respective input strings (i.e., the MWDSM problem). This is a natural variant of the problem, as considering the position of the preserved features in the strings provides solutions with better quality in many applications, such as in comparative genomics in which more weight could indicate more synteny between the corresponding preserved features. We gave a polynomial-time 6-approximation algorithm for the MWDSM problem using the local ratio technique. Although the approximation factor of our algorithm is a bit large in compare to that of algorithms for the unweighted version of the problem, our algorithm is much simpler as it benefits from the simplicity of the local ratio technique. Giving approximation algorithms with better approximation factors for the MWDSM problem remains open for future work.

References

1. Bar-Yehuda, R., Even, S.: A local-ratio theorem for approximating the weighted vertex cover problem. In: Ausiello, G., Lucertini, M. (eds.) Analysis and Design of Algorithms for Combinatorial Problems, vol. 109, pp. 27–45. North-Holland (1985)
2. Beretta, S., Castelli, M., Dondi, R.: Parameterized tractability of the maximum-duo preservation string mapping problem. Theor. Comput. Sci. **646**, 16–25 (2016)
3. Boria, N., Cabodi, G., Camurati, P., Palena, M., Pasini, P., Quer, S.: A 7/2-approximation algorithm for the maximum duo-preservation string mapping problem. In: Proceedings of the 27th Annual Symposium on Combinatorial Pattern Matching (CPM 2016), Tel Aviv, Israel, pp. 11:1–11:8 (2016)

4. Boria, N., Kurpisz, A., Leppänen, S., Mastrolilli, M.: Improved approximation for the maximum duo-preservation string mapping problem. In: Brown, D., Morgenstern, B. (eds.) WABI 2014. LNCS, vol. 8701, pp. 14–25. Springer, Heidelberg (2014). doi:10.1007/978-3-662-44753-6_2

5. Brubach, B.: Further improvement in approximating the maximum duo-preservation string mapping problem. In: Frith, M., Storm Pedersen, C.N. (eds.) WABI 2016. LNCS, vol. 9838, pp. 52–64. Springer, Cham (2016). doi:10.1007/978-3-319-43681-4_5

6. Bulteau, L., Fertin, G., Komusiewicz, C., Rusu, I.: A fixed-parameter algorithm for minimum common string partition with few duplications. In: Darling, A., Stoye, J. (eds.) WABI 2013. LNCS, vol. 8126, pp. 244–258. Springer, Heidelberg (2013). doi:10.1007/978-3-642-40453-5_19

7. Bulteau, L., Komusiewicz, C.: Minimum common string partition parameterized by partition size is fixed-parameter tractable. In: Proceedings of the Twenty-Fifth Annual ACM-SIAM Symposium on Discrete Algorithms (SODA 2014), Portland, Oregon, USA, pp. 102–121 (2014)

8. Chan, T.M., Har-Peled, S.: Approximation algorithms for maximum independent set of pseudo-disks. Discrete Comput. Geometry 48(2), 373–392 (2012)

9. Chen, W., Chen, Z., Samatova, N.F., Peng, L., Wang, J., Tang, M.: Solving the maximum duo-preservation string mapping problem with linear programming. Theor. Comput. Sci. 530, 1–11 (2014)

10. Chen, X., Zheng, J., Zheng, F., Nan, P., Zhong, Y., Lonardi, S., Jiang, T.: Assignment of orthologous genes via genome rearrangement. IEEE/ACM Trans. Comput. Biology Bioinform. 2(4), 302–315 (2005)

11. Chrobak, M., Kolman, P., Sgall, J.: The greedy algorithm for the minimum common string partition problem. In: Jansen, K., Khanna, S., Rolim, J.D.P., Ron, D. (eds.) APPROX/RANDOM -2004. LNCS, vol. 3122, pp. 84–95. Springer, Heidelberg (2004). doi:10.1007/978-3-540-27821-4_8

12. Cormode, G., Muthukrishnan, S.: The string edit distance matching problem with moves. ACM Trans. Algorithms 3(1), 2:1–2:19 (2007)

13. Dudek, B., Gawrychowski, P., Ostropolski-Nalewaja, P.: A family of approximation algorithms for the maximum duo-preservation string mapping problem. CoRR, abs/1702.02405 (2017)

14. Goldstein, A., Kolman, P., Zheng, J.: Minimum common string partition problem: hardness and approximations. Electr. J. Comb. 12 (2005)

15. Hardison, R.C.: Comparative genomics. PLoS Biol. 1(2), e58 (2003)

16. Jiang, H., Zhu, B., Zhu, D., Zhu, H.: Minimum common string partition revisited. J. Comb. Optim. 23(4), 519–527 (2012)

17. Kolman, P., Walen, T.: Reversal distance for strings with duplicates: linear time approximation using hitting set. Electr. J. Comb. 14(1) (2007)

18. Mushegian, A.R.: Foundations of Comparative Genomics. Academic Press (AP), Cambridge (2007)

19. Mustafa, N.H., Ray, S.: Improved results on geometric hitting set problems. Discrete Comput. Geometry 44(4), 883–895 (2010)

20. Swenson, K.M., Marron, M., Earnest-DeYoung, J.V., Moret, B.M.E.: Approximating the true evolutionary distance between two genomes. ACM J. Experimental Algorithmics 12, 3.5:1–3.5:17 (2008)

An Incentive Compatible, Efficient Market for Air Traffic Flow Management

Ruta Mehta[1]([✉]) and Vijay V. Vazirani[2]

[1] Department of Computer Science, University of Illinois at Urbana-Champaign,
Champaign, USA
rutamehta@cs.illinois.edu
[2] College of Computing, Georgia Tech, Atlanta, Georgia

Abstract. We present a market-based approach to the Air Traffic Flow Management (ATFM) problem. The goods in our market are delays and buyers are airline companies; the latter pay money to the Federal Aviation Administration (FAA) to buy away the desired amount of delay on a per flight basis. We give a notion of equilibrium for this market and an LP whose every optimal solution gives an equilibrium allocation of flights to landing slots as well as equilibrium prices for the landing slots. Via a reduction to matching, we show that this equilibrium can be computed combinatorially in strongly polynomial time. Moreover, there is a special set of equilibrium prices, which can be computed easily, that is identical to the VCG solution, and therefore the market is incentive compatible in dominant strategy.

1 Introduction

Air Traffic Flow Management (ATFM) is a challenging operations research problem whose importance keeps escalating with the growth of the airline industry. In the presence of inclement weather, the problem becomes particularly serious and leads to substantial monetary losses and delays[1], Yet, despite massive efforts on the part of the U.S. Federal Aviation Administration (FAA), airline companies, and even the academia, the problem remains largely unsolved.

In a nutshell, the reason for this is that any viable solution needs to satisfy several conflicting requirements, e.g., in addition to ensuring efficiency the solution also needs to be viewed as "fair" by all parties involved. Indeed, [8] state that " ... While this work points at the possibility of dramatically reducing delay costs to the airline industry vis-a-vis current practice, the vast majority of these proposals remain unimplemented. The ostensible reason for this is fairness ... ." It also needs to be computationally efficient – even moderate sized airports today handle hundreds of flights per day, with the 30 busiest ones handling anywhere from 1000

---

[1] According to [7], the U.S. Congress Joint Economic Committee estimated that in 2007, the loss to the U.S. economy was \$25.7 billion, due to 2.75 million hours of flight delays. In contrast, the total profit of U.S. airlines in that year was \$5 billion. Also, see [4] for another perspective.

© Springer International Publishing AG 2017
Y. Cao and J. Chen (Eds.): COCOON 2017, LNCS 10392, pp. 407–419, 2017.
DOI: 10.1007/978-3-319-62389-4_34

to 3000 flights per day. The full problem involves scheduling flight-landings simultaneously for multiple airports over a large period of time, taking into consideration inter-airport constraints. Yet, according to [7], current research has mostly remained at the level of a single airport because of computational tractability reasons.

Building on a sequence of recent ideas that were steeped in sound economic theory, and drawing on ideas from game theory and the theory of algorithms, we present a solution that has a number of desirable properties. Our solution for allocating flights to landing slots at a single airport is based on the principle of a free market, which is known to be fair and a remarkably efficient method for allocating scarce resources among alternative uses (sometimes stated in the colorful language of the "invisible hand of the market" [27]). We define a market in which goods are delays and buyers are airline companies; the latter pay money to the FAA to buy away the desired amount of delay on a per flight basis and we give a notion of equilibrium for this market. W.r.t. equilibrium prices, the total cost (price paid and cost of delay) of each agent, i.e., flight, is minimized.

This involves a multi-objective optimization, one for each agent, just like all market equilibrium problems. Yet, for some markets an equilibrium can be found by optimizing only one function. As an example, consider the linear case of Fisher's market [10] for which an optimal solution to the remarkable Eisenberg-Gale [16] convex formulation gives equilibrium prices and allocations. For our market, we give a special LP whose optimal solution gives an equilibrium.

Using results from matching theory, we show how to find equilibrium allocations and prices in strongly polynomial time. Moreover, using [21] it turns out that our solution is incentive compatible in dominant strategy, i.e., the players will not be able to game the final allocation to their advantage by misreporting their private information.

We note that the ATFM problem involves several issues that are not of a game-theoretic or algorithmic nature, e.g., the relationship between long term access rights (slot ownership or leasing) and short term access rights on a given day of operations, e.g., see [6]. Our intention in this paper is not to address the myriad of such issues. Instead, we have attempted to identify a mathematically clean, core problem that is amenable to the powerful tools developed in the theories stated above, and whose solution could form the core around which a practical scheme can be built.

Within academia, research on this problem started with the pioneering work of Odoni [24] and it flourished with the extensive work of Bertsimas et. al.; we refer the reader to [7,9] for thorough literature overviews and references to important papers. These were centralized solutions in which the FAA decides a schedule that is efficient, e.g., it decides which flights most critically need to be served first in order to minimize cascading delays in the entire system.

A conceptual breakthrough came with the realization that *the airlines themselves are the best judge of how to achieve efficiency*[2], thus moving away from centralized solutions. This observation led to a solutions based on collaborative decision making (CDM) which is used in practice [5, 28, 29].

More recently, a market based approach was proposed by Castelli, Pesenti and Ranieri [11]. Although their formulation is somewhat complicated, the strength of their approach lies in that it not only leads to efficiency but at the same time, it finesses away the sticky issue of fairness – whoever pays gets smaller delays, much the same way as whoever pays gets to fly comfortably in Business Class! Paper [11] also gave a tatonnement-based implementation of their market. Each iteration starts with FAA announcing prices for landing slots. Then, airlines pick their most preferred slots followed by FAA adjusting prices, to bring parity between supply and demand, for the next iteration. However, they are unable to show convergence of this process and instead propose running it a pre-specified number of times, and in case of failure, resorting to FAA's usual solution. They also give an example for which incentive compatibility does not hold.

Our market formulation is quite different and achieves both efficient running time and incentive compatibility. We believe that the simplicity of our solution for this important problem, and the fact that it draws on fundamental ideas from combinatorial optimization and game theory, should be viewed as a strength rather than a weakness.

1.1 Salient Features of Our Solution

In Sect. 2 we give details of our basic market model for allocating a set of flights to landing slots for one airport. This set of flights is picked in such a way that their actual arrival times lie in a window of a couple of hours; the reason for the latter will be clarified in Sect. 4. The goods in our market are delays and buyers are airline companies; the latter pay money to the FAA to buy away the desired amount of delay on a per flight basis. Typically flights have a myriad interdependencies with other flights – because of the use of the same aircraft for subsequent flights, passengers connecting with other flights, crew connecting with other flights, etc. The airline companies, and not FAA, are keenly aware of these and are therefore in a better position to decide which flights to avoid delay for. The information provided by airline companies for each flight is the dollar value of delay as perceived by them.

For finding equilibrium allocations and prices in our market, we give a special LP in which parameters can be set according to the prevailing conditions at the airport and the delay costs declared by airline companies. We arrived at this LP as follows. Consider a traffic network in which users selfishly choose paths from

---

[2] e.g., they know best if a certain flight needs to be served first because it is carrying CEOs of important companies who have paid a premium in order to reach their destination on time or if delaying a certain flight by 30 min will not have dire consequences, however delaying it longer would propagate delays through their entire system and result in a huge loss.

their source to destination. One way of avoiding congestion is to impose tolls on roads. [13] showed the existence of such tolls for minimizing the total delay for the very special case of one source and one destination, using Kakutani's fixed point theorem. Clearly, their result was highly non-constructive. In a followup work, [18] gave a remarkable LP whose optimal solution yields such tolls for the problem of arbitrary sources and destinations and moreover, this results in a polynomial time algorithm. Their LP, which was meant for a multi-commodity flow setting, was the starting point of our work. One essential difference between the two settings is that whereas they sought a Nash equilibrium, we seek a market equilibrium; in particular, the latter requires the condition of market clearing.

We observe that the underlying matrix of our LP is totally unimodular and hence it admits an integral optimal solution. Such a solution yields an equilibrium schedule for the set of flights under consideration and the dual of this LP yields equilibrium price for each landing slot. Equilibrium entails that each flight is scheduled in such a way that the sum of the delay price and landing price is minimum possible. We further show that an equilibrium can be found via an algorithm for the minimum weight perfect b-matching problem and hence can be computed combinatorially in strongly polynomial time. In hindsight, our LP resembles the b-matching LP, but there are some differences.

Since the b-matching problem reduces to the maximum matching problem, our market is essentially a matching market. Leonard [21] showed that the set of equilibrium prices of a matching market with minimum sum corresponds precisely to VCG payments [23], thereby showing that the market is incentive compatible in dominant strategy. Since equilibrium prices form a lattice [1,15,26], the one minimizing sum has to be simultaneously minimum for all goods. For our market, we give a simple, linear-time procedure that converts arbitrary equilibrium prices to ones that are simultaneously minimum for all slots. Incentive compatibility with these prices follows. An issue worth mentioning is that the total revenue, or the total cost, of VCG-based incentive compatible mechanisms has been studied extensively, mostly with negative results [2,14,17,19,20]. In contrast, since the prices in our natural market model happened to be VCG prices, we have no overhead for making our mechanism incentive compatible.

The next question is how to address the scheduling of landing slots over longer periods at multiple airports, taking into consideration inter-airport constraints. Airlines can and do anticipate future congestion and delay issues and take these into consideration to make complex decisions. However, sometimes unexpected events happening even at a few places are likely to have profound cascading effects at geographically distant airports, making it necessary to make changes dynamically. For such situations, in Sect. 4, we propose a dynamic solution by decomposing this entire problem into many small problems, each of which will be solved by the method proposed above. The key to this decomposition is the robustness of our solution for a single set of flights at one airport: we have not imposed any constraints on delay costs, not even monotonicity. Therefore, airline companies can utilize this flexibility to encode a wide variety of inter-airport constraints.

We note that this approach opens up the possibility of making diverse types of travelers happy through the following mechanism: the additional revenues generated by FAA via our market gives it the ability to subsidize landing fees for low budget airlines. As a result, both types of travelers can achieve an end that is most desirable to them, business travelers and casual/vacation travelers. The former, in inclement weather, will not be made to suffer delays that ruin their important meetings and latter will get to fly for a lower price (and perhaps sip coffee for an additional hour on the tarmac, in inclement weather, while thinking about their upcoming vacation).

To the best of our knowledge, ours is the first work to give a simple LP-based efficient solution for the ATFM problem. We note that an LP similar to ours is also given in [1]. This paper considers two-sided matching markets with payments and non-quasilinear utilities. They show that the lowest priced competitive equilibria are group strategy proof, which induces VCG payments for the case of quasilinear utilities. Another related paper is [12], which considers a Shapley-Shubik assignment model for unit-demand buyers and sellers with one indivisible item each. Buyers have budget constraint for every item. This sometimes prevents a competitive equilibrium from existing. They give a strongly polynomial-time algorithm to check if an equilibrium exists or not, and if it does exist, then it computes the one with lowest prices. However, they do not ensure incentive compatibility.

2   The Market Model

In this section we will consider the problem of scheduling landings at one airport only. Let A be the set of all flights, operated by various airlines, that land in this airport in a given period of time. We assume that the given period of time is partitioned into a set of landing time slots, in a manner that is most convenient for this airport; let S denote this set. Each slot s has a capacity $cap(s) \in \mathbf{Z}^+$ specifying the number of flights that can land in this time slot. As mentioned in [6] the arrival of each aircraft consumes approximately the same amount of airport capacity, therefore justifying the slot capacities as the number of flights while ignoring their types. We will assume that $cap(s)$ is adjusted according to the prevailing weather condition.

For $i \in A$, the airline of this flight decides the *landing window* for flight i, denoted by $W(i)$. This gives the set of time slots in which this flight should land as per prevailing conditions, e.g., if there are no delays, the earliest time slot in $W(i)$ will be the scheduled arrival time[3] of flight i. For each slot $s \in W(i)$, the airline also decides its *delay cost*, denoted by $c_{is} \geq 0$. Thus, if slot s is the scheduled arrival time of flight i, then $c_{is} = 0$[4] and in general c_{is} is the dollar value of the cost perceived by the airline, for delay due to landing in slot s.

---

[3] We will assume that if the flight arrives before this time, it will have to wait on the tarmac for some time. This appears to be a standard practice for the majority of times, in case gates are not available.

[4] All the results of this paper hold even if $c_{is} \neq 0$.

A *landing schedule* is an assignment of flights to time slots, respecting capacity constraints. Each time slot will be assigned a *landing price* which is the amount charged by FAA from the airline company if its flight lands in this time slot. We will define the *total cost* incurred by a flight to be the sum of the price paid for landing and the cost of the delay.

We say that a given schedule and prices are *an equilibrium landing schedule and prices* if:

1. W.r.t. these prices, each flight incurs a minimum total cost.
2. The landing price of any time slot that is not filled to capacity is zero. This condition is justified by observing that if at equilibrium, a slot with zero price is not filled to capacity, then clearly its price cannot be made positive. This is a standard condition in equilibrium economics.

2.1   LP Formulation

In this section, we will give an LP that yields an equilibrium schedule; its dual will yield equilibrium landing prices. Section 3 shows how they can be computed in strongly polynomial time.

For $s \in S$, x_{is} will be the indicator variable that indicates whether flight i is scheduled in time slot s; naturally, in the LP formulation, this variable will be allowed to take fractional values. The LP given below obtains a scheduling where a flight may be assigned partially to a slot (fractional scheduling), that minimizes the total dollar value of the delays incurred by all flights, subject to capacity constraints of the time slots. (Note that the inequality in the first constraint will be satisfied with equality since the objective is being minimized; the formulation below was chosen for reasons of convenience).

$$
\begin{aligned}
\text{minimize} \quad & \textstyle\sum_{i \in A, s \in S} c_{is} x_{is} \\
\text{subject to} \quad & \forall i \in A : \textstyle\sum_{s \in W(i)} x_{is} \geq 1 \\
& \forall s \in S : \textstyle\sum_{i \in A, s \in W(i)} x_{is} \leq cap(s) \\
& \forall i \in A, \ s \in W(i) : \ x_{is} \geq 0
\end{aligned} \tag{1}
$$

Let p_s denote the dual variable corresponding to the second set of inequalities. We will interpret p_s as the price of landing in time slot s. Thus if flight i lands in time slot s, the total cost incurred by it is $p_s + c_{is}$. Let t_i denote the dual variable corresponding to the first set of inequalities. In Lemma 1 we will prove that t_i is the total cost incurred by flight i w.r.t. the prices found by the dual; moreover, each flight incurs minimum total cost.

The dual LP is the following.

$$
\begin{aligned}
\text{maximize} \quad & \textstyle\sum_{i \in A} t_i \ - \ \textstyle\sum_{s \in S} cap(s) \cdot p_s \\
\text{subject to} \quad & \forall i \in A, \ \forall s \in W(i) : \ t_i \leq p_s + c_{is} \\
& \forall i \in A : t_i \geq 0 \\
& \forall s \in S : p_s \geq 0
\end{aligned} \tag{2}
$$

Lemma 1. *W.r.t. the prices found by the dual LP (2), each flight i incurs minimum total cost and it is given by t_i.*

Proof. Apply complementary slackness conditions to the primal variables we get

$$\forall i \in A, \ \forall s \in W(i): \quad x_{is} > 0 \quad \Rightarrow \quad t_i = p_s + c_{is}.$$

Moreover, for time slots $s \in S$ which are not used by flight i, i.e., for which $x_{is} = 0$, by the dual constraint, the total cost of using this slot can only be higher than t_i. The lemma follows.

The second condition required for equilibrium is satisfied because of complementarity applied to the variables p_s: If $\sum_{i \in A, s \in W(i)} x_{is} < cap(s)$, then $p_s = 0$.

At this point, we can provide an intuitive understanding of how the actual slot assigned to flight i by LP (1) is influenced by the delay costs declared for flight i and how LP (2) sets prices of slots. Assume that time slot s is the scheduled arrival time of flight i, i.e., $c_{is} = 0$ and s' is a later slot. Then by Lemma 1, slot s will be preferred to slot s' only if $p_s - p_{s'} \leq c_{is'}$. Thus $c_{is'}$ places an upper bound on the extra money that can be charged for buying away the delay incurred by landing in s instead of s'. Clearly, flight i will incur a smaller delay, at the cost of paying more, if its airline declares large delay costs for late landing. Furthermore, by standard LP theory, the dual variables, p_s, will adjust according to the demand of each time slot, i.e., a time slot s that is demanded by a large number of flights that have declared large delay costs will have a high price. In particular, if a slot is not allocated to capacity, its price will be zero as shown above.

It is easy to see that the matrix underlying LP (1) is totally unimodular, since it corresponds to the incidence matrix of a bipartite graph [3]. Therefore, it has an integral optimal solution. Further, minimization ensures that for every flight i at most one of the x_{is}s is one and the rest are zero. Hence we get:

Theorem 1. *Solution of LP (1) and its dual (2) give an (optimal) equilibrium schedule and equilibrium prices.*

3   Strongly Polynomial Implementation

As discussed in the previous section, LP (1) has an integral optimal solution as its underlying matrix is totally unimodular. In this section, we show that the problem of obtaining such a solution can be reduced to a minimum weight perfect b-matching problem[5], and hence can be found in strongly polynomial time; see [25] Volume A. The equilibrium prices, i.e., solution of (2), can be obtained from the dual variables of the matching. Furthermore, we show that there exist equilibrium prices that induce VCG payments, and hence is incentive compatible in dominant strategy. Finally, we give a strongly polynomial time procedure to compute such prices.

Consider the edge-weighted bipartite graph (A', S, E), with bipartition $A' = A \cup \{v\}$, where A is the set of flights and v is a special vertex, and S is the set of

---

[5] The instance we construct can also be reduced to a minimum weight perfect matching problem with quadratic increase in number of nodes.

time slots. The set of edges E and weights are as follows: for $i \in A$, $s \in W(i)$, (i, s) is an edge with weight c_{is}, and for each $s \in S$, there are $cap(s)$ many (v, s) edges[6], each with unit weight (a multi-graph).

The matching requirements are: $b_i = 1$ for each $i \in A$, $b_s = cap(s)$ for each $s \in S$, and $b_v = \sum_{s \in S} cap(s) - |A|$ for v. Clearly, the last quantity is non-negative, or else LP (1) is infeasible. The following lemmas show that the equilibrium landing schedule and prices can be computed using minimum weight perfect b-matching of graph (A', S, E).

Lemma 2. *Let $F^* \subset E$ be a perfect b-matching in (A', S, E) and x^* be a schedule where $x_{is}^* = 1$ if $(i, s) \in F^*$. F^* is a minimum weight perfect b-matching if and only if x^* is an optimal solution of LP (1).*

Proof. To the contrary suppose x' and not x^* is the optimal solution of LP (1). Let $F' = \{(i, s) \in E \mid x_{is}' = 1\} \cup \{(cap(s) - \sum_{i; s \in W(i)} x_{is}') \text{ many } (v, s) \mid s \in S\}$ be the set of edges corresponding to schedule x'. Clearly, F' is a perfect b-matching. Note that the matching edges incident on v contribute cost b_v in any perfect b-matching. Since, x' and not x^* is an optimal solution of LP (1), we have,

$$\sum_{i \in A, s \in W(i)} c_{is} x_{is}' + b_v < \sum_{i \in A, s \in W(i)} c_{is} x_{is}^* + b_v \Rightarrow \sum_{(i,j) \in F'} c_{ij} < \sum_{(i,j) \in F^*} c_{ij}$$

Contradicting F^* being the minimum weight perfect matching. The reverse implication follows by similar argument in the reverse order.

Using Lemma 2, next we show that the dual variables of the b-matching LP give an equilibrium price vector. In the b-matching LP there is an equality for each node to ensure its matching requirement. Let u_v, u_i and q_s be the dual variables corresponding to the equalities of nodes v, $i \in A$ and $s \in S$. Then the dual LP for minimum weight perfect b-matching in graph (A', S, E) is as follows.

$$\max : \sum_{i \in A} u_i + \sum_{s \in S} cap(s) q_s + u_v b_v$$
$$\text{s.t.} \quad \forall i \in A, \ s \in W(i) : u_i \leq -q_s + c_{is} \quad (3)$$
$$\forall s \in S : \qquad u_v \leq -q_s + 1$$

There are no non-negativity constraints on the dual variables since the corresponding primal constraints are equality.

Lemma 3. *There exists a dual solution (u^*, q^*) of (3) with $u_v^* = 1$, and given that, $-q^*$ yields a solution of LP. (2).*

Proof. If (u^*, q^*) is a dual solution then so is $v = (u^* + \delta, q^* - \delta)$ for any $\delta \in \mathbb{R}$. This is because, clearly v is feasible. Further, since $|A| + b_v = \sum_s cap(s)$ the value of objective function at v is same as that at (u^*, q^*).

Therefore given any solution of the dual, we can obtain one with $u_v^* = 1$ by an additive scaling. Replacing u_v with 1 and q_s with $-p_s$ in (3) gives $max\{\sum_i u_i - \sum_s cap(s) p_s + b_v \mid u_i \leq p_s + c_{is}, \ p_s \geq 0\}$, which is exactly (2).

---

[6] This is not going to affect strong polynomiality, because we can assume that $cap(s) \leq |A|, \forall s$ without loss of generality.

Since a primal and a dual solution of a minimum weight perfect b-matching can be computed in strongly polynomial time [25], the next theorem follows using Lemmas 2 and 3, and Theorem 1.

Theorem 2. *There is a combinatorial, strongly polynomial algorithm for computing an equilibrium landing schedule and equilibrium prices.*

3.1   Incentive Compatible in Dominant Strategy

Since equilibrium price vectors of the market is in one-to-one correspondence with the solutions of the dual matching LP with $u_v = 1$ (Lemma 3), they need not be unique, and in fact form a convex set. In this section we show that one of them induces VCG payments, and therefore is incentive compatible in dominant strategy. Further, we will design a method to compute such VCG prices in strongly polynomial time.

An instance of the perfect b-matching problem can be reduced to the perfect matching problem by duplicating node n, b_n times. Therefore, if we convert the costs c_{is} on edge (i, s) to payoffs $H - c_{is}$ for a big enough constant H, the market becomes an equivalent matching market (also known as assignment game) [26] where the costs of producing goods, the slots in our case, are zero. It is not difficult to check that equilibrium allocations and prices of our original market and the transformed matching market exactly match.

For such a market, Leonard [21] showed that the set of equilibrium prices of a matching market with minimum sum correspond precisely to VCG payments [23], thereby showing that the market is incentive compatible in dominant strategy at such a price vector. Since the proof in [21] is not formal, we have provided a complete proof in the full version [22]. Since equilibrium prices form a lattice [1,15,26], the one minimizing sum has to be simultaneously minimum for all goods.[7] Clearly, such a price vector has to be unique. Next we give a procedure to compute the minimum equilibrium price vector, starting from any equilibrium price vector p^* and corresponding equilibrium schedule x^*.

The procedure is based on the following observation: Given equilibrium prices p^* and corresponding schedule x^*, construct graph $G(x^*, p^*)$ where slots form the node set. Put a directed edge from slot s to slot s' if there exists a flight, say i, scheduled in s at x^*, and it is indifferent between s and s' in terms of total cost, i.e. $x_{is}^* = 1$ and $p_s^* + c_{is} = p_{s'}^* + c_{is'}$. An edge in graph $G(x^*, p^*)$ indicates that if the price of slot s' is decreased then i would prefer s' over s. Therefore, in order to maintain x^* as an equilibrium schedule the price of s also has to be decreased by the same amount.

Lemma 4. *Prices p^{*m} give the minimum equilibrium prices if and only if every node in $G(x^*, p^{*m})$ has a directed path from a zero priced node, where x^* is the corresponding equilibrium schedule.*

---

[7] Equilibrium prices p are minimum if for any other equilibrium prices p' we have $p_s \leq p_s'$, $\forall s \in S$.

Proof. Suppose slot s does not have a path from a zero priced node. Consider the set D of nodes which can reach s in $G^* = G(x^*, p^{*m})$; these have positive prices. $\exists \epsilon > 0$ s.t. the prices of all the slots in D can be lowered by ϵ without violating the equilibrium condition (1), contradicting minimality of p^{*m}.

For the other direction, the intuition is that if every node is connected to a zero priced node in $G(x^*, p^{*m})$, then price of any slot can not be reduced without enforcing price of some other slot go negative, in order to get the corresponding equilibrium schedule. The formal proof is as follows:

To the contrary suppose every node is connected to a zero priced node in G^* and there are equilibrium prices $p' \leq p^{*m}$ such that for some $s \in S, p_s^{*m} > p_s' \geq 0$. Consider, one such s with the smallest path from a zero-priced node in G^*. Since, $p_s' \geq 0$, we have $p_s^{*m} > 0$, and therefore s is filled to its capacity at prices p^{*m} (using equilibrium condition (2) of Sect. 2). Let x' be the equilibrium schedule corresponding to prices p'.

Let there be a directed edge from s' to s in G^*. By choice of s we have that $p_{s'}' = p_{s'}^{*m}$. In that case, a flight, say i', allocated to s' at p^* will move to s at p'. This implies that $\sum_i x_{is'}' < \sum_i x_{is'}^* \leq cap(s')$. Hence $p_{s'}' = 0 \Rightarrow p_{s'}^{*m} = 0$ (using equilibrium condition (2)). Let $Z = \{s \mid p_s^{*m} = 0\}$. There are two cases:

Case I - Flights in slot s at x^* remain in s at x', i.e., $\{i \mid x_{is}^* = 1\} \subseteq \{i \mid x_{is}' = 1\}$: Since, $x_{i's}' = 1$ and $x_{i's}^* = 0$, implying $\sum_i x_{is}' > \sum_i x_{is}^* = cap(s)$, a contradiction.

Case II - $\exists i, x_{is}^* = 1, x_{is}' = 0$: Wlog let x' be an optimal allocation at p' that is at minimum distance from x^* in l_1-norm. In other words, a flight that is allocated to slot u at prices p^{*m} moves to slot v at prices p' only if price of slot v has decreased.

Construct a graph H, where slots are nodes, and there is an edge from u to v if $\exists i, x_{iu}^* = 1, x_{iv}' = 1$, i.e., flight i moved from u to v when prices change from p^{*m} to p', with weight being number of edges moved. Note that price of every node with an incoming edge should have decreased while going from p^{*m} to p'. Therefore, nodes of Z have no incoming edges. Further, nodes with incoming edges are filled to capacity at p^{*m} since their prices are non-zero. Hence, total out going weight of such a node should be at least total incoming weight in H.

If there is a cycle in H, then subtract weight of one from all its edges, and remove zero-weight edges. Repeat this until there are no cycles. Since, $s' \in Z$, it had no incoming edge, but had an edge to s. Therefore, there is a path in remaining H starting at s'. Consider the other end of this path. Clearly, it has to be filled beyond its capacity at x', a contradiction.

Using the fact established by Lemma 4 next we design a procedure to compute the minimum equilibrium prices in Table 1, given any equilibrium (p^*, x^*).

Lemma 5. *Given an equilibrium* (x^*, p^*), *MinimumPrices*(x^*, p^*) *outputs minimum prices in time* $O(|A||S|^2)$.

Table 1. Procedure for computing minimum optimal prices

MinimumPrices(x^*, p^*)
1. $Z \leftarrow$ Nodes reachable from zero-priced nodes in $G(x^*, p^*)$.
2. Pick a $d \in S \setminus Z$
3. $D \leftarrow$ {Nodes that can reach d in $G(x^*, p^*)$}, $\delta \leftarrow 0$,
   and $p_s^* \leftarrow p_s^* - \delta, \forall s \in D$
4. Increase δ until one of the following happen
   - If price of a slot in D becomes zero, then go to 1.
   - If a new edge appears in $G(x^*, p^*)$, then recompute Z.
     If $d \in Z$ then go to 2 else go to 3.
5. Output p^* as the minimum prices.

Proof. Note that the size of Z and edges in $G(x^*, p^*)$ are increasing. Therefore, Step 3 is executed $O(|S|)$ many times in total. Step 4 may need $O(|A||S|)$ time to compute the threshold δ. Therefore the running time of the procedure MinimumPrices is $O(|A||S|^2)$. Let the output price vector be p^{*m}. The lemma follows from the fact that (x^*, p^{*m}) still satisfy both the equilibrium conditions, and every slot is reachable from a zero priced node in $G(x^*, p^{*m})$ (Lemma 4).

Theorems 1 and 2, Lemma 5, together with [21] give:

Theorem 3. *There exists an incentive compatible (in dominant strategy) market mechanism for scheduling a set of flight landings at a single airport; moreover, it is computable combinatorially in strongly polynomial time.*

4   Dealing with Multiple Airports

In this section, we suggest how to use the above-stated solution to deal with unexpected events that result in global, cascading delays. Our proposal is to decompose the problem of scheduling landing slots over a period of a day at multiple airports into many small problems, each dealing with a set of flights whose arrival times lie in a window of a couple of hours – the window being chosen in such a way that all flights would already be in the air and their actual arrival times, assuming no further delays, would be known to the airline companies and to FAA. At this point, an airline company has much crucial information about all the other flights associated with its current flight due to connections, crew availability, etc. It is therefore in a good position to determine how much delay it needs to buy away for its flight and how much it is willing to pay, by setting c_{is}s accordingly. This information is used by FAA to arrive at a landing schedule. The process is repeated every couple of hours at each airport.

References

1. Alaei, S., Jain, K., Malekian, A.: Competitive equilibria in two sided matching markets with non-transferable utilities (2012). arxiv:1006.4696
2. Archer, A., Tardos, E.: Frugal path mechanisms. In: ACM-SIAM Annual Symposium on Discrete Algorithms, pp. 991–999 (2002)
3. Bapat, B.R.: Incidence matrix. In: Bapat, B.R. (ed.) Graphs and Matrices. Springer, London (2010). doi:10.1007/978-1-84882-981-7
4. Ball, M., Barnhart, C., Dresner, M., Hansen, M., Neels, K., Odoni, A., Peterson, E., Sherry, L., Trani, A., Zou, B., Britto, R.: Total delay impact study. In: NEXTOR Research Symposium, Washington DC (2010)
5. Ball, M., Barnhart, C., Nemhauser, G., Odoni, A.: Air transportation: irregular operations and control. In: Barnhart, C., Laporte, G. (eds.) Handbook of Operations Research and Management Science: Transportation (2006)
6. Ball, M.O., Donohue, G., Hoffman, K.: Auctions for the safe, efficient and equitable allocation of airspace system resources. In: Cramton, P., Shoham, Y., Steinberg, R. (eds.) Combinatorial Auctions, pp. 507–538. MIT Press, Cambridge (2005)
7. Barnhart, C., Bertsimas, D., Caramanis, C., Fearing, D.: Equitable and efficient coordination in traffic flow management. Transp. Sci. **42**(2), 262–280 (2012)
8. Bertsimas, D., Farias, V., Trichakis, N.: The price of fairness. Oper. Res. **59**(1), 17–31 (2011)
9. Bertsimas, D., Gupta, S.: A proposal for network air traffic flow management incorporating fairness and airline collaboration. Oper. Res. (2011)
10. Brainard, W.C., Scarf, H.E.: How to compute equilibrium prices in 1891. Cowles Foundation Discussion Paper 1270 (2000)
11. Castelli, E., Pesenti, R., Ranieri, A.: The design of a market mechanism to allocate air traffic flow management slots. Trans. Res. Part C **19**, 931–943 (2011)
12. Chen, N., Deng, X., Ghosh, A.: Competitive equilibria in matching markets with budgets. SIGecom Exch. **9**(1), 5:1–5:5 (2010)
13. Cole, R., Dodis, Y., Roughgarden, T.: Pricing network edges for heterogeneous selfish users. In: STOC, pp. 521–530 (2003)
14. Conitzer, V., Sandholm, T.: Failures of the VCG mechanism in combinatorial auctions and exchanges. In: AAMAS, pp. 521–528 (2006)
15. Demange, G., Gale, D.: The strategy structure of two-sided matching markets. Econometrica **53**(4), 873–888 (1985)
16. Eisenberg, E., Gale, D.: Consensus of subjective probabilities: the Pari-Mutuel method. Ann. Math. Stat. **30**, 165–168 (1959)
17. Elkind, E., Sahai, A., Steiglitz, K.: Frugality in path auctions. In: ACM-SIAM Annual Symposium on Discrete Algorithms, pp. 701–709 (2004)
18. Fleischer, L., Jain, K., Mahdian, M.: Tolls for heterogeneous selfish users in multicommodity networks and generalized congestion games. In: STOC, pp. 277–285 (2004)
19. Hartline, J.D., Roughgarden, T.: Simple versus optimal mechanisms. In: EC (2009)
20. Karlin, A.R., Kempe, D., Tamir, T.: Beyond VCG: frugality of truthful mechanisms. In: FOCS, pp. 615–624 (2005)
21. Leonard, H.B.: Elicitation of honest preferences for the assignment of individuals to positions. J. Polit. Econ. **91**(3), 461–479 (1983)
22. Mehta, R., Vazirani, V.V.: An incentive compatible, efficient market for air traffic flow management (2017). arxiv:1305.3241

23. Nisan, N.: Introduction to mechanism design (for computer scientists). In: Nisan, N., Roughgarden, T., Tardos, E., Vazirani, V.V. (eds.) Algorithmic Game Theory, pp. 209–241. Cambridge University Press (2007)
24. Odoni, A.: The flow management problem in air traffic control. In: Odoni, A., Szego, G. (eds.) Flow Control of Congested Networks. Springer, Berlin (1987)
25. Schrijver, A.: Combinatorial Optimization. Springer, Heidelberg (2003)
26. Shapley, L.S., Shubik, M.: The assignment game I: The core. Int. Game Theor. **1**(2), 111–130 (1972)
27. Smith, A.: The Wealth of Nations. Forgotten Books, London (1776)
28. Vossen, T., Ball, M.: Slot trading opportunities in collaborative ground delay programs. Transp. Sci. **40**, 29–43 (2006)
29. Wambsganss, M.: Collaborative decision making through dynamic information transfer. Air Traffic Control Q. **4**, 107–123 (1996)

Linear Representation of Transversal Matroids and Gammoids Parameterized by Rank

Pranabendu Misra[1(✉)], Fahad Panolan[1],
M.S. Ramanujan[2], and Saket Saurabh[1,3]

[1] Department of Informatics, University of Bergen, Bergen, Norway
{pranabendu.misra,fahad.panolan}@ii.uib.no
[2] TU Wien, Vienna, Austria
ramanujan@ac.tuwien.ac.at
[3] The Institute of Mathematical Sciences, HBNI, Chennai, India
saket@imsc.res.in

Abstract. Given a bipartite graph $G = (U \uplus V, E)$, a linear representation of the transversal matroid associated with G on the ground set U, can be constructed in randomized polynomial time. In fact one can get a linear representation deterministically in time $2^{\mathcal{O}(m^2 n)}$, where $m = |U|$ and $n = |V|$, by looping through all the choices made in the randomized algorithm. Other important matroids for which one can obtain linear representation deterministically in time similar to the one for transversal matroids include gammoids and strict gammoids. Strict gammoids are duals of transversal matroids and gammoids are restrictions of strict gammoids. We give faster deterministic algorithms to construct linear representations of transversal matroids, gammoids and strict gammoids. All our algorithms run in time $\binom{m}{r} m^{\mathcal{O}(1)}$, where m is the cardinality of the ground set and r is the rank of the matroid. In the language of parameterized complexity, we give an XP algorithm for finding linear representations of transversal matroids, gammoids and strict gammoids parameterized by the rank of the given matroid.

1 Introduction

Matroids are important mathematical objects in the theory of algorithms and combinatorial optimization. Often an algorithm for a class of matroids gives us an algorithmic meta theorem, which gives a unified solution to several problems. For example, it is known that any problem which admits a greedy algorithm can be embedded into a matroid and finding minimum (maximum) weighted independent set in this matroid corresponds to finding a solution to the problem. Other important examples are the MATROID INTERSECTION and MATROID PARITY problems, which encompass several combinatorial optimization problems such as BIPARTITE MATCHING, 2-EDGE DISJOINT SPANNING TREES and ARBORESCENCE. A matroid M is defined as a pair $(E, \mathcal{I})$, where E is called the ground set and $\mathcal{I}$ is a family of subsets of E, called independent sets, with following three properties: (i) $\emptyset \in \mathcal{I}$, (ii) if $A \in \mathcal{I}$ and $A' \subseteq A$, then $A' \in \mathcal{I}$ and

© Springer International Publishing AG 2017
Y. Cao and J. Chen (Eds.): COCOON 2017, LNCS 10392, pp. 420–432, 2017.
DOI: 10.1007/978-3-319-62389-4_35

(iii) If $A, B \in \mathcal{I}$ and $|A| < |B|$, then there is a $e \in B \setminus A$ such that $A \cup \{e\} \in \mathcal{I}$. As the cardinality of $\mathcal{I}$ could be exponential in $|E|$, as it is in many applications, explicitly listing $\mathcal{I}$ for algorithms is highly inefficient both in terms of time complexity as well as space complexity. Thus, during the early works of algorithms using matroids, algorithms were designed using the *oracle* model. An *independence oracle* for a matroid is a black box algorithm which takes as input a subset of the ground set and returns YES if the set is independent in the matroid and NO otherwise. Many algorithms are designed using these oracles. These oracle based algorithms lead to efficient algorithms for problems where we have good algorithms that can act as an oracles. A few matroids for which we have efficient oracles include, but are not limited to, graphic matroids, co-graphic matroids, transversal matroids and linear matroids.

Another way of representing a matroid succinctly is by encoding the information about the family of independent sets in a matrix. A matrix A over a field $\mathbb{F}$ is called a linear representation of a matroid $M = (E, \mathcal{I})$, if there is a bijection between the columns of A and E and a subset $S \subseteq E$ is independent in M if and only if the corresponding columns in A are linearly independent over the field $\mathbb{F}$. Note that while not all matroids admit a linear representation, a number of important classes of matroids do. Recently, several algorithmic results have been obtained in the fields of Parameterized Complexity and Exact Algorithms, which require a linear representations of certain classes of matroids [2–6,10,11,13,15]. This naturally motivates the question of constructing linear representations for various classes of matroids *efficiently*. Deterministic polynomial time algorithms were known for linear representations of many important classes of matroids such as uniform matroids, partition matroids, graphic matroids and co-graphic matroids. In all these algorithms the running time is polynomial in the size of the ground set. However, for transversal matroids and gammoids, only randomized polynomial time algorithms are known for constructing its linear representations. These matroids feature in many of the results mentioned above, and deterministic algorithms to compute linear representations for them will derandomize several algorithms in literature. In this paper we give a modest improvement over the naïve algorithm for constructing deterministic linear representations for transversal matroids and gammoids.

Let $G = (U \uplus V, E)$ be a bipartite graph, the *transversal matroid* M_G on the ground set U has the following family of independent sets: $U' \subseteq U$ is independent in M_G if and only if there is a matching in G saturating U'. Furthermore assume that $|U| = |V| = m$ and G has a perfect matching. A natural question in this direction is as follows.

Question 1: Could we exploit the fact that G has a perfect matching to design a deterministic polynomial time algorithm to find a linear representation for M_G, where we can test whether a subset is independent or not in deterministic polynomial time?

The answer to this question is of course, Yes! An $m \times m$ identity matrix is a linear representation of M_G. This naturally leads to the following question.

Question 2: Suppose G has a matching of size $m-\ell$, where ℓ is a constant. Can we design a deterministic polynomial time algorithm to find a linear representation for M_G, where we can test whether a subset is independent or not in deterministic polynomial time?

This question is the starting point of the present work. As mentioned earlier, there is a randomized polynomial time algorithm to obtain a linear representation of a transversal matroid for any bipartite graph. Let $G = (U \uplus V, E)$ be a bipartite graph, where $U = \{u_1, \ldots, u_m\}$ and $V = \{v_1, \ldots, v_n\}$. Let $X = \{x_{i,j} \mid i \in [n], j \in [m]\}$. Define an $n \times m$ matrix A as follows: for each $i \in [n], j \in [m], A[i,j] = 0$ if $v_i u_j \notin E$ and $x_{i,j}$ otherwise. Then for any $R \subseteq [n], C \subseteq [m], |R| = |C|, \det(A[R,C]) \not\equiv 0$ if and only if there is a perfect matching in $G[\{v_i \mid i \in R\} \cup \{u_j \mid j \in C\}]$. This implies that A is in fact a linear representation of the transversal matroid M_G on the ground set U over the field of fractions $\mathbb{F}(\mathbf{X})$, where $\mathbb{F}$ is any field of size at least 2. Notice that the above construction can be done in deterministic polynomial time.

However, in the representation above, to check whether a set is linearly independent we need to test if the corresponding determinant polynomial, which is a multivariate polynomial, is identically non-zero. This is a case of the well known polynomial identity testing (PIT) problem, and we do not know of a deterministic polynomial time algorithm for this problem. Hence, this representation is difficult to use in deterministic algorithms for many applications [10,13]. Furthermore, it rather difficult to carry out field operations over the field of fractions $\mathbb{F}(\mathbf{X})$ in polynomial time. As a result, even though we get the above linear representation in deterministic polynomial time, we are not able to use it for deterministic algorithms *efficiently*. We can obtain another representation, by substituting random values for each $x_{i,j} \in X$ from a field of size at least $2^p m 2^m$, where $p \in \mathbb{N}$, and succeed with probability at least $\left(1 - \frac{1}{2^p}\right)$. This leads to a randomized polynomial time algorithm [13], to obtain a representation over a finite field or a field such as $\mathbb{R}$, where field operations can be carried out efficiently. It appears that derandomizing the above approach has some obstacles, as this will have some important consequences on lower bounds in complexity theory [9]. Observe that the above approach implies a deterministic algorithm of running time $2^{\mathcal{O}(m^2 n)}$, that tests all possible assignments of at most mn variables from a field of size $m 2^m$ (setting $p = \mathcal{O}(1)$), since one of them will certainly be a linear representation of M_G. Although we have not been able to obtain *polynomial time* deterministic algorithms for computing linear representations of transversal matroids and gammoids, our results do imply an affirmative answer to Question 2. Our main theorem is the following.

Theorem 1. *There is a deterministic algorithm that, given a bipartite graph $G = (U \uplus V, E)$ with a maximum matching of size r, outputs a linear representation of the transversal matroid $M_G = (U, \mathcal{I})$ over a field $\mathbb{F}$ of size $> \binom{|U|}{r}$, in time $\mathcal{O}(\binom{|U|}{r} \cdot |E| \sqrt{|V|} + N)$, where N is the time required to perform $\mathcal{O}(\binom{|U|}{r}|V| \cdot |U|)$ operations over $\mathbb{F}$.*

An XP algorithm for a parameterized problem $\Pi \subseteq \Sigma^* \times \mathbb{N}$, takes as input $(I,k) \in \Sigma^* \times \mathbb{N}$ and decides whether $(I,k) \in \Pi$ in time $|I|^{g(k)}$, where $g(k)$ is a computable function in k alone. Thus, in the language of parameterized complexity, Theorem 1 gives an XP algorithm for finding a linear representation of transversal matroids parameterized by the rank of the given matroid. Observe that if r is the rank of the matroid then the maximum matching of the graph is r. That is, $r = |U| - \ell$ and hence $\ell = |U| - r$. This together with the fact that $\binom{m}{a} = \binom{m}{m-a}$ implies that Theorem 1 gives a polynomial time algorithm for Question 2, whenever ℓ is a constant.

Transversal matroids are closely related to the class of *gammoids*. A gammoid is defined by a digraph D and two vertex subsets S and T of $V(D)$. Here, T is the ground set and a subset X of T is independent if and only if X is reachable from S by a collection of $|X|$ vertex disjoint paths. It was shown by Ingleton and Piff [8], that a subclass of gammoids, called *strict gammoids* where $T = V(D)$, are the duals of transversal matroids. Thus, one can also view gammoids as matroids obtained from strict gammoids by deleting some of the elements from the ground set. Therefore the task of designing an algorithm to construct a linear representation for gammoids is at least as hard as constructing one for transversal matroids. In this work we prove the following theorem.

Theorem 2. *There is a deterministic algorithm that, given an n-vertex digraph D and $S, T \subseteq V(D)$ such that $|S| = r$ and $|T| = n'$, outputs a linear representation of the gammoid defined in D with ground set T, over a field $\mathbb{F}$ of size strictly greater than $\binom{n'}{r}$ in time $\mathcal{O}(\binom{n'}{r}n^3 + N)$, where N is the time required to perform $\mathcal{O}(\binom{n'}{r}n^3)$ operations over $\mathbb{F}$.*

2   Preliminaries

For $n \in \mathbb{N}$, we use $[n]$ to denote the set $\{1, 2, \ldots, n\}$. Let U be a set. We use $|U|$ to denote the cardinality of U and 2^U to denote the set of subsets of U. For $i \in [|U|]$, $\binom{U}{i}$ denotes the set $\{S \subseteq U \mid |S| = i\}$.

Graphs. We use G to denote a graph and D to denote a digraph. The vertex set and edge (arc) set of a graph (digraph) are denoted as $V(G)$ ($V(D)$) and $E(G)$ ($A(D)$) respectively. We also use $G = (V, E)$ (or $D = (V, A)$) to denote a graph (digraph) with vertex set V and edge set E (arc set A). We use standard notations from graph theory [1]. For a graph G, and $u, v \in V(G)$ we use uv to denote the edge with end vertices u and v. An induced subgraph of G on the vertex set $V' \subseteq V(G)$ is written as $G[V']$. A *matching* in a graph G is a collection of edges such that no two edges share a common end vertices. We say a matching M in a graph G *saturates* $V' \subseteq V(G)$, if all the vertices in V' are end vertices of edges in M. For a digraph D and $u, v \in V(D)$, a *path* from u to v is a sequence of vertices $u_1, u_2, \ldots, u_k$ such that $u_1 = u, u_k = v$ and for all $i \in [k-1]$, $u_i u_{i+1} \in A(D)$.

Matrices and Linear algebra. For a finite field $\mathbb{F}$, $\mathbb{F}[X]$ denotes the ring of polynomials in X over $\mathbb{F}$ and $\mathbb{F}(X)$ denotes the *field of fractions* of $\mathbb{F}[X]$. A vector

v over a field $\mathbb{F}$ is an array of elements from $\mathbb{F}$. A set of vectors $\{v_1, v_2, \ldots, v_k\}$ are said to be linearly dependent if there exist elements $a_1, a_2, \ldots, a_k \in \mathbb{F}$, not all zero, such that $\sum_{i=1}^{k} a_i v_i = 0$. Otherwise these vectors are called linearly independent. We say a matrix A has dimension $n \times m$ (or $n \times m$ matrix A) if A has n rows and m columns. For a matrix A (or a vector v) we use A^T (or v^T) to denote its *transpose*. The determinant of an $n \times n$ matrix A is denoted by $\det(A)$. For a matrix A, with rows indexed by elements from the set R and columns indexed by elements from the set C, $R' \subseteq R$ and $C' \subseteq C$, we use $A[R', C']$ to denote the matrix A restricted to rows R' and columns C'. The rank of a matrix is the cardinality of a maximum sized set of columns which are linearly independent. Equivalently, the rank of a matrix A is the maximum number r such that there is a $r \times r$ submatrix whose determinant is non-zero.

Matroids. We recall a few definitions related to matroids. For a broader overview on matroids, we refer to [14].

Definition 1. *A matroid M is a pair $(E, \mathcal{I})$, where E is a set, called ground set and $\mathcal{I}$ is a family of subsets of E, called independent sets, with the following three properties: (i) $\emptyset \in \mathcal{I}$, (ii) if $I \in \mathcal{I}$ and $I' \subseteq I$, then $I' \in \mathcal{I}$, and (iii) if $I, J \in \mathcal{I}$ and $|I| < |J|$, then there is $e \in J \setminus I$ such that $I \cup \{e\} \in \mathcal{I}$.*

Any set $F \subseteq E$, $F \notin \mathcal{I}$, is called a dependent set and an inclusion wise maximal set B such that $B \in \mathcal{I}$ is called a basis. The cardinality of a basis in a matroid M is called the rank of M and is denoted by $\mathsf{rank}(M)$. The rank function of M, denoted by r_M, is a function from 2^E to $\mathbb{N} \cup \{0\}$ and is defined as, $r_M(S) = \max_{S' \subseteq S, S' \in \mathcal{I}} |S'|$ for any $S \subseteq E$.

Proposition 1. *Let $M = (E, \mathcal{I})$ be a matroid. If $A \in \mathcal{I}$, $B \subseteq E$ and $r_M (A \cup B) = \ell$ then there is a subset $B' \subseteq B$, $|B'| = \ell - |A|$ such that $A \cup B' \in \mathcal{I}$.*

Definition 2. *Let A be a matrix over a field $\mathbb{F}$ and E be the set of columns of A. The pair $(E, \mathcal{I})$, where $\mathcal{I}$ defined as follows, is a matroid, called linear matroid and is denoted by $M[A]$. For any $X \subseteq E$, $X \in \mathcal{I}$ if and only if the columns of A corresponding to X are linearly independent over $\mathbb{F}$. If a matroid M can be defined by a matrix A over a field $\mathbb{F}$, then we say that the matroid is representable over $\mathbb{F}$ and A is a linear representation of M.*

A matroid $M = (E, \mathcal{I})$ is called *representable* or *linear* if it is representable over some field $\mathbb{F}$.

Definition 3. *The dual of a matroid $M = (E, \mathcal{I})$, is a matroid, denoted by M^*, on the same ground set E, in which a set S is independent if and only if there is basis of M fully contained in $E \setminus S$.*

Definition 4. *Let $M = (E, \mathcal{I})$ be a matroid and $F \subseteq E$. Then the deletion of F in M, is a matroid, denoted by $M \setminus F$, whose independent sets are the independent sets of M that are contained in $E \setminus F$.*

Definition 5. *Let $M = (E, \mathcal{I})$ be a matroid and $F \subseteq E$. Then the contraction of M by F, is a matroid, denoted by M/F, on the ground set $E \setminus F$, whose rank function is defined as, $r_{M/F}(A) = r_M(A \cup F) - r_M(F)$ for any $A \subseteq E \setminus F$.*

Proposition 2 ([14]). *Let $M = (E, \mathcal{I})$ be a matroid and $F \subseteq E$. Then $M^* \setminus F = (M/F)^*$.*

Proposition 3 ([14]). *Given a matroid $M = (E, \mathcal{I})$ with an $n \times m$ matrix A over a field $\mathbb{F}$ as its linear representation and $F \subseteq E$. Then linear representations of M^*, $M \setminus F$ and M/F over the same field $\mathbb{F}$ can be computed by using only polynomially (in n and m) many operations over $\mathbb{F}$.*

Definition 6 (Truncation of a matroid). *Let $M = (E, \mathcal{I})$ be a matroid and $k \in \mathbb{N}$. Then the k-truncation of M is a matroid $M' = (E', \mathcal{I}')$ such that for any $F \subseteq E$, $F \in \mathcal{I}'$ if and only if $|F| \leq k$ and $F \in \mathcal{I}$.*

The following theorem proved in [12] implies that there is an algorithm which given a linear representation of a matroid M over a field $\mathbb{F}$, outputs a linear representation of a truncation of M over the field $\mathbb{F}(Y)$ using only polynomially many operations over $\mathbb{F}$.

Theorem 3 ([12]). *There is an algorithm that, given an $n \times m$ matrix M of rank n over a field $\mathbb{F}$ and a number $k \leq n$, uses $\mathcal{O}(mnk)$ field operations over $\mathbb{F}$ and computes a matrix M_k over the field $\mathbb{F}(Y)$, which represents the k-truncation of M. Furthermore, given M_k and a set of columns in M_k, we can test whether they are linearly independent using $\mathcal{O}(n^2 k^3)$ field operations over $\mathbb{F}$.*

Note that, in the above theorem Y is a single variable, rather than a set of variables. Hence testing if a determinant is non-zero over this field is simply a matter of checking if a univariate polynomial in the variable Y is non-zero. Furthermore, in many cases the output of the above theorem lies in a finite degree extension of $\mathbb{F}$. The two important matroids we are interested in this work are transversal matroids and gammoids.

Definition 7. *Let $G = (U \uplus V, E)$ be a bipartite graph. The transversal matroid associated with G, denoted by M_G, has ground set U and a set $U' \subseteq U$ is independent in M_G if and only if there is a matching in G saturating U'.*

Definition 8. *Let D be a digraph and $S, T \subseteq V(D)$. Then a gammoid with respect to D and S on ground set T is a matroid $(T, \mathcal{I})$, where $\mathcal{I}$ is defined as follows. For any $T' \subseteq T$, $T' \in \mathcal{I}$, if and only if there are $|T'|$ vertex disjoint paths whose start vertices are in S and end vertices are in T'. When $T = V(D)$, the gammoid is called a strict gammoid. In other words, for any $T \subseteq V(D)$ a gammoid on the ground set T is obtained from the strict gammoid by deleting $V(D) \setminus T$.*

As mentioned earlier, there is a randomized polynomial time algorithm to construct a linear representation of transversal matroids [13]. The following useful result is proved by Ingleton and Piff in 1972 [8] (see also [13,14]), shows that strict gammoids and transversal matroids are duals.

Lemma 1. *Let D be a digraph and $S \subseteq V(D)$. Let $T = V(D)$ and $V' = V(D)\backslash S$ be two copies of vertex subsets. Then there is bipartite graph $G = (T \uplus V', E)$ such that the strict gammoid with respect to D and S is the dual of the transversal matroid M_G on the ground set T. Moreover, there is a polynomial time algorithm that, given D and S, outputs the graph G.*

3   The Algorithm for Representing Transversal Matroids

In this section, we first give a deterministic algorithm to compute a linear representation of transversal matroids. That is, we give a deterministic algorithm to construct a linear representation of transversal matroids, and moreover, one can test whether a subset of the ground set is independent or not, in polynomial time. In the next section, we show that this algorithm may be modified to obtain more efficient algorithms for other classes of matroids that are related to transversal matroids.

Let $G = (U \uplus V, E)$ be a bipartite graph such that U and V contain m and n vertices, respectively. Let $U = \{u_1, \ldots, u_m\}$ and $V = \{v_1, \ldots, v_n\}$. Let $M_G = (U, \mathcal{I})$ be the transversal matroid associated with G where the ground set is U. That is, $S \subseteq U$ is independent in M_G if and only if there is a matching in G, saturating S. Let A' be the bipartite adjacency matrix of G. That is, the rows of A' are indexed with elements from V, columns of A' are indexed with elements from U, and for any $v \in V, u \in U$, $A'[v, u] = 1$ if and only if $vu \in E$. Let A be an $n \times m$ matrix defined as follows. For any $v_i \in V, u_j \in U$, $A[v_i, u_j] = A'[v_i, u_j] \cdot x_{i,j}$, where $X = \{x_{i,j} \mid i \in [n], j \in [m]\}$ is a set of variables. Notice that for any $v \in V, u \in U$, $A[v, u] = 0$ if and only if $vu \notin E$. Also, note that each variable $x_{i,j}$ appears at most once in the matrix A.

Now we describe our algorithm to find a linear representation of transversal matroids. We call this algorithm, Algorithm $\mathcal{A}$. For each $j \in [m]$, we define a set $X_j = \{x_{i,j} \mid i \in [n]\}$. Our algorithm is an iterative algorithm that produces values for $X_1, X_2, \ldots, X_m$ in order, by solving a system of linear inequations in the variables in X_i in the i-th iteration. Let r be the size of a maximum matching in G. Now we define a family of subsets of U as $\mathcal{B} = \{S \in \binom{U}{r} \mid$ there is a matching in G saturating $S\}$. That is $\mathcal{B}$ is the set of bases in M_G and let $\mathcal{B} = \{S_1, \ldots, S_t\}$. Notice that $t = |\mathcal{B}|$, is the number of bases in M_G. Now, for any $S \in \mathcal{B}$, we fix a matching $\mathsf{M}(S)$ saturating S. For any $S \in \mathcal{B}$, let $R(S)$ be the set of vertices from V, saturated by $\mathsf{M}(S)$. Note that $R(S) \in \binom{V}{r}$, and $G[S \cup R(S)]$ has a perfect matching (a matching of size r). Our goal is to assign values to all the variables in X from a field such that for any $S \in \mathcal{B}$, $\det(A[R(S), S]) \neq 0$. We will then show that this is enough to produce a linear representation of M_G (the details may be found in the correctness proof presented later in this section).

Let us now describe the steps of our algorithm. Recall that for each $j \in [m]$, $X_j = \{x_{i,j} \mid i \in [n]\}$. Let $X_{<j} = \bigcup_{j'=1}^{j-1} X_{j'}$ for any $j \in [m] \setminus \{1\}$ and $X_{<1} = \emptyset$. For $S \subseteq U$, and $j \in [m]$, we define $S(j) = \{u_{j'} \in S \mid j' \in [j]\}$. Let $\mathbb{F}$ be a field of size strictly more than $\binom{m}{r}$. Our algorithm assigns values for the variables in

X_j in the increasing order of j. In the j-th iteration, for $j \geq 1$, the algorithm will assign values for variables in X_j as follows. Note that at this stage, all the variables in $X_{<j}$ have been assigned values already. We denote by A_{j-1} the matrix A instantiated with the values for $X_{<j}$.

For any $S_i \in \mathcal{B}$, recall that $\mathsf{M}(S_i)$ is a fixed matching. Let M_{ij} be the set of edges in $\mathsf{M}(S_i)$ which saturate the vertices in $S_i(j)$. Let R_{ij} be the subset of V saturated by M_{ij}. Notice that the matching M_{ij} saturates the vertices $S_i(j) \cup R_{ij}$. Now the algorithm obtains values for X_j by solving the following inequalities.

$$\det(A_{j-1}[R_{ij}, S_i(j)]) \neq 0 \tag{1}$$

Observe that for each $i \in [t]$, $\det(A_{j-1}[R_{ij}, S_i(j)])$ is a linear function of the variables in X_j, since all the entries in $A_{j-1}[R_{ij}, S_i(j)]$ are elements from $\mathbb{F}$ except the entries in the last column which are either 0 or variables from X_j. Therefore, (1) can be stated as a system of linear inequations in the variables in X_j.

$$DX_j \neq \mathbf{0} \tag{2}$$

where D is a $t' \times n$ matrix for some $t' \leq t$ and $\mathbf{0}$ is a zero column vector of length t'. The entries in DX_j are linear functions, and note that the constraints require that every one of these linear functions is non-zero. We also know that $|\mathbb{F}| > t'$. Now our algorithm will execute the following algorithm (Lemma 2) to find a solution to the system (2).

Lemma 2. *Let $\mathbb{F}$ be a field of size $> t$ and let D be a $t \times n$ matrix over $\mathbb{F}$ such that no row vector in D contains only zeros. Then there is an n-length column vector Y^* such that $DY^* \neq \mathbf{0}$. Moreover such a vector can be computed using $\mathcal{O}(t \cdot n)$ operations over the field $\mathbb{F}$.*

Proof. Let $Y = [y_1, \ldots, y_n]^T$ be a column vector containing n variables. We will directly give an n step iterative process to compute Y^*, an instantiation of Y satisfying the system of linear inequations $DY \neq \mathbf{0}$. Initially, set all the variables in Y to be 0 and we use $Y(0)$ to denote this assignment. In other words, $Y(0)$ is a n length zero column vector. At each step we find out new assignment for Y. In step i we find out an assignment $Y(i)$ for Y and prove that indeed $D \cdot Y(i) \neq \mathbf{0}$. Now for any $i \in [n]$, at step i, $Y(i)$ is computed as follows. Note that at this step we have the assignment $Y(i-1)$ for Y. In other words, $Y(i-1)$ is an n-length column vector such that the jth entry is same as the value assigned for y_j in step $i-1$. Let Z_i be an n-length column vector where all entries are same as the entries in $Y(i-1)$, except the ith entry which is the variable y_i. That is, in Z_i, we did not assign any value to y_i, but all other variables are assigned the same value as in the assignment $Y(i-1)$. Now consider the entries in the column vector DZ_i. Some entries are elements in the field $\mathbb{F}$ and some are linear functions on variable y_i. Let $P_1(y_i) = p_1 y_i + q_1, \ldots, P_{t'}(y_i) = p_{t'} y_i + q_{t'}$ be the entries in DZ_j which are linear functions. Notice that $t' \leq t$. Since the size of the field $\mathbb{F}$ is strictly larger than $t \geq t'$, there is an element $a \in \mathbb{F}$ such that for all $j \in [t']$, $P_j(y_i) \neq 0$, because for any linear function $P_j(y_i)$, $P_j(y_i) = 0$ only when $y_i = -q_j/p_j$. Now we set $Y(i)$ as follows. Each entry in $Y(i)$ is same as each entry in $Y(i-1)$, except the ith entry which is set to a. Our algorithm will output $Y(n)$ as the required column vector. The correctness of the algorithm follows from the claim below.

Claim (⋆¹). For all $i \in \{0, 1, \ldots, n\}$, $Y(i)$ satisfies the set of inequalities in $DY \neq \mathbf{0}$, containing variables only from $\{y_1, \ldots, y_i\}$.

At step i, the algorithm computes DZ_i and chooses a value for y_i by looking at a linear function of y_i in the t-length column vector DZ_i. Since Z_{i-1} and Z_i differ only in two entries, DZ_i can be computed from DZ_{i-1}, and the $(i-1)^{st}$ and i^{th} columns of D, using $\mathcal{O}(t)$ operations over $\mathbb{F}$. Since algorithm has n iterations, the number of field operations performed is $\mathcal{O}(t \cdot n)$. $\qquad \square$

Our algorithm iterates over all values of j from $1, 2, \ldots m$, and produces an assignment of values from the field $\mathbb{F}$ for the variables in X_j in the j-th iteration. After m iterations an assignment for all variables will have been computed, and we let A_m be the instantiation of the matrix A with this assignment. Our algorithm outputs A_m as the representation of the transversal matroid M_G. This completes the description of Algorithm $\mathcal{A}$. The following two lemmata are required for proving the correctness of Algorithm $\mathcal{A}$.

Lemma 3 (⋆). *For any $j \in [m]$ and $S_i \in \mathcal{B}$, let M_{ij} be the set of edges in $\mathsf{M}(S_i)$ which saturates $S_i(j)$. Let $R_{ij} \subseteq V$ be the set of vertices in V saturated by M_{ij}. Then $\det(A_j[R_{ij}, S_i(j)]) \neq 0$.*

Lemma 4 (⋆). *Let $S' \subseteq U$ such that there is no matching saturating S' (or in other words $S' \notin \mathcal{I}$). Then the columns corresponding to S' in A_m are linearly dependent.*

Lemma 5. *The matrix A_m is a linear representation of the matroid $M_G = (U, \mathcal{I})$.*

Proof. We need to show that for any $U' \subseteq U$, $U' \in \mathcal{I}$ if and only if the columns in A_m indexed with elements from U' are linearly independent. If $U' \notin \mathcal{I}$, then by Lemma 4, the columns in A_m indexed with elements from U' are linearly dependent. Now we need to show that if $U' \in \mathcal{I}$, then the corresponding columns in A_m are linearly independent. Since $(U, \mathcal{I})$ is a matroid, there is a set $S \in \mathcal{B}$ such that $U' \subseteq S$. Note that $S(m) = S$ and $\mathsf{M}(S)$ is a fixed matching. Let R be the subset of V, saturated by the matching $\mathsf{M}(S)$. By Lemma 3, $\det(A_m[R, S]) \neq 0$. This implies that the columns of A_m corresponding to S are linearly independent and since $U' \subseteq S$, the columns of A_m corresponding to U' are also linearly independent. This completes the proof of the lemma. $\qquad \square$

Lemma 6. *Algorithm $\mathcal{A}$ runs in time $\mathcal{O}(\binom{m}{r} \cdot |E|\sqrt{n} + N)$, where N is the time required to perform $\mathcal{O}(\binom{m}{r}n \cdot m)$ operations over $\mathbb{F}$.*

---

¹ Proof of results marked (⋆) have been omitted due to space constraints.

Proof. Algorithm $\mathcal{A}$ first computes $\mathcal{B}$ and for each $S \in \mathcal{B}$ a matching $\mathsf{M}(S)$. This can be done by executing the bipartite maximum matching algorithm for each r-vertex subset of U. Since there are $\binom{m}{r}$ such subsets and each execution of the bipartite maximum matching algorithm is on a bipartite graph having at most $r + n \leq 2n$ vertices, this step takes time $\mathcal{O}(\binom{m}{r} \cdot |E|\sqrt{n})$ [7]. Following this, Algorithm $\mathcal{A}$ executes the algorithm of Lemma 2 once for each X_i, where $i \in [m]$. Since each execution of this algorithm takes $\mathcal{O}(\binom{m}{r}n)$ field operations, the total number of field operations required for the second phase of Algorithm $\mathcal{A}$ is $\mathcal{O}(\binom{m}{r}mn)$. This completes the proof of the lemma. $\qquad\square$

The correctness and running time bounds we have obtained for Algorithm $\mathcal{A}$ imply Theorem 1.

4   Representing Matroids Related to Transversal Matroids

In this section, we give deterministic algorithms for constructing linear representations of gammoids and strict gammoids. These algorithms utilize the algorithm for constructing linear representation of transversal matroids.

Truncations of transversal matroids. Several algorithmic applications require a linear representation of the k-truncation of matroids [2,5,12]. While, we can obtain a representation of the k-truncation of a transversal matroid M_G, by applying Theorem 3 to a representation of M_G, it very inefficient when $k \ll n$ which is usually the case in many applications. We can get a faster algorithm for this problem by slightly modifying Algorithm $\mathcal{A}$ and using Theorem 3. In the new algorithm, call it Algorithm $\mathcal{A}'$, we define $\mathcal{B}$ as $\mathcal{B} = \{S \in \binom{U}{k} \mid$ *there is a matching in G saturating S*$\}$. That is, $\mathcal{B}$ is directly defined to be the set of bases in the k-truncation of M_G. Then we follow the steps of algorithm $\mathcal{A}$. This algorithm will output an $n \times m$ matrix $\hat{A}$ over a field of size strictly more than $\binom{|U|}{k}$. Lemmata 3 and 4 are clearly true for Algorithm $\mathcal{A}'$ as well. That is, all the columns corresponds to a basis in k-truncation of M_G form a set of linearly independent vectors. But there may be a set of columns of size strictly greater than k which are also linearly independent. To get rid of this, we apply Theorem 3 to obtain the k-truncation of $\hat{A}$, which gives us a linear representation of the k-truncation of M_G.

Theorem 4. *There is a deterministic algorithm that, given a bipartite graph $G = (U \uplus V, E)$ and $k \in \mathbb{N}$, outputs a linear representation of the k-truncation of the transversal matroid $M_G = (U, \mathcal{I})$ over a field $\mathbb{F}(Y)$ where $\mathbb{F}$ has size strictly greater than $\binom{|U|}{k}$, in time $\mathcal{O}(\binom{|U|}{k}|V| \cdot |E| + N)$, where N is the time required to perform $\mathcal{O}(\binom{|U|}{k}|V| \cdot |U|)$ operations over $\mathbb{F}$.*

Contractions of transversal matroids. Here, we give a faster algorithm to compute a linear representation for a contracted transversal matroid by modifying Algorithm $\mathcal{A}$. We will use this linear representation to obtain a linear representation of gammoids. Let $M_G = (U, \mathcal{I})$ be the transversal matroid associated with the graph $G = (U \uplus V, E)$ on ground set U. Let $F \subseteq U$. One way of getting a representation of M_G/F is to find a linear representation of M_G and then find a linear representation of M_G/F by applying Proposition 3, but we can do better. Let r be the size of a maximum matching in G, that is $r_{M_G}(U) = r$. Let $r_{M_G}(F) = \ell$ and $k = r_{M_G}(U) - r_{M_G}(F)$. Note that the rank of M_G/F is k. Now we explain how to modify Algorithm $\mathcal{A}$ to get an algorithm, $\mathcal{A}''$, for computing a linear representation of M_G/F. Towards that we first define $\mathcal{B}$ as $\mathcal{B} = \{S \in \binom{U}{r} \mid$ there is a matching in G saturating S and $|S \setminus F| = k\}$. Now, Algorithm $\mathcal{A}''$ follows the steps of Algorithm $\mathcal{A}$ and it constructs an $n \times m$ matrix A_m. Lemmata 3 and 4 are true in this case as well. Let $M[A_m] = (E, \mathcal{I}'')$ be the matroid represented by the matrix A_m. Now Algorithm $\mathcal{A}''$ run the algorithm mentioned in Proposition 3 to compute a linear representation C of $M[A_m]/F$.

Lemma 7 ($\star$). *The matrix C is a linear representation of M_G/F.*

Now consider the running time of Algorithm $\mathcal{A}''$. Let M be a maximum matching in $G[F \cup V]$ and the F' be the set of vertices from F saturated by M. By Proposition 1, for any $S' \subseteq U \setminus F$ of cardinality k, there is a matching of size r in $G[F \cup S']$ if and only if there is a matching of size r in $G[F' \cup S']$. This implies that algorithm $\mathcal{A}''$ can construct $\mathcal{B}$, by running the bipartite maximum matching algorithm at most $\binom{n-|F|}{k}$ times. Thus, we get the following theorem.

Theorem 5. *There is a deterministic algorithm that, given a bipartite graph $G = (U \uplus V, E)$ and a vertex set $F \subseteq U$, outputs a linear representation of M_G/F over a field $\mathbb{F}$ of size strictly greater than $\binom{|U|-|F|}{k}$, in time $\mathcal{O}((\binom{|U|}{k})|V| \cdot |E| + N)$, where $k = \mathsf{rank}(M_G/F)$ and N is the time required to perform $\mathcal{O}((\binom{|U|-|F|}{k})|V| \cdot |U|)$ operations over $\mathbb{F}$.*

Gammoids. Now we give an algorithm to compute a linear representation of a gammoid efficiently. By Lemma 1, we know that there is a polynomial time algorithm which given a digraph D and $S \subseteq V(D)$, outputs a bipartite graph $G = (T \uplus V', E)$, where $T = V(D)$ and $V' = V(D) \setminus S$, such that the strict gammoid with respect to D and S is the dual of the transversal matroid M_G on the ground set T. Thus by Lemma 1, Theorem 1 and Proposition 3, we get the following theorem.

Theorem 6. *Let D be an n-vertex digraph, $S \subseteq V(D)$, $|S| = r$ and $\mathbb{F}$ be a field of size strictly greater than $\binom{n}{n-r}$. Then there is a deterministic algorithm which outputs a linear representation of the strict gammoid with respect to D and S, over $\mathbb{F}$ in time $\mathcal{O}((\binom{n}{n-r})n^3 + N)$, where N is the time required to perform $\mathcal{O}((\binom{n}{n-r})n^2)$ operations over $\mathbb{F}$.*

One way to get a representation of a gammoid is to first construct a representation of the *strict* gammoid in the graph and then delete some elements from the strict gammoid. However, observe that if we compute the representation of the strict gammoid via the algorithm of Theorem 6, the running time depends on the total number of vertices in the graph. We can obtain a much faster algorithm as follows. Let D be a digraph and let S and W be subsets of $V(D)$. Let M be the gammoid in D with ground set $W \subseteq V(D)$, with respect to $S \subseteq V(D)$ of rank r. We may assume that $r = |S|$. Otherwise, we construct the graph D' obtained from D by adding S', a set of r new vertices, and all possible edges from S' to S. Now consider the gammoid in D' with ground set W with respect to S'. It is easy to see that, for any subset of X of W, there are $|X|$ vertex disjoint paths from S' to X in D' if and only if there are $|X|$ vertex disjoint paths from S to X in D. So these two gammoids are the same and our assumption holds. Now let M_S be the strict gammoid in D with respect to S and note that the rank of M_S and M are same. Let M_S^* be the transversal matroid which is the dual of M_S and it is defined on the bipartite graph $G = (V(D) \uplus V', E)$, where $V' = V(D) \setminus S$. Now let N be the matroid obtained from M_S^* by contracting $V(D) \setminus W$. It is easy to see the following lemma.

Lemma 8. $M = N^*$.

Proof. Since N is a matroid obtained by contracting $V(D) \setminus W$ in M_S^*, Proposition 2 implies that N^* is the matroid $M_S \setminus (V(D) \setminus W)$. That is $N^* = M$. $\square$

Combining Lemma 8, Theorem 5 and Proposition 3 we obtain Theorem 2.

References

1. Diestel, R.: Graph Theory. Graduate Texts in Mathematics, vol. 173, 3rd edn. Springer, Berlin (2005)
2. Fomin, F.V., Golovach, P.A., Panolan, F., Saurabh, S.: Editing to connected f-degree graph. In: 33rd Symposium on Theoretical Aspects of Computer Science, STACS 2016, 17–20 February 2016, Orléans, France, pp. 36:1–36:14 (2016)
3. Fomin, F.V., Lokshtanov, D., Panolan, F., Saurabh, S.: Representative sets of product families. In: Schulz, A.S., Wagner, D. (eds.) ESA 2014. LNCS, vol. 8737, pp. 443–454. Springer, Heidelberg (2014). doi:10.1007/978-3-662-44777-2_37
4. Fomin, F.V., Lokshtanov, D., Panolan, F., Saurabh, S.: Efficient computation of representative families with applications in parameterized and exact algorithms. J. ACM **63**(4), 29:1–29:60 (2016)
5. Goyal, P., Misra, P., Panolan, F., Philip, G., Saurabh, S.: Finding even subgraphs even faster. In: 35th IARCS Annual Conference on Foundation of Software Technology and Theoretical Computer Science, FSTTCS 2015, 16–18 December 2015, Bangalore, India, pp. 434–447 (2015)
6. Hols, E.C., Kratsch, S.: A randomized polynomial kernel for subset feedback vertex set. In: 33rd Symposium on Theoretical Aspects of Computer Science, STACS 2016, 17–20 February 2016, Orléans, France, pp. 43:1–43:14 (2016)
7. Hopcroft, J.E., Karp, R.M.: An $n^{5/2}$ algorithm for maximum matchings in bipartite graphs. SIAM J. Comput. **2**, 225–231 (1973)
8. Ingleton, A., Piff, M.: Gammoids and transversal matroids. J. Comb. Theory Ser. B **15**(1), 51–68 (1973)
9. Kabanets, V., Impagliazzo, R.: Derandomizing polynomial identity tests means proving circuit lower bounds. Comput. Complex. **13**(1–2), 1–46 (2004)

10. Kratsch, S., Wahlström, M.: Representative sets and irrelevant vertices: new tools for kernelization. In: Proceedings of the 53rd Annual Symposium on Foundations of Computer Science (FOCS), pp. 450–459 (2012)
11. Kratsch, S., Wahlström, M.: Compression via matroids: a randomized polynomial kernel for odd cycle transversal. ACM Trans. Algorithms **10**(4), 20:1–20:15 (2014)
12. Lokshtanov, D., Misra, P., Panolan, F., Saurabh, S.: Deterministic truncation of linear matroids. In: Halldórsson, M.M., Iwama, K., Kobayashi, N., Speckmann, B. (eds.) ICALP 2015. LNCS, vol. 9134, pp. 922–934. Springer, Heidelberg (2015). doi:10.1007/978-3-662-47672-7_75
13. Marx, D.: A parameterized view on matroid optimization problems. Theor. Comput. Sci. **410**(44), 4471–4479 (2009)
14. Oxley, J.G.: Matroid Theory. Oxford Graduate Texts in Mathematics, vol. 21, 2nd edn. Oxford University Press, Cambridge (2010)
15. Shachnai, H., Zehavi, M.: Representative families: a unified tradeoff-based approach. In: Schulz, A.S., Wagner, D. (eds.) ESA 2014. LNCS, vol. 8737, pp. 786–797. Springer, Heidelberg (2014). doi:10.1007/978-3-662-44777-2_65

Dynamic Rank-Maximal Matchings

Prajakta Nimbhorkar[1]([✉]) and Arvind Rameshwar V.[2]

[1] Chennai Mathematical Institute, Chennai, India
prajakta@cmi.ac.in
[2] Birla Institute of Technology and Science Pilani, Hyderabad Campus,
Hyderabad, India
arvind.rameshwar@gmail.com

Abstract. We consider the problem of matching applicants to posts where applicants have preferences over posts. Thus the input to our problem is a bipartite graph $G = (\mathcal{A} \cup \mathcal{P}, E)$, where $\mathcal{A}$ denotes a set of applicants, $\mathcal{P}$ is a set of posts, and there are ranks on edges which denote the preferences of applicants over posts. A matching M in G is called *rank-maximal* if it matches the maximum number of applicants to their rank 1 posts, subject to this the maximum number of applicants to their rank 2 posts, and so on.

We consider this problem in a dynamic setting, where vertices and edges can be added and deleted at any point. Let n and m be the number of vertices and edges in an instance G, and r be the maximum rank used by any rank-maximal matching in G. We give a simple $O(r(m+n))$-time algorithm to update an existing rank-maximal matching under each of these changes. When $r = o(n)$, this is faster than recomputing a rank-maximal matching completely using a known algorithm like that of Irving et al. [13], which takes time $O(\min((r+n, r\sqrt{n})m)$.

1 Introduction

We consider matchings under one-sided preferences. The problem can be modeled as that of matching applicants to posts where applicants have preferences over posts. This problem has several important practical applications like allocation of graduates to training positions [11] and families to government housing [19]. The input to the problem consists of a bipartite graph $G = (\mathcal{A} \cup \mathcal{P}, E)$, where $\mathcal{A}$ is a set of applicants, $\mathcal{P}$ is a set of posts. Each applicant has a subset of posts ranked in an order of preference. This is referred to as the *preference list* of the applicant. An edge (a, p) has rank i if p is an ith choice of a. An applicant can have any number of posts at rank i, including zero. Thus the edge-set E can be partitioned as $E = E_1 \dot\cup \ldots \dot\cup E_r$, where E_i contains the edges of rank i.

This problem has received lot of attention and there exist several notions of optimality like pareto-optimality [1], rank-maximality [13], popularity [2], and fairness. The notion of *rank-maximality* has been first studied by Irving [12], who

R.V. Arvind—Part of the work was done when the author was a summer intern at Chennai Mathematical Institute.

Y. Cao and J. Chen (Eds.): COCOON 2017, LNCS 10392, pp. 433–444, 2017.
DOI: 10.1007/978-3-319-62389-4_36

called it *greedy matchings* and also gave an algorithm for computing such matchings in case of strict lists. A rank-maximal matching matches maximum number
of applicants to their rank 1 posts, subject to that, maximum number of applicants to their rank 2 posts and so on. Irving et al. [13] gave an $O(\min(n +
r, r\sqrt{n})m)$-time algorithm to compute a rank-maximal matching. Here $n =
|\mathcal{A}| + |\mathcal{P}|$, $m = |E|$, and r denotes the maximum rank on any edge in a rank-
maximal matching. The weighted and capacitated versions of this problem have
been studied in [14] and [17] respectively.

We consider the rank-maximal matching problem in a dynamic setting where
vertices and edges are added and deleted over time. The requirement of dynamic
updates in matchings has been well-studied in literature, with the motivation
of updating an existing optimal matching without recomputing it completely.
Dynamic updates are important in real-world applications as applicants matched
to posts can leave their jobs, or new applicants can apply for a job, or an applicant
can acquire new skills and hence becomes eligible for more posts.

Related work: Bipartite matchings as well as popular matchings have been
extensively studied in a dynamic setting [3–7,10,16]. The algorithms for maintaining maximum matchings in dynamic bipartite graphs maintain a matching
under addition and deletion of edges that closely approximates the maximum
cardinality matching, and the update time is small i.e. sub-linear or even poly-
logarithmic in the size of the graph. The algorithm of [7] maintains a matching
that has an unpopularity factor of $(\Delta + k)$ with $O(\Delta + \Delta^2/k)$ amortized changes
per round for addition or deletion of an edge, and $O(\Delta^2 + \Delta^3/k)$ changes per
round for addition and deletion of a vertex for any $k > 0$. In contrast to this, our
algorithm maintains rank-maximal matchings exactly but needs $O(r(m + n))$
time for each update. We describe our contribution below.

Recently, independent of our work, [8] give an $O(m)$ algorithm for updating
rank-maximal matchings under addition and deletion of vertices using techniques
similar to ours.

1.1 Our Contribution

We consider the problem of updating an existing rank-maximal matching when
a vertex or edge is added or deleted. We show the following in this paper:

Theorem 1. *Given an instance of the rank-maximal matching problem with n
vertices and m edges, there is an $O(r(m + n))$-time algorithm for updating a
rank-maximal matching when a vertex or edge is added to or deleted from the
instance. Here r is the maximum rank used in any rank-maximal matching in
the instance.*

When $r = o(n)$, this is faster than recomputing a rank-maximal matching using
the fastest known algorithm by Irving et al. [13].

Our algorithm crucially uses Irving et al.'s algorithm and the graphs it creates
for each stage. In Irving et al.'s algorithm, at stage i, edges of rank i are added
to the instance and some edges which can not belong to any rank-maximal
matching are deleted. We show that addition or deletion of a vertex or edge can

lead to addition and deletion of several edges at each stage, however, at most one augmenting path is created at each stage. This helps us update each stage in time $O(m+n)$, thus total time taken is $O(r(m+n))$ where r is the maximum rank on any edge in a rank-maximal matching.

It is important to note that addition or deletion of even one edge can change an existing rank-maximal matching by as much as $\Omega(n)$ edges. Also, addition or deletion of a vertex can potentially lead to addition or deletion of $\Omega(n)$ edges. In light of this, it is an interesting aspect of our algorithm that it avoids a complete recomputation of a rank-maximal matching. Also, in the instances that arise in practice, where there is a large number of applicants and posts, typically each applicant ranks only a small subset of posts. Therefore our algorithm is useful for updating a rank-maximal matching in such instances substantially faster than recomputing it completely. We refer the reader to the full version [15] for omitted details.

1.2 Organization of the Paper

In Sect. 2, we give some definitions and recall the algorithm of Irving et al. [13] for computing a rank-maximal matching along with some of its properties. The preprocessing and an overview of the algorithm appear in Sect. 3. The description and analysis of the algorithm is given in Sect. 4. We discuss some related questions in Sect. 5.

2 Preliminaries

We recall some well-known definitions and terminology (see e.g. [9]). A matching M in a graph G is a subset of edges, such that no two of them share a vertex. For a matched vertex u, we denote by $M(u)$ its partner in M.

Properties of maximum matchings in bipartite graphs: Let $G = (\mathcal{A} \cup \mathcal{P}, E)$ be a bipartite graph and let M be a maximum matching in G. The matching M defines a partition of the vertex set $\mathcal{A} \cup \mathcal{P}$ into three disjoint sets, defined below:

Definition 1 (Even, odd, unreachable vertices). *A vertex $v \in \mathcal{A} \cup \mathcal{P}$ is even (resp. odd) if there is an even (resp. odd) length alternating path with respect to M from an unmatched vertex to v. A vertex v is unreachable if there is no alternating path from an unmatched vertex to v.*

The following lemma is well-known in matching theory; see [18] or [13] for a proof.

Lemma 1 ([18]). *Let $\mathcal{E}$, $\mathcal{O}$, and $\mathcal{U}$ be the sets of even, odd, and unreachable vertices defined by a maximum matching M in G. Then,*

(a) $\mathcal{E}$, $\mathcal{O}$, and $\mathcal{U}$ are disjoint, and are the same for all the maximum matchings in G.

(b) In any maximum matching of G, every vertex in $\mathcal{O}$ is matched with a vertex in $\mathcal{E}$, and every vertex in $\mathcal{U}$ is matched with another vertex in $\mathcal{U}$. The size of a maximum matching is $|\mathcal{O}| + |\mathcal{U}|/2$.

(c) No maximum matching of G contains an edge with one end-point in $\mathcal{O}$ and the other in $\mathcal{O} \cup \mathcal{U}$. Also, G contains no edge with one end-point in $\mathcal{E}$ and the other in $\mathcal{E} \cup \mathcal{U}$.

Rank-maximal matchings: An instance of the rank-maximal matchings problem consists of a bipartite graph $G = (\mathcal{A} \cup \mathcal{P}, E)$, where $\mathcal{A}$ is a set of applicants, $\mathcal{P}$ is a set of posts, and applicants rank posts in order of their preference. That is the input is a bipartite graph $G = (\mathcal{A} \cup \mathcal{P}, E)$ where the edges in E can be partitioned as $E_1 \cup E_2 \cup \ldots \cup E_r$. Here E_i denotes the edges of rank i, and r denotes the maximum rank any applicant assigns to a post. An edge (a, p) has rank i if p is an ith choice of a.

Definition 2 (Signature). *The signature of a matching M is defined as an r-tuple $\rho(M) = (x_1, \ldots, x_r)$ where, for each $1 \leq i \leq r$, x_i is the number of applicants who are matched to their ith rank post in M.*

Let M, M' be two matchings in G, with signatures $\rho(M) = (x_1, \ldots, x_r)$ and $\rho(M') = (y_1, \ldots, y_r)$. Define $M \succ M'$ if $x_i = y_i$ for $1 \leq i < k \leq r$ and $x_k > y_k$.

Definition 3 (Rank-maximal matching). *A matching M in G is rank-maximal if M has the maximum signature under the above ordering $\succ$.*

Observe that all the rank-maximal matchings in an instance have the same cardinality and the same signature.

Construction of Rank-Maximal Matchings: Now we recall Irving et al.'s algorithm [13] for computing a rank-maximal matching in a given instance $G = (\mathcal{A} \cup \mathcal{P}, E_1 \cup \ldots \cup E_r)$. The pseudocode of the algorithm is given in Algorithm 1, we refer the reader to [13] for details. Recall that E_i is the set of edges of rank i.

Algorithm 1. An algorithm to compute a rank-maximal matching from [13].

Input: $G = (\mathcal{A} \cup \mathcal{P}, E_1 \cup E_2 \cup \cdots \cup E_r)$.
Output: A rank maximal matching M in G.
1: Let $G_i = (\mathcal{A} \cup \mathcal{P}, E_1 \cup E_2 \cup \cdots \cup E_i)$
2: Construct $G_1' = G_1$. Let M_1 be a maximum matching in G_1'.
3: **for** $i = 1 \ldots r$ **do**
4:     Partition $\mathcal{A} \cup \mathcal{P}$ as $\mathcal{O}_i, \mathcal{E}_i, \mathcal{U}_i$ with respect to M_i in G_i'.
5:     Delete all edges of rank $j > i$ incident on vertices in $\mathcal{O}_i \cup \mathcal{U}_i$.
6:     Delete all edges from G_i' between a node in $\mathcal{O}_i$ and a node in $\mathcal{O}_i \cup \mathcal{U}_i$.
7:     Add edges in E_{i+1} to G_i'; denote the resulting graph G_{i+1}'.
8:     Compute a maximum matching M_{i+1} in G_{i+1}' by augmenting M_i.
9: **end for**
10: Delete all edges from G_{r+1}' between a node in $\mathcal{O}_{r+1}$ and a node in $\mathcal{U}_{r+1}$.
11: Denote the graph G_{r+1}' as G'.
12: Return a rank-maximal matching $M = M_{r+1}$.

We note the following properties of Irving et al.'s algorithm:

(*I1*) For every $1 \leq i \leq r$, every rank-maximal matching in G_i is contained in G_i'.
(*I2*) The matching M_i is rank-maximal in G_i, and is a maximum matching in G_i'.

(*I3*) If a rank-maximal matching in G has signature $(s_1, \ldots, s_i, \ldots s_r)$ then M_i has signature $(s_1, \ldots, s_i)$.

(*I4*) The graphs $G'_i, 1 \leq i \leq r$ constructed at the end of iteration i of Irving et al.'s algorithm, and G' are independent of the rank-maximal matching computed by the algorithm. This follows from Lemma 1 and the above invariants.

3   Preprocessing and Overview

In the preprocessing stage, we store the information necessary to perform an update in $O(r(m + n))$ time. The preprocessing time is asymptotically same as that of computing a rank-maximal matching in a given instance by Irving et al.'s algorithm viz. $O(\min((r + n, r\sqrt{n})m))$ and uses $O((m + n) \log n)$ storage.

3.1   Preprocessing

Given an instance of the rank-maximal matching problem, $G = (\mathcal{A} \cup \mathcal{P}, E)$ and ranks on edges, we execute Irving et al.'s algorithm on G. (Algorithm 1 from Sect. 2.) Recall that n is the number of vertices and m is the number of edges.

We use the reduced graphs G'_i for $1 \leq i \leq r$, where $G'_i = (\mathcal{A} \cup \mathcal{P}, E'_i)$, computed by Algorithm 1 for updating a rank-maximal matching in G on addition or deletion of an edge or a vertex. If M is a rank-maximal matching in G, then in each G'_i, we consider the matching $M_i = M \cap E'_i$. By Invariant (*I2*) from Sect. 2, M_i is rank-maximal in G_i. When a vertex or an edge is added to or deleted from G, the goal is to emulate Algorithm 1 on the new instance H using the reduced graphs G'_i for each i.

We prove in Lemma 2 below that we do not need to store the reduced graphs explicitly. The storage can be achieved by storing the original graph G along with some extra information for each stage. If a vertex becomes odd (respectively unreachable) at stage i of Algorithm 1, we store the number i and one bit 0 (respectively 1) indicating that, at stage i, the vertex became odd (respectively unreachable). For each edge, we store the stage at which it gets deleted, if at all. This takes $O((m + n) \log n)$ bits of extra storage.

Lemma 2. *A reduced graph G'_i of any stage i of Algorithm 1 can be completely reconstructed from the stored information as described above. Moreover, this reconstruction can be done in $O(m + n)$ time.*

Proof. Edge-set E'_i of G'_i is a subset of $E_1 \cup \ldots \cup E_i$. We go over all the edges in $E_1 \cup \ldots \cup E_i$ and keep those edges in E'_i which have not been deleted up to stage i. This is precisely the information we have stored for each edge. As we go over each edge exactly once, we need $O(m + n)$ time.

3.2   An Overview of the Algorithm

Let G be a given instance and let H be the updated instance obtained by addition or deletion of an edge or a vertex. As stated earlier, the goal of our algorithm is to emulate Algorithm 1 on H using stored in the preprocessing step described above. Thus our algorithm constructs the reduced graphs H_i' for H by updating the reduced graphs G_i', and also a rank-maximal matching M' in H by updating a rank-maximal matching M in G. We prove that the graphs H_i' are same as the reduced graphs that would be obtained by executing Algorithm 1 on H.

The reduced graph H_i' can be significantly different from the reduced graph G_i' for a stage i. However, we show that there is at most one augmenting path in H_i' for any stage i. Thus each H_{i+1}' and M' can be obtained from H_i', G_{i+1}', and M_i in linear time i.e. $O(m+n)$ time. Also, we need to recompute the labels $\mathcal{E}, \mathcal{O}, \mathcal{U}$. This can be done in $O(m+n)$ time by connecting all the unmatched vertices to a new vertex s and performing BFS from s.

We note that, in Irving et al.'s algorithm, an applicant is allowed to have any number of posts of a rank i, including zero. Also, because of the one-sided preferences model, each edge has a unique rank associated with it. Thus addition of an applicant is analogous to addition of a post. In both the cases, a new vertex is added to the instance, along with the edges incident on it, and along with the ranks on these edges. The ranks can be viewed from either applicants' side or posts' side. Therefore, we describe our algorithm for addition of an applicant, but the same can be used for addition of a post. The same is true for deletion of a vertex. Deletion of an applicant or post involves deleting a vertex, along with its incident edges. Hence the same algorithm applies to deletion of both applicants and posts.

4   The Algorithm

We describe the update algorithm here. Throughout this discussion, we assume that G is an instance of the rank-maximal matching problem and H is an updated instance, where an update could be addition or deletion of an edge or a vertex. We discuss each of these updates separately.

As described in Sect. 3, we first run Algorithm 1 on G and compute a rank-maximal matching M in G. We also store the information regarding each vertex and edge as described in Sect. 3. In the subsequent discussion, we assume that we have the reduced graphs G_i' for each rank $1 \leq i \leq r$, which can be obtained in linear-time from the stored information as proved in Lemma 2.

4.1   Addition of a Vertex:

We describe the procedure for addition of a vertex in terms of addition of an applicant. Addition of a post is analogous as explained in Sect. 3. A description of the vertex-addition algorithm is given below and then we prove its correctness.

Description of vertex-addition algorithm: Let a be a new applicant to be added to the instance G. Let E_a be the set of edges along with their ranks, that correspond to the preference list of a. Thus the new instance is $H = ((\mathcal{A} \cup \{a\}) \cup \mathcal{P}, E \cup E_a)$. The update algorithm starts from G_1', adds edges of rank 1 from E_a to G_1' to get H_1' and then updates M and H_1' as follows:

Initialization: $S, T = \emptyset$. These sets are used later as described below.

The following cases arise while updating H'_1:

Case 1: Each rank 1 post p of a is odd in G'_1: Then H'_1 is same as G'_1, along with a and its rank 1 edges added.

Case 2: No rank 1 post of a is even but some post is unreachable in G'_1: Update the labels $\mathcal{E}, \mathcal{O}, \mathcal{U}$.[1] Add those applicants whose label changes from $\mathcal{U}$ to $\mathcal{E}$ to the set S, as they need to get higher rank edges in subsequent stages. Note that their higher rank edges are deleted by Algorithm 1 as they become unreachable in G'_1. Thus S always stores the vertices which need to get higher rank edges in subsequent iterations.

Case 3: A rank 1 post p of a is even in G'_1: Then there is an augmenting path starting with (a, p) in H'_1. Find it and augment M_1 to get a rank-maximal matching M'_1 in H'_1. Recompute the $\mathcal{E}, \mathcal{O}, \mathcal{U}$ labels. Delete higher rank edges on those vertices whose labels change from $\mathcal{E}$ in G'_1 to $\mathcal{U}$ in H'_1.

Delete $\mathcal{OO}$ and $\mathcal{OU}$ edges if present. Add those vertices to T which are odd or unreachable in H'_1. These are precisely those vertices that will not get higher rank edges in any subsequent iteration even if they become even in one such iteration.

For each subsequent stage $i > 1$, the algorithm proceeds as follows:

1. Start with $H'_i = G'_i$. Add a and its undeleted edges up to rank i to H'_i.
2. If there are applicants in the set S as described in Case 2 above, add edges of rank i incident on them to H'_i.
3. Start with a matching M'_i in H'_i such that M'_i has all the edges of M'_{i-1} and those rank i edges of M_i which are not incident on any vertex matched in M'_{i-1}.
4. Check if there is an augmenting path in H'_i with respect to M'_i. If so, augment M'_i.
5. Recompute the labels $\mathcal{E}, \mathcal{O}, \mathcal{U}$.
6. Delete higher rank edges on those vertices whose labels change from $\mathcal{E}$ to $\mathcal{U}$ or $\mathcal{O}$. Remove such vertices from S if they are present in S.
7. Delete $\mathcal{OO}$ or $\mathcal{OU}$ edges, if present. Now we have the final updated reduced graph H'_i.
8. Add those vertices from $V \setminus T$ to S whose labels change from $\mathcal{U}$ or $\mathcal{O}$ to $\mathcal{E}$. Add those vertices to T which are odd or unreachable in H'_i.

The algorithm stops when there are no more edges left in H. Figure 1 shows an example of the various cases considered above.

Analysis of the Vertex-Addition Algorithm: Recall the notation that G is the given instance and H is the instance obtained by adding an applicant a along with its incident edges. Moreover, G_i and H_i are subgraphs of G and H

---

[1] In Irving et al.'s algorithm, these labels are called $\mathcal{E}_1, \mathcal{O}_1, \mathcal{U}_1$. We omit the subscripts for the sake of bravity. The subscripts are clear from the stage under consideration.

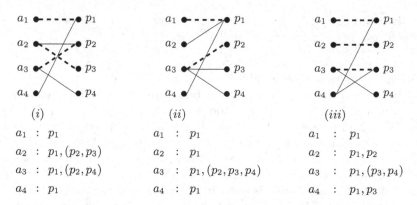

$$
\begin{array}{lll}
a_1 : p_1 & a_1 : p_1 & a_1 : p_1 \\
a_2 : p_1, (p_2, p_3) & a_2 : p_1 & a_2 : p_1, p_2 \\
a_3 : p_1, (p_2, p_4) & a_3 : p_1, (p_2, p_3, p_4) & a_3 : p_1, (p_3, p_4) \\
a_4 : p_1 & a_4 : p_1 & a_4 : p_1, p_3
\end{array}
$$

Fig. 1. Example of status change of nodes after addition of applicant a_4. Dashed lines indicate a rank-maximal matching before addition of a_4. In (i), a_1, p_1 are unreachable before adding a_4. After adding a_4, p_1 becomes odd while a_1 becomes even. In (ii), there is no status change after adding a_4. In (iii), there is an augmenting path a_4, p_3, a_3, p_4 after adding a_4. Augmentation makes all the nodes unreachable. Preference list for each figure is shown below the figure. Note that some edges on p_1 are deleted because they are $\mathcal{OO}$ or $\mathcal{OU}$ edges.

respectively, consisting of edges up to rank i respectively from G and H. Also G'_i is the reduced graph corresponding to stage i of an execution of Algorithm 1 on G whereas H'_i is the graph of stage i for H constructed by the vertex-addition algorithm. In Theorem 2, we prove that H'_i is indeed the reduced graph that would be constructed by an execution of Algorithm 1 on H.

The following Lemma is useful in analyzing the running time of the algorithm. It proves that there can be at most one new augmenting path at any stage i with respect to M_i in H'_i. Recall that M is a rank-maximal matching in G and M_i is the subset of M consisting of edges of rank only up to i. Also, M'_i is a rank-maximal matching in H_i.

Lemma 3. *At each stage i, $|M_i| \leq |M'_i| \leq |M_i| + 1$. Thus, for any stage i, there can be at most one augmenting path with respect to M_i in H_i.*

Proof. Recall from invariant $(I3)$ of Algorithm 1 mentioned in Sect. 2 that M_i and M'_i are the rank-maximal matchings in G_i and H_i respectively. Here G_i and H_i are the instances G and H with only the edges of ranks 1 to i present.

Consider $M_i \oplus M'_i$, which is the set of edges present in exactly one of the two matchings. This is a collection of vertex-disjoint paths and cycles. Each path on which the new applicant a does not appear, and each cycle must have the same number of edges of each rank from M_i and M'_i. Otherwise we can obtain a matching which has a better signature than either M_i or M'_i, which contradicts the rank-maximality of both the matchings in G_i and H_i respectively. At most one path can have the new applicant a as one end-point. This path can contain at most one more edge of M'_i than that of M_i. This proves the first part.

To see that there can be an augmenting path at multiple stages, consider the case where a post p gets matched to the new applicant a at stage i. If p is matched to an applicant b in M and the edge (b, p) has rank j such that $j > i$, then b is matched in G'_j but not in H'_j. Hence there can possibly be a new augmenting path in H'_j starting at b. □

Correctness of the algorithm is given by the following theorem.

Theorem 2. *The vertex-addition amunl43098165928lgorithm correctly updates the rank-maximal matching and the reduced graphs. Moreover, it runs in time* $O(r(m + n))$.

Proof. We prove this by induction on ranks. Thus we prove that, if the stage-wise reduced graphs are updated correctly up to stage $i - 1$, then the algorithm correctly constructs H'_i, and gives a rank-maximal matching M'_i in H_i. The base case is straightforward, we give the induction step here:

Assume that the algorithm has correctly computed H'_j for $1 \leq j < i$. We show that the algorithm then correctly computes H'_i and M'_i.

Initialization: The algorithm starts from $H'_i = G'_i$ and the matching M'_i in H_i is initialized to $M'_{i-1} \cup$ set of those edges in M_i that are vertex-disjoint from edges in M'_{i-1}. Note that there could be a vertex that is matched in M'_{i-1} but not in M_{i-1}, and possibly matched in M_i. Thus the initial matching M'_i is same as M_i except for the updates performed in earlier stages.

Recall that S is the set of vertices which do not have rank i edges in G'_i but need to get rank i edges in H'_i. The algorithm adds rank i edges on such vertices. It also adds applicant a and its undeleted edges to H'_i.

Checking for augmenting path: If a is still unmatched, then there could be an augmenting path in H'_i starting at a. Even if a is matched in M'_{i-1}, there could still be an augmenting path in H'_i with respect to M'_i, as explained below:

If a is matched in M'_{i-1}, say by a rank $j \leq i-1$ edge, then there is a post that is matched in M'_j but not in M_j, say q. This is because augmentation along an augmenting path always matches an additional applicant (in this case, a) and an additional post (in this case q). However, in M, i.e. prior to addition of a, q may have been matched to some applicant b at a rank $k > j$. Now q is matched to a, so b loses its matched edge at stage k. This needs updating labels $\mathcal{O}, \mathcal{U}, \mathcal{E}$ at subsequent stages. Also, we need to find an augmenting path from b, if any, at a later stage.

Note that there can be at most one augmenting path according to Lemma 3.

Recomputation of labels: Thus, at each stage, the algorithm looks for an augmenting path and augments M'_i, if such a path is found. The augmentation can lead to change of labels, and deletion of edges on those vertices whose labels change from $\mathcal{E}$ to $\mathcal{U}$ due to the augmentation. Also, even if there is no augmentation, there could still be a change of labels due to addition of edges on vertices in S and also due to addition of edges incident on a. Thus the labels need to be recomputed anyway. The sets S and T are updated as mentioned in the base case above.

As all the possible differences between G'_i to H'_i are considered above, H'_i is the correct reduced graph of stage i. Further, by Lemma 3, M'_i is a maximum matching in H_i obtained by augmenting a rank-maximal matching M'_{i-1} from H'_{i-1}. Thus M'_i is rank-maximal in H_i by correctness of Algorithm 1.

Each of the three operations described above need $O(m+n)$ time. Whenever the label on a vertex changes, or an edge is deleted, the stored information can be updated in $O(1)$ time. Thus each stage can be updated in time $O(m+n)$, so total update time is $O(r(m+n))$.   □

4.2   Deletion of a Vertex

Let an applicant a be deleted from the instance. The case of deletion of a post p is analogous, as explained in Sect. 3. Let G be the given instance and H be the updated instance. Thus $H = (\mathcal{A} \setminus \{a\} \cup \mathcal{P}, E \setminus E_a)$ where E_a is the set of edges incident on a. Let M be a rank-maximal matching in G. Also assume that the preprocessing step is executed on G and the information as mentioned in Sect. 3 is stored.

Description of the Vertex-Deletion Algorithm. If a is not matched in M, then M clearly remains rank-maximal in H, although the reduced graphs H'_i could differ a lot from the corresponding reduced graphs G'_i for each i. We describe the algorithm below.

Initialization: $S, T = \emptyset$. These sets will be used later, as given in the following description.

Case (I): a is matched in M: Let j be the rank of the matched edge in M incident on a and Let $M(a) = p$. Thus a remains even in the execution of Algorithm 1 on G at least for j iterations. The algorithm now works as follows: For each rank i from 1 to $j-1$, initialize $H'_i = G'_i$ and $M'_i = M_i$. Delete edges of rank up to i incident on a from H'_i. Recompute the labels $\mathcal{E}, \mathcal{O}, \mathcal{U}$. Delete from H the edges of rank $> i$ on those applicants whose label changes from $\mathcal{E}$ to $\mathcal{U}$. This is the final reduced graph H'_i. Add odd and unreachable vertices from H'_i to T. The set T contains those vertices that will not get higher rank edges at later stages even if their label changes to $\mathcal{E}$.

Now we come to the rank j at which a is matched in M. Initialize $H'_j = G'_j$ and $M'_j = M_j \setminus \{(a, p)\}$. Delete edges incident on a from H'_j. The following cases arise:

Case 1: p is odd in G'_j: Find an augmenting path in H'_j with respect to M'_j starting at p. Augment M'_j along this path. Recompute the labels. Delete from H the edges of rank $> j$ incident on those applicants whose labels change from $\mathcal{E}$ in G'_j to $\mathcal{U}$ in H'_j.

Case 2: p is unreachable in G'_j: Recompute the labels $\mathcal{E}, \mathcal{O}, \mathcal{U}$. Include those posts to S whose label changes from $\mathcal{U}$ to $\mathcal{E}$. These posts need to get edges of rank $> j$ in subsequent iterations.

Case 3: p is even in G'_j: Recompute the labels $\mathcal{E}, \mathcal{O}, \mathcal{U}$ in H'_j. Remove higher rank edges on those posts whose labels change from $\mathcal{E}$ to $\mathcal{U}$.

Add the odd and unreachable vertices from H_j' to T. Remove such vertices from S, if they are present in S. These are the vertices that will not get higher rank edges even if they get the label $\mathcal{E}$ at a later stage.

For each rank i from $j + 1$ to r, initialize $H_i' = G_i'$, except for a and its incident edges. Add edges of rank i on posts in S. Initialize $M_i' = M_{i-1}' \cup$ set of those edges in M_i which are disjoint from the edges in M_{i-1}'. Look for an augmenting path, and augment M_i' if an augmenting path is found. Recompute the labels $\mathcal{E}, \mathcal{O}, \mathcal{U}$. Update S and T as mentioned above.

Case (II): a **is unmatched in** M: The algorithm involves iterating over $i = 1$ to r and computing the reduced graphs H_i' as follows: Start with $H_i' = G_i'$, deleting a and its incident edges from H_i', add rank i edges on vertices in S, recompute the labels $\mathcal{E}, \mathcal{O}, \mathcal{U}$, include those vertices from $V \setminus T$ into set S whose labels change from $\mathcal{U}$ to $\mathcal{E}$. Add vertices with $\mathcal{O}$ or $\mathcal{U}$ labels to T. Delete higher rank edges on the vertices whose labels are $\mathcal{O}$ or $\mathcal{U}$.

The correctness of the vertex-deletion algorithm is given by the theorem below. The proof and an example appear in the full version [15].

Theorem 3. *The vertex-deletion algorithm correctly updates the rank-maximal matching M on deletion of an applicant. Moreover, it takes time $O(r(m + n))$.*

4.3 Addition and Deletion of an Edge

Modules similar to those for vertex-addition and vertex-deletion can be written for addition and deletion of an edge, which would have time complexity $O(r(m + n))$ each. However, both edge-addition and edge-deletion can be performed as a vertex-deletion followed by vertex-addition, achieving the same running time $O(r(m + n))$. We explain this here. To add an edge (a, p), one can first delete applicant a using the vertex-deletion algorithm thereby deleting all the edges E_a incident on a, and then the applicant a is added back along with the edge-set $E_a \cup \{(a, p)\}$. The case of edge-deletion is analogous. It is clear that both edge-addition and edge-deletion can thus be carried out in $O(r(m + n))$ time.

5 Discussion

In this paper, we give an $O(r(m + n))$ algorithm to update a rank-maximal matching when vertices or edges are added and deleted over time. Independent of our work, [8] give an algorithm for vertex addition and deletion that runs in $O(m)$ time using similar techniques.

In [9], a switching graph characterization of rank-maximal matchings has been developed, which has found several applications. It is an interesting question to explore whether this characterization can be used for dynamic updates.

Acknowledgement. We thank anonymous reviewers for their comments on an earlier version of this paper. We thank Meghana Nasre for helpful discussions.

References

1. Abraham, D.J., Cechlárová, K., Manlove, D.F., Mehlhorn, K.: Pareto-optimality in house allocation problems. In: Proceedings of 15th ISAAC, pp. 3–15 (2004)
2. Abraham, D.J., Irving, R.W., Kavitha, T., Mehlhorn, K.: Popular matchings. SIAM J. Comput. **37**(4), 1030–1045 (2007)
3. Abraham, D.J., Kavitha, T.: Dynamic matching markets and voting paths. In: Arge, L., Freivalds, R. (eds.) SWAT 2006. LNCS, vol. 4059, pp. 65–76. Springer, Heidelberg (2006). doi:10.1007/11785293_9
4. Baswana, S., Gupta, M., Sen, S.: Fully dynamic maximal matching in o(log n) update time. SIAM J. Comput. **44**(1), 88–113 (2015)
5. Bhattacharya, S., Henzinger, M., Italiano, G.F.: Deterministic fully dynamic data structures for vertex cover and matching. In: Proceedings of the Twenty-Sixth Annual ACM-SIAM Symposium on Discrete Algorithms, SODA 2015, pp. 785–804 (2015)
6. Bhattacharya, S., Henzinger, M., Nanongkai, D.: New deterministic approximation algorithms for fully dynamic matching. In: Proceedings of the 48th Annual ACM SIGACT Symposium on Theory of Computing, STOC 2016, pp. 398–411 (2016)
7. Bhattacharya, S., Hoefer, M., Huang, C.-C., Kavitha, T., Wagner, L.: Maintaining near-popular matchings. In: Halldórsson, M.M., Iwama, K., Kobayashi, N., Speckmann, B. (eds.) ICALP 2015. LNCS, vol. 9135, pp. 504–515. Springer, Heidelberg (2015). doi:10.1007/978-3-662-47666-6_40
8. Ghosal, P., Kunysz, A., Paluch, K.: The dynamics of rank-maximal and popular matchings. CoRR, abs/1703.10594 (2017)
9. Ghosal, P., Nasre, M., Nimbhorkar, P.: Rank-maximal matchings – structure and algorithms. In: Ahn, H.-K., Shin, C.-S. (eds.) ISAAC 2014. LNCS, vol. 8889, pp. 593–605. Springer, Cham (2014). doi:10.1007/978-3-319-13075-0_47
10. Gupta, M., Peng, R.: Fully dynamic (1+ e)-approximate matchings. In: Proceedings of the 54th Annual IEEE Symposium on Foundations of Computer Science, FOCS 2013, pp. 548–557 (2013)
11. Hylland, A., Zeckhauser, R.: The efficient allocation of individuals to positions. J. Polit. Econ. **87**(2), 293–314 (1979)
12. Irving, R.W.: Greedy matchings. Technical report, University of Glasgow, TR-2003-136 (2003)
13. Irving, R.W., Kavitha, T., Mehlhorn, K., Michail, D., Paluch, K.E.: Rank-maximal matchings. ACM Trans. Algorithms **2**(4), 602–610 (2006)
14. Kavitha, T., Shah, C.D.: Efficient algorithms for weighted rank-maximal matchings and related problems. In: Asano, T. (ed.) ISAAC 2006. LNCS, vol. 4288, pp. 153–162. Springer, Heidelberg (2006). doi:10.1007/11940128_17
15. Nimbhorkar, P., Arvind, R.V.: Dynamic rank-maximal matchings. CoRR, abs/1704.00899 (2017)
16. Onak, K., Rubinfeld, R.: Maintaining a large matching and a small vertex cover. In: Proceedings of the Forty-Second ACM Symposium on Theory of Computing, STOC 2010, pp. 457–464 (2010)
17. Paluch, K.: Capacitated rank-maximal matchings. In: Spirakis, P.G., Serna, M. (eds.) CIAC 2013. LNCS, vol. 7878, pp. 324–335. Springer, Heidelberg (2013). doi:10.1007/978-3-642-38233-8_27
18. Pulleyblank, W.R.: Matchings and extensions. In: Handbook of Combinatorics, vol. 1, pp. 179–232. MIT Press, Cambridge (1995)
19. Yuan, Y.: Residence exchange wanted: a stable residence exchange problem. Eur. J. Oper. Res. **90**(3), 536–546 (1996)

Bend Complexity and Hamiltonian Cycles in Grid Graphs

Rahnuma Islam Nishat[(✉)] and Sue Whitesides

Department of Computer Science, University of Victoria, Victoria, BC, Canada
{rnishat,sue}@uvic.ca

Abstract. Let G be an $m \times n$ rectangular grid graph. We study the problem of transforming Hamiltonian cycles on G under two basic operations we call *flip* and *transpose*. We introduce a new complexity measure, the *bend complexity*, for Hamiltonian cycles. Given any two Hamiltonian cycles C_1 and C_2 of bend complexity 1, we show that C_1 can be transformed to C_2 using only a linear number of flips and transposes.

1 Introduction

A *grid graph* is a finite, embedded, vertex-induced subgraph of the infinite two dimensional integer grid. Its vertex set is finite and its vertices have integer coordinates. Two vertices are adjacent if and only if they are distance one apart. A *solid grid graph* is a grid graph without holes, i.e., each bounded face of the graph is a unit square. An $m \times n$ *rectangular grid graph* is a solid grid graph of area $(m - 1) \times (n - 1)$ such that the boundary of the outer face is rectangular. This graph has m rows and n columns.

The Hamiltonian path and cycle problems have been studied in detail in the context of grid graphs as initiated by Itai *et al.* in 1982 [8]. They showed that it is NP-complete to decide whether a grid graph has a Hamiltonian path or a Hamiltonian cycle. They also gave necessary and sufficient conditions for a rectangular grid graph to have a Hamiltonian cycle. They left the problem of deciding whether a Hamiltonian cycle exists in a solid grid graph open. Later Umans and Lenhart [9] gave a polynomial time algorithm to find a Hamiltonian cycle (if it exists) in a solid grid graph. Afrati [1] gave a linear time algorithm for finding Hamiltonian cycles in *restricted grids*. Cho and Zelikovsky [4] studied *spanning closed trails* containing all the vertices of a grid graph. They showed that the problem of finding such a trail is NP-complete for grid graphs, but solvable for a subclass of grid graphs that they called *polyminos*. Researchers have explored other grids as well. In 2009, Arkin *et al.* [2] studied the existence of Hamiltonian cycles in grid graphs and proved several complexity results for a variety of types of grids.

In this paper, we define two simple "local" operations, *flip* and *transpose*, and study transformation of one Hamiltonian cycle of a rectangular grid graph into another Hamiltonian cycle under those two operations. The idea of applying local operations to transform one cycle into another was inspired by the study of

© Springer International Publishing AG 2017
Y. Cao and J. Chen (Eds.): COCOON 2017, LNCS 10392, pp. 445–456, 2017.
DOI: 10.1007/978-3-319-62389-4_37

transforming triangulations by "flipping" one edge at a time. In 1936, Wagner showed that any triangulation can be transformed into a canonical form, and hence can be transformed into any other triangulation [10]. Researchers since have studied different aspects of this problem (see [3] for a survey).

Our work is motivated also by applications of Hamiltonian paths and cycles in grid graphs. Grid graphs are used in path planning problems [7]. Suppose a vacuum cleaning robot is cleaning an indoor environment mapped as a grid graph, and to prevent damage to the floor, the robot must visit each vertex exactly once. The problem gives rise to a Hamiltonian path or cycle problem [6]. Another application is exploring an area with a given map. By overlaying the map with a grid graph, exploration is reduced to a Hamiltonian cycle problem. Enumeration of Hamiltonian paths and cycles in grid graphs has found application in polymer science [5].

From now on, unless specified otherwise, the term $m \times n$ *grid graph*, or simply *grid graph*, refers to a *rectangular grid graph*.

The rest of this paper is organized as follows. In Sect. 2, we introduce two operations, *flip* and *transpose*, and a complexity measure for Hamiltonian cycles in rectangular grid graphs that we call *bend complexity*. Section 3 gives an algorithm that, given any two Hamiltonian cycles C_1 and C_2 with bend complexity 1, transforms C_1 into C_2 using only a linear number of flips and transposes. Section 4 concludes the paper with some open problems.

2   Preliminaries

In this section, we introduce two local operations on Hamiltonian cycles on grid graphs that we call *flip* and *transpose*, and a complexity measure for such cycles that we call the *bend complexity*.

Let G be an $m \times n$ grid graph. We assume that G is embedded on the integer grid such that the top left corner vertex is at $(0,0)$. We also assume that the positive y direction is downward. A vertex of G with integer coordinates (x, y) is denoted as $v_{x,y}$. Vertices with the same x-coordinates or y-coordinates form a *column* or a *row*, respectively. All the vertices of *column* a have x-coordinate a, and all the vertices of *row* b have y-coordinate b. We call a subgraph G' of G a *subgrid* of G if G' is a grid graph, i.e., G' is an $m' \times n'$ rectangular grid graph.

We now define, by way of concrete examples, two "local" operations on any Hamiltonian cycle on G, where by *local*, we mean that the changes to the cycle are made within a small subgrid G' of G.

Flip. Let G' be a 3×2 or a 2×3 subgrid of a grid graph G, where the vertices a, b, d, f, e, c appear on the boundary of the outer face of G' in clockwise order, and the corner vertices of G' are a, b, f, e. Let C be a Hamiltonian cycle of G such that only the edges $(a, c), (c, d), (d, b)$ and (e, f) of G' are included in C. A *flip* operation on C in G' is performed in two steps. First, we remove the edges $(a, c), (c, d)$ and (d, b) and connect a to b, which gives a Hamiltonian cycle on $G - \{c, d\}$. We then remove the edge (e, f) and add the edges $(e, c), (c, d)$

and (d, f), which gives a Hamiltonian cycle C' on G. See Fig. 1(a). Note that applying flip to C' in G' gives back the original Hamiltonian cycle C.

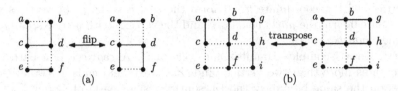

Fig. 1. Two local operations on a Hamiltonian cycle on a grid graph.

Transpose. Let G'' be a 3×3 subgrid of a grid graph G, where the vertices a, b, g, h, i, f, e, c appear on the boundary of the outer face of G'' in clockwise order, and the corner vertices of G'' are a, g, i, e. The vertex d of G'' is the only vertex that does not appear on the boundary of the outer face of G''. Let C be a Hamiltonian cycle on G such that the edges $(a, c), (f, d), (d, b), (b, g), (g, h), (h, i)$ are included in C. Either or both or none of the edges (e, c) and (e, f) might be on C. A *transpose* operation on C in G'' is performed in two steps. First, we remove the edges $(f, d), (d, b), (b, g), (g, h), (h, i)$ and add the edge (f, i). This gives a Hamiltonian cycle in $G - \{d, b, g, h\}$. We then remove the edge (a, c) and add the edges $(a, b), (b, g), (g, h), (h, d), (d, c)$, which gives a Hamiltonian cycle C' in G. See Fig. 1(b). The edges incident to e are not affected by the transpose operation. We call e the *pin vertex* of G''. Note that C' can be transformed back to C by applying transpose on C' in G''.

We introduce a complexity measure for Hamiltonian cycles in G.

Bend Complexity. Let C be a Hamiltonian cycle of G. Let v be a vertex of G and let e_1 and e_2 be the two edges incident to v that are also included in C. We say that v is a *bend* on C if one of e_1, e_2 is horizontal and the other is vertical. We define the *level of a bend* on C as follows. If the bend is on the boundary of the outer face of G, then its level is 0. Otherwise, if the bend is connected to a bend of level k by a straight line of edges of C then the level of this bend is $k + 1$. If the bend is connected to two bends, its level is one more than the minimum of the levels of the neighboring bends.

The *bend complexity* $B(C)$ of Hamiltonian cycle C is the highest level of any bend on C. We call a Hamiltonian cycle with bend complexity k a *k-complex Hamiltonian cycle*. The 1-complex Hamiltonian cycles have simple structures that look like "combs"; cycles with higher bend complexity are more complex.

3   1-Complex Hamiltonian Cycles

In this section, we study the structure of Hamiltonian cycles with bend complexity one, i.e., the 1-complex Hamiltonian cycles. We show that any 1-complex Hamiltonian cycle can be transformed to any other 1-complex Hamiltonian cycle using a linear number of flips and transposes.

We first define some terminology. Let G be an $m \times n$ rectangular grid graph. Throughout this section, we assume that $m, n \geq 3$ as otherwise G has at most one Hamiltonian cycle. The vertices of G with x-coordinate 0 form the *west boundary*, the vertices with x-coordinate $n - 1$ form the *east boundary*, the vertices with y-coordinate 0 form the *north boundary*, and the vertices with y-coordinate $m - 1$ form the *south boundary*.

Let C be a 1-complex Hamiltonian cycle of G. A *cookie* c is a path in C that contains more than two vertices and that starts and ends at two adjacent vertices on the same boundary (north, south, west or east) of G but does not contain any boundary edges; thus the path travels along a row or column, bends before reaching the opposite side, then immediately bends again to return to the same boundary, ending at a vertex adjacent to the start vertex. Thus, C consists of a set of cookies, and a set of sequences of vertices on C that form straight line segments on the boundary connecting two cookies or a cookie to a corner vertex. We call the length of the parallel sides of c the *height* of c. Observe that the height of a cookie is always greater than zero. Based on which one of the four boundaries contains the endpoints of a cookie, we have four *types* of cookies: north, south, west and east.

A cookie c lies in a pair of adjacent rows (e.g., rows y and $y + 1$, for $0 < y < m - 2$, when c is a west or east cookie) or adjacent columns (e.g., columns x and $x + 1$, for $0 < x < n - 2$, when c is a north or a south cookie). We call a pair of adjacent rows of G a *horizontal track* and denote it by t_x, and we call a pair of adjacent columns a *vertical track* and denote it by t_y, where x and y are the smaller indices of the pairs.

If C has only one type of cookie, then we call C a *canonical Hamiltonian cycle*. Note that if G has a Hamiltonian cycle, at least one of m and n must be even [8]. If only one of them is even, then G has two canonical Hamiltonian cycles; otherwise G has four canonical Hamiltonian cycles. Let C_1 and C_2 be any two 1-complex Hamiltonian cycles of G. Let $\mathbb{C}_1$ and $\mathbb{C}_2$ be two canonical Hamiltonian cycles of G. We show that C_1 can be transformed to C_2 using only flip and transpose operations in four steps as shown in Fig. 2: (a) we first transform C_1 to $\mathbb{C}_1$, (b) we transform C_2 to $\mathbb{C}_2$, (c) if $\mathbb{C}_1 \neq \mathbb{C}_2$, we then transform $\mathbb{C}_1$ to $\mathbb{C}_2$, and (d) finally, we reverse the steps of (b) to transform $\mathbb{C}_2$ to C_2. Effectively, we transform C_1 to $\mathbb{C}_1$ to $\mathbb{C}_2$ to C_2 as shown in Fig. 2.

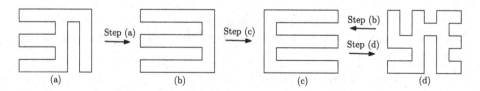

Fig. 2. (a) C_1, (b) $\mathbb{C}_1$, (c) $\mathbb{C}_2$, and (d) C_2. Steps (a) and (b): Transforming 1-complex Hamiltonian cycles C_1 and C_2 to canonical Hamiltonian cycles $\mathbb{C}_1$ and $\mathbb{C}_2$, respectively. Step (c): Transforming $\mathbb{C}_1$ to $\mathbb{C}_2$. Step (d): Reversing Step (b).

We first assume that C_1 has only three types of cookies and give an algorithm to transform C_1 to a canonical Hamiltonian cycle in Sect. 3.1. In Sect. 3.2, we use the algorithm in Sect. 3.1 to design an algorithm for the general case, where C_1 can have four types of cookies (Steps (a), (b) and (d)). In Sect. 3.3, we give an algorithm for transforming any canonical Hamiltonian cycle to another canonical Hamiltonian cycle (Step (c)). Finally, in Sect. 3.4, we analyze the time complexity of the algorithms.

3.1   1-Complex Hamiltonian Cycles with Three Types of Cookies

In this section, we give an algorithm, called *Algorithm* 1, to transform a 1-complex Hamiltonian cycle with at most three types of cookies to a canonical Hamiltonian cycle. We first give an overview of Algorithm 1.

Let C be a 1-complex Hamiltonian cycle on an $m \times n$ grid graph G such that C has at most three types of cookies. Since we can rotate G by multiples of $90°$, we can assume that C has no east cookies. Thus all the edges on the east boundary, which is also column $n - 1$, must be on C. The only bends are at the corner vertices $v_{n-1,0}$ and $v_{n-1,m-1}$. We therefore start from column $n - 2$ and scan the vertices from top (y-coordinate 0) to bottom (y-coordinate $m-1$) in that column. Scanning down column $n-2$ from the first internal vertex $v_{n-2,1}$ toward the last one $v_{n-2,m-2}$ we note that these internal vertices must be covered by a north cookie until the first encounter with a west or south cookie. After a west cookie is met, subsequent internal vertices must be covered by west cookies until either a south cookie is met or the south boundary is reached. If a south cookie is met, it contains all remaining internal vertices. Note that any north and south cookies covering the internal vertices of column $n - 2$ must be on the vertical track t_{n-3}. Let the height of the north and south cookies in vertical track t_{n-3} be y_N and y_S, respectively, where $y_N = 0$ if there is no north cookie in this track and $y_S = 0$ if there is no south cookie in this track. Then, the west cookies we encounter in column $n - 2$ must be in horizontal tracks $t_{y_N+1}, t_{y_N+3}, \ldots, t_{m-3-y_S}$.

We apply transpose operations on the west cookies until there are no west cookies in the vertical track t_{n-3} and the height of the north cookie is $m-2-y_S$. Then we apply flip on the south cookie until the height of the north cookie is maximum (i.e., $m-2$). An example is shown in Fig. 3. We then move two columns to the left and continue the same process until we reach column 1 or column 0 (the west boundary). Observe that at any time, the internal vertices in the vertical tracks that have already been processed are covered by north cookies.

If n is even, we reach column 0 eventually. Then the algorithm ends as the Hamiltonian cycle now has only north cookies, which means that we have obtained a canonical Hamiltonian cycle. Otherwise, we reach column 1. In this case the internal vertices in column 1 must be covered by only west cookies, since a north or south cookie requires two columns. We then apply $(n - 3)/2$ transpose operations on each west cookie, starting from the bottommost (in the horizontal track t_{m-3}) west cookie, and get a canonical Hamiltonian cycle with only west cookies. We now prove the correctness of Algorithm 1.

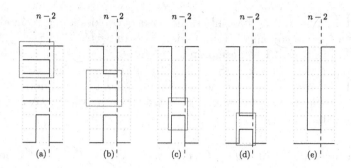

Fig. 3. (a) Column $n-2$ intersects two west cookies and a south cookie. Apply transpose in the 3×3 subgrid inside the red box, where the top left corner of the subgrid is the pin vertex, (b) apply another transpose, (c) then a flip in the 3×2 subgrid inside the red box, (d) apply another flip, and reach (e) final state. (Color figure online)

Theorem 1. *Let C be a 1-complex Hamiltonian cycle on an $m \times n$ grid graph G with no east cookies. Then Algorithm 1 transforms C to a canonical Hamiltonian cycle $\mathbb{C}$ of G, where $\mathbb{C}$ has either north cookies or west cookies.*

Proof. Since C is a Hamiltonian cycle, at least one of m and n must be even [8]. If n is even, we eventually reach column 0, the west boundary, and all the internal vertices in the columns between the east and west boundaries are covered by north cookies only. We now assume that n is odd, and hence m must be even. When we reach column 1, all the internal vertices in columns 2 through $n-2$ are covered by only north cookies, so the internal vertices on column 1 must be covered by only west cookies. We then apply $(n-3)/2$ transposes on each west cookie. Since m is even and we are moving two rows up after each step, we will reach row 0, at which stage we will have a canonical Hamiltonian cycle with only west cookies.   □

3.2   1-Complex Hamiltonian Cycles with Four Types of Cookies

In this section, we give an algorithm to transform any 1-complex Hamiltonian cycle to a canonical Hamiltonian cycle.

Let C be a 1-complex Hamiltonian cycle on a grid graph G with as many as four types of cookies. A *subpath* P' of C is a sequence of consecutive vertices on C such that P' is a Hamiltonian path in an $m' \times n'$ subgrid G' of G, where the two endpoints of P' are corner vertices of G' on the same row or same column. Without loss of generality, assume that the endpoints of P' are on row 0 of G'. We create a *subproblem* C' for C by adding another row above row 0 of G' and connecting the row's endpoints to the endpoints of P' to form a Hamiltonian cycle in that augmentation of G'. An example is shown in Fig. 4.

We first observe some properties of C.

Properties. Assume there exists a column x that intersects an east and a west cookie. Then there cannot exist any row that intersects both a north cookie

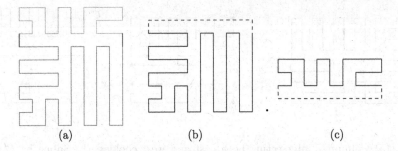

Fig. 4. (a) A 1-complex Hamiltonian cycle C. Two subpaths of C are shown in red and gray. (b) A subproblem for C created from the red subpath, and (c) a subproblem for C created from the gray subpath. The dashed lines show the edges that were added to the subpaths to create the subproblems. (Color figure online)

and a south cookie. Moreover, there must exist a pair of east and west cookies intersecting column x such that if the east cookie is on horizontal track t_y, then the west cookie must be on horizontal track t_{y+2} or t_{y-2}.

Let $C_1, C_2, \ldots, C_p$ be suproblems for C and let $P_1, P_2, \ldots, P_p$ be the corresponding subpaths. We say that C is *partitioned* into subproblems $C_1, C_2, \ldots, C_p$ if each vertex of C appears on exactly one P_i, $1 \leq i \leq p$. We now use the above properties to show that C can be partitioned into at most two subproblems, each with at most three types of cookies.

Lemma 1. *Let C be a 1-complex Hamiltonian cycle. Then C can be partitioned into at most two disjoint subproblems such that each subproblem has at most three types of cookies.*

Proof. We start scanning from the east boundary. If all the edges on that boundary are on C, then there are no east cookies. Therefore, C is a subproblem in itself with at most three types of cookies. Otherwise, there is at least one east cookie. We move one column to the left and continue scanning until one of the two following events occurs.

1. **A column intersects both east and west cookies.** Then by the properties of Hamiltonian cycles, there is no row that intersects both a north and a south cookie, and there exists a pair of east west cookies such that if the east cookie is in horizontal track t_y then the west cookie must be in horizontal track t_{y+2} or horizontal track t_{y-2}. Without loss of generality, we assume that the west cookie is in horizontal track t_{y-2} as shown in Fig. 5(a). Then there is no north cookie that goes below row $y - 1$ and no south cookie that goes above row y. Let P' be the subpath of C from vertex $v_{0,y}$ to $v_{n-1,y}$ that includes the vertex $v_{0,m-1}$, and let P'' be the subpath of C from vertex $v_{0,y-1}$ to $v_{n-1,y-1}$ that includes the vertex $v_{0,0}$ as shown in Fig. 5(b). We then create two subproblems C' and C'' as shown in Fig. 5(c).

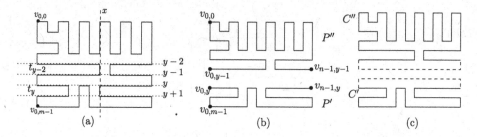

Fig. 5. (a) Column x intersecting both east and west cookies. (b) Subpaths P' and P'', (c) the two subproblems C' and C'' created from P' and P'', respectively.

2. **A column has no east cookies.** Let that column be x. There can be at most one south cookie and at most one north cookie intersecting column x. We now have the following cases to consider.

(a) Any north and south cookies that x intersects are in vertical track t_{x-1} as shown in Fig. 6(a). Let P' be the subpath of C from vertex $v_{x,0}$ to $v_{x,m-1}$ that includes the vertex $v_{0,0}$, and let P'' be the subpath of C from vertex $v_{x+1,0}$ to $v_{x+1,m-1}$ that includes the vertex $v_{n-1,0}$ (Fig. 6(b)). Then the subproblem created from P' will have no east cookies and the subproblem created from P'' will have no west cookies.

(b) Any north and south cookies are in vertical track t_x as shown in Fig. 6(c). Since this is the first column that intersects no east cookies when scanning to the left from the east boundary, there is at least one east cookie intersecting column $x + 1$. Let the first east cookie from the top be in horizontal track t_y, where $0 < y \le m - 3$. Then there must be a west cookie in the same track, since the vertex $v_{x,y}$ cannot be covered by any north or south cookies. Then no north cookie goes below row $y - 1$ and no south cookie comes above row $y + 2$. Let P' be the subpath of C from vertex $v_{0,y-1}$ to $v_{n-1,y-1}$ that

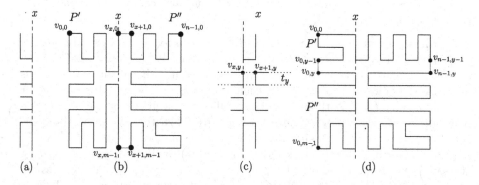

Fig. 6. (a) Vertical track t_{x-1} contains north or south cookies intersected by column x, (b) example of this case showing subpaths P' and P''. (c) Vertical track t_x contains north or south cookies intersected by column x, (d) example of this case.

includes the vertex $v_{0,0}$, and let P'' be the subpath of C from vertex $v_{0,y}$ to $v_{n-1,y}$ that includes the vertex $v_{0,m-1}$ (Fig. 6(d)). Then the subproblem created from P' will have no south cookies and the subproblem created from P'' will have no north cookies.

(c) Column x intersects north and south cookies on different vertical tracks. Let the height of the cookie in vertical track t_x be y. We assume that the north cookie is in vertical track t_x as shown in Fig. 7(a), where possibly column x intersects some west cookies. The case when the south cookie is in vertical track t_x is similar. Column $x + 1$ must intersect an east cookie but no west cookies, so vertex $v_{x+1,y+1}$ must be covered by an east cookie. From x we move to the left column by column until we find a west cookie which must occur at some column $x' > 0$. There may be three cases as described below; see Figs. 7(b)–(d). If there is a west cookie in column x (Fig. 7(b)), then there must be one in horizontal track t_{y+1}, as the height of the north cookie is y. If there is no west cookie in column x, then since x does not intersect any east cookie, vertices $v_{x,y+1}$ to $v_{x,m-1}$ must lie in a south cookie or on the south boundary. Thus no column left of x intersects an east cookie, and thus the vertex $v_{x-1,y}$ is covered by a north cookie or a west cookie. Vertex $v_{x-1,y+1}$ must be covered by a south cookie.

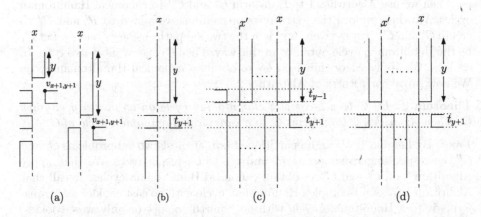

Fig. 7. (a) The north cookie is in vertical track t_x and the south cookie is in vertical track t_{x-1}. (b) Column x intersects a west cookie in horizontal track t_{y+1}. (c) The first west cookie to the left of column x is in horizontal track t_{y-1}. (d) The first west cookie to the left of column x is in horizontal track t_{y+1}.

We keep moving to the left until we arrive at a column x' such that there is a west cookie in column $x' - 1$. Observe that all the north cookies between (and including) columns x' and x have height y and all the south cookies between (and including) these columns have height $m - 2 - y$. If there is a north cookie in vertical track $t_{x'}$, then the west cookie will be in horizontal track t_{y-1} as shown in Fig. 7(c). Otherwise, there is a north cookie in vertical track $t_{x'-1}$, and the west cookie will be in horizontal track t_{y+1} as shown in Fig. 7(d). Since we have a west cookie in either horizontal track t_{y-1} or t_{y+1} with height $x' - 1$ and an east cookie in horizontal track t_{y+1} with height $n - 2 - x$, and all the north cookies in columns x' to x have height y, there is no north cookie in C with height greater than y. Similarly, there is no south cookie in C that has height greater than $m - 2 - y$.

We then create two subproblems in a similar way to that shown in Figs. 6(c)–(d). Let P' be the subpath of C from vertex $v_{0,y}$ to $v_{n-1,y}$ that includes the vertex $v_{0,0}$, and let P'' be the subpath of C from vertex $v_{0,y+1}$ to $v_{n-1,y+1}$ that include the vertex $v_{0,m-1}$. Then the subproblem C' created from P' will have no south cookies and the subproblem C'' created from P'' will have no north cookies.

Therefore, C can always be partitioned into at most two subproblems, each with at most three types of cookies. Observe that in each case above, the subpaths are connected on C by two edges on the boundary.   □

We now give an algorithm to transform any 1-complex Hamiltonian cycle C to a canonical Hamiltonian cycle. We call it Algorithm 2.

We first partition C into at most two subproblems C' and C'' using Lemma 1. Let C' and C'' correspond to the subpaths P' and P'' of C, respectively. By the proof of Lemma 1, P' and P'' must be connected by two boundary edges e_1 and e_2. Then we use Algorithm 1 to transform C' and C'' to canonical Hamiltonian cycles. We then remove the extra rows or columns we added to P' and P'' to create C' and C'', respectively, and join the two subpaths using e_1 and e_2. Let C_1 be the Hamiltonian cycle obtained in this way. Then C_1 has at most two types of cookies. We apply Algorithm 1 on C_1 to obtain a canonical Hamiltonian cycle. We now prove correctness of Algorithm 2.

Theorem 2. *Let C be a 1-complex Hamiltonian cycle on an $m \times n$ grid graph G. Then Algorithm 2 transforms C to a canonical Hamiltonian cycle of G.*

Proof. By Lemma 1, C can be partitioned into at most two subproblems C' and C'', corresponding to subpaths P' and P'' of C, respectively. We then apply Algorithm 1 on C' and C'' to obtain canonical Hamiltonian cycles. Recall that Algorithm 1 takes a 1-complex Hamiltonian cycle with no east cookies and transforms it to a Hamiltonian cycle with only north cookies or only west cookies. Thus the canonical Hamiltonian cycles obtained from C' and C'' cannot have cookies from the boundaries that were added to P' and P''; removing those boundaries gives two subpaths that can be joined to obtain another 1-complex Hamiltonian cycle C_1 on G. Since C was partitioned into at most two subproblems, C_1 can have at most two types of cookies. Now we can apply Algorithm 1 on C_1 again to obtain a canonical Hamiltonian cycle of G.   □

3.3 Transformation Between Canonical Hamiltonian Cycles

In this section we give an algorithm to transform a canonical Hamiltonian cycle $\mathbb{C}_1$ of an $m \times n$ grid graph G to another canonical Hamiltonian cycle $\mathbb{C}_2$ of G. We call it Algorithm 3.

Without loss of generality, assume that $\mathbb{C}_1$ has west cookies. If $\mathbb{C}_2$ has east cookies, we apply $n - 2$ flips on the west cookie in horizontal track t_y, $1 \leq y \leq m - 3$, starting from horizontal track t_1 until we have only an east cookie in that track. If $\mathbb{C}_2$ has north (or south) cookies, we apply $(m - 2)/2$ transposes in the vertical tracks t_x, $1 \leq x \leq n - 3$, such that the pin vertex is in column $x - 1$, starting from vertical track t_{n-3} until that track is covered by only one north (south) cookie.

3.4   Complexity Analysis

In this section we analyze the time complexity of Algorithms 1, 2 and 3.

Let C be a 1-complex Hamiltonian cycle in an $m \times n$ grid graph G. In Algorithm 1, we scan every column of G from column $n - 2$ to column 1 or column 0. The scanning takes linear time $O(mn)$. We apply at most $m - 2$ transposes and flips in each vertical track of C. Since there are $(n-2)/2$ vertical tracks, at most $(m-2)(n-2)/4$ operations are performed on the vertical tracks. In case n is odd, $(n - 3)/2$ transposes are applied on each horizontal track. Since there are $(m - 2)/2$ horizontal tracks, exactly $(m - 2)(n - 3)/4$ additional transpose operations are performed in this case. Therefore, the algorithm takes $O(mn)$ time in total, since each flip or transpose operation takes $O(1)$ time.

In Algorithm 2, partitioning C into two subproblems and merging the solutions to the subproblems take $O(mn)$ time. We apply Algorithm 1 on the two subproblems and finally on the cycle obtained from merging the solutions to the subproblems. Therefore, the total running time of Algorithm 2 is $O(mn)$.

In Algorithm 3, we apply $O(mn)$ flips or transposes, which takes $O(mn)$ time. Since Algorithm 2 (Steps (a) and (c)) takes $O(mn)$ time and Algorithm 3 (Step (b)) takes $O(mn)$ time, we can transform any 1-complex Hamiltonian cycle on G to any other 1-complex Hamiltonian cycle on G in $O(mn)$ time with $O(mn)$ flips and transpose operations, in other words, in linear time (and space).

4   Conclusions

In this paper, we have studied Hamiltonian cycles on grid graphs, by which we mean solid rectangular grid graphs. We defined a complexity measure for Hamiltonian cycles on rectangular grid graphs, and two local operations, flip and transpose. We showed that any 1-complex Hamiltonian cycle can be transformed to any other 1-complex Hamiltonian cycle using only a linear number of flips and transposes, thereby initiating a study of k-complex Hamiltonian cycles in grid graphs as we next describe.

We define *Hamiltonian cycle graph* $\mathcal{G}_{m,n}$ to be the graph whose vertices are Hamiltonian cycles on an $m \times n$ grid graph G. Two vertices u, v of $\mathcal{G}_{m,n}$ have an edge between them if one can obtain u from v and vice versa by applying a single flip or transpose operation. The subgraph $\mathcal{G}_{m,n,k}$ of $\mathcal{G}_{m,n}$ contains exactly the Hamiltonian cycles with bend complexity k. Our result shows that $\mathcal{G}_{m,n,1}$ is a connected graph and that the diameter of $\mathcal{G}_{m,n,1}$ is at most $O(mn)$. We pose the question whether $\mathcal{G}_{m,n,k}$ is a connected graph, where $k > 1$. We also ask whether similar results can be achieved for grid graphs with non-rectangular outer face boundary or grid graphs with holes. It would also be interesting to find out whether our results can be extended to Hamiltonian paths.

References

1. Afrati, F.: The hamilton circuit problem on grids. RAIRO Theor. Inform. Appl. **28**(6), 567–582 (1994)
2. Arkin, E.M., Fekete, S.P., Islam, K., Meijer, H., Mitchell, J.S., Rodríguez, Y.N., Polishchuk, V., Rappaport, D., Xiao, H.: Not being (super)thin or solid is hard: a study of grid hamiltonicity. Comput. Geom. **42**(6–7), 582–605 (2009)
3. Bose, P.: Flips. In: Didimo, W., Patrignani, M. (eds.) GD 2012. LNCS, vol. 7704, pp. 1–1. Springer, Heidelberg (2013). doi:10.1007/978-3-642-36763-2_1
4. Cho, H.-G., Zelikovsky, A.: Spanning closed trail and hamiltonian cycle in grid graphs. In: Staples, J., Eades, P., Katoh, N., Moffat, A. (eds.) ISAAC 1995. LNCS, vol. 1004, pp. 342–351. Springer, Heidelberg (1995). doi:10.1007/BFb0015440
5. des Cloizeaux, J., Jannik, G.: Polymers in Solution: Their Modelling and Structure. Clarendon Press, Oxford (1987)
6. Gorbenko, A., Popov, V.: On hamilton paths in grid graphs. Adv. Stud. Theor. Phys. **7**(3), 127–130 (2013)
7. Gorbenko, A., Popov, V., Sheka, A.: Localization on discrete grid graphs. In: He, X., Hua, E., Lin, Y., Liu, X. (eds.) CICA 2011. Lecture Notes in Electrical Engineering, vol. 107, pp. 971–978. Springer, Dordrecht (2012). doi:10.1007/978-94-007-1839-5_105
8. Itai, A., Papadimitriou, C.H., Szwarcfiter, J.L.: Hamilton paths in grid graphs. SIAM J. Comput. **11**(4), 676–686 (1982)
9. Umans, C., Lenhart, W.: Hamiltonian cycles in solid grid graphs. In: Proceedings of the 38th Annual Symposium on Foundations of Computer Science, FOCS 1997, pp. 496–505. IEEE Computer Society (1997)
10. Wagner, K.: Bemerkungen zum vierfarbenproblem. Jahresbericht der Deutschen Mathematiker-Vereinigung **46**, 26–32 (1936)

Optimal Covering and Hitting of Line Segments by Two Axis-Parallel Squares

Sanjib Sadhu[1]($\boxtimes$), Sasanka Roy[2], Subhas C. Nandy[2], and Suchismita Roy[1]

[1] Department of CSE, National Institute of Technology Durgapur, Durgapur, India
sanjibsadhu411@gmail.com
[2] Indian Statistical Institute, Kolkata, India

Abstract. This paper discusses the problem of covering and hitting a set of line segments $\mathcal{L}$ in $\mathbb{R}^2$ by a pair of axis-parallel squares such that the side length of the larger of the two squares is minimized. We also discuss the restricted version of covering, where each line segment in $\mathcal{L}$ is to be covered completely by at least one square. The proposed algorithm for the covering problem reports the optimum result by executing only two passes of reading the input data sequentially. The algorithm proposed for the hitting and restricted covering problems produces optimum result in $O(n \log n)$ time. All the proposed algorithms are in-place, and they use only $O(1)$ extra space. The solution of these problems also give a $\sqrt{2}$ approximation for covering and hitting those line segments $\mathcal{L}$ by two congruent disks of minimum radius with same computational complexity.

Keywords: Two-center problem · Covering line segments by squares · Two pass algorithm · Computational geometry

1 Introduction

Covering a point set by squares/disks has drawn interest to the researchers due to its applications in sensor network. Covering a given point set by k congruent disks of minimum radius, known as k-center problem, is NP-Hard [12]. For $k = 2$, this problem is referred to as the *two center problem* [3,5,6,8,9,13].

A line segment ℓ_i is said to be covered (resp. hit) by two squares if every point (resp. at least one point) of ℓ_i lies inside one or both of the squares. For a given set $\mathcal{L}$ of line segments, the objective is to find two axis-parallel congruent squares such that each line segment in $\mathcal{L}$ is covered (resp. hit) by the union of these two squares, and the size of the squares is as small as possible. There are mainly two variations of the covering problem: standard version and discrete version. In discrete version, the center of the squares must be on some specified points, whereas there are no such restriction in standard version. In this paper, we focus our study on the standard version of covering and hitting a set $\mathcal{L}$ of line segments in $\mathbb{R}^2$ by two axis-parallel congruent squares of minimum size.

As an application, consider a sensor network, where each mobile sensor is moving to and fro along different line segment. The objective is to place two

© Springer International Publishing AG 2017
Y. Cao and J. Chen (Eds.): COCOON 2017, LNCS 10392, pp. 457–468, 2017.
DOI: 10.1007/978-3-319-62389-4_38

base stations of minimum transmission range so that each of mobile sensors are always (resp. intermittently) connected to any of the base stations. This problem is exactly same as to cover (resp. hit) the line segments by two congruent disks (in our case axis-parallel congruent squares) of minimum radius.

Most of the works on the *two center problem* deal with covering a given point set. Kim and Shin [11] provided an optimal solution for the *two center problem* of a convex polygon where the covering objects are two disks. As mentioned in [11], the major differences between the *two-center problem for a convex polygon P* and the *two-center problem for a point set S* are (i) points covered by the two disks in the former problem are *in convex positions* (instead of arbitrary positions), and (ii) the union of two disks should also cover the edges of the polygon P. The feature (i) indicates the problem may be easier than the standard two-center problem for points, but feature (ii) says that it might be more difficult. To the best of our knowledge, there are no works on covering or hitting a set of line segments by two congruent squares of minimum size.

Related Work: Drenzer [4] covered a given point set S by two axis-parallel squares of minimum size in $O(n)$ time, where where $n = |S|$. Ahn and Bae [10] proposed an $O(n^2 \log n)$ time algorithm for covering a given point set S by two disjoint rectangles where one of the rectangles is axis parallel and other one is of arbitrary orientation, and the area of the larger rectangle is minimized. Two congruent squares of minimum size covering all the points in S, where each one is of arbitrary orientation, can be computed in $O(n^4 \log n)$ time [1]. The best known deterministic algorithm for the standard version of two-center problem for a point set S is given by Sharir [13] that runs in $O(n \log^9 n)$ time. Eppstein [5] proposed a randomized algorithm for the same problem with expected time complexity $O(n \log^2 n)$. The standard and discrete versions of the two-center problem for a convex polygon P was first solved by Kim and Shin [11] in $O(n \log^3 n \log \log n)$ and $O(n \log^2 n)$ time respectively. Hoffmann [7] solved the rectilinear 3-center problem for a point set in $O(n)$ time. However none of the algorithms in [1,4,7] can handle the line segments.

Our Work: We propose in-place algorithms for covering and hitting n line segments in $\mathbb{R}^2$ by two axis-parallel congruent squares of minimum size. We also study the restricted version of the covering problem where each object needs to be completely covered by at least one of the reported squares. The time complexities of our proposed algorithms for these three problems are $O(n)$, $O(n \log n)$ and $O(n \log n)$ respectively, and these work using $O(1)$ extra work-space. The same algorithms work for covering/hitting a polygon, or a set of polygons by two axis-parallel congruent squares of minimum size. We show that the result of this algorithm can produce a solution for the problem of covering (resp. hitting) these line segments by two congruent disks of minimum radius in $O(n)$ (resp. $O(n \log n)$) time with an approximation factor $\sqrt{2}$.

1.1   Notations and Terminologies Used

Throughout this paper, unless otherwise stated a *square* is used to imply an axis-parallel square. We will use the following notations and definition.

Symbols used	Meaning
$\overline{pq}$ and $\lvert pq \rvert$	the line segment joining two points p and q, and its length
$x(p)$ (resp. $y(p)$)	x- (resp. y-) coordinates of the point p
$\lvert x(p) - x(q) \rvert$	*horizontal distance* between a pair of points p and q
$\lvert y(p) - y(q) \rvert$	*vertical distance* between a pair of points p and q
$s \in \overline{pq}$	the point s lies on the line segment $\overline{pq}$
$\square efgh$	an axis-parallel rectangle with vertices at e, f, g and h
$size(\mathcal{S})$	size of square $\mathcal{S}$; it is the length of its one side
$LS(\mathcal{S})$, $RS(\mathcal{S})$	Left-side of square $\mathcal{S}$ and right-side of square $\mathcal{S}$
$TS(\mathcal{S})$, $BS(\mathcal{S})$	Top-side of square $\mathcal{S}$ and bottom-side of square $\mathcal{S}$

Definition 1. *A square is said to be* **anchored** *with a vertex of a rectangle* $\mathcal{R} = \square efgh$, *if one of the corners of the square coincides with that vertex of* $\mathcal{R}$.

2   Covering Line Segments by Two Congruent Squares

LCOVER problem: Given a set $\mathcal{L} = \{\ell_1, \ell_2, \ldots, \ell_n\}$ of n line segments (possibly intersecting) in $\mathbb{R}^2$, the objective is to compute two congruent squares $\mathcal{S}_1$ and $\mathcal{S}_2$ of minimum size whose union covers all the members in $\mathcal{L}$.

In the first pass, a linear scan is performed among the objects in $\mathcal{L}$, and four points a, b, c and d are identified with minimum x-, maximum y-, maximum x- and minimum y-coordinate respectively among the end-points of $\mathcal{L}$. This defines an axis-parallel rectangle $\mathcal{R} = \square efgh$ of minimum size that covers $\mathcal{L}$, where $a \in \overline{he}$, $b \in \overline{ef}$, $c \in \overline{fg}$ and $d \in \overline{gh}$. We use $L = \lvert x(c) - x(a) \rvert$ and $W = \lvert y(b) - y(d) \rvert$ as the length and width respectively of the rectangle $\mathcal{R} = \square efgh$, and we assume that $L \geq W$. We assume that $\mathcal{S}_1$ lies to the left of $\mathcal{S}_2$. $\mathcal{S}_1$ and $\mathcal{S}_2$ may or may not overlap (see Fig. 1). We use $\sigma = size(\mathcal{S}_1) = size(\mathcal{S}_2)$.

Lemma 1

(a) *There exists an optimal solution of the problem where $LS(\mathcal{S}_1)$ and $RS(\mathcal{S}_2)$ pass through the points a and c respectively.*

(b) *The top side of at least one of $\mathcal{S}_1$ and $\mathcal{S}_2$ pass through the point b, and the bottom side of at least one of $\mathcal{S}_1$ and $\mathcal{S}_2$ pass through the point d.*

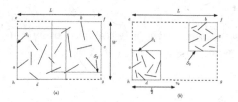

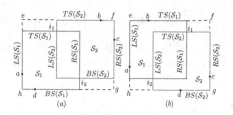

Fig. 1. Squares S_1 and S_2 are (a) over-lapping, (b) disjoint

Fig. 2. (a) **Configuration 1** and (b) **Configuration 2** of squares S_1 and S_2

Thus in an optimal solution of the **LCOVER** problem, $a \in LS(S_1)$ and $c \in RS(S_2)$. We need to consider two possible configurations of an optimum solution (i) $b \in TS(S_2)$ and $d \in BS(S_1)$, and (ii) $b \in TS(S_1)$ and $d \in BS(S_2)$. These are named as **Configuration 1** and **Configuration 2** respectively (see Fig. 2).

Observation 1

(a) If the optimal solution of LCOVER problem satisfies **Configuration 1**, then the bottom-left corner of S_1 will be anchored at the point h, and the top-right corner of S_2 will be anchored at the point f.

(b) If the optimal solution of LCOVER problem satisfies **Configuration 2**, then the top-left corner of S_1 will be anchored at the point e, and the bottom-right corner of S_2 will be anchored at the point g.

We consider each of the configurations separately, and compute the two axis-parallel congruent squares S_1 and S_2 of minimum size whose union covers the given set of line segments $\mathcal{L}$. If σ_1 and σ_2 are the sizes obtained for **Configuration 1** and **Configuration 2** respectively, then we report $\min(\sigma_1, \sigma_2)$.

Consider the rectangle $\mathcal{R} = \Box efgh$ covering $\mathcal{L}$, and take six points k_1, k_2, k_3, k_4, v_1 and v_2 on the boundary of $\mathcal{R}$ satisfying $|k_1 f| = |ek_3| = |hk_4| = |k_2 g| = W$ and $|ev_1| = |hv_2| = \frac{L}{2}$ (see Fig. 3). Throughout the paper we assume h as the origin in the co-ordinate system, i.e. $h = (0,0)$.

Observation 2

(i) The Voronoi partitioning line λ_1 of the corners f and h of $\mathcal{R} = \Box efgh$ with respect to the L_∞ norm[1] is the polyline $k_1 z_1 z_2 k_4$, where the coordinates of its defining points are $k_1 = (L - W, W)$, $z_1 = (L/2, L/2)$, $z_2 = (L/2, W - L/2)$ and $k_4 = (W, 0)$ (see Fig. 3(a)).

(ii) The Voronoi partitioning line λ_2 of e and g of $\mathcal{R} = \Box efgh$ in L_∞ norm is the polyline $k_3 z_1 z_2 k_2$ where $k_3 = (W, W)$ and $k_2 = (L - W, 0)$ (see Fig. 3(b)).

---

[1] L_∞ distance between two points a and b is given by $\max(|x(a) - x(b)|, |y(a) - y(b)|)$.

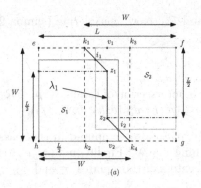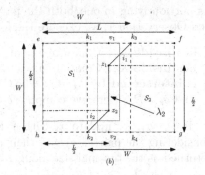

Fig. 3. Voronoi partitioning line (a) $\lambda_1 = k_1 z_1 z_2 k_4$ of f and h in **Configuration 1** (b) $\lambda_2 = k_3 z_1 z_2 k_2$ of e and g in **Configuration 2**

Note that, if $W \le \frac{L}{2}$, then the voronoi partitioning lines λ_1 and λ_2 for both the pairs (f, h) and (e, g) will be same, i.e., $\lambda_1 = \lambda_2 = \overline{v_1 v_2}$, where $v_1 = (\frac{L}{2}, 0)$ and $v_2 = (\frac{L}{2}, W)$.

Lemma 2

(a) *For* **Configuration 1**, *All the points p inside the polygonal region $e k_1 z_1 z_2 k_4 h$ satisfy $d_\infty(p, h) < d_\infty(p, f)$, and all points p inside the polygonal region $k_1 f g k_4 z_2 z_1$ satisfy $d_\infty(p, f) < d_\infty(p, h)$ (see Fig. 3(a)).*

(b) *Similarly for* **Configuration 2**, *all points p inside polygonal region $e k_3 z_1 z_2 k_2 h$, satisfy $d_\infty(p, e) < d_\infty(p, g)$, and all points p that lie inside the polygonal region $k_3 f g k_2 z_2 z_1$, satisfy $d_\infty(p, g) < d_\infty(p, e)$ (see Fig. 3(b)).*

Lemma 3. *If S_1 and S_2 intersect, then the points of intersection i_1 and i_2 will always lie on voronoi partitioning line $\lambda_1 = k_1 z_1 z_2 k_4$ (resp. $\lambda_2 = k_3 z_1 z_2 k_2$) depending on whether S_1 and S_2 satisfy* **Configuration 1** *or* **Configuration 2**.

Our algorithm consists of two passes. In each pass we sequentially read each element of the input array $\mathcal{L}$ exactly once. We consider $W > \frac{L}{2}$ only. The other case i.e. $W \le \frac{L}{2}$ can be handled in the similar way.

Pass-1: We compute the rectangle $\mathcal{R} = \Box efgh$, and the voronoi partitioning lines λ_1 and λ_2 (see Fig. 3) for handling **Configuration 1** and **Configuration 2**. We now discuss Pass 2 for **Configuration 1**. The same method works for **Configuration 2**, and for both the configurations, the execution run simultaneously keeping a $O(1)$ working storage.

Pass-2: λ_1 splits $\mathcal{R}$ into two disjoint parts, namely $\mathcal{R}_1$ = region $e k_1 z_1 z_2 k_4 h$ and $\mathcal{R}_2$ = region $f k_1 z_1 z_2 k_4 g$. We initialize $\sigma_1 = 0$. Next, we read elements in the input array $\mathcal{L}$ in sequential manner. For each element $\ell_i = [p_i, q_i]$, we identify its portion lying in one/both the parts $\mathcal{R}_1$ and $\mathcal{R}_2$. Now, considering Lemma 2 and Observation 1, we execute the following:

ℓ_i **lies inside** $\mathcal{R}_1$: Compute $\delta = \max(d_\infty(p_i, h), d_\infty(q_i, h))$.
ℓ_i **lies inside** $\mathcal{R}_2$: Compute $\delta = \max(d_\infty(p_i, f), d_\infty(q_i, f))$.
ℓ_i **is intersected by** λ_1: Let θ be the point of intersection of ℓ_i and λ_1, $p_i \in \mathcal{R}_1$ and $q_i \in \mathcal{R}_2$. Here, we compute $\delta = \max(d_\infty(p_i, h), d_\infty(\theta, h), d_\infty(q_i, f))$.

If $\delta > \sigma_1$, we update σ_1 with δ. Similarly, σ_2 is also computed in this pass considering the pair(e, g) and their partitioning line λ_2. Finally, $\min(\sigma_1, \sigma_2)$ is returned as the optimal size along with the centers of the squares $\mathcal{S}_1$ and $\mathcal{S}_2$.

Theorem 1. *Given a set of line segments $\mathcal{L}$ in $\mathbb{R}^2$ in an array, one can compute two axis-parallel congruent squares of minimum size whose union covers $\mathcal{L}$ by reading the input array only twice in sequential manner, and maintaining $O(1)$ extra work-space.*

3   Hitting Line Segments by Two Congruent Squares

Definition 2. *A geometric object Q is said to be* hit *by a square $\mathcal{S}$ if at least one point of Q lies inside (or on the boundary of) $\mathcal{S}$.*

Line segment hitting (LHIT) problem: Given a set $\mathcal{L} = \{\ell_1, \ell_2, \ldots, \ell_n\}$ of n line segments in $\mathbb{R}^2$, compute two axis-parallel congruent squares $\mathcal{S}_1$ and $\mathcal{S}_2$ of minimum size whose union hits all the line segments in $\mathcal{L}$.

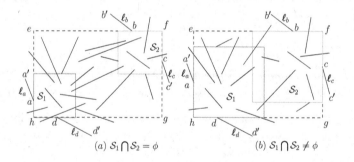

(a) $\mathcal{S}_1 \bigcap \mathcal{S}_2 = \phi$ (b) $\mathcal{S}_1 \bigcap \mathcal{S}_2 \neq \phi$

Fig. 4. Two axis-parallel congruent squares $\mathcal{S}_1$ and $\mathcal{S}_2$ hit line segments in $\mathcal{L}$

The squares $\mathcal{S}_1$ and $\mathcal{S}_2$ may or may not be disjoint (see Fig. 4). We now describe the algorithm for this **LHIT** problem.

For each line segment ℓ_i, we use $LP(\ell_i)$, $RP(\ell_i)$, $TP(\ell_i)$ and $BP(\ell_i)$ to denote its left end-point, right end-point, top end-point and bottom end-point using the relations $x(LP(\ell_i)) \leq x(RP(\ell_i))$ and $y(BP(\ell_i)) \leq y(TP(\ell_i))$. Now we compute four line segments ℓ_a, ℓ_b, ℓ_c, and $\ell_d \in \mathcal{L}$ such that one of their end-points a, b, c and d, respectively satisfy the following

$$a = \min_{\forall \ell_i \in \mathcal{L}} x(RP(\ell_i)), b = \max_{\forall \ell_i \in \mathcal{L}} y(BP(\ell_i)), c = \max_{\forall \ell_i \in \mathcal{L}} x(LP(\ell_i)), d = \min_{\forall \ell_i \in \mathcal{L}} y(TP(\ell_i))$$

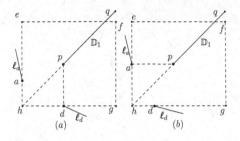
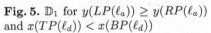
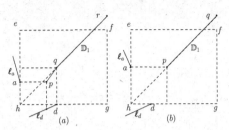

Fig. 5. $\mathbb{D}_1$ for $y(LP(\ell_a)) \geq y(RP(\ell_a))$ and $x(TP(\ell_d)) < x(BP(\ell_d))$

Fig. 6. $\mathbb{D}_1$ for $y(LP(\ell_a)) \geq y(RP(\ell_a))$ and $x(TP(\ell_d)) \geq x(BP(\ell_d))$

We denote the other end point of ℓ_a, ℓ_b, ℓ_c and ℓ_d by a', b', c' and d', respectively. The four points a, b, c, d define an axis-parallel rectangle $\mathcal{R} = \Box efgh$ of minimum size that hits all the members of $\mathcal{L}$ (as per Definition 2), where $a \in \overline{he}$, $b \in \overline{ef}$, $c \in \overline{fg}$ and $d \in \overline{gh}$ (see Fig. 4). We use $L = |x(c) - x(a)|$ and $W = |y(b) - y(d)|$ as the length and width of the rectangle $\mathcal{R}$, and assume $L \geq W$. Let $\mathcal{S}_1$ and $\mathcal{S}_2$ be the two axis-parallel congruent squares that hit the given line segments $\mathcal{L}$ optimally, where $\mathcal{S}_1$ lies to the left of $\mathcal{S}_2$.

Observation 3. *(a) The left side of $\mathcal{S}_1$ (resp. right side of $\mathcal{S}_2$) must not lie to the right of (resp. left of) the point a (resp. c), and (b) the top side (resp. bottom side) of both $\mathcal{S}_1$ and $\mathcal{S}_2$ cannot lie below (resp. above) the point b (resp. d).*

For the **LHIT** problem, we say $\mathcal{S}_1$ and $\mathcal{S}_2$ are in **Configuration 1**, if $\mathcal{S}_1$ hits both ℓ_a and ℓ_d, and $\mathcal{S}_2$ hits both ℓ_b and ℓ_c. Similarly, $\mathcal{S}_1$ and $\mathcal{S}_2$ are said to be in **Configuration 2**, if $\mathcal{S}_1$ hits both ℓ_a and ℓ_b, and $\mathcal{S}_2$ hits both ℓ_c and ℓ_d.

Without loss of generality, we assume that $\mathcal{S}_1$ and $\mathcal{S}_2$ are in **Configuration 1**. We compute the reference (poly) line $\mathbb{D}_1$ (resp. $\mathbb{D}_2$) on which the top-right corner of $\mathcal{S}_1$ (resp. bottom-left corner of $\mathcal{S}_2$) will lie. Let $\mathbb{T}_1$ (resp. $\mathbb{T}_2$) be the line passing through h (resp. f) with slope 1. Our algorithm consists of the following phases:

1. Computation of the reference lines $\mathbb{D}_1$ and $\mathbb{D}_2$.
2. Computation of event points for the top-right (resp. bottom-left) corner of $\mathcal{S}_{.1}$ (resp. $\mathcal{S}_2$) on $\mathbb{D}_1$ (resp. $\mathbb{D}_2$).
3. Searching for pair $(\mathcal{S}_1, \mathcal{S}_2)$ that hit all the line segments in $\mathcal{L}$ and $\max(size(\mathcal{S}_1), size(\mathcal{S}_2))$ is minimized.

Computation of the reference lines $\mathbb{D}_1$ and $\mathbb{D}_2$: The reference line $\mathbb{D}_1$ is computed based on the following four possible orientations of ℓ_a and ℓ_d

(i) $y(LP(\ell_a)) \geq y(RP(\ell_a))$ and $x(TP(\ell_d)) < x(BP(\ell_d))$: Here $\mathbb{D}_1$ is the segment $\overline{pq}$ on $\mathbb{T}_1$ where p is determined (i) by its x-coordinate i.e. $x(p) = x(d)$, if $|ha| < |hd|$ (see Fig. 5(a)), (ii) by its y-coordinate i.e. $y(p) = y(a)$, if $|ha| \geq |hd|$ (see Fig. 5(b)). The point q on $\mathbb{T}_1$ satisfy $x(q) = x(f)$.

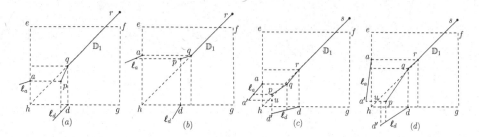

Fig. 7. $\mathbb{D}_1$ for $y(LP(\ell_a)) < y(RP(\ell_a))$ and $x(TP(\ell_d)) > x(BP(\ell_d))$

(ii) $y(LP(\ell_a)) \geq y(RP(\ell_a))$ and $x(TP(\ell_d)) \geq x(BP(\ell_d))$: Here,
if $|ha| < |hd|$ (see Fig. 6(a)), then the reference line $\mathbb{D}_1$ is a polyline $\overline{pqr}$,
where (i) $y(p) = y(a)$ and $x(p)$ satisfies $|x(p) - x(a)| =$ vertical distance of
p from the line segment ℓ_d, (ii) the point q lies on $\mathbb{T}_1$ satisfying $x(q) = x(d)$
and (iii) the point r lies on $\mathbb{T}_1$ satisfying $x(r) = x(f)$.
If $|ha| \geq |hd|$ (see Fig. 6(b)), then the reference line $\mathbb{D}_1$ is a line segment $\overline{pq}$,
where p, q lies on $\mathbb{T}_1$, and p satisfies $y(p) = y(a)$ and q satisfies $x(q) = x(f)$.
(iii) $y(LP(\ell_a)) < y(RP(\ell_a))$ and $x(TP(\ell_d)) \leq x(BP(\ell_d))$: This case is similar
to case (ii), and we can compute the respective reference lines.
(iv) $y(LP(\ell_a)) < y(RP(\ell_a))$ and $x(TP(\ell_d)) > x(BP(\ell_d))$: There are two
possible subcases:
(A) If ℓ_a and ℓ_d are parallel or intersect (after extension) at a point to the
right of $\overline{he}$ (Fig. 7(a,b)), then the reference line $\mathbb{D}_1$ is a polyline $\overline{pqr}$, where
(a) if $|ha| < |hd|$ (Fig. 7(a)), then (1) $y(p) = y(a)$ and $|x(p) - x(a)| =$ the
vertical distance of p from ℓ_d, (2) the points q and r lie on $\mathbb{T}_1$ satisfying
$x(q) = x(d)$ and $x(r) = x(f)$, **(b) if $|ha| > |hd|$** (Fig. 7(b)), then (1) $x(p) =$
$x(d)$ and $|y(p) - y(d)| =$ the horizontal distance of p from ℓ_a, (2) the points
q and r lie on $\mathbb{T}_1$ satisfying $y(q) = y(a)$ and $x(r) = x(f)$.

(B) If extended ℓ_a and ℓ_d intersect at a point to the left of $\overline{he}$ (Fig. 7(c,d)),
then $\mathbb{D}_1$ is a polyline $\overline{pqrs}$, where
(i) the line segment $\overline{pq}$ is such that for every point $\theta \in \overline{pq}$, the horizontal
distance of θ from ℓ_a and the vertical distance of θ from ℓ_d are same.
(ii) the line segment $\overline{qr}$ is such that for every point $\theta \in \overline{qr}$, we have
if $|ha| < |hd|$ then $|x(\theta) - x(a)| =$ vertical distance of θ from ℓ_d (Fig. 7(c)),
else $|y(\theta) - x(d)| =$ horizontal distance of θ from ℓ_a, (Fig. 7(d))
(iii) the point s lies on $\mathbb{T}_1$ satisfying $x(s) = x(f)$.

In the same way, we can compute the reference line $\mathbb{D}_2$ based on the four
possible orientations of ℓ_b and ℓ_c. The break points/end points of $\mathbb{D}_2$ will be
referred to as p', q', r', s' depending on the appropriate cases. From now onwards,
we state the position of square $\mathcal{S}_1$ (resp. $\mathcal{S}_2$) in terms of the position of its top-
right corner (resp. bottom-left corner).

Observation 4. *The point $p \in \mathbb{D}_1$ (resp. $p' \in \mathbb{D}_2$) gives the position of minimum sized axis-parallel square $\mathcal{S}_1$ (resp. $\mathcal{S}_2$) that hit ℓ_a and ℓ_d (resp. ℓ_b and ℓ_c).*

Computation of discrete event points on $\mathbb{D}_1$ and $\mathbb{D}_2$: Observe that the line segments in $\mathcal{L}$ that hits the vertical half-line below the point $p \in \mathbb{D}_1$ (resp. above the point $p' \in \mathbb{D}_2$), or the horizontal half-line to the left of the point $p \in \mathbb{D}_1$ (resp. to the right of the point $p' \in \mathbb{D}_2$) will be hit by any square that hits ℓ_a and ℓ_d (resp. ℓ_b and ℓ_c), and these line segments need not contribute any event point on $\mathbb{D}_1$ (resp. $\mathbb{D}_2$). For each of the other segments $\ell_i \in \mathcal{L}$, we create an event point e_i^1 (resp. e_i^2) on $\mathbb{D}_1$ (resp. $\mathbb{D}_2$) as follows:

(i) p is an event point on $\mathbb{D}_1$ and p' is an event point on $\mathbb{D}_2$ (see Observation 4)

(ii) If ℓ_i lies completely above $\mathbb{D}_1$ (resp. $\mathbb{D}_2$), then we compute an event point $e_i^1 = (x_{i_1}, y_{i_1})$ on $\mathbb{D}_1$ (resp. $e_i^2 = (x_{i_2}, y_{i_2})$ on $\mathbb{D}_2$) where $y_{i_1} = y(BP(\ell_i))$ (resp. $x_{i_2} = x(RP(\ell_i))$). (e.g. e_1^1 for ℓ_1 and e_4^2 for ℓ_4 in Fig. 8).

(iii) If ℓ_i lies completely below $\mathbb{D}_1$ (resp. $\mathbb{D}_2$), we compute an event point $e_i^1 = (x_{i_1}, y_{i_1})$ on $\mathbb{D}_1$ (resp. $e_i^2 = (x_{i_2}, y_{i_2})$ on $\mathbb{D}_2$) where $x_{i_1} = x(LP(\ell_i))$ (resp. $y_{i_2} = y(TP(\ell_i))$). (e.g. e_3^1 for ℓ_3 and e_6^2 for ℓ_6 in Fig. 8).

(iv) If ℓ_i intersects with $\mathbb{D}_1$ (resp. $\mathbb{D}_2$) at point p_1 (resp. q_1), then we create the event point e_i^1 on $\mathbb{D}_1$ (resp. e_i^2 on $\mathbb{D}_2$) according to the following rule:

  (a) If the $x(BP(\ell_i)) > x(p_1)$ (resp. $x(TP(\ell_i)) < x(q_1)$), then we take p_1 (resp. q_1) as the event point e_1^i (resp. e_2^i). (e.g. e_4^1 for ℓ_4 in Fig. 8).

  (b) If $x(BP(\ell_i)) < x(p_1)$ then if $BP(\ell_i)$ lies below $\mathbb{D}_1$ then we consider the point of intersection by $\mathbb{D}_1$ with the vertical line passing through the $BP(\ell_i)$ as the event point e_i^1 (see e_2^1 for ℓ_2 in Fig. 8), and if $BP(\ell_i)$ lies above $\mathbb{D}_1$ then we consider the point of intersection $\mathbb{D}_1$ with the horizontal line passing through $BP(\ell_i)$ as the event point e_i^1 (see e_5^1 for ℓ_5 in Fig. 8).

  (c) If $x(TP(\ell_i)) > x(q_1)$ then if $TP(\ell_i)$ lies above $\mathbb{D}_2$ then we consider the point of intersection by $\mathbb{D}_2$ with the vertical line passing through $TP(\ell_i)$ as the event point e_i^2, and if $TP(\ell_i)$ lies below $\mathbb{D}_2$ then we consider the point of intersection $\mathbb{D}_2$ with the horizontal line passing through $TP(\ell_i)$ as the event point e_i^2.

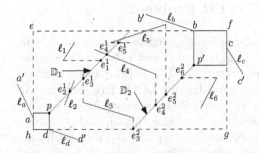

Fig. 8. Event points for **LHIT** problem under **Configuration 1**

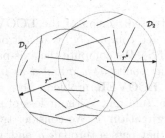

Fig. 9. Covering $\mathcal{L}$ by two disks $\mathcal{D}_1$ & $\mathcal{D}_2$

Observation 5

(i) An event e_i^1 on $\mathbb{D}_1$ shows the position of the top-right corner of the minimum sized square $\mathcal{S}_1$ that hits ℓ_a, ℓ_d and ℓ_i, and an event e_i^2 on $\mathbb{D}_2$ shows the position of the bottom-left corner of the minimum sized square $\mathcal{S}_2$ that hits ℓ_b, ℓ_c and ℓ_i.

(ii) The square $\mathcal{S}_1$ whose top-right corner is at e_i^1 on $\mathbb{D}_1$ hits all those line segments ℓ_j whose corresponding event points e_j^1 on $\mathbb{D}_1$ satisfies $x(h) \leq x(e_j^1) \leq x(e_i^1)$. Similarly, the square $\mathcal{S}_2$ whose bottom-left corner is at e_i^2 on $\mathbb{D}_2$ hits all those line segments ℓ_j whose corresponding event point e_j^2 on $\mathbb{D}_2$ satisfies $x(e_i^1) \leq x(e_j^1) \leq x(f)$.

Let us consider an event point e_i^1, and the corresponding square $\mathcal{S}_1$. We can identify the size and position of the other square $\mathcal{S}_2$ that hit the line segments ℓ_j which were not hit by $\mathcal{S}_1$ in linear time. Observe that, as the size of $\mathcal{S}_1$ increases, the size of $\mathcal{S}_2$ either decreases or remains same. Thus $\max(size(\mathcal{S}_1), size(\mathcal{S}_2))$ is a convex function, and we can compute the minimum value of this function in $O(\log n)$ iterations.

We initially set $\alpha = 1$ and $\beta = n$. In each iteration of our *in-place algorithm*, we compute $\mu = \lfloor \frac{\alpha+\beta}{2} \rfloor$, and compute the μ-th smallest element e^* among $\{e_i^1, i = 1, 2, \ldots, n\}$, and define $\mathcal{S}_1$ in $O(n)$ time [2]. Next, in a linear pass, we compute $\mathcal{S}_2$ for hitting the line segments ℓ_j that are not hit by $\mathcal{S}_1$. If $size(\mathcal{S}_1) < size(\mathcal{S}_2)$ then we set $\alpha = \mu$, otherwise we set $\beta = \mu$ to execute the next iteration. If in two consecutive iterations we get the same μ, the process terminates.

Similarly, we can determine the optimal size of the congruent squares $\mathcal{S}_1$ and $\mathcal{S}_2$ in **Configuration 2**. Finally we consider that configuration for which the size of the congruent squares is minimized. Thus we get the following result:

Theorem 2. *The* **LHIT** *problem can be solved optimally in* $O(n \log n)$ *time using* $O(1)$ *extra work-space.*

4   Restricted Version of LCOVER Problem

In restricted version of the **LCOVER** problem, each line segment in $\mathcal{L}$ is to be covered completely by atleast one of the two congruent axis-parallel squares $\mathcal{S}_1$ and $\mathcal{S}_2$. We compute the axis-parallel rectangle $\mathcal{R} = \Box efgh$ passing through the four points a, b, c and d as in our algorithm for **LCOVER** problem. As in the **LCOVER** problem, here also we have two possible configurations for optimal solution. Without loss of generality, we assume that $\mathcal{S}_1$ and $\mathcal{S}_2$ satisfy **Configuration 1**. We consider two reference lines $\mathbb{D}_1$ and $\mathbb{D}_2$, each with unit slope that passes through h and f, respectively. These reference lines $\mathbb{D}_1$ and $\mathbb{D}_2$ are the locus of the top-right corner of $\mathcal{S}_1$ and bottom-left corner of $\mathcal{S}_2$, respectively. For each line segment ℓ_i, we create an event point $e_i^1 = (x_{i_1}, y_{i_1})$ on $\mathbb{D}_1$ (resp. $e_i^2 = (x_{i_2}, y_{i_2})$ on $\mathbb{D}_2$) as follows:

(i) If ℓ_i lies completely above $\mathbb{D}_1$ (resp. $\mathbb{D}_2$), then the event point e_i^1 on $\mathbb{D}_1$ (resp. e_i^2 on $\mathbb{D}_2$) will satisfy $y_{i_1} = y(TP(\ell_i))$ (resp. $x_{i_2} = x(LP(\ell_i))$).

(ii) If ℓ_i lies completely below $\mathbb{D}_1$ (resp. $\mathbb{D}_2$) then the event point e_i^1 on $\mathbb{D}_1$ (resp. e_i^2 on $\mathbb{D}_2$) will satisfy $x_{i_1} = x(RP(\ell_i))$ (resp. $y_{i_2} = y(BP(\ell_i))$).

(iii) If ℓ_i intersects with $\mathbb{D}_1$ then we create the event point e_i^1 on $\mathbb{D}_1$ as follows: Let the horizontal line through $TP(\ell_i)$ intersect with $\mathbb{D}_1$ at point p, and the vertical line through $BP(\ell_i)$ intersect with $\mathbb{D}_1$ at point q. If $x(p) > x(q)$, then we take p (else q) as the event point on $\mathbb{D}_1$.

(iv) If ℓ_i intersects with $\mathbb{D}_2$, then we create the event point e_i^2 on $\mathbb{D}_2$ as follows: Let the vertical line through $BP(\ell_i)$ intersect with $\mathbb{D}_2$ at point p, and the horizontal line through $TP(\ell_i)$ intersect with $\mathbb{D}_2$ at point q. If $x(p) > x(q)$, then we take q (else p) as the event point on $\mathbb{D}_2$.

Observation similar to Observation 5 in LHIT problem also holds for this problem where $\mathcal{S}_1$ and $\mathcal{S}_2$ cover $\mathcal{L}$ with restriction. Thus, here we can follow the same technique as in LHIT problem to obtain the following result:

Theorem 3. *The restricted version of* **LCOVER** *problem can be solved optimally in $O(n \log n)$ time using $O(1)$ extra work-space.*

5   Covering/Hitting Line Segments by Two Congruent Disks

In this section, we consider problems related to **LCOVER**, **LHIT** and **restricted LCOVER** problem, called *two center problem*, where the objective is to cover, hit or restricted-cover the given line segments in $\mathcal{L}$ by two congruent disks so that their (common) radius is minimized. Figure 9 demonstrates a covering instance of this *two center problem*. Here, we first compute two axis-parallel squares $\mathcal{S}_1$ and $\mathcal{S}_2$ whose union covers/ hits all the members of $\mathcal{L}$ optimally as described in the previous section. Then we report the circum-circles $\mathcal{D}_1$ and $\mathcal{D}_2$ of $\mathcal{S}_1$ and $\mathcal{S}_2$ respectively as an approximate solution of the *two center problem*.

Lemma 4. *A lower bound for the optimal radius of two center problem for $\mathcal{L}$ is the radius r' of in-circle of the two congruent squares $\mathcal{S}_1$ and $\mathcal{S}_2$ of minimum size that cover/ hit/ restricted-cover $\mathcal{L}$; i.e. $r' \leq r^*$.*

The radius r of the circum-circle $\mathcal{D}_1$ and $\mathcal{D}_2$ of the squares $\mathcal{S}_1$ and $\mathcal{S}_2$ is $\sqrt{2}$ times of the radius r' of their in-circles. Lemma 4 says that $r' \leq r^*$. Thus, we have

Theorem 4. *Algorithm* **Two center** *generates a $\sqrt{2}$ approximation result for* LCOVER, LHIT *and* restricted LCOVER *problems for the line segments in $\mathcal{L}$.*

References

1. Bhattacharya, B., Das, S., Kameda, T., Sinha Mahapatra, P.R., Song, Z.: Optimizing squares covering a set of points. In: Zhang, Z., Wu, L., Xu, W., Du, D.-Z. (eds.) COCOA 2014. LNCS, vol. 8881, pp. 37–52. Springer, Cham (2014). doi:10.1007/978-3-319-12691-3_4

2. Carlsson, S., Sundström, M.: Linear-time in-place selection in less than $3n$ comparisons. In: Staples, J., Eades, P., Katoh, N., Moffat, A. (eds.) ISAAC 1995. LNCS, vol. 1004, pp. 244–253. Springer, Heidelberg (1995). doi:10.1007/BFb0015429

3. Chan, T.M.: More planar two-center algorithms. Comput. Geom. Theory Appl. **13**(3), 189–198 (1999)

4. Drezner, Z.: On the rectangular p-center problem. Naval Res. Log. **34**(2), 229–234 (1987)

5. Eppstein, D.: Faster construction of planar two-centers. In: 8th ACM-SIAM Symposium On Discrete Algorithms (SODA), pp. 131–138 (1997)

6. Hershberger, J.: A fast algorithm for the two-Center decision Problem. Inf. Process. Lett. (Elsevier) **47**(1), 23–29 (1993)

7. Hoffmann, M.: A simple linear algorithm for computing rectilinear 3-centers. Comput. Geom. **31**(3), 150–165 (2005)

8. Jaromczyk, J.W., Kowaluk, M.: An efficient algorithm for the euclidean two-center problem. In: Mehlhorn, K. (ed.) Symposium on Computational Geometry, pp. 303–311. ACM (1994)

9. Katz, M.J., Kedem, K., Segal, M.: Discrete rectilinear 2-center problems. Comput. Geom. **15**(4), 203–214 (2000)

10. Kim, S.S., Bae, S.W., Ahn, H.K.: Covering a point set by two disjoint rectangles. Int. J. Comput. Geometry Appl. **21**(3), 313–330 (2011)

11. Kim, S.K., Shin, C.-S.: Efficient algorithms for two-center problems for a convex polygon. In: Du, D.-Z.-Z., Eades, P., Estivill-Castro, V., Lin, X., Sharma, A. (eds.) COCOON 2000. LNCS, vol. 1858, pp. 299–309. Springer, Heidelberg (2000). doi:10.1007/3-540-44968-X_30

12. Marchetti-Spaccamela, A.: The p center problem in the plane is NP complete. In: Proceedings of the 19th Allerton Conference on Communication, Control and Computing, pp. 31–40 (1981)

13. Sharir, M.: A near-linear algorithm for the planar 2-center problem. Discrete Comput. Geom. **18**(2), 125–134 (1997)

Complexity and Algorithms for Finding a Subset of Vectors with the Longest Sum

Vladimir Shenmaier[✉]

Sobolev Institute of Mathematics, 4 Koptyug Avenue,
630090 Novosibirsk, Russia
shenmaier@mail.ru

Abstract. The problem is, given a set of n vectors in a d-dimensional normed space, find a subset with the largest length of the sum vector. We prove that the problem is APX-hard for any ℓ_p norm, $p \in [1, \infty)$. For the general problem, we propose an algorithm with running time $O(n^{d-1}(d + \log n))$, improving previously known algorithms. In particular, the two-dimensional problem can be solved in a nearly linear time. We also present an improved algorithm for the cardinality-constrained version of the problem.

Keywords: Computational geometry · Vector sum · Normed space · APX-hardness · Optimal solution

1 Introduction

We consider the following problem:

Longest vector sum (LVS). Given a set X of n vectors in a normed space $(\mathbb{R}^d, \|.\|)$, find a subset $S \subseteq X$ with the maximum value of

$$f(S) = \left\| \sum_{x \in S} x \right\|.$$

The variation where the subset S is required to have a given cardinality $k \in [1, n]$ will be called the *longest k-vector sum* problem (Lk-VS).

These problems have applications in such fields as geophysics, radiolocation, signal filtering. As an example, suppose that we are given a set of measurements of the direction to some signal source (an acoustic or radio wave source, the magnetic field), each measurement is a unit vector in $\mathbb{R}^3$. The measurements contain an error and some of them may be received from outliers (other random sources, reflected waves). How to determine the true direction? A reasonable answer is to find a subset of measurements with the maximum aggregate correlation, which equals to the Euclidean length of the sum vector.

The cardinality-constrained version also arises in the context of searching a quasiperiodically repeating fragment in a noisy number sequence [5] and detecting the active and passive states of an observed object [3].

Y. Cao and J. Chen (Eds.): COCOON 2017, LNCS 10392, pp. 469–480, 2017.
DOI: 10.1007/978-3-319-62389-4_39

Related Work. The LVS and Lk-VS problems are strongly NP-hard if the Euclidean norm is used [1,13] and polynomially solvable in the case of the ℓ_∞ norm and other polyhedral norms with a polynomial number of facets [2]. For an arbitrary norm, both problems become polynomial if the dimension of space is fixed [8]. The best known algorithm for the LVS finds an optimal solution in time $O(n^d)$ [11][1]. The Lk-VS problem can be solved in time $O(n^{4d})$ [8], but the algorithm from [14] solves the Euclidean problem in time $O(n^{d+1})$.

Our Contributions. First, we prove a stronger hardness result: for any ℓ_p norm, $p \in [1, \infty)$, the LVS and Lk-VS problems are NP-hard to approximate within a factor better than $\alpha^{1/p}$, where α is the inapproximability bound for Max-Cut, $\alpha = 16/17$. This holds even if coordinates of the input vectors belong to $\{0, \pm 1\}$.

On the other hand, for an arbitrary norm, we propose an algorithm solving the LVS problem in time $O(n^{d-1}(d + \log n))$ if $d \geq 2$. The algorithm is based on counting regions in hyperplane arrangements in $\mathbb{R}^d$. In particular, we have a simple nearly linear-time algorithm for the two-dimensional case. Finally, for the general Lk-VS problem, we give a Voronoi-based algorithm with running time $O(dn^{d+1})$, generalizing the result of [14].

2   The Hardness Result

In this section, we prove that the longest vector sum problem is APX-hard. The proof is done by a simple PTAS-reduction from the following problem:

Max-Cut. Given an undirected graph $G = (V, E)$, find a subset of vertices $S \subseteq V$ for which the value of $cut(S)$ is maximized, where $cut(S)$ is the number of edges in E with one endpoint in S and the other in $V \setminus S$.

The complexity of this problem can be described by the following facts:

Fact 1. [6] *Max-Cut is NP-hard to approximate within a factor better than* $16/17$.

Fact 2. [9] *Assuming the Unique Games Conjecture, there is no polynomial-time algorithm which approximates Max-Cut within a factor of* 0.879.

Consider the LVS problem for the ℓ_p norm, $p \in [1, \infty)$. In this case,

$$\|x\| = \Big(\sum_{i=1}^{d} |x(i)|^p \Big)^{1/p},$$

where $x(i)$ is the ith coordinate of x. Denote this special case by LVS$_p$.

Theorem 1. *For any $p \in [1, \infty)$ and $\alpha \in (0, 1]$, if there exists an α-approximation polynomial-time algorithm for the LVS$_p$, then there exists an α^p-approximation polynomial-time algorithm for Max-Cut.*

---

[1] According to [12], the running time of this algorithm is as pointed above and not $O(n^{d-1})$ as is asserted in [11].

Proof. Given a graph $G = (V, E)$, we construct a set X, an instance of the LVS, in the following way. Let A be the incidence matrix of G, i.e., the $|V| \times |E|$-matrix whose (v, e)-entry is 1 if vertex v is incident to edge e and 0 otherwise. Then, for each $e \in E$, the eth column of A contains exactly two 1s, at the rows indexed by the endpoints of e. Denote one of these endpoints by $a(e)$, the other by $b(e)$, and replace 1 at the row $b(e)$ with -1. Let x_v, $v \in V$, be the vth row of the matrix obtained in this way and $X = \{x_v | v \in V\}$.

Given a subset of vertices $S \subseteq V$, compute the vector $\xi = \sum_{v \in S} x_v$. For each $e \in E$, we have

$$\xi(e) = \begin{cases} 1 & \text{if } a(e) \in S \text{ and } b(e) \notin S, \\ -1 & \text{if } a(e) \notin S \text{ and } b(e) \in S, \\ 0 & \text{otherwise.} \end{cases}$$

So

$$\|\xi\|^p = \sum_{e \in E} |\xi(e)|^p = cut(S).$$

But $\|\xi\| = f(X_S)$, where $X_S = \{x_v | v \in S\}$. It follows that X_S is an α-approximate solution of the LVS problem on the input set X if and only if S is an α^p-approximate solution of the Max-Cut problem on the graph G. The theorem is proved. □

Corollary 1. *For any $p \in [1, \infty)$, the LVS_p problem is NP-hard to approximate within a factor better than $(16/17)^{1/p}$. Assuming the Unique Games Conjecture, there is no polynomial-time algorithm which approximates this problem within a factor of $0.879^{1/p}$. Both statements hold even if $X \subseteq \{0, \pm 1\}^d$.*

Clearly, all inapproximability factors for the LVS problem are also valid for its cardinality-constrained version. Moreover, by using the above idea, we can get a similar reduction to the Lk-VS from the Maximum Bisection problem which yields an inapproximability factor of $(15/16)^{1/p}$ under the assumption $NP \not\subseteq \cap_{\epsilon > 0} DTIME(2^{n^\epsilon})$ [7].

3   Algorithms

For a fixed space dimension, both LVS and Lk-VS problems are polynomially solvable. However, the computational time of $O(n^d)$ for the first of them and that of $O(n^{4d})$ for the second seem to be excessive. In this section, we show how to improve these time bounds to $O(n^{d-1} \log n)$ and $O(n^{d+1})$ respectively.

3.1   Finding an Optimal Solution for the LVS Problem

Make some notations. Given a set $S \subseteq X$, put $\Sigma S = \sum_{x \in S} x$. For any vectors $x, y \in \mathbb{R}^d$, denote by xy their dot product. For any $v \in \mathbb{R}^d$, define the set $S(v, X) = \{x \in X | xv > 0\}$ and let $D(X) = \{v \in \mathbb{R}^d | xv \neq 0 \text{ for all } x \in X \setminus \{0\}\}$, where $\mathbf{0}$ is the origin.

Lemma 1. (Folklore) *Let P be a convex polytope in $\mathbb{R}^d$ and $u \in P$. Then u is a vertex of P if and only if there exists a vector $v \in \mathbb{R}^d$ for which u is the unique maximum point of xv over $x \in P$.*

Proof. By a definition, u is a vertex of P if there exists a supporting hyperplane h such that u is the unique intersection of h and P (e.g., see [10]). Clearly, the outer normal of h is the required vector v.   □

Lemma 2. *Let $\mathcal{F}(X) = \{\Sigma S \,|\, S \subseteq X\}$, $Z(X)$ be the convex hull of $\mathcal{F}(X)$, and $u \in \mathbb{R}^d$. Then u is a vertex of the polytope $Z(X)$ if and only if $u = \Sigma S(v, X)$ for some $v \in D(X)$.*

Proof. Suppose that u is a vertex of $Z(X)$. Then $u \in \mathcal{F}(X)$, i.e., $u = \Sigma U$ for some $U \subseteq X$. By Lemma 1, there exists a vector $v \in \mathbb{R}^d$ for which u is the unique maximum point of zv over $z \in Z(X)$. Then $v \in D(X)$ since otherwise $xv = 0$ for some $x \in X \setminus \{\mathbf{0}\}$ that gives another maximum point, either $u - x$ if $x \in U$ or $u + x$ if $x \notin U$. On the other hand, by the construction of $Z(X)$, we have $(\Sigma U) v > (\Sigma S) v$ for all $S \subseteq X$ such that $S \neq U$. Since $xv \neq 0$ for all $x \in X$, it follows that U consists exactly of those $x \in X$ for which $xv > 0$. This means that $U = S(v, X)$.

Suppose that $u = \Sigma S(v, X)$ for some $v \in D(X)$. By the construction of the sets $S(v, X)$ and $D(X)$, we have $(\Sigma S(v, X)) v > (\Sigma S) v$ for all $S \subseteq X$ such that $S \neq S(v, X)$. Therefore, u is the unique maximum point of zv over $z \in Z(X)$. By Lemma 1, it follows that u is a vertex of $Z(X)$. Lemma 2 is proved.   □

Theorem 2. *There exists an algorithm which computes all the vectors from the set $\mathcal{S}(X) = \{\Sigma S(v, X) \,|\, v \in D(X)\}$ in time $O(n^{d-1}(d + \log n))$ if $d \geq 2$. For any output vector $u \in \mathcal{S}(X)$, the algorithm allows to restore a set $S \subseteq X$ such that $\Sigma S = u$ in time $O(dn)$.*

Theorem 2 will be proved in Sect. 4.

Theorem 3. *There is an algorithm which finds an optimal solution of the LVS problem in time $O(n^{d-1}(d + \log n))$ if $d \geq 2$.*

Proof. Since any norm is a convex function, the maximum of $\|.\|$ over the polytope $Z(X)$ is reached at its vertices. On the other hand, all these vertices belong to the set $\mathcal{F}(X)$ and $\mathcal{F}(X) \subset Z(X)$. Therefore, we can find an optimal solution of the LVS problem among all the sets $S \subseteq X$ for which ΣS is a vertex of $Z(X)$. By Lemma 2, it is equivalent to finding a vector

$$u^* = \arg \max_{u \in \mathcal{S}(X)} \|u\|$$

and a set $S \subseteq X$ such that $\Sigma S = u^*$. By Theorem 2, we can do it in time $O(n^{d-1}(d + \log n))$. Theorem 3 is proved.   □

Remark 1. It is easy to see that Theorem 3 is also true for the more general problem of maximizing any convex function on the set $\mathcal{F}(X)$.

3.2   Finding Optimal Solutions for the Lk-VS Problem

Given a point $v \in \mathbb{R}^d$, define a set $S_k(v, X)$ of k vectors from X with the maximum dot products with v and let $N_k(v, X)$ be a set of k vectors from X nearest to v (if such sets are not unique, break the ties arbitrarily). Denote by $D_k(X)$ the set of vectors $v \in \mathbb{R}^d$ for which $S_k(v, X)$ is uniquely defined and let $E_k(X)$ be the similar set for $N_k(., X)$.

Lemma 3. *Let $\mathcal{F}_k(X) = \{\Sigma S \,|\, S \subseteq X, |S| = k\}$, $Z_k(X)$ be the convex hull of $\mathcal{F}_k(X)$, and $u \in \mathbb{R}^d$. Then the following holds:*

(a) u is a vertex of the polytope $Z_k(X)$ if and only if $u = \Sigma S_k(v, X)$ for some $v \in D_k(X)$;

(b) if u is a vertex of $Z_k(X)$, then $u = \Sigma N_k(v, X)$ for some $v \in E_k(X)$.

Proof. (a) Suppose that u is a vertex of $Z_k(X)$. Then $u \in \mathcal{F}_k(X)$, i.e., $u = \Sigma U$ for some k-element set $U \subseteq X$. By Lemma 1, there exists a vector $v \in \mathbb{R}^d$ for which u is the unique maximum point of zv over $z \in Z_k(X)$. By the construction of $Z_k(X)$, we have $(\Sigma U) v > (\Sigma S) v$ for all the k-element sets $S \subseteq X$ such that $S \neq U$. It follows that U is the only set of k vectors from X with the maximum dot products with v. This means that $v \in D_k(X)$ and $U = S_k(v, X)$.

Suppose that $u = \Sigma S_k(v, X)$ for some $v \in D_k(X)$. By the construction of the sets $S_k(v, X)$ and $D_k(X)$, we have $(\Sigma S_k(v, X)) v > (\Sigma S) v$ for all the k-element sets $S \subseteq X$ such that $S \neq S_k(v, X)$. Therefore, u is the unique maximum point of zv over $z \in Z_k(X)$. By Lemma 1, it follows that u is a vertex of $Z_k(X)$.

(b) According to (a), if u is a vertex of $Z_k(X)$, then $u = \Sigma S_k(v, X)$ for some $v \in D_k(X)$. By Lemma 2 of [14], there exists a vector $v' \in \mathbb{R}^d$ such that, for all $x, y \in X$, if $xv > yv$, then $\|x - v'\| < \|y - v'\|$. It follows that $u = \Sigma N_k(v', X)$ and $v' \in E_k(X)$. Lemma 3 is proved. □

Theorem 4. *There exists an algorithm which computes all the sets $\mathcal{N}_k(X) = \{\Sigma N_k(v, X) \,|\, v \in E_k(X)\}$, $k = 1, \ldots, n$, in time $O(dn^{d+1})$. For any output vector $u \in \mathcal{N}_k(X)$, the algorithm allows to restore a k-element set $S \subseteq X$ such that $\Sigma S = u$ in time $O(dn)$.*

Theorem 4 will be proved in Sect. 5.

Theorem 5. *There is an algorithm which finds optimal solutions of the Lk-VS problem for all $k = 1, \ldots, n$ in time $O(dn^{d+1})$.*

Proof. Since any norm is a convex function, the maximum of $\|.\|$ over the polytope $Z_k(X)$ is reached at its vertices. On the other hand, all these vertices belong to the set $\mathcal{F}_k(X)$ and $\mathcal{F}_k(X) \subset Z_k(X)$. Therefore, we can find an optimal solution of the Lk-VS problem among all the k-element sets $S \subseteq X$ for which ΣS is a vertex of $Z_k(X)$. By Lemma 3(b), it is sufficient to find a vector

$$u^* = \arg \max_{u \in \mathcal{N}_k(X)} \|u\|$$

and a k-element set $S \subseteq X$ such that $\Sigma S = u^*$. By Theorem 4, we can do it in time $O(dn^{d+1})$. Theorem 5 is proved. □

Remark 2. It is easy to see that Theorem 5 is also true for the more general problem of maximizing any convex function on the set $\mathcal{F}_k(X)$.

4   Proof of Theorem 2

In this section, we present an algorithm which computes all the vectors from the set $\mathcal{S}(X)$ and, for any output vector u, allows to restore a set $S \subseteq X$ such that $\Sigma S = u$. The idea of this algorithm is based on the following concept.

Definition 1. *The hyperplane arrangement for the set X in space $\mathbb{R}^d$ is the collection $A(X)$ of all the open d-dimensional regions in the partition of this space by the hyperplanes $H(x) = \{v \in \mathbb{R}^d \mid xv = 0\}$, $x \in X \setminus \{\mathbf{0}\}$.*

It is easy to see that, for all vectors v from the same region $C \in A(X)$, the sets $S(v, X)$ coincide. We will denote these sets by $S(C, X)$. On the other hand, the set $D(X)$ from the definition of $\mathcal{S}(X)$ is exactly the union of all the regions in the arrangement $A(X)$. Thus, we have

$$\mathcal{S}(X) = \{\Sigma S(C, X) \mid C \in A(X)\} = \{\Sigma S(v, X) \mid v \in \mathcal{M}\},$$

where $\mathcal{M}$ is any collection of points in $\mathbb{R}^d$ such that $\mathcal{M} \cap C \neq \emptyset$ for each region $C \in A(X)$. Such a collection will be referred to as a *member collection* for the arrangement $A(X)$.

4.1   Algorithm for the Plane

First, describe a quick way to construct the set $\mathcal{S}(X)$ when the vectors of X lie in some two-dimensional subspace G of space $\mathbb{R}^d$. This case is one of the recursion base cases for the future high-dimensional algorithm.

Let E be an orthonormal basis of the linear span of X, which can be constructed by using the rules of linear algebra. If $|E| = 1$, the vectors of X are collinear and $\mathcal{S}(X) = \{\Sigma S(\pm e_1, X)\}$, where e_1 is the only vector of E. Henceforth, we assume that E consists of two vectors, say, e_1 and e_2. We will measure the angles between vectors in G in the direction from e_1 to e_2. In particular, the angle from e_1 to e_2 is $\pi/2$, the angle from e_2 to e_1 is $-\pi/2$.

Clearly, the regions of the two-dimensional arrangement $A(X)$ in the plane G are bounded by the rays starting at the origin and orthogonal to the non-zero vectors from X. For each $x \in X \setminus \{\mathbf{0}\}$, we have two such rays. These pass through the vectors $l(x)$ and $r(x)$ obtained by rotating x through angles of $\pi/2$ and $-\pi/2$ respectively. Denote these rays by $[\mathbf{0}, l(x))$ and $[\mathbf{0}, r(x))$. Put $L(\lambda) = \{x \in X \mid x \neq \mathbf{0}, l(x) \in \lambda\}$ and $R(\lambda) = \{x \in X \mid x \neq \mathbf{0}, r(x) \in \lambda\}$.

Lemma 4. *Suppose that C_1, C_2 are adjacent regions in $A(X)$, λ is a ray between them, and C_2 lies in the direction of negative angles from λ. Then*

$$S(C_2, X) = (S(C_1, X) \cup L(\lambda)) \setminus R(\lambda). \tag{1}$$

Proof. Let $v_1 \in C_1$, $v_2 \in C_2$, and $x \in X \setminus \{\mathbf{0}\}$. Then we have three following cases. If $l(x) \in \lambda$, then $v_2 x > 0$ and $v_1 x < 0$. If $r(x) \in \lambda$, then $v_2 x < 0$ and $v_1 x > 0$. Finally, if $l(x) \notin \lambda$ and $r(x) \notin \lambda$, then the sign of $v_2 x$ equals to that of $v_1 x$. It follows that $S(v_2, X) = (S(v_1, X) \cup L(\lambda)) \setminus R(\lambda)$ which is equivalent to (1). The lemma is proved. □

Lemma 4 allows to enumerate all the sets $S(C, X)$, $C \in A(X)$, by sequential turning around the origin from ray to ray and updating the set $S(C, X)$ according to Eq. (1). Therefore, all the vectors from $\mathcal{S}(X)$ can be computed with the following algorithm:

Algorithm 1

Step 1. Construct an orthonormal basis E of the linear span of X.

Step 2. If $|E| = 1$, return the set $\{\Sigma S(\pm e_1, X)\}$.

Step 3. For each $x \in X \setminus \{\mathbf{0}\}$, construct the vectors $l(x)$ and $r(x)$.

Step 4. Sort the rays $[\mathbf{0}, l(x))$, $[\mathbf{0}, r(x))$, $x \in X \setminus \{\mathbf{0}\}$, by decreasing the angle from e_1; remove coincided; denote the resulting sequence of rays by $\lambda_1, \ldots, \lambda_t$; determine the sets $L(\lambda_i)$, $R(\lambda_i)$, $i = 1, \ldots, t$.

Step 5. Calculate the vector $u_1 = \Sigma S(v, X)$, where v is any non-zero vector on the bisector of the rays λ_1 and λ_2; for each $i = 2, \ldots, t$, compute the vector $u_i = u_{i-1} + \Sigma L(\lambda_i) - \Sigma R(\lambda_i)$.

Lemma 5. *Algorithm 1 computes all the vectors from the set $\mathcal{S}(X)$ in time $O(n(d + \log n))$. For each output vector u, the algorithm allows to restore a set $S \subseteq X$ such that $\Sigma S = u$ in time $O(dn)$.*

Proof. Lemma 4 yields that $\mathcal{S}(X) = \{u_1, \ldots, u_t\}$. Estimate the running time of the algorithm. Sorting the rays $[\mathbf{0}, p)$, where $p = l(x), r(x)$, $x \in X \setminus \{\mathbf{0}\}$, takes time $O(dn + n \log n)$ since it is reduced to computing and sorting the angles from e_1 to the corresponding vectors p. The other operations can be realized in time $O(dn)$. For each u_i, the set S for which $\Sigma S = u_i$ equals to the set $S(v, X)$, where v is any non-zero vector on the bisector of λ_i and λ_{i+1} (here $\lambda_{t+1} = \lambda_1$). Therefore, this set can be computed in a linear time. The lemma is proved. □

4.2   Algorithm for Higher Dimensions

Suppose that a space G is either $\mathbb{R}^d$ or the intersection of some hyperplanes $H(x)$, $x \in X \setminus \{\mathbf{0}\}$. Let $x_1, \ldots, x_n$ be the vectors of the input set X and $g_1, \ldots, g_n$ be their orthogonal projections into the space G. Given an integer $m \in [1, n]$, put $X(G, m) = \{g_1, \ldots, g_m\}$ and denote by $A(G, m)$ the arrangement for the set $X(G, m)$ in the space G.

Definition 2. *A member collection $\mathcal{M}$ for the arrangement $A(G, m)$ is called normal if $\mathcal{M} \subset D(X(G, n))$.*

Let $P(G, m)$ be the problem of computing a set of vectors which can be represented as $\{\Sigma S(v, X) \mid v \in \mathcal{M}\}$, where $\mathcal{M}$ is any normal member collection for the arrangement $A(G, m)$.

Lemma 6. *Suppose that the space G is k-dimensional, $k, d \geq 2$, and the set $X(G, m)$ is given explicitly. Then the problem $P(G, m)$ can be solved in time*

$$\text{Time}(k, m) \leq cn(d + \log n) \min\{k, 3\} m^{k-2},$$

where c is a universal constant independent of k, d, m, n.

Proof. We will use induction on k and m.

Base case $k = 2$: Construct an orthonormal basis E of the linear span of $X(G, n)$. If E consists of a single element e_1, the set $\{\Sigma S(\pm e_1, X)\}$ is a solution of the problem $P(G, m)$. Otherwise, perform Steps 3 and 4 of Algorithm 1 to the set $X(G, n)$ and obtain a sequence $\lambda_1, \ldots, \lambda_t$ of the rays bounding the regions of the two-dimensional arrangement $A(G, n)$. For each $i = 1, \ldots, t$, choose any non-zero vector v_i from the bisector of the rays λ_i and λ_{i+1} (where $\lambda_{t+1} = \lambda_1$). Then the vectors $v_1, \ldots, v_t$ form a normal member collection for $A(G, n)$.

It remains to compute the vectors $\Sigma S(v_i, X)$. Note that, for each $i = 1, \ldots, t$ and $j = 1, \ldots, n$, we have $v_i x_j = v_i g_j$ since $v_i \in G$. Then, by the observations similar to Lemma 4, the vectors $\Sigma S(v_i, X)$ can be computed as follows:

Step 5'. Calculate the vector $\Sigma S(v_1, X)$; for each $i = 2, \ldots, t$, compute the vector $\Sigma S(v_i, X)$ as $\Sigma S(v_{i-1}, X) + \Sigma\{x_j | g_j \in L(\lambda_i)\} - \Sigma\{x_j | g_j \in R(\lambda_i)\}$.

Thus, we obtain a solution of the problem $P(G, n)$. Show that, for an arbitrary $m \leq n$, we can obtain a solution of $P(G, m)$ which contains at most $2m$ vectors. Define the set I_m of indices i such that λ_i is one of the rays $[\mathbf{0}, l(x))$, $[\mathbf{0}, r(x))$, $x \in X(G, m) \setminus \{\mathbf{0}\}$. Then $|I_m| \leq 2m$ and the vectors v_i, $i \in I_m$, form a normal member collection for the arrangement $A(G, m)$. Therefore, the set $\{\Sigma S(v_i, X) | i \in I_m\}$ is a desired solution of $P(G, m)$.

Constructing the basis E takes time $O(dn)$. Steps 3 and 4 of Algorithm 1 and also Step 5' can be performed in time $O(n(d + \log n))$. The set I_m can be determined in a linear time. Thus, the problem $P(G, m)$ can be solved in time at most $cn(d + \log n)$, where c is a universal constant.

Base case $m = 1$: Given a vector $v = (v(1), \ldots, v(d))$, let $\alpha(v)$ be the value of the first non-zero coordinate of v and $\beta(v) = \max_i |v(i)|/\alpha(v)$ (put $\alpha(\mathbf{0}) = 0$, $\beta(\mathbf{0}) = 1$). Define the vector $p = (b^{d-1}, b^{d-2}, \ldots, 1)$, where $b = \max_j |\beta(g_j)| + 1$. Then, for each $j = 1, \ldots, n$, the sign of pg_j equals to that of $\alpha(g_j)$. On the other hand, $pg_j = qg_j$, where q is the projection of p into the space G. Therefore, $q \in D(X(G, n))$, so the vectors $\pm q$ form a normal member collection for the arrangement $A(G, 1)$. At the same time, we have $qg_j = qx_j$, which follows that $S(q, X) = \{x_j \mid \alpha(g_j) > 0\}$ and $S(-q, X) = \{x_j \mid \alpha(g_j) < 0\}$. Thus, the problem $P(G, 1)$ can be solved in time $O(dn)$.

Induction step, $k \geq 3$, $m \geq 2$: If $g_m = \mathbf{0}$ or g_m is collinear with one of the non-zero vectors from $X(G, m - 1)$, then the arrangement $A(G, m)$ equals to $A(G, m - 1)$, so the solution of the problem $P(G, m - 1)$ is also a solution for $P(G, m)$. Suppose that $g_m \neq \mathbf{0}$ and g_m is non-collinear with the non-zero vectors from $X(G, m - 1)$.

Denote by H the $(k-1)$-dimensional subspace $G \cap H(g_m)$. Note that the arrangement $A(G, m)$ can be obtained from $A(G, m-1)$ by dividing an each region $C \in A(G, m-1)$ crossed by H into two parts, C^+ and C^-. In the first of them, the dot products with g_m are positive; in the second, these are negative. At the same time, the intersection of the region C with H is some region C_H in the arrangement $A(H, m-1)$.

Consider any solution of the problem $P(H, m-1)$ and let $\Sigma S(v_H, X)$ be a vector from this solution for which $v_H \in C_H$. Put $v^+ = v_H + \varepsilon g_m$ for some sufficiently small $\varepsilon > 0$. Then $v^+ \in C^+$ and, since $v_H \in D(X(H, n))$, we have $v^+ \in D(X(G, n))$ and $S(v^+, X) = S(v_H, X) \cup Dir(g_m)$, where $Dir(v) = \{x_j \mid g_j \neq \mathbf{0}, g_j \text{ is codirected with } v, j = 1, \ldots, n\}$. On the other hand, the sets $S(v_H, X)$ and $Dir(g_m)$ are disjoint since g_m is orthogonal to v_H. Therefore,

$$\Sigma S(v^+, X) = \Sigma S(v_H, X) + \Sigma Dir(g_m). \tag{2}$$

Similarly, for some $v^- \in C^- \cap D(X(G, n))$, we have

$$\Sigma S(v^-, X) = \Sigma S(v_H, X) + \Sigma Dir(-g_m). \tag{3}$$

Then, after we append the vectors of the type $\Sigma S(v^\pm, X)$ to a solution of the problem $P(G, m-1)$, we get that of $P(G, m)$. Thus, we reduce the problem $P(G, m)$ to the problems $P(G, m-1)$ and $P(H, m-1)$. To apply induction, it remains to determine the vectors of the set $X(H, n)$. We can calculate them by using the equation

$$h_j = g_j - z_m(z_m g_j), \tag{4}$$

where h_j is the projection of x_j into the space H, $j = 1, \ldots, n$, $z_m = g_m / \|g_m\|$. Estimate the running time of the resulting algorithm.

Lemma 7. *Let $t(k, m)$ be the number of output vectors in the obtained solution of the problem $P(G, m)$. Then $t(k, m) \leq 2m^{k-1}$.*

Proof. We use induction on k and m. If $k = 2$ or $m = 1$, the obtained solutions are at most $2m$-element. Suppose that $k \geq 3$ and $m \geq 2$. Then $t(k, m) \leq t(k, m-1) + 2t(k-1, m-1)$ which is at most $2(m-1)^{k-1} + 4(m-1)^{k-2}$ by induction hypothesis. Next, for any positive integers a and b, a binomial inequality $(a+1)^b \geq a^b + ba^{b-1}$ holds. Replacing a by $a-1$ yields

$$(a-1)^b \leq a^b - b(a-1)^{b-1}. \tag{5}$$

Therefore, $t(k, m) \leq 2(m^{k-1} - (k-1)(m-1)^{k-2}) + 4(m-1)^{k-2} \leq 2m^{k-1}$. The lemma is proved. $\square$

By the construction of the obtained solution, $\text{Time}(k, m)$ is at most

$$\text{Time}(k, m-1) + \text{Time}(k-1, m-1) + c_1 dn + c_2 d\, t(k-1, m-1), \tag{6}$$

where $c_1 dn$ is a time bound for checking the non-collinearity of g_m, computing two sets $Dir(\pm g_m)$ and their sums, and constructing the input of the problem

$P(H, m-1)$ by using Eq. (4); $c_2 d$ is a time bound for calculating a pair of vectors $\Sigma S(v^{\pm}, X)$ by using Eqs. (2, 3).

By induction hypothesis and Lemma 7, expression (6) does not exceed

$$cn(d + \log n)\left[\min\{k, 3\}(m - 1)^{k-2} + \min\{k - 1, 3\}(m - 1)^{k-3}\right]$$
$$+ c_1 dn + 2c_2 d(m - 1)^{k-2}.$$

By (5), the first term in the square brackets can be estimated as

$$\min\{k, 3\}\left(m^{k-2} - (k - 2)(m - 1)^{k-3}\right).$$

Then, since $-\min\{k, 3\}(k - 2) + \min\{k - 1, 3\} \leq -1$ for $k \geq 3$, we obtain

$$\text{Time}(k, m) \leq cn(d + \log n)\left[\min\{k, 3\}\, m^{k-2} - (m - 1)^{k-3}\right]$$
$$+ c_1 dn + 2c_2 d(m - 1)^{k-2}$$

which is at most $cn(d + \log n)\min\{k, 3\}\, m^{k-2}$ whenever $c \geq c_1 + 2c_2$. Lemma 6 is proved.   □

To compute all the vectors from the set $\mathcal{S}(X)$, it is sufficient to solve the problem $P(\mathbb{R}^d, n)$, which can be performed in time $O(n^{d-1}(d + \log n))$ by Lemma 6. To enable restoring the set S of summands of any output vector u, we equip each vector $\Sigma S(v^{\pm}, X)$ obtained by using Eqs. (2, 3) in the above algorithm with pointers into the vector $\Sigma S(v_H, X)$ and the corresponding set $Dir(\pm g_m)$. This allows to represent the set S as $Dir(y_1) \cup \cdots \cup Dir(y_s) \cup S(v, X)$, where each y_i is a vector of the type $\pm g_m$, $s \leq \min\{d - 2, n - 1\}$, and $S(v, X)$ is a set whose sum is an output vector for one of the base cases, $k = 2$ or $m = 1$. To restore the set $S(v, X)$ in the case $k = 2$, we equip each output vector $\Sigma S(v_i, X)$, $i = 1, \ldots, t$, with the corresponding vector v_i, which is computed explicitly. In the case $m = 1$, we equip each output vector $\Sigma S(\pm q, X)$ with a pointer to the corresponding set $S(\pm q, X)$. Thus, since $s \leq n$, the required set S can be restored in time $O(dn)$. Theorem 2 is proved.

5   Proof of Theorem 4

In this section, we present an algorithm which computes all the sets $\mathcal{N}_k(X)$, $k = 1, \ldots, n$, and, for any output vector $u \in \mathcal{N}_k(X)$, allows to restore a k-element set $S \subseteq X$ such that $\Sigma S = u$. The idea of this algorithm is based on using higher-order Voronoi diagrams.

Definition 3. *The Voronoi cell for a subset $S \subseteq X$ is the set*

$$V(S, X) = \{z \in \mathbb{R}^d \mid \|z - x\| < \|z - y\| \ \text{for all} \ x \in S, y \in X \setminus S\}.$$

Given an integer $k \in [1, n]$, the k-order Voronoi diagram for the set X is the collection $\dot{V}_k(X)$ of all the non-empty Voronoi cells $V(S, X)$, $S \subseteq X$, $|S| = k$.

A Voronoi cell $V(S, X)$ consists of the points of $\mathbb{R}^d$ for which the elements of S are closer than the other elements of X. Clearly, for each cell $C \in V_k(X)$, the k-element subset $S \subseteq X$ such that $C = V(S, X)$ is exactly the set $N_k(z, X)$ for any $z \in C$. We denote this subset by $N_k(C, X)$.

Fact 3. [4] *The total number of cells in all the Voronoi diagrams of order* 1 *to* n *is* $O(n^{d+1})$.

Edelsbrunner et al. [4] proposed an algorithm for computing higher-order Voronoi diagrams. We will denote this algorithm by $\mathcal{V}$. Its output is a data structure which describes the construction (faces and cells) of all $V_k(X)$, $k = 1, \ldots, n$. Particularly, each cell $C \in V_k(X)$, $k > 1$, is equipped with a pointer into a cell $C' \in V_{k-1}(X)$ and a vector $s(C) \in N_k(C, X) \setminus N_{k-1}(C', X)$ for which $N_k(C, X) = N_{k-1}(C', X) \cup \{s(C)\}$; each cell of $V_1(X)$ is equipped with the vector $s(C)$ for which $N_1(C, X) = \{s(C)\}$. It follows that

$$N_k(C, X) = \{s(C), s(C'), s(C''), \ldots\}. \tag{7}$$

Fact 4. [4] *The running time of the algorithm* $\mathcal{V}$ *is* $O(n^{d+1})$.

On the other hand, for each $k = 1, \ldots, n$, the set $E_k(X)$ from the definition of $\mathcal{N}_k(X)$ is exactly the union of all the cells from $V_k(X)$. Therefore, we can compute the vectors from all the sets $\mathcal{N}_k(X)$ with the following algorithm:

Algorithm 2

Step 1. By using algorithm $\mathcal{V}$, construct all the Voronoi diagrams $V_k(X)$, $k = 1, \ldots, n$.

Step 2. For each $k = 1, \ldots, n$ and each cell $C \in V_k(X)$, calculate the vector $\Sigma N_k(C, X)$ that equals to either $\Sigma N_{k-1}(C', X) + s(C)$ if $k > 1$ or $s(C)$ if $k = 1$.

By Facts 3 and 4, the running time of Algorithm 2 is $O(dn^{d+1})$. For any output vector u, we may restore a k-element set $S \subseteq X$ such that $\Sigma S = u$ by using Eq. (7) in time $O(dn)$. Theorem 4 is proved.

6   Conclusion

We present improved hardness and algorithmic results for the problem of finding subset of vectors with the longest sum. We prove its APX-hardness in the case of any ℓ_p norm, $p \in [1, \infty)$, and reduce the computational time for an arbitrary norm. A similar result for the cardinality-constrained version is given.

An interesting open question is the existence of a polynomial-time constant-factor approximation algorithm for the LVS and Lk-VS problems. In other words, are these problems APX-complete? Another open question is their parameterized complexity with respect to the parameter d in the Euclidean case.

Acknowledgments. This work is supported by the Russian Science Foundation under grant 16-11-10041.

References

1. Baburin, A.E., Gimadi, E.K., Glebov, N.I., Pyatkin, A.V.: The problem of finding a subset of vectors with the maximum total weight. J. Appl. Industr. Math. **2**(1), 32–38 (2008)
2. Baburin, A.E., Pyatkin, A.V.: Polynomial algorithms for solving the vector sum problem. J. Appl. Industr. Math. **1**(3), 268–272 (2007)
3. Dolgushev, A.V., Kel'manov, A.V., Shenmaier, V.V.: Polynomial-time approximation scheme for a problem of partitioning a finite set into two clusters. Proc. Steklov Inst. Math. **295**(Suppl 1), 47–56 (2016)
4. Edelsbrunner, H., O'Rourke, J., Seidel, R.: Constructing arrangements of lines and hyperplanes with applications. SIAM J. Comput. **15**(2), 341–363 (1986)
5. Gimadi, E.K., Kel'manov, A.V., Kel'manova, M.A., Khamidullin, S.A.: A posteriori detecting a quasiperiodic fragment in a numerical sequence. Pattern Recogn. Image Anal. **18**(1), 30–42 (2008)
6. Håstad, J.: Some optimal inapproximability results. J. ACM **48**(4), 798–859 (2001)
7. Holmerin, J., Khot, S.: A new PCP outer verifier with applications to homogeneous linear equations and max-bisection. In: 36th Annual ACM Symposium on Theory of Computing, pp. 11–20. ACM, New York (2004)
8. Hwang, F.K., Onn, S., Rothblum, U.G.: A polynomial time algorithm for shaped partition problems. SIAM J. Optim. **10**(1), 70–81 (1999)
9. Khot, S., Kindler, G., Mossel, E., O'Donnell, R.: Optimal inapproximability results for MAX-CUT and other 2-variable CSPs? SIAM J. Comput. **37**(1), 319–357 (2007)
10. Matoušek, J.: Lectures on Discrete Geometry. Springer, New York (2002)
11. Onn, S., Schulman, L.J.: The vector partition problem for convex objective functions. Math. Oper. Res. **26**(3), 583–590 (2001)
12. Onn, S.: Personal communication, November 2016
13. Pyatkin, A.V.: On the complexity of the maximum sum length vectors subset choice problem. J. Appl. Industr. Math. **4**(4), 549–552 (2010)
14. Shenmaier, V.V.: Solving some vector subset problems by Voronoi diagrams. J. Appl. Industr. Math. **10**(4), 560–566 (2016)

Simple $O(n\ log^2\ n)$ Algorithms for the Planar 2-Center Problem

Xuehou Tan[1,2(✉)] and Bo Jiang[1]

[1] Dalian Maritime University, Linghai Road 1, Dalian, China
[2] Tokai University, 4-1-1 Kitakaname, Hiratsuka 259-1292, Japan
tan@wing.ncc.u-tokai.ac.jp

Abstract. The *planar 2-center* problem for a set S of points in the plane asks for two congruent circular disks of the minimum radius, whose union covers all points of S. In this paper, we present a simple $O(n \log^2 n)$ time algorithm for the planar 2-center problem, improving upon the previously known bound by a factor of $(\log \log n)^2$. We first describe an $O(n \log^2 n)$ time solution to a restricted 2-center problem in which the given points are in convex position (i.e., they are the vertices of a convex polygon), and then extend it to the case of arbitrarily given points. The novelty of our algorithms is their simplicity: our algorithms use only binary search and the known algorithms for computing the smallest enclosing disk of a point set, avoiding the use of relatively complicated parametric search, which is the base for most planar 2-center algorithms. Our work sheds more light on the (open) problem of developing an $O(n \log n)$ time algorithm for the planar 2-center problem.

Keywords: Computational geometry · Planar 2-center problem · The smallest enclosing circle problem · Convex polygons

1 Introduction

Let S denote a set of n points in the plane. The planar *p-center problem* asks for p congruent closed disks of the minimum radius, whose union covers S. The problem is NP-complete if p is part of input [10], and can be solved in $O(n^{2p-1} \log n)$ time for any fixed p [6]. On the other hand, efficient algorithms are known for small values of p. The 1-center problem, known as the *smallest enclosing disk* problem, is widely studied and can be solved in $O(n)$ time [3,4,9].

The planar 2-center problem has also been investigated extensively. After several near-quadratic time algorithms were given, Sharir [11] was the first to present a near-linear algorithm with running time $O(n \log^9 n)$, using the powerful (but complicated) parametric search paradigm. Later, Eppstein [5] presented a randomized algorithm with $O(n \log^2 n)$ expected time, and Chan [2]

The work by Tan was partially supported by JSPS KAKENHI Grant Number 15K00023, and the work by Jiang was partially supported by National Natural Science Foundation of China under grant 61173034.

© Springer International Publishing AG 2017
Y. Cao and J. Chen (Eds.): COCOON 2017, LNCS 10392, pp. 481–491, 2017.
DOI: 10.1007/978-3-319-62389-4_40

proposed an $O(n \log^2 n)$ time randomized algorithm with high probability, and an $O(n(\log n \log \log n)^2)$ time deterministic algorithm. Both Eppstein and Chan made some further refinements of Sharir's work. Whether an $o(n \log^2 n)$ time algorithm for the planar 2-center problem can be developed is left as an open problem [2].

Kim and Shin [8] have also studied the 2-center problem for a set of points in convex position. A set of points is said to be in *convex position* if it is the set of vertices of a convex polygon. Let P be a convex polygon of n vertices. We want to find two congruent closed disks of the minimum radius, whose union covers the whole polygon P, or only the vertices of P. Kim and Shin [8] announced an $O(n \log^3 n \log \log n)$ algorithm for covering the whole polygon P and an $O(n \log^2 n)$ time algorithm for covering all vertices of P. But, there is an error in their analysis on the time complexity of their algorithm for covering all vertices of P. A detailed discussion on it is provided in the appendix.

Suppose that r^* is the solution (optimum radius) for the planar 2-center problem for a given point set S. The idea of parametric search is to use a data structure for the *decision problem* (i.e., deciding whether S can be covered by two disks of a given radius r) to test whether r^* is larger than or smaller than some given value r. By making sure that the test for $r = r^*$ actually happens, one can eventually obtain the solution to the 2-center problem. In the known near-linear algorithms for the planar 2-center problem, $O(n)$ tests are usually used and each test takes $O(polylog\ n)$ time. Since parallel algorithms are usually used, the parametric search paradigm is complicated and not easy to be implemented [5].

Note that the p-center problem has an important application in modern wireless communication systems. It is a natural requirement that some transmissions should be set up so that mobile phones in a city (or region) can be serviced. Usually, the city can be represented by a simple or roughly convex polygon, and the region covered by a transmission is represented by a closed disk with radius r. The solution to the p-center problems helps give an optimal placement of transmissions.

Our work. The general approach employed in most previous work for the planar 2-center problem is first to solve the decision problem, and then apply an optimization scheme, such as the matrix search, parametric search, or expander-based technique [2,5,7,11]. Hence, the resulting algorithms are relatively complicated.

In this paper, we present a simple approach to the following two restricted 2-center problems: One is to cover the set of vertices of a given convex polygon, and the other is that the asked two congruent disks of the minimum radius are assumed (or previously known) to have a non-empty overlap and a point in the overlap is also given. Our new idea is to partition (or organize) the given point set S into two subsets such that a sequence of locally optimal two disks for either subset can be found; the union of two optimal disks covers S and the radii of locally optimal disks in the sequence are monotonically decreasing and then increasing. So, a binary search on either sequence of locally optimal solutions can be used to find the overall optimum solution.

We will develop two $O(n \log^2 n)$ time algorithms for the restricted 2-center problems described above. Our second result, together with a known result of Eppstein [5], directly leads to an $O(n \log^2 n)$ time solution to the planar 2-center problem. This improves upon the previously known bound $O(n(\log n \log \log n)^2)$ [2]. The novelty of our algorithms is their simplicity: our algorithms use only binary search and the known algorithms for computing the smallest enclosing disk of a point set, avoiding the use of complicated parametric search, which is the base for most planar 2-center algorithms. Although our improvement upon the previous results is small, it sheds more light on the (open) problem of developing an optimal $O(n \log n)$ time algorithm for the planar 2-center problem.

2   Covering a Set of Points in Convex Position

Let P be a convex polygon in the plane. In order to solve the 2-center problem for the set of vertices of P, we make a substantial use of the smallest disk enclosing P as well as the convexity of P. Let S denote the set of vertices of P. It is well known that the smallest disk covering the point set S, which is determined either by the diameter of S or by three points of S on its boundary, can be found in $O(n)$ time [9]. Denote by s_1, s_2 and s_3 (if it exists), in clockwise order, the points on the boundary of the smallest disk enclosing S. Without loss of generality, assume that the x-coordinate of s_1 is smaller than those of s_2 and s_3.

Denote by P_1 the polygonal chain of P from s_1 to s_2 clockwise, and P_2 the chain of P from s_3 (or s_2 if point s_3 does not exist) to s_1 clockwise. For two vertices x and y of P (P_1, or P_2), denote by $P[x, y]$ ($P_1[x, y]$, or $P_2[x, y]$) the polygonal chain of P (P_1, or P_2) from x to y *clockwise*. Also, denote by $P(x, y)$ the *open* polygonal chain of $P[x, y]$, in which x and y are excluded.

Suppose that D_1 and D_2 are two optimum disks for the 2-center problem for the point set S. Clearly, three points s_1, s_2 and s_3 cannot be contained in either optimum disk. Following from the pigeon-hole principle, two of s_1, s_2 and s_3 are contained in one of the optimum disks.

Assume that D_1 contains s_1, and D_2 contains s_2 and s_3 (if it exists). Since P is convex, all the vertices of $P[s_2, s_3]$ have to be contained in D_2; otherwise, either s_2 or s_3 is contained in D_1, contradicting our assumption. (If point s_3 does not exist, then $P[s_2, s_3]$ is a single point s_2.) Suppose that the number of the vertices of P_1 is larger than or equal to that of P_2. For ease of presentation, we assume below that both P_1 and P_2 consist of exactly m vertices. (To this end, one can simply add some "pseudo-vertices" to an edge of P_2.) Note that the vertices of the polygonal chain $P(s_2, s_3)$ are excluded from those $2m$ vertices.

Let us number the points on P_1 clockwise by $0, 1, \ldots, m - 1$, and the points on P_2 by $m, m + 1, \ldots, 2m - 1$. (So, s_1 and s_2 are numbered as 0 and $m - 1$ in P_1, respectively. And, s_1 and s_3 are numbered as $2m - 1$ and m in P_2, respectively.) For two points i and j, $1 \leq i \leq m - 1$ and $m+1 \leq j \leq 2m-1$, denote by $D(j, i-1)$ the smallest disk enclosing the polygonal chain $P[j, i-1]$, and $D(i, j-1)$ the smallest disk enclosing $P[i, j-1]$. Also, denote by $R(j, i - 1)$ and $R(i, j - 1)$ the radii of $D(j, i - 1)$ and $D(i, j - 1)$, respectively.

Let $R^*(i,j) = \max\{R(j,i-1), R(i,j-1)\}$. We can consider $R^*(i,j)$ as a (not an optimum) solution to the 2-center problem for the vertices of P.

For a point i $(1 \leq i \leq m-1)$, let $r_i^* = \min\{R^*(i,m+1), R^*(i,m+2),\ldots,R^*(i,2m-1)\}$. We will refer r_i^* to as a *locally* optimal solution to the 2-center problem for the vertices of P, because it is required in the solution r_i^* that two points $i-1$ and i be in different disks. See Fig. 1. Two adjacent solutions r_p^* and r_{p+1}^* can be identical; in this case, two disks determining the solution r_p^*, say, $R(q,p-1)$ and $R(p,q-1)$ $(m+1 \leq q \leq 2m-1)$, are *identical* to those determining r_{p+1}^*.

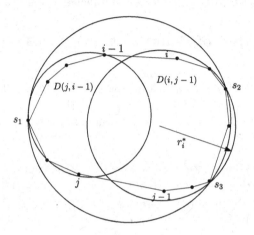

Fig. 1. The locally optimal radius r_i^*.

From the definition of radius $R^*(i,j)$, the following observation can be made.

Observation 1. *Suppose that* $r_i^* = R^*(i,q)$ *holds, for some pair of points i and q, $1 \leq i \leq m-1$ and $m+1 \leq q \leq 2m-1$. Then,* $R^*(i,m+1) \geq R^*(i,m+2) \geq \ldots \geq R^*(i,q-1) \geq R^*(i,q)$, *and* $R^*(i,q) \leq R^*(i,q+1) \leq \ldots \leq R^*(i,2m-1)$.

Analogously, let $r_j^* = \min\{R^*(1,j), R^*(2,j),\ldots,R^*(m-1,j)\}$, $m+1 \leq j \leq 2m-1$. Also, r_j is a locally optimal solution to the 2-center problem for the vertices of P.

Lemma 1. *If* $R(j,i-1) > R(i,j-1)$ *and* $R(j,i-2) < R(i-1,j-1)$ *hold, then* $r_j^* = \min\{R(j,i-1), R(i-1,j-1)\}$, *where* $2 \leq i \leq m-1$ *and* $m+1 \leq j \leq 2m-1$.

Proof. Since $R(j,i-1) > R(i,j-1)$ holds, we have $R(j,i) > R(i+1, j-1), R(j,i+1) > R(i+2,j-1),\ldots,R(j,m-2) > R(m-1,j-1)$. This is because $R(j,i-1) \leq R(j,i) \leq \ldots \leq R(j,m-2)$ and $R(i,j-1) \geq R(i+1,j-1) \geq \ldots \geq R(m-1,j-1)$. Hence, $R^*(i,j) = R(j,i-1), R^*(i+1,j) = R(j,i)\ldots, R^*(m-1,j) = R(j,m-2)$.

Also, since $R(j, i-2) < R(i-1, j-1)$ holds, we have $R(j, i-3) < R(i-2, j-1), \ldots, R(j, 0) < R(1, j-1)$. This is because $R(j, i-2) \geq R(j, i-3) \geq \ldots \geq R(j, 0)$ and $R(i-1, j-1) \leq R(i-2, j-1) \leq \ldots \leq R(1, j-1)$. Hence, $R^*(i-1, j) = R(i-1, j-1), R^*(i-2, j) = R(i-2, j-1) \ldots, R^*(1, j) = R(1, j-1)$.

Finally, since $R(j, i-1) \leq R(j, i) \leq \ldots \leq R(j, m-2)$ and $R(i-1, j-1) \leq R(i-2, j-1) \leq \ldots \leq R(1, j-1)$, we obtain $r_j^* = \min\{R(j, i-1), R(i-1, j-1)\}$. $\square$

Denote by r^* the optimum solution to the 2-center problem for the vertices of P. Following from the convexity of P and our assumption that point s_1 is contained in D_1 and two points s_2 and s_3 are in D_2,

$$r^* = \min_{1 \leq i \leq m-1, m+1 \leq j \leq 2m-1} \max\{R(j-1, i), R(i-1, j)\}.$$

Equivalently, we have $r^* = \min\{r_1^*, r_2^*, \ldots, r_{m-1}^*\} = \min\{r_{m+1}^*, r_{m+2}^*, \ldots, r_{2m-1}^*\}$. So, there are two points p and q, $1 \leq p \leq m-1$ and $m+1 \leq q \leq 2m-1$, such that $r^* = r_p^* = r_q^*$.

Lemma 2. *Suppose that $r_i^* = R^*(i, j)$ holds for some pair of points i and j, $1 \leq i \leq m-1$ and $m+1 \leq j \leq 2m-1$. If $r_j^* = R^*(i, j)$ also holds, then $r^* = r_i^*$ $(= r_j^*)$.*

Proof. Suppose that (i, j) is such a pair of points that $r_i^* = R^*(i, j) = r_j^*$. Assume by contradiction that $r^* < r_i^*$. If $R^*(i^-, j) < R^*(i, j)$ or $R^*(i^+, j) < R^*(i, j)$ holds for any $1 \leq i^- \leq i-1$ or $i+1 \leq i^+ \leq m-1$, then r_j^* is not locally optimal, a contradiction. Also, if $R^*(i, j^-) < R^*(i, j)$ or $R^*(i, j^+) < R^*(i, j)$ holds for any $m+1 \leq j^- \leq j-1$ or $j+1 \leq j^+ \leq 2m-1$, then r_i^* is not locally optimal, a contradiction. In the case that none of $R^*(i^-, j) < R^*(i, j)$, $R^*(i^+, j) < R^*(i, j)$, $R^*(i, j^-) < R^*(i, j)$ and $R^*(i, j^+) < R^*(i, j)$ holds, r_i^* (or r_j^*) is overall optimum, contradicting the assumption that $r^* < r_i^*$. The proof is complete. $\square$

Let us now introduce an important concept to locally optimal solutions. For a point i $(1 \leq i \leq m-1)$, the solution r_i^* is said to be *adjustable* on P_1 if there exists another solution $r_{i'}^*$, $1 \leq i' \leq m-1$, such that $r_{i'}^* < r_i^*$.[1]

Lemma 3. *Suppose that the solution r_i^* is adjustable on P_1, and $r_i^* = R^*(i, j)$ holds for some point $j \in P_2$. If $R(j, i-1) > R(i, j-1)$ then $r_{i-1}^* \leq r_i^*$, $2 \leq i \leq m-1$. Also, if $R(j, i-1) < R(i, j-1)$ then $r_{i+1}^* \leq r_i^*$, $1 \leq i \leq m-2$.*

Proof. We first claim that $R(j, i-1) \neq R(i, j-1)$. Assume that $R(j, i-1) = R(i, j-1)$. Then, $r_j^* = R^*(i, j)$, because any solution $R^*(i', j)$, $i' \neq i$, has to have a disk that contains the chain $P[j, i-1]$ or $P[i, j-1]$ and is thus of radius at least $R(j, i-1)$. It follows from Lemma 1 that $r_i^* = r^*$, a contradiction. Our claim is proved.

---

[1] The idea of defining the adjustablity of locally optimal solutions originates from the previous work on the well known *watchman route problem* in computational geometry [12].

By symmetry, we prove only that if $R(j, i-1) > R(i, j-1)$ then $r_{i-1}^* \le r_i^*$, $2 \le i \le m-1$. First, we have $R^*(i,j) = R(j, i-1)$ by the definition of $R^*(i,j)$. If $R(i-1, j-1) > R(j, i-2)$ and $R(i-1, j-1) > R(j, i-1)$ hold, then from our assumption that $R(j, i-1) > R(i, j-1)$, we have $r_j^* = \min\{R(j, i-1), R(i-1, j-1)\}$ (Lemma 1) $= R(j, i-1) = R^*(i,j)$. From the assumption of the lemma that $r_i^* = R^*(i,j)$, we have $r^* = r_i^*$ (Lemma 2), a contradiction. Hence, if $R(i-1, j-1) > R(j, i-2)$ holds, then $R(i-1, j-1) \le R(j, i-1)$; in this case, $R^*(i-1, j) = R(i-1, j-1) \le R(j, i-1) = R^*(i,j)$. Therefore, $r_{i-1}^* \le R^*(i-1, j) \le R^*(i,j) = r_i^*$. If $R(i-1, j-1) < R(j, i-2)$ holds, then $R^*(i-1, j) = R(j, i-2) \le R(j, i-1) = R^*(i,j)$. Again, we have $r_{i-1}^* \le R^*(i-1, j) \le R^*(i,j) = r_i^*$. $\square$

From Lemma 3, we say the *adjustable direction* of r_i^* is leftwards on P_1, if $R(j, i-1) > R(i, j-1)$. Also, if $R(j, i-1) < R(i, j-1)$, the adjustable direction of r_i^* is rightwards on P_1. Thus, the adjustable direction of r_i^* on P_1 is unique.

Lemma 4. *A locally optimal solution r_i^*, $1 \le i \le m-1$, gives r^* if and only if it is not adjustable on P_1.*

Proof. Assume first that $r^* = r_i^*$ holds for some point i. From the optimality of r^*, the solution r_i^* is not adjustable.

On the other hand, if r_i^* is not adjustable, then there are no other solutions, which are strictly smaller than r_i^*. Thus, $r_i^* = r^*$ holds. $\square$

Theorem 1. *Suppose that $r^* = r_i^*$ holds for some point i, $1 \le i \le m-1$. Then, $r_1^* \ge r_2^* \ge \dots \ge r_i^*$ and $r_i^* \le r_{i+1}^* \le \dots \le r_{m-1}^*$.*

Proof. Assume first that $r^* = r_i^*$ holds, for some point $i \ge 2$. Then, r_1^* is adjustable on P_1 (Lemma 4). Since r_1^* is the very first locally optimal solution on P_1, the adjustable direction of r_1^* (by the definition of adjustable solutions) is rightwards. Thus, $r_1^* \ge r_2^*$ (Lemma 3). So, if $r_2^* = r_i^*$ (probably, $i > 2$), we obtain $r_1^* > r_2^* = \dots = r_i^*$.

Consider now the situation in which $r_2^* > r_i^*$. Again, it follows from Lemma 4 that r_2^* is adjustable. If the adjustable direction of r_2^* is leftwards, then as the adjustable direction of r_1^* is rightwards, $r_1^* = r_2^*$ (Lemma 3). This implies that two disks determining r_1^* are identical to those determining r_2^*, contradicting the claim that their adjustable directions are different. Hence, the adjustable direction of r_2^* has to be rightwards, and thus $r_2^* \ge r_3^*$. In this way, we can eventually obtain $r_1^* \ge r_2^* \ge \dots \ge r_i^*$.

Analogously, if $i \le m-2$, we have $r_i^* \le r_{i+1}^* \le \dots \le r_{m-1}^*$. $\square$

By a similar argument, we can also show that there exists a point j in P_2, $m+1 \le j \le 2m-1$, such that $r_{m+1}^* \ge r_{m+2}^* \ge \dots \ge r_j^*$ and $r_j^* \le r_{j+1}^* \le \dots \le r_{2m-1}^*$.

Theorem 2. *Given a set S of n points in convex position, one can compute in $O(n \log^2 n)$ time the minimum radius r^* such that the union of two disks of radius r^* covers S.*

Proof. Let P be the convex polygon whose vertices coincide with the points of S. First, we compute the smallest enclosing disk of S. Denote by s_1, s_2 and s_3 the points of S on the boundary of the enclosing disk. (The special situation in which only two points are on the boundary of the enclosing disk can be dealt with analogously.)

Suppose that D_1 and D_2 are two optimum disks for the 2-center problem for the set of the vertices of P, and r^* is the radius of them. Then, two of s_1, s_2 and s_3 are contained in one optimum disk, and the third is in the other disk. There are three such combinations of s_1, s_2 and s_3. The algorithm described below is performed for each of these combinations, and the smallest among their outputted radii then gives the overall optimum radius r^*.

Suppose below that D_1 contains s_1 and D_2 contains s_2 and s_3. Assume also that the x-coordinate of s_1 is smaller than those of s_2 and s_3. If the locally optimal solution giving r^* is unique, it can simply be found by a binary search on the sequence $r_1^*, r_2^*, \dots r_{m-1}^*$ (Theorem 1). In the case that there are multiple optimal solutions r_k^* of the same value r^*, these points k are consecutive on the polygonal chain P_1 (Theorem 1). With a little more caution, r^* can also be found by a binary search on the sequence $r_1^*, r_2^*, \dots, r_{m-1}^*$.

Finally, a locally optimal solution r_i^* ($1 \le i \le m - 1$) can be found by a binary search on the sequence $R^*(i, m + 1), R^*(i, m + 2), \dots, R^*(i, 2m - 1)$ (Observation 1). Since the radius of the smallest disk enclosing a given point set can be computed in linear time, r_i^* can be found in $O(n \log n)$ time. Therefore, the optimum radius r^* can be computed in $O(n \log^2 n)$ time.   □

Remark. It is the first time to make use of three points s_1, s_2 and s_3 on the smallest enclosing disk of a point set in a solution to the planar 2-center problem. This helps to reduce the 2-center problem into at most three subproblems, which have an important property as described by Theorem 1.

3   Extension to the set of arbitrarily given points

Let us briefly review the known results on the 2-center problem for a set S of points in the plane. Sharir [11] was the first to present a near-linear algorithm with running time $O(n \log^9 n)$, using the powerful (but complicated) parametric search paradigm. Subsequently, Eppstein [5] presented a randomized algorithm with $O(n \log^2 n)$ expected time. The best known deterministic algorithm with $O(n(\log n \log \log n)^2)$ running time was due to Chan [2], who refined the deterministic and randomized algorithms of Sharir and Eppstein.

We first outline the basic strategy employed in these near-linear time algorithms. Depending on the ratio of the distance between two centers of optimum disks to r^*, two different algorithms are designed: one is for the case that two optimum disks are well separated, and the other for the case that they are close to each other, or to be exact, they have a non-empty common area. To avoid the work of determining which case the input data falls into, one can perform the algorithms for both cases and then return or report the smaller of the outputs. (The smaller output tells us which case actually happens.)

Two optimum disks D_1 and D_2 are said to be *well separated* if the distance between two centers of D_1 and D_2 is at least ϵr^*, for some constant $\epsilon > 0$. In this case, Eppstein [5] has given an $O(n \log^2 n)$ time algorithm to compute r^*, using a parametric search scheme.

Two disks D_1 and D_2 are said to be *close to each other* if the distance between two centers of D_1 and D_2 is strictly smaller than $(2 - \epsilon)r^*$, for some constant $\epsilon > 0$. In this case, one can generate a constant number of points such that at least one of them belongs to $D_1 \cap D_2$. We don't know which of these points is the one in $D_1 \cap D_2$. Again, we can try all of them, as the number of these points is a constant. The 2-center problem is then reduced to a constant number of the restricted problems, in which a known point is assumed to be in $D_1 \cap D_2$.

From the discussion made above, what we need to do is to solve the following restricted 2-center problem: for a given point set S and a given point t, find two congruent disks D_1 and D_2 of the minimum radius such that $S \subseteq D_1 \cup D_2$ and $t \in D_1 \cap D_2$ [2,5]. (It is not necessary for t to be a point of S.) Again, assume that three points s_1, s_2 and s_3 in clockwise order are on the smallest enclosing disk of S, and the x-coordinate of s_1 is smaller than those of s_2 and s_3. Assume also that D_1 contains s_1 and D_2 contains s_2 and s_3 (if it exists). We first compute (in $O(n \log n)$ time) the convex hull of S. Still, denote by $P[s_2, s_3]$ the convex chain between s_2 and s_3, which does not contain s_1, and $P(s_2, s_3)$ the open chain of $P[s_2, s_3]$.

Denote by E_1 the smallest disk that encloses s_1 and t, and E_2 the smallest disk that encloses the vertices of $P[s_2, s_3]$ and t. Denote by $S(E_1)$ $(S(E_2))$ the set of the points contained in E_1 (E_2). All the points of $S(E_1)$ $(S(E_2))$, by its definition, are contained in D_1 (D_2), and can thus be omitted in the following discussion.

Let γ_1 and γ_2 be the rays that emanate from t and pass through s_1 and s_2, respectively. Two rays γ_1 and γ_2 together partition the points of $S - S(E_1) \cup S(E_2)$ into two families. Let P_1 (P_2) denote the family of the points of $S - S(E_1) \cup S(E_2)$, which are vertically above (below) either γ_1 or γ_2. Notice that $S - S(E_1) \cup S(E_2)$ does not contain any point that is vertically below γ_2 and above the ray, which emanates from t and passes through s_3.

For ease of presentation, assume that s_1 and s_2 (s_3) also belong to P_1 (P_2), and both P_1 and P_2 have k points. Let us sort the points of P_1 (P_2) radially around point t in clockwise order. That is, the points of P_1 (P_2) are encountered in order if one rotates γ_1 (γ_2) around t clockwise. We number the sorted points of P_1 by $0, 1, \ldots, k - 1$, and the sorted points of P_2 by $k, k + 1, \ldots, 2k - 1$. For the example given in Fig. 2, the points of P_1 and P_2 are numbered as $0, 1, \ldots,$ 13 (point 13 is identical to point 0). Again, s_1 and s_2 are numbered as 0 and $k - 1$ in P_1 respectively, and s_1 and s_3 are numbered as $2k - 1$ and k in P_2 respectively.

Since we have assumed that both optimum disks D_1 and D_2 (of the same radius r^*) contain t, the boundaries of D_1 and D_2 have two intersection points; one is vertically above γ_1 or γ_2, and the other is below γ_1 or γ_2. Denote by γ_3 and γ_4 two rays, which emanate from t and pass through these points, see Fig. 2.

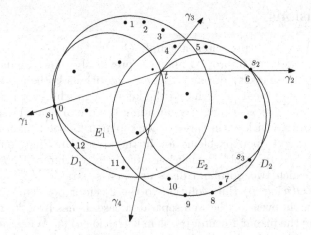

Fig. 2. Illustration for the sorted points of P_1 and P_2.

The following observation directly follows from the definitions of γ_3, γ_4, and the sorted points of P_1 and P_2.

Observation 2. γ_3 (γ_4) partitions the sorted points of P_1 (P_2) into two subsets: one has all its points in D_1 and the other has its points in D_2.

We use the same notation as described in Sect. 2. From Observation 2,

$$r^* = \min_{1 \le i \le k-1, k+1 \le j \le 2k-1} \max\{R(j-1, i), R(i-1, j)\}.$$

Thus, the argument given in Sect. 2 can directly be applied to the point sequences P_1 and P_2. Let $r_i^* = \min\{R^*(i, k+1), R^*(i, k+2), \ldots, R^*(i, 2k-1)\}$, $1 \le i \le k-1$. Again, r_i^* is the locally optimal solution to the 2-center problem for the points of S, under the requirement that two points $i-1$ and i be in different disks. Hence, we have the following result.

Theorem 3. Suppose that $r^* = r_i^*$ holds for some point i of P_1, $1 \le i \le k-1$. Then, $r_1^* \ge r_2^* \ge \ldots \ge r_i^*$ and $r_i^* \le r_{i+1}^* \le \ldots \le r_{k-1}^*$.

The method described in Sect. 2 can then be used to compute r^*. What is new in computing a locally optimal solution is that the points in P_1 (P_2) are now not in convex position. Since the smallest enclosing disk of a point set can be computed in linear time, it does not affect the time complexity of our algorithm at all. Still, the optimum radius r^* can be computed in $O(n \log^2 n)$ time.

Combining with Eppstein's result for the well separated case [5], we obtain the main result of this paper.

Theorem 4. For a set S of n given points, we can compute in $O(n \log^2 n)$ time the minimum radius r^* such that the union of two disks of radius r^* covers S.

4   Conclusions

In this paper, we have presented a simple $O(n \log^2 n)$ time algorithm for the planar 2-center problem, improving upon the previously known bound by a factor of $(\log \log n)^2$. The novelty of our presented algorithms is their simplicity: our algorithms use only binary search and the known algorithms for computing the smallest enclosing disk of a point set, avoiding the use of relatively complicated parametric search, which is the base for most planar 2-center algorithms.

We pose several open problems for further research. First, our algorithm for the planar 2-center problem makes use of the algorithm of Eppstein for the first case, in which two optimum disks are well separated [5]. We would like to ask for an $O(n \log^2 n)$ time solution to the 2-center problem without using parametric search, even for the well separated case. It has been known that the lower bound on the planar 2-center problem is $\Omega(n \log n)$ [5]. Whether an optimal $O(n \log n)$ time algorithm can be developed is a challenging open problem. From the simplicity of our presented algorithms, we believe that the method employed in this work can be used to develop an $O(n \log n)$ time algorithm for the planar 2-center problem.

Finally, Agarwal *et al.* have also considered the *discrete* 2-center problem, where the centers of two disks are restricted to be at points of S, and given an $O(n^{4/3} \log^5 n)$ time algorithm [1]. Whether or not a near-linear time algorithm can be developed is an interesting open problem.

Appendix: Note on the Kim and Shin algorithm for covering a set of points in convex position

In this appendix, we point out a mistake that occurs in the previously announced $O(n \log^2 n)$ time algorithm, by Kim and Shin, for covering a set A of points in convex position [8].

We use the same notation as given in [8]. For any $i, j \in A$, let $r_{i,j}^1$ ($r_{i,j}^2$) be the radius of the smallest disk containing $\langle i, j-1 \rangle$ ($\langle j, i-1 \rangle$). Then, the optimum solution $r^* = \min_{i,j \in A} \max\{r_{i,j}^1, r_{i,j}^2\}$.

Let $r_0^* = \min_{j \in A} \max\{r_{0,j}^1, r_{0,j}^2\}$, and let k be the point such that $r_0^* = \max\{r_{0,k}^1, r_{0,k}^2\}$. Then, it was stated in Lemma 7 of [8] that "For any i, j such that $i, j \in \langle 0, k-1 \rangle$ or $i, j \in \langle k, n-1 \rangle$, $\max\{r_{i,j}^1, r_{i,j}^2\} > r_0^{*}$".

By Lemma 7 of [8], all the pairs of $i, j \in \langle 0, k-1 \rangle$ or $i, j \in \langle k, n-1 \rangle$ needn't be considered. Kim and Shin then described a divide-and-conquer procedure to compute all the pairs of i and j such that $i \in \langle 0, k-1 \rangle$ and $j \in \langle k, n-1 \rangle$, in the paragraph immediately following Lemma 7 of [8]. Note that the number of the pairs of i and j, which are found by their divide-and-conquer procedure, is $O(n)$ (other than $O(\log n)$). To compute r^*, these $O(n)$ pairs of i and j have further to be handled (e.g., using parametric search to compute the smallest enclosing disks for them). However, it is NOT considered in [8]. Therefore, the $O(n \log^2 n)$ time solution to the planar 2-center problem for a set of points in convex position was not obtained in [8].

References

1. Agarwal, P.K., Sharir, M., Welzl, E.: The discrete 2-center problem. Discret. Comput. Geom. **20**, 287–305 (1998)
2. Chan, T.M.: More planar two-center algorithms. Comput. Geom. Theor. Appl. **13**, 189–198 (1999)
3. Chazelle, B., Matoušek, J.: On linear-time deterministic algorithms for optimization problems in fixed dimension. J. Algorithms **21**, 579–597 (1996)
4. Dyer, M.E.: On a multidimensional search technique and its application to the Euclidean one-center problem. SIAM J. Comput. **15**, 725–738 (1986)
5. Eppstein, D.: Faster construction of planar two-centers. In: Proceedings of 8th ACM-SIAM Symposium on Discrete Algorithms, pp. 131–138 (1997)
6. Hwang, R.Z., Lee, R.C.T., Chang, R.C.: The slab dividing approach to the euclidean P-center problem. Algorithmica **9**, 1–22 (1993)
7. Katz, M.J., Sharir, M.: An expander-based approach to geometric optimization. SIAM J. Comput. **26**, 1384–1408 (1997)
8. Kim, S.K., Shin, C.-S.: Efficient algorithms for two-center problems for a convex polygon. In: Du, D.-Z.-Z., Eades, P., Estivill-Castro, V., Lin, X., Sharma, A. (eds.) COCOON 2000. LNCS, vol. 1858, pp. 299–309. Springer, Heidelberg (2000). doi:10.1007/3-540-44968-X_30
9. Megiddo, N.: Linear time algorithms for linear programming in R^3 and related problems. SIAM J. Comput. **12**, 759–776 (1983)
10. Megiddo, N., Supowit, K.: On the complexity of some common geometric location problems. SIAM J. Comput. **13**, 1182–1196 (1984)
11. Sharir, M.: A near-linear time algorithm for the planar 2-center problem. Discret. Comput. Geom. **18**, 125–134 (1997)
12. Tan, X.: Fast computation of shortest watchman routes in simple polygons. Inform. Process. Lett. **87**, 27–33 (2001)

Stable Matchings in Trees

Satoshi Tayu$^{(\boxtimes)}$ and Shuichi Ueno

Department of Information and Communications Engineering,
Tokyo Institute of Technology, S3-57, Tokyo 152-8550, Japan
tayu@eda.ict.e.titech.ac.jp

Abstract. The maximum stable matching problem (Max-SMP) and the minimum stable matching problem (Min-SMP) have been known to be NP-hard for subcubic bipartite graphs, while Max-SMP can be solved in polynomal time for a bipartite graph G with a bipartition (X, Y) such that $\deg_G(v) \leq 2$ for any $v \in X$. This paper shows that both Max-SMP and Min-SMP can be solved in linear time for trees. This is the first polynomially solvable case for Min-SMP, as far as the authors know. We also consider some extensions to the case when G is a general/bipartite graph with edge weights.

1 Introduction

Let G be a simple bipartite graph (bigraph) with vertex set $V(G)$ and edge set $E(G)$. For each vertex $v \in V(G)$, let $I_G(v)$ be the set of all edges incident with v, and $\deg_G(v) = |I_G(v)|$. For each $v \in V(G)$, $\preceq_v$ is a total preorder (a binary relation with transitivity, totality, and hence reflexivity) on $I(v)$, and $\preceq_G = \{\preceq_v | \ v \in V(G)\}$. A total preorder $\preceq_v$ is said to be *strict* if $e \preceq_v f$ and $e \neq f$ imply $f \not\preceq_v e$. We say that $\preceq_G$ is *strict* if $\preceq_v$ is strict for every $v \in V(G)$. It should be noted that a strict total preorder is just a linear order. A pair $(G, \preceq_G)$ is called a *preference system*. A preference system $(G, \preceq_G)$ is said to be *strict* if $\preceq_G$ is strict. We say that an edge e *dominates* f at vertex v if $e \preceq_v f$. A matching M of G is said to be *stable* if each edge of G is dominated by some edge in M. The stable matching problem (SMP) is to find a stable matching of a preference system $(G, \preceq_G)$. It is well-known that any preference system $(G, \preceq_G)$ has a stable matching, and SMP can be solved in linear time by using the Gale/Shapley algorithm [3]. It is also well-known that every stable matching for a strict preference system has the same size and spans the same set of vertices, while a general preference system can have stable matchings of different sizes [3]. This leads us to the following two problems. The maximum stable matching problem (Max-SMP) is to find a stable matching with the maximum cardinality, and the minimum stable matching problem (Min-SMP) is to find a stable matching with the minimum cardinality. Manlove, Irving, Iwama, Miyazaki, and Morita showed that Max-SMP and Min-SMP are both NP-hard [9].

Let (X, Y) be a bipartition of a bigraph G. A bigraph G is called a (p, q)-*graph* if $\deg_x(\leq)p$ for every $x \in X$, and $\deg_y(\leq)q$ for every $y \in Y$. Irving, Manlove, and O'Malley showed that Max-SMP is NP-hard even for $(3, 3)$-graphs, while

© Springer International Publishing AG 2017
Y. Cao and J. Chen (Eds.): COCOON 2017, LNCS 10392, pp. 492–503, 2017.
DOI: 10.1007/978-3-319-62389-4_41

Max-SMP can be solved in polynomial time for $(2, \infty)$-graphs [7]. Some indepth consideration on the approximation for both problems can be found in [5].

The purpose of the paper is to show that Max-SMP and Min-SMP can be solved in linear time if G is a tree. This is the first polynomially solvable case for Min-SMP, as far as the authors know. We also consider some extensions to the case when G is a general/bipartite graph with edge weights.

The rest of the paper is organized as follows. Section 2 gives a foundation for our algorithms. Section 3.1 shows a linear time algorithm based on a dynamic programming to compute the size of a maximum stable matching in a tree. Section 3.2 shows a linear time algorithm, a modification of the algorithm in Sect. 3.1, to compute a maximum stable matching in a tree. Section 3.3 mentions an extension of the algorithm in Sect. 3.2 to compute a maximum-weight stable matching in linear time for trees with edge weights. Section 4.1 mentions that minimum stable matchings can be computed in linear time for trees (with edge weights) by modifying algorithms in Sect. 3. Section 4.2 mentions some extensions of our results to the case when G is a general graph with edge weights.

2   Stable Matchings in Trees

We need preliminaries to describe our algorithms.

Let T be a tree, and $(T, \preceq_T)$ be a preference system, which is called a tree preference system. A stable matching of $(T, \preceq_T)$ is called a *stable matching* of T, for simplicity. We use the following notations:

- we write $u \preceq_v w$ (or $w \succeq_v u$) if $(v, u) \preceq_v (v, w)$,
- we write $u \equiv_v w$ if $(v, u) \preceq_v (v, w)$ and $(v, w) \preceq_v (v, u)$,
- we write $u \prec_v w$ (or $w \succ_v u$) if $u \preceq_v w$ and $u \not\equiv_v w$.

It should be noted that if $\preceq_v$ is strict, then $u \equiv_v w$ if and only if $u = w$. It should be also noted that $\equiv_v$ is an equivalence relation on $I_G(v)$.

We consider T as a rooted tree with the root r, which is a leaf (a vertex of degree one) of T. For each vertex $v \in V(T) - \{r\}$, $p(v)$ is the parent of v, and $D(v)$ is the set of descendants of v. For any $v \in V(T) - \{r\}$, we denote by $T(v)$ the subtree induced by $D(v) \cup \{p(v)\}$. A matching M of $T(v)$ is said to be *v-stable* if every edge of $E(T(v)) - \{(v, p(v))\}$ is dominated by some edge in M. A vertex v is said to be *matched* with u in M if $(u, v) \in M$.

We define five sets of v-stable matchings of $T(v)$ as follows.

- $\mathcal{M}_v^P$ is the set of v-stable matchings of $T(v)$ in which v is matched with $p(v)$.
- $\mathcal{M}_v^H$ is the set of v-stable matchings of $T(v)$ in which v is matched with a child c such that $c \preceq_v p(v)$.
- $\mathcal{M}_v^L$ is the set of v-stable matchings of $T(v)$ in which v is matched with a child c such that $c \succ_v p(v)$.
- $\mathcal{M}_v^F$ is the set of v-stable matchings of $T(v)$ in which v is matched with no other vertices of $T(v)$.
- $\mathcal{M}_v^{\overline{P}}$ is the set of v-stable matchings of $T(v)$ in which v is not matched with $p(v)$.

If $v(\neq r)$ is a leaf, $T(v)$ is a tree with $E(T(v)) = \{(v, p(v))\}$, and we have $E(T(v)) - \{(v, p(v))\} = \emptyset$. Thus, we have the following.

Lemma 1. *If $v(\neq r)$ is a leaf, then $\mathcal{M}_v^P = \{\{(v, p(v))\}\}$, $\mathcal{M}_v^H = \mathcal{M}_v^L = \emptyset$, and $\mathcal{M}_v^{\overline{P}} = \mathcal{M}_v^F = \emptyset$.* □

It should be noted that for any $v \in V(T) - \{r\}$, $\mathcal{M}_v^{\overline{P}} = \mathcal{M}_v^H \cup \mathcal{M}_v^L \cup \mathcal{M}_v^F$, $\mathcal{M}_v^{\overline{P}} \cap \mathcal{M}_v^P = \emptyset$, and every v-stable matching of $T(v)$ is in $\mathcal{M}_v^P \cup \mathcal{M}_v^{\overline{P}}$.

Let r' be the child of r. Since r' is the only child of r, we obtain the following.

Lemma 2. *A set $M \subseteq E(T)$ is a stable matching of T if and only if $M \in \mathcal{M}_{r'}^P \cup \mathcal{M}_{r'}^H$.*

Proof. Suppose M is a stable matching of $T = T(r')$. Since M is an r'-stable matching, $M \in \mathcal{M}_{r'}^S$ for some $S \in \{P, H, L, F\}$. If $(r', r) \in M$ then $M \in \mathcal{M}_{r'}^P$. If $(r', r) \notin M$ then (r', r) must be dominated by an edge in M, and thus, there is a child c of r' such that $(r', c) \in M$ and $c \preceq_{r'} r$, which means that $M \in \mathcal{M}_{r'}^H$. Therefore, we conclude that $M \in \mathcal{M}_{r'}^P \cup \mathcal{M}_{r'}^H$.

Conversely, suppose $M \in \mathcal{M}_{r'}^P \cup \mathcal{M}_{r'}^H$. Since M is an r'-stable matching of $T = T(r')$, every edge in $E(T(v)) - \{(r', r)\}$ is dominated by an edge in M. If $M \in \mathcal{M}_{r'}^P$ then $(r', r) \in M$. If $M \in \mathcal{M}_{r'}^H$, then there exists a child c of r' such that $(r', c) \in M$ and $c \preceq_{r'} r$, which means that (r', r) is dominated by $(r', c) \in M$. Thus, we conclude that M is a stable matching of T, and we have the lemma. □

For a vertex $v \in V(T)$, let $\mathcal{C}(v)$ be the set of children of v. For a set $M \subseteq E(T)$ and $v \in V(T) - \{r\}$, we define $M(v) = E(T(v)) \cap M$.

Lemma 3. *If M is a v-stable matching of $T(v)$ then $M(c)$ is a c-stable matching of $T(c)$ for any $c \in \mathcal{C}(v)$.*

Proof. Since M is a matching of $T(v)$, $M(c)$ is a matching of $T(c)$. Since M is a v-stable matching of $T(v)$, every edge in $E(T(c)) - \{(c, v)\}$ is dominated by an edge in $M(c)$. Thus, $M(c)$ is a c-stable matching of $T(c)$. □

Lemma 4. *For any $v \in V(T) - \{r\}$ and set $M \subseteq E(T(v))$, $M \in \mathcal{M}_v^P$ if and only if the following conditions are satisfied:*

(i) $(v, p(v)) \in M$,
(ii) $M(c) \in \mathcal{M}_c^{\overline{P}}$ for every $c \in \mathcal{C}(v)$, and
(iii) $M(c) \in \mathcal{M}_c^H$ for every $c \in \mathcal{C}(v)$ with $c \prec_v p(v)$.

Proof. Suppose $M \in \mathcal{M}_v^P$. Then, (i) follows from the definition of $\mathcal{M}_v^P$. Since M is a v-stable matching of $T(v)$, $M(c)$ is a c-stable matching of $T(c)$ for every $c \in \mathcal{C}(v)$ by Lemma 3. Since v is matched with $p(v)$, $(v, c) \notin M$ for any $c \in \mathcal{C}(v)$, that is, $M(c) \in \mathcal{M}_c^{\overline{P}}$. Thus, we have (ii). For any $c \in \mathcal{C}(v)$ with $c \prec_v p(v)$, (v, c) is not dominated by $(v, p(v))$. Thus, there exists $g \in \mathcal{C}(c)$ such that (c, g) dominates $(c, v) = (v, c)$. Since $g \preceq_c v$, $M(c) \in \mathcal{M}_c^H$, and we have (iii).

Conversely, suppose a set $M \subseteq E(T(v))$ satisfies (i), (ii), and (iii). For any $c \in C(v)$, $M(c)$ is a matching such that v is matched with no other vertex by (ii). Thus, M is a matching of $T(v)$. For any $c \in C(v)$, each edge of $E(T(c)) - \{(v,c)\}$ is dominated by an edge in M, since $M(c)$ is a c-stable matching by (i), (ii), and (iii). For any $c \in C(v)$, if $c \succeq_v p(v)$ then (v,c) is dominated by $(v,p(v))$, which is in M by (i). If $c \prec_v p(v)$ then (v,c) is dominated by an edge in M by (iii). Thus, M is a v-stable matching of $T(v)$, and we conclude that $M \in \mathcal{M}_v^P$ by (i).   □

Lemma 5. *For any $v \in V(T) - \{r\}$ and set $M \subseteq E(T(v))$, $M \in \mathcal{M}_v^H$ if and only if the following conditions are satisfied:*

(i) $(v,p(v)) \notin M$, and
(ii) there exists $c' \in C(v)$ such that the following conditions are satisied:
  (ii-1) $c' \preceq_v p(v)$,
  (ii-2) $M(c') \in \mathcal{M}_{c'}^P$,
  (ii-3) $M(c) \in \mathcal{M}_c^{\overline{P}}$ for every $c \in C(v) - \{c'\}$, and
  (ii-4) $M(c) \in \mathcal{M}_c^H$ for every $c \in C(v)$ with $c \prec_v c'$.

Proof. Suppose $M \in \mathcal{M}_v^H$. Then, (i) and (ii-1) follow from the definition of $\mathcal{M}_v^H$. Since M is a v-stable matching of $T(v)$, $M(c)$ is a c-stable matching of $T(c)$ for every $c \in C(v)$ by Lemma 3. Since $M \in \mathcal{M}_v^H$, there exists $c' \in C(v)$ such that $(v,c') \in M$ and $c' \preceq_v p(v)$, that is, $M(c') \in \mathcal{M}_{c'}^P$. Thus, we have (ii-2). Since M is a matching and $(v,c') \in M$, $(v,c) \notin M(c)$ for every $c \in C(v) - \{c'\}$. Thus, $M(c) \in \mathcal{M}_c^{\overline{P}}$ for every $c \in C(v) - \{c'\}$, and we have (ii-3). For any $c \in C(v)$ with $c \prec_v c'$, (v,c) is not dominated by (v,c'). Therefore, there exists $g \in C(c)$ such that (c,g) dominates (c,v). Since $g \preceq_c v$, $M(c) \in \mathcal{M}_c^H$, and we have (ii-4).

Conversely, suppose a set $M \subseteq E(T(v))$ satisfies (i) and (ii). For any $c \in C(v) - \{c'\}$, $M(c)$ is a matching such that v is matched with no other vertex by (ii-3). Also, $M(c')$ is a matching by (ii-2). Thus M is a matching of $T(v)$. For any $c \in C(v) - \{c'\}$, each edge in $E(T(c)) - \{(v,c)\}$ is dominated by an edge in M, since $M(c)$ is a c-stable matching by (ii-3). For any $c \in C(v)$, if $c \succeq_v c'$ then (v,c) is dominated by (v,c'), which is in M by (ii-2). If $c \prec_v c'$ then (v,c) is dominated by an edge in M by (ii-4). Thus, M is a v-stable matching of $T(v)$ by (i) and (ii), and we conclude that $M \in \mathcal{M}_v^H$ by (ii-1) and (ii-2).   □

Lemma 6. *For any $v \in V(T) - \{r\}$ and set $M \subseteq E(T(v))$, $M \in \mathcal{M}_v^L$ if and only if the following conditions are satisfied:*

(i) $(v,p(v)) \notin M$, and
(ii) there exists $c' \in C(v)$ such that the following conditions are satisfied:
  (ii-1) $c' \succ_v p(v)$,
  (ii-2) $M(c') \in \mathcal{M}_{c'}^L$,
  (ii-3) $M(c) \in \mathcal{M}_c^{\overline{P}}$ for every $c \in C(v) - \{c'\}$, and
  (ii-4) $M(c) \in \mathcal{M}_c^H$ for every $c \in C(v)$ with $c \prec_v c'$.

Proof. Suppose $M \in \mathcal{M}_v^L$. Then, (i) and (ii-1) follow from the definition of $\mathcal{M}_v^L$. Since M is a v-stable matching of $T(v)$, $M(c)$ is a c-stable matching of $T(c)$ for every $c \in \mathcal{C}(v)$ by Lemma 3. Since $M \in \mathcal{M}_v^L$, there exists $c' \in \mathcal{C}(v)$ such that $(v, c') \in M$ and $c' \succ_v p(v)$, that is, $M(c') \in \mathcal{M}_{c'}^L$. Thus, we have (ii-2). Since M is a matching and $(v, c') \in M$, $(v, c) \notin M(c)$ for every $c \in \mathcal{C}(v) - \{c'\}$. Thus, $M(c) \in \mathcal{M}_c^{\overline{P}}$ for every $c \in \mathcal{C}(v) - \{c'\}$, and we have (ii-3). For any $c \in \mathcal{C}(v)$ with $c \prec_v c'$, (v, c) is not dominated by (v, c'). Therefore, there exists $g \in \mathcal{C}(c)$ such that (c, g) dominates (c, v). Since $g \preceq_c v$, $M(c) \in \mathcal{M}_c^H$, and we have (ii-4).

Conversely, suppose a set $M \subseteq E(T(v))$ satisfies (i) and (ii). For any $c \in \mathcal{C}(v) - \{c'\}$, $M(c)$ is a matching such that v is matched with no other vertex by (ii-3). Also, $M(c')$ is a matching by (ii-2). Thus M is a matching of $T(v)$. For any $c \in \mathcal{C}(v) - \{c'\}$, each edge in $E(T(c)) - \{(v, c)\}$ is dominated by an edge in M, since $M(c)$ is a c-stable matching by (ii-3). For any $c \in \mathcal{C}(v)$, if $c \succeq_v c'$ then (v, c) is dominated by (v, c'), which is in M by (ii-2). If $c \prec_v c'$ then (v, c) is dominated by an edge in M by (ii-4). Thus, M is a v-stable matching of $T(v)$ by (i) and (ii), and we conclude that $M \in \mathcal{M}_v^L$ by (ii-1) and (ii-2).   □

Lemma 7. *For any* $v \in V(T) - \{r\}$ *and set* $M \subseteq E(T(v))$, $M \in \mathcal{M}_v^F$ *if and only if the following conditions are satisfied:*

(i) $(v, p(v)) \notin M$, *and*
(ii) $M(c) \in \mathcal{M}_c^H$ *for any* $c \in \mathcal{C}(v)$.

Proof. Suppose $M \in \mathcal{M}_v^F$. Then, (i) follows from the definition of $\mathcal{M}_v^F$. Since M is a v-stable matching of $T(v)$, $M(c)$ is a c-stable matching of $T(c)$ for every $c \in \mathcal{C}(v)$ by Lemma 3. Since $M \in \mathcal{M}_v^F$, for any $c \in \mathcal{C}(v)$, there exists $g \in \mathcal{C}(c)$ such that (c, g) dominates (c, v). Since $g \preceq_c v$, $M(c) \in \mathcal{M}_c^H$ for any $c \in \mathcal{C}(v)$, and we have (ii).

Conversely, suppose a set $M \subseteq E(T(v))$ satisfies (i) and (ii). For any $c \in \mathcal{C}(v)$, $M(c)$ is a matching such that v is matched with no other vertex by (ii). Thus, M is a matching of $T(v)$. For any $c \in \mathcal{C}(v)$, each edge in $E(T(c)) - \{(c, v)\}$ is dominated by an edge in M, since $M(c)$ is a c-stable matching by (ii). For any $c \in \mathcal{C}(v)$, (c, v) is dominated by an edge in M by (ii). Thus, M is a v-stable matching of $T(v)$, and we conclude that $M \in \mathcal{M}_v^F$ by (i).   □

3   Linear Time Algorithms for Trees

3.1   Computing the Size of Maximum Stable Matchings

Now, we are ready to show a linear time algorithm to compute the size of a maximum stable matching for a tree preference system. Our algorithm applies a dynamic programming scheme based on the results in the previous section.

Let $(T, \preceq_T)$ be a tree preference system. We consider T as a rooted tree with root r, which is a leaf of T. For any $v \in V(T) - \{r\}$ and $S \in \{P, H, L, F, \overline{P}\}$, we define that

$$\mu_v^S = \max_{M \in \mathcal{M}_v^S} |M|. \tag{1}$$

That is, μ_v^S is the maximum number of edges of a v-stable matching in $\mathcal{M}_v^S$. We define that $\mu_v^S = -\infty$ if $\mathcal{M}_v^S = \emptyset$.

From Lemma 1, we have the following.

Lemma 8. *If $v(\neq r)$ is a leaf of T,*

$$\mu_v^H = \mu_v^L = -\infty, \tag{2}$$

$$\mu_v^P = 1, and \tag{3}$$

$$\mu_v^F = 0. \tag*{$\square$}$$

Define that for any $v \in V(T) - \{r\}$ and $c \in \mathcal{C}(v)$,

$$\mu_{v,c} = \mu_c^P + \sum_{c \in \mathcal{C}(v),\ c' \prec_v c} \mu_{c'}^H + \sum_{c \in \mathcal{C}(v),\ c' \succeq_v c,\ c' \neq c} \mu_{c'}^{\overline{P}}. \tag{4}$$

From Lemmas 4 –7, we have the following.

Lemma 9. *For any $v \in V(T) - \{r\}$,*

$$\mu_v^P = \sum_{c \in \mathcal{C}(v),\ c \prec_v p(v)} \mu_c^H + \sum_{c \in \mathcal{C}(v),\ c \succeq_v p(v)} \mu_c^{\overline{P}} + 1, \tag{5}$$

$$\mu_v^H = \max\{\mu_{v,c} \mid c \in \mathcal{C}(v), c \preceq_v p(v)\}, \tag{6}$$

$$\mu_v^L = \max\{\mu_{v,c} \mid c \in \mathcal{C}(v), c \succ_v p(v)\}, \tag{7}$$

$$\mu_v^F = \sum_{c \in \mathcal{C}(v)} \mu_c^H, and \tag{8}$$

$$\mu_v^{\overline{P}} = \max\{\mu_v^H, \mu_v^L, \mu_v^F\}. \tag*{(9)\quad$\square$}$$

From Lemmas 8 and 9, we have the following.

Lemma 10. *Procedure* COMP_$\mu(v)$ *shown in Fig. 1 computes μ_v^S for all $v \in V(T) - \{r\}$ and $S \in \{P, H, L, F, \overline{P}\}$.* $\qquad\square$

Now, we are ready to show the following.

Theorem 1. *Algorithm* MAX-SIZE$(T, \preceq_T, r)$ *shown in Fig. 2 solves Max-SMP for a tree preference system $(T, \preceq_T)$ in linear time.*

Proof. The validity of the algorithm follows from Lemmas 2 and 10. We shall show that the time complexity of the algorithm is $O(n)$, where $n = |V(T)|$.

Lemma 11. *Given μ_c^S for every $c \in \mathcal{C}(v)$ and $S \in \{P, H, L, F, \overline{P}\}$, $\{\mu_{v,c} \mid c \in \mathcal{C}(v)\}$ can be computed in $O(|\mathcal{C}(v)|)$ time,*

Input $v \in V(T) - \{r\}$.
begin
  for all $c \in \mathcal{C}(v)$
    $\text{COMP-}\mu(c)$.
  endfor
  for all $c \in \mathcal{C}(v)$
$$\mu_{v,c} = \mu_c^P + \sum_{c' \in \mathcal{C}(v),\ c' \prec_v c} \mu_{c'}^H + \sum_{c' \in \mathcal{C}(v),\ c' \succeq_v c, c' \neq c} \mu_{c'}^{\overline{P}}.$$
  endfor
$$\mu_v^P = \sum_{c \in \mathcal{C}(v),\ c \prec_v p(v)} \mu_c^H + \sum_{c \in \mathcal{C}(v),\ c \succeq_v p(v)} \mu_c^{\overline{P}} + 1, \text{ where } \mu_v^P = 1 \text{ if } v \text{ is a leaf.}$$
  (See Lemma 8.)
$$\mu_v^F = \sum_{c \in \mathcal{C}(v)} \mu_c^H, \text{ where } \mu_v^F = 0 \text{ if } v \text{ is a leaf.}$$
$$\mu_v^H = \max_{c \in \mathcal{C}(v), c \preceq_v p(v)} \mu_{v,c}, \text{ where } \mu_v^H = -\infty \text{ if } v \text{ is a leaf.}$$
$$\mu_v^L = \max_{c \in \mathcal{C}(v), c \prec_v p(v)} \mu_{v,c}, \text{ where } \mu_v^L = -\infty \text{ if } v \text{ is a leaf.}$$
  if $\mu_v^H \geq \max\{\mu_v^L, \mu_v^F\}$ **then**
    $\mu_v^{\overline{P}} = \mu_v^H$.
  endif
end

Fig. 1. Procedure $\text{COMP-}\mu(v)$.

Proof of Lemma 11. If v is a leaf, $\{\mu_{v,c} \mid c \in \mathcal{C}(v)\}$ can be computed in $O(1)$ time, by definition. Let $v \in V(T)$ be a vertex of degree at least 2, $\delta = |\mathcal{C}(v)|$, and $c_1, c_2, \ldots, c_\delta$ be a sequence of the children of v such that $c_i \preceq_v c_{i+1}$ for any $i \in [\delta - 1]$, where $c_1, c_2, \ldots, c_\delta$ are sorted in the problem instance. Define that for any $v \in V(T) - \{r\}$ and $c \in \mathcal{C}(v)$,

$$[c]_v = \{c' \mid c' \equiv_v c\},$$

which is an equivalence class for an equivalence relation $\equiv_v$ on $I_G(v)$. Then, $\mathcal{C}(v)$ is partitioned into equivalence classes for $\equiv_v$. Let $x \geq 2$. If $c_x \equiv_v c_{x-1}$ then

$$\mu_{v,c_x} = \mu_{v,c_{x-1}} - (\mu_{c_{x-1}}^P + \mu_{c_x}^{\overline{P}}) + (\mu_{c_{x-1}}^{\overline{P}} + \mu_{c_x}^P).$$

Thus, by (4), μ_{v,c_x} can be computed in $O(1)$ time by using $\mu_{v,c_{x-1}}$. If $c_{x-1} \prec_v c_x$, let y be the integer satisfying

$$[c_{x-1}]_v = \{c_y, c_{y+1}, \ldots, c_{x-1}\}.$$

Since $c_{x-1} \not\equiv_v c_x$, c_{x-1} and c_x are contained in different equivalnce classes for $\equiv_v$. Thus from (4), we have

$$\mu_{v,c_x} = \mu_{v,c_{x-1}} - \left(\sum_{i=y}^{x-2} \mu_{c_i}^{\overline{P}} + \mu_{c_{x-1}}^P + \mu_{c_x}^{\overline{P}} \right) + \left(\sum_{i=y}^{x-1} \mu_{c_i}^H + \mu_{c_x}^P \right). \qquad (10)$$

Therefore, μ_{v,c_x} can be computed in $O(|[c_{x-1}]_v|)$ time by using $\mu_{v,c_{x-1}}$.

Thus, $\{\mu_{v,c} \mid c \in \mathcal{C}(v)\}$ can be computed in $O(|\mathcal{C}(v)|)$ time, since the computation shown in (10) is executed once for each x with $c_x \succ_v c_{x-1}$.

Once $\mu_{v,c}$ are obtained for every $c \in \mathcal{C}(v)$, $\{\mu_v^S \mid S \in \{H, L, P, F, \overline{P}\}\}$ can be computed in $O(|\mathcal{C}(v)|) = O(\deg Tv)$ time, by Eqs. (5)–(9). From Lemma 10, except for the recursive calls, $\mathrm{COMP\text{-}}\mu(v)$ for each vertex v can be done in $O(\deg Tv)$ time. Moreover, in the execution of $\mathrm{MAX\text{-}SIZE}(T, \preceq_T, r)$ shown in Fig. 2, $\mathrm{COMP\text{-}}\mu(v)$ is called once for every $v \in V(T)$, and thus, $\mathrm{MAX\text{-}SIZE}(T, \preceq_T, r)$ can be executed in $\sum_{v \in V(T)} \deg Tv = O(|V(T)|)$ time. Since $\mathrm{MAX\text{-}SIZE}(T, \preceq_T, r)$ computes $\max\{\mu_{r'}^H, \mu_{r'}^P\}$, we have the theorem, by Lemma 2. □

Input tree preference system $(T, \preceq_T)$, a root $r \in V(T)$ of T.
Output the size of a maximum stable matching.
begin
    $M = \emptyset$.
    let r' be the child of r.
    $\mathrm{COMP\text{-}}\mu(r')$.
    return $\max\{\mu_{r'}^H, \mu_{r'}^P\}$.
end

Fig. 2. Algorithm $\mathrm{MAX\text{-}SIZE}(T, \preceq_T, r)$.

3.2 Computing Maximum Stable Matchings

Before describing an algorithm for computing a maximum stable matching, we modify $\mathrm{COMP\text{-}}\mu(v)$ to store an edge (u, v) in a matching with μ_v^S edges for any $S \in \{H, \overline{P}\}$ when μ_v^S is computed. We use two variables $\gamma(H, v)$ and $\gamma(\overline{P}, v)$ to store a child c of v. $\gamma(H, v)$ stores a child c with $(c, v) \in M$ for any $M \in \mathcal{M}_v^H$ corresponding to μ_v^H. $\gamma(\overline{P}, v)$ stores a child c with $(c, v) \in M$ for any $M \in \mathcal{M}_v^H$ corresponding to μ_v^P.

Figure 3 shows a recursive procedure $\mathrm{COMP\text{-}}\gamma(v)$, which is obtained from $\mathrm{COMP\text{-}}\mu(v)$ shown in Fig. 1 by adding some instructions for $\gamma(S, v)$.

Lemma 12. *For any $S \in \{H, \overline{P}\}$ and $v \in V(T) - \{r\}$, $\gamma(S, v)$ stores the edge of M incident with v such that $M \in \mathcal{M}_v^S$ with $|M| = \mu_v^S$, if any.* □

We show an algorithm for computing a maximum stable matching of a tree preference system $(T, \preceq_T)$ in Fig. 4, where Procedure $\mathrm{ADD\text{-}EDGES}(v, S, M)$ is shown in Fig. 5.

Procedure $\mathrm{ADD\text{-}EDGES}(v, S, M)$ traverses vertices of T in DFS order, and we have the following.

Lemma 13. *For any $S \in \{P, H, \overline{P}\}$ and $M \subseteq E(T)$ with $M \cap E(T(v)) = \emptyset$, $\mathrm{ADD\text{-}EDGES}(v, S, M)$ adds edges in M' to M for some $M' \in \mathcal{M}_v^S$ satisfying $|M'| = \mu_v^S$.* □

Input $v \in V(T) - \{r\}$.
begin
  for all $c \in \mathcal{C}(v)$
    COMP_$\gamma(c)$.
  endfor
  for all $c \in \mathcal{C}(v)$

$$\mu_{v,c} = \sum_{c' \in \mathcal{C}(v),\ c' \prec_v c} \mu_{c'}^H + \mu_c^P + \sum_{c' \in \mathcal{C}(v),\ c' \succeq_v c, c' \neq c} \mu_{c'}^{\overline{P}}.$$

  endfor

$$\mu_v^P = \sum_{c \in \mathcal{C}(v),\ c \prec_v p(v)} \mu_c^H + \sum_{c \in \mathcal{C}(v),\ c \succeq_v p(v)} \mu_c^{\overline{P}} + 1, \text{ where } \mu_v^P = 1 \text{ if } v \text{ is a leaf}.$$

  (See Lemma 8.)

$$\mu_v^F = \sum_{c \in \mathcal{C}(v)} \mu_c^H, \text{ where } \mu_v^F = 0 \text{ if } v \text{ is a leaf}.$$

$$\mu_v^H = \max_{c \in \mathcal{C}(v), c \preceq_v p(v)} \mu_{v,c}, \text{ where } \mu_v^H = -\infty \text{ if } v \text{ is a leaf}.$$

$$\mu_v^L = \max_{c \in \mathcal{C}(v), c \prec_v p(v)} \mu_{v,c}, \text{ where } \mu_v^L = -\infty \text{ if } v \text{ is a leaf}.$$

  let $\gamma(H, v)$ be a child c of v
    such that $\mu_v^H = \mu_{v,c}$ and $c \preceq_v p(v)$.
  if $\mu_v^H \geq \max\{\mu_v^L, \mu_v^F\}$ **then**
    $\mu_v^{\overline{P}} = \mu_v^H$.
    $\gamma(\overline{P}, v) = \gamma(H, v)$.
  elseif $\mu_v^L \geq \mu_v^F$ **then**
    $\mu_v^{\overline{P}} = \mu_v^L$.
    let $\gamma(\overline{P}, v)$ be a child c of v
      such that $\mu_v^L = \mu_{v,c}$ and $c \succ_v p(v)$.
  else   /* $\mu_v^F > \max\{\mu_v^H, \mu_v^L\}$ */
    $\mu_v^{\overline{P}} = \mu_v^F$.
    $\gamma(\overline{P}, v) = $ NULL.
  endif
end

Fig. 3. Procedure COMP_$\gamma(S, v)$.

From Lemmas 12 and 13, we have the following.

Theorem 2. *Algorithm* MAX-SMP$(T, \preceq_T, r)$ *solves Max-SMP in linear time for a tree preference system* $(T, \preceq_T)$.   □

3.3 Computing Maximum-Weight Stable Matchings

A *weighted preference system* $(G, \preceq_G, w)$ is a preference system $(G, \preceq_G)$ with a weight function $w : E(G) \to \mathbb{Z}^+$. For a matching M of G, $w(M) = \sum_{e \in M} w(e)$ is a weight of M. The maximum-weight stable matching problem (Max-WSMP) is to find a stable matching with maximum weight for a weighted preference system. It is easy to see that we can solve Max-WSMP for weighted tree preference systems by the algorithm in Sect. 3.2 with a slight modification of μ_v^S. For any

```
Input tree preference system (T, ⪯_T), a root r ∈ V(T) of T.
Output a maximum stable matching M and μ = |M|.
begin
    M = ∅.
    let r′ be the child of r.
    COMP_γ(r′).
    if μ_{r′}^H ≥ μ_{r′}^P.
        μ = μ_{r′}^H.
        ADD_EDGES(r′, H, M)
    else
        μ = μ_{r′}^P.
        ADD_EDGES(r′, P, M)
    endif
    return M and μ.
end
```

Fig. 4. Algorithm MAX-SMP$(T, ⪯_T, r)$.

$S \in \{H, L, P, F, \overline{P}\}$, we define that

$$\mu_v^S = \max_{M \in \mathcal{M}_v^S} w(M) \tag{11}$$

instead of (1). We also replace (3) and (5) with

$$\mu_v^P = w(v, p(v)) \text{ and} \tag{12}$$

$$\mu_v^P = \sum_{c \in \mathcal{C}(v),\ c \prec_v p(v)} \mu_c^H + \sum_{c \in \mathcal{C}(v),\ c \succeq_v p(v)} \mu_c^{\overline{P}} + w(v, p(v)), \tag{13}$$

respectively, and let MAX-WSMP$(T, ⪯_T, r, w)$ be the algorithm obtained by the modifications above. Thus, we have the following.

Theorem 3. *Algorithm* MAX-WSMP$(T, ⪯_T, r, w)$ *solves Max-WSMP in linear time for a weighted tree preference system* $(T, ⪯_T, w)$. □

4   Concluding Remarks

4.1   Min-SMP and Min-WSMP

The minimum-weight stable matching problem (Min-WSMP) is to find a stable matching with minimum weight for a weighted preference system. We can compute minimum stable matchings in a similar way. We define $\mu_v^S = +\infty$ if $\mathcal{M}_v^S = \emptyset$. Let MIN-SMP$(T, ⪯_T, r)$ be the algorithm obtained from MAX-SMP$(T, ⪯_T, r)$ by replacing (2), (6), (7), and (9) with

$$\mu_v^H = \mu_v^L = +\infty$$

$$\mu_v^H = \min\{\mu_{v,c} \mid c \in \mathcal{C}(v), c \preceq_v p(v), c′ \neq p(v)\},$$

$$\mu_v^L = \min\{c \in \mathcal{C}(v),\ \mu_{v,c} \mid c \succ_v p(v)\}, \text{ and}$$

$$\mu_v^{\overline{P}} = \min\{\mu_v^H, \mu_v^L, \mu_v^F\},$$

Input $v \in V(T) - \{r\}$, $S \in \{P, H, \overline{P}\}$, $M \subseteq E(T)$.
begin
  if $S = P$ **then**
    $c_0 = p(v)$.
  elseif $S = H$ **then**
    $c_0 = \gamma(v, H)$.
  else
    $c_0 = \gamma(v, \overline{P})$.
  endif
  for all $c \in \mathcal{C}(v)$ with $c \prec_v c_0$, /* $c \prec_v c_0$ if $c_0 = \text{NULL}$ */
    ADD_EDGES(c, H, M).
  endfor
  for all $c \in \mathcal{C}(v)$ with $c \succeq_v c_0$ and $c \neq c_0$
    ADD_EDGES$(c, \overline{P}, M)$.
  endfor
  if $S = P$ **then** /* $c_0 = p(v)$ */
    add $(v, p(v))$ to M.
  elseif $c_0 \neq \text{NULL}$ **then** /* $S = H$ or $\overline{P}$ */
    ADD_EDGES(c_0, P, M).
  endif
end

Fig. 5. Procedure ADD_EDGES(v, S, M).

respectively. Then, we have the following.

Theorem 4. *Algorithm* MIN-SMP$(T, \preceq_T, r)$ *solves Min-WSMP in linear time for a tree preference system* $(T, \preceq_T)$. $\qquad\qquad\square$

Moreover, we can also compute the size of a minimum weighted stable matching for a weighted preference system by replacing (3) and (5) with (12) and (13), respectively. Let MIN-WSMP$(T, \preceq_T, r, w)$ be the algorithm obtained from MIN-SMP$(T, \preceq_T, r)$ by the modifications. Then, we have the following.

Theorem 5. *Algorithm* MIN-WSMP$(T, \preceq_T, r, w)$ *solves Min-WSMP in linear time for a weighted tree preference system* $(T, \preceq_T, w)$. $\qquad\qquad\square$

4.2   Extensions

The (weighted) preference system can be defined for general graphs without any modification. The general stable matching problem (GSMP) is to find a stable matching of a general preference system $(G, \preceq_G)$, where G is a general graph. It has been known that there exists a general preference system which has no stable matching, and GSMP is NP-hard [6,10]. If G is a bipartite graph, $(G, \preceq_G)$ is referred to as a bipartite preference system in this section.

The maximum-weight general stable matching problem (Max-WGSMP) is to find a stable matching with maximum weight for a weighted general preference system. The minimum-weight general stable matching problem (Min-WGSMP) is to find a stable matching with minimum weight for a weighted general preference system.

It has been known that both Max-WGSMP and Min-WGSMP are NP-hard even for weighted strict general preference systems [2]. It is also known that both problems are sovable in $O(m^2 \log m)$ time for a weighted strict bipartite preference system $(G, \preceq_G, w)$, where $m = |E(G)|$ [1,4,8]. An extension can be found in [1].

It is interesting to note that both Max-WGSMP and Min-WGSMP can be solved in plynomial time if the treewidth of G is bounded by a constant. We can prove the following by extending our resutlts on trees, although the proof is complicated.

Theorem 6. *Both Max-WGSMP and Min-WGSMP can be solved in $O(n\Delta^{k+1})$ time if the treewidth of G is bounded by k, where $n = |V(G)|$, and Δ is the maximum degree of a vertex of G.* $\square$

It should be noted that if $k = 1$, both problems can be solved in linear time as shown in Theorems 3 and 5.

Acknowledgements. The research was partially supported by JSPS KAKENHI Grant Number 26330007.

References

1. Chen, X., Ding, G., Hu, X., Zang, W.: The maximum-weight stable matching problem: duality and efficiency. SIAM J. Discret. Math. **26**, 1346–1360 (2012)
2. Feder, T.: A new fixed point approach for stable networks and stable marriages. J. Comput. Syst. Sci. **45**, 233–284 (1992)
3. Gale, D., Shapley, L.S.: College admissions and the stability of marriage. Am. Math. Mon. **69**, 9–15 (1962)
4. Gusfield, D., Irving, R.W.: The Stable Marriage Problem - Structure and Algorithms. Foundations of computing series. MIT Press, Cambridge (1989)
5. Halldórsson, M., Irving, R., Iwama, K., Manlove, D., Miyazaki, S., Morita, Y., Scott, S.: Approximability results for stable marriage problems with ties. Theor. Comput. Sci. **306**, 431–447 (2003)
6. Irving, R.W., Manlove, D.: The stable roommates problem with ties. J. Algorithms **43**, 85–105 (2002)
7. Irving, R.W., Manlove, D., O'Malley, G.: Stable marriage with ties and bounded length preference lists. J. Discret. Algorithms **1**, 213–219 (2009)
8. Király, T., Pap, J.: Total dual integrality of rothblum's description of the stable-marriage polyhedron. Math. Oper. Res. **33**, 283–290 (2008)
9. Manlove, D., Irving, R., Iwama, K., Miyazaki, S., Morita, Y.: Hard variants of stable marriage. Theor. Comput. Sci. **276**, 261–279 (2002)
10. Ronn, E.: NP-complete stable matching problems. J. Algorithms **11**, 285–304 (1990)

Maximum Matching on Trees
in the Online Preemptive
and the Incremental Dynamic Graph Models

Sumedh Tirodkar[1]([✉]) and Sundar Vishwanathan[2]

[1] School of Technology an Computer Science, TIFR, Mumbai, India
sumedh.tirodkar@tifr.res.in
[2] Department of Computer Science and Engineering, IIT Bombay, Mumbai, India
sundar@cse.iitb.ac.in

Abstract. We study the Maximum Cardinality Matching (MCM) and
the Maximum Weight Matching (MWM) problems, on trees and on some
special classes of graphs, in the Online Preemptive and the Incremental
Dynamic Graph models. In the *Online Preemptive* model, the edges of
a graph are revealed one by one and the algorithm is required to always
maintain a valid matching. On seeing an edge, the algorithm has to either
accept or reject the edge. If accepted, then the adjacent edges are discarded, and all rejections are permanent. In this model, the complexity
of the problems is settled for deterministic algorithms [11,15]. Epstein
et al. [5] gave a 5.356-competitive randomized algorithm for MWM, and
also proved a lower bound of 1.693 for MCM. The same lower bound
applies for MWM.

In this paper we show that some of the results can be improved in the
case of trees and some special classes of graphs. In the online preemptive model, we present a 64/33-competitive (in expectation) randomized
algorithm (which uses only two bits of randomness) for MCM on trees.

Inspired by the above mentioned algorithm for MCM, we present the
main result of the paper, a randomized algorithm for MCM with a "worst
case" update time of $O(1)$, in the incremental dynamic graph model,
which is 3/2-approximate (in expectation) on trees, and 1.8-approximate
(in expectation) on general graphs with maximum degree 3.

We also present a minor result for MWM in the online preemptive
model, a 3-competitive (in expectation) randomized algorithm (that uses
only $O(1)$ bits of randomness) on growing trees (where the input revealed
upto any stage is always a tree, i.e. a new edge never connects two disconnected trees).

1 Introduction

The *Maximum (Cardinality/Weight) Matching* problem is one of the most extensively studied problems in Combinatorial Optimization. See Schrijver's book [12]

S. Tirodkar—A part of this work was done when the author was a student in the
Department of Computer Science and Engineering at IIT Bombay.

Y. Cao and J. Chen (Eds.): COCOON 2017, LNCS 10392, pp. 504–515, 2017.
DOI: 10.1007/978-3-319-62389-4_42

and references therein for a comprehensive overview of classic work. A *matching* $M \subseteq E$ is a set of edges such that at most one edge is incident on any vertex. Traditionally the problem was studied in the offline setting where the entire input is available to the algorithm beforehand. But over the last few decades it has been extensively studied in various other models where the input is revealed in pieces, like the vertex arrival model (adversarial and random), the edge arrival model (adversarial and random), streaming and semi-streaming models, the online preemptive model, etc. [4–6,9–11]. In this paper, we study the Maximum Cardinality Matching (MCM) and the Maximum Weight Matching (MWM) problems, on trees and on some special classes of graphs, in the *Online Preemptive* model, and in the *Incremental Dynamic Graph* model. (Refer Appendix D in [14] for a comparison between the two models.)

In the online preemptive model, the edges arrive in an online manner, and the algorithm is supposed to accept or reject an edge on arrival. If accepted, the algorithm can reject it later, and all rejections are permanent. The algorithm is supposed to always maintain a valid matching. There is a 5.828-competitive deterministic algorithm due to McGregor [11] for MWM, and a tight lower bound on deterministic algorithms due to Varadaraja [15]. Epstein et al. [5] gave a 5.356-competitive randomized algorithm for MWM, and also proved a 1.693 lower bound on the competitive ratio achievable by any randomized algorithm for MCM. No better lower bound is known for MWM.

In [3], the authors gave the first randomized algorithm with competitive ratio (28/15 in expectation) less than 2 for MCM in the online preemptive model, on growing trees (defined in Sect. 1.1). In Sect. 2, we extend their algorithm to give a 64/33-competitive (in expectation) randomized (which uses only two bits of randomness) algorithm for trees. Although the algorithm is an extension of the one for growing trees in [3], it serves as a basis of the algorithm (described in Sect. 3) for MCM in the incremental dynamic graph model.

Note that the adversary presenting the edges in the online preemptive model is oblivious, and does not have access to the random choices made by the algorithm.

In recent years, algorithms for approximate MCM in dynamic graphs have been the focus of many studies due to their wide range of applications. Here [1,2,8,13] is a non-exhaustive list some of the studies. The objective of these dynamic graph algorithms is to efficiently process an online sequence of update operations, such as edge insertions and deletions. It has to quickly maintain an approximate maximum matching despite an adversarial order of edge insertions and deletions. Dynamic graph problems are usually classified according to the types of updates allowed: incremental models allow only insertions, decremental models allow only deletions, and fully dynamic models allow both. We study MCM in the incremental model. Gupta [7] proved that for any $\epsilon \leq 1/2$, there exists an algorithm that maintains a $(1+\epsilon)$-approximate MCM on bipartite graphs in the incremental model in an "amortized" update time of $O\left(\frac{\log^2 n}{\epsilon^4}\right)$. We present a randomized algorithm for MCM in the incremental model with a "worst case" update time of $O(1)$, which is 3/2-approximate (in expectation) on trees, and 1.8-approximate (in expectation) on general graphs with maximum degree 3. Note that the algorithm of Gupta [7] is based on multiplicative weights update, and it therefore seems unlikely that a better running time analysis for special classes of graphs is possible.

We present a minor result in Sect. 4, a 3-competitive (in expectation) randomized algorithm (which uses only $O(1)$ bits of randomness) for MWM on growing trees in the online preemptive model. Although, growing trees is a very restricted class of graphs, there are a couple of reasons to study the performance of the algorithm on this class of input. Firstly, almost all lower bounds, including the one due to Varadaraja [15] for MWM are on growing trees. Secondly, even for this restricted class, the analysis is involved. We use the primal-dual technique for analyzing the performance of this algorithm, and show that this analysis is indeed tight by giving an example, for which the algorithm achieves the competitive ratio 3. We describe the algorithm for general graphs, but were only able to analyze it for growing trees, and new ideas are needed to prove a better bound for general graphs.

1.1 Preliminaries

We use primal-dual techniques to analyze the performance of all the randomized algorithms described in this paper. Here are the well known Primal and Dual formulations of the matching problem.

Primal LP	Dual LP
$\max \sum_e w_e x_e$	$\min \sum_v y_v$
$\forall v : \sum_{v \in e} x_e \leq 1$	$\forall e : y_u + y_v \geq w_e$
$x_e \geq 0$	$y_v \geq 0$

For MCM, $w_e = 1$ for any edge. Any matching M implicitly defines a feasible primal solution. If an edge $e \in M$, then $x_e = 1$, otherwise $x_e = 0$.

Suppose an algorithm outputs a matching M, then let P be the corresponding primal feasible solution. Let D denote some feasible dual solution. The following claim can be easily proved using weak duality.

Claim 1. *If $D \leq \alpha \cdot P$, then the algorithm is α-competitive.*

If M is any matching, then for an edge e, $X(M, e)$ denotes edges in M which share a vertex with the edge e. We will say that a vertex(/an edge) is *covered* by a matching M if there is an edge in M which is incident on(/adjacent to) the vertex(/edge). We also say that an edge is *covered* by a matching M if it belongs to M.

In the online preemptive model, growing trees are trees, such that a new edge has exactly one vertex common with already revealed edges.

2   MCM in the Online Preemptive Model

In this section, we present a randomized algorithm (that uses only 2 bits of randomness) for MCM on trees in the online preemptive model.

The algorithm maintains four matchings M_1, M_2, M_3, M_4, and it tries to ensure that a large number of input edges are covered by some or other matchings. (Here, the term "large number" is used vaguely. Suppose more than four edges are incident on a vertex, then at most four of them will belong to matchings, one to each.) One of the four matchings is output uniformly at random. A more formal description of the algorithm follows.

Algorithm 1. Randomized Algorithm for MCM on Trees

1. Pick $l \in_{u.a.r.} \{1, 2, 3, 4\}$.
2. The algorithm maintains four matchings: M_1, M_2, M_3, and M_4.
3. On arrival of an edge e, the processing happens in two phases.
   (a) **The augment phase.** The new edge e is added to each M_i in which there are no edges adjacent to e.
   (b) **The switching phase.** For $i = 2, 3, 4$, in order, $M_i \leftarrow M_i \setminus X(M_i, e) \cup \{e\}$, provided it decreases the quantity $\sum_{j \in [4], i \neq j, |X(M_i \cap M_j, e)| = |X(M_i, e)|} |M_i \cap M_j|$.
4. Output M_l.

Although, l is picked randomly at the beginning of the algorithm, this value is not known to the adversary.

Note that in the switching phase, the expected size of the matching stored by the algorithm might decrease. For example, consider two disjoint edges e_1 and e_2 that have been revealed. Each of them will belong to all four matchings. So the expected size of the matching stored by the algorithm is 2. Now, if an edge e is revealed between e_1 and e_2, then e will be added to M_2 and M_3. The expected size of the matching is now 1.5. The important thing to notice here is that the decrease is not too much, and we are able to prove that the competitive ratio of the algorithm still remains below 2.

We begin with the following observations.

- After an edge is revealed, its end points are covered by all four matchings.
- An edge e that does not belong to any matching has four edges incident on its end points such that each of these edges belongs to a distinct matching. This holds when the edge is revealed, and does not change subsequently.
- Every edge is covered by at least three matchings.

An edge is called *internal* if there are edges incident on both its end points which belong to some matching. An edge is called a *leaf edge* either if one of its end point is a leaf or if all the edges incident on one of its end points do not belong to any matching. An edge is called *bad* if its end points are covered by only three matchings.

We begin by proving some properties about the algorithm. The key structural lemma that keeps "influences" of bad edges local is given below.

Lemma 1. *At most five consecutive vertices on a path can have bad edges incident on them.*

According to Lemma 1, there can be at most four consecutive internal bad edges or at most five bad leaf edges incident on five consecutive vertices of a path. Lemma 1 is proved in Appendix A of [14].

Once all edges have been seen, we distribute the primal charge among the dual variables, and use the primal-dual framework to infer the competitive ratio. If end points of every edge are covered with four matchings, then the distribution of dual charge is easy. However we do have bad edges, and would like the edges in matchings to contribute more to the end-points of these edges. Then, the charge on the other end-point would be less and we need to balance this through other edges. Details follow.

Lemma 2. *There exists an assignment of the primal charge to the dual variables such that the dual constraint for each edge $e \equiv (u, v)$ is satisfied at least $\frac{33}{64}$ in expectation, i.e. $\mathbb{E}[y_u + y_v] \geq \frac{33}{64}$.*

Proof. Root the tree at an arbitrary vertex. For any edge $e \equiv (u, v)$, let v be the parent vertex, and u be the child vertex. The dual variable assignment is done after the entire input is seen, as follows.

Dual Variable Management: An edge e will distribute its primal weight between its end-points. The exact values are discussed below. In general, we look to transfer all of the primal charge to the parent vertex. But this does not work and we need a finer strategy. This is detailed below.

- If e does not belong to any matching, then it does not contribute to the value of dual variables.
- If e belongs to a single matching then, depending on the situation, one of 0, ϵ, 2ϵ, 3ϵ, 4ϵ, or 5ϵ of its primal charge will be assigned to u and the rest will be assigned to v.
- If e belongs to two matchings, then at most 6ϵ of its primal charge will be assigned to u as required. The rest is assigned to v.
- If e belongs to three or four matchings, then its entire primal charge is assigned to v.

We will show that $y_u + y_v \geq 2 + \epsilon$ for such an edge, when summed over all four matchings. The value of ϵ is chosen later.

For the sake of analysis, if there are bad leaf edges incident on both end points of an internal edge, then we analyze it as a bad internal edge. We need to do this because a bad leaf edge might need to transfer its entire primal charge to the vertex on which there are edges which do not belong to any matching. Note that the end points of the internal edge would still be covered by three matchings, even if we consider that the bad leaf edges do not exist on its end points. The analysis breaks up into eight cases.

Case 1. Suppose e does not belong to any matching. There must be a total of at least 4 edges incident on u and v besides e, each belonging to a distinct matching. Of these 4, at least a total of 3, say e_1, e_2, and e_3, must be between some children of u and v, to u and v respectively. The edges e_1, e_2, and e_3, each assign a charge of at least $1 - 5\epsilon$ to y_u and y_v, respectively. Therefore, $y_u + y_v \geq 3 - 15\epsilon \geq 2 + \epsilon$.

Case 2. Suppose e is a bad leaf edge that belongs to a single matching, and internal edges are incident on v. This implies that there is an edge e_1 from a child vertex of v to v, which belongs to single matching, and another edge e_2, also belonging to single matching from v to its parent vertex. The edge e assigns a charge of 1 to y_v. If e_1 assigns a charge of 1 or $1 - \epsilon$ or $1 - 2\epsilon$ or $1 - 3\epsilon$ or $1 - 4\epsilon$ to y_v, then e_2 assigns ϵ or 2ϵ or 3ϵ or 4ϵ or 5ϵ respectively to y_v. In either case, $y_u + y_v = 2 + \epsilon$. The key fact is that e_1 could not have assigned 5ϵ to its child vertex. Since, then, by Lemma 1, e cannot be a bad edge.

Case 3. Suppose e is a bad leaf edge that belongs to a single matching, and internal edges are incident on u. This implies that there are two edges e_1 and e_2 from children of u to u, each belonging to a single distinct matching. The edge e assigns a charge of 1 to y_v. Both e_1 and e_2 assign a charge of at least $1 - 4\epsilon$ to y_u. In either case, $y_u + y_v \geq 3 - 8\epsilon \geq 2 + \epsilon$. The key fact is that neither e_1 nor e_2 could have assigned more than 4ϵ to their corresponding child vertices. Since, then, by Lemma 1, e cannot be a bad edge.

Case 4. Suppose e is an internal bad edge. This implies (by Lemma 1) that there is an edge e_1 from a child vertex of u to u, which belongs to a single matching. Also, there is an edge e_2, from v to its parent vertex (or from a child vertex v to v), which also belongs to a single matching. The edge e assigns its remaining charge (1 or $1 - \epsilon$ or $1 - 2\epsilon$ or $1 - 3\epsilon$ or $1 - 4\epsilon$) to y_v. If e_1 assigns a charge of 1 or $1 - \epsilon$ or $1 - 2\epsilon$ or $1 - 3\epsilon$ or $1 - 4\epsilon$ to y_u, then e_2 assigns ϵ or 2ϵ or 3ϵ or 4ϵ or 5ϵ respectively to y_v. In either case, $y_u + y_v = 2 + \epsilon$. The key fact is that e_1 could not have assigned 5ϵ to its child vertex. Since, then, by Lemma 1, e cannot be a bad edge.

Case 5. Suppose e is not a bad edge, and it belongs to a single matching. Then either there are at least two edges e_1 and e_2 from child vertices of u or v to u or v respectively, or e_1 on u and e_2 on v, each belonging to a single matching, or one edge e_3 from a child vertex of u or v to u or v, respectively, which belongs to two matchings, or one edge e_4 from a child vertex of u or v to u or v, respectively, which belongs to single matching, and one edge e_5 from v to its parent vertex which belongs to two matchings. In either case, $y_u + y_v \geq 3 - 10\epsilon \geq 2 + \epsilon$.

Case 6. Suppose e is a bad edge that belongs to two matchings, and internal edge is incident on u or v. This implies that there is an edge e_1, from a child vertex of u to u or from v to its parent vertex which belongs to a single matching. The edge e assigns a charge of 2 to y_v, and the edge e_1 assigns a charge of ϵ to y_u or y_v respectively. Thus, $y_u + y_v = 2 + \epsilon$.

Case 7. Suppose e is not a bad edge and it belongs to two matchings. This means that either there is an edge e_1 from a child vertex of u to u, which belongs to at least one matching, or there is an edge from child vertex of v to v that belongs to at least one matching, or there is an edge from v to its parent vertex which belongs to two matchings. The edge e assigns a charge of 2 among y_u and y_v. The neighboring edges assign a charge of ϵ to y_u or y_v (depending on which vertex it is incident to), yielding $y_u + y_v \geq 2 + \epsilon$.

Case 8. Suppose, e belongs to 3 or 4 matchings, then trivially $y_u + y_v \geq 2 + \epsilon$.

From the above cases, $y_v + y_v \geq 3 - 15\epsilon$ and $y_u + y_v \geq 2 + \epsilon$. The best value for the competitive ratio is obtained when $\epsilon = \frac{1}{16}$, yielding $\mathbb{E}[y_u + y_v] \geq \frac{33}{64}$. □

Lemma 2 immediately implies Theorem 2 using Claim 1.

Theorem 2. *Algorithm 1 is a $\frac{64}{33}$-competitive randomized algorithm for finding MCM on trees.*

3   MCM in the Incremental Dynamic Graph Model

In this section, we present our main result, a randomized algorithm (that uses only $O(1)$ bits of randomness) for MCM in the incremental dynamic graph model, which is 3/2-approximate (in expectation) on trees, and is 1.8-approximate (in expectation) on general graphs with maximum degree 3. It is inspired by the randomized algorithm for MCM on trees described in Sect. 2. In the online preemptive model, we cannot add edges in the matching which were discarded earlier, which results in the existence of bad edges. But in the incremental dynamic graph model, there is no such restriction. For some $i \in [3]$, let $e \equiv (u, v) \in M_i$ be switched out by some edge $e' \equiv (u, u')$, i.e. $M_i \leftarrow M_i \setminus \{e\} \cup \{e'\}$. If there is an edge $e'' \equiv (v, v') \in M_j$ for $i \neq j$, then we can add e'' to M_i if possible. Using this simple trick, we get a better approximation ratio in this model, and also, the analysis becomes significantly simpler. Details follow.

Algorithm 2. Randomized Algorithm for MCM

---
1. Pick $l \in_{u.a.r.} \{1, 2, 3\}$.
2. The algorithm maintains three matchings: M_1, M_2, and M_3.
3. When an edge e is inserted, the processing happens in two phases.
   (a) **The augment phase.** The new edge e is added to each M_i in which there are no edges adjacent to e.
   (b) **The switching phase.** For $i = 2, 3$, in order, $M_i \leftarrow M_i \setminus X(M_i, e) \cup \{e\}$, provided it decreases the quantity $\sum_{j \in [3], i \neq j, |X(M_i \cap M_j, e)| = |X(M_i, e)|} |M_i \cap M_j|$. For every edge e' discarded from M_i, add edges on the other end point of e' in M_j ($\forall j \neq i$) to M_i if possible.
4. Output the matching M_l.

Note that the end points of every edge will be covered by all three matchings.

We again use the primal-dual technique to analyze the performance of this algorithm on trees.

Lemma 3. *There exists an assignment of the primal charge amongst the dual variables such that the dual constraint for each edge $e \equiv (u, v)$ is satisfied at least $\frac{2}{3}$rd in expectation.*

Proof. Root the tree at an arbitrary vertex. For any edge $e \equiv (u, v)$, let v be the parent vertex, and u be the child vertex. The dual variable assignment is done at the end of input, as follows.

- If e does not belong to any matching, then it does not contribute to the value of dual variables.
- If e belongs to a single matching, then its entire primal charge is assigned to v as $y_v = 1$.
- If e belongs to two matchings, then its entire primal charge is assigned equally amongst u and v, as $y_u = 1$ and $y_v = 1$.
- If e belongs to three matchings, then its entire primal charge is assigned to v as $y_v = 3$.

The analysis breaks up into three cases.

Case 1. Suppose e does not belong to any matching. There must be a total of at least 2 edges incident amongst u and v besides e, each belonging to a distinct matchings, from their respective children. Therefore, $y_u + y_v \geq 2$.

Case 2. Suppose e belongs to a single matching. Then either there is an edge e' incident on u or v which belongs to a single matching, from their respective children, or there is an edge e'' incident on u or v which belongs to two matchings. In either case, $y_u + y_v \geq 2$.

Case 3. Suppose e belongs to two or three matchings, then $y_u + y_v \geq 2$ trivially.   □

Lemma 3 immediately implies Theorem 3 using Claim 1.

Theorem 3. *Algorithm 2 is a $\frac{3}{2}$-approximate (in expectation) randomized algorithm for MCM on trees, with a worst case update time of $O(1)$.*

We also analyze Algorithm 2 for general graphs with maximum degree 3, and prove the following Theorem.

Theorem 4. *Algorithm 2 is a 1.8-approximate (in expectation) randomized algorithm for MCM on general graphs with maximum degree 3, with a worst case update time of $O(1)$.*

The proof of Theorem 4 is presented in Appendix E of [14].

4 MWM in the Online Preemptive Model

In this section, we present a randomized algorithm (that uses only $O(1)$ bits of randomness) for MWM in the online preemptive model, and analyze its performance for growing trees. We describe the algorithm next. (An intuition for the algorithm is presented in Appendix B of [14].)

Algorithm 3. Randomized Algorithm for MWM

---

1. Maintain two matchings M_1 and M_2. Let $j = 1$ with probability p, and $j = 2$ otherwise.
2. On receipt of an edge e:
   For $i = 1, 2$, if $w(e) > (1 + \gamma_i)w(X(M_i, e))$, then $M_i = M_i \setminus X(M_i, e) \cup \{e\}$.
3. Output M_j.

Note that we cannot just output the best of two matchings because that could violate the constraints of the online preemptive model.

4.1 Analysis

We use the primal-dual technique to analyze the performance of this algorithm. The primal-dual technique used to analyze McGregor's deterministic algorithm for MWM described in [3] is fairly straightforward. However the management becomes complicated with the introduction of randomness, and we are only able to analyze the algorithm in a very restricted class of graphs, which are growing trees.

Theorem 5. *The expected competitive ratio of Algorithm 3 on growing trees is*

$$\max \left\{ \frac{1 + \gamma_1}{p}, \frac{1 + \gamma_2}{1 - p}, \frac{(1 + \gamma_1)(1 + \gamma_2)(1 + 2\gamma_1)}{p \cdot \gamma_1 + (1 - p)\gamma_2 + \gamma_1 \gamma_2} \right\},$$

where p is the probability to output M_1.

We maintain both primal and dual variables along with the run of the algorithm. Consider a round in which an edge $e \equiv (u, v)$ is revealed, where v is the new vertex. Before e is revealed, let e_1 and e_2 be the edges incident on u which belong to M_1 and M_2 respectively. If such an e_i does not exist, then we may assume $w(e_i) = 0$. The primal and dual variables are updated as follows.

- e is rejected by both matchings, we set the primal variable $x_e = 0$, and the dual variable $y_v = 0$.
- e is added to M_1 only, then we set the primal variable $x_e = p$, and the dual variable $y_u = \max(y_u, \min((1 + \gamma_1)w(e), (1 + \gamma_2)w(e_2)))$, and $y_v = 0$;.
- e is added to M_2 only, then we set the primal variable $x_e = 1 - p$, and the dual variable $y_u = \max(y_u, \min((1 + \gamma_1)w(e_1), (1 + \gamma_2)w(e)))$, and $y_v = 0$.

- e is added to both the matchings, then we set the primal variable $x_e = 1$, and the dual variables $y_u = \max(y_u, (1 + \gamma_1)w(e))$ and $y_v = (1 + \gamma_1)w(e)$.
- When an edge e' is evicted from M_1 (or M_2), we decrease its primal variable $x_{e'}$ by p (or $(1 - p)$ respectively), and the corresponding dual variables are unchanged.

We begin with three simple observations.

1. The cost of the primal solution is equal to the expected weight of the matching maintained by the algorithm.
2. The dual variables never decrease. Hence, if a dual constraint is feasible once, it remains so.
3. $y_u \geq \min((1 + \gamma_1)w(e_1), (1 + \gamma_2)w(e_2))$.

The idea behind the analysis is to prove a bound on the ratio of the dual cost and the primal cost while maintaining dual feasibility. By Observation 2, to ensure dual feasibility, it is sufficient to ensure feasibility of the dual constraint of the new edge. If the new edge e is not accepted in any M_i, then $w(e) \leq \min((1 + \gamma_1)w(e_1), (1 + \gamma_2)w(e_2))$. Hence, the dual constraint is satisfied by Observation 3. Else, it can be seen that the dual constraint is satisfied by the updates performed on the dual variables.

The following lemma implies Theorem 5 using Claim 1.

Lemma 4. $\frac{\Delta Dual}{\Delta Primal} \leq \max\left\{ \frac{1+\gamma_1}{p}, \frac{1+\gamma_2}{1-p}, \frac{(1+\gamma_1)(1+\gamma_2)(1+2\gamma_1)}{p \cdot \gamma_1 + (1-p)\gamma_2 + \gamma_1\gamma_2} \right\}$ *after every round.*

We will use the following simple technical lemma to prove Lemma 4.

Lemma 5. $\frac{ax+b}{cx+d}$ *increases with x iff $ad - bc \geq 0$.*

Proof (of Lemma 4). There are four cases to be considered.

1. If edge e is accepted in M_1, but not in M_2. Then $(1 + \gamma_1)w(e_1) < w(e) \leq (1 + \gamma_2)w(e_2)$. By Observation 3, before e was revealed, $y_u \geq (1 + \gamma_1)w(e_1)$. After e is accepted in M_1, $\Delta Primal = p(w(e) - w(e_1))$, and $\Delta Dual \leq (1 + \gamma_1)(w(e) - w(e_1))$. Hence,

$$\frac{\Delta Dual}{\Delta Primal} \leq \frac{(1 + \gamma_1)}{p}.$$

2. If edge e is accepted in M_2, but not in M_1. Then $(1 + \gamma_2)w(e_2) < w(e) \leq (1 + \gamma_1)w(e_1)$. By Observation 3, before e was revealed, $y_u \geq (1 + \gamma_2)w(e_2)$. After e is accepted in M_2, $\Delta Primal = (1 - p)(w(e) - w(e_2))$, and $\Delta Dual \leq (1 + \gamma_2)(w(e) - w(e_2))$. Hence,

$$\frac{\Delta Dual}{\Delta Primal} \leq \frac{(1 + \gamma_2)}{1 - p}.$$

3. If edge e is accepted in both the matchings, and $(1+\gamma_1)w(e_1) \le (1+\gamma_2)w(e_2)$ $< w(e)$. By Observation 3, before e was revealed, $y_u \ge (1+\gamma_1)w(e_1)$. After e is accepted in both the matchings, $\Delta\text{Dual} \le (1+\gamma_1)(2w(e) - w(e_1))$. The change in primal cost is

$$\Delta\text{Primal} \ge w(e) - p \cdot w(e_1) - (1-p) \cdot w(e_2)$$

$$\ge w(e) - p \cdot w(e_1) - (1-p) \cdot \frac{w(e)}{1+\gamma_2}$$

$$= \frac{p+\gamma_2}{1+\gamma_2} w(e) - p \cdot w(e_1).$$

$$\frac{\Delta\text{Dual}}{\Delta\text{Primal}} \le (1+\gamma_1)\frac{2w(e) - w(e_1)}{\frac{p+\gamma_2}{1+\gamma_2}w(e) - p \cdot w(e_1)}.$$

By Lemma 5, this value increases, for a fixed $w(e)$, with $w(e_1)$ if $\gamma_2 \le \frac{p}{1-2p}$, and its worst case value is achieved when $(1+\gamma_1)w(e_1) = w(e)$. Thus,

$$\frac{\Delta\text{Dual}}{\Delta\text{Primal}} \le (1+\gamma_1)\frac{2(1+\gamma_1)(1+\gamma_2) - (1+\gamma_2)}{(p+\gamma_2)(1+\gamma_1) - p(1+\gamma_2)}$$

$$= (1+\gamma_1)(1+\gamma_2)\frac{1+2\gamma_1}{p \cdot \gamma_1 + (1-p)\gamma_2 + \gamma_1\gamma_2}.$$

4. If e is accepted in both the matchings, and $(1+\gamma_2)w(e_2) \le (1+\gamma_1)w(e_1) < w(e)$. By Observation 3, before e was revealed, $y_u \ge (1+\gamma_2)w(e_2)$. The following bound can be proved similarly.

$$\frac{\Delta\text{Dual}}{\Delta\text{Primal}} \le (1+\gamma_1)(1+\gamma_2)\frac{1+2\gamma_1}{p \cdot \gamma_1 + (1-p)\gamma_2 + \gamma_1\gamma_2}.$$

□

The following theorem is an immediate consequence of Theorem 5.

Theorem 6. *Algorithm 3 is a 3-competitive (in expectation) randomized algorithm for MWM on growing trees, when $p = 1/3$, $\gamma_1 = 0$, and $\gamma_2 = 1$; and the analysis is tight.*

In Appendix C of [14], an input is presented for which both M_1 and M_2 are simultaneously of 1/3rd the weight of the optimum.

Note. In the analysis of Algorithm 3 for growing trees, we crucially use the following fact in the dual variable assignment. If an edge $e \notin M_i$ for some i, then a new edge incident on its leaf vertex will definitely be added to M_i, and it suffices to assign a zero charge to the corresponding dual variable. This is not necessarily true for more general classes of graphs, and new ideas are needed to analyze the performance for those classes.

Acknowledgements. The first author would like to thank Ashish Chiplunkar for helpful suggestions to improve the competitive ratio of Algorithm 3, and also to improve the presentation of Sect. 4.

References

1. Bernstein, A., Stein, C.: Faster fully dynamic matchings with small approximation ratios. In: Proceedings of the Twenty-Seventh Annual ACM-SIAM Symposium on Discrete Algorithms, SODA 2016, Arlington, VA, USA, 10–12 January 2016, pp. 692–711 (2016)
2. Bhattacharya, S., Henzinger, M., Nanongkai, D.: Fully dynamic approximate maximum matching and minimum vertex cover in $O(\log^3 n)$ worst case update time. In: Proceedings of the Twenty-Eighth Annual ACM-SIAM Symposium on Discrete Algorithms, SODA 2017, Barcelona, Spain, Hotel Porta Fira, 16–19 January, pp. 470–489 (2017)
3. Chiplunkar, A., Tirodkar, S., Vishwanathan, S.: On randomized algorithms for matching in the online preemptive model. In: Bansal, N., Finocchi, I. (eds.) ESA 2015. LNCS, vol. 9294, pp. 325–336. Springer, Heidelberg (2015). doi:10.1007/978-3-662-48350-3_28
4. Epstein, L., Levin, A., Mestre, J., Segev, D.: Improved approximation guarantees for weighted matching in the semi-streaming model. SIAM J. Discrete Math. **25**(3), 1251–1265 (2011)
5. Epstein, L., Levin, A., Segev, D., Weimann, O.: Improved bounds for online preemptive matching. In: Proceedings of the 30th International Symposium on Theoretical Aspects of Computer Science, STACS 2013, Kiel, Germany, pp. 389–399 (2013)
6. Feigenbaum, J., Kannan, S., McGregor, A., Suri, S., Zhang, J.: On graph problems in a semi-streaming model. Theor. Comput. Sci. **348**(2), 207–216 (2005)
7. Gupta, M.: Maintaining approximate maximum matching in an incremental bipartite graph in polylogarithmic update time. In: Proceedings of the 34th International Conference on Foundation of Software Technology and Theoretical Computer Science, FSTTCS 2014, New Delhi, India, 15–17 December 2014, pp. 227–239 (2014) (2014)
8. Gupta, M., Peng, R.: Fully dynamic $(1+\epsilon)$-approximate matchings. In: Proceedings of the 2013 IEEE 54th Annual Symposium on Foundations of Computer Science, FOCS 2013, pp. 548–557. IEEE Computer Society, Washington, DC (2013)
9. Kalyanasundaram, B., Pruhs, K.: Online weighted matching. J. Algorithms **14**(3), 478–488 (1993)
10. Karp, R.M., Vazirani, U.V., Vazirani, V.V.: An optimal algorithm for on-line bipartite matching. In: Proceedings of the Twenty-Second Annual ACM Symposium on Theory of Computing, STOC 1990, pp. 352–358. ACM, New York (1990)
11. McGregor, A.: Finding graph matchings in data streams. In: Chekuri, C., Jansen, K., Rolim, J.D.P., Trevisan, L. (eds.) APPROX/RANDOM -2005. LNCS, vol. 3624, pp. 170–181. Springer, Heidelberg (2005). doi:10.1007/11538462_15
12. Schrijver, A.: Combinatorial Optimization - Polyhedra and Efficiency. Springer, Heidelberg (2003)
13. Solomon, S.: Fully dynamic maximal matching in constant update time. In: IEEE 57th Annual Symposium on Foundations of Computer Science, FOCS 2016, Hyatt Regency, New Brunswick, New Jersey, USA, 9–11 October 2016, pp. 325–334 (2016)
14. Tirodkar, S., Vishwanathan, S.: Maximum matching on trees in the online preemptive and the incremental dynamic graph models. CoRR abs/1612.05419 (2016). http://arxiv.org/abs/1612.05419
15. Badanidiyuru Varadaraja, A.: Buyback problem - approximate matroid intersection with cancellation costs. In: Aceto, L., Henzinger, M., Sgall, J. (eds.) ICALP 2011. LNCS, vol. 6755, pp. 379–390. Springer, Heidelberg (2011). doi:10.1007/978-3-642-22006-7_32

Approximation Algorithms for Scheduling Multiple Two-Stage Flowshops

Guangwei Wu[1,2] and Jianxin Wang[1(✉)]

[1] School of Information Science and Engineering, Central South University,
Changsha, People's Republic of China
jxwang@csu.edu.cn
[2] College of Computer and Information Engineering,
Central South University of Forestry and Technology,
Changsha, People's Republic of China

Abstract. This paper studies the problem that schedules n two-stage jobs on m multiple two-stage flowshops, with the objective of minimizing the makespan. The problem is NP-hard even when m is a fixed constant, and becomes strongly NP-hard when m is a part of input. A 17/6-approximation algorithm along with its analysis is presented for arbitrary $m \geq 2$. This is the first approximation algorithm for multiple flowshops when the number m of flowshops is a part of input. The arbitrary m and the time complexity $O(n \log n + mn)$ of the algorithm demonstrate that the problem, which plays an important role in the current research in cloud computing and data centers, can be solved efficiently with a reasonable level of satisfaction.

Keywords: Scheduling · Multiple two-stage flowshops · Approximation algorithm · Cloud computing

1 Introduction

Motivated by the current research in data centers and cloud computing, this paper studies the scheduling problem for two-stage jobs on multiple two-stage flowshops. A job is a *two-stage job* if it consists of an R-operation and a T-operation. A flowshop is a *two-stage flowshop* if it contains an R-processor and a T-processor. When a job is assigned to a flowshop, its R-operation and T-operation are executed by R-processor and T-processor of the same flowshop, respectively, and the T-operation cannot start until the R-operation is finished. Therefore in this problem, a schedule is to assign the jobs to flowshops, and then for each flowshop, to determine the executing order of the R- and T-operations of the jobs assigned to that flowshop. The problem is formally stated as follows:

This work is supported by the National Natural Science Foundation of China under grants 61420106009, 61672536, 61232001, and 61472449, Scientific Research Fund of Hunan Provincial Education Department under grant 16C1660.

Y. Cao and J. Chen (Eds.): COCOON 2017, LNCS 10392, pp. 516–528, 2017.
DOI: 10.1007/978-3-319-62389-4_43

Construct a schedule for n given two-stage jobs on m identical two-stage flowshops that minimizes the makespan, i.e., the completion time of the last job.

The drive of our paper on the multiple two-stage flowshops problem came from the recent research on data centers and cloud computing. Cloud computing refers to both the applications delivered as services over the Internet and the hardware and system software in the data centers that provide these services [3]. Today's data center contains tens of thousands or more servers consisting of CPUs, memory, network interface, and local high-speed I/O (disk or flash), where applications and data are stored as resources [1,9]. In this kind of context, when clients request services in cloud computing, for a corresponding resource, the server needs first to read the resource from local I/O to memory – this is the R-operation, and then to send it through network interface over Internet to the clients – this is the T-operation. Therefore, given a set of resources requested from clients, scheduling them on multiple servers in data centers to make the completion time of the last resource minimum is meaningful in practical applications, which is exactly the scheduling problem studied in this paper. The throughput of server is limited by the bandwidth of disk (or flash) and of network [13]. Typical settings of server are shown in Table 1 [2], which shows that the R-operation and the T-operation in a typical server are comparable in general. The two operations need not to have a linear relation due to the impact of cache system, therefore, neither of them can be simply ignored when considering this scheduling problem. There are other characteristics to be considered. For example, the number of servers in data centers is usually very large and may vary frequently due to the issues such as economic factors [9]. Thus, it is natural to consider the number of servers as a part of input rather than a fixed integer.

Table 1. Typical server settings in data centers

Media	Capacity (TB)	Bandwidth (GB/sec)
Disk(x18)	12–144	0.2–2
Flash(x4)	1–4	1–4
Network	N/A	1.25–5

When the number m of flowshops is 1, this scheduling problem becomes the classic two-stage flowshop problem, which can be solved in $O(n \log n)$ time by the classical Johnson's algorithm where n is the number of jobs [12]. For a general case where $m \geq 2$, unfortunately, even in the special case when the R-operation of each job costs 0, the problem becomes the classic MAKESPAN problem, which is NP-hard when $m \geq 2$ is a fixed number and becomes strongly NP-hard when m is an arbitrary integer [5]. As a consequence, the scheduling problem studied in the current paper, where m is a part of input, is strongly NP-hard.

The current paper focuses on approximation algorithms for scheduling on multiple two-stage flowshops, with the objective of minimizing the scheduling makespan. An α-*approximation algorithm*, that achieves the *approximation ratio* α, is a polynomial-time algorithm that for all instances of the problem produces a schedule whose makespan is within α times the minimum makespan [16].

We review some important results related to the problem studied in this paper. For the classic MAKESPAN problem, which can be viewed as scheduling one-stage jobs on one-stage flowshops, the well-known *ListRanking* algorithm is an important technique when the number m of flowshops is arbitrary. The ListRanking algorithm was due to Graham who also proved that the algorithm achieves an approximation ratio $(2 - 1/m)$ for arbitrary m [6]. Graham further improved the approximation ratio to $(4 - 1/m)/3$ by sorting the jobs in non-increasing order before *ListRanking* [7]. Hochbaum and Shmoys gave a polynomial-time approximation scheme using dual approximation algorithm [11]. When m is a fixed integer, Sahni proposed a fully polynomial-time approximation scheme [14].

Scheduling multiple two-stage flowshops had not been studied thoroughly until very recently, and most studies focused on the problem when the number m of flowshops is a fixed constant. He *et al.* [10] was the first group who studied this problem, motivated by applications in glass manufacturing. A mixed-integer programming formulation was proposed and a heuristic algorithm was developed. Vairaktarakis *et al.* [15] studied the problem in order to cope with the hybrid flowshop problem. In particular, when the number m of flowshops equals 2, they proposed a formulation that leads to a pseudo-polynomial time exact algorithm. Zhang *et al.* [19] developed a 3/2-approximation algorithm for $m = 2$ and a 12/7-approximation algorithm for $m = 3$ for the problem. Both algorithms running in time $O(n \log n)$ first sort the jobs by Johnson's algorithm, then split the jobs into two (or three) subsequences for two (or three) flowshops. Using a similar formulation to that for $m = 2$ in [15], Dong *et al.* [4] developed a pseudo-polynomial time exact algorithm, and proposed a fully polynomial-time approximation scheme based on the exact algorithm for a fixed constant $m \geq 2$. Very recently, Wu *et al.* [17] proposed a new formulation for the problem that leads directly to improvements to the complexity of scheduling algorithms. They further proposed a new approach that significantly improves scheduling algorithms when the costs of the two stages are different significantly. Further improvements for the approximation algorithms were also studied.

When the number of flowshops is unbounded, the multiple two-stage flowshops problem is strongly NP-hard and had almost been untouched. Wu *et al.* [18] considered two restricted versions of the problem: one restricts the R-operation to cost no less time than the T-operation for each job, while the other assumes that the T-operation is more time-consuming than the R-operation for each job. For the first case, an online 2-competitive algorithm and an offline 11/6-approximation algorithm were developed. For the second case, an online 5/2-competitive algorithm and an offline 11/6-approximation algorithm were given.

To the best of our knowledge, no approximation results for the problem of scheduling multiple flowshops have been known when m is not a fixed constant. On the other hand, as explained earlier in this section, for applications in cloud computing and data centers, scheduling two-stage jobs on unbounded number of two-stage flowshops seems a common practice. This motivated the research of the current paper. In this paper, we present a 17/6-approximation algorithm for the multiple two-stage flowshop scheduling problem for arbitrary $m \geq 2$, which runs in time $O(n \log n + nm)$. This is the first approximation algorithm for the problem when the number of flowshops is not a fixed constant. The arbitrary m and the time complexity of the algorithm demonstrate that the problem, which plays an important role in the current research in cloud computing and data centers, can be solved efficiently with a reasonable level of satisfaction. Moreover, practically, for the case where the number of flowshops is a fixed constant, our algorithm is more efficient when compared with those approximation algorithms constructed from pseudo-polynomial time exact algorithms [4,17].

2 Preliminary

For n two-stage jobs $G = \{J_1, \ldots, J_n\}$ to be scheduled on m identical two-stage flowshops $M = \{M_1, \ldots, M_m\}$, we make the following assumptions:

1. each job consists of an R-operation and a T-operation;
2. each flowshop has an R-processor and a T-processor that can run in parallel and can process the R-operations and the T-operations, respectively, of the assigned jobs;
3. the R-operation and T-operation of a job must be executed in the R-processor and T-processor, respectively, of the same flowshop, in such a way that the T-operation cannot start unless the R-operation is completed;
4. there is no precedence constraints among the jobs; and
5. preemption is not allowed.

Under this model, each job J_i can be given as a pair (r_i, t_i) of integers, where r_i, the R-time, is the time for processing the R-operation of J_i by an R-processor, and t_i, the T-time, is the time for processing the T-operation of J_i by a T-processor. A *schedule* S of a set of jobs $\{J_1, \ldots, J_n\}$ on m flowshops $M_1, \ldots,$ M_m consists of an *assignment* that assigns each job to a flowshop, and, for each flowshop, the execution orders of the R- and T-operations of the jobs assigned to that flowshop. The *completion time* of a flowshop M_j under the schedule S is the time when M_j finishes the last T-operation of the assigned jobs. The *makespan* $C_{\max}$ of S is the largest flowshop completion time under the schedule S over all flowshops. Following the three-field notation $\alpha|\beta|\gamma$ suggested by Graham et al. [8], we refer to the scheduling model studied in this paper as $P|2\mathrm{FL}|C_{\max}$, or as $P_m|2\mathrm{FL}|C_{\max}$ if the number m of flowshops is a fixed constant.

We will use an ordered sequence $\langle J_1, J_2, \ldots, J_t \rangle$ of two-stage jobs to denote the schedule S of the jobs on a single two-stage flowshop, in which both the executions of the R- and T-operations of the jobs, by the R- and T-processor of the flowshop, respectively, strictly follow the given order. If our objective is to minimize the makespan of schedules, then we can make the following assumption, which was proved in [17].

Lemma 1. *Let $S = \langle J_1, J_2, \ldots, J_t \rangle$ be a two-stage job schedule on a single two-stage flowshop, where $J_i = (r_i, t_i)$, for $1 \leq i \leq t$. Let $\bar{\rho}_i$ and $\bar{\tau}_i$, respectively, be the times at which the R-operation and the T-operation of job J_i are started. Then for all i, $1 \leq i \leq t$, we can assume:*

(1) $\bar{\rho}_i = \sum_{k=1}^{i-1} r_k$; and
(2) $\bar{\tau}_i = \max\{\bar{\rho}_i + r_i, \bar{\tau}_{i-1} + t_{i-1}\}$.

We give a brief description of this lemma. For an ordered job sequence on a two-stage flowshop, the R- and T-operation of each job on the flowshop always start once they can, without further wait if the objective is to minimize the completion time of the flowshop. Therefore, the execution of the R-processor is continuous: the R-operation of a job starts immediately when the R-operations of the previous jobs in the sequence are finished. On the other hand, the execution of the T-operation of a job waits for the time not only when the T-operations of the previous jobs are finished but also when its corresponding R-operation is finished. Each "gap" in the execution of the T-processor implies that there is a T-operation of a job waiting for its R-operation to finish.

Therefore, the status of a flowshop M_q at any moment can be represented by a *configuration* (ρ_q, τ_q) for a corresponding schedule, where ρ_q and τ_q are the completion times of the R-processor and the T-processor, respectively, of the flowshop M_q. The status (ρ_q, τ_q) of the flowshop M_q can be easily updated when a new job $J_i = (r_i, t_i)$ is added to it. We give the procedure of assigning a next two-stage job J_i to a two-stage flowshop M_q as Fig. 1.

Algorithm AssignJob

INPUT: a job $J_i = (r_i, t_i)$, a flowshop M_q with configuration (ρ_q, τ_q).
OUTPUT: assign the job J_i to the flowshop M_q and update the configuration
of the flowshop M_q.

1. $\rho_q = \rho_q + r_i$;
2. **if** $\rho_q \leq \tau_q$ **then** $\tau_q = \tau_q + t_i$ **else** $\tau_q = \rho_q + t_i$;

Fig. 1. Assigning a new job J_i to a flowshop M_q

3   An Approximation Algorithm for $P|2FL|C_{\max}$

In this section, we consider approximation algorithms for the multiple two-stage flowshops problem, that is: schedule n two-stage jobs $G = \{J_1, \ldots, J_n\}$ on m two-stage flowshops $M = \{M_1, \ldots, M_m\}$ with the objective of minimizing the makespan, where m is a part of input. An approximation algorithm along with its analysis is given in this section.

We start this section with the description of the approximation algorithm. Before scheduling a job set G on m flowshops, without loss of generality, we first sort G into a job sequence $\langle J_1, ..., J_{d-1}, J_d, ..., J_n \rangle$, where the subsequence $\langle J_1, ..., J_{d-1} \rangle$, denoted by G_1, contains the jobs whose R-times are no larger than their corresponding T-times and which are sorted in non-increasing order by their T-times, and $\langle J_d, ..., J_n \rangle$, denoted by G_2, contains the rest of the jobs sorted in an arbitrary order. We introduce a new array $\psi = \{\psi_1, ..., \psi_m\}$ where ψ_q, for all $1 \le q \le m$, records the sum of the T-times of the jobs assigned to the flowshop M_q for now. Now we are ready for our algorithm given in Fig. 2.

Algorithm Approx

INPUT: a job sequence $G = \langle J_1, ..., J_{d-1}, J_d, ..., J_n \rangle$ of n two-stage jobs, which
        has been sorted into G_1 and G_2. A integer m which represents the
        number of two-stage flowshops.
OUTPUT: a makespan of scheduling the two-stage job set G on m two-stage
          flowshops $M = \{M_1, ..., M_m\}$.

1. **for** $q = 1$ **to** m **do**
1.1 $\rho_q = 0, \tau_q = 0, \psi_q = 0$;
2. **for** $i = 1$ **to** $d - 1$ **do**
2.1   find the flowshop M_q whose ψ_q is minimum, for all $1 \le q \le m$;
2.2   call the algorithm **AssignJob** to assign J_i to M_q;
2.3 $\psi_q = \psi_q + t_i$;
3. **for** $i = d$ **to** n **do**
3.1   find the flowshop M_q whose ρ_q is minimum, for all $1 \le q \le m$;
3.2   call the algorithm **AssignJob** to assign J_i to M_q;
4. return the value $\max_{1 \le q \le m}\{\tau_q\}$;

Fig. 2. An approximation algorithm for $P|2FL|C_{\max}$

The algorithm **Approx** contains two main steps: first to schedule the jobs in G_1, which follows this ordered job sequence G_1, and then to schedule the rest of the jobs following the job sequence G_2. When scheduling a job in G_1, this algorithm always picks the flowshop M_q, whose ψ_q is minimum for all $1 \le q \le m$, to assign the job to, i.e., the flowshop which has the minimum sum of the T-times of the jobs on it over all flowshops. Then, at the second step when scheduling a job in G_2, the algorithm picks M_q with the minimum ρ_q instead of the minimum ψ_q, that is the flowshop which has the minimum completion time of the R-processor over all flowshops at that time. Notice that there may be more than one flowshops, which have the same minimum cost of the T-operations of the assigned jobs (or the same minimum completion time of the R-processor) when scheduling a job in G_1 (or G_2). We mention that this algorithm always selects the flowshop with the minimum flowshop index from those flowshops. The time complexity of the algorithm is given in the following theorem.

Theorem 1. *The algorithm* **Approx** *runs in time* $O(n \log n + nm)$.

We define some notations which will be used in the rest of this paper. Given a job set G of n two-stage jobs and a integer m, let $\text{Opt}(G)$ be the minimum makespan of scheduling G on m two-stage flowshops. It is supposed that M_h is the flowshop which achieves the makespan after scheduling the job set G by the algorithm **Approx**. The completion time of M_h is denoted by τ^* instead of $C_{\max}$ for convenience. Without loss of generality, denote by $\langle J_1, ..., J_c \rangle$ the job sequence scheduled on the flowshop M_h. Let k be the minimum job index from which the T-operations of the following jobs on M_h are executed continuously by M_h. In this paper, the analysis of the approximation ratio for the algorithm **Approx** is divided into three cases, based on which job set J_c and J_k belong to:

Case 1. J_c belongs to the job set G_1;
Case 2. J_c and J_k belong to the job set G_2;
Case 3. J_c belongs to the job set G_2, and J_k belongs to the job set G_1.

According to Fig. 2, the algorithm **Approx** does not start scheduling the jobs in G_2 unless the jobs in G_1 are finished, thus the case, where the job J_k belongs to G_2 while the last job J_c on M_h belongs to G_1, cannot exist.

We start the analysis with Case 1. The condition in this case, that J_c belongs to the job set G_1, means that all the jobs in $\{J_1, ..., J_c\}$ belong to G_1, i.e., for each job J_i scheduled on M_h, we have $r_i \leq t_i$. Furthermore, we can discard the jobs following J_c in the job sequence G to get a new instance. Such action does not affect our analysis for the approximation ratio: first, the completion time of the flowshop M_h dose not change, because J_c is the last job on M_h thus the following jobs discarded are scheduled on other flowshops; second, $\text{Opt}(G)$ cannot increase by discarding jobs in G. Thus in Case 1, we can make an assumption that all the jobs in G belong to G_1, and have been sorted in non-increasing order by their T-times before scheduling by the algorithm **Approx**.

For a job $J_i = (r_i, t_i)$, the T-*partial job* of J_i is $J_i^T = (0, t_i)$, i.e., the T-partial job J_i^T is constructed from the original job J_i by setting its R-time to 0. Given a job set $G = \{J_1, ..., J_n\}$, the T-partial job set G^T of G is $\{J_1^T, ..., J_n^T\}$, where J_i^T is the T-partial job of J_i, for $1 \leq i \leq n$. It is obvious that $\text{Opt}(G) \geq \text{Opt}(G^T)$. We also have further observation as the following lemma.

Lemma 2. *Given a job set G of n two-stage jobs, construct its T-partial job set G^T. On m two-stage flowshops, if all jobs in G belong to G_1, then a job J_i is assigned to a flowshop M_q when scheduling the job set G by the algorithm* **Approx** *if and only if its T-partial job J_i^T is assigned to the same flowshop M_q when scheduling the corresponding T-partial job set G^T by the same algorithm.*

The algorithm **Approx** on the job set G^T, where each job with its R-time equaling 0, can be regarded as a one-stage job, is actually the well-known *ListRanking* algorithm based on ψ, that sets all ψ_q in ψ to 0 at the initial step, finds the flowshop M_q whose ψ_q is minimum in ψ over all flowshops for the next job J_i^T, increases ψ_q by t_i after placing J_i^T to M_q, and returns the maximum value in ψ over all flowshops as the makespan after finishing the scheduling of all jobs. As supposed, G^T has been sorted in non-increasing order by T-time before scheduling. Therefore the makespan of scheduling the job set G^T on m flowshops by the algorithm **Approx** is bounded by $\frac{4}{3}\text{Opt}(G^T)$ for arbitrary $m \geq 2$ [7], thus further by $\frac{4}{3}\text{Opt}(G)$. Based on Lemma 2, the following theorem holds.

Theorem 2. *When scheduling a job set G of n two-stage jobs on m two-stage flowshops by the algorithm **Approx**, if all the jobs in G belong to G_1, then ψ_q, for all $1 \leq q \leq m$, is bounded by $\frac{4}{3}Opt(G)$ for all $m \geq 2$.*

Notice that in Case 1, we make an assumption that all jobs in the job set G belong to G_1, which does not affect the analysis of the approximation ratio for the algorithm **Approx**. Thus the above theorem can also be written as:

Theorem 3. *In Case 1, ψ_q of scheduling a job set G of n two-stage jobs on m two-stage flowshops by the algorithm **Approx**, for all $1 \leq q \leq m$, is bounded by $\frac{4}{3}Opt(G)$ for all $m \geq 2$.*

As supposed before, M_h is the flowshop achieving the makespan τ^* of scheduling a job set G of n two-stage jobs on m two-stage flowshops by the algorithm **Approx**, and the job sequence $\langle J_1, ..., J_c \rangle$ is the schedule on M_h. Let $a_0 = \sum_{i=1}^{k-1} r_i$ represent the sum of the R-times of the assigned jobs on M_h from J_1 to J_{k-1}, and $b_0 = \sum_{i=1}^{k-1} t_i$ denote the sum of the T-times of these jobs correspondingly. Similarly, $a_1 = \sum_{i=k+1}^{c-1} r_i$ and $b_1 = \sum_{i=k+1}^{c-1} t_i$. It is straightforward that $b_0 \geq a_0$ and $b_1 \geq a_1$, because we have $r_i \leq t_i$ for all jobs J_i where $1 \leq i \leq n$ in Case 1. Figure 3 illustrates the state of M_h. Note that for convenience, we denote the sum of the T-times of the jobs on M_h before J_k by b_0, though the execution of the T-operations of these jobs may not be continuous. Now we begin our analysis for the algorithm **Approx** in this case.

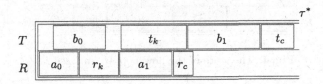

Fig. 3. The state of the flowshop M_h in Case 1

Theorem 4. *In Case 1, the algorithm **Approx** is $\frac{11}{6}$-approximation for all $m \geq 2$.*

Proof. The makespan of scheduling n two-stage jobs on m two-stage flowshops by the algorithm **Approx**, that is the completion time τ^* of the flowshop M_h, is shown as:

$$\tau^* = a_0 + r_k + t_k + b_1 + t_c \leq b_0 + r_k + t_k + b_1 + t_c \tag{1}$$

$$\leq \frac{4}{3}\text{Opt}(G) + \frac{1}{2}\text{Opt}(G) = \frac{11}{6}\text{Opt}(G). \tag{2}$$

We explain the derivations in (1) and (2). The execution of the R-processor of any flowshop is continuous by Lemma 1. From the definition of k, the T-operations of the jobs on M_h from the job J_k to J_c are executed continuously. Moreover, note that the T-operation of J_k starts right after its R-operation is finished: otherwise there would be no execution gap between the T-operations of the jobs J_{k-1} and J_k, contradicting the assumption of the minimality of the index k. These explain the equation in (1). As discussed above, a_0 is no larger than b_0, that is why the inequation in (1) holds. The inequation in (2) holds for two reasons: first, by Theorem 3, all ψ_q where $1 \leq q \leq m$ are bounded by $\frac{4}{3}\mathrm{Opt}(G)$, which implies $b_0 + t_k + b_1 + t_c \leq \frac{4}{3}\mathrm{Opt}(G)$; second, it is clear that $\mathrm{Opt}(G)$ must be no smaller than the time of completing any job, that is $r_i + t_i \leq \mathrm{Opt}(G)$ for $1 \leq i \leq n$. Combining the assumption that r_i is no larger than t_i for all jobs in this case, we have $r_k \leq \frac{1}{2}\mathrm{Opt}(G)$. The equation in (2) gives the conclusion that τ^* is no larger than $\frac{11}{6}\mathrm{Opt}(G)$. □

Now consider Case 2 where the jobs J_c and J_k belong to the job set G_2, in which each job has its R-time larger than its T-time. We use the same notations defined in the previous case: $a_0 = \sum_{i=1}^{k-1} r_i$, $b_0 = \sum_{i=1}^{k-1} t_i$, $a_1 = \sum_{i=k+1}^{c-1} r_i$ and $b_1 = \sum_{i=k+1}^{c-1} t_i$. The state of the flowshop M_h, which achieves the makespan τ^*, is shown in Fig. 4. According to Fig. 2, the job set G has been sorted before scheduling and the algorithm **Approx** does not start scheduling the jobs in G_2 unless it finishes the scheduling of all the jobs in G_1. Therefore the assumption in this case, that J_k belongs to G_2, implies that the jobs on M_h following J_k are also in G_2 and hence $b_1 < a_1$. We also have the following theorem.

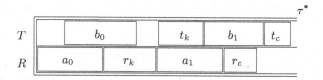

Fig. 4. The state of the flowshop M_h in Case 2

Theorem 5. *In Case 2, $a_0 + r_k + a_1$ is bounded by $\mathrm{Opt}(G)$.*

Proof. This case assumes that the job J_c belongs to the job set G_2. According to the algorithm **Approx**, the jobs in G_2 are scheduled by step 3 of this algorithm. Therefore the flowshop M_h must have the minimum ρ_h when J_c arrives, which means that the completion times of the R-processors of all flowshops are no smaller than ρ_h before J_c is assigned, which is expressed as $a_0 + r_k + a_1$ here. By Lemma 1, the execution of the R-processor of any flowshop is continuous, thus the makespan on any schedule must be no smaller than the minimum completion time of the R-processor over all flowshops, that is $\mathrm{Opt}(G) \geq a_0 + r_k + a_1$. □

The approximation ratio of the algorithm in Case 2 is shown as follows.

Theorem 6. *In Case 2, the algorithm* **Approx** *is 2-approximation for all* $m \geq 2$.

Proof. The makespan τ^* achieved on the flowshop M_h is expressed as:

$$\tau^* = a_0 + r_k + t_k + b_1 + t_c < a_0 + r_k + t_k + a_1 + t_c \tag{3}$$

$$< \mathrm{Opt}(G) + \frac{1}{2}\mathrm{Opt}(G) + \frac{1}{2}\mathrm{Opt}(G) = 2\mathrm{Opt}(G). \tag{4}$$

The reason for the equation in (3) is the same as that for the equation in (1). The inequation in (3) holds obviously: $b_1 < a_1$ in this case. By Theorem 5, $a_0 + r_k + a_1$ is no larger than $\mathrm{Opt}(G)$. This case assumes that J_k and J_c belong to the job set G_2, thus both jobs have their R-times larger than their corresponding T-times. Due to a similar reason to that explained in the inequation in (2), the fact that $\mathrm{Opt}(G)$ is no smaller than the time of completing any job J_i which equals $r_i + t_i$, implies that t_k and t_c are smaller than $\frac{1}{2}\mathrm{Opt}(G)$. These explain the inequation in (4). The equation in (4) completes the proof of the approximation ratio for the algorithm in Case 2. □

Finally we consider Case 3 where the job J_c belongs to the job set G_2, while J_k belongs to G_1. As supposed in the previous cases, the job sequence $\langle J_1, ..., J_c \rangle$ is scheduled on the flowshop M_h achieving the makespan τ^*. It is obvious that there must be a job index d ($k < d \leq c$) such that the assigned jobs on M_h whose job index are smaller than d are in the job set G_1, and the rest of these jobs on M_h belong to G_2. We define the notations slightly different from those in the previous cases: $a_0 = \sum_{i=1}^{k-1} r_i$, $b_0 = \sum_{i=1}^{k-1} t_i$, $a_1 = \sum_{i=k+1}^{d-1} r_i$, $b_1 = \sum_{i=k+1}^{d-1} t_i$, $a_2 = \sum_{i=d}^{c-1} r_i$, and $b_2 = \sum_{i=d}^{c-1} t_i$. From the definition of d, it is clear that $b_2 < a_2$. The state of the flowshop M_h after scheduling is shown in Fig. 5. Just as in Case 2, M_h has the minimum ρ_h over all flowshops when scheduling J_c in this case, thus the following theorem similar to Theorem 5 holds.

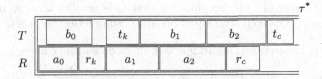

Fig. 5. The state of the flowshop M_h in Case 3

Theorem 7. *In Case 3,* $a_0 + r_k + a_1 + a_2$ *is bounded by* $\mathrm{Opt}(G)$.

The following theorem also holds.

Theorem 8. *In Case 3,* $b_0 + t_k + b_1$ *is bounded by* $\frac{4}{3}\mathrm{Opt}(G)$.

Proof. The algorithm **Approx** can be considered as two main steps: the first step is scheduling the jobs in the job set G_1, where the R-operation takes no more time than the T-operation for each job and which is sorted in non-increasing order by T-time; the second step is scheduling the jobs in G_2, where the R-operation costs more time than the T-operation for each job. As discussed in Theorem 2, when finishing the first step of scheduling all jobs in G_1, ψ_q, which records the sum of the T-times of the jobs on the flowshop M_q at that time for all $1 \leq q \leq m$, is no larger than $\frac{4}{3}\text{Opt}(G_1)$. It is clear that $\text{Opt}(G) \geq \text{Opt}(G_1)$, because G_1 is a subsequence of the job sequence G. Thus the ψ_h, which equals $b_0 + t_k + b_1$ after finishing the scheduling of all jobs in G_1, is bounded by $\frac{4}{3}\text{Opt}(G)$. □

We mention that there is a subcase where the T-operation of the job J_d may start after the time when the R-operation of the job J_{c-1} is finished, which is not shown in Fig. 5 and obviously is not against Theorems 7 and 8. Actually, it is easy to see that the other subcase shown in Fig. 5, where the T-operation of J_d starts before the time when the R-operation of J_{c-1} is finished, means that $b_0 + t_k + b_1$ is no larger than $a_0 + r_k + a_1 + a_2$ thus further $\text{Opt}(G)$, which is tighter than Theorem 8. As a consequence, the following analysis is still based on Theorem 8, which holds in both subcases.

Theorem 9. *In Case 3, the algorithm* **Approx** *is $\frac{17}{6}$-approximation for all $m \geq 2$.*

Proof. The makespan τ^* is expressed as:

$$\tau^* = a_0 + r_k + t_k + b_1 + b_2 + t_c < a_0 + r_k + t_k + b_1 + a_2 + t_c \qquad (5)$$

$$< \text{Opt}(G) + \frac{4}{3}\text{Opt}(G) + \frac{1}{2}\text{Opt}(G) = \frac{17}{6}\text{Opt}(G). \qquad (6)$$

The equation in (5) is straightforward from Fig. 5, whose reason is similar to that for the equation in (1). The fact $b_2 < a_2$ in this case proves the inequation in (5). The inequation in (6) holds for three reasons: first, Theorem 7 implies $a_0 + r_k + a_2 \leq \text{Opt}(G)$ because a_1 is obviously no smaller than 0; second, by Theorem 8, we have $t_k + b_1 \leq \frac{4}{3}\text{Opt}(G)$ due to that $b_0 \geq 0$; third, as discussed in the inequation in (4), the assumption that t_c is smaller than r_c in this case means that $t_c < \frac{1}{2}\text{Opt}(G)$. The equation in (6) completes the proof. □

Combining Theorems 4, 6 and 9 about three cases, the conclusion follows.

Theorem 10. *The algorithm* **Approx** *is $\frac{17}{6}$-approximation for all $m \geq 2$.*

Let $\mathcal{S} = \{\mathcal{S}_1, ..., \mathcal{S}_m\}$ be the schedule for the job set G of n two-stage jobs on m two-stage flowshops by the algorithm **Approx**, where especially the job sequence $\mathcal{S}_q$ is the schedule on the flowshop M_q, for $1 \leq q \leq m$. Notice that $\mathcal{S}_q$ must be a subsequence of the job sequence G, which has been sorted into two parts before scheduling and does not follow the order by Johnson's algorithm, thus may not be the optimal schedule for M_q when aiming at minimizing the makespan. We can sort all the $\mathcal{S}_q$ separately by Johnson's algorithm after scheduling by the algorithm **Approx**. It is easy to see that the additional sorting step can get a new schedule whose makespan is no larger than that of the original one, and can be done in time $O(n \log n)$. Therefore such step will not change the approximation ratio and the time complexity of the algorithm **Approx**.

4   Conclusion

This paper studied the multiple two-stage flowshops problem for scheduling n two-stage jobs on m identical two-stage flowshops. In particular, to meet the practical demands in our research on data centers and cloud computing, we addressed this problem in the situation where the number m of flowshops is a part of input, which makes the problem become strongly NP-hard. We proposed an efficient algorithm which runs in time $O(n \log n + mn)$. The subsequent analysis showed that this algorithm achieves an approximation ratio $17/6$. To the best of our knowledge, it is the first approximation algorithm for the multiple two-stage flowshops problem when the number m of flowshops is a part of input.

The approximation algorithms given in [4,17], which are based on different pseudo-polynomial time exact algorithms for a fixed constant $m \geq 2$, produce schedules with makespan bounded by $\text{Opt}(G)(1 + \epsilon)$, but run in time $O(2^{2m-1} n^{2m} m^{2m+1} / \epsilon^{2m-1})$ and $O(n^{2m-1} m^{2m} / \epsilon^{2m-2})$ respectively. Compared to these time complexities, which become unacceptable high when considering the enormous number m of servers in data centers even when ϵ is bigger than 1, the time complexity $O(n \log n + mn)$ for arbitrary m makes our algorithm much more efficient in practical applications with a reasonable level of satisfaction.

References

1. Abts, D., Felderman, B.: A guided tour through data-center networking. Queue **10**(5), 10 (2012)
2. Dell. http://www.dell.com/us/business/p/servers
3. Armbrust, M., Fox, A., Griffith, R., Joseph, A.D., Katz, R., Konwinski, A., Lee, G., Patterson, D., Rabkin, A., Stoica, I., et al.: A view of cloud computing. Commun. ACM **53**(4), 50–58 (2010)
4. Dong, J., Tong, W., Luo, T., Wang, X., Hu, J., Xu, Y., Lin, G.: An FPTAS for the parallel two-stage flowshop problem. Theoret. Comput. Sci. **657**, 64–72 (2017)
5. Garey, M.R., Johnson, D.S.: Computers and Intractability: A Guide to the Theory of NP-completeness. Freeman, San Francisco (1979)
6. Graham, R.L.: Bounds for certain multiprocessing anomalies. Bell Labs Tech. J. **45**(9), 1563–1581 (1966)
7. Graham, R.L.: Bounds on multiprocessing timing anomalies. SIAM J. Appl. Math. **17**(2), 416–429 (1969)
8. Graham, R.L., Lawler, E.L., Lenstra, J.K., Kan, A.R.: Optimization and approximation in deterministic sequencing and scheduling: a survey. Ann. Discret. Math. **5**, 287–326 (1979)
9. Greenberg, A., Hamilton, J., Maltz, D.A., Patel, P.: The cost of a cloud: research problems in data center networks. ACM SIGCOMM Comput. Commun. Rev. **39**(1), 68–73 (2008)

10. He, D.W., Kusiak, A., Artiba, A.: A scheduling problem in glass manufacturing. IIE Trans. **28**(2), 129–139 (1996)
11. Hochbaum, D.S., Shmoys, D.B.: Using dual approximation algorithms for scheduling problems theoretical and practical results. J. ACM (JACM) **34**(1), 144–162 (1987)
12. Johnson, S.M.: Optimal two-and three-stage production schedules with setup times included. Nav. Res. Logist. (NRL) **1**(1), 61–68 (1954)
13. Li, H., Ghodsi, A., Zaharia, M., Shenker, S., Stoica, I.: Tachyon: reliable, memory speed storage for cluster computing frameworks. In: Proceedings of the ACM Symposium on Cloud Computing, pp. 1–15. ACM (2014)
14. Sahni, S.K.: Algorithms for scheduling independent tasks. J. ACM (JACM) **23**(1), 116–127 (1976)
15. Vairaktarakis, G., Elhafsi, M.: The use of flowlines to simplify routing complexity in two-stage flowshops. IIE Trans. **32**(8), 687–699 (2000)
16. Vazirani, V.V.: Approximation Algorithms. Springer Science & Business Media, New york (2013)
17. Wu, G., Chen, J., Wang, J.: On scheduling two-stage jobs on multiple two-stage flowshops. Technical report, School of Information Science and Engineering Central South University (2016)
18. Wu, G., Chen, J., Wang, J.: On approximation algorithms for two-stage scheduling problems. In: Xiao, M., Rosamond, F. (eds.) Frontiers in Algorithmics, FAW 2017. LNCS, vol. 10336, pp. 241–253. Springer, Cham (2017)
19. Zhang, X., van de Velde, S.: Approximation algorithms for the parallel flow shop problem. Eur. J. Oper. Res. **216**(3), 544–552 (2012)

The Existence of Universally Agreed Fairest Semi-matchings in Any Given Bipartite Graph

Jian Xu[✉], Soumya Banerjee, and Wenjing Rao[✉]

Department of Electrical and Computer Engineering,
University of Illinois at Chicago, Chicago, IL 60607, USA
{jxu39,sbaner8,wenjing}@uic.edu

Abstract. In a bipartite graph $G = (U \cup V, E)$ where $E \subseteq U \times V$, a *semi-matching* is defined as a set of edges $M \subseteq E$, such that each vertex in U is incident with exactly one edge in M. Many previous works focus on the problem of fairest semi-matchings: ones that assign U-vertices with V-vertices as fairly as possible. In these works, fairness is usually measured according to a specific index. In fact, there exist many different fairness measures, and they often disagree on the fairness comparison of some semi-matching pairs. In this paper, we prove that for any given bipartite graph, there always exists a (set of equally) fairest semi-matching(s) universally agreed by all the fairness measures. In other words, given that fairness measures disagree on many comparisons between semi-matchings, they nonetheless are all in agreement on the (set of) fairest semi-matching(s), for any given bipartite graph. To prove this, we propose a partial order relationship (Transfer-based Comparison) among the semi-matchings, showing that the greatest elements always exist in such a partially ordered set. We then show that such greatest elements can guarantee to be the fairest ones under the criteria of Majorization [10]. We further show that all widely used fairness measures are in agreement on such a (set of equally) fairest semi-matching(s).

Keywords: Bipartite graph · Semi-matching · Fairness measure · Partial order · Majorization · Load balancing · Resource allocation

1 Introduction

This paper focuses on the problem of a *Fairest Semi-matching*. A *semi-matching* in a bipartite graph $G = (U \cup V, E)$ where $E \subseteq U \times V$, is defined as a set of edges $M \subseteq E$ such that each vertex in U is incident with exactly one edge in M. A vertex in V might be incident with more than one edge in M. Note that in general, valid semi-matchings can be easily obtained by matching each vertex $u \in U$ with an arbitrary vertex $v \in V$ for which $(u, v) \in E$. The problem of finding a *fairest* semi-matching is related to the load-balancing problem in a system, where discrete resources (set of U) need to be assigned to the agents

This work is supported by NSF Grant CNS-1149661.

Y. Cao and J. Chen (Eds.): COCOON 2017, LNCS 10392, pp. 529–541, 2017.
DOI: 10.1007/978-3-319-62389-4_44

(set of V) in the fairest way. Particularly, the given bipartite graph corresponds to a system under some assignment constraints, where the edge set E indicate the possible assignment options. This is representative for a large-scale system, where resource assignment is constrained to a limited set of agents in the local area. In such cases, the optimal solution of a load-balance problem is given by a fairest semi-matching indicating how the resources (of U-vertices) should be assigned to the agents (of V-vertices), under assignment constraints (of edge set E).

Given a semi-matching M, the *quota* of vertex $v \in V$, is defined as the number of U-vertices matched with v. The corresponding *quotas vector* of M, denoted as Q_M, is the vector of which each element $Q_M(v)$ represents the *quota* of a vertex $v \in V$. The fairness of a semi-matching is usually defined based on its quotas vector, with regard to some fairness measure. An example of the fairest semi-matching according to Jain's index [6] is shown in Fig. 1.

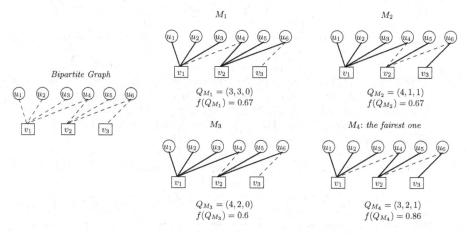

Fig. 1. An example of the fairest semi-matching, where $f(Q_M) = \frac{(\sum_{i=1}^{m} Q_M(i))^2}{m \sum_{i=1}^{m} Q_M(i)^2}$ (where $m = |Q_M|$) represents the score of Q_M from Jain's index [6].

Many fairness measures have been proposed to compare quotas vectors, such as Jain's index [6], Entropy [11], Atkinson's index [1], α-fairness [9], Max-min index [2], convex cost functions [5], L_p-norm [5], lexicographically minimum fairness [3]. Lan et al. [8] constructed a family of fairness measures which satisfies several axioms. Note that, for all the fairness measures, any pair of semi-matchings with the same *sorted* quotas vector are considered equal in fairness. Nonetheless, the existence of numerous fairness measures indicates that they do not agree on the fairness comparison between many semi-matchings (or the quotas vectors of them). For this reason, most approaches choose a specific fairness measure to work with.

Harvey et al. [5] proposed a number of algorithms to achieve a semi-matching which is fairest with regard to Max-min index and L_2-norm, and showed that the obtained semi-matching is also the fairest with regard to any convex cost functions and any L_p-norm where $p \in R$ and $p > 1$. Bokal et al. [3] showed a semi-matching is the fairest with regard to L_2-norm if and only if the sorted quotas vector of this semi-matching is lexicographically minimum.

In this paper, we prove that for any given bipartite graph, there always exists a set of equally fair semi-matchings, which are uniformly considered *the fairest* by all the fairness measures. This paper is organized as follows. Section 2 discusses the consensus among fairness measures with regard to a partially ordered measure of Majorization. Section 3 defines a partial order (namely Transfer-based Comparison) between any pair of semi-matchings, which is shown to be more rigorous than Majorization. Section 4 proves there exists a set of the greatest semi-matchings in the proposed partially ordered set based on the Transfer-based Comparison. Moreover, those greatest semi-matchings are shown to be agreed by all existing fairness measures. Section 5 provides a discussion on various issues related to the Transfer-based Comparison, and some algorithms for achieving the fairest semi-matchings. This paper is concluded in Sect. 6.

2   Consensus Among Fairness Measures

Most of the existing and widely used fairness measures are index-based. An index-based fairness measure is a function f which maps a sorted quotas vector into a real number, so that it can be compared against any other ones. For instance, Jain's index [6] was defined as the modified "variance" among elements in a quotas vector. Using an index-based fairness measure, any arbitrary pair of quotas vectors are comparable (either one is considered fairer, or both considered equally fair). However, different index-based fairness measures often disagree on the comparison of some pairs of quotas vectors. For example, for the comparison of $Q_1 = (0, 3, 3)$ and $Q_2 = (4, 1, 1)$, the following fairness measures disagree, and their comparison results are shown in Table 1. Suppose quotas vector $Q = (q_1, \cdots, q_i, \cdots, q_m)$, Q_1 is considered fairer than Q_2, if $f(Q_1) > f(Q_2)$, where $f(Q)$ represent the score of Q from a fairness measure f.

- *Jain's index* [6]: $J(Q) = \frac{(\sum_{i=1}^{m} q_i)^2}{m \sum_{i=1}^{m} q_i^2}$.
- *Max-Min index* [2]: $M(Q) = min_{i=1,\cdots,m}\{q_i\}$.
- An index proposed in [8]: $G(Q) = \{\sum_{i=1}^{m} (\frac{q_i}{s})^{1-\beta}\}^{\frac{1}{\beta}}$, where $\beta \in (-1, -\infty)$ and $s = \sum_{i=1}^{m} q_i$.

Majorization [10] is a partial order over quotas vectors, which by allowing some pairs to be "incomparable", offers a "stricter" way of fairness comparison.

Definition 1 [10] *(**Majorization**). For $x, y \in R^n$, x is majorized by y (denoted as $x \preceq_{Maj} y$), if $\sum_{i=1}^{n} x_i = \sum_{i=1}^{n} y_i$, and $\sum_{i=1}^{d} x_i^\uparrow \leqslant \sum_{i=1}^{d} y_i^\uparrow$ for $d = 1, \ldots, n$, where $x_i^\uparrow$ and $y_i^\uparrow$ are the ith elements of $x^\uparrow$ and $y^\uparrow$, which are sorted in ascending order.*

Table 1. Comparison among *Jain's index*, *Max-min index*, and an index in [8] ($\beta = -2$)

	Jain's index [6]	Max-min index [2]	An index in [8] ($\beta = -2$)
$Q_1 = (0, 3, 3)$	0.67	0	2
$Q_2 = (4, 1, 1)$	0.67	1	1.809
Fairness Comparison	$Q_1 = Q_2$	$Q_1 < Q_2$	$Q_1 > Q_2$

For example, $x = (0, 3, 3)$ is majorized by $y = (3, 1, 2)$. The sorted vectors are $x^\uparrow = (0, 3, 3)$ and $y^\uparrow = (1, 2, 3)$ respectively. Let $S_x^d = \sum_{i=1}^{d} x_i^\uparrow$ for $d = 1, 2, 3$. Then we have $S_x = (S_x^1, S_x^2, S_x^3) = (0, 0+3, 0+3+3) = (0, 3, 6)$ and $S_y = (S_y^1, S_y^2, S_y^3) = (1, 1+2, 1+2+3) = (1, 3, 6)$. Therefore, it meets $x \preceq_{Maj} y$. An example of incomparable pair is $x = (0, 3, 3)$ and $y = (4, 1, 1)$, such that $S_x = (S_x^1, S_x^2, S_x^3) = (0, 3, 6)$ and $S_y = (S_y^1, S_y^2, S_y^3) = (1, 2, 6)$. Because $S_x^1 < S_y^1$ and $S_x^2 > S_y^2$, x and y are incomparable with regard to Majorization.

A fairness measure f is called "compatible" with Majorization,[1] if it satisfies that $f(x) < f(y)$ when $x \preceq_{Maj} y$. Lan et al. [8] studied various fairness measures in networking field, and constructed a unique family of fairness measures satisfying five axioms (continuity, homogeneity, saturation, partition, and starvation), which includes all the popularly used fairness measures such as Atkinson's index, α-fairness, Jain's index, Entropy function. This family of fairness measures has been proved to be compatible with Majorization. To our best knowledge, all existing fairness measures in the literature are compatible with Majorization.

3   Transfer-Based Comparison for Semi-matching

For a given semi-matching M in a bipartite graph $G = (U \cup V, E)$, we define a *Transfer* t on M to be a sequence of alternating edges ($\{v_1, u_1\}, \{u_1, v_2\}, \{v_2, u_3\}, \ldots, \{u_{k-1}, v_k\}$) with $v_i \in V$, $u_i \in U$, and $\{v_i, u_i\} \in M$ for each $1 \leq i \leq k - 1$. Note that according to the definition of semi-matching, when $\{v_i, u_i\} \in M$, then $\{u_i, v_{i+1}\} \notin M$. Essentially, a Transfer t on M is a path beginning and ending in V-vertices in M, consisting of alternating edges in and out of M. The *application of Transfer* t to semi-matching M is defined as switching the matching and non-matching edges in M along Transfer t. The result of the application of t to M will change M to a different semi-matching M', which includes all the $\{u_i, v_{i+1}\}$ edges in t, but excludes all the $\{v_i, u_i\}$ edges in t. For example, in Fig. 1, a sequence of edges ($\{v_1, u_4\}, \{u_4, v_2\}, \{v_2, u_6\}, \{u_6, v_3\}$) is one Transfer t on semi-matching M_3, and the application of t to M_3 yields M_4 which is also shown in Fig. 1.

For a Transfer t on semi-matching M, $t = (\{v_1, u_1\}, \{u_1, v_2\}, \{v_2, u_3\}, \ldots, \{u_{k-1}, v_k\})$, we call v_1 and v_k to be the *source* and *destination* vertex of t, denoted as $v_s(t)$ and $v_d(t)$ respectively. The application of t to M decreases

---

[1] A function compatible with Majorization is also known as Schur-convex function.

the quota of the source vertex $v_s(t)$ by 1, and increases the quota of the sink vertex $v_d(t)$ by 1, without changing the quotas of any other vertices. Depending on the original quota difference between the source and destination vertices, we can define the Transfer t as one of the three types shown below. Let $Q_M(v_s(t))$ and $Q_M(v_d(t))$ represent the quota of vertex $v_s(t)$ and $v_d(t)$ in M respectively. Suppose $v_s(t) \neq v_d(t)$ (source and destination are not the same vertex).

- If $Q_M(v_s(t)) > Q_M(v_d(t)) + 1$, then t is called an *Improving Transfer (IT)*
- If $Q_M(v_s(t)) = Q_M(v_d(t)) + 1$, then t is called a *Neutral Transfer (NT)*
- If $Q_M(v_s(t)) < Q_M(v_d(t)) + 1$, then t is called a *Deteriorating Transfer (DT)*

The special case of $v_s(t) = v_d(t)$ is denoted as a *Cyclic Neutral Transfer*, and can be classified as *Neutral Transfer*, as it does not change the quota of any V-vertex. Overall, the application of any Transfer t to a semi-matching M will at most change the quotas of two vertices $v_s(t)$ and $v_d(t)$. If t is an Improving Transfer, it will reduce the absolute quotas difference between $v_s(t)$ and $v_d(t)$. Thus, an Improving Transfer always changes a semi-matching to a fairer one. Similarly, a Deteriorating Transfer always changes a semi-matching to a less fair one; a Neutral Transfer changes a semi-matching to an equally fair one.

A comparison relationship between semi-matchings is proposed based on the properties of a sequence of Transfers that can be used to change one semi-matching into the other one, shown as in Definition 2.

Definition 2 (Transfer-based Comparison). *For any two semi-matchings M_x and M_y, if there exists a Transfer Sequence $(t_1, t_2, \cdots, t_i, \cdots, t_n)$ such that $M_x \xrightarrow{t_1} M_1 \xrightarrow{t_2} M_2 \cdots \rightarrow M_{i-1} \xrightarrow{t_i} M_i \rightarrow \cdots \xrightarrow{t_n} M_y$, and for each $1 \leq i \leq n$, t_i is IT or NT, then M_y is defined as not less fair than M_x in terms of a Transfer-based Comparison, denoted as $M_x \preceq_T M_y$. Specifically, if all t_i are NT, then M_x is defined as equally fair with M_y, denoted as $M_x \approx_T M_y$; if all t_i are either IT or NT, with at least one t_i being IT, then M_y is defined as fairer than M_x, denoted as $M_x \prec_T M_y$.*

If for every possible sequences $(t_1, t_2, \cdots, t_i, \cdots, t_n)$ such that $M_x \xrightarrow{t_1} M_1 \xrightarrow{t_2} M_2 \cdots \rightarrow M_{i-1} \xrightarrow{t_i} M_i \rightarrow \cdots \xrightarrow{t_n} M_y$, there always exist some t_i, t_j with $1 \leq i, j \leq n$ such that t_i is IT and t_j is DT, then M_x and M_y are incomparable, denoted as $M_x \prec\succ_T M_y$.

Such a Transfer-based Comparison aims at reserving the "strictly comparable" relationships between semi-matchings via identifiable Improving or Neutral Transfer Sequence, while leaving out the "incomparable" ones with mixtures of both Improving and Deteriorating Transfers. It is easy to prove this defined comparison meets the properties of *Reflexivity, Antisymmetry, Transitivity*. Thus it is a partial order on the set of all semi-matchings in a given bipartite graph.

Lemma 1 (Partially Ordered Set (poset)). *The set of all semi-matchings of a bipartite graph is a partially ordered set with regard to the Transfer-based Comparison defined above.*

Different from the partial order of Majorization, which is defined on the quotas vectors, the proposed Transfer-based Comparison is a partial order defined over the set of semi-matchings. Moreover, we prove that the Transfer-based Comparison implies the Majorization relationship between the corresponding quotas vectors. In other words, the proposed Transfer-based Comparison is a "stricter" fairness measure than Majorization, by leaving out some pairs that can be compared under Majorization as incomparable.

Lemma 2 *(Improving Transfer $\Rightarrow$ Majorization).* *If $M_x \preceq_T M_y$, then $Q_{M_x} \preceq_{Maj} Q_{M_y}$.*

Proof. Let $Q^{\uparrow}_{M_x} = (q_1^x, q_2^x, \cdots, q_i^x, \cdots, q_n^x)$ be the sorted quotas vectors of M_x in ascending order and the partial sums $S_x^d = \sum_{i=1}^{d} q_i^x$ for $d = 1, \cdots, n$. Assume the application of an Improving Transfer t to M_x yields another semi-matching M_x', which decreases the quota of source vertex of t (suppose q_j^x) by 1, and increases the quota of destination vertex of t (suppose q_i^x) by 1. Note that Transfer t is IT, therefore the quota of its source vertex must be larger than the quota of its source vertex in M_x, thus $i < j$. Let $Q^{\uparrow}_{M_{x'}}$ represents the quotas vector of $M_{x'}$ sorted in ascending order, and $S_{x'}^d$ represents the partial sums of $Q^{\uparrow}_{M_{x'}}$ for $d = 1, \cdots, n$. It can be easily derived that $S_{x'}^d = S_x^d$ when $d < i$, that $S_{x'}^d > S_x^d$ when $i \le d < j$, and that $S_{x'}^d = S_x^d$ when $j \le d \le n$. Likewise, if a Neutral Transfer is applied, then $S_{x'}^d = S_x^d$ when $1 \le d \le n$. Thus, it meets $Q_{M_x} \preceq_{Maj} Q_{M_{x'}}$ if M_x' can be derived by applying an Improving or Neutral Transfer to M_x.

If $M_x \preceq_T M_y$, then M_y can be derived by applying a sequence of Improving or Neutral Transfers to M_x. Thus, it meets $Q_{M_x} \preceq_{Maj} Q_{M_y}$ due to the Transitivity of the partial order "$\preceq_{Maj}$". $\square$

Corollary 1. *Any fairest semi-matching with regard to "$\preceq_T$" is the fairest with regard to "$\preceq_{Maj}$".*

Corollary 2. *Any fairest semi-matching with regard to "$\preceq_T$" is the fairest with regard to any fairness measure compatible with Majorization.*

4   The Existence of the Fairest Semi-matchings Under Transfer-Based Comparison

In this section, we prove the poset of semi-matchings under Transfer-based Comparison has the *greatest elements*, i.e., the fairest semi-matchings. This is done by showing there always exists a fairer semi-matching for any two incomparable ones. The proof is done in two steps. First, we show that if one semi-matching M_x can be changed to another one M_y via a Bitonic Transfer Sequence (essentially a Transfer Sequence where all $IT's$ are before all $DT's$), then a fairer semi-matching can be (straightforwardly) found. Then, we show how to construct a Bitonic Transfer Sequence between any two incomparable semi-matchings by modifying a Transfer Sequence between them.

Definition 3 (Bitonic Transfer Sequence). *For a Transfer Sequence* $(t_1, t_2, \cdots, t_i, \cdots, t_n)$ *such that* $M_x \xrightarrow{t_1} M_1 \xrightarrow{t_2} M_2 \cdots \to M_{i-1} \xrightarrow{t_i} M_i \to \cdots \xrightarrow{t_n} M_y$, *if there exists* $1 \leq k < n$, *such that for all* t_i *with* $i \leq k$ *are either IT or NT, with at least one* t_i *being IT, and for all* t_j *with* $k < j \leq n$ *are either DT or NT, with at least one* t_j *being DT, then* $(t_1, t_2, \cdots, t_i, \cdots, t_n)$ *is called a Bitonic Transfer Sequence.*

Lemma 3. *For a pair of semi-matchings* M_x *and* M_y, *if there exists a Bitonic Transfer Sequence* $(t_1, t_2, \cdots, t_i, \cdots, t_n)$ *such that* $M_x \xrightarrow{t_1} M_1 \xrightarrow{t_2} M_2 \cdots \to M_{i-1} \xrightarrow{t_i} M_i \to \cdots \xrightarrow{t_n} M_y$, *then there exists a semi-matching* M_k *satisfying that* $M_x \preceq_T M_k$ *and* $M_y \preceq_T M_k$.

Proof. A Bitonic Transfer Sequence $(t_1, t_2, \cdots, t_i, \cdots, t_n)$ can be divided into two Sub-Sequences $T_{front} = (t_1, t_2, \cdots, t_i, \cdots, t_k)$ and $T_{back} = (t_{k+1}, t_{k+2}, \cdots, t_j, \cdots, t_n)$, where each $t_i \in T_{front}$ is either IT or NT, and each $t_j \in T_{back}$ is either DT or NT.

Let M_k be the semi-matching obtained by applying T_{front} to M_x. Then M_k is not less fair than M_x ($M_x \preceq_T M_k$). Besides, it is obvious that M_y can be derived by applying T_{back} to M_k, that is $M_k \xrightarrow{t_{k+1}} M_{k+1} \xrightarrow{t_{k+2}} M_{k+2} \cdots \to M_{j-1} \xrightarrow{t_j} M_j \to \cdots \xrightarrow{t_n} M_y$. Then it meets $M_y \preceq_T M_k$. $\square$

Lemma 4. *For a pair of incomparable semi-matchings* M_x *and* M_y ($M_x \prec\succ_T M_y$), *there always exists a Bitonic Transfer Sequence which changes* M_x *to* M_y.

Proof. Assume there exists a "Simplest" Transfer Sequence changing M_x to M_y ($M_x \prec\succ_T M_y$), of which the source vertex of one Transfer cannot be the sink vertex of another Transfer (otherwise, the Sequence can be further "Simplified" by merging such two Transfers into one larger Transfer). We will prove any Simplest Transfer Sequence can be re-organized to form a Bitonic Transfer Sequence which also changes M_x to M_y.

For two adjacent Transfers t_a and t_b in a Simplest Transfer Sequence (assume t_a precedes t_b) such that $M_i \xrightarrow{t_a} M_a \xrightarrow{t_b} M_j$, swapping the order of t_a and t_b (such that $M_i \xrightarrow{t_b} M_b \xrightarrow{t_a} M_j$) will not change the beginning and ending semi-matchings M_i and M_j, but will yield a different intermediate one M_b. The following proves that if t_b is Improving Transfer before the swapping (i.e., t_b is IT in $M_a \xrightarrow{t_b} M_j$), then t_b remains to be Improving Transfer after the swapping (i.e., t_b is IT in $M_i \xrightarrow{t_b} M_b$).

Suppose the source and sink vertex of t_a are a and a' respectively, and the source and sink vertex of t_b are b and b' respectively. If t_b is an Improving Transfer before the swapping, then $Q_{M_a}(b) - Q_{M_a}(b') > 1$. The following shows it is also true that $Q_{M_i}(b) - Q_{M_i}(b') > 1$, by examining how the application of t_a to M_i changes the quotas of b and b'. In a Simplest Transfer Sequence, it guarantees that $a \neq b'$ and $a' \neq b$. Thus, one of the following four cases should apply:

– *Case 1:* t_a and t_b have no common source vertex and no common sink vertex ($a \neq b$ and $a' \neq b'$). The application of t_a to M_i will not change the quotas of b and b', thus $Q_{M_i}(b) = Q_{M_a}(b)$, $Q_{M_i}(b') = Q_{M_a}(b')$. It indicates $Q_{M_i}(b) - Q_{M_i}(b') = Q_{M_a}(b) - Q_{M_a}(b') > 1$.

– *Case 2:* t_a and t_b have common source vertex but no common sink vertex ($a = b$ and $a' \neq b'$). The application of t_a to M_i will not change the quota of b', but will decrease the quota of b by 1, thus $Q_{M_i}(b) = Q_{M_a}(b) + 1$, $Q_{M_i}(b') = Q_{M_a}(b')$. It indicates $Q_{M_i}(b) - Q_{M_i}(b') = Q_{M_a}(b) - Q_{M_a}(b') + 1 > 1$.
– *Case 3:* t_a and t_b have no common source vertex but common sink vertex ($a \neq b$ and $a' = b'$). The application of t_a to M_i will not change the quota of b, but will increase the quota of b' by 1, thus $Q_{M_i}(b) = Q_{M_a}(b)$, $Q_{M_i}(b') = Q_{M_a}(b') - 1$. It indicates $Q_{M_i}(b) - Q_{M_i}(b') = Q_{M_a}(b) - Q_{M_a}(b') + 1 > 1$.
– *Case 4:* t_a and t_b have common source vertex and common sink vertex ($a = b$ and $a' = b'$). The application of t_a to M_i will decrease the quota of b by 1 and increase the quota of b' by 1, thus $Q_{M_i}(b) = Q_{M_a}(b) + 1$, $Q_{M_i}(b') = Q_{M_a}(b') - 1$. It indicates $Q_{M_i}(b) - Q_{M_i}(b') = Q_{M_a}(b) - Q_{M_a}(b') + 2 > 1$.

The above-proven property of "IT conservation when swapped ahead" ensures that it is possible to re-organize any Simplest Transfer Sequence with a mixture of $IT's$ and $DT's$ into a Bitonic Transfer Sequence, by swapping all the $IT's$ to be ahead of all the $DT's$. As the swapping operations do not change the individual Transfers, the original Simplest Transfer Sequence and the resultant Bitonic Transfer Sequence have the same set of Transfers, thus perform the same operation of changing M_x to M_y.

In fact, for any pair of semi-matchings, one can always find a Simplest Transfer Sequence and a sequence of Cyclic Neutral Transfers, which change one semi-matching to another. The construction process is shown in the Appendix. For any pair of incomparable semi-matchings M_x and M_y, let S_{Simp} and S_{cyclic} represent the Simplest Transfer Sequence and the sequence of Cyclic Neutral Transfers respectively, the concatenation of which changes M_x to M_y. Obviously, S_{Simp} contains both $IT's$ and $DT's$, and can be re-organized into a Bitonic Transfer Sequence. Then, adding S_{cyclic} into the front or back of this Bitonic Transfer Sequence, results in yet another Bitonic Transfer Sequence changing M_x to M_y.   □

Corollary 3. *Let poset$_G$ represents a partially ordered set defined by "$\preceq_T$" which contains all semi-matchings of a bipartite Graph G. For any pair of semi-matchings $M_x, M_y \in$ poset$_G$, if $M_x \prec\succ_T M_y$, then there exists at least one semi-matching $M_z \in$ poset$_G$ satisfying that $M_x \preceq_T M_z$ and $M_y \preceq_T M_z$.*

From Corollary 3 and the properties of a poset, Theorem 1 can be derived.

Theorem 1 (Existence of a set of fairest semi-matchings with regard to Transfer-based Comparison for any bipartite graph). *For any bipartite graph G, let poset$_G$ represents the partially ordered set defined by "$\preceq_T$" which contains all semi-matchings of G. There always exists a set of equally fair semi-matchings that constitute the greatest elements in poset$_G$, and all semi-matchings in this set have the same sorted quotas vector.*

From Theorem 1 and Corollary 2, we can furthermore derive Theorem 2.

Theorem 2. *For any bipartite graph, there always exists a set of equally fair semi-matchings (with the same sorted quotas vector), which are uniformly considered the fairest by all the fairness measures which are compatible with Majorization.*

5 Discussions

A. The poset from Transfer-based Comparison might be a semi-lattice
For the poset derived from Transfer-based comparison, we did not prove it to be a join-semilattice defined as the poset which has a least upper bound for any nonempty finite subset [4]. In other words, we proved that the poset itself has a greatest element which is the least upper bound, but this might not be true for all of its subsets. There might exist a counter example, such as a pair of semi-matchings M_1, M_2 having three common upper bounds M_3, M_4, M_5 which meet $M_3 \prec\succ_T M_4$ and $M_3 \prec_T M_5$, $M_4 \prec_T M_5$. That being said, we also did not find such counter examples for any given bipartite graph setting. Consequently, whether the poset is a semi-lattice remains a hypothesis.

B. The existence of the least elements with regard to Transfer-based Comparison is not always true
This paper has proved the existence of the greatest elements in the partially ordered set with regard to Transfer-based Comparison. However, the least elements in this partially ordered set do not always exist. This means there does not exist a uniformly agreed upon "most unfair" semi-matching set among all the fairness measures. For example, in Fig. 2, the two worst semi-matchings M_1 and M_2 are incomparable with regard to Transfer-based Comparison. Fairness measures disagree on the comparison of their quotas vectors $Q_{M_1} = (4, 1, 1)$ and $Q_{M_2} = (0, 3, 3)$. The details of conflicting comparisons between Q_{M_1} and Q_{M_2} have been shown in Table 1.

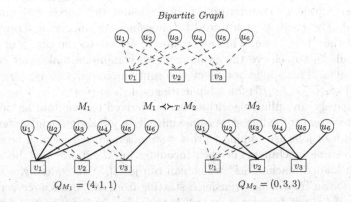

Fig. 2. An example of two incomparable semi-matchings both as the most unfair ones regarding "$\preceq_T$".

The insight behind the existence of the greatest elements but not the least elements in this partially ordered set, is in the asymmetry of "swapping IT" ahead: we have proved that an IT can be swapped ahead without losing its attribute of being an Improving Transfer, but this is not true for a DT. A Deteriorating Transfer, when swapped ahead, does not always maintain to be a DT. As a result, for any two incomparable semi-matchings M_x and M_y, it is guaranteed that there always exists a fairer one M_z that $M_x \preceq_T M_z$, $M_x \preceq_T M_z$, but not necessarily an "unfairer" one $M_{z'}$ that $M_{z'} \preceq_T M_x$, $M_{z'} \preceq_T M_y$.

C. Algorithms to achieve the fairest semi-matching in applications

The problem addressed in this paper has many applications such as load balancing, fair allocation, especially for their online scenario. Kleinberg et al. [7] studied on the load balancing problem which is concerned with assigning uniform jobs J to machines M. For each job J_i, there is a set $S_i \subseteq M$ on which job J_i can run. The aim is to assign each job J_i to a machine $M \in S_i$ in a way that the assignment has the fairest loads among all machines. An iterative network flow algorithm was proposed by Kleinberg et al. [7] to achieve an assignment of jobs to machines, which is the fairest with regard to lexicographically minimum fairness. This algorithm builds the assignment step by step starting with an empty assignment. In each step, it applies a max-flow algorithm to find a "partial" assignment, until eventually a legitimate assignment is reached which will turn out to be the fairest under lexicographical measure. However, this algorithm cannot be applied in the scenario of having to start from an existing assignment, and iteratively improving to achieve a fairest one. This limits its applicability in the online scenario, where a new assignment is always based on modifying an existing one.

Based on the conclusions achieved from this paper, two algorithms can be derived to achieve the fairest assignment, for the above job assignment system described in [7]. One algorithm is to iteratively improve an arbitrary assignment, according to the Transfer-based Comparison, to eventually achieve one of the fairest assignments. Starting from an arbitrary assignment, by iteratively applying an Improving Transfer until no more found, one fairest assignment will be achieved. The result is guaranteed by choosing any Improving Transfer, not necessarily the "best" one. This algorithm is similar to the one proposed by Harvey et al. [5] to achieve the assignment with minimal makespan and minimal flow time. The upper bound of the runtime complexity is shown to be $O(min\{|U|^{3/2}, |U||V|\} \cdot |E|)$ for a bipartite graph $G = (U \cup V, E)$.

Alternatively, an online algorithm can be derived to maintain an online job assignment system to be fairest. Assume initially a job assignment system is the fairest, and new jobs are coming to this system sequentially, which needs to be assigned to some machines. For each incoming job J_i, it will first be assigned to the least loaded machine M^* which can run job J_i. Subsequently, a search is conducted for an Improving Transfer t, starting from M^* as a source vertex. We claim that, if t is not found, the system is then already the fairest; if t is found, then by applying t which is a chain of jobs re-assignments to the system, the system can be successfully updated to be a fairest one with the new job. This claim can be easily proved based on the conclusions achieved from this paper.

6   Conclusion

This paper proves there always exist the universally agreed fairest semi-matchings in any given bipartite graph. To prove this, we define a poset of semi-matchings based on the way to transfer one into another. This poset (Transfer-based Comparison) is shown to always have the greatest elements, as the fairest semi-matchings. Subsequently, we show that the proposed Transfer-based Comparison can strictly imply the Majorization order. In other words, the fairest semi-matchings under the proposed Transfer-based Comparison are always regarded as fairest under Majorization. To our best knowledge, all existing fairness measures in the literature are compatible with Majorization. In conclusion, for any bipartite graph, there always exists a set of equally fair semi-matchings, which are universally regarded as the fairest ones, by all existing fairness measures, even though they may disagree on the comparisons among the ones that are not the fairest.

Acknowledgments. We are grateful to Kai Da Zhao, Professor Miloš Žefran, and Professor Bhaskar DasGupta for their helpful insights and discussion.

Appendix

Claim. For any pair of semi-matchings M_x and M_y, there always exists a sequence containing of a Simplest Transfer Sequence and a sequence of Cyclic Neutral Transfer, which changes M_x to M_y.

Proof. The notion $M_x \oplus M_y$ denotes the symmetric difference of edges set M_x and M_y, that is, $M_x \oplus M_y = (M_x \setminus M_y) \cup (M_y \setminus M_x)$. Let S represents the set of all Cyclic Neutral Transfers in $M_x \oplus M_y$. Suppose $M_{x'}$ can be derived by applying all Cyclic Neutral Transfers of S to M_x.

We assume all the edges of $M_{x'} \setminus M_y$ are colored by green, and all the edges of $M_y \setminus M_{x'}$ are colored by red. An observation on $M_{x'} \oplus M_y$ is that there exist one or more V-vertices which are endpoints of only green edges, but not red edges. We call those V-vertices as *Starting Vertices*. We build a Transfer which is an alternating green-red sequence of edges, as follows. (1) Find a green edge of which the V-endpoint is one arbitrary *Starting Vertex*, and set its U-endpoint as *Current Vertex*. Then repeat the following two steps. (2) Find a red edge of which the U-endpoint is *Current Vertex*. Set its V-endpoint as *Current Vertex*. (3) Find a green edge of which the V-endpoint is *Current Vertex*. Set its U-endpoint as *Current Vertex*. Continue until we cannot find any green edges. Delete all chosen edges from $M_{x'} \oplus M_y$, and then repeat above procedures to build more Transfers until $M_{x'} \oplus M_y$ becomes empty.

Throughout this process, we maintain that among all the obtained Transfers, the source vertex of one Transfer cannot be the sink vertex of another Transfer. Then, an arbitrarily ordered sequence of all the obtained Transfers constructs a Simplest Transfer Sequence from $M_{x'}$ to M_y. An illustration of the Transfers Construction is shown in Fig. 3.   □

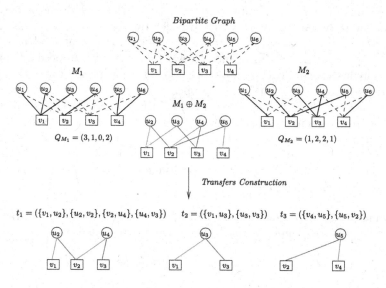

Fig. 3. An illustration of Transfers Construction.

References

1. Atkinson, A.B.: On the measurement of inequality. J. Econ. Theory **2**(3), 244–263 (1970)
2. Bansal, N., Sviridenko, M.: The santa claus problem. In: Proceedings of the Thirty-Eighth Annual ACM Symposium on Theory of Computing, pp. 31–40. ACM (2006)
3. Bokal, D., Brešar, B., Jerebic, J.: A generalization of hungarian method and hall's theorem with applications in wireless sensor networks. Discret. Appl. Math. **160**(4), 460–470 (2012)
4. Davey, B.A., Priestley, H.A.: Introduction to Lattices and Order. Cambridge University Press, Cambridge (2002)
5. Harvey, N.J., Ladner, R.E., Lovász, L., Tamir, T.: Semi-matchings for bipartite graphs and load balancing. J. Algorithms **59**(1), 53–78 (2006)
6. Jain, R., Chiu, D.M., Hawe, W.R.: A Quantitative Measure of Fairness and Discrimination for Resource Allocation in Shared Computer System, vol. 38. Eastern Research Laboratory, Digital Equipment Corporation Hudson, MA (1984)
7. Kleinberg, J., Rabani, Y., Tardos, É.: Fairness in routing and load balancing. In: 40th Annual Symposium on FOCS, 1999, pp. 568–578. IEEE (1999)

8. Lan, T., Kao, D., Chiang, M., Sabharwal, A.: An axiomatic theory of fairness in network resource allocation. IEEE (2010)
9. Mo, J., Walrand, J.: Fair end-to-end window-based congestion control. IEEE/ACM Trans. Netw. (ToN) **8**(5), 556–567 (2000)
10. Olkin, I., Marshall, A.W.: Inequalities: Theory of Majorization and Its Applications, vol. 143. Academic Press, New York (2016)
11. Tse, D., Viswanath, P.: Fundamentals of Wireless Communication. Cambridge University Press, New York (2005)

Better Inapproximability Bounds and Approximation Algorithms for Min-Max Tree/Cycle/Path Cover Problems

Wei Yu and Zhaohui Liu$^{(\boxtimes)}$

Department of Mathematics, East China University of Science and Technology,
Shanghai 200237, China
{yuwei,zhliu}@ecust.edu.cn

Abstract. We study the problem of covering the vertices of an undirected weighted graph with a given number of trees (cycles, paths) to minimize the weight of the maximum weight tree (cycle, path). Improved inapproximability lower bounds are proved and better approximation algorithms are designed for several variants of this problem.

Keywords: Approximation hardness · Approximation algorithm · Tree cover · Cycle cover · Path cover · Traveling salesman problem

1 Introduction

Given an undirected graph with nonnegative edge weights and a positive integer k, the objective is to find k trees (cycles, paths) to cover all the vertices such that the weight of the maximum weight tree (cycle, path) is minimized. This fundamental optimization problem, known as Min-Max Tree/Cycle/Path Cover Problem, and its variants have attracted considerable research attention in recent decades. This is mainly due to its wide range of application in both operations research and computer science, such as mail and newspaper delivery [10], nurse station location [8], disaster relief efforts routing [5], data gathering and wireless recharging in wireless sensor networks [22], multi-vehicle scheduling problem [4,15], political districting [14], and so on. For more practical examples that can be modeled by min-max tree/cycle/path cover problems, we refer to [1,21,22, 24,25].

Given the NP-hardness of the min-max tree/cycle/path cover problem [8], most researchers have concentrated on designing approximation algorithms that generate near-optimal solutions in polynomial time as well as on deducing inapproximability lower bounds.

1.1 Previous Results

For the Min-Max Tree Cover Problem (MMTCP), Even et al. [8] and Arkin et al. [1] devised independently 4-approximation algorithms, which was improved

© Springer International Publishing AG 2017
Y. Cao and J. Chen (Eds.): COCOON 2017, LNCS 10392, pp. 542–554, 2017.
DOI: 10.1007/978-3-319-62389-4_45

to a 3-approximation algorithm by Khani and Salavatipour [16]. The above-mentioned 4-approximation algorithm by Arkin et al. [1] is actually developed for the Min-Max Path Cover Problem (MMPCP), which is still the best available algorithm. For the Min-Max Cycle Cover Problem (MMCCP), Xu et al. [25] gave a 6-approximation algorithm, which can also be achieved by a combination of the 3-approximation algorithm for MMTCP with standard edge-doubling strategy. Xu et al. [22] and Jorati [13] proposed independently 16/3-approximation algorithms. Recently, Yu, Liu [26] developed a min$\{4\rho, 5\}$-approximation algorithm provided an algorithm for TSP with approximation ratio ρ is available.

In Rooted MMTCP/MMCCP/MMPCP, which is an important variant treated in the literature, a depot set D is prescribed and the aim is to cover all the vertices not in D with at most k trees/cycles/paths each of which contains at least one vertex from D. If each vertex in D has a unit-capacity, i.e., it can be included in at most one tree/cycle/path, we obtain a capacitated rooted problem. Otherwise, if each vertex in D is allowed to be present in an arbitrary number of trees/cycles/paths, we have an uncapacitated rooted problem. As shown in [22, 24, 26], an $(\alpha + 1)$-approximation algorithm for the uncapacitated rooted problem can be derived from an α-approximation algorithm for the corresponding unrooted problem. For Single-Depot Rooted MMTCP/MMCCP/MMPCP, i.e., $|D| = 1$, better algorithms with approximation ratios 3, $\rho+1$, 3 were obtained by Nagamochi [18], Frederickson et al. [10], Xu et al. [24], respectively (ρ defined as before).

For Capacitated Rooted MMCCP, Xu et al. [25] developed the first approximation algorithm with ratio 13 and obtained a 7-approximation algorithm for the case $|D| = k$. The latter result was shown independently by Jorati [13]. Subsequently, Xu et al. [22] improved the ratio to 7 for the general problem. For Capacitated Rooted MMTCP with $|D| \doteq k$, Even et al. [8] provided a 4-approximation algorithm.

In addition, the above results can be further improved if either the graph has a special structure, such as a tree [19], is embedded into a Euclidean space [14] or if k is fixed [9].

As for the approximation hardness, Xu, Wen [23] proved an inapproximability lower bound of 3/2 for both MMTCP and MMPCP, which applies to the capacitated rooted problems even with the restriction $|D| = k$. Xu et al. [24] obtained the same result for Uncapacitated Rooted MMTCP/MMPCP. Xu et al. [25] gave a lower bound of 4/3 for both MMCCP and its rooted version. Xu, Wen [23] also showed a lower bound of 10/9 (20/17) for Single-Depot Rooted MMTCP (MMCCP), complemented by a lower bound of 4/3 for Single-Depot Rooted MMPCP in [24].

1.2   Our Results

We obtain both improved approximation algorithms and better inapproximability lower bounds for several variants of the min-max tree/cycle/path cover problem. Firstly, we devise a 6-approximation algorithm for Capacitated Rooted MMCCP, improving on the algorithm in [22] with an approximation ratio of 7.

Second, we propose the first approximation algorithms for Capacitated Rooted MMTCP and Capacitated Rooted MMPCP, both of which have a constant ratio of 7. Third, we provide further improvement on the approximation algorithms for some interesting special cases, i.e., planar problem, graphic problem (see Sect. 2 for definitions). Last, we prove better inapproximability lower bounds for (Rooted) MMCCP, Single-Depot Rooted MMCCP, Single-Depot Rooted MMTCP, as illustrated in Table 1, and show that the reductions used to prove the lower bounds are essentially best possible.

Table 1. Previous and new lower bounds on min-max tree/cycle/path cover problems

	Unrooted	Uncapacitated rooted $|D| = 1$	$|D| = k$	Capacitated rooted $|D| = k$
MMTCP	4/3	10/9 4/3 (Theorem 2)	3/2	3/2
MMCCP	4/3	20/17	4/3	4/3
	3/2 (Theorem 1)	5/4 (Theorem 3)	3/2 (Theorem 1)	3/2 (Theorem 1)
MMPCP	3/2	4/3	3/2	3/2

The remainder of the paper is organized as follows. We explain some notations and formally describe the problems in Sect. 2. The inapproximablility lower bounds are proved in Sect. 3. This is followed by approximation algorithms for the capacitated rooted min-max tree/cycle/path cover problem in Sect. 4.

2   Preliminaries

We denote by $G = (V, E)$ an undirected weighted complete graph with the vertex set V and the edge set E. Let $w(e)$ denote the weight or length of edge e. If $e = (u, v)$, we also use $w(u, v)$ to denote the weight of e. The weights are metric, i.e., nonnegative, symmetric and satisfy the triangle inequality. For $B > 0$, $G[B]$ denotes the subgraph of G obtained by removing all the edges in E with weights greater than B. For a subgraph H (e.g. tree, cycle, path) of G, let $V(H), E(H)$ denote the vertex set and edge set of H, respectively. The weight of H is defined as $w(H) = \sum_{e \in E(H)} w(e)$. For $u \in V$, $d(u, H) = \min_{v \in V(H)} w(u, v)$ denotes the distance from u to H. The weight of a path or cycle is also called its length. A tree (cycle, path) containing only one vertex and no edges is a trivial tree (cycle, path) and its weight is defined as zero.

For a subset V' of V, a set $\{T_1, T_2, \ldots, T_k\}$ of trees (some of them may be trivial trees) is called a *tree cover* of V' if $V' \subseteq \cup_{i=1}^k V(T_i)$. The cost of this tree cover is defined as $\max_{1 \leq i \leq k} w(T_i)$, i.e., the maximum weight of the trees. Particularly, a tree cover of V is simply called a tree cover. By replacing trees with cycles (paths) we can define *cycle cover* (*path cover*) and its cost similarly. Note that a cycle containing exactly two vertices u, v consists of two copies of edge (u, v) and has a length of $2w(u, v)$.

Now we can formally state the following problems.

Definition 1. *In the Min-Max Tree Cover Problem (MMTCP), we are given an undirected complete graph $G = (V, E)$ with a metric weight function w on E and a positive integer k, the goal is to find a minimum cost tree cover with at most k trees.*

Definition 2. *In the Rooted Min-Max Tree Cover Problem (RMMTCP), we are given an undirected complete graph $G = (V, E)$ with a metric weight function w on E, a depot set $D \subseteq V$, and a positive integer k, the objective is to find a minimum cost tree cover of $V \setminus D$ with at most k trees such that each tree contains exactly one vertex in D.*

Definition 3. *In the Capacitated Rooted Min-Max Tree Cover Problem (CRMMTCP), we are given an undirected complete graph $G = (V, E)$ with a metric weight function w on E, a depot set $D \subseteq V$, and a positive integer k, the objective is to find a minimum cost tree cover of $V \setminus D$ with at most k trees such that each tree contains exactly one vertex in D and each vertex in D is contained in at most one tree.*

The first problem is called unrooted problem as before, since no depot set is given. The second problem is referred to as uncapacitated rooted problem while the last problem is called capacitated rooted problem.

By replacing trees with cycles or paths in the above definitions we have another six problems, which are denoted by similar acronyms. For example, CRMMPCP stands for the Capacitated Rooted Min-Max Path Cover Problem while RMMCCP refers to the Rooted Min-Max Cycle Cover Problem. Note that in a rooted path cover, not only should each path contain one vertex in D but also this vertex has to be an end vertex of the path. This is natural if one interprets paths as the routing of a fleet of k vehicles that service the customers located at the vertices from given depots.

In our problem definitions we assume that the graphs are complete. This involves no loss of generality since we can take the metric closure of a connected graph if it is not complete but connected. For a disconnected graph, we first add some edges of sufficiently large length to obtain a connected graph, and then consider the problems defined on the metric closure of this connected graph. In MMTCP, by restricting the graph to be the metric closure of a connected (weighted) planar graph we obtain (Weighted) Planar MMTCP. In Graphic MMTCP, the graph is required to be the metric closure of a connected unweighted graph. The same special case of the min-max cycle/path cover problems is defined similarly. We also suppose w.l.o.g. that the weights of the edges are integers.

Given an instance of some min-max tree/cycle/path cover problem, OPT indicates its optimal value as well as the corresponding optimal solution. Each tree (cycle, path) in OPT is called an optimum tree (cycle, path). By the triangle inequality, except for the uncapacitated rooted problem, we can assume w.l.o.g that any two optimum cycles (paths) are vertex-disjoint, since one can eliminate the repeated vertices without increasing the cost. However, the optimum trees are not necessarily vertex-disjoint since there may exist vertices of degree greater than 2 and eliminating these vertices may increase the cost.

3   Inapproximability Lower Bounds

In this section we show that even Planar Min-Max Tree/Cycle/Path Cover Problems are hard to approximate and give improved lower bounds for some single-depot, i.e., $|D| = 1$, rooted problems. However, Weighted Planar TSP, the special case of Weighted Planar MMCCP with $k = 1$, admits PTAS (see [3,17]).

Our results are based on reductions from a restriction of the following well-known NP-Complete problem (see [11]).

3-Dimensional Matching Problem (3DM). Given disjoint sets X, Y, Z that have the same number q of elements and a set $\mathcal{M} \subseteq X \times Y \times Z$ of triples with $|\mathcal{M}| = m$, is there an exact matching $\mathcal{M}' \subseteq \mathcal{M}$ such that $|\mathcal{M}'| = q$ and any two triples in $\mathcal{M}'$ do not have any common coordinate?

Associated with an instance of 3DM there is a bipartite graph $B = (W \cup (X \cup Y \cup Z), E)$, where each element of W represents a triple of $\mathcal{M}$. There is an edge connecting $w \in W$ to $v \in X \cup Y \cup Z$ if and only if v is a coordinate of the triple represented by w. By restricting the associated bipartite graph to be planar we obtain Planar 3-Dimensional Matching Problem (Planar 3DM), which is still NP-Complete (see [7]).

Xu and Wen [23], Xu et al. [24], Xu et al. [25] derived the lower bounds for min-max tree/cycle/path cover problems summarized in Table 1. All these bounds are deduced by gap-reductions from 3DM. In what follows, we show that the a lower bound of 3/2 holds even for the planar version of all the problems in Table 1 except for the single-depot rooted problems. First, we observe that restricting the planar bipartite graph to be connected in Planar 3DM preserves NP-Completeness. This special case is called Connected Planar 3DM.

Lemma 1. *Connected Planar 3DM is NP-Complete.*

Now we are ready to show the inapproximablity lower bounds.

Theorem 1. *It is NP-hard to approximate the planar version of the min-max tree/cycle/path cover problem and its rooted version within a performance ratio less than 3/2.*

Remark 1. Replacing 3DM with Planar 3DM in the reductions of Xu, Wen [23] and redefining the edge length as the shortest path length are not sufficient to obtain our results. In this case we may derive a disconnected planar graph. To make it connected we have to add some long edges. As a result, we have a connected weighted planar graph, instead of a connected unweighted planar graph.

In Single-Depot RMMTCP/RMMCCP/RMMPCP, the depot set D contains a single vertex. Xu, Xu, Li [24] derived a lower bound of 4/3 for Graphic Single-Depot RMMPCP using the following reduction. Given an instance I of 3DM, each triple $w = (x, y, z) \in \mathcal{M}$ corresponds to a subgraph of G as illustrated in Fig. 1. They proved that I has an exact matching if and only if the metric closure of G has a path cover with $k = q + 2m$ paths of cost at most 3. We note that the same reduction also implies a lower bound of 4/3 for Graphic Single-Depot RMMTCP. A tree cover of the metric closure of G with $k = q + 2m$ trees of cost at most 3 has to be a path cover with $q + 2m$ paths of cost at most 3.

Theorem 2. *It is NP-hard to approximate Graphic Single-Depot RMMTCP within a performance ratio less than 4/3.*

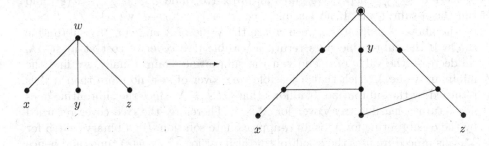

Fig. 1. The constructions in proof of Theorem 2 (The circled vertex is the only depot.)

For the single-depot cycle cover problem we have the following result.

Theorem 3. *It is NP-hard to approximate Graphic Single-Depot RMMCCP within a performance ratio less than 5/4.*

We end this section by showing that the results in Theorems 1, 2 and 3 are the best possible if we are confined to give a gap-reduction from an NP-Complete problem to derive inapproximability lower bounds for graphic tree/cycle/path cover problems.

Lemma 2. *There is a polynomial time algorithm to decide if (i) the optimal value of an instance of Graphic (Rooted) MMTCP/MMPCP equals 1; (ii) the optimum value of an instance of Graphic (Rooted) MMCCP equals 2.*

Lemma 3. *There is a polynomial time algorithm to decide if (i) the optimal value of an instance of Graphic Single-Depot RMMTCP/RMMPCP is not greater than 2; (ii) the optimal value of an instance of Graphic Single-Depot RMMCCP is not greater than 3.*

4   Approximation Algorithms

In this section we show how to obtain approximation algorithms for Capacitated Rooted Min-Max Tree/Cycle/Path Cover Problem by using the approximation algorithm for the unrooted problem.

To design an approximation algorithm for CRMMTCP, we adopt a two-stage process. The first is to obtain an α-subroutine defined as below.

Definition 4. *A polynomial time algorithm is called an α-subroutine for CRMMTCP, if for any instance consisting of $G = (V, E)$ and $k > 0$, and any nonnegative integer $\lambda \leq \sum_{e \in E} w(e)$, the algorithm always returns a feasible tree cover of cost at most $\alpha\lambda$, as long as $OPT \leq \lambda$.*

Given an instance of CRMMTCP consisting of $G = (V, E)$ and $k > 0$, clearly $0 \leq OPT \leq \sum_{e \in E} w(e)$. We guess an objective value $\lambda \in [0, \sum_{e \in E} w(e)]$ and run the α-subroutine. If no feasible tree cover is returned, we have $OPT > \lambda$ by the above definition. So we increase the value of λ and run the subroutine again. If the subroutine does return a feasible tree cover of cost at most $\alpha\lambda$, we decrease the value of λ and run the subroutine again. Finally, we find the minimum value λ^* such that a feasible tree cover of cost no more than $\alpha\lambda^*$ is returned by the subroutine. It follows that $OPT \geq \lambda^*$ since the subroutine does not return a feasible tree cover for $\lambda^* - 1$. Therefore, the tree cover returned by the α-subroutine for λ^* is an α-approximate solution. By a binary search for λ^*, this procedure uses the α-subroutine at most $\log \sum_{e \in E} w(e)$ times and hence runs in polynomial time.

One can see that the above procedure applies to other min-max tree/cycle/path cover problems.

Lemma 4. *A $T(n)$ time α-subroutine for MMTCP (MMCCP, MMPCP) or its rooted version implies an α-approximation algorithm for the same problem that runs in $O(T(n) \log \sum_{e \in E} w(e))$ time.*

4.1   Cycle Cover

Suppose there is a ρ_c-subroutine for MMCCP. We give the following algorithm for CRMMCCP. The input is an instance of CRMMCCP consisting of $G = (V, E)$ and $k > 0$ and a value $\lambda \in [0, \sum_{e \in E} w(e)]$.

Algorithm $CRMMCCP(\rho_c)$

Step 1. Delete all the edges with weight greater than $\frac{\lambda}{2}$ in G. If $G[\frac{\lambda}{2}]$ has some connected component containing no vertices in D, stop and return failure. Otherwise, remove the connected components containing no vertices in $V \setminus D$ and let the remaining p connected components be $F_1, F_2, \ldots, F_p$.

Step 2. For each $i = 1, 2, \ldots, p$,

Step 2.1. Let $D_i = D \cap V(F_i) = \{r_1, r_2, \ldots, r_{|D_i|}\}$ and find the minimum integer k_i with $1 \leq k_i \leq |D_i|$ such that the ρ_c-subroutine for MMCCP outputs a cycle cover $\mathcal{C}_i = \{C_{i,1}, C_{i,2}, \ldots, C_{i,k_i}\}$ of $V(F_i) \setminus D_i$ of cost at most $\rho_c \lambda$, while using λ, the complete subgraph G_i of G induced by $V(F_i) \setminus D_i$ and k_i as inputs.

Step 2.2. Construct a bipartite graph $H_i = (U \cup D_i, E_i')$, where $U = \{t_1, \ldots, t_{k_i}\}$ represents the k_i cycles in $\mathcal{C}_i$ (t_j corresponds to $C_{i,j}$) and $(t_j, r_l) \in E_i'$ if and only if $d(r_l, C_{i,j}) \leq \frac{\lambda}{2}$. Find a maximum matching M_i on H_i.

Step 2.3. If M_i matches all the vertices in U, i.e., $|M_i| = k_i$, then we can obtain a feasible cycle cover of $V(F_i) \setminus D_i$ as follows. For $(t_j, r_l) \in M_i$ we add two copies of the edge in the original graph G that corresponds to $d(r_l, C_{i,j})$ and make a shortcut to derive a cycle $C_{i,j}'$ on $V(C_{i,j}) \cup \{r_l\}$.

Step 2.4. If M_i does not match all the vertices in U, i.e., $|M_i| < k_i$, choose any unmatched vertex $u \in U$ and determine the set S of vertices that can be reached by an alternating path starting from u. Let $U' = S \cap U$ and $D' = S \cap D_i$. Run the ρ_c-subroutine for λ and the instance consisting of the complete subgraph induced by $\cup_{t_j \in U'} V(C_{i,j})$ and $|D'|$ to obtain a cycle cover of $\cup_{t_j \in U'} V(C_{i,j})$ of cost at most $\rho_c \lambda$. Update $\mathcal{C}_i$ by replacing the cycles corresponding to the vertices in U' by the newly generated cycles. For simplicity the cycles in $\mathcal{C}_i$ are still denoted by $C_{i,1}, C_{i,2}, \ldots, C_{i,k_i-1}$. Set $k_i := k_i - 1$ and go to Step 2.2.

Step 3. If $\sum_{i=1}^p k_i \leq k$, return the cycle cover $C_{1,1}', \ldots, C_{1,k_1}', \ldots \ldots,$ $C_{p,1}', \ldots, C_{p,k_p}'$ of $V \setminus D$; otherwise, return failure.

Let $C_1^*, C_2^*, \ldots, C_k^*$ be all the vertex-disjoint optimum cycles. By the triangle inequality, it is easy to verify that

Observation 1. *If $OPT \leq \lambda$, then $w(e) \leq \frac{OPT}{2} \leq \frac{\lambda}{2}$ for each $e \in \cup_{i=1}^k E(C_i^*)$.*

By this observation the vertex set of each optimum cycle is contained entirely in exactly one of $V(F_1), V(F_2), \ldots, V(F_p)$. As a consequence, the optimum cycles whose vertex sets are contained in $V(F_i)$ constitute a cycle cover of $V(F_i) \setminus D_i$. Moreover, the cost of this cycle cover of $V(F_i) \setminus D_i$ is at most OPT since the length of each optimum cycle is no more than OPT. Therefore, we have

Observation 2. *If $OPT \leq \lambda$, the optimum cycles can be partitioned into p groups such that the ith ($i = 1, 2, \ldots, p$) group consisting of $k_i^* \geq 1$ optimum cycles is a cycle cover of $V(F_i) \setminus D_i$ with cost at most λ.*

The performance of Algorithm $CRMMCCP(\rho_c)$ can be derived by the following lemma.

Lemma 5. *If $OPT \leq \lambda$, then*

(i) *Any connected component of $G[\frac{\lambda}{2}]$ contains at least one vertex in D;*

(ii) *For each $i = 1, 2, \ldots, p$, Step 2.1. returns a cycle cover of $V(F_i) \setminus D_i$ of cost at most $\rho_c\lambda$ and $k_i \leq k_i^*$;*

(iii) *For each $i = 1, 2, \ldots, p$, whenever Step 2.4. is executed for $|M_i| < k_i$ it holds that $|D'| = |U'| - 1 \geq 1$ and the ρ_c-subroutine returns a cycle cover of $\cup_{t_j \in U'} V(C_{i,j})$ of cost at most $\rho_c\lambda$;*

(iv) *Algorithm $CRMMCCP(\rho_c)$ returns a feasible cycle cover of cost at most $(\rho_c + 1)\lambda$.*

This lemma shows that Algorithm $CRMMCCP(\rho_c)$ is a $(\rho_c + 1)$-subroutine for CRMMCCP. Combining this with Lemma 4, we have the following result.

Theorem 4. *Given a ρ_c-subroutine for MMCCP, there is a $(\rho_c + 1)$-approximation algorithm for CRMMCCP.*

In [26], we have obtained a 5-subroutine for MMCCP running in $O(n^3)$ time for instances with n-vertex graphs. Plugging in this subroutine and using the $O(n^{2.5})$ time matching algorithm of Hopcroft, Karp [12] for n-vertex bipartite graphs we deduce that the Algorithm $CRMMCCP(\rho_c)$ has a time complexity of $O(n^3k)$ since the possible running of k subroutines dominates the complexity. Therefore we conclude that

Theorem 5. *There is an $O(n^3k \log \sum_{e \in E} w(e))$ time 6-approximation algorithm for CRMMCCP.*

Given a ρ-approximation algorithm for TSP, we also devised a 4ρ-subroutine for MMCCP in [26] that runs in polynomial time. Since Weighted Planar TSP admits a PTAS [3, 17], we have a $(4+\epsilon)$-subroutine for Weighted Planar MMCCP. For Graphic MMCCP, using the fact that a set of k connected subgraphs can be turned into a single connected subgraphs by adding $k - 1$ unit-length edge one can show that the same subroutine in [26] always returns a cycle cover of cost at most $2\rho_c(\lambda + 2)$ whenever $OPT \leq \lambda$. By Lemma 2, we can assume $\lambda \geq 3$. This implies $2\rho(\lambda + 2) \leq \frac{10}{3}\rho\lambda$ and we have a $\frac{10}{3}\rho_c$-subroutine for Graphic MMCCP. Using Theorem 4, we have improved the results for weighted planar or graphic cycle cover problem.

Corollary 1. *For any $\epsilon > 0$, there is a $(5 + \epsilon)$-approximation algorithm for Weighted Planar CRMMCCP and a $(\frac{13}{3} + \epsilon)$-approximation algorithm for Planar CRMMCCP.*

4.2 Path Cover

The framework to design approximation algorithms for CRMMPCP is similar to the cycle cover problem. However, in Step 1., we delete edges of length greater than λ instead of $\frac{\lambda}{2}$. And in other steps, we construct a path (cover) instead of a cycle (cover). Given a ρ_p-subroutine for MMPCP, the algorithm accepts an instance of CRMMPCP consisting of the graph $G = (V, E)$ and $k > 0$ and a value $\lambda \in [0, \sum_{e \in E} w(e)]$ as inputs.

Algorithm $CRMMPCP(\rho_p)$

Step 1. Delete all the edges with weight greater than λ in G. If $G[\lambda]$ has some connected component containing no vertices in D, stop and return failure. Otherwise, remove the connected components containing no vertices in $V \setminus D$ and the remaining p connected components are $F_1, F_2, \ldots, F_p$.

Step 2. For each $i = 1, 2, \ldots, p$,

Step 2.1. Let $D_i = V \cap V(F_i) = \{r_1, r_2, \ldots, r_{|D_i|}\}$ and find the minimum integer k_i with $1 \leq k_i \leq |D_i|$ such that the ρ_p-subroutine for MMPCP outputs a path cover $\mathcal{P}_i = \{P_{i,1}, P_{i,2}, \ldots, P_{i,k_i}\}$ of $V(F_i) \setminus D_i$ of cost at most $\rho_p \lambda$ using λ, the complete subgraph G_i of G induced by $V(F_i) \setminus D_i$ and k_i as inputs.

Step 2.2. Construct a bipartite graph $H_i = (U \cup D_i, E_i')$, where $U = \{t_1, \ldots, t_{k_i}\}$ represents the k_i paths in $\mathcal{P}_i$ (t_j corresponds to $P_{i,j}$) and $(t_j, r_l) \in E_i'$ if and only if $d(r_l, P_{i,j}) \leq \lambda$. Find a maximum matching M_i on H_i.

Step 2.3. If M_i matches all the vertices in U, i.e., $|M_i| = k_i$, then we can obtain a feasible path cover of $V(F_i) \setminus D_i$ as follows. For each $(t_j, r_l) \in M_i$, let (u_j, r_l) be the corresponding edge in the original graph. Denote by v_j, w_j the two end vertices of $P_{i,j}$ such that $w(u_j, v_j) \leq w(u_j, w_j)$. After adding edge (u_j, r_l) to connect r_l with $P_{i,j}$ and edge (u_j, v_j), we can derive an Eulerian path from r_l to w_j that goes through $\{r_l\} \cup V(P_{i,j})$. By shortcutting this path, we obtain a path with one end vertex $r_l \in D$ covering all the vertices in $V(P_{i,j})$.

Step 2.4. If M_i does not match all the vertices in U, i.e., $|M_i| < k_i$, choose any unmatched vertex $u \in U$ and determine the set S of vertices in H_i that can be reached by an alternating path starting from u. Let $U' = S \cap U$ and $D' = S \cap D_i$. Run the ρ_p-subroutine for λ and the instance consisting of the complete subgraph induced by $\cup_{t_j \in U'} V(P_{i,j})$ and $|D'|$ to obtain a path cover of $\cup_{t_j \in U'} V(P_{i,j})$ of cost at most $\rho_p \lambda$. Update $\mathcal{P}_i$ by replacing the paths corresponding to the vertices in U' by the newly generated paths. For simplicity the paths in $\mathcal{P}_i$ are still denoted by $P_{i,1}, P_{i,2}, \ldots, P_{i,k_i-1}$. Set $k_i := k_i - 1$ and go to Step 2.2.

Step 3. If $\sum_{i=1}^{p} k_i \leq k$, return the path cover $P_{1,1}, \ldots, P_{1,k_1}, \ldots \ldots, P_{p,1} \ldots P_{p,k_p}$ of $V \setminus D$; otherwise, return failure.

Let $P_1^*, P_2^*, \ldots, P_k^*$ be all the vertex-disjoint optimum paths. We have two observations similar to the cycle cover problem.

Observation 3. *If $OPT \leq \lambda$, then $w(e) \leq OPT \leq \lambda$ for each $e \in \cup_{i=1}^{k} E(P_i^*)$.*

Observation 4. *If $OPT \leq \lambda$, the optimum paths can be partitioned into p groups such that the ith ($i = 1, 2, \ldots, p$) group consisting of $k_i^* \geq 1$ optimum paths is a path cover of $V(F_i) \setminus D_i$ with cost at most λ.*

Based on these facts and an analogous proof to Lemma 5 we have

Lemma 6. *If $OPT \leq \lambda$, then Algorithm $CRMMCCP(\rho_p)$ returns a feasible path cover of cost at most $(\frac{3}{2}\rho_p + 1)\lambda$.*

By the results of Akin, Hassin, Levin [1], there is an $O(n^2)$ time ρ_p-subroutine with $\rho_p = 4$. This implies

Theorem 6. *There is an $O(n^{2.5}k \log \sum_{e \in E} w(e))$ time 7-approximation algorithm for CRMMPCP.*

Again one can show that for graphic instances, the subroutine in [1] returns a path cover of cost at most $2(\lambda + 1)$ whenever $OPT \leq \lambda$. By Lemma 2, we may assume that $\lambda \geq 2$ and hence $2(\lambda + 1) \leq 3\lambda$. So we have a 3-subroutine for Graphic MMPCP, which implies by Theorem 6 that

Corollary 2. *There is an 4-approximation algorithm for Graphic CRMMCCP.*

4.3   Tree Cover

For any $V' \subseteq V$, a cycle (path) cover of cost at most λ can be transformed into a cycle (path) cover of V' of cost at most λ by shortcuting the vertices of $V \setminus V'$, since the vertices in a cycle (path) are of degree no more than 2. This transformation cannot be conducted directly for a tree cover since the degree of the vertices of a tree can be greater than 2. However, by doubling the edges of a tree cover of cost at most λ we first obtain a cycle cover of cost at most 2λ. After that we can shortcut the vertices of $V \setminus V'$ and delete one edge from each cycle to obtain a path cover of V' (and hence a tree cover of V') of cost no greater than 2λ.

Due to this subtle difference, we obtain an algorithm for CRMMTCP proceeding as follows. First, we replace paths by trees in Algorithm $CRMMPCP(\rho_p)$ and change ρ_p-subroutine for MMPCP to ρ_t-subroutine for MMTCP. In Step 2.1. and Step 2.4. whenever running the subroutine for MMTCP, we also replace the value λ with 2λ. As a consequence, we obtain a tree cover of cost at most $2\rho_t\lambda$. In addition, in Step 2.3. the construction of a tree cover is simplified to connect each matched depot vertex to the corresponding tree by the edge in M_i. One can verify that this procedure is a $(2\rho_t + 1)$-subroutine for CRMMTCP. Since Khani, Salavatipour [16] derived an $O(n^5)$ time 3-subroutine for MMTCP, we have the following result.

Theorem 7. *There is an $O(n^5 k \log \sum_{e \in E} w(e))$ time 7-approximation algorithm for CRMMTCP.*

Acknowledgements. This research is supported in part by the National Natural Science Foundation of China under grants number 11671135, 11301184.

References

1. Arkin, E.M., Hassin, R., Levin, A.: Approximations for minimum and min-max vehicle routing problems. J. Algorithms **59**, 1–18 (2006)
2. Arora, S.: Polynomial time approximation schemes for euclidean traveling salesman and other geometric problems. J. ACM **45**, 753–782 (1998)
3. Arora, S., Grigni, M., Karger, D.R., Klein, P., Woloszyn, A.: A polynomial-time approximation scheme for weighted planar graph TSP. In: the Proceedings of the 9th Annual ACM-SIAM Symposium on Discrete Algorithms, pp. 33–41 (1998)
4. Bhattacharya, B., Hu, Y.: Approximation algorithms for the multi-vehicle scheduling problem. In: Cheong, O., Chwa, K.-Y., Park, K. (eds.) ISAAC 2010. LNCS, vol. 6507, pp. 192–205. Springer, Heidelberg (2010). doi:10.1007/978-3-642-17514-5_17
5. Campbell, A.M., Vandenbussche, D., Hermann, W.: Routing for relief efforts. Transportation Sci. **42**, 127–145 (2008)
6. Christofides, N.: Worst-case analysis of a new heuristic for the traveling salesman problem. Technical report, Graduate School of Industrial Administration, Carnegie-Mellon University, Pittsburgh, PA (1976)
7. Dyer, M., Frieze, A.: Planar 3DM is NP-complete. J. Algroithms **7**, 174–184 (1986)
8. Even, G., Garg, N., Koemann, J., Ravi, R., Sinha, A.: Min-max tree covers of graphs. Operations Res. Letters **32**, 309–315 (2004)
9. Farbstein, B., Levin, A.: Min-max cover of a graph with a small number of parts. Discrete Optimiz. **16**, 51–61 (2015)
10. Frederickson, G.N., Hecht, M.S., Kim, C.E.: Approximation algorithms for some routing problems. SIAM J. Comput. **7**(2), 178–193 (1978)
11. Garey, M., Johnson, D.: Computers and Intractability. Freeman, San Francisco (1979)
12. Hopcroft, J., Karp, R.: An $n^{2.5}$ algorithm for maximum matchings in bipartite graphs. SIAM J. Comput. **2**, 225–231 (1973)
13. Jorati, A.: Approximation algorithms for some min-max vehicle routing problems. Master thesis, University of Alberta (2013)
14. Karakawa, S., Morsy, E., Nagamochi, H.: Minmax tree cover in the euclidean space. J. Graph Algorithms Appl. **15**, 345–371 (2011)
15. Karuno, Y., Nagamochi, H.: 2-Approximation algorithms for the multi-vehicle scheduling problem on a path with release and handling times. Discrete Appl. Mathe. **129**, 433–447 (2003)
16. Khani, M.R., Salavatipour, M.R.: Approximation algorithms for min-max tree cover and bounded tree cover problems. Algorithmica **69**, 443–460 (2014)
17. Klein, P.: A linear-time approximation scheme for TSP in undirected planar graphs with edge-weights. SIAM J. Comput. **37**, 1926–1952 (2008)
18. Nagamochi, H.: Approximating the minmax rooted-subtree cover problem. IEICE Trans. Fund. Electron. **E88–A**, 1335–1338 (2005)
19. Nagamochi, H., Okada, K.: Polynomial time 2-approximation algorithms for the minmax subtree cover problem. In: Ibaraki, T., Katoh, N., Ono, H. (eds.) ISAAC 2003. LNCS, vol. 2906, pp. 138–147. Springer, Heidelberg (2003). doi:10.1007/978-3-540-24587-2_16
20. Nagamochi, H., Okada, K.: Approximating the minmax rooted-tree cover in a tree. Inf. Process. Lett. **104**, 173–178 (2007)
21. Nagarajan, V., Ravi, R.: Approximation algorithms for distance constrained vehicle routing problems. Networks **59**(2), 209–214 (2012)

22. Xu, W., Liang, W., Lin, X.: Approximation algorithms for min-max cycle cover problems. IEEE Trans. Comput. **64**, 600–613 (2015)
23. Xu, Z., Wen, Q.: Approximation hardness of min-max tree covers. Oper. Res. Lett. **38**, 408–416 (2010)
24. Xu, Z., Xu, L., Li, C.-L.: Approximation results for min-max path cover problems in vehicle routing. Nav. Res. Log. **57**, 728–748 (2010)
25. Xu, Z., Xu, L., Zhu, W.: Approximation results for a min-max location-routing problem. Discrete Appl. Mathe. **160**, 306–320 (2012)
26. Yu, W., Liu, Z.: Improved approximation algorithms for some min-max and minimum cycle cover problems. Theor. Comput. Sci. **654**, 45–58 (2016)

On the Complexity of k-Metric Antidimension Problem and the Size of k-Antiresolving Sets in Random Graphs

Congsong Zhang[✉] and Yong Gao

Department of Computer Science, University of British Columbia Okanagan,
Kelowna, BC V1V 1V7, Canada
congsong.zhang@alumni.ubc.ca, yong.gao@ubc.ca

Abstract. Network analysis has benefited greatly from published data of social networks. However, the privacy of users may be compromised even if the data are released after applying anonymization techniques. To measure the resistance against privacy attacks in an anonymous network, Trujillo-Rasua R. et al. introduce the concepts of k-antiresolving set and k-metric antidimension [1]. In this paper, we prove that the problem of k-metric antidimension is NP-hard. We also study the size of k-antiresolving sets in random graphs. Specifically, we establish three bounds on the size of k-antiresolving sets in Erdős-Rényi random graphs.

1 Introduction

The study of privacy in social networks has attracted much recent interest [2]. A common approach to data privacy is to use anonymization techniques when publishing social network data [3–5]. However, an adversary may compromise the privacy of users in anonymous social networks by active and passive attacks [6–9]. For example, Peng, W. et. al introduce a two-stage deanonymization attack in [8]. Firstly, an adversary can register new users with connections to the targeted users in a social network, and then creates edges between the newly registered users to construct a special subgraph. After that, the adversary identifies the subgraph in the anonymized social network that is released. Then the adversary identifies the targeted users. To measure the resistance against such attacks, Trujillo-Rasua R. et al. introduce the following concepts: metric representation, k-antiresolving set, k-metric antidimension, k-antiresolving basis, and (k, l)-anonymity [1].

Let $G = (V, E)$ be a simple connected graph, S be a subset of V, and v be a vertex in $V \setminus S$. The metric representation of v with respect to S is a tuple formed by the shortest-path distances from v to vertices in S. The set S is a k-antiresolving set if k is the greatest integer such that for any vertex $v \in V \setminus S$ there are at least $k - 1$ different vertices in $V \setminus S$ having the same metric representation with respect to S as v. A k-antiresolving basis is defined to be a k-antiresolving set of minimum cardinality. The k-metric antidimension is the cardinality of a k-antiresolving basis. We denote the k-metric antidimension

© Springer International Publishing AG 2017
Y. Cao and J. Chen (Eds.): COCOON 2017, LNCS 10392, pp. 555–567, 2017.
DOI: 10.1007/978-3-319-62389-4_46

by $\text{adim}_k(G)$. The graph G meets (k,l)-anonymity if k is the smallest positive integer such that $\text{adim}_k(G)$ is less than or equal to l.

It has been shown that the probability of identifying an anonymous vertex not in a k-antiresolving set S by exploiting the variation of metric representations with respect to S is at most $\frac{1}{k}$ [1]. The $\text{adim}_k(G)$ is the lower bound on the number of vertices controlled by an adversary to approach this probability.

In this paper, we prove that the problem of computing $\text{adim}_k(G)$ is NP-hard[1]. We also establish three bounds on the size of k-antiresolving sets in the Erdős-Rényi random graphs $G(n,p)$ with constant p.

2   Preliminary and Main Results

The following definitions of metric representation, k-antiresolving set, k-antiresolving basis, k-metric antidimension, and (k,l)-anonymity are from [1].

Definition 1 (Metric representation). *Let $G = (V,E)$ be a simple connected graph and $d_G(u,v)$ be the length of a shortest path between the vertices u and v in G. For a set $S = \{u_1, ..., u_t\}$ of vertices in V and a vertex v, we call the t-tuple $r(v|S) := (d_G(v, u_1), ..., d_G(v, u_t))$ the metric representation of v with respect to S.*

Definition 2 (*k-antiresolving set*). *Let $G = (V,E)$ be a simple connected graph and let $S = \{u_1, ..., u_t\}$ be a subset of vertices of G. The set S is called a k-antiresolving set if k is the greatest positive integer such that for every vertex $v \in V \setminus S$ there exist at least k - 1 different vertices $v_1, ..., v_{k-1} \in V \setminus S$ with $r(v|S) = r(v_1|S) = ... = r(v_{k-1}|S)$, i.e., v and v_1 , ..., v_{k-1} have the same metric representation with respect to S.*

Definition 3 (*k-metric antidimension and k-antiresolving basis*). *The k-metric antidimension of a simple connected graph $G = (V,E)$ is the minimum cardinality amongst the k-antiresolving sets in G and is denoted by $\text{adim}_k(G)$. A k-antiresolving set of cardinality $\text{adim}_k(G)$ is called a k-antiresolving basis for G.*

Definition 4 (*(k,l)-anonymity*). *A graph G meets (k,l)-anonymity with respect to active attacks if k is the smallest positive integer such that the k-metric antidimension of G is lower than or equal to l.*

Below are the notations used in this paper.

1. Let $G = (V, E)$ be a simple connected graph, S be a subset of V, and v be a vertex in $V \setminus S$. We define $N_s(v) = \{u | u \in S, u \text{ and } v \text{ are neighbors}\}$. For two vertices u, v in $V \setminus S$, we use $u =_s v$ to mean $r(u|S) = r(v|S)$. Moreover, we denote $\{v | v \in V \setminus S, r(v|S) = r'\}$ by $M_s[r']$ for a metric representation r' with respect to S.

---

[1] After finishing our proof, we found that Chatterjee T. et al. had proved that the problem of computing $\text{adim}_k(G)$ is NP-hard by a different reduction independently [10].

2. Let $G(n,p)$ be an Erdős-Rényi random graph and m be an integer. We define Γ_m as the family of all m-element subsets of vertices in $G(n,p)$. Let S be a vertex set in $G(n,p)$. We define R_S as the family of metric representations of vertices not in S with respect to S. Moreover, we use **whp** as the abbreviation of *with high probability* to mean that the occurrence probability of an event tends to 1 when n tends to infinity.

3. We define $\beta_u(S) = |\{e|e \in S, e = u\}|$ where S is a tuple of integers and u is an integer. We denote an n-tuple where all elements are valued x by $(x)_n$.

Our main results are: (1) a reduction from the exact cover by 3-sets problem $(X3C)^2$ to prove that the problem of computing $\text{adim}_k(G)$ is NP-hard; (2) three bounds on the size of k-antiresolving sets in $G(n,p)$ with constant p. The first bound is an upper bound such that **whp** there is no k-antiresolving set where k is constant. The second one is a lower bound such that **whp** there is no k-antiresolving set where k is not constant. The last one is a lower bound such that **whp** there is at least one k-antiresolving set where k is constant.

3   Computational Complexity of Computing $\text{Adim}_k(G)$

Before proving the computational complexity of computing $\text{adim}_k(G)$, we show two properties of k-antiresolving sets in simple connected graphs.

Proposition 1. *Let S be a subset of a k-antiresolving set S_k in a simple connected graph G. If there is a metric representation r with respect to S such that $0 < |M_s[r]| < k$, then $M_s[r] \subset S_k$.*

By the definition of k-antiresolving set, we know that this proposition is true.

Proposition 2. *Let G be a simple connected graph, S be a k-antiresolving set in G with $k \geq 3$, and $P_n = \{v_{p_1}, ..., v_{p_n}\}$ be a path in G satisfying: (1) the degree of the vertex v_{p_n} is equal to 1; (2) the degree of the vertex v_{p_i} is equal to 2 where $i \in [2, n-1]$. Then $P_n \subseteq S$ if any $v_{p_i} \in S$ where $i \in [2, n]$.*

Proof. By the definition of k-antiresolving set, we know that for a vertex v in S $|N_{V \setminus S}(v)| = 0$ or $|N_{V \setminus S}(v)| \geq k$. Therefore, if a vertex v_{p_i} is in S where $i \in [2, n]$, its neighborhood $\{v_{p_{i-1}}, v_{p_{i+1}}\}$ is also in S. By applying this observation again, we have that the neighborhood of $v_{p_{i-1}}$ is in S if $i-1 > 1$, and the neighborhood of $v_{p_{i+1}}$ is in S if $i+1 < n$. By repeating this observation, we have $P_n \subseteq S$. □

Theorem 1. *The problem of computing $\text{adim}_k(G)$ is NP-hard.*

Proof. We prove that the following decision version of the problem of computing $\text{adim}_k(G)$ is NP-complete: given two integers k and m, is there a k-antiresolving set of the cardinality is less than or equal to m in a simple connected graph $G = (V, E)$? Given a set of vertices in V, in polynomial time, we can verify that whether the cardinality of the set is less than or equal to m and whether the set is a k-antiresolving set. Therefore, this decision problem belongs to the class NP.

---

[2] X3C is a well known NP-complete problem [13].

We prove the NP-completeness by a reduction from the X3C problem : given a set $B = \{e_1, ..., e_{3q}\}$ and a family $\mathcal{S} = \{S_1, ..., S_p\}$ of 3-element subsets of B, does $\mathcal{S}$ contain a subfamily such that every element in B occurs in exactly one member of the subfamily. The X3C problem is NP-complete [13]. We may suppose that $p - q \geq 12$. If not, we create a set $B' = \{e_{3q+1}, ..., e_{3q+36}\}$ and a family $\mathcal{S}' = \{S_{p+1}, ..., S_{p+24}\}$ of 3-element subsets of B' such that $\mathcal{S}'$ contains an exact cover for B'. Then we get a new instance of X3C with $B \cup B'$ and $\mathcal{S} \cup \mathcal{S}'$ that satisfies our assumption. Apparently, $\mathcal{S} \cup \mathcal{S}'$ contains an exact cover for $B \cup B'$ iff $\mathcal{S}$ contains an exact cover for B. Let n be an integer such that $\lfloor \frac{p-q}{3} \rfloor < n < \lfloor \frac{p-q}{2} \rfloor$. We construct a simple connected graph $G = (V, E)$ with $|V| = 3qn(q+2) + p$ from an instance of X3C with $3q$ elements and p subsets as follows. Note that $n < \lfloor \frac{p-q}{2} \rfloor$. Then we can get G in polynomial time. Figure 1 shows the gadget corresponding to a subset $S_i = \{e_a, e_b, e_c\}$.

1. Vertices
    (a) For each S_i, we create a vertex v_{S_i}.
    (b) For each e_i, we create n vertices $v_{e_{i,1}}, ..., v_{e_{i,n}}$.
    (c) For each $v_{e_{i,j}}$, we create $q + 1$ vertices $v_{p_{i,j,1}}, ..., v_{p_{i,j,q+1}}$.
2. Edges
    (a) We create edges $\{v_{S_i}, v_{S_j}\}$ where $i \neq j$.
    (b) We create edges $\{v_{e_{i,j}}, v_{e_{i',j'}}\}$ where $i \neq i'$ or $j \neq j'$.
    (c) For each $v_{p_{i,j,1}}, ..., v_{p_{i,j,q+1}}$, we create a path $= \{v_{p_{i,j,1}}, ..., v_{p_{i,j,q+1}}\}$.
    (d) For each $v_{e_{i,j}}$, we create edges $\{v_{e_{i,j}}, v_{p_{i',j',1}}\}$ where $i \neq i'$ or $j \neq j'$.
    (e) If a subset $S_i = \{e_a, e_b, e_c\}$, we create edges $\{v_{S_i}, v_{e_{a,j}}\}$, $\{v_{S_i}, v_{e_{b,j}}\}$, and $\{v_{S_i}, v_{e_{c,j}}\}$ for all $j \in [1, n]$.

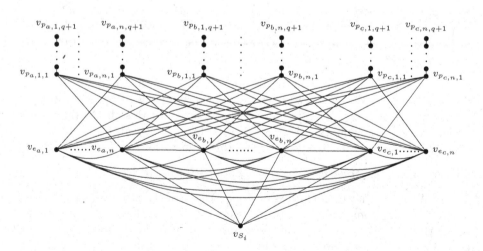

Fig. 1. A gadget for a subset $S_i = \{e_a, e_b, e_c\}$

Lemma 1. *If X3C has an exact cover $\mathcal{S}_c$, there is a $(p-q)$-antiresolving set S in G such that $|S| \leq q$.*

Proof. Let $V_{\mathcal{S}_c}$ and $V_{\mathcal{S}}$ be the sets of vertices corresponding to the subsets in $\mathcal{S}_c$ and $\mathcal{S}$. As $\mathcal{S}_c$ is an exact cover, we know $|V_{\mathcal{S}_c}| = q$ and $|V_{\mathcal{S}} \setminus V_{\mathcal{S}_c}| = p-q$. Let v be a vertex in $V_{\mathcal{S}} \setminus V_{\mathcal{S}_c}$. Then we have $r(v|V_{\mathcal{S}_c}) = (1)_q$. Moreover, for a vertex $v_{e_{i,j}}$, $\beta_1(r(v_{e_{i,j}}|V_{\mathcal{S}_c})) = 1$ and $\beta_2(r(v_{e_{i,j}}|V_{\mathcal{S}_c})) = q - 1$. Futhermore, there are $3n - 1$ different vertices $v_{e_{i',j'}}$ such that $v_{e_{i',j'}} =_{V_{\mathcal{S}_c}} v_{e_{i,j}}$. Similarly, for a vertex $v_{p_{i,j,l}}$ we can see that $r(v_{p_{i,j,l}}|V_{\mathcal{S}_c}) = (l+1)_q$, and there are $3qn - 1$ different vertices $v_{p_{i',j',l}}$ where $v_{p_{i',j',l}} =_{V_{\mathcal{S}_c}} v_{p_{i,j,l}}$. Because $3qn > 3n > p - q$, we have that $V_{\mathcal{S}_c}$ is a $(p - q)$-antiresolving set. $\qquad \square$

Lemma 2. *If there is a $(p-q)$-antiresolving set S in G such that $|S| \leq q$, X3C has an exact cover.*

Proof. We show that there are no vertices $v_{e_{i,j}}$ or $v_{p_{i,j,l}}$ in S. As the length of a path $\{v_{p_{i,j,1}}, ..., v_{p_{i,j,q+1}}\}$ is $q+1$, we know that the vertex $v_{p_{i,j,l}}$ where $l \in [2, q+1]$ is not in S by Proposition 2. Moreover, as $M_{\{v_{e_{i,j}}\}}[(q+2)] = \{v_{p_{i,j,q+1}}\}$, we get that the vertex $v_{e_{i,j}}$ is not in S by Proposition 1. For vertices $v_{p_{i,j,1}}$, we consider two cases: (1) more than two vertices $v_{p_{i,j,1}}, v_{p_{i',j',1}}$ are in S; (2) only one vertex $v_{p_{i,j,1}}$ is in S. In Case 1, we get a contradiction that the vertex $v_{p_{i,j,2}}$ should be in S by Proposition 1 as $M_{\{v_{p_{i,j,1}}, v_{p_{i',j',1}}\}}[(1,3)] = \{v_{p_{i,j,2}}\}$. In Case 2, because $\{v_{p_{i,j,1}}\}$ is not a $(p-q)$-antiresolving set, there is at least one vertex v_{S_i} in S. Then we get the same contradiction by Proposition 1 as $M_{\{v_{p_{i,j,1}}, v_{S_i}\}}[(1,3)] = \{v_{p_{i,j,2}}\}$.

We now prove $|S| \geq q$. If not, more than $p - q$ vertices v_{S_i} not in S have $r(v_{S_i}|S) = (1)_{|S|}$. Moreover, $3qn$ vertices $v_{p_{i,j,l}}$ where $l \in [1, q+1]$ have $r(v_{p_{i,j,l}}|S) = (l+1)_{|S|}$. Therefore, only vertices $v_{e_{i,j}}$ may have metric representations with respect to S different from $(l)_{|S|}$ where $l \in [1, q+2]$. For a metric representation r with respect to S different from $(l)_{|S|}$ where $l \in [1, q+2]$, we have $|M_S[r]|$ is n, $2n$, or $3n$. Note that $\lfloor \frac{p-q}{3} \rfloor < n < \lfloor \frac{p-q}{2} \rfloor$. Hence we get a contradiction that S is not a $(p - q)$-antiresolving set.

In the following, we show $\beta_1(r(v_{e_{i,j}}|S)) = 1$ for any vertex $v_{e_{i,j}}$. As the discussion in the previous paragraph, we know that $|M_S[(1)_q]|$ should be $p - q$ which means $r(v_{e_{i,j}}|S) \neq (1)_q$. Moreover, if $r(v_{e_{i,j}}|S) \neq (2)_q$, we have $\beta_1(r(v_{e_{i,j}}|S)) = 1$. If not, there are three vertices v_{S_a}, v_{S_b}, and v_{S_c} in S such that $v_{e_{i,j}} \in M_{\{v_{S_a}, v_{S_b}, v_{S_c}\}}[(1,1,2)]$. Because S is a $(p-q)$-antiresolving set, there should be $3n - 1$ different vertices $v_{e_{i',j'}}$ such that $v_{e_{i',j'}} =_S v_{e_{i,j}}$. Let S_a and S_b be the subsets in $\mathcal{S}$ corresponding to v_{S_a} and v_{S_b}. By the construction of G, we have $S_a = S_b$ which contradicts that no subsets in $\mathcal{S}$ are equal. Next, we prove $r(v_{e_{i,j}}|S) \neq (2)_q$. If not, let V_e be the set of vertices $v_{e_{i,j}}$ which are adjacent to at least one vertex in S. Then, for a $v \in V_e$ we have $\beta_1(r(v|S)) = 1$, and there are $3n - 1$ different vertices v' in V_e such that $v' =_S v$. As $|V_e| < 3qn$, we know that at least one vertex in S is not adjacent to any vertex $v_{e_{i,j}}$. Let S' be the 3-element subset in $\mathcal{S}$ corresponding to the vertex. We have a contradiction that $S' = \emptyset$.

We have proved that for any vertex $v_{e_{i,j}}$ $\beta_1(r(v_{e_{i,j}}|S)) = 1$, and there are $3n - 1$ different vertices $v_{e_{i',j'}}$ where $v_{e_{i',j'}} =_s v_{e_{i,j}}$. Let $\mathcal{S}'$ be the subfamily of 3-element subsets in $\mathcal{S}$ corresponding to the vertices in S. Then we know that $\mathcal{S}'$ is an exact cover for B. $\square$

Lemmas 1 and 2 complete the reduction from the X3C problem to the decision version of the problem of computing $\text{adim}_k(G)$. Therefore the problem of computing $\text{adim}_k(G)$ is NP-hard. $\square$

4   Antiresolving Sets in Random Graphs

Since the problem of computing $\text{adim}_k(G)$ is NP-hard, the (k, l)-anonymity problem is also NP-hard. To understand the relationship between the two parameters k and l, we study the size of k-antiresolving sets in random graphs which, we hope, will help characterize the trade-off between the level of anonymity and the cost of achieving such level of anonymity. For constant k, we establish three bounds on the size of k-antiresolving sets in $G(n, p)$ where the edge probability p is constant.

As shortest-path distances between different pairs of vertices are not mutually independent in $G(n, p)$, the analysis of the size of k-antiresolving sets is much more difficult than the analysis of the size of cliques and independent sets studied in the literature [12]. To overcome this difficulty due to the correlation among distances, we introduce the concept of a relaxed metric representation and prove bounds on the size of k-antiresolving sets under the relaxed metric representation. These bounds are then converted to the bounds on the size of k-antiresolving sets under the standard metric representation by taking into the consideration of an observation on the diameter of the random graph $G(n, p)$.

In the relaxed metric representation, we use the relaxed shortest-path distance instead of the shortest-path distance. Given two vertices v_i, v_j in $G(n, p)$, we define the relaxed shortest-path distance d_{ij} as

$$d_{ij} = \begin{cases} 1: & v_i \text{ and } v_j \text{ are adjacent,} \\ *: & \text{otherwise.} \end{cases}$$

For a set S of vertices and a vertex $v \notin S$, we denote the relaxed metric representation of v with respect to S by $r^*(v|S)$, and denote the family of relaxed metric representations with respect to S by R_S^*. Let

$$C_\alpha = \min \left[\frac{\alpha^2}{2}, (1+\alpha)\ln(1+\alpha) - \alpha\right] \text{ and } \epsilon = \ln\left[\frac{(2+\beta)}{C_\alpha \ln(\frac{1}{p_m})} \ln^2(n)\right] \cdot [\ln(n)]^{-1}$$

where α, β are arbitrary small constants, and $p_m = \min(p, 1 - p)$. We have the following theorem.

Theorem 2. *Given a random graph $G(n, p)$ where p is constant, **whp** there is no k-antiresolving set S such that k is constant and $|S| \leq (1-\epsilon)\log_{\frac{1}{p_m}}(n)$ where $p_m = \min(p, 1 - p)$.*

Proof. Let S be a subset in Γ_i with $i \leq (1 - \epsilon) \log_{\frac{1}{p_m}}(n)$, v be a vertex not in S, and r'_* be a relaxed metric representation in R^*_S. We define $I_v(r'_*, S)$ as the indicator function of the event $r^*(v|S) = r'_*$ and $X_{r'_*}(S) = \sum_{v \notin S} I_v(r'_*, S)$. As the relaxed shortest-path distances between v and vertices in S are mutually independent, we have

$$\mathbb{E}(I_v(r'_*, S)) = p^{\beta_1(r'_*)} \cdot (1 - p)^{|S| - \beta_1(r'_*)} \text{ and } \mathbb{E}(X_{r'_*}(S)) \geq (n - |S|) \cdot p_m^{|S|}.$$

By the assumptions of α and ϵ, we get

$$(n - |S|) \cdot p_m^{|S|} \geq n^{\epsilon} - |S| \cdot n^{\epsilon - 1} \text{ where } n^{\epsilon} = \frac{(2 + \beta)}{C_\alpha \ln(\frac{1}{p_m})} \ln^2(n)$$

which means $\min(\mathbb{E}(X_{r'_*}(S))) \in \Omega(\ln^2(n))$. Let $B(r'_*, S)$ be the event

$$|X_{r'_*}(S) - \mathbb{E}(X_{r'_*}(S))| \leq \alpha\mathbb{E}(X_{r'_*}(S)).$$

By Chernoff Bound, see Corollary A.1.14 in [11], we know

$$\Pr\{^\sim B(r'_*, S)\} = \Pr\{|X_{r'_*}(S) - \mathbb{E}(X_{r'_*}(S))| \geq \alpha\mathbb{E}(X_{r'_*}(S))\} \leq 2e^{-C_\alpha\mathbb{E}(X_{r'_*}(S))}.$$

Let r' be a metric representation with respect to S corresponding to r'_*. We define $I_v(r', S)$ as the indicator function of the event $r(v|S) = r'$ and $X_{r'}(S) = \sum_{v \notin S} I_v(r', S)$. Let $B(r', S)$ be the event

$$|X_{r'}(S) - \mathbb{E}(X_{r'_*}(S))| \leq \alpha\mathbb{E}(X_{r'_*}(S)).$$

Let D be the event that the diameter of $G(n, p)$ is less than or equal to 2. Then $^\sim B(r', S) = (^\sim B(r', S) \cap D) \cup (^\sim B(r', S) \cap^\sim D)$. Note that the event $^\sim B(r', S) \cap D$ is the event $^\sim B(r'_*, S) \cap D$. Therefore we have

$$\Pr\{^\sim B(r', S)\} = \Pr\{^\sim B(r'_*, S) \cap D\} + \Pr\{^\sim B(r', S) \cap^\sim D\} \leq \Pr\{^\sim B(r'_*, S)\} + \Pr\{^\sim D\}.$$

Note that two vertices have the shortest-path distance greater than 2 **iff** they have no common neighbors. Let X be the number of pairs of vertices in $G(n, p)$ that they have no common neighbors. By Markov's inequality, we have

$$\Pr\{^\sim D\} = \Pr\{X > 0\} \leq \mathbb{E}(X) = \binom{n}{2}(1 - p^2)^{n-2}.$$

Let $B(S)$ be the event $\bigcap_{r' \in R_S} B(r', S)$. By Boole's inequality, also well known by the union bound, we have

$$\Pr\{B(S)\} \geq 1 - n^{|S|} \cdot \left[2e^{-C_\alpha(n - |S|) \cdot p_m^{|S|}} + \binom{n}{2}(1 - p^2)^{n-2}\right].$$

Let $A = |\{S | S \in \Gamma_i, \sim B(S)\}|$. By the inequality $\binom{n}{k} \leq (\frac{en}{k})^k$, we have

$$\lim_{n \to \infty} \mathbb{E}(A) \leq \lim_{n \to \infty} 2 \cdot \left\{ \frac{en^2}{|S|} \cdot \left[\frac{e^{C_\alpha n^{\epsilon-1}}}{n^{2+\beta}} + \left(\binom{n}{2}(1-p^2)^{n-2} \right)^{\frac{1}{|S|}} \right] \right\}^{|S|} = 0.$$

By Markov's inequality, we get $\lim_{n \to \infty} \Pr\{A = 0\} = 1$.   □

Theorem 3. *Given a random graph $G(n,p)$ with constant p,* **whp** *there is no k-antiresolving set S such that k is not constant and $|S| \geq \log_{\frac{1}{2p^2-2p+1}}(n)$.*

Proof. Let S be a subset in Γ_i with $i \geq \log_{\frac{1}{2p^2-2p+1}}(n)$. We prove Theorem 3 by considering two cases: (1) $i \in \Theta(n)$; (2) $i \in o(n)$. In Case 1, we use $I_v(r', S)$ as the definition in Theorem 2. Let $p_m = \min(p, 1-p)$. Then we have

$$\Pr\{I_v(r', S) = 1\} \leq (1 - p_m)^{|S|}.$$

Let $B(r', S)$ be the event $\sum_{v \notin S} I_v(r', S) \in \omega(1)$.
Thus we know

$$\Pr\{B(r', S)\} \leq \binom{n}{\omega(1)}(1 - p_m)^{|S|\omega(1)}.$$

Let $B(S)$ be the event that S is a k-antiresolving set where $k \in \Theta(1)$. By Fréhet inequalities, also known as Boole-Fréhet inequalities, we get

$$\Pr\{B(S)\} = 1 - \Pr\{\bigcap_{r' \in R_S} B(r', S)\} \geq 1 - \binom{n}{\omega(1)}(1 - p_m)^{|S|\omega(1)}.$$

Let $A = |\{S | S \in \Gamma_i, \sim B(S)\}|$. Hence we have

$$\lim_{n \to \infty} \mathbb{E}(A) \leq \lim_{n \to \infty} \left[\frac{ne}{\omega(1)(\frac{1}{1-p_m})^{\frac{|S|}{2}}} \right]^{\omega(1)} \cdot \left[\frac{ne}{|S|(\frac{1}{1-p_m})^{\frac{\omega(1)}{2}}} \right]^{|S|}$$

$$\leq \lim_{n \to \infty} \left[\frac{\frac{1}{c_1}e}{(\frac{1}{1-p_m})^{\frac{\omega(1)}{2}}} \right]^{|S|} = 0.$$

By the notation of Θ, we know that there are a n_0 and a constant c_1, such that $n > n_0$, $c_1 n \leq |S|$. Then we have $\lim_{n \to \infty} \Pr\{A = 0\} = 1$.

In Case 2, we firstly prove that if $|R_S| \in \Theta(n - |S|)$, S is a k-antiresolving set of constant k. If not, we have that for any $r' \in R_S$ $\sum_{v \notin S} I_v(r', S) \in \omega(1)$. Therefore, we know that the number of vertices not in S is in $\omega(1) \cdot \Theta(n-|S|) \in \omega(n-|S|)$. This contradicts that the number of vertices not in S is exactly $n-|S|$.

In the following, we prove that $\mathbb{E}(|R_S^*|) \in \Theta(n - |S|))$ in Case 2. Given a $r'_* \in R_S^*$, let $I_S(r'_*)$ be the indicator function as

$$I_S(r'_*) = \begin{cases} 1 : \exists v \notin S \text{ such that } r^*(v|S) = r'_*, \\ 0 : \text{otherwise.} \end{cases}$$

Then we have

$$\Pr\{I_S(r'_*) = 1\} = 1 - \left[1 - p^{\beta_1(r'_*)}(1-p)^{|S|-\beta_1(r'_*)}\right]^{n-|S|}.$$

Let $\Phi = p^{\beta_1(r'_*)}(1-p)^{|S|-\beta_1(r'_*)}$. By the following inequality $1 - (1-x)^n \geq nx - \frac{(nx)^2}{2}$ for $x, n > 0$, we get

$$\Pr\{I_S(r'_*) = 1\} \geq [\Phi \cdot (n - |S|)] - \frac{[\Phi \cdot (n - |S|)]^2}{2}$$

which means

$$\mathbb{E}(|R_S^*|) \geq \sum_{\beta_1(r'_*)=0}^{|S|} \binom{|S|}{\beta_1(r'_*)} \left\{ [\Phi \cdot (n - |S|)] - \frac{[\Phi \cdot (n - |S|)]^2}{2} \right\}$$

$$\geq (n - |S|) - \frac{1}{2}n^2(2p^2 - 2p + 1)^{|S|} - \frac{1}{2}|S|^2(2p^2 - 2p + 1)^{|S|}.$$

As $|S| \geq \log_{\frac{1}{2p^2-2p+1}}(n)$, we know $\mathbb{E}(|R_S^*|) \in \Omega(n - |S|)$. Clearly, $|R_S^*| \leq n - |S|$. Then we have $\mathbb{E}(|R_S^*|) \in \Theta(n - |S|)$.

For a vertex $v \notin S$, we know that $r^*(v|S)$ takes values from the Cartesian product $\Lambda_v = \{1, *\}^{|S|}$. For $v_1, ..., v_{n-|S|} \notin S$, we define a function f such that $f(r^*(v_1|S), ..., r^*(v_{n-|S|}|S)) = |R_S^*|$. If two vectors $x, x' \in \prod_{j=1}^{n-|S|} \Lambda_{v_j}$ are different with only one coordinate, we have $|f(x) - f(x')| \leq 1$. Let α be an arbitary small constant. By the Azuma-Hoeffding inequality, see Corollary 2.27 in [12], we get

$$\Pr\{|R_S^*| - \mathbb{E}(|R_S^*|)| \geq \alpha\mathbb{E}(|R_S^*|)\} \leq 2e^{-\frac{(\alpha\mathbb{E}(|R_S^*|))^2}{2(n-|S|)}}.$$

As $\mathbb{E}(|R_S^*|) \in \Theta(n - |S|)$, there are a n_0 and a constant c_1, such that $n > n_0$, $c_1(n - |S|) \leq \mathbb{E}(|R_S^*|)$. Let $B(S)$ be the event $|R_S| \in \Theta(n - |S|)$ and $B^*(S)$ be the event $|R_S^*| \in \Theta(n - |S|)$. As $|R_S| \geq |R_S^*|$, we have

$$\Pr\{{\sim}B(S)\} \leq \Pr\{{\sim}B^*(S)\}$$

which means $\lim_{n\to\infty} \Pr\{{\sim}B(S)\} \leq 2e^{-C_\alpha(n-|S|)}$ where $C_\alpha = \frac{(\alpha c_1)^2}{2}$. Let $A = |\{S|S \in \Gamma_i, {\sim}B(S)\}|$. By $|S| \in o(n)$ and the inequality $\binom{n}{k} \leq (\frac{en}{k})^k$, we have

$$\lim_{n\to\infty} \mathbb{E}(A) \leq \lim_{n\to\infty} 2 \cdot \left(\frac{ne}{|S|}\right)^{|S|} e^{-C_\alpha(n-|S|)} = \lim_{n\to\infty} 2 \cdot \left[\frac{e \cdot \frac{n}{|S|}}{e^{C_\alpha(\frac{n}{|S|}-1)}}\right]^{|S|} = 0.$$

By Markov's inequality, we know $\lim_{n\to\infty} \Pr\{A = 0\} = 1$.   □

Theorem 4. *Given a random graph $G(n, p)$ with constant p,* **whp** *there is at least one k-antiresolving set S such that k is constant and $|S| \geq \log_{\frac{1}{p_m}}(n)$ where $p_m = \min(p, 1 - p)$.*

Proof. Because the conclusion of Theorem 3, we only need to consider the case $i \in \Theta(\log_{\frac{1}{p_m}}(n))$ in here. We use $I_v(r'_*, S)$ as the definition in Theorem 3. Let S be a subset in Γ_i and v be a vertex not in S. We define r^S_* as the relaxed metric representation with respect to S such that $\Pr\{I_v(r^S_*, S) = 1\} = p_m^{|S|}$. Let $B(S)$ be the event that S is a k-antiresolving set of constant k and $C(S)$ be the event that $\sum_{v \notin S} I_v(r^S_*, S)$ is a constant c. Clearly, $\Pr\{B(S)\} \geq \Pr\{C(S)\}$. Let $A = |\{S|S \in \Gamma_i, C(S)\}|$. Then we have

$$\mathbb{E}(A) = \binom{n}{|S|}\Pr\{C(S)\} = \binom{n}{|S|+c}p_m^{|S|c}(1 - p_m^{|S|})^{n-|S|-c}.$$

By the inequality $\binom{n}{k} \geq (\frac{n}{k})^k$ and $\lim_{n\to\infty}(1 - p_m^{|S|})^{n-|S|-c} \geq e^{-1}$ for $|S| \geq \log_{\frac{1}{p_m}}(n)$, we know

$$\lim_{n\to\infty} \mathbb{E}(A) = \infty.$$

By Chap. 4.1, Corollaries 4.3.3, and 4.3.4 in [11], we have the following inequality

$$\mathrm{Var}(A) \leq \mathbb{E}(A) + \sum_{S,S'\in\Gamma_i,S\neq S'} \mathrm{Cov}(C(S), C(S'))$$

where

$$\sum_{S,S'\in\Gamma_i,S\neq S'} \mathrm{Cov}(C(S), C(S')) \leq \mathbb{E}(A) \cdot \sum_{S,S'\in\Gamma_i,S\neq S'} \Pr\{C(S')|C(S)\}.$$

In the following, we show that $\mathrm{Var}(A) \in o(\mathbb{E}(A)^2)$ by considering two cases: (1) $S \cap S' \neq \emptyset$; (2) $S \cap S' = \emptyset$. Let $d_m = 1$ if $p = p_m$; otherwise let $d_m = *$. In Case 1, let $y = |S \cap S'|$. Then, given a vertex $v_u \notin S' \setminus S$ and a vertex $v_t \in S' \setminus S$, we have

$$\Pr\{d_{ut} = d_m|C(S)\} = \Pr\{d_{ut} = d_m, v_u \in S|C(S)\} + \Pr\{d_{ut} = d_m, v_u \notin S|C(S)\}$$

$$\leq \frac{|S|}{n - |S| + y} + p_m$$

which means

$$\Pr\{r^*(v|S') = r^{S'}_*|C(S)\} \leq \Pr\{r^*(v|S' \setminus S) = r^{S'\setminus S}_*|C(S)\}$$

$$\leq (\frac{|S|}{n - |S| + y} + p_m)^{|S'|-y}.$$

Let

$$p_m(y) = \frac{|S|}{n - |S| + y} + p_m \tag{1}$$

and z be the number of vertices $v \notin S \cup S'$ such that $r^*(v|S) = r^S_*$ and $r^*(v|S') = r^{S'}_*$. Then we have

$$\Pr\{C(S')|C(S)\} \le \binom{n}{c-z}[p_m(y)]^{c(|S'|-y)}$$

which means

$$\sum_{S,S'\in\Gamma_i, S\ne S'} \Pr\{C(S')|C(S)\} \le \sum_{z=0}^{c} \sum_{y=1}^{|S|-1} \binom{n}{c-z}\binom{n}{|S|-y}[p_m(y)]^{c(|S|-y)}.$$

By changing variables z to $c - z$, y to $|S| - y$, we have

$$p_m(y) = \frac{|S|}{n-y} + p_m$$

$$\sum_{S,S'\in\Gamma_i, S\ne S'} \Pr\{C(S')|C(S)\} \le \sum_{z=0}^{c} \sum_{y=1}^{|S|-1} \binom{n}{z}\binom{n}{y}[p_m(y)]^{cy}.$$

By $y \in O(\log_{\frac{1}{p_m}}(n))$, we know

$$\lim_{n\to\infty} \binom{n}{y+1}[p_m(y+1)]^{c(y+1)} \ge \lim_{n\to\infty} \binom{n}{y}[p_m(y)]^{cy}$$

which means

$$\lim_{n\to\infty} \max_{y\in[1,|S|-1]} \left[\binom{n}{y}[p_m(y)]^{cy}\right] = \lim_{n\to\infty} \binom{n}{|S|-1}[p_m(|S|-1)]^{c(|S|-1)}. \quad (2)$$

By the inequality $\binom{n}{k} \ge \frac{n^k}{4k!}$ for $n^{\frac{1}{2}} \ge k$, $\lim_{n\to\infty}(1-p_m^{|S|})^{n-|S|-c} \ge e^{-1}$, and (2), we have

$$\lim_{n\to\infty} \frac{\text{Var}(A)}{\mathbb{E}(A)^2} \le \lim_{n\to\infty} \frac{c \cdot n^c \cdot |S| \cdot \binom{n}{|S|-1} \cdot [p_m(|S|-1)]^{c(|S|-1)}}{\frac{n^{|S|+c}}{4(|S|+c)!} \cdot e^{-1} \cdot p_m^{c|S|}}$$

$$\le \lim_{n\to\infty} \frac{4e \cdot c \cdot |S| \cdot (|S|+c)^{1+c} \cdot (1 + \frac{|S|}{n-|S|+1} \cdot \frac{1}{p_m})^{c(|S|-1)}}{p_m^c \cdot n}$$

$$= 0.$$

Now we consider Case 2. Let V' be the set of vertices $v \notin S'$ where $r^*(v|S') = r^{S'}_*$. If $V' \cap S = \emptyset$, $C(S)$ and $C(S')$ are independent. Otherwise, let $z = |V' \cap S|$. Then we have

$$\Pr\{C(S')|C(S)\} \le \sum_{z=1}^{c} \binom{|S|}{z}\binom{n-2|S|}{c-z}[p_m(0)]^{c|S|}$$

where $p_m(0)$ is the formula by plugging $y = 0$ into (1). Then we have

$$\lim_{n \to \infty} \frac{\text{Var}(A)}{\mathbb{E}(A)^2} \leq \lim_{n \to \infty} \frac{c \cdot |S|^c \cdot \binom{n}{|S|} \cdot n^{c-1} \cdot [p_m(0)]^{c|S|}}{\frac{n^{|S|+c}}{4(|S|+c)!} \cdot e^{-1} \cdot p_m^{c|S|}}$$

$$\leq \lim_{n \to \infty} \frac{4ec \cdot |S|^c \cdot (|S|+c)^c \cdot [1 + \frac{|S|}{n-|S|} \cdot \frac{1}{p_m}]^{c|S|}}{n}$$

$$= 0.$$

Now we have proved that $\text{Var}(A) \in o(\mathbb{E}(A)^2)$. By the inequality

$$\Pr\{A = 0\} \leq \frac{\text{Var}(A)}{\mathbb{E}(A)^2},$$

see Theorem 4.3.1 in [11], we have $\lim_{n \to \infty} \Pr\{A = 0\} = 0$. As $\Pr\{B(S)\} \geq \Pr\{C(S)\}$, by the coupling method, see Example 7.1 in [14], we know

$$\lim_{n \to \infty} \Pr\{\bigcup_{S \in \Gamma_i} B(S)\} \geq \lim_{n \to \infty} \Pr\{\bigcup_{S \in \Gamma_i} C(S)\} = \lim_{n \to \infty} \Pr\{A > 0\} = 1. \qquad \square$$

5   Future Discussion

The time complexity of brute force algorithm to find a k-antiresolving basis in a simple conntected graph $G = (V, E)$ is in $O(2^{|V|} \cdot |V|^2)$. We are looking for an exact algorithm to find a k-antiresolving basis which performs better than the brutal one. For $G(n, p)$ with constant p, there still is a gap of $\Theta(\ln(\ln(n)))$ between the first bound and the third bound. We conjecture there is a greater value for the first bound, or $\log_{\frac{1}{p_m}}(n)$ is the sharp threshold for the appearance of k-antiresolving sets with constant k. We also conjecture there is a lower value for the second bound. Moreover, we are interest in how to add noises into anonymized social networks to increase the resistance against privacy attacks.

References

1. Trujillo-Rasua, R., Yero, I.G.: k-Metric antidimension: a privacy measure for social graphs. Inf. Sci. **328**, 403–417 (2016)
2. Netter, M., Herbst, S., Pernul, G.: Analyzing privacy in social networks - an interdisciplinary approach. In: IEEE 3rd International Conference on Social Computing, pp. 1327–1334 (2011)
3. Zhou, B., Pei, J., Luk, W.: A brief survey on anonymization techniques for privacy preserving publishing of social network data. ACM SIGKDD Explor. **10**(2), 1222 (2008)
4. Meyerson, A., Williams, R.: On the complexity of optimal K-anonymity. In: ACM Symposium on the Principles of Database Systems, pp. 223–228 (2004)

5. Wang, S.-L., Tsai, Z.-Z., Hong, T.-P., Ting, I.-H.: Anonymizing shortest paths on social network graphs. In: Nguyen, N.T., Kim, C.-G., Janiak, A. (eds.) ACIIDS 2011. LNCS, vol. 6591, pp. 129–136. Springer, Heidelberg (2011). doi:10.1007/978-3-642-20039-7_13
6. Liu, K., Terzi, E.: Towards identity anonymization on graphs. In: ACM SIGMOD International Conference on Management of Data, pp. 93–106 (2008)
7. Backstrom, L., Dwork, C., Kleinberg, J.: Wherefore art thou R3579X? anonymized social networks, hidden patterns, and structural steganography. In: 16th International Conference on World Wide Web, pp. 181–190 (2007)
8. Peng, W., Li, F., Zou, X., Wu, J.: A two-stage deanonymization attack against anonymized social networks. IEEE Trans. Comput. **63**(2), 290–303 (2014)
9. Narayanan, A., Shmatikov, V.: De-anonymizing social networks. In: 30th IEEE Symposium on Security and Privacy, pp. 173–187 (2009)
10. Chatterjee, T., DasGupta, B., Mobasheri, N., Srinivasan, V., Yero, I.G.: On the computational complexities of three privacy measures for large networks under active attackde-anonymizing social networks (2016). arxiv:1607.01438
11. Alon, N., Spencer, J.H.: The Probabilistic Method, 2nd edn. A Wiley-Interscience Publication, New York (2000)
12. Janson, S., Łuczak, T., Ruciński, A.: Random Graphs. A Wiley-Interscience Publication, New York (2000)
13. Garey, M., Johnson, D.: Computers and Intractability: A Guide to the Theory of NP-Completeness. W.H. Freeman and Company, New York (1979)
14. Dubhashi, D., Panconesi, A.: Concentration of Measure for the Analysis of Randomised Algorithms, 1st edn. Cambridge University Press, New York (2009). ISBN 978-0-521-88427-3

A Local Search Approximation Algorithm for the k-means Problem with Penalties

Dongmei Zhang[1], Chunlin Hao[2], Chenchen Wu[3], Dachuan Xu[2(✉)], and Zhenning Zhang[2]

[1] School of Computer Science and Technology, Shandong Jianzhu University, Jinan 250101, People's Republic of China
[2] Department of Information and Operations Research, College of Applied Sciences, Beijing University of Technology, Beijing 100124, People's Republic of China
xudc@bjut.edu.cn
[3] College of Science, Tianjin University of Technology, Tianjin 300384, People's Republic of China

Abstract. In this paper, we study the k-means problem with (nonuniform) penalties (k-MPWP) which is a natural generalization of the classic k-means problem. In the k-MPWP, we are given an n-client set $\mathcal{D} \subset \mathbb{R}^d$, a penalty cost $p_j > 0$ for each $j \in \mathcal{D}$, and an integer $k \leq n$. The goal is to open a center subset $F \subset \mathbb{R}^d$ with $|F| \leq k$ and to choose a client subset $P \subseteq \mathcal{D}$ as the penalized client set such that the total cost (including the sum of squares of distance for each client in $\mathcal{D} \setminus P$ to the nearest open center and the sum of penalty cost for each client in P) is minimized. We offer a local search $(81 + \varepsilon)$-approximation algorithm for the k-MPWP by using single-swap operation. We further improve the above approximation ratio to $(25 + \varepsilon)$ by using multi-swap operation.

Keywords: Approximation algorithm · k-means · Penalty · Local search

1 Introduction

The k-means problem is a classic problem in the Machine Learning area. The problem is defined formally as follows. Given an n-client set $\mathcal{D}$ in $\mathbb{R}^d$ and an integer k, the goal is to partition $\mathcal{D}$ into k clusters $X_1, ..., X_k$ and assign each cluster a center such that the sum of squares of distances from each client to its center is minimized.

Since the k-means problem is NP-hard [1,8,15], one natural way is to design efficient heuristic algorithm. Among all the heuristic algorithms, the Lloyd's algorithm is the most popular one [13,14]. Another important way is to design approximation algorithm. Most approximation algorithms for the k-means problem are based on the important observation of Matoušek [18] which points out the connection between k-means and k-median problems via constructing an approximate centroid set.

© Springer International Publishing AG 2017
Y. Cao and J. Chen (Eds.): COCOON 2017, LNCS 10392, pp. 568–574, 2017.
DOI: 10.1007/978-3-319-62389-4_47

For the metric k-median problem, Charikar et al. [6] give the first constant approximation algorithm. This ratio is further improved by [2,4,12]. Zhang [20] gives a local search $(2+\sqrt{3}+\varepsilon)$-approximation algorithm for the general k-facility location problem.

For the k-means problem, Kanungo et al. [10] give the first constant local search $(9 + \varepsilon)$-approximation algorithm. Makarychev et al. [16] present a bi-criteria approximation algorithm. There are many practical and interesting variants for the k-means problem (cf. [3,9,17]).

In order to handle outliers in clustering applications, we introduce a k-means problem with (nonuniform) penalties (k-MPWP) which is a natural generalization of the classic k-means problem. In the k-MPWP, we are given a n-client set $\mathcal{D} \subset \mathbb{R}^d$, a penalty cost $p_j > 0$ for each $j \in \mathcal{D}$, and an integer $k \leq n$. The goal is to open a center subset $F \subset \mathbb{R}^d$ with $|F| \leq k$ and to choose a client subset $P \subseteq \mathcal{D}$ as the penalized client set such that the total cost (including the sum of squares of distance for each client in $\mathcal{D} \setminus P$ to the nearest open center and the sum of penalty cost for each client in P) is minimized. Following the approaches of Kanungo et al. [10] for the k-means problem and Arya et al. [2] for the k-median problem, we present two local search constant approximation algorithms for the k-MPWP.

We remark that Tseng [17] introduces the penalized and weighted k-means problem which is similar to the k-MPWP but with uniform penalties. The penalty version for the classic facility location problem and k-median problem is proposed by Charikar et al. [7]. The currently best approximation ratios for the facility location problems with linear/submodular penalties are given by [11].

The contributions of our paper are summarized as follows.

- Introduce firstly the k-means problem with (nonuniform) penalties which generalizes the k-means problem;
- Offer $(81 + \varepsilon)$- and $(25 + \varepsilon)$- approximation algorithms for the k-MPWP by using single-swap and multi-swap local search techniques respectively. To the best of our knowledge, no constant approximation algorithm is known for this problem even for the uniform penalties version.

The organization of this paper is as follows. In Sect. 2, we present a single-swap local search $(81+\varepsilon)$-approximation algorithm for the k-MPWP. We further improve this approximation ratio to $25 + \varepsilon$ by using multi-swap in Sect. 3. We give some discussions in Sect. 4.

All the proofs are deferred to the journal version.

2   A Local Search $(81 + \varepsilon)$-approximation Algorithm

2.1   Notations

The square of distance between any two points $a, b \in \mathbb{R}^d$, is defined as follows,

$$\Delta(a,b) := ||a - b||^2.$$

Given a client subset $U \subseteq \mathcal{D}$, a point $c \in \mathbb{R}^d$, we define the total sum of squares distances of U with respect to c and the centroid of U as follows,

$$\Delta(c, U) := \sum_{j \in U} \Delta(c, j) = \sum_{j \in U} ||c - j||^2,$$

$$\mathrm{cen}(U) := \frac{1}{|U|} \sum_{j \in U} j.$$

The following lemma gives the important property of centroidal solution.

Lemma 1 [10]. *For any subset $U \subseteq \mathcal{D}$ and a point $c \in \mathbb{R}^d$, we have*

$$\Delta(c, U) = \Delta(\mathrm{cen}(U), U) + |U|\Delta(\mathrm{cen}(U), c). \tag{1}$$

Let S and O be a feasible solution and a global optimal solution to the k-MPWP with penalized sets P and P^*, respectively. For each client $j \in \mathcal{D} \setminus P$, we denote $S_j := \min_{s \in S} \Delta(j, s)$ and $s_j := \arg\min_{s \in S} \Delta(j, s)$. For each client $j \in \mathcal{D} \setminus P^*$, we denote $O_j := \min_{o \in O} \Delta(j, o)$ and $o_j := \arg\min_{o \in O} \Delta(j, o)$. For each center $o \in O$, we denote $s_o := \arg\min_{s \in S} \Delta(o, s)$ and say that s_o captures o. Furthermore, we denote

- $N_S(s) := \{j \in \mathcal{D} \setminus P | s_j = s\}, \forall s \in S$;
- $C_s := \sum_{j \in \mathcal{D} \setminus P} S_j = \sum_{s \in S} \Delta(s, N_S(s))$, $C_p := \sum_{j \in P} p_j$, $\mathrm{cost}(S) := C_s + C_p$;
- $N_O(o) := \{j \in \mathcal{D} \setminus P^* | o_j = o\}, \forall o \in O$;
- $C_s^* := \sum_{j \in \mathcal{D} \setminus P^*} O_j = \sum_{o \in O} \Delta(o, N_O(o))$, $C_p^* := \sum_{j \in P^*} p_j$, $\mathrm{cost}(O) := C_s^* + C_p^*$.

We remark that $o = \mathrm{cen}(N_O(o))$ for each $o \in O$.

Matoušek [18] introduces the following concept of approximate centroid set.

Definition 2. *A set of $\mathcal{C} \subset \mathbb{R}^d$ is an $\hat{\varepsilon}$-approximate centroid set for $\mathcal{D} \subset \mathbb{R}^d$ if for any set $U \subseteq \mathcal{D}$, we have*

$$\min_{c \in \mathcal{C}} \sum_{j \in U} \Delta(c, j) \le (1 + \hat{\varepsilon}) \min_{c \in \mathbb{R}^d} \sum_{j \in U} \Delta(c, j).$$

One can compute a $\hat{\varepsilon}$-approximate centroid set $\mathcal{C}$ for $\mathcal{D}$ of polynomial size in polynomial time [16,18]. Equipped with this result, we assume that all candidate centers are chosen from $\mathcal{C}$.

2.2 Local Search Algorithm

For any feasible solution S, we define the single-swap operation $\mathrm{swap}(a, b)$ as follows. Given $a \in S$ and $b \in \mathcal{C} \setminus S$, the center a is deleted from S and the center b is added to S.

We define the neighborhood of S associated with the above operation as follows,

$$\mathrm{Ngh}_1(S) := \{S \setminus \{a\} \cup \{b\} | a \in S, b \in \mathcal{C} \setminus S, \}.$$

For any given $\varepsilon > 0$, we present our single-swap local search algorithm as follows.

Algorithm 1 (ALG1(ε))

Step 0. **(Initialization)** *Setting*

$$\hat{\varepsilon} := 72 + \varepsilon - \sqrt{72^2 + 64\varepsilon},$$

construct $\hat{\varepsilon}$-approximate centroid set $\mathcal{C}$. Arbitrarily choose a feasible solution S.

Step 1. **(Local search)** *Compute*

$$S_{\min} := \arg\min_{S' \in \mathrm{Ngh}_1(S)} \mathrm{cost}(S').$$

Step 2. **(Stop criterion)** *If* $\mathrm{cost}(S_{\min}) \geq \mathrm{cost}(S)$, *output S. Otherwise, set $S := S_{\min}$ and go to Step 1.*

2.3 Analysis

To proceed the analysis, we need the following technical lemma.

Lemma 3. *Let S and O be a local optimal solution and a global optimal solution to the k-MPWP, respectively. We have,*

$$\sum_{j \in \mathcal{D} \backslash (P \cup P^*)} \sqrt{S_j O_j} \leq \sqrt{\sum_{j \in \mathcal{D} \backslash (P \cup P^*)} S_j} \cdot \sqrt{\sum_{j \in \mathcal{D} \backslash (P \cup P^*)} O_j}, \tag{2}$$

$$\sqrt{\Delta(s_{o_j}, j)} \leq \sqrt{S_j} + 2\sqrt{O_j}, \qquad \forall j \in \mathcal{D} \backslash (P \cup P^*). \tag{3}$$

In order to proceed the analysis, we need to introduce a center $\hat{o} \in \mathcal{C}$ associated with each $o \in O$. For each $o \in O$, let us define

$$\hat{o} := \arg\min_{c \in \mathcal{C}} \Delta(c, N_O(o)). \tag{4}$$

From Definitions 2 and 4, we have

$$\begin{aligned}
\sum_{j \in N_O(o)} \Delta(\hat{o}, j) &= \Delta(\hat{o}, N_O(o)) \\
&= \min_{c \in \mathcal{C}} \Delta(c, N_O(o)) \\
&\leq (1 + \hat{\varepsilon}) \min_{c \in \mathbb{R}^d} \Delta(c, N_O(o)) \\
&= (1 + \hat{\varepsilon}) \Delta(o, N_O(o)) \\
&= (1 + \hat{\varepsilon}) \sum_{j \in N_O(o)} \Delta(o, j). \tag{5}
\end{aligned}$$

We remark that the idea of choosing $\hat{o}$ satisfying inequality 5 is essentially due to Ward [19].

Then, we estimate the cost of S in the following theorem.

Theorem 4. *For any given $\varepsilon > 0$, ALG1(ε) produces a local optimal solution S satisfying*

$$C_s + C_p \leq (81 + \varepsilon)C_s^* + 18C_p^*.$$

Using standard technique of [2,5], we can obtain a polynomial time local search algorithm which sacrifices any given $\varepsilon' > 0$ in the approximation ratio.

3   Improved Local Search $(25 + \varepsilon)$-approximation Algorithm

For any feasible solution S, we define the so-called multi-swap operation swap(A, B) as follows. In the operation, we are given two subsets $A \subseteq S$ and $B \subseteq C \setminus S$ with $|A| = |B| \leq p$, where p is a fixed integer. All centers in A are deleted from S. Meanwhile, all centers in B are added to S.

We define the neighborhood of S with respect to the above multi-swap operation as follows,

$$\text{Ngh}_p(S) := \{S \setminus A \cup B | A \subseteq S, B \subseteq C \setminus S, |A| = |B| \leq p\}.$$

For any given $\varepsilon > 0$, we present our multi-swap local search algorithm as follows.

Algorithm 2 (ALG2(ε))

Step 0. **(Initialization)** *Setting*

$$p := \left\lceil \frac{5}{\sqrt{25 + \varepsilon} - 5} \right\rceil,$$

$$\hat{\varepsilon} := \frac{6}{p} + \frac{5}{p^2},$$

construct $\hat{\varepsilon}$-approximate centroid set C. Arbitrarily choose a feasible solution S.

Step 1. **(Local search)** *Compute*

$$S_{\min} := \arg \min_{S' \in \text{Ngh}_p(S)} \text{cost}(S').$$

Step 2. **(Stop criterion)** *If $\text{cost}(S_{\min}) \geq \text{cost}(S)$, output S. Otherwise, set $S := S_{\min}$ and go to Step 1.*

Theorem 5. *For any given $\varepsilon > 0$, ALG2(ε) produces a local optimal solution S satisfying*

$$C_s + C_p \leq (25 + \varepsilon)\,C_s^* + \left(5 + \frac{\varepsilon}{5}\right)C_p^*.$$

4   Discussions

In this paper, we study the k-MPWP and present $(81 + \varepsilon)$- and $(25 + \varepsilon)$- approximation algorithms using single-swap and multi-swap respectively. Since there is a local search $(9 + \varepsilon)$-approximation algorithm for the classic k-means problem [10], it is natural to ask whether our $(25 + \varepsilon)$-approximation can be further improved. Another research direction is to design approximation algorithm for this problem using LP rounding technique.

Acknowledgements. The research of the first author is supported by Higher Educational Science and Technology Program of Shandong Province (No. J15LN23). The second author is supported by Ri-Xin Talents Project of Beijing University of Technology. The third author is supported by Natural Science Foundation of China (No. 11501412). The fourth author is supported by Natural Science Foundation of China (No. 11531014). The fifth author is supported by Beijing Excellent Talents Funding (No. 2014000020124G046).

References

1. Aloise, D., Deshpande, A., Hansen, P., Popat, P.: NP-hardness of Euclidean sum-of-squares clustering. Mach. Learn. **75**, 245–249 (2009)
2. Arya, V., Garg, N., Khandekar, R., Meyerson, A., Munagala, K., Pandit, V.: Local search heuristics for k-median and facility location problems. SIAM J. Comput. **33**, 544–562 (2004)
3. Bandyapadhyay, S., Varadarajan, K.: On variants of k-means clustering. In: Proceedings of SoCG, Article No. 14, pp. 14:1–14:15 (2016)
4. Byrka, J., Pensyl, T., Rybicki, B., Srinivasan, A., Trinh, K.: An improved approximation for k-median, and positive correlation in budgeted optimization. In: Proceedings of SODA, pp. 737–756 (2014)
5. Charikar, M., Guha, S.: Improved combinatorial algorithms for the facility location and k-median problems. In: Proceedings of FOCS, pp. 378–388 (1999)
6. Charikar, M., Guha, S., Tardos, É., Shmoys, D.B. A constant-factor approximation algorithm for the k-median problem. In: Proceedings of STOC, pp. 1–10 (1999)
7. Charikar, M., Khuller, S., Mount, D.M., Narasimhan, G.: Algorithms for facility location problems with outliers. In: Proceedings of SODA, pp. 642–651 (2001)
8. Dasgupta, S. The hardness of k-means clustering. Technical Report CS2007-0890, University of California, San Diego (2007)
9. Georgogiannis, A.: Robust k-means: a theoretical revisit. In: Proceedings of NIPS, pp. 2883–2891 (2016)
10. Kanungo, T., Mount, D.M., Netanyahu, N.S., Piatko, C.D., Silverman, R., Wu, A.Y.: A local search approximation algorithm for k-means clustering. Comput. Geom. Theory Appl. **28**, 89–112 (2004)
11. Li, Y., Du, D., Xiu, N., Xu, D.: Improved approximation algorithms for the facility location problems with linear/submodular penalties. Algorithmica **73**, 460–482 (2015)
12. Li, S., Svensson, O.: Approximating k-median via pseudo-approximation. In: Proceedings of STOC, pp. 901–910 (2013)
13. Lloyd, S.: Least squares quantization in PCM. Technical report, Bell Laboratories (1957)

14. Lloyd, S.: Least squares quantization in PCM. IEEE Trans. Inf. Theory **28**, 129–137 (1982)
15. Mahajan, M., Nimbhorkar, P., Varadarajan, K.: The planar k-means problem is NP-hard. In: Proceedings of WALCOM, pp. 274–285 (2009)
16. Makarychev, K., Makarychev, Y., Sviridenko, M., Ward, J.: A bi-criteria approximation algorithm for k-means. In: Proceedings of APPROX/RONDOM, Article No. 14, pp. 14:1–14:20 (2016)
17. Tseng, G.C.: Penalized and weighted k-means for clustering with scattered objects and prior information in high-throughput biological data. Bioinformatics **23**, 2247–2255 (2007)
18. Matoušek, J.: On approximate geometric k-clustering. Discrete Comput. Geom. **24**, 61–84 (2000)
19. Ward, J. Private Communication (2017)
20. Zhang, P.: A new approximation algorithm for the k-facility location problem. Theoret. Comput. Sci. **384**, 126–135 (2007)

Improved Approximation Algorithm for the Maximum Base Pair Stackings Problem in RNA Secondary Structures Prediction

Aizhong Zhou[1], Haitao Jiang[1(✉)], Jiong Guo[1], Haodi Feng[1],
Nan Liu[2], and Binhai Zhu[3]

[1] School of Computer Science and Technology, Shandong University,
Jinan, Shandong, China
398239146@qq.com, {htjiang,jguo,fenghaodi}@sdu.edu.cn
[2] School of Computer Science and Technology, Shandong Jianzhu University,
Jinan, Shandong, China
belovedmilk@126.com
[3] Gianforte School of Computing, Montana State University,
Bozeman, MT 59717-3880, USA
bhz@montana.edu

Abstract. We investigate the maximum base pair stackings problem from RNA Secondary Structures prediction in this paper. Previously, Ieong *et al.* defined a basic version of this maximum base pair stackings problem as: given an RNA sequence, finding a set of base pairs to constitute a maximum number of stackings, and proved it to be NP-hard, where the base pairs are default under some biology principle and are given implicitly. Jiang proposed a generalized version of this problem, where the candidate base pairs are given explicitly as input and presented an approximation algorithm with a factor 8/3. In this paper, we present a new approximation algorithm for the generalized maximum base pair stackings problem by a two-stage local search method, improving the approximation factor from $8/3+\varepsilon$ to $5/2$. Since we adopt only two basic local operations, 1-substitutions and 2-substitutions, during the local improvement stage, the time complexity can be bounded by $O(n^7)$, much faster than the previous approximation algorithms.

1 Introduction

According to the central dogma of biology, Ribonucleic acids (RNAs) play an important role in regulating genetic and metabolic activities. Moreover, as new RNA sequences are constantly being discovered, in order to understand the biological functions of RNAs elaborately, we need to first know their structures.

An RNA is single-stranded chain and can be viewed as a sequence of nucleotides (also known as bases, denoted by A, C, G and U). The order of A, C, G, U's on the sequence form the *primary* structure of an RNA strand. An RNA folds into a three-dimensional structure by forming hydrogen bonds between nonconsecutive bases that are complementary, such as the Watson-Crick pairs

© Springer International Publishing AG 2017
Y. Cao and J. Chen (Eds.): COCOON 2017, LNCS 10392, pp. 575–587, 2017.
DOI: 10.1007/978-3-319-62389-4_48

A-U and C-G and the wobble pair G-U. The three-dimensional arrangement of the atoms in the folded RNA molecule forms the *tertiary* structure; the collection of base pairs in the tertiary structure forms the *secondary* structure. The secondary structure can in fact tell us where there are additional connections between the bases, and where the RNA molecule could be folded. In [13], the authors claimed that "the folding of RNA is hierarchical, since secondary structure is much more stable than tertiary folding", which implies that the tertiary folding would mostly obey the secondary structure. Since the three-dimensional structure determines the function of the RNA to some extent, predicting the secondary structure of RNA becomes a key problem to study RNA in a larger and deeper scope.

In 1978, Nussinov *et al.* [9] initiated the computational study of RNA secondary structures prediction, but this problem is still not well-solved yet. The biggest impediment is the existing of *pseudoknots*, which is composed of two interleaving base pairs provided that we arrange the RNA sequence in a linear order.

In the case where there is no pseudoknot, there have been a lot of positive results. Almost all of them use a dynamic programming method [7–9,11,15, 16]. As a consequence, the optimal RNA secondary structure can be computed roughly in $O(n^3)$ time and $O(n^2)$ space.

When pseudoknots do exist in some RNAs, the secondary structures prediction problem is harder. Lyngsø and Pedersen [6] proved that determining the optimal secondary structure possibly with pseudoknots is NP-hard under some special energy functions. Akutsu [1] showed that it remains NP-hard, even if the secondary structure requires to be planar. For limited types of pseudoknots, polynomial-time algorithms have been presented [1,10,14].

According to Tinoco's energy model [12], an RNA structure can be decomposed recursively into loops with independent free energy, the stacking loops formed by two adjacent base pairs have negative energy, which stabilizes the RNA structure. Hence Ieong *et al.* [3] initiated the study for the maximum base pair stackings problem with arbitrary pseudoknots. They proved that it is NP-hard to compute the planar secondary structure with the largest number of stackings, and proposed a 2-approximation for the planar version and a 3-approximation for the general version of this problem. Later, Lyngsø [5] proved that the maximum base pair stacking loops problem without the planar restriction remains NP-hard, even for binary sequences with 0–1 base pairs. He also devised a polynomial-time approximation scheme (PTAS) for this problem, with bases over a fixed-size alphabet Σ and the base pairs being a subset of $\Sigma \times \Sigma$, which runs in $O(n^{|\Sigma|^{\frac{1}{\varepsilon}}})$ time. Unfortunately, this PTAS is impractical even for $|\Sigma| = 4$ (e.g., $\Sigma = \{A, C, G, U\}$), and $\varepsilon = 1/2$.

Among all the above results, the base pairs are given implicitly, that is, under some fix biology principle, e.g., Watson-Crick base pairs: A-U and C-G, where any two such bases can form a base pair. As an alternative, the set of candidate base pairs may be given explicitly as input, because there could be additional conditions from comparative analysis which prevents two bases from forming a pair. This generalizes the maximum base pair stacking problem with implicit base pairs, hence the problem remains NP-hard. Jiang [4] improved the approximation factor for the maximum base pair stackings problem with explicit base pairs to $8/3+\varepsilon$. Jiang's algorithm combines the greedy strategy of Ieong's approximation algorithm and Berman's [2] approximation algorithm for computing a Maximum Weight Independent Set in $(d+1)$-claw-free graphs; to be more precise, its approximation factor is $8/3+\varepsilon$, and the time complexity is $O(n^{\log_d \frac{1}{\varepsilon}})$.

In this paper, we devise a new approximation algorithm for the maximum base pair stacking problems with explicit base pairs. Our method is based on local search. The new approximation factor is $5/2$, and the time complexity is $O(n^7)$.

2 Preliminaries

Let $S = s_1 s_2 \cdots s_n$ be an RNA sequence of n bases on $\{A, C, G, U\}$. We say that two bases s_i and s_{i+1} ($1 \leq i \leq n-1$) are *continuous* on S. A secondary structure of S is a set of base pairs (s_{i_1}, s_{j_1}), (s_{i_2}, s_{j_2}), ..., (s_{i_r}, s_{j_r}), where $i_k + 2 \leq j_k$ for all $k = 1, \ldots, r$ and no two base pairs share a base. Two base pairs, such as (s_i, s_j) and (s_{i+1}, s_{j-1}) with $i + 4 \leq j$, are said to be *adjacent*. A *stacking* is a loop formed by two adjacent base pairs (s_i, s_j) and (s_{i+1}, s_{j-1}), denoted by $(s_i, s_{i+1}; s_{j-1}, s_j)$. A *helix* H of length q is composed of $q + 1$ consecutive base pairs (s_i, s_j), (s_{i+1}, s_{j-1}), ..., (s_{i+q}, s_{j-q}), denoted by $(s_i, s_{i+1}, \ldots, s_{i+q}; s_{j-q}, s_{j-q+1}, \ldots, s_j)$. (s_i, s_j) and (s_{i+q}, s_{j-q}) are called *ending* base pairs of the helix H. We refer the segment of bases $s_i, s_{i+1}, \ldots, s_{i+q}$ as the α-side of the helix, and $s_{j-q}, s_{j-q+1}, \ldots, s_j$ as the β-side of the helix, denoted by H_α and H_β respectively. Note that there are exactly q stackings in a helix of length q. A helix contains at least two stackings is called a *long* helix, and a stacking is also called a *short* helix.

Now we present the formal definition of the problem to be studied in this paper. An example is shown in Fig. 1.

Problem Description: Maximum Base Pair Stackings
Input: An RNA sequence S, and a set of candidate base pairs BP.
Output: A set of chosen base pairs to constitute a maximum number of stackings.

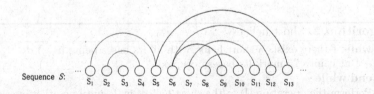

Fig. 1. The optimal base pairs found: (s_1, s_5), (s_2, s_4); and (s_6, s_{12}), (s_7, s_{11}), (s_8, s_{10}) and the maximum base pair stackings is 3. We cannot choose the base pair (s_5, s_{13}), as the base s_5 has been chosen in (s_1, s_5).

3   Our Algorithm

In this section, we depict our algorithm in detail, where the main idea of our algorithm is a two-stage local search. We firstly search for long helices by 1-substitutions and some special 2-substitutions, without considering the short helices. Then, we search for the remaining short helices, with the long helices previously found unchanged. A base is *free* if it is not involved in any stacking, otherwise, it is *occupied*. The stackings in set T are the output of our algorithm, where T is initialized as the empty set. In the algorithm, we will perform the following 3 operations of local search to obtain more stackings.

- *Operation* ① (1-substitution local improvement $\langle\, q\, \rangle$): For a helix H of length q $(q \geq 2)$ in T, replace it by other long helices with a total length of $q'(q' > q)$; for a short helix in T, replace it by other short helices with a total length of $q'(q' > 1)$.
- *Operation* ② (2-substitution local improvement $\langle\, 2, p\, \rangle$ $(p \geq 3)$): For a helix H of length 2 and another helix H' of length p $(p \geq 3)$ in T, replace H and an ending base pair of H' by other long helices with a total length of $p'(p' > 3)$.
- *Operation* ③ (2-substitution local improvement $\langle\, 2, 2\, \rangle$): For two helices H and H' both of length 2 in T, Replace H and H' by other long helices with a total length of $p'(p' > 4)$.

Algorithm 1. Long helices

1: **for** $(q$ from 8 to 2) **do**
2:     **while** (there exists a helix of length q, with all its bases being free) **do**
3:         Put it into T; mark its bases occupied.
4:     **end while**
5: **end for**
6: Perform the operation ① to the long helices in T until no other operation ① can not be performed.
7: Perform the operation ②,③ until no other operation ② and ③ can not be performed.

The following Algorithm 2 shows how to search short helices and locally improve them by 1-substitutions.

Algorithm 2. Short helices

1: **while** (there exists a short helix, with all its bases being free) **do**
2:     Put it into T; mark its bases occupied.
3: **end while**
4: Perform the operation ① to the short helices in T, until no other operation ① can not be performed.

Theorem 1. *The time complexity of Algorithms 1 and 2 is $O(n^7)$.*

Proof. To generate an initial feasible solution, we search for long helices of length at most 8, it takes $O(n)$ time to find such a helix, and there are at most $O(n)$ such helices. In total, it takes $O(n^2)$ time to obtain an initial feasible solution.

During the 1-substitution local improvement process in Algorithm 1, each long helix of length q $(2 \leq q \leq 7)$ occupies $2q$ bases, while we possibly make use of the 4 bases adjacent to them, so we search for long helices from these (at most) $2q + 4$ bases, which means we can obtain at most $2q + 4$ base pairs. These base pairs can constitute at most $(2q + 4)/3$ long helices. To fix a helix, it suffices to fix its starting base pair. Hence we have to search for at most $(2q+4)/3$ starting base pairs, while fixing each starting base pair takes $O(n)$ time. Consequently, for each iteration of the 1-substitution local improvement process, it takes $O(n^6)$ time.

During the 2-substitution local improvement process in Algorithm 1, we choose a helix of length 2 and a single base pair from a helix of length greater than 2, or two helices of length 2. In total they occupy 8 bases, while there are at most 8 bases adjacent to them, hence we search for long helices from these (at most) 16 bases. Similar to the above argument, it takes $O(n^5)$ time. By a similar analysis, the time complexity of the 1-substitution local improvement process in Algorithm 2 is $O(n^4)$.

Since the value of our solution is at most n, and the value of our solution would increase by at least 1 during each local improvement step, the algorithm executes at most n local improvements.

Finally, to check whether our solution cannot be further improved, we have to check all the $O(n^2)$ pairs of long helices (at least one of which must be of length 2), and check all the long helices of length less than 8 individually. In summary, the time complexity of Algorithm 1 and Algorithm 2 is $O(n^7)$.   □

4   Approximation Factor Analysis

To analyze the performance of our algorithm, we should compare our solution with the optimal solution. At the termination of our algorithm, there would not be any stacking with its four bases being free, then, all the stacking in the optimal solution would either be found by our algorithm or at least one of its bases be occupied by stackings in our solution. For a stacking $T^* = (s_i, s_{i+1}; s_{j-1}, s_j)$ in the optimal solution, we say that it is *destroyed* by these helices in our solution using s_i, s_{i+1}, s_{j-1} or s_j (even if T^* is also in a stacking in our solution); moreover, it can be destroyed by H_α (or H_β), where H is helix and some bases of s_i, s_{i+1}, s_{j-1} or s_j belong to H_α (or H_β). The following lemma shows an upper bound of the number of stackings in the optimal solution which is destroyed by some helices in our solution.

Lemma 1. *A helix of length q in our solution can destroyed at most $2q + 4$ stackings in the optimal solution.*

Proof. Let $H = (s_i, s_{i+1}, \ldots, s_{i+q}; s_{j-q}, s_{j-q+1}, \ldots, s_j)$ be a helix of length q. It contains $q+1$ base pairs, as well as $2(q+1)$ bases. Each stacking in the optimal solution is composed of two adjacent bases in one segment. The segment of bases $s_i, s_{i+1}, \ldots, s_{i+q}$ can constitute at most $q+2$ adjacent bases together with another two bases: s_{i-1}, s_{i+q+1}. Similarly, the segment of bases $s_{j-q}, s_{j-q+1}, \ldots, s_j$ can constitute at most $q+2$ adjacent bases together with another two bases: s_{j-q-1}, s_{j+1}. In total, at most $2q+4$ stackings in the optimal solution, using these bases, could be destroyed.   □

A stacking in the optimal solution is *singly* destroyed, if only one helix in our solution using its bases, otherwise, it is *multiply* destroyed.

Lemma 2. *At the termination of Algorithm 1, the number of stackings, which are singly destroyed by a long helix H of length q $(2 \le q \le 7)$ in our solution and all of which can constitute long helices without H, is at most q.*

Proof. Since otherwise, by replacing H, we would obtain some long helices with more stackings, which means the Algorithm 1 would not terminate.   □

To analyze the performance of our algorithm, we divide the stackings in the optimal solution into two parts: (1) stackings that are singly destroyed by some helices in our solution; (2) stackings that are multiply destroyed by at least two long helices in our solution. Then we assign the stackings in the optimal solution to destroyed-sets of helices by our solution. For a helix of length 8 or more in our solution, its destroyed-set DS_H contains the stackings in the optimal solution that are destroyed by it, and these stackings could appear in any other destroyed-sets. For a long helix H of length q $(2 \le q \le 7)$ in our solution, its destroyed-set DS_H contains the stackings in the optimal solution that are destroyed by it. For a short helix H' in our solution, its destroyed-set $DS_{H'}$ contains the stackings in the optimal solution that are destroyed by it but not by any long helix. As each stacking contributes a weight of 1 to the value in the optimal solution, we assign the weight for destroyed-sets as follows:

- a singly destroyed stacking contributes a weight of 1 to the destroyed-set containing it;
- a multiply destroyed stacking contributes a weight of $1/2$ to each destroyed-set containing it (see Fig. 2 for an example).

Obviously, the total weight of all the destroyed-sets would be greater than or equal to the optimal solution value, since all the stackings in the optimal solution are destroyed. To guarantee a 2.5 approximation ratio, it is sufficient to show that, given each helix H of length q, with the weight of its destroyed-set being $W(DS_H)$, it satisfies $W(DS_H)/q \le 2.5$. Henceforth, we say a helix is *safe* if the above condition is fulfilled.

Let $H = (s_i, s_{i+1}, \ldots, s_{i+q}; s_{j-q}, s_{j-q+1}, \ldots, s_j')$ be a helix of length q in our solution, two consecutive bases s_k, s_{k+1} $(i \le k \le i+q-1$ or $j-q \le k \le j-1)$ form a *gap*, if they are not in a common stacking in the optimal solution. Define $l(H_\alpha)$ and $l(H_\beta)$ to be the total length of long helices in the optimal solution,

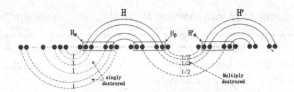

Fig. 2. The stackings in the H_α side are singly destroyed and the stackings in the other side are multiply destroyed by H and H'.

that are singly destroyed by H, and use the bases from the α-side and the β-side of H respectively. Define $s(H_\alpha)$ and $s(H_\beta)$ to be the number of short helices in the optimal solution, that are singly destroyed by H, and also use the bases from the α-side and the β-side of H respectively. Define $g(H_\alpha)$ and $g(H_\beta)$ to be the number of gaps, in the α-side and the β-side of H respectively.

Lemma 3. *At the termination of Algorithm 1, let H be a helix of length q in our solution, $s(H_\alpha) \leq g(H_\alpha) + 1 + \lfloor \frac{q+2-\max\{1,l(H_\alpha)+g(H_\alpha)\}}{5} \rfloor$, and $s(H_\beta) \leq g(H_\beta) + 1 + \lfloor \frac{q+2-\max\{1,l(H_\beta)+g(H_\beta)\}}{5} \rfloor$.*

Proof. We just prove the former inequality, since the other is exactly the same. From the definition of $s(H_\alpha)$, the short helices cannot share common base pairs, otherwise, they would form long helices. Therefore, the short helices and long helices must be spaced by gaps or multiply destroyed stackings. At the termination of Algorithm 1, our solution only contains long helices, and each long helix of length p $(p \geq 2)$ can occupy $p+1$ bases, and destroy $p+2$ stackings of a helix in the optimal solution. Hence, between two short helices which are singly destroyed by H_α, if there is no gap, there should be at least four stackings which are multiply destroyed. In other words, without gaps, every 5 consecutive stacking can contain a singly destroyed short helix. In case there is no gap and no long helix, the singly destroyed short helices are spaced by segments of multiply destroyed stackings, there would be one more singly destroyed short helix. So we have an extra one in the inequality, meanwhile this short helix occupies two consecutive bases. □

In fact, it always holds that $l(H_\alpha) + s(H_\alpha) + g(H_\alpha) \leq q + 2$. Hence, when $q = 3$ and $l(H_\alpha) = 3$, we have $s(H_\alpha) \leq g(H_\alpha) \leq 1$.

Now we show that most helices are safe, except a specific case, which we would analyze separately.

Lemma 4. *A helix H of length q, where $q \geq 8$, is safe.*

Proof. From Lemma 1, H can destroy at most $2q + 4$ stackings in the optimal solution, all of which are in the destroyed-set of H. Then, we have $(2q+4)/q \leq 2.5$, provided that $q \geq 8$. □

Lemma 5. *At the termination of Algorithm 1, a helix H of length q, where $3 \leq q \leq 7$, is safe.*

Proof. From Lemma 1, H can destroy at most $2q + 4$ stackings in the optimal solution. At the termination of Algorithm 1, from Lemma 2, $l(H_\alpha) + l(H_\beta) \leq q$. Each singly destroyed stacking contributes a weight of 1 to the destroyed-set of H, and each multiply destroyed stacking contributes a weight of $1/2$ to the destroyed-set of H, totally, we can show,

$$\frac{W(DS_H)}{q} \leq \frac{3q + 4 + \rho}{2q} \tag{1}$$

From Lemma 3, we conclude that $\rho \leq 1$, when $q = 3$; $\rho \leq 2$, when $q = 4$; and $\rho \leq 3$, when $5 \leq q \leq 7$. Therefore,

$$\frac{W(DS_H)}{q} \leq 5/2 \tag{2}$$

In fact, even if we add a weight $1/2$ to the numerator, the inequality still holds. □

It remains to deal with the helices of length 2.

Lemma 6. *Let* $H = (s_i, s_{i+1}, s_{i+2}; s_{j-2}, s_{j-1}, s_j)$ *be a helix of length 2 in our solution, if* $l(H_\alpha) = 0$*, then the total weight of stackings in the optimal solution assigned to* H *by* H_α *is at most 2.5.*

Proof. From Lemma 3, we have, $s(H_\alpha) \leq g(H_\alpha) + 1 + \lfloor \frac{2+2-\max\{1, g(H_\alpha)\}}{5} \rfloor$. That means $s(H_\alpha) - g(H_\alpha) \leq 1$. As there are at most 4 possible stackings in the optimal solution using bases of H_α, besides the gaps and short helices, all the other possible stackings can contribute a weight of $1/2$ to the destroyed-set of H, so we have $s(H_\alpha) + \frac{4 - s(H_\alpha) - g(H_\alpha)}{2} \leq 2.5$. (An example is shown in Fig. 3(a)). □

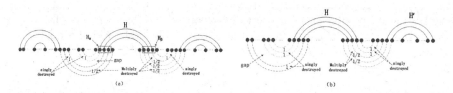

Fig. 3. (a) Case 1: the H_α side is a singly destroyed stacking with a gap, and the H_β side is a singly destroyed stacking together with three multiply destroyed stackings, and the total weight is at most be 5/2.(b) Case 2: the H_α and H_β sides decide that H gets a weight 3 when $l(H_\alpha) = 2$ or $l(H_\beta) = 2$.

Lemma 7. *Let* $H = (s_i, s_{i+1}, s_{i+2}; s_{j-2}, s_{j-1}, s_j)$ *be a helix of length 2 in our solution, if* $l(H_\alpha) = 2$*, then the total weight of stackings in the optimal solution assigned to* H *by* H_α *is at most 3.*

Proof. From Lemma 3, we have, $s(H_\alpha) \leq g(H_\alpha) + 1 + \lfloor \frac{2+2-\max\{1, l(H_\alpha) + g(H_\alpha)\}}{5} \rfloor$. As there are at most 4 possible stackings in the optimal solution using bases of H_α, if $s(H_\alpha) = 1$, to split this short helix with the long helix, then $g(H_\alpha)=1$, and there would be no other possible stackings left. If $s(H_\alpha) = 0$, besides the gaps and short helices, all the other possible stackings can contribute a weight of $1/2$ to the destroyed-set of H, so we have $l(H_\alpha) + s(H_\alpha) + \frac{4-l(H_\alpha)-s(H_\alpha)-g(H_\alpha)}{2} \leq 3$. (An example is shown in Fig. 3(b).) $\square$

The following lemma can be obtained directly from Lemma 6.

Lemma 8. *Let $H = (s_i, s_{i+1}, s_{i+2}; s_{j-2}, s_{j-1}, s_j)$ be a helix of length 2 in our solution, if $l(H_\alpha) = l(H_\beta) = 0$, then H is safe.*

Obviously, there could also be helices of length 2 which is not safe. Suppose $H = (s_i, s_{i+1}, s_{i+2}; s_{j-2}, s_{j-1}, s_j)$ is such an unsafe helix, in this case, one of $l(H_\alpha)$ and $l(H_\beta)$ could not be zero. Without loss of generality, we assume that $l(H_\alpha) = 2$ and $l(H_\beta) = 0$. From Lemma 2, we have $l(H_\alpha) + l(H_\beta) \leq 2$. Then the total weight of stackings in the optimal solution assigned to H by H_α is at most 3, and the total weight of stackings in the optimal solution assigned to H by H_β is at most 2.5. There are four possible stackings in the optimal solution using the bases of H_β, we define a weight-vector to record the weight of these four stackings contributing to H,

$$V(H_\beta) = \langle W(s_{i-1}s_i), W(s_i s_{i+1}), W(s_{i+1}s_{i+2}), W(s_{i+2}s_{i+3}) \rangle,$$

where $W(s_{k-1}s_k) \in \{1, 0.5, 0\}, i \leq k \leq i+3$. To make $W(s_{i-1}s_i) + W(s_i s_{i+1}) + W(s_{i+1}s_{i+2}) + W(s_{i+2}s_{i+3}) \leq 2.5$, since there could not be two continuous 1's, $V(H_\beta)$ has two choices: either $V(H_\beta)$ contains exactly one 1 and three 0.5's; or $V(H_\beta)$ contains exactly two 1's, one 0.5, and one zero. More specifically, $V(H_\beta)$ has the following configurations: (a) $\langle 1, 0.5, 0.5, 0.5 \rangle$ or symmetrically $\langle 0.5, 0.5, 0.5, 1 \rangle$, (b) $\langle 0.5, 1, 0.5, 0.5 \rangle$ or symmetrically $\langle 0.5, 0.5, 1, 0.5 \rangle$, (c) $\langle 1, 0, 1, 0.5 \rangle$ or symmetrically $\langle 0.5, 1, 0, 1 \rangle$. (See Fig. 4 for examples.) We have the following lemmas.

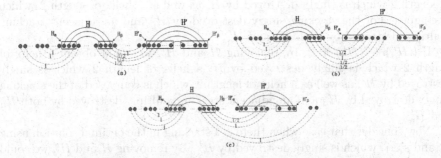

Fig. 4. (a),(b) and (c) are the examples of configuration (a),configuration (b) and configuration (c). They are all unsafe.

Lemma 9. *At the termination of Algorithm 1, configuration (a) could not exist.*

Proof. Omitted due to space constraint. □

We can observe that either configuration (b) or configuration (c) contains two continuous weight $\langle 1, 0.5 \rangle$ with the weight of $1/2$ being from a stacking using s_j and s_{j+1}, and the weight of 1 being from a stacking using s_{j-1} and s_j. Without loss of generality, let the stacking $S_H^{\frac{1}{2}}$ using s_{j-1} and s_j be the stacking singly destroyed by H_β, and the stacking $S_H^{\frac{1}{2}} = (s_j, s_{j+1}; s_r, s_{r+1})$ be a multiply destroyed stacking, which is destroyed by both H_β and $H' = (H'_\alpha, H'_\beta)$ where $H'_\alpha = (s_{x-p}, s_{x-p+1}, \ldots, s_x)$ and $H'_\beta = (s_{y-p}, s_{y-p+1}, \ldots, s_y)$ and where $x = r$. But s_{j+1} could be possible occupied by another long helix H'' in our solution, which means the stacking $S_H^{\frac{1}{2}}$ is commonly destroyed by H, H' and H''. Then we reassign the weight of $S_H^{\frac{1}{2}}$ as follows:

- If $S_H^{\frac{1}{2}}$ is destroyed by only H and H', we assign the total weight 1 of $S_H^{\frac{1}{2}}$ to the destroyed-set of H'.
- If $S_H^{\frac{1}{2}}$ is destroyed by H, H' and H'', we assign a weight of $1/2$ to each of the destroyed-set of H' and H''.

It is obviously that, under this weight assignment, H is always safe, no matter whether $V(H_\beta)$ is in configuration (b) or configuration (c). We still have to prove that under this new weight assignment, H' and H'' are safe.

Lemma 10. *At the termination of Algorithm 1, if $S_H^{\frac{1}{2}}$ is destroyed by only H_β and H'_α, then $p = 2$ and $l(H'_\alpha) = l(H'_\beta) = 0$; moreover, there cannot be a stacking in the optimal solution using s_x and s_{x-1}, which is singly destroyed by H'_α.*

Proof. We show that if any of the three consequences in the lemma does not hold, there would exist feasible local improvement for our solution, which contradicts the assumption that Algorithm 1 has terminated.

If $p \geq 3$, by removing H and the base pair (s_x, s_y), we could obtain a helix of length 2 which is singly destroyed by H, as well as a helix of length 2, which is composed of the stacking singly destroyed by H_β and an adjacent stacking multiply destroyed by both H_β and H'_α.

If $l(H'_\alpha) + l(H'_\beta) = 2$, by removing H and H', we could obtain a helix of length 2 which is singly destroyed by H, a helix of length 2 which is singly destroyed by H', as well as a helix of length 2, which is composed of the stacking singly destroyed by H_β and an adjacent stacking multiply destroyed by both H_β and H'_α.

Now consider that case when there is a stacking in the optimal solution using s_x and s_{x-1}, which is singly destroyed by H'_α. By removing H and H', we could obtain the helix of length 2 which is singly destroyed by H, as well as a helix of length 3, which is composed of the stacking singly destroyed by H_β, an adjacent stacking multiply destroyed by both H_β and H'_α, and the stacking using s_x and s_{x-1} singly destroyed by H'_α. In all these cases, a local improvement is possible. □

Lemma 11. *The weight reassignment cannot generate two new continuous stackings, both of which contributing a weight of one to the destroyed-set of H'.*

Proof. The weight assignment only involves the weight of $S_H^{\frac{1}{2}}$. If $S_H^{\frac{1}{2}}$ is destroyed by only H and H', from Lemma 10, $S_H^{\frac{1}{2}}$ cannot be adjacent to a stacking which contributes a weight of one to H'. If $S_H^{\frac{1}{2}}$ is destroyed by H, H' and H'', the weight assignment cannot generate a new stacking which contributes a weight of 1 to H' or H''.   □

Lemma 11 indicates that, if H' cannot be replaced by 1-substitution after the termination of Algorithm 1, then it still cannot be replaced by any 1-substitution local improvement, even if $S_H^{\frac{1}{2}}$ is singly destroyed by H'.

Now, we show that, after the weight reassignment, H' remains safe. It can be shown similarly that H'' would be safe as well, since viewing from H, the role of H' and H'' is equivalent.

Lemma 12. *If $p \geq 3$, then H' is safe.*

Proof. From Lemma 11, no matter how much weight H' contributes from the weight reassignment, there would not be two new continuous stackings, both of which contributes a weight of one to the destroyed-set of H'. Since H' cannot be replaced by any 1-substitution local improvement, Lemma 1 still holds. Lemma 3 holds since we only keep long helix in our solution at the termination of Algorithm 1. By the same argument as in Lemma 5, we can conclude that the lemma holds.   □

Lemma 13. *If $p = 2$ and $S_H^{\frac{1}{2}}$ is destroyed by only H and H', then H' is safe.*

Proof. From Lemma 11, no matter how much weight H' contributes from the weight reassignment, there would not be two continuous stackings, both of which contributes a weight of one to the destroyed-set of H'. Then, following Lemma 8, H' is safe.   □

Lemma 14. *If $p = 2$ and if $S_H^{\frac{1}{2}}$ is destroyed by H, H' and H'', then H' is safe.*

Proof. Omitted due to space constraint.   □

Lemma 15. *At the termination of Algorithm 2, every short helix H is safe.*

Proof. There would not be any long helix left after Algorithm 1, hence a short helix can only destroy some other short helices. A short helix occupies four bases, then it can destroy at most 4 short helices in the optimal solution. By the 1-substitution local improvement process, among the destroyed short helices, at most one of them could be singly destroyed. All the other short helices must be multiply destroyed, each of which contributing a weight of 1/2 to the destroy set of H. Hence, we have,

$$W(DS_H) \leq 1 + 3/2 = 2.5,$$

and we are done.   □

Theorem 2. *Our algorithm approximates the maximum base pair stackings within a factor 5/2.*

Proof. At the termination of Algorithm 2, all the stackings in the optimal solution would be destroyed. (Recall that if some stackings are also found by our algorithm, we consider them to be destroyed by themselves.) If a stacking in the optimal solution is singly destroyed by some helix H, it contributes a weight of 1 to $W(DS_H)$; if a stacking in the optimal solution is multiply destroyed, it contributes a weight of 1/2 to two of the helices destroying it; if the weight of some stacking in the optimal solution is reassigned, its total contribution remains 1. Consequently, the total weight assigned to the destroyed-set of all the helices in our solution is exactly the value of the optimal solution. From Lemmas 12, 13, 14, 15, the weight contributed by each helix is at most 2.5 times of its own length. Since the total length of helices is exactly the value in our solution, then we are done. □

5   Concluding Remarks

In this paper, we investigate the maximum base pair stackings problem, which is a well-defined combinatorial problem from RNA secondary structures prediction. We obtain a 5/2-approximation by a two-stage local search method together with an amortization analysis. A direction for future research is to further improve the approximation factor. In our algorithm, we use only 1-substitutions and 2-substitutions. In fact, as we discussed in this paper, using 1-substitutions alone cannot reach this ratio.

Acknowledgments. This research is partially supported by NSF of China under grant 61472222 and 61202014. Haitao Jiang is supported by Young Scholars Program of Shandong University. Nan Liu is supported by he Foundation for Outstanding Young Scientists in Shandong Province (project no. BS2014DX017), by the Doctoral Foundation of Shandong Jianzhu University (project no. 0000601512). Haodi Feng is supported by NSF of China under grant 61672325. Binhai Zhu is supported by NSF of China under grant 61628207.

References

1. Akutsu, T.: Dynamic programming algorithms for RNA secondary structure prediction with pseudoknots. Discrete Appl. Mathe. **104**(1–3), 45–62 (2000)
2. Berman, P.: A $d/2$ approximation for maximum weight independent set in d-claw free graphs. Nordic J. Comput. **7**, 178–184 (2000)
3. Ieong, S., Kao, M.-Y., Lam, T.-W., Sung, W.-K., Yiu, S.-M.: Predicting RNA secondary structure with arbitrary pseudoknots by maximizing the number of stacking pairs. J. Comput. Biol. **10**, 981–995 (2003)
4. Jiang, M.: Approximation algorithms for predicting RNA secondary structures with arbitrary pseudoknots. IEEE/ACM Trans. Comput. Biol. Bioinform. **7**(2), 323–332 (2010)

5. Lyngsø, R.B.: Complexity of pseudoknot prediction in simple models. In: Díaz, J., Karhumäki, J., Lepistö, A., Sannella, D. (eds.) ICALP 2004. LNCS, vol. 3142, pp. 919–931. Springer, Heidelberg (2004). doi:10.1007/978-3-540-27836-8_77
6. Lyngsø, R.B., Pedersen, C.N.S.: RNA pseudoknot prediction in energy based models. J. Comput. Biol. **7**(3/4), 409–428 (2000)
7. Lyngsø, R.B., Zuker, M., Pedersen, C.N.S.: Fast evaluation of interval loops in rna secondary structure prediction. Bioinformatics **15**, 440–445 (1999)
8. Nussinov, R., Jacobson, A.B.: Fast algorithm for predicting the secondary structure of single-stranded RNA. Proc. Nat. Acad. Sci. USA **77**, 6309–6313 (1980)
9. Nussinov, R., Pieczenik, G., Griggs, J.R., Kleitman, D.J.: Algorithms for loop matchings. SIAM J. Appl. Mathe. **35**(1), 68–82 (1978)
10. Rivas, E., Eddy, S.R.: A dynamic programming algorithm for RNA structure prediction including pseudoknots. J. Mol. Biol. **285**(5), 2053–2068 (1999)
11. Sankoff, D.: Simultaneous solution of the RNA folding, alignment and protosequence problems. SIAM J. Appl. Mathmatics **45**, 810–825 (1985)
12. Tinoco, I., Borer, P.N., Dengler, B., Levine, M.D., Uhlenbeck, O.C., Crothers, D.M., Gralla, J.: Improved estimation of secondary structure in ribonucleic acids. Nat. New Biol. **246**, 40–42 (1973)
13. Tinoco, I., Bustamante, C.: How RNA folds. J. Mol. Biol. **293**, 271–281 (1999)
14. Uemura, Y., Hasegawa, A., Kobayashi, S., Yokomori, T.: Tree adjoining grammars for RNA structure prediction. Theore. Comput. Sci. **210**(2), 277–303 (1999)
15. Zuker, M., Sankoff, D.: RNA secondary structures and their prediction. Bull. Math. Biol. **46**, 591–621 (1984)
16. Zuker, M., Stiegler, P.: Optimal computer folding of large RNA sequences using thermodynamics and auxiliary information. Nucleic Acids Res. **9**, 133–148 (1981)

CSoNet Papers

Cooperative Game Theory Approaches for Network Partitioning

Konstantin E. Avrachenkov[1](✉), Aleksei Yu. Kondratev[2], and Vladimir V. Mazalov[2]

[1] INRIA, 2004 Route des Lucioles, Sophia-Antipolis, France
k.avrachenkov@sophia.inria.fr
[2] Karelian Research Center, Institute of Applied Mathematical Research, Russian Academy of Sciences, 11, Pushkinskaya Street, Petrozavodsk 185910, Russia
vmazalov@krc.karelia.ru

Abstract. The paper is devoted to game-theoretic methods for community detection in networks. The traditional methods for detecting community structure are based on selecting denser subgraphs inside the network. Here we propose to use the methods of cooperative game theory that highlight not only the link density but also the mechanisms of cluster formation. Specifically, we suggest two approaches from cooperative game theory: the first approach is based on the Myerson value, whereas the second approach is based on hedonic games. Both approaches allow to detect clusters with various resolution. However, the tuning of the resolution parameter in the hedonic games approach is particularly intuitive. Furthermore, the modularity based approach and its generalizations can be viewed as particular cases of the hedonic games.

Keywords: Network partitioning · Community detection · Cooperative games · Myerson value · Hedonic games

1 Introduction

Community detection in networks or network partitioning is a very important topic which attracted the effort of many researchers. Let us just mention several main classes of methods for network partitioning. The first very large class is based on spectral elements of the network matrices such as adjacency matrix and Laplacian (see e.g., the survey [23] and references therein). The second class of methods, which is somehow related to the first class, is based on the use of random walks (see e.g., [1,2,6,16,18,20] for the most representative works in this research direction.) The third class of approaches to network partitioning is based on methods from statistical physics [3,21,22]. The fourth class, which is probably most related to our approach, is based on the concept of modularity and its various generalizations [4,9,19,24]. For a very thorough overview of the community detection methods we recommend the survey [7].

© Springer International Publishing AG 2017
Y. Cao and J. Chen (Eds.): COCOON 2017, LNCS 10392, pp. 591–602, 2017.
DOI: 10.1007/978-3-319-62389-4_49

In essence, all the above methods (may be with some exception of the statistical physics methods), try to detect denser subgraphs inside the network and do not address the question: what are the natural forces and dynamics behind the formation of network clusters. We feel that the game theory, and in particular, cooperative game theory is the right tool to explain the formation of network clusters.

In the present work, we explore two cooperative game theory approaches to explain possible mechanisms behind cluster formation. Our first approach is based on the Myerson value in cooperative game theory, which particularly emphasizes the value allocation in the context of games with interactions between players constrained by a network. The advantage of the Myerson value is in taking into account the impact of all coalitions. We use the efficient method developed in [14,15] based on characteristic functions to calculate quickly the Myerson value in the network. We would like to mention that in [14] a network centrality measure based on the Myerson value was proposed. It might be interesting to combine node ranking and clustering based on the same approach such as the Myerson value to analyze the network structure.

The second approach is based on hedonic games, which are games explaining the mechanism behind the formation of coalitions. Both our approaches allow to detect clusters with varying resolution and thus avoiding the problem of the resolution limit [8,12]. The hedonic game approach is especially well suited to adjust the level of resolution as the limiting cases are given by the grand coalition and maximal clique decomposition, two very natural extreme cases of network partitioning. Furthermore, the modularity based approaches can be cast in the setting of hedonic games. We find that this gives one more, very interesting, interpretation of the modularity based methods.

Some hierarchical network partitioning methods based on tree hierarchy, such as [9], cannot produce a clustering on one resolution level with the number of clusters different from the predefined tree shape. Furthermore, the majority of clustering methods require the number of clusters as an input parameter. In contrast, in our approaches we specify the value of the resolution parameter and the method gives a natural number of clusters corresponding to the given resolution parameter.

In addition, our approach easily works with multi-graphs, where several edges (links) are possible between two nodes. A multi-edge has several natural interpretations in the context of social networks. A multi-edge can represent: a number of telephone calls; a number of exchanged messages; a number of common friends; or a number of co-occurrences in some social event.

The paper is structured as follows: in the following section we formally define network partitioning as a cooperative game. Then, in Sect. 3 we present our first approach based on the Myerson value. The second approach based on the hedonic games is presented in Sect. 4. In both Sects. 3 and 4 we provide illustrative examples which explain the essence of the methods. Finally, Sect. 5 concludes the paper with directions for future research.

2   Network Partitioning as a Cooperative Game

Let $g = (N, E)$ denote an undirected multi-graph consisting of the set of nodes N and the set of edges E. We denote a link between node i and node j as ij. The interpretation is that if $ij \in E$, then the nodes $i \in N$ and $j \in N$ have a connection in network g, while $ij \notin E$, then nodes i and j are not directly connected. Since we generally consider a multi-graph, there could be several edges between a pair of nodes. Multiple edges can be interpreted for instance as a number of telephone calls or as a number of message exchanges in the context of social networks.

We view the nodes of the network as players in a cooperative game. Let $N(g) = \{i : \exists j \text{ such that } ij \in g\}$. For a graph g, a sequence of different nodes $\{i_1, i_2, \dots, i_k\}$, $k \geq 2$, is a path connecting i_1 and i_k if for all $h = 1, \dots, k-1$, $i_h i_{h+1} \in g$. The length of the path l is the number of links in the path, i.e. $l = k - 1$. The length of the shortest path connecting i and j is distance between i and j. Graph g on the set N is connected graph if for any two nodes i and j there exists a path in g connecting i and j.

We refer to a subset of nodes $S \subset N$ as a coalition. The coalition S is connected if any two nodes in S are connected by a path which consists of nodes from S. The graph g' is a component of g, if for all $i \in N(g')$ and $j \in N(g')$, there exists a path in g' connecting i and j, and for any $i \in N(g')$ and $j \in N(g)$, $ij \in g$ implies that $ij \in g'$. Let $N|g$ is the set of all components in g and let $g|S$ is the subgraph with the nodes in S.

Let $g - ij$ denote the graph obtained by deleting link ij from the graph g and $g + ij$ denote the graph obtained by adding link ij to the graph g.

The result of community detection is a partition of the network (N, E) into subsets (coalitions) $\{S_1, \dots, S_K\}$ such that $S_k \cap S_l = \emptyset, \forall k, l$ and $S_1 \cup \dots \cup S_K = N$. This partition is *internally stable* or *Nash stable* if for any player from coalition S_k it is not profitable to join another (possibly empty) coalition S_l. We also say that the partition is *externally stable* if for any player $i \in S_l$ for whom it is benefitial to join a coalition S_k there exists a player $j \in S_k$ for whom it is not profitable to include there player i. The payoff definition and distribution will be discussed in the following two sections.

3   Myerson Cooperative Game Approach

In general, cooperative game of n players is a pair $< N, v >$ where $N = \{1, 2, \dots, n\}$ is the set of players and $v: 2^N \to R$ is a map prescribing for a coalition $S \in 2^N$ some value $v(S)$ such that $v(\emptyset) = 0$. This function $v(S)$ is the total utility that members of S can jointly attain. Such a function is called the characteristic function of cooperative game [13].

Characteristic function (payoff of coalition S) can be determined in different ways. Here we use the approach of [10,11,14,15], which is based on discounting directed paths. The payoff to an individual player is called an imputation. The imputation in this cooperative game will be Myerson value [14,15,17].

Let $< N, v >$ be a cooperative game with partial cooperation presented by graph g and characteristic function v. An allocation rule Y describes how the value associated with the network is distributed to the individual players. Denote by $Y_i(v, g)$ the value allocated to player i from graph g under the characteristic function v.

Myerson proposed in [17] the allocation rule

$$Y(v, g) = (Y_1(v, g), \ldots, Y_n(v, g)),$$

which is uniquely determined by the following two axioms:

A1. If S is a component of g then the members of the coalition S ought to allocate to themselves the total value $v(S)$ available to them, i.e. $\forall S \in N|g$

$$\sum_{i \in S} Y_i(v, g) = v(S). \tag{1}$$

A2. $\forall g$, $\forall ij \in g$ both players i and j obtain equal payoffs after adding or deleting a link ij,

$$Y_i(v, g) - Y_i(v, g - ij) = Y_j(v, g) - Y_j(v, g - ij). \tag{2}$$

Let us determine the characteristic function by the following way

$$v_g(S) = \sum_{K \in S|g} v(K).$$

Then the Myerson value can be calculated by the formula

$$Y_i(v, g) = \sum_{S \subset N \setminus \{i\}} (v_g(S \cup i) - v_g(S)) \frac{s!(n - s - 1)!}{n!}, \tag{3}$$

where $s = |S|$ and $n = |N|$.

Let us determine the characteristic function which is determined by the scheme proposed by Jackson [11]: every direct connection gives to coalition S the impact r, where $0 \le r \le 1$. Moreover, players obtain an impact from indirect connections. Each path of length 2 gives to coalition S the impact r^2, a path of length 3 gives to coalition the impact r^3, etc. So, for any coalition S we obtain

$$v(S) = a_1 r + a_2 r^2 + \cdots + a_k r^k + \cdots + a_L r^L = \sum_{k=1}^{L} a_k r^k, \tag{4}$$

where L is a maximal distance between two nodes in the coalition; a_k is the number of paths of length k in this coalition. Set

$$v(i) = 0, \quad \forall i \in N.$$

In [15] it was proven that the Myerson value can be found by the following simple procedure of allocation the general gain $v(N)$ to each player $i \in N$:

Stage 1. Two direct connected players together obtain r. Individually, they would receive nothing. So, each of them receives at least $r/2$. If player i has some direct connections then she receives the value $r/2$ times the number of paths of length 1 which contain the node i.

Stage 2. Three connected players obtain r^2, so each of them must receive $r^2/3$, and so on.

Arguing this way, we obtain the allocation rule of the following form:

$$Y_i(v,g) = \frac{a_1^i}{2}r + \frac{a_2^i}{3}r^2 + \cdots + \frac{a_L^i}{L+1}r^L = \sum_{k=1}^{L} \frac{a_k^i}{k+1}r^k, \qquad (5)$$

where a_k^i is the number of all paths of length k which contain the node i.

Example 1. Consider network of six nodes presented in Fig. 1 Below we show how to calculate characteristic function for different coalitions.

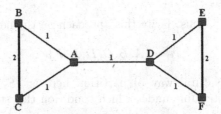

Fig. 1. Network of six nodes.

For the network $N = \{A, B, C, D, E, F\}$ we find $L = 3$, $a_1 = 9$, $a_2 = 4$, $a_3 = 4$. Consequently, the value of grand-coalition is

$$v(N) = 9r + 4r^2 + 4r^3.$$

For coalition $S = \{A, B, C, D\}$ we have $L = 2$, $a_1 = 5$, $a_2 = 2$ and we obtain

$$v(S) = 5r + 2r^2.$$

This way we can calculate the values of characteristic function for all coalitions $S \subset N$. After that we can find the Myerson vector.

Example 1 (ctnd). Let us calculate the Myerson value for player A in Example 1 using the allocation rule (5). Mark all paths which contain node A. The paths of length 1 are: {A,B}, {A,C}, {A,D}, hence $a_1^A = 3$. The paths of length 2 are: {B,A,C}, {B,A,D}, {C,A,D}, {A,D,E}, {A,D,F}, so $a_2^A = 5$. The paths of length 3: {B,A,D,E}, {B,A,D,F}, {C,A,D,E}, {C,A,D,F}, so $a_3^A = 4$. Consequently,

$$Y_A = \frac{3}{2}r + \frac{5}{3}r^2 + r^3.$$

Thus, we can propose the following algorithm for network partitioning based on the Myerson value: Start with any partition of the network $N = \{S_1, \ldots, S_K\}$. Consider a coalition S_l and a player $i \in S_l$. In cooperative game with partial cooperation presented by the graph $g|S_l$ we find the Myerson value for player i, $Y_i(g|S_l)$. That is reward of player i in coalition S_l. Suppose that player i decides to join the coalition S_k. In the new cooperative game with partial cooperation presented by the graph $g|S_k \cup i$ we find the Myerson value $Y_i(g|S_k \cup i)$. So, if for the player $i \in S_l : Y_i(g|S_l) \geq Y_i(g|S_k \cup i)$ then player i has no incentive to join to new coalition S_k, otherwise the player changes the coalition. The partition $N = \{S_1, \ldots, S_K\}$ is the Nash stable if for any player there is no incentive to move from her coalition. Notice that for unweighted graphs the definition of the Myerson value implies that for any coalition it is always beneficial to accept a new player (of course, for the player herself it might not be profitable to join that coalition), the Nash stability (internal stability) in this game coincides with the external stability.

Example 1 (ctnd). Let us clarify this approach on the network

$$N = \{A, B, C, D, E, F\}$$

presented in Fig. 1 Natural way of partition here is $\{S_1 = (A, B, C), S_2 = (D, E, F)\}$. Let us determine under which condition this structure will present the stable partition.

Suppose that characteristic function is determined by (4). For coalition S_1 the payoff $v(S_1) = 4r$. The payoff of player A is $Y_A(g|S_1) = r$. Imagine that player A decides to join the coalition S_2.

Coalition $S_2 \cup A$ has payoff $v(S_2 \cup A) = 5r + 2r^2$. The imputation in this coalition is $Y_A(g|S_2 \cup A) = r/2 + 2r^2/3, Y_D(g|S_2 \cup A) = 3r/2 + 2r^2/3, Y_E(g|S_2 \cup A) = Y_F(g|S_2 \cup A) = 3r/2 + r^2/3$. We see that for player A it is profitable to join this new coalition if $r/2 + 2r^2/3 > r$, or $r > 3/4$. Otherwise, the coalitional structure is stable.

Thus, for the network in Fig. 1 the Myerson value approach will give the partition $\{S_1 = (A, B, C), S_2 = (D, E, F)\}$ if $r < 3/4$ and, otherwise, it leads to the grand coalition. This example already gives a feeling that the parameter r can be used to tune the resolution of network partitioning. Such tuning will be even more natural in the ensuing approach.

4   Hedonic Coalition Game Approach

There is another game-theoretic approach for the partitioning of a society into coalitions based on the ground-breaking work [5]. We apply the framework of

Hedonic games [5] to network partitioning problem, particularly, specifying the preference function.

Assume that the set of players $N = \{1, \ldots, n\}$ is divided into K coalitions: $\Pi = \{S_1, \ldots, S_K\}$. Let $S_\Pi(i)$ denote the coalition $S_k \in \Pi$ such that $i \in S_k$. A player i preferences are represented by a complete, reflexive and transitive binary relation $\succeq_i$ over the set $\{S \subset N : i \in S\}$. The preferences are additively separable [5] if there exists a value function $v_i : N \to \mathbb{R}$ such that $v_i(i) = 0$ and

$$S_1 \succeq_i S_2 \Leftrightarrow \sum_{j \in S_1} v_i(j) \geq \sum_{j \in S_2} v_i(j).$$

The preferences $\{v_i, i \in N\}$ are symmetric, if $v_i(j) = v_j(i) = v_{ij} = v_{ji}$ for all $i, j \in N$. The symmetry property defines a very important class of Hedonic games.

As in the previous section, the network partition Π is *Nash stable*, if $S_\Pi(i) \succeq_i S_k \cup \{i\}$ for all $i \in N, S_k \in \Pi \cup \{\emptyset\}$. In the Nash-stable partition, there is no player who wants to leave her coalition.

A potential of a coalition partition $\Pi = \{S_1, \ldots, S_K\}$ (see [5]) is

$$P(\Pi) = \sum_{k=1}^{K} P(S_k) = \sum_{k=1}^{K} \sum_{i,j \in S_k} v_{ij}. \tag{6}$$

Our method for detecting a stable community structure is based on the following better response type dynamics:

Start with any partition of the network $N = \{S_1, \ldots, S_K\}$. Choose any player i and any coalition S_k different from $S_\Pi(i)$. If $S_k \cup \{i\} \succeq_i S_\Pi(i)$, assign node i to the coalition S_k; otherwise, keep the partition unchanged and choose another pair of node-coalition, etc.

Since the game has the potential (6), the above algorithm is guaranteed to converge in a finite number of steps.

Proposition 1. *If players' preferences are additively separable and symmetric ($v_{ii} = 0, v_{ij} = v_{ji}$ for all $i, j \in N$), then the coalition partition Π giving a local maximum of the potential $P(\Pi)$ is the Nash-stable partition.*

One natural way to define a symmetric value function v with a parameter $\alpha \in [0, 1]$ is as follows:

$$v_{ij} = \begin{cases} 1 - \alpha, & (i, j) \in E, \\ -\alpha, & (i, j) \notin E, \\ 0, & i = j. \end{cases} \tag{7}$$

For any subgraph $(S, E|S)$, $S \subseteq N$, denote $n(S)$ as the number of nodes in S, and $m(S)$ as the number of edges in S. Then, for the value function (7), the potential (6) takes the form

$$P(\Pi) = \sum_{k=1}^{K} \left(m(S_k) - \frac{n(S_k)(n(S_k) - 1)\alpha}{2} \right). \tag{8}$$

We can completely characterize the limiting cases $\alpha \to 0$ and $\alpha \to 1$.

Proposition 2. *If $\alpha = 0$, the grand coalition partition $\Pi_N = \{N\}$ gives the maximum of the potential (8). Whereas if $\alpha \to 1$, the maximum of (8) corresponds to a network decomposition into maximal cliques. (A maximal clique is a clique which is not contained in another clique.)*

Proof. It is immediate to check that for $\alpha = 0$ the grand coalition partition $\Pi_N = \{N\}$ gives the maximum of the potential (8), and $P(\Pi_N) = m(N)$.

For values of α close to 1, the partition into maximal cliques $\Pi = \{S_1, \ldots, S_K\}$ gives the maximum of (8). Indeed, assume that a player i from the clique $S_\Pi(i)$ of the size m_1 moves to a clique S_j of the size $m_2 < m_1$. The player $i \in S_\Pi(i)$ and S_j are connected by at most m_2 links. The impact on $P(\Pi)$ of this movement is not higher than

$$m_2(1 - \alpha) - (m_1 - 1)(1 - \alpha) \le 0.$$

Now, suppose that player i from the clique $S_\Pi(i)$ moves to a clique S_j of the size $m_2 \ge m_1$. The player $i \in S_\Pi(i)$ is connected with the clique S_j by at most $m_2 - 1$ links. Otherwise, it contradicts the fact that Π is maximal clique cover and the clique S_j can be increased by adding of i. If i has an incentive to move from $S_\Pi(i)$ to the clique S_j, then for new partition the sum (8) would be not higher than for partition Π by

$$m_2 - 1 - m_2\alpha - (m_1 - 1)(1 - \alpha) = m_2 - m_1 - \alpha(m_2 - m_1 + 1).$$

For α close to 1, this impact is negative, so there is no incentive to join the coalition S_j.

The grand coalition and the maximal clique decomposition are two extreme partitions into communities. By varying the parameter α we can easily tune the resolution of the community detection algorithm.

Example 2. Consider graph $G = G_1 \cup G_2 \cup G_3 \cup G_4$, which consists of $n = 26$ nodes and $m = 78$ edges (see Fig. 2) This graph includes 4 fully connected subgraphes: $(G_1, 8, 28)$ with 8 vertices connected by 28 links, $(G_2, 5, 10)$, $(G_3, 6, 15)$ and $(G_4, 7, 21)$. Subgraph G_1 is connected with G_2 by 1 edge, G_2 with G_3 by 2 edges, and G_3 with G_4 by 1 edge.

Firstly, find the potentials (8) for large-scale decompositions of G for any parameter $\alpha \in [0, 1]$. It is easy to check, that $P(G) = 78 - 325\alpha$, $P(\{G_1, G_2 \cup G_3 \cup G_4\}) = 77 - 181\alpha$, $P(\{G_1, G_2 \cup G_3, G_4\}) = 76 - 104\alpha$, $P(\{G_1, G_2, G_3, G_4\}) = 74 - 74\alpha$.

Other coalition partitions give smaller potentials: $P(\{G_1 \cup G_2, G_3 \cup G_4\}) = 76 - 156\alpha < 76 - 104\alpha$, $P(\{G_1 \cup G_2 \cup G_3, G_4\}) = 77 - 192\alpha < 77 - 181\alpha$, $P(\{G_1, G_2, G_3 \cup G_4\}) = 75 - 116\alpha < 76 - 104\alpha$, $P(\{G_1 \cup G_2, G_3, G_4\}) = 75 - 114\alpha < 76 - 104\alpha$.

We solve a sequence of linear inequalities in order to find maximum of the potential for all $\alpha \in [0, 1]$. The result is presented in the table below.

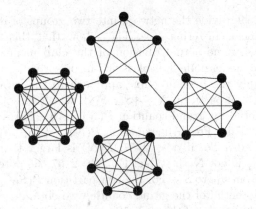

Fig. 2. Graph with four fully connected subgraphs.

Nash-stable coalition partitions in Example 2.

α	Coalition partition	potential
$[0, 1/144]$	$G_1 \cup G_2 \cup G_3 \cup G_4$	$78 - 325\alpha$
$[1/144, 1/77]$	$G_1, G_2 \cup G_3 \cup G_4$	$77 - 181\alpha$
$[1/77, 1/15]$	$G_1, G_2 \cup G_3, G_4$	$76 - 104\alpha$
$[1/15, 1]$	G_1, G_2, G_3, G_4	$74 - 74\alpha$

Example 1 (ctnd). Note that for the unweighted version of the network example presented in Fig. 1, there are only two stable partitions: $\Pi = N$ for small values of $\alpha \leq 1/9$ and $\Pi = \{\{A, B, C\}, \{D, E, F\}\}$ for $\alpha > 1/9$.

Example 3. Consider the popular example of the social network from Zachary karate club (see Fig. 3). In his study [25], Zachary observed 34 members of a karate club over a period of two years. Due to a disagreement developed between the administrator of the club and the club's instructor there appeared two new clubs associated with the instructor (node 1) and administrator (node 34) of sizes 16 and 18, respectively.

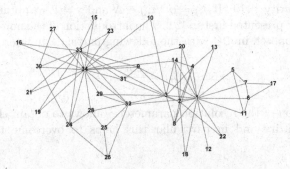

Fig. 3. Zachary karate club network.

The authors of [9] divide the network into two groups of roughly equal size using the hierarchical clustering tree. They show that this split corresponds almost perfectly with the actual division of the club members following the break-up. Only one node, node 3, is classified incorrectly.

Let us now apply the hedonic game approach to the karate club network. We start from the final partition $N = \{S_{15}, S_{19}\}$, which was obtained in [9]. We calculate the potential for grand-coalition $P(N) = 78 - 561\alpha$ and for partition $P(S_{15}, S_{19}) = 68 - 276\alpha$. From the equation $P(N) = P(S_{15}, S_{19})$ we obtain the cutoff point $\alpha = 2/57$. So, if $\alpha < 2/57$, $P(N)$ is larger than $P(S_{15}, S_{19})$, so partition $\{S_{15}, S_{19}\}$ is not Nash-stable. For $\alpha = 2/57$ the potential increases if the node 3 moves from S_{19} to S_{15}. For the new partition $P(S_{16}, S_{18}) = 68 - 273\alpha$. Comparing with potential of the grand coalition we obtain $\alpha = 5/144$. For $\alpha = 5/144$ the potential increases if the node 10 moves to S_{16}. Now $P(S_{17}, N \setminus S_{17}) = 68 - 272\alpha$ and the new cutoff point is $\alpha = 10/289$. Finally, in order to find the upper bound of the resolution parameter, we have to check that for any player there is no incentive to move from her coalition to the empty coalition.

Thus, for $1/16 \geq \alpha \geq 10/289$ the Nash-stable partition is

$$S_{17} = \{1, 2, 3, 4, 5, 6, 7, 8, 10, 11, 12, 13, 14, 17, 18, 20, 22\} \cup \{N \setminus S_{17}\}.$$

Notice that in this new partition the node 3 belongs to the "right" coaltion.

Another natural approach to define a symmetric value function is, roughly speaking, to compare the network under investigation with the configuration random graph model. The configuration random graph model can be viewed as a null model for a network with no community structure. Namely, the following value function can be considered:

$$v_{ij} = \beta_{ij} \left(A_{ij} - \gamma \frac{d_i d_j}{2m} \right), \tag{9}$$

where A_{ij} is a number of links between nodes i and j, d_i and d_j are the degrees of the nodes i and j, respectively, $m = \frac{1}{2} \sum_{l \in N} d_l$ is the total number of links in the network, and $\beta_{ij} = \beta_{ji}$ and γ are some parameters.

Note that if $\beta_{ij} = \beta, \forall i, j \in N$ and $\gamma = 1$, the potential (8) coincides with the network modularity [9,19]. If $\beta_{ij} = \beta, \forall i, j \in N$ and $\gamma \neq 1$, we obtain the generalized modularity presented first in [22]. The introduction of the non-homogeneous weights was proposed in [24] with the following particularly interesting choice:

$$\beta_{ij} = \frac{2m}{d_i d_j}.$$

The introduction of the resolution parameter γ allows to obtain clustering with varying granularity and in particular this helps to overcome the resolution limit [8].

Thus, we have now a game-theoretic interpretation of the modularity function. Namely, the coalition partition $\Pi = \{S_1, \ldots, S_K\}$ which maximises the modularity

$$P(\Pi) = \sum_{k=1}^{K} \sum_{i,j \in S_k, i \neq j} \left(A_{ij} - \frac{d_i d_j}{2m} \right) \tag{10}$$

gives the Nash-stable partition of the network in the Hedonic game with the value function defined by (9), where $\gamma = 1$ and $\beta_{ij} = \beta$.

Example 1 (ctnd). For the network example presented in Fig. 1 we calculate $P(N) = 3/2, P(\{B, C\} \cup \{A, D\} \cup \{E, F\}) = P(\{A, B, C, D\} \cup \{E, F\}) = 7/2$ and $P(\{A, B, C\} \cup \{D, E, F\}) = 5$. Thus, according to the value function (9) with $\gamma = 1$ and $\beta_{ij} = \beta$ (modularity value function), $\Pi = \{\{A, B, C\}, \{D, E, F\}\}$ is the unique Nash-stable coalition.

Example 3 (ctnd). Numerical calculations show that the partition $S_{17} \cup \{N \setminus S_{17}\}$ gives the maximum of potential function (10). It means that this partition is Nash stable.

5 Conclusion and Future Research

We have presented two cooperative game theory based approaches for network partitioning. The first approach is based on the Myerson value for graph constrained cooperative game, whereas the second approach is based on hedonic games which explain coalition formation. We find the second approach especially interesting as it gives a very natural way to tune the clustering resolution and generalizes the modularity based approaches. Our near term research plans are to test our methods on more social networks and to develop efficient computational Monte Carlo type methods.

Acknowledgements. This research is supported by Russian Humanitarian Science Foundation (project 15-02-00352), Russian Fund for Basic Research (projects 16-51-55006 and 17-11-01079), EU Project Congas FP7-ICT-2011-8-317672 and Campus France.

References

1. Avrachenkov, K., Dobrynin, V., Nemirovsky, D., Pham, S.K., Smirnova, E.: Pagerank based clustering of hypertext document collections. In: Proceedings of ACM SIGIR 2008, pp. 873–874 (2008)
2. Avrachenkov, K., El Chamie, M., Neglia, G.: Graph clustering based on mixing time of random walks. In: Proceedings of IEEE ICC 2014, pp. 4089–4094 (2014)
3. Blatt, M., Wiseman, S., Domany, E.: Clustering data through an analogy to the Potts model. In: Proceedings of NIPS 1996, pp. 416–422 (1996)
4. Blondel, V.D., Guillaume, J.L., Lambiotte, R., Lefebvre, E.: Fast unfolding of communities in large networks. J. Stat. Mech. Theory Exp. **10**, P10008 (2008)

5. Bogomolnaia, A., Jackson, M.O.: The stability of hedonic coalition structures. Games Econ. Behav. **38**(2), 201–230 (2002)
6. Dongen, S.: Performance criteria for graph clustering and Markov cluster experiments. CWI Technical report (2000)
7. Fortunato, S.: Community detection in graphs. Phys. Rep. **486**(3), 75–174 (2010)
8. Fortunato, S., Barthelemy, M.: Resolution limit in community detection. Proc. Nat. Acad. Sci. **104**(1), 36–41 (2007)
9. Girvan, M., Newman, M.E.J.: Community structure in social and biological networks. Proc. Nat. Acad. Sci. USA **99**(12), 7821–7826 (2002)
10. Jackson, M.O.: Allocation rules for network games. Games Econ. Behav. **51**(1), 128–154 (2005)
11. Jackson, M.O.: Social and Economic Networks. Princeton University Press, Princeton (2008)
12. Leskovec, J., Lang, K.J., Dasgupta, A., Mahoney, M.W.: Community structure in large networks: Natural cluster sizes and the absence of large well-defined clusters. Internet Math. **6**(1), 29–123 (2009)
13. Mazalov, V.: Mathematical Game Theory and Applications. Wiley, Hoboken (2014)
14. Mazalov, V., Avrachenkov, K., Trukhina, I.: Game-theoretic centrality measures for weighted graphs. Fundamenta Informaticae **145**(3), 341–358 (2016)
15. Mazalov, V.V., Trukhina, L.I.: Generating functions and the Myerson vector in communication networks. Disc. Math. Appl. **24**(5), 295–303 (2014)
16. Meila, M., Shi, J.: A random walks view of spectral segmentation. In: Proceedings of AISTATS 2001
17. Myerson, R.B.: Graphs and cooperation in games. Math. Oper. Res. **2**, 225–229 (1977)
18. Newman, M.E.J.: A measure of betweenness centrality based on random walks. Proc. Nat. Acad. Sci. USA **27**, 39–54 (2005)
19. Newman, M.E.J.: Modularity and community structure in networks. Soc. Netw. **103**(23), 8577–8582 (2006)
20. Pons, P., Latapy, M.: Computing communities in large networks using random walks. J. Graph Algorithms Appl. **10**(2), 191–218 (2006)
21. Raghavan, U.N., Albert, R., Kumara, S.: Near linear time algorithm to detect community structures in large-scale networks. Phys. Rev. E **76**(3), 036106 (2007)
22. Reichardt, J., Bornholdt, S.: Statistical mechanics of community detection. Phys. Rev. E **74**(1), 016110 (2006)
23. von Luxburg, U.: A tutorial on spectral clustering. Stat. Comput. **17**(4), 395–416 (2007)
24. Waltman, L., van Eck, N.J., Noyons, E.C.: A unified approach to mapping and clustering of bibliometric networks. J. Inform. **4**(4), 629–635 (2010)
25. Zachary, W.W.: An information flow model for conflict and fission in small groups. J. Anthropol. Res. **33**(4), 452–473 (1977)

Chain of Influencers: Multipartite Intra-community Ranking

Pavla Drazdilova$^{(\boxtimes)}$, Jan Konecny, and Milos Kudelka

Department of Computer Science, Faculty of Electrical Engineering
and Computer Science, VŠB - Technical University of Ostrava,
17. listopadu 15/2172, 708 33 Ostrava, Czech Republic
{pavla.drazdilova,jan.konecny.st,milos.kudelka}@vsb.cz

Abstract. Ranking of vertices is an important part of social network analysis. However, thanks to the enormous growth of real-world networks, the global ranking of vertices on a large scale does not provide easily comparable results. On the other hand, the ranking can provide clear results on a local scale and also in heterogeneous networks where we need to work with vertices of different types. In this paper, we present a method of ranking objects in a community which is closely related to the analysis of heterogeneous information networks. Our method assumes that the community is a set of several groups of objects of different types where each group, so-called object pool, contains objects of the same type. These community object pools can be connected and ordered to the chain of influencers, and ranking can be applied to this structure. Based on the chain of influencers, the heterogeneous network can be converted to a multipartite graph. In our approach, we show how to rank vertices of the community using the mutual influence of community object pools. In our experiments, we worked with a computer science research community. Objects of this domain contain authors, papers (articles), topics (keywords), and years of publications.

Keywords: Ranking · Multipartite graph · Heteregonous network · Community

1   Introduction

Ranking of vertices is a task historically associated with the social network analysis (SNA). The first methods of measurement of vertex importance were based on so-called centrality. Freeman [2] formalized three different measures of vertex importance: degree, closeness and betweenness centralities. With the enormous growth of networks and, in particular, thanks to the Internet, the original centralities did not provide reliable results on a global scale, and new approaches were investigated. As a key method, Page et al. published the so-called PageRank [11]. This method is based on a simple recursive principle that defines the important vertex as a vertex which is a neighbor of many other important vertices. The idea of Pagerank was followed by other similar approaches focusing on specific properties of vertices.

© Springer International Publishing AG 2017
Y. Cao and J. Chen (Eds.): COCOON 2017, LNCS 10392, pp. 603–614, 2017.
DOI: 10.1007/978-3-319-62389-4_50

With the growth of social and information networks, the ranking of vertices loses, in part, its importance for the analysis of network structure. In addition to the other, the original purpose of the ranking was to sort vertices by their importance; it is impractical in networks with billions of vertices. This importance is, in particular, dependent on the perspective from which the network vertices are assessed.

Networks are inherently heterogeneous. We can have a network of authors and their papers, topics, and affiliation. We can have customers who buy various products from different categories at a different time. Similarly, we can have users who watch movies of various genres, with different actors and which are created by different directors. In each of these examples, a group of vertices of different types can be found (authors, products, genres, etc.). The well-known approach of analysis of heterogeneous information networks (HIN) was presented by Han et al. [3].

The aim of our approach is to evaluate a community in a specific domain (e.g., publication activities, e-shops, the Internet movie database). As the community, we understand several groups of vertices of different types. Moreover, these individual groups of vertices (pools containing objects of the same type) are connected in pairs (see Fig. 1). By the evaluation the community we understand the answer to a simple question: "What value has a selected vertex from a selected group in the community?" As we show in this paper, this question leads to a more precise formulation of the problem. Its solution is based on organizing pools to *Chain of Influencers* and the subsequent analysis of the multipartite graph. In our experiment with a community from the publication domain, we demonstrate how the different pools in the chain of influencers affect the evaluation of the vertices.

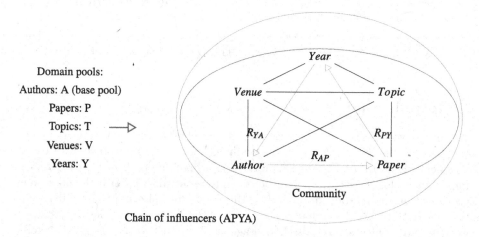

Domain pools:

Authors: A (base pool)

Papers: P

Topics: T

Venues: V

Years: Y

Chain of influencers (APYA)

Fig. 1. Selected domain pools based on DBLP data organized, as a community, to complete graph of pools. A base pool is authors from one computer science department. Vertices represent domain pools and the edges between vertices represent relations between pools. Closed chain of infuencers is represented by a sequence APYA.

2   Related Work

Ranking of vertices is an important data mining task in network analysis. Many ranking methods have been proposed for homogeneous networks. For example, PageRank [11] evaluates the importance of vertices through a random walk process in the graph of the Web. HITS (Hyperlink-Induced Topic Search) [6] ranks objects in the directed network using authority and hub scores. Authority score estimates the value of the content of the object, whereas hub scores measure the value of its links to other objects. SimRank [5] measures the similarity of the structural context in which objects occur, based on their relationships with other objects. Authors compute a measure that says "two objects are similar if they are related to similar objects."

These approaches consider the same type of objects in homogeneous networks, and they cannot be applied in heterogeneous networks with different types of objects. To rank tweets effectively by capturing the semantics and importance of different linkages, Huang et al. [4] propose the Tri-HITS model. This model makes use of heterogeneous networks composed of tweets, users, and web documents to rank tweets.

Some researchers focused on the co-ranking of multiple types of objects. Zhou et al. [20] co-rank authors and their publications by coupling two random walk processes. Sun et al. [17] address the problem of generating clusters for a specified type of objects, as well as ranking information for all types of objects based on these clusters in a heterogeneous information network. They performed experiments on the bi-type information network extracted from DBLP. Soulier et al. [15] propose a bi-type entity ranking algorithm to rank documents and authors in a bibliographic network jointly by combining content-based and network-based features. These methods can rank different types of objects existing in heterogeneous information networks; however, it is restricted to bipartite graphs. MultiRank [7] determines the importance of both objects and relations simultaneously for multi-relational data.

In recent years many articles with analysis of HIN have been published [1,9,18,19]. As a feature of HIN, the links connect different types of objects containing semantics. The meta path [16], connecting object types via a sequence of relations, has been widely used to capture relation semantics. Liu et al. [8] propose a ranking method with pseudo-relevance feedback by investigating some hypothesis-driven meta-paths on the scholarly heterogeneous graph. HeteSim is proposed by Shi et al. [12] to measure the relevance scores of heterogeneous objects in HIN. HRank [13] is based on a path-constrained random walk process and it can simultaneously determine the importance of objects and constrained meta-paths by applying the tensor analysis. A survey of models and methods for analysis of heterogeneous information networks is in [14].

A new network data model is proposed in [10] - a Network of Networks (NoN), where each vertex of the main network itself can be further represented as another (domain-specific) network. Authors formulate ranking tasks in this model as a regularized optimization problem.

2.1   Terminology

We have designed our terminology differently from the terminology used by Han et al. in [3]. Our terminology is concentrated on the heterogeneous networks where the relation between groups of objects is symmetric. The following definitions make it easier to navigate through the rest of the text.

Definition 1. Domain *is a set of objects of different types.*

Definition 2. Domain Pool *is a set of objects of the same type which are independent of each other.*

Definition 3. Community Pool *(or only* Pool*) is a non-empty subset of the domain pool.*

Definition 4. Community *is a set of community pools where:*

– *One pool is selected as a* Base Pool.
– *Each of the other community pools contains all objects from the corresponding domain pool which are adjacent to at least one object of the base pool.*

The community can be described as a complete graph where vertices are the pools and edges are symmetrical relations between them.

Definition 5. Object Rank *is an evaluation of an object regarding its importance in the community.*

Definition 6. Ranked Pool *is a pool in which each object has its object rank.*

Definition 7. *Two different pools are in* Influencer-Influencee relation *when object ranks of objects in the pool Influencee depend on object ranks of objects in the pool Influencer.*

Definition 8. Chain of Influencers *is a sequence of at least three pools such that:*

1. *Consecutive pools are different.*
2. *Two consecutive pools are in the relation Influencer-Influencee.*
3. *The first and the last pools are the same.*

Example 1. Let the above-defined domain is a research area. As a base pool, for instance, we can select people of an academic or research department. Other pools can be papers, topics, and years; all objects of all of the pools are adjacent to at least one person from the department. The community contains all authors of the department, all papers/articles in which these authors participated in, all topics which they deal with, and all years when they published results. We can set a chain of influencers as this sequence of pools: authors → papers → topics → years → authors. The influencer-influencee relations are between authors and papers, papers and topics, topics and years, and years and authors. Now, we are able to calculate individual rankings for all of these pools (authors, papers, topics, years).

2.2 Multipartite Graph

We use a heterogeneous network to describe a domain. A heterogeneous network contains different types of objects and different types of links. We focus on heterogeneous networks where objects in different pools can be connected via different types of links, and objects in the same pool are not connected.

Let a heterogeneous network is represented by multipartite undirected graph $G = (\{V_1, \ldots, V_n\}, \{E_{12} \ldots, E_{(n-1)n}\})$ where V_i are pools, and edges E_{ij} between objects of two pools (V_i, V_j) have a different type than edges E_{kl} between objects belonging to another pair of pools (V_k, V_l). A biadjacency matrix of a subgraph (V_i, V_j, E_{ij}) corresponds to heterogeneous relation $R_{ij} \subseteq V_i \cdot V_j$.

An example of a heterogeneous network based on DBLP data[1] with selected domain pools is shown in Fig. 1. Figure 2 contains a network representing a community extracted from this data and the APYA chain of influencers.

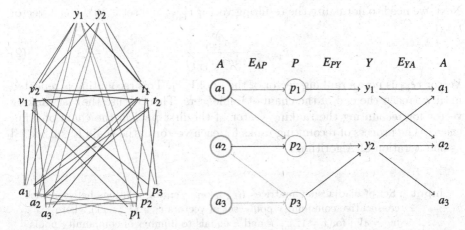

Fig. 2. On the left is an example of a heterogeneous network with heterogeneous vertices and edges. A multipartite graph on the right represents the APYA chain of influencers. A vertex a_1 is influencer of vertices p_1, p_2 and one vertex p_3 is influencee of vertices a_2, a_3.

3 Ranking Algorithm

The proposed ranking method is based on composing relations between domain pools. This composition of relations corresponds to a certain chain of influencers on the complete graph representing a heterogeneous network. Each specific chain of influencers over the heterogeneous network is projected into a specific way of ranking of all the community pools belonging to the chain. Each pool is represented by a vector of the dimension corresponding to the number of objects

---

[1] This data is freely available at http://dblp.org/xml/release/

in the pool. In the beginning, the rank of each object is set to any initial value; this value is 1 in our case.

Ranked pools are a result of the iteration process described in Algorithm 1. This process recalculates pool (vector) ranking in each step of the algorithm.

3.1   Mathematical Formulation

Let us have a selected chain of influencers $V_1 \xrightarrow{R_{1,2}} V_2 \xrightarrow{R_{2,3}} \ldots \xrightarrow{R_{k-1,k}} V_k$ where $V_1, \ldots, V_k$ are the community pools and $R_{1,2}, \ldots, R_{k-1,k}$ are biadjacency matrices corresponding to relations between adjacent pools. We can define ranking vector r_i for each community pool V_i in the chain of influencers where $|V_i| = |r_i|$.

Now we can count for vector $\mathbf{r_i}$ the corresponding vector $\mathbf{r'_{i+1}}$:

$$\mathbf{r'_{i+1}} = \mathbf{r_i} \cdot R_{i,i+1} \tag{1}$$

Next, we need to normalize the resulting vector $\mathbf{r'_{i+1}}$, and set it as the new vector $\mathbf{r_{i+1}}$:

$$\mathbf{r'_{i+1}} = \frac{\mathbf{r'_{i+1}}}{max(\mathbf{r'_{i+1}})} \tag{2}$$

Vector $\mathbf{r_{i+1}}$ is now a ranking vector of the pool V_{i+1}. This recounting is repeated until we reach the end of the chain of influencers. Then we use the last ranking vector for recounting the ranking vector of the first pool in the chain of influencers. The process of recounting ranks for each vector in the chain is considered as one iteration of Algorithm 1.

---

input : Set of biadjacency matrices $R_{i,i+1}$ representing relations between each
consecutive community pools, set of vectors $\mathbf{r_i} = (1, \ldots, 1)$ where
$|\mathbf{r_i}| = |V_i|$ for $i = 1, \ldots, k$ and k equals to number of community pools
and θ is a accuracy.
output: Ranked pools V_i
while *Consecutive iteration of* $\mathbf{r_i}$ *differ more than* θ **do**
  for $i = 1, 2, \ldots k$ **do**
    $\mathbf{r_{i+1}} = \mathbf{r_i} \cdot R_{i,i+1}$
    Normalize($\mathbf{r_{i+1}}$)
    if *(i+1) mod k = 1* **then**
    | $\mathbf{r_1} = \mathbf{r_{i+1}}$
    endif
  end
end

Algorithm 1. Pseudo-code of the ranking algorithm

4   Experiments

We extracted the experimental dataset from the DBLP data. The dataset contains authors, their publications, topics and, years; the base community pool is 53 authors from the Department of Computer Science and Electrical Engineering at Stanford University, Stanford, California. Other community pools - 792 topics, 4231 papers/articles, and 42 years -were chosen regarding objects relationship with at least one author. If the paper/article had no topic, the special topic was used for this case ("no topic" keyword). Most popular keywords representing topics in this paper can be found at FacetedDBLP[2]; topics were detected in titles of papers/articles from the experimental dataset. We extracted relations from the dataset for each pair of the community pools.

Next, we had to select the chains of influencers for ranking. There were many possibilities. We could e.g. define ranking methods like APA (authors → papers → authors) that rank authors only by their publications and publications by their authors. We could also create more, e.g. APYTA (authors → papers → years → topics → authors) which have more aspects influencing the ranking of authors, publications, years, and topics. In the end, we ordered authors by the number of publications and could compare this simple ordering with the results of our methods.

A list of further used chains of influencers: AYA (authors → years → authors), ATA (authors → topics → authors), ATPA (authors → topics → papers → authors), APTYA (authors → papers → topics → years → authors), and the longest chain of influencers AYAPATA (authors → years → authors → papers → authors → topics → authors).

4.1   Experiment: AYA Ranking

We can see results for AYA ranking in Table 1. There are top 5 authors, top 5 years, r_A is the ranking of authors, and r_Y is the ranking of years. The AYA ranking adds more importance to authors publishing in many highly ranked years. Since each year is ranked, an author with a smaller number of years can have a higher ranking than an author who was active in numerous years with a lower ranking. It is important to say that there is no notion of how many times an author did publish in the specific year since this method uses a binary matrix of heterogeneous relation. However, in more complex chains with more pools, the composition of relations results in a more complex multipartite graph in which this quantitative information is also indirectly taken into account.

4.2   Experiment: ATPA Ranking

In Table 2, there are results for a more complex method ATPA. This method considers authors as more important if they use more popular topics in their

---

[2] http://dblp.l3s.de/browse.php?browse=mostPopularKeywords.

Table 1. Ranking of AYA chain of influencers – Top 5

Authors	r_A	Year	r_Y
H. Garcia-Molina	1.00	2011	1.00
M. Horowitz	0.98	2015	0.99
L. J. Guibas	0.97	2013	0.96
P. Hanrahan	0.97	2010	0.95
J. Widom	0.97	2008	0.94

publications. The finding is that top 5 authors were using almost all top 300 best-ranked topics. It is interesting that, except for prof. J. Guibas who is the author of the second best-ranked publication, top 5 publications were not written by top 5 authors. This is because an author with a smaller number of topics can have top-ranked publications if she/he uses interesting topics in these publications but can still be considered less influential in comparison with other authors who used much more topics in their publications.

Table 2. Ranking of ATPA chain of influencers – Top 5

Author	r_A	Topics	r_T	Publications	r_P
L. J. Guibas	1.00	no topic	1.00	Efficient randomized web-cache ...	1.00
H. Garcia-Molina	0.91	optimization	0.86	Learning Large-Scale Dynamic ...	0.98
J. Widom	0.56	applications	0.83	Time and Cost-Efficient Modeling ...	0.92
O. Khatib	0.49	efficiency	0.8	Compressed Sensing and ...	0.91
D. Boneh	0.48	algorithms	0.79	A scalable parallel framework ...	0.83

Table 3 contains a summary of top 5 authors, topics, and papers. In the left part, there are numbers for the authors in top 5. $\#A_T$ is the number of topics which the author used, $\#A_P$ is the number of papers which the author published, and $\#A_Y$ is the number of years in which the author published. In the middle part, there are numbers for the topic in the top 5. $\#T_A$ means how many authors used the topic and $\#T_P$ means how many papers deal with the topic. In the right part, there are the numbers for the papers in the top 5. $\#P_A$ means how many co-authors created the paper and $\#P_T$ means how many topics were used in the title of the paper.

4.3   Experiment: APTYA Ranking

Results for method APTYA are shown in Table 4. In this method, authors are ranked by the importance of their publications, topics and, years. We can say that an author is more important if she/he published in important years where many important topics were published in many important publications.

Table 3. Summary of quantitative indicators of ATPA – Top 5

$\#A_T$	$\#A_P$	$\# A_Y$	$\#T_A$	$\# T_P$	$\#P_A$	$\#P_T$
198	382	40	44	349	1	7
206	360	35	34	116	1	5
135	178	31	34	100	1	5
89	154	28	32	130	1	6
127	221	23	32	108	1	5

Table 4. Ranking for APTYA chain of influencers – Top 5

Author	r_A	Publication	r_P	Topic	r_T	Year	r_Y
H. Garcia-Molina	1.00	The case for RAMClouds...	1.00	no topic	1.00	2013	1.00
L. J. Guibas	0.98	The case for RAMCloud	1.00	systems	0.41	2011	0.98
M. Horowitz	0.97	CSEL: Securing a Mote for 20...	0.7	network	0.38	2007	0.95
J. Widom	0.96	The Performance Impact...	0.59	efficiency	0.36	2012	0.95
P. Hanrahan	0.96	The Stanford FLASH Multi...	0.59	algorithms	0.3	2008	0.93

4.4 Experiment: AYAPATA Ranking of Authors

An interesting aspect of the chain of influencers is the possibility of repeating pools. If we choose chain AYAPATA, then we have three different author's ratings. Each of these rankings is most affected by one of the Y-P-T pools. Moreover, each of the pools Y-P-T is influenced by another ranking of the authors. When we use this chain of influencers, we will get more varied intra-community ranking. Rankings of authors for the AYAPATA chain is in Table 5. Columns A1.TA and r_{TA} contain the names of authors and their first ranking by the chain of influencers AYAPATA, columns A2.YA and r_{YA} contain the names of authors and their second ranking by the chain of influencers AYAPATA and columns A3.PA and r_{PA} contain the names of authors and their third ranking in the same chain of influencers.

Table 5. Rankings of authors by AYAPATA chain of influencers – Top 5

A1.TA	r_{TA}	A2.YA	r_{YA}	A3.PA	r_{PA}
H. Garcia-Molina	1.00	H. Garcia-Molina	1.00	H. Garcia-Molina	1.00
L. J. Guibas	0.96	L.J. Guibas	0.98	L. Guibas	0.98
J. Widom	0.73	M. Horowitz	0.97	J. Widom	0.51
M. Horowitz	0.67	P. Hanrahan	0.96	D. Boneh	0.50
P. Hanrahan	0.66	J. Widom	0.95	M. Horowitz	0.38

4.5 Correlations Between Rankings of Authors

In this section, we show information about the similarity between the presented methods. This similarity is expressed by Spearman's correlations coefficient which was evaluated on an author's ranking created by different methods. Spearman's rank correlation coefficient ρ (rho) is a nonparametric measure of rank correlation (statistical dependence between the ranking of two variables - for example between the ranking of authors by the method AYA and the method ATPA).

Table 6. Spearmans correlation coefficient between presented methods for authors ranking.

C. Coeff	ATA	A1.TA	APTYA	AYA	A2.YA	APA	ATPA	A3.PA	P.Count
ATA	1.00	0.99	0.84	0.83	0.83	0.68	0.94	0.92	0.9
A1.TA	0.99	1.00	0.86	0.85	0.85	0.67	0.95	0.94	0.92
APTYA	0.84	0.86	1.00	0.99	0.99	0.57	0.86	0.94	0.88
AYA	0.83	0.85	0.99	1.00	0.99	0.57	0.86	0.94	0.87
A2.YA	0.83	0.85	0.99	0.99	1.00	0.57	0.86	0.94	0.88
APA	0.68	0.67	0.57	0.57	0.57	1.00	0.64	0.65	0.63
ATPA	0.94	0.95	0.86	0.86	0.86	0.64	1.00	0.96	0.96
A3.PA	0.92	0.94	0.94	0.94	0.94	0.65	0.96	1.00	0.97
P.Count	0.9	0.92	0.88	0.87	0.88	0.63	0.96	0.97	1.00

We compared methods presented in the experiments section in Table 6. P.Count represents authors sorted by the number of their publications. A1.TA represents the first ranking of authors in the chain of influencers AYAPATA, A2.YA represents the second ranking of authors and A3.PA represents the third ranking of authors in the same chain of influencers. The blue highlighted submatrices correspond to the strongly correlated rankings of authors by different methods. We can see that AYA and APTYA and A2.YA have a high correlation. This can be explained because, in all of these chain of influencers, the pool of years is the influencer of the pool of authors. We can also see that the ordering by P.Count best correlates with A1.TA and ATPA. Both these methods depend on the number of publications. The ranking by APA least correlates with all other methods. The reason is that some authors do not have co-authors, and the graph corresponding to composite relation $R_{AP} \circ R_{PA}$ is not connected. Moreover, the proposed method ranks, simultaneously, sets of authors who do not have common publications. A better solution is to use more pools in the chain of influencers to ensure a connected graph and obtain a more complex ranking for all the pools in the chain.

5 Conclusion

In the paper, we introduced ranking of vertices of a multipartite graph representing a heterogeneous network. We understand this multipartite graph as a representation of the community which is a composition of objects of different types. Groups of objects of the same type, pools, are organized to the complete graph. Edges of this graph represent symmetrical relationships between pools. For the purpose of ranking of vertices, we defined "chain of influencers" represented by the sequence (graph walk) of pools. Ranking of vertices in one pool in the chain is most influenced by the ranking of vertices of the previous pool. We described the theoretical background of our approach and performed experiments with the community from the publication domain. The experimental community we used contained authors, topics, papers/articles, and a temporal aspect. Our method ranks all the community pools in the chain of influencers. We compared the ranking authors resulting from various chains of influencers with a simple quantitative evaluation based on numbers of published papers/articles. As a major contribution, we showed that the chain of influencers provides a straightforward and understandable alternative to this quantitative evaluation. This alternative takes into account other aspects of the intra-community evaluation.

Further research will be focused on the application of the presented approach in other domains and on the analysis of the impact of ranking on the relationships inside the community.

Acknowledgement. This work was supported by the Czech Science Foundation under the grant no. GA15-06700S, and by the projects SP2017/100 and SP2017/85 of the Student Grant System, VŠB-Technical University of Ostrava.

References

1. Chen, J., Dai, W., Sun, Y., Dy, J.: Clustering and ranking in heterogeneous information networks via gamma-poisson model. In: Proceedings of the 2015 SIAM International Conference on Data Mining, pp. 424–432. SIAM (2015)
2. Freeman, L.C.: Centrality in social networks conceptual clarification. Soc. Netw. **1**(3), 215–239 (1978)
3. Han, J., Sun, Y., Yan, X., Yu, P.S.: Mining heterogeneous information networks. In: Tutorial at the 2010 ACM SIGKDD Conference on Knowledge Discovery and Data Mining (KDD 2010), Washington, DC (2010)
4. Huang, H., Zubiaga, A., Ji, H., Deng, H., Wang, D., Le, H.K., Abdelzaher, T.F., Han, J., Leung, A., Hancock, J.P., et al.: Tweet ranking based on heterogeneous networks. In: COLING, pp. 1239–1256 (2012)
5. Jeh, G., Widom, J.: Simrank: a measure of structural-context similarity. In: Proceedings of the Eighth ACM SIGKDD International Conference on Knowledge Discovery And Data Mining, pp. 538–543. ACM (2002)
6. Kleinberg, J.M.: Authoritative sources in a hyperlinked environment. J. ACM (JACM) **46**(5), 604–632 (1999)
7. Li, X., Ng, M., Ye, Y.: Multirank: co-ranking for objects and relations in multi-relational data. In: 17th ACM SIGKDD Conference on Knowledge Discovery and Data Mining (KDD-2011), pp. 1217–1225 (2011)

8. Liu, X., Yu, Y., Guo, C., Sun, Y.: Meta-path-based ranking with pseudo relevance feedback on heterogeneous graph for citation recommendation. In: Proceedings of the 23rd ACM International Conference on Conference on Information and Knowledge Management, pp. 121–130. ACM (2014)

9. Liu, X., Yu, Y., Guo, C., Sun, Y., Gao, L.: Full-text based context-rich heterogeneous network mining approach for citation recommendation. In: Proceedings of the 14th ACM/IEEE-CS Joint Conference on Digital Libraries, pp. 361–370. IEEE Press (2014)

10. Ni, J., Tong, H., Fan, W., Zhang, X.: Inside the atoms: ranking on a network of networks. In: Proceedings of the 20th ACM SIGKDD International Conference on Knowledge Discovery and Data Mining, pp. 1356–1365. ACM (2014)

11. Page, L., Brin, S., Motwani, R., Winograd, T.: The pagerank citation ranking: bringing order to the web. Technical report, Stanford InfoLab (1999)

12. Shi, C., Kong, X., Huang, Y., Philip, S.Y., Wu, B.: Hetesim: a general framework for relevance measure in heterogeneous networks. IEEE Trans. Knowl. Data Eng. **26**(10), 2479–2492 (2014)

13. Shi, C., Li, Y., Philip, S.Y., Wu, B.: Constrained-meta-path-based ranking in heterogeneous information network. Knowl. Inf. Syst. **49**(2), 719–747 (2016)

14. Shi, C., Li, Y., Zhang, J., Sun, Y., Philip, S.Y.: A survey of heterogeneous information network analysis. IEEE Trans. Knowl. Data Eng. **29**(1), 17–37 (2017)

15. Soulier, L., Jabeur, L.B., Tamine, L., Bahsoun, W.: On ranking relevant entities in heterogeneous networks using a language-based model. J. Am. Soc. Inf. Sci. Technol. **64**(3), 500–515 (2013)

16. Sun, Y., Han, J., Yan, X., Yu, P.S., Wu, T.: Pathsim: Meta path-based top-k similarity search in heterogeneous information networks. Proc. VLDB Endow. **4**(11), 992–1003 (2011)

17. Sun, Y., Han, J., Zhao, P., Yin, Z., Cheng, H., Rankclus, T.: Integrating clustering with ranking for heterogeneous information network analysis. In: Proceedings of the 12th International Conference on Extending Database Technology: Advances in Database Technology, pp. 565–576. ACM (2009)

18. Tsai, M.-H., Aggarwal, C., Huang, T.: Ranking in heterogeneous social media. In: Proceedings of the 7th ACM International Conference on Web Search and Data Mining, pp. 613–622. ACM (2014)

19. Yu, X., Ren, X., Sun, Y., Gu, Q., Sturt, B., Khandelwal, U., Norick, B., Han, J.: Personalized entity recommendation: a heterogeneous information network approach. In: Proceedings of the 7th ACM International Conference on Web Search and Data Mining, pp. 283–292. ACM (2014)

20. Zhou, D., Orshanskiy, S.A., Zha, H., Giles, C.L.: Co-ranking authors and documents in a heterogeneous network. In: Seventh IEEE International Conference on Data Mining, ICDM 2007, pp. 739–744. IEEE (2007)

Influence Spread in Social Networks with both Positive and Negative Influences

Jing (Selena) He[1], Ying Xie[1], Tianyu Du[3], Shouling Ji[2(✉)], and Zhao Li[3]

[1] Department of Computer Science, Kennesaw State University,
Marietta, GA 30060, USA
{jhe4,yxie2}@kennesaw.edu
[2] College of Computer Science and Technology, Zhejiang University,
Hangzhou, China
sji@zju.edu.cn
[3] Alibaba Group, Hangzhou, China
{tianyu.dty,lizhao.lz}@alibaba-inc.com

Abstract. Social networks are important mediums for spreading information, ideas, and influences among individuals. Most of existing research works of social networks focus on understanding the characteristics of social networks and spreading information through the "word of mouth" effect. However, most of them ignore negative influences among individuals and groups. Motivated by alleviating social problems, such as drinking, smoking, gambling, and influence spreading problems such as promoting new products, we take both positive and negative influences into consideration and propose a new optimization problem, named the Minimum-sized Positive Influential Node Set (MPINS) selection, to identify the minimum set of influential nodes, such that every node in the network can be positively influenced by these selected nodes no less than a threshold θ. Our contributions are threefold. First, we prove that, under the independent cascade model considering both positive and negative influences, MPINS is APX-hard. Subsequently, we present a greedy approximation algorithm to address the MPINS selection problem. Finally, to validate the proposed greedy algorithm, extensive simulations and experiments are conducted on random Graphs and seven different real-world data sets representing small, medium, and large scale networks.

Keywords: Influence spread · Social networks · Positive influential node set · Greedy algorithm · Positive and negative influences

1 Introduction

A social network (e.g., Facebook, Google+, and MySpace) is composed of a set of nodes that share the similar interest or purpose. The network provides a powerful medium of communication for sharing, exchanging, and disseminating information. With the emergence of social applications (such as Flickr, Wikis,

© Springer International Publishing AG 2017
Y. Cao and J. Chen (Eds.): COCOON 2017, LNCS 10392, pp. 615–629, 2017.
DOI: 10.1007/978-3-319-62389-4_51

Netflix, and Twitter, *etc.*), there has been tremendous interests in how to effectively utilize social networks to spread ideas or information within a community [1–8]. In a social network, individuals may have both positive and negative influence on each other. For example, within the context of gambling, a gambling insulator has positive influence on his friends/neighbors. Moreover, if many of an individual's friends are gambling insulators, the aggregated positive influence is exacerbated. However, an individual might turn into a gambler, who brings negative impact on his friends/neighbors.

One application of MPINS is described as follows. A community wants to implement a smoking intervention program. To be cost effective and get the maximum effect, the community wishes to select a small number of influential individuals in the community to attend a quit-smoking campaign. The goal is that all other individuals in the community will be positively influenced by the selected users. Constructing an MPINS is helpful to alleviate the aforementioned social problem, and it is also helpful to promote new products in the social network. Consider the following scenario as another motivation example. A small company wants to market a new product in a community. To be cost effective and get maximum profit, the company would like to distribute sample products to a small number of initially chosen influential users in the community. The company wishes that these initial users would like the product and positively influence their friends in the community. The goal is to have other users in the community be positively influenced by the selected users no less than θ eventually. To sum up, the specific problem we investigate in this work is the following: given a social network and a threshold θ, identify a minimum-sized subset of individuals in the network such that the subset can result in a positive influence on every individual in the network no less than θ.

Hence, we explore the MPINS selection problem under the *independent cascade model* considering both positive and negative influences, where individuals can positively or negatively influence their neighbors with certain probabilities.

In this paper, first we formally define the MPINS problem and then propose a greedy approximation algorithm to solve it. Particularly, the main contributions of this work are summarized as follows:

- Taking both positive and negative influences into consideration, we introduce a new optimization problem, named the Minimum-sized Positive Influential Node Set (MPINS) selection problem, for social networks, which is to identify the minimum-sized set of influential nodes, that could positively influence every node in the network no less than a pre-defined threshold θ. We prove that it is an APX-hard problem under the independent cascade model.
- We define a contribution function, which suggests us a greedy approximation algorithm called MPINS-GREEDY to address the MPINS selection problem. The correctness of the proposed algorithm is analyzed in the paper as well.
- We also conduct extensive simulations and experiments to validate our proposed algorithm. The simulation and experiment results show that the proposed greedy algorithm works well to solve the MPINS selection problem.

More importantly, the solutions obtained by the greedy algorithm is very close to the optimal solution of MPINS in small scale networks.

The rest of this paper is organized as follows: in Sect. 2, we review some related literatures with remarking the difference. In Sect. 3, we first introduce the network model and then we formally define the MPINS selection problem and prove its APX-hardness. The greedy algorithm and theoretical analysis on the correctness of the algorithm are presented in Sect. 4. The simulation and experimental results are presented in Sect. 5 to validate our proposed algorithm. Finally, the paper is concluded in Sect. 6.

2   Related Work

In this section, we first briefly review the related works of social influence analysis. Subsequently, we summarize some related literatures of the PIDS problem and the influence maximization problem.

Influence maximization, initially proposed by Kempe et al. [1], is targeting at selecting a set of users in a social network to maximize the expected number of influenced users through several steps of information propagation [9]. A series of empirical studies have been performed on influence learning [10,11], algorithm optimizing [12,13], scalability promoting [14,15], and influence of group conformity [4,16]. Saito et al. predicted the information diffusion probabilities in social networks under the independent cascade model in [17]. Tang et al. argued that the effect of the social influence from different angles (topics) may be different. Hence, they introduced Topical Affinity Propagation (TAP) to model topic-related social influence on large social networks in [18,19]. Later, Tang et al. [20] proposed a Dynamic Factor Graph (DFG) model to incorporate the time information to analyze dynamic social influences.

Wang et al. first proposed the PIDS problem under the deterministic linear threshold model in [21], which is to find a set of nodes D such that every node in the network has at least half of its neighbor nodes in D. Subsequently, Zhu et al. proved that PIDS is APX-hard and proposed two greedy algorithms with approximation ratio analysis in [22,23]. He et al. [24,25] proposed a new optimization problem named the Minimum-sized Influential Node Set (MINS) selection problem, which is to identify the minimum-sized set of influential nodes. But they neglected the existence of negative influences.

To address the scalability problem of the algorithms in [1,26], Leskovec et al. [27] presented a "lazy-forward" optimization scheme on selecting initial nodes, which greatly reduces the number of influence spread evaluations. Laterly, Chen et al. [28] showed that the problem of computing exact influence in social networks under both models are #P-Hard. They also proposed scalable algorithms under both models, which are much faster than the greedy algorithms in [1,26]. Most recently, consider the data from both cyber-physical world and online social network, [29,30] proposed methods to solve the problem of influence maximization comprehensively.

However, all the aforementioned works did not consider negative influence when they model the social networks. Besides taking both positive and negative influences into consideration, our work try to find a minimum-sized set of individuals that guarantees the positive influences on every node in the network no less than a threshold θ, while the influence maximization problem focuses on choosing a subset of a pre-defined size k that maximizes the expected number of influenced individuals. Since we study the MPINS selection problem under the independent cascade model and take both positive and negative influences into consideration, our problem is more practical. In addition, PIDS is investigated under the deterministic linear threshold model.

3   Problem Definition and Hardness Analysis

In this section, we first introduce the network model. Subsequently, we formally define the MPINS selection problem and make some remarks on the proposed problem. Finally, we analyze the hardness of the MPINS selection problem.

3.1   Network Model

We model a social network by an undirected graph $\mathcal{G}(\mathcal{V}, \mathcal{E}, \mathcal{P}(\mathcal{E}))$, where $\mathcal{V}$ is the set of n nodes, denoted by u_i, and $0 \leq i < n$. i is called the node ID of u_i. An undirected edge $(u_i, u_j) \in \mathcal{E}$ represents a social tie between the pair of nodes. $\mathcal{P}(\mathcal{E}) = \{p_{ij} \mid \text{if } (u_i, u_j) \in \mathcal{E}, 0 < p_{ij} \leq 1, \text{ else } p_{ij} = 0\}$, where p_{ij} indicates the social influence between nodes u_i and u_j[1]. It is worth to mention that the social influence can be categorized into two groups: positive influence and negative influence. For simplicity, we assume the links are undirected (bidirectional), which means two linked nodes have the same social influence (*i.e.*, p_{ij} value) on each other.

3.2   Problem Definition

The objective of the MPINS selection problem is to identify a subset of influential nodes as the initialized nodes. Such that, all the other nodes in a social network can be positively influenced by these nodes no less than a threshold θ. For convenient, we call the initial nodes been selected as *active nodes*, otherwise, *inactive nodes*. Therefore, how to define *positive influence* is critical to solve the MPINS selection problem. In the following, we first formally define some terminologies, and then give the definition of the MPINS selection problem.

Definition 1. *Positive Influential Node Set (*$\mathcal{I}$*).* For social network $\mathcal{G}(\mathcal{V}, \mathcal{E}, \mathcal{P}(\mathcal{E}))$, the positive influential node set is a subset $\mathcal{I} \subseteq \mathcal{V}$, such that all the nodes in $\mathcal{I}$ are initially selected to be the active nodes.

---

[1] This model is reasonable since many empirical studies have analyzed the social influence probabilities between nodes [10, 17, 20].

Definition 2. *Neighboring Set ($\mathcal{B}(u_i)$).* For social network $\mathcal{G}(\mathcal{V}, \mathcal{E}, \mathcal{P}(\mathcal{E}))$, $\forall u_i \in \mathcal{V}$, the neighboring set of u_i is defined as: $\mathcal{B}(u_i) = \{u_j \mid (u_i, u_j) \in \mathcal{E}, p_{ij} > 0\}$.

Definition 3. *Active Neighboring Set ($\mathcal{A}^{\mathcal{I}}(u_i)$).* For social network $\mathcal{G}(\mathcal{V}, \mathcal{E}, \mathcal{P}(\mathcal{E}))$, $\forall u_i \in \mathcal{V}$, the active neighboring set of u_i is defined as: $\mathcal{A}^{\mathcal{I}}(u_i) = \{u_j \mid u_j \in \mathcal{B}(u_i), u_j \in \mathcal{I}\}$.

Definition 4. *Non-active Neighboring Set ($\mathcal{N}^{\mathcal{I}}(u_i)$).* For social network $\mathcal{G}(\mathcal{V}, \mathcal{E}, \mathcal{P}(\mathcal{E}))$, $\forall u_i \in \mathcal{V}$, the non-active neighboring set of u_i is defined as: $\mathcal{N}^{\mathcal{I}}(u_i) = \{u_j \mid u_j \in \mathcal{B}(u_i), u_j \notin \mathcal{I}\}$.

Definition 5. *Positive Influence ($p_{u_i}(\mathcal{A}^{\mathcal{I}}(u_i))$).* For social network $\mathcal{G}(\mathcal{V}, \mathcal{E}, \mathcal{P}(\mathcal{E}))$, a node $u_i \in \mathcal{V}$, and a positive influential node set $\mathcal{I}$, we define a joint influence probability of $\mathcal{A}^{\mathcal{I}}(u_i)$ on u_i, denoted by $p_{u_i}(\mathcal{A}^{\mathcal{I}}(u_i))$ as $p_{u_i}(\mathcal{A}^{\mathcal{I}}(u_i)) = 1 - \prod\limits_{u_j \in \mathcal{A}^{\mathcal{I}}(u_i)} (1 - p_{ij})$.

Definition 6. *Negative Influence ($p_{u_i}(\mathcal{N}^{\mathcal{I}}(u_i))$).* For social network $\mathcal{G}(\mathcal{V}, \mathcal{E}, \mathcal{P}(\mathcal{E}))$, a node $u_i \in \mathcal{V}$, and a positive influential node set $\mathcal{I}$, we define a joint influence probability of $\mathcal{N}^{\mathcal{I}}(u_i)$ on u_i, denoted by $p_{u_i}(\mathcal{N}^{\mathcal{I}}(u_i))$ as $p_{u_i}(\mathcal{N}^{\mathcal{I}}(u_i)) = 1 - \prod\limits_{u_j \in \mathcal{N}^{\mathcal{I}}(u_i)} (1 - p_{ij})$.

Definition 7. *Ultimate Influence ($\varrho^{\mathcal{I}}(u_i)$).* For social network $\mathcal{G}(\mathcal{V}, \mathcal{E}, \mathcal{P}(\mathcal{E}))$, a node $u_i \in \mathcal{V}$, and a positive influential node set $\mathcal{I}$, we define an ultimate influence of $\mathcal{B}(u_i)$ on u_i, denoted by $\varrho^{\mathcal{I}}(u_i)$ as $\varrho^{\mathcal{I}}(u_i) = p_{u_i}(\mathcal{A}^{\mathcal{I}}(u_i)) - p_{u_i}(\mathcal{N}^{\mathcal{I}}(u_i))$. Moreover, if $\varrho^{\mathcal{I}}(u_i) < 0$, we set $\varrho^{\mathcal{I}}(u_i) = 0$. If $\varrho^{\mathcal{I}}(u_i) \geq \theta$, where $0 < \theta < 1$ is a pre-defined threshold, then u_i is said been positively influenced. Otherwise, u_i is not been positively influenced.

Definition 8. *Minimum-sized Positive Influential Node Set (MPINS).* For social network $\mathcal{G}(\mathcal{V}, \mathcal{E}, \mathcal{P}(\mathcal{E}))$, the MPINS selection problem is to find a minimum-sized positive influential node set $\mathcal{I} \subseteq \mathcal{V}$, such that $\forall u_i \in \mathcal{V} \setminus \mathcal{I}$, u_i is positively influenced, *i.e.*, $\varrho^{\mathcal{I}}(u_i) = p_{u_i}(\mathcal{A}^{\mathcal{I}}(u_i)) - p_{u_i}(\mathcal{N}^{\mathcal{I}}(u_i)) \geq \theta$, where $0 < \theta < 1$.

3.3 Problem Hardness Analysis

In general, given an arbitrary threshold θ, the MPINS selection problem is APX-hard. We prove the APX-hardness of MPINS by constructing a *L-reduction* from Vertex Cover problem in Cubic Graph (denoted by VCCG) to the MPINS selection problem. The decision problem of VCCG is APX-hard which is proven in [31]. A cubic graph is a graph with every vertex's degree of exactly three. Given a cubic graph, VCCG is to find a minimum-sized vertex cover[2].

First, consider a cubic graph $\mathcal{G}(\mathcal{V}, \mathcal{E}, \mathcal{P}(\mathcal{E}))$, where $\mathcal{P}(\mathcal{E}) = \{1 \mid (u_i, u_j) \in \mathcal{E}; u_i, u_j \in \mathcal{V}\}$, as an instance of VCCG. We construct a new graph $\widehat{\mathcal{G}}$ as follows:

---

[2] A vertex cover is defined as a subset of nodes in a graph $\mathcal{G}$ such that each edge of the graph is incident to at least one vertex of the set.

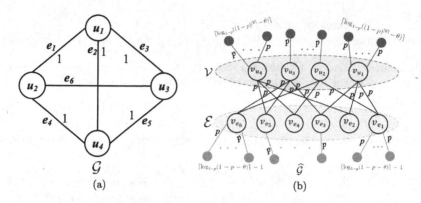

Fig. 1. Illustration of the construction from $\mathcal{G}$ to $\widehat{\mathcal{G}}$.

(1) We create $|\mathcal{V}| + |\mathcal{E}|$ nodes with $|\mathcal{V}|$ nodes $\mathbf{v}_{u_i} = \{v_{u_1}, v_{u_2}, \cdots, v_{u_{|\mathcal{V}|}}\}$ representing the nodes in $\mathcal{G}$ and $|\mathcal{E}|$ nodes $\mathbf{v}_{e_i} = \{v_{e_1}, v_{e_2}, \cdots, v_{e_{|\mathcal{E}|}}\}$ representing the edges in $\mathcal{G}$. (2) We add an edge with influence weight p between nodes v_{u_i} and v_{e_j} if and only if node u_i is an endpoint of edge e_j. (3) We attach additional $\lceil \log_{1-p}((1-p)^{|\mathcal{V}|} - \theta) \rceil$ active nodes to each node v_{u_i}, denoted by set $\mathbf{v}_{u_i}^{\mathcal{A}} = \{v_{u_i}^j \mid 1 \le j \le \lceil \log_{1-p}((1-p)^{|\mathcal{V}|} - \theta) \rceil\}$. Obviously, $|\mathbf{v}_{u_i}^{\mathcal{A}}| = \lceil \log_{1-p}((1-p)^{|\mathcal{V}|} - \theta) \rceil$. (4) We attach additional $\lceil \log_{1-p}(1 - p - \theta) \rceil - 1$ active nodes to each node v_{e_j}, denoted by set $\mathbf{v}_{e_j}^{\mathcal{A}} = \{v_{e_j}^j \mid 1 \le j \le \lceil \log_{1-p}(1 - p - \theta) \rceil - 1\}$. Obviously, $|\mathbf{v}_{e_j}^{\mathcal{A}}| = \lceil \log_{1-p}(1 - p - \theta) \rceil - 1$. (5) $\widehat{\mathcal{G}} = \{\widehat{\mathcal{V}}, \widehat{\mathcal{E}}\}$, where $\widehat{\mathcal{V}} = \{v_{u_1}, \cdots, v_{u_{|\mathcal{V}|}}\} \cup \{v_{e_1}, \cdots, v_{e_{|\mathcal{E}|}}\} \cup \bigcup_{i=1}^{|\mathcal{V}|} \mathbf{v}_{u_i}^{\mathcal{A}} \cup \bigcup_{i=1}^{|\mathcal{E}|} \mathbf{v}_{e_i}^{\mathcal{A}}$, $\widehat{\mathcal{E}}$ is the set of all the edges associated with the nodes in $\widehat{\mathcal{V}}$, and $\mathcal{P}(\widehat{\mathcal{E}}) = \{p \mid$ for every edge in $\widehat{\mathcal{E}}\}$.

Taking the cubic graph shown in Fig. 1(a) as an example to illustrate the construction procedure from $\mathcal{G}$ to $\widehat{\mathcal{G}}$. There are 4 nodes and 6 edges in $\mathcal{G}$. Therefore, we first create $\{v_{u_i}\}_{i=1}^4$ and $\{v_{e_i}\}_{i=1}^6$ nodes in $\widehat{\mathcal{G}}$. Then we add edges with influence weight p between nodes v_{u_i} and v_{e_j} based on the topology shown in $\mathcal{G}$. Subsequently, we add additional $\mathbf{v}_{u_i}^{\mathcal{A}} = \{v_{u_i}^j \mid 1 \le j \le \lceil \log_{1-p}((1-p)^{|\mathcal{V}|} - \theta) \rceil\}$ active nodes to each node v_{u_i} (marked by upper shaded nodes in Fig. 1(b)). Similarly, we add additional $\mathbf{v}_{e_j}^{\mathcal{A}} = \{v_{e_j}^j \mid 1 \le j \le \lceil \log_{1-p}(1 - p - \theta) \rceil - 1\}$ active nodes to each node v_{e_j} (marked by bottom shaded nodes in Fig. 1(b)). The influence weights on all the additional edges are p. Finally, the new graph $\widehat{\mathcal{G}}$ is constructed as shown in Fig. 1(b).

Lemma 1. $\mathcal{G}$ has a VCCG $\mathcal{D}$ of size at most d if and only if $\widehat{\mathcal{G}}$ has a positive influential node set $\mathcal{I}$ of size at most k by setting $k = |\mathcal{V}|\lceil \log_{1-p}((1-p)^{|\mathcal{V}|} - \theta) \rceil + |\mathcal{E}|(\lceil \log_{1-p}(1 - p - \theta) \rceil - 1) + d$.

Due to limited space in this paper, for a comprehensive proof of Lemma 1 we refer the reader to our technical report in [32].

Theorem 1. The MPINS selection problem is APX-hard.

Proof. An immediate conclusion of Lemma 1 is that $\mathcal{G}$ has a minimum-sized vertex cover of size $OPT_{VCCG}(\mathcal{G})$ if and only if $\widehat{\mathcal{G}}$ has a minimum-sized positive influential node set of size

$$OPT_{MPINS}(\widehat{\mathcal{G}})$$
$$= |\mathcal{V}|\lceil \log_{1-p}((1-p)^{|\mathcal{V}|} - \theta)\rceil + |\mathcal{E}|(\lceil \log_{1-p}(1-p-\theta)\rceil - 1) + OPT_{VCCG}(\mathcal{G}).$$
(1)

Note that in a cubic graph $\mathcal{G}$, $|\mathcal{E}| = \frac{3|\mathcal{V}|}{2}$. Hence, we have

$$\frac{|\mathcal{V}|}{2} = \frac{|\mathcal{E}|}{3} \le OPT_{VCCG}(\mathcal{G}).$$
(2)

Based on Lemma 1, plugging

$$|\mathcal{V}| = \frac{OPT_{MPINS}(\widehat{\mathcal{G}}) - OPT_{VCCG}(\mathcal{G})}{\lceil \log_{1-p}((1-p)^{|\mathcal{V}|} - \theta)\rceil + \frac{3}{2}(\lceil \log_{1-p}(1-p-\theta)\rceil - 1)}$$
(3)

into the inequality 2, we have

$$OPT_{MPINS}(\widehat{\mathcal{G}})$$
$$\le [2\lceil \log_{1-p}((1-p)^{|\mathcal{V}|} - \theta)\rceil + 3\lceil \log_{1-p}(1-p-\theta)\rceil - \tfrac{1}{2}]OPT_{VCCG}(\mathcal{G}).$$
(4)

This means that VCCG is L-reducible to MPINS. In conclusion, we proved that a specific case of the MPINS selection problem is APX-hard, since the VCCG problem is APX-hard. Consequently, the general MPINS selection problem is also at least APX-hard.

Based on Theorem 1, we conclude that MPINS cannot be solved in polynomial time. Therefore, we propose a greedy algorithm to solve the problem in the next section.

4   Greedy Algorithm and Performance Analysis

Since MPINS is APX-hard, we propose a greedy algorithm to solve it named MPINS-GREEDY. Before introducing MPINS-GREEDY, we first define a useful contribution function as follows:

Definition 9. *Contribution function* $(f(\mathcal{I}))$. For a social network represented by graph $\mathcal{G}(\mathcal{V}, \mathcal{E}, \mathcal{P}(\mathcal{E}))$, and a positive influential node set $\mathcal{I}$, the contribution function of $\mathcal{I}$ to $\mathcal{G}$ is defined as: $f(\mathcal{I}) = \sum_{i=1}^{|\mathcal{V}|} \max\{\min(\varrho^{\mathcal{I}}(u_i), \theta), 0\}$.

Based on the defined contribution function, we propose a heuristic algorithm, which has two phases. First, we find the node u_i with the maximum $f(\mathcal{I})$, where $\mathcal{I} = \{u_i\}$; and after that, we select a Maximal Independent Set (MIS)[3] induced

---

[3] MIS can be defined formally as follows: given a graph $G = (V, E)$, an Independent Set (IS) is a subset $I \subset V$ such that for any two vertex $v_1, v_2 \in I$, they are not adjacent, *i.e.*, $(v_1, v_2) \notin E$. An IS is called an MIS if we add one more arbitrary node to this subset, the new subset will not be an IS any more.

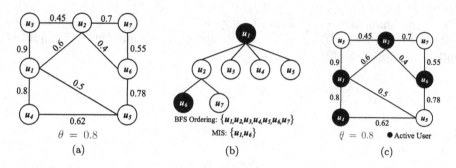

Fig. 2. Illustration of MPINS-Greedy algorithm.

by a breadth-first-search (BFS) ordering starting from u_i. Second, employ the pre-selected MIS denoted by $\mathcal{M}$ as the initial active node set to perform the greedy algorithm called MPINS-GREEDY as shown in Algorithm 1. MPINS-GREEDY starts from $\mathcal{I} = \mathcal{M}$. Each time, it adds the node having the maximum $f(\cdot)$ value into $\mathcal{I}$. The algorithm terminates when $f(\mathcal{I}) = |\mathcal{V}|\theta$.

Algorithm 1. MPINS-GREEDY Algorithm

Require: A social network represented by graph $\mathcal{G}(\mathcal{V}, \mathcal{E}, \mathcal{P}(\mathcal{E}))$; a pre-defined threshold θ.
1: Initialize $\mathcal{I} = \mathcal{M}$
2: **while** $f(\mathcal{I}) < |\mathcal{V}|\theta$ **do**
3:    choose $u \in \mathcal{V} \setminus \mathcal{I}$ to maximize $f(\mathcal{I} \bigcup \{u\})$
4:    $\mathcal{I} = \mathcal{I} \bigcup \{u\}$
5: **end while**
6: **return** $\mathcal{I}$

To better understand the proposed algorithm, we use the social network represented by the graph shown in Fig. 2(a) to illustrate the selection procedure as follows. In the example, $\theta = 0.8$. Since u_1 has the maximum $f(\{u_1\})$ value, we construct a BFS tree rooted at u_1, as shown in Fig. 2(b) with the help of the BFS ordering, we find the MIS set which is $\mathcal{M} = \{u_1, u_6\}$. Next, we go to the second phase to perform Algorithm 1. (1) First round: $\mathcal{I} = \mathcal{M} = \{u_1, u_6\}$. (2) Second round: we first compute $f(\mathcal{I} = \{u_1, u_2, u_6\}) = 4.45$, $f(\mathcal{I} = \{u_1, u_3, u_6\}) = 3.018$, $f(\mathcal{I} = \{u_1, u_4, u_6\}) = 3.65$, $f(\mathcal{I} = \{u_1, u_5, u_6\}) = 3.65$, and $f(\mathcal{I} = \{u_1, u_6, u_7\}) = 3.778$. Therefore, we have $\mathcal{I} = \{u_1, u_2, u_6\}$, which has the maximum $f(\mathcal{I})$ value. However, $f(\mathcal{I} = \{u_1, u_2, u_6\}) = 4.45 < 7 * 0.8 = 5.6$. Consequently, the selection procedure continues. (3) Third round: we first computer $f(\mathcal{I} = \{u_1, u_2, u_3, u_6\}) = 4.45$, $f(\mathcal{I} = \{u_1, u_2, u_4, u_6\}) = 5.6$, $f(\mathcal{I} = \{u_1, u_2, u_5, u_6\}) = 5.6$, and $f(\mathcal{I} = \{u_1, u_2, u_6, u_7\}) = 4.45$. Therefore, we have $\mathcal{I} = \{u_1, u_2, u_4, u_6\}$[4]. Since $f(\mathcal{I} = \{u_1, u_2, u_4, u_6\}) = 7 * 0.8 = 5.6$, algorithm

---

[4] If there is a tie on the $f(\mathcal{I})$ value, we use the node ID to break the tie.

terminates and outputs set $\mathcal{I} = \{u_1, u_2, u_4, u_6\}$ as shown in Fig. 2(c), where black nodes represent the selected influential nodes.

Based on Algorithm 1, in each iteration, only one node is selected to be added into the output set $\mathcal{I}$. In the worst case, all nodes are added into $\mathcal{I}$ in the $|\mathcal{V}|$-th iteration. Then, $f(\mathcal{I}) = f(\mathcal{V}) = |\mathcal{V}|\theta$ and Algorithm 1 terminates and outputs $\mathcal{I} = \mathcal{V}$. Therefore, Algorithm 1 terminates for sure. Also, if $f(\mathcal{I}) = |\mathcal{V}|\theta$, then $\forall u_i \in \mathcal{V}, \varrho^{\mathcal{I}}(u_i) \geq \theta$ followed by Definition 9. Therefore, all nodes in the network are positively influenced. In another side, if $\forall u_i \in \mathcal{V}, \varrho^{\mathcal{I}}(u_i) \geq \theta$, then we obtain $\forall u_i \in \mathcal{V}, \min(\varrho^{\mathcal{I}}(u_i), \theta) = \theta$. Therefore, Algorithm 1 must produce a feasible solution of the MPINS selection problem.

5   Performance Evaluation

Since there is no existing work studying the MPINS problem under the independent cascade model currently, in the real data experiments, the results of MPINS-GREEDY (MPINS) are compared with the most related work [22] (PIDS), and the optimal solution of MPINS (OPTIMAL) which is obtained by exhausting searching. To ensure the fairness of comparison, the condition of termination to the algorithm proposed in [22] is changed to find a PIDS, such that every node in the network is positively influenced no less than the same threshold θ in MPINS. All experiments were performed on a desktop computer equipped with Inter(R) Core(TM) 2 Quad CPU 2.83 GHz and 6 GB RAM.

5.1   Experimental Setting

We also implement experiments run on different kinds of real-world data sets. The first group of data sets are shown in Table 1 come from SNAP[5]. The network statistics are summarized by the number of nodes and edges, and the diameter (i.e., longest shortest path). The data collected in Table 1 is based on the *Customers Who Bought This Item Also Bought* feature of the Amazon website. Four different networks are composed of the data collected from March to May in 2003 in Amazon. In each network, for a pair of nodes (products) i and j, there is an edge between them if and only if a product i is frequently co-purchased with product j [33]. Besides the Amazon product co-purchasing data sets shown in Table 1, we also evaluate our algorithm in the additional real data sets listed as follows:

1. *WikiVote*: a data set obtained from [34], which contains the vote history data of Wikipedia. The data set includes 7115 vertices and 103689 edges which contains the voting data of Wikipedia from the inception till January 2008. If user i voted on user j for the administrator election, there will be an edge between i to j.

---

[5] http://snap.stanford.edu/data/.

Table 1. Data Set 1 in Our Experiment

Data	Nodes	Edges	Diameter
A1	262111	1234877	29
A2	400727	3200440	18
A3	410236	3356824	21
A4	403394	3387388	21

2. *Coauthor*: a data set obtained from [35], which hold the coauthors information maintained by ArnetMiner.

   We chosen the subset which include 53442 vertices and 127968 edges. When the author i has a relationship with author j, there will be one edge between i to j.
3. *Twitter*: a data set obtained from [36,37], which stores the information collected from Twitter. We picked the subset with 92180 vertices and 188971 edges, which represent the user account and their relationships.

Moreover, the social influence on each edge (i, j) is calculated by $\frac{1}{deg(j)}$ [38], where $deg(j)$ is the degree of node j. Similarly, if one node is selected as the active node, it has positively influence on all its neighbors. Otherwise, it only has negative influence on its neighbors.

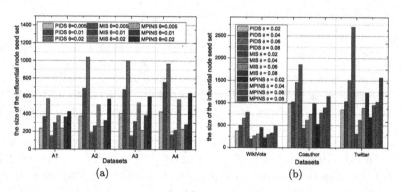

(a)   (b)

Fig. 3. Size of influential nodes in (a) Amazon, (b) WikiVote, Coauthor, and Twitter.

Experimental Results. The impacts of θ on the size of MIS, the solutions of MPINS, and the solution of PIDS on Amazon co-purchase data sets, when θ change from 0.005 to 0.02, are shown in Fig. 3(a). As shown in Fig. 3(a), the solution sizes of PIDS and MINS increase when θ increases. This is because, when the pre-set threshold becomes large, more influential nodes are required to be chosen to influence the whole network. On average, the difference between the size of PIDS and MPINS solutions is 37.23%. This is because that MPINS chooses the most influential node first instead of the node with the largest degree first. Moreover, the growth rate of the solution size of PIDS is higher than that of MPINS.

Similarly, the impacts of θ on the size of MIS, the solutions of MPINS, and the solution of PIDS on WikiVote, Coauthor, and Twitter, when θ change from 0.02 to 0.08, are shown in Fig. 3(b). The solution sizes of PIDS and MINS increase when θ increases as well. For the Twitter data set, MPINS outperforms PIDS significantly, *i.e.*, MPINS selects 45.45% less influential nodes than that of PIDS. On average, the difference between the sizes of PIDS and MPINS solutions is 36.37%.

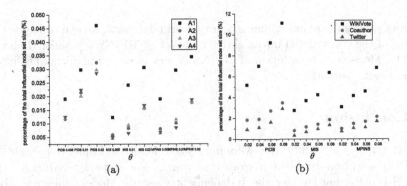

(a)   (b)

Fig. 4. % of influential nodes in (a) Amazon (b) WikiVote, Coauthor, and Twitter.

Figure 4 shows how many nodes are selected as the influential nodes represented by the ratio over the total number of nodes in the network. Figure 4 (a) shows the impacts of θ on the ratio of MIS, MPINS, and PIDS on Amazon co-purchase data sets. While, Fig. 4 (b) shows the the impacts of θ on the ratio of MIS, MPINS, and PIDS on WikiVote, Coauthor, and Twitter data sets. One interesting observation here is that much less nodes are selected as the influential nodes for Amazon co-purchase data sets compared to the WikiVote, Coauthor, and Twitter data sets.

Finally, we compare the performance of our proposed method MPINS with PIDS and the method denoted by "Random", which randomly chooses a node as the influential node. The impacts of θ on the sizes of the solutions of MPINS, PIDS, and Random, when θ change from 0.02 to 0.08, are shown in Fig. 5 for the WikiVote, Coauthor data set, and Twitter data sets. As shown in Fig. 5, the solution sizes of Random, PIDS and MPINS increase when n increases. Moreover, for a specific θ, MPINS produces a smaller influential node set than PIDS. This is consistent with the simulation results and previous experimental results. Furthermore, both PIDS and MPINS produce much smaller influential node sets than Random for a specific θ. This is because Random picks node randomly without any selection criterion. However, PIDS's selection process is based on degree and our MPINS greedy criterion is based on social influence.

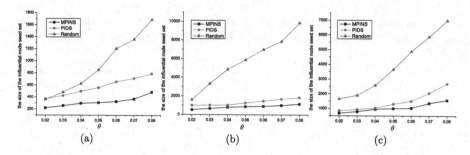

Fig. 5. MPINS VS. PIDS VS. Random in (a) WikiVote (b) Coauthor (c) Twitter

From the results of experiments on real-world data sets, we can conclude that the size of the constructed initial active node set of MPINS is smaller than that of PIDS. Moreover, the solution of MPINS is very close to the optimal solutions in small scale networks.

6   Conclusion

In this paper, we study the Minimum-sized Positive Influential Node Set (MPINS) selection problem in social networks. We show by reduction that MPINS is APX-hard under the Independent Cascade Model. Subsequently, a greedy algorithm is proposed to solve the problem. Furthermore, we validate our proposed algorithm through simulations on random graphs and experiments on seven different real-world data sets. The simulation and experimental results indicate that MPINS-GREEDY can construct smaller sized satisfied initial active node sets than the latest related work PIDS. Moreover, for small scale network, MPINS-GREEDY has very similar performance as the optimal solution of MPINS. Furthermore, the simulation and experimental results indicate that MPINS-GREEDY considerably outperforms PIDS in medium and large scale networks, sparse networks, and for high threshold θ.

Acknowledgment. This research is funded in part by the Kennesaw State University College of Science and Mathematics Interdisciplinary Research Opportunities (IDROP) Program, the Provincial Key Research and Development Program of Zhejiang, China under No. 2016C01G2010916, the Fundamental Research Funds for the Central Universities, the Alibaba-Zhejiang University Joint Research Institute for Frontier Technologies (A.Z.F.T.) under Program No. XT622017000118, and the CCF-Tencent Open Research Fund under No. AGR20160109.

References

1. Kempe, D., Kleinberg, J., Tardos, É.: Maximizing the spread of influence through a social network. In: Proceedings of the Ninth ACM SIGKDD International Conference on Knowledge Discovery and Data Mining, pp. 137–146. ACM (2003)

2. Saito, K., Kimura, M., Motoda, H.: Discovering influential nodes for sis models in social networks. In: Gama, J., Costa, V.S., Jorge, A.M., Brazdil, P.B. (eds.) DS 2009. LNCS, vol. 5808, pp. 302–316. Springer, Heidelberg (2009). doi:10.1007/978-3-642-04747-3_24

3. Li, Y., Chen, W., Wang, Y., Zhang, Z.-L.: Influence diffusion dynamics and influence maximization in social networks with friend and foe relationships. In: Proceedings of the sixth ACM International Conference on Web Search and Data Mining, pp. 657–666. ACM (2013)

4. Han, M., Yan, M., Cai, Z., Li, Y., Cai, X., Yu, J.: Influence maximization by probing partial communities in dynamic online social networks. Trans. Emerging Telecommun. Technol

5. He, X., Song, G., Chen, W., Jiang, Q.: Influence blocking maximization in social networks under the competitive linear threshold model. In: Proceedings of the 2012 SIAM International Conference on Data Mining, pp. 463–474. SIAM (2012)

6. Lu, W., Bonchi, F., Goyal, A., Lakshmanan, L.V.: The bang for the buck: fair competitive viral marketing from the host perspective. In: Proceedings of the 19th SIGKDD International Conference on Knowledge Discovery and Data Mining, pp. 928–936. ACM (2013)

7. Selcuk Uluagac, A., Beyah, R., Ji, S., He, J. (Selena), Li, Y.: Cell-based snapshot and continuous data collection in wireless sensor networks. ACM Trans. Sensor Networks (TOSN) **9**(4)

8. He, J. (Selena), Ji, S., Li, Y.: Genetic-algorithm-based construction of load-balanced cdss in wireless sensor networks. MILCOM **9**(4)

9. Albinali, H., Han, M., Wang, J., Gao, H., Li, Y.: The roles of social network mavens. In: The 12th International Conference on Mobile Ad-hoc and Sensor Networks

10. Goyal, A., Bonchi, F., Lakshmanan, L.V.: Learning influence probabilities in social networks. In: Proceedings of the third ACM International Conference on Web Search and Data Mining, pp. 241–250. ACM (2010)

11. Han, M., Yan, M., Li, J., Ji, S., Li, Y.: Generating uncertain networks based on historical network snapshots. In: Du, D.-Z., Zhang, G. (eds.) COCOON 2013. LNCS, vol. 7936, pp. 747–758. Springer, Heidelberg (2013). doi:10.1007/978-3-642-38768-5_68

12. Goyal, A., Lu, W., Lakshmanan, L.V.: Celf++: optimizing the greedy algorithm for influence maximization in social networks. In: Proceedings of the 20th International Conference Companion on World Wide Web, pp. 47–48. ACM (2011)

13. Han, M., Yan, M., Cai, Z., Li, Y.: An exploration of broader influence maximization in timeliness networks with opportunistic selection. J. Netw. Comput. Appl. **63**, 39–49 (2016)

14. Wang, C., Chen, W., Wang, Y.: Scalable influence maximization for independent cascade model in large-scale social networks. Data Mining Know. Discovery **25**(3), 545 (2012)

15. Han, M., Duan, Z., Ai, C., Lybarger, F.W., Li, Y., Bourgeois, A.G.: Time constraint influence maximization algorithm in the age of big data. Int. J. Comput. Sci. Eng

16. Tang, J., Wu, S., Sun, J.: Confluence: conformity influence in large social networks. In: Proceedings of the 19th ACM SIGKDD International Conference on Knowledge Discovery and Data Mining, pp. 347–355. ACM (2013)

17. Saito, K., Nakano, R., Kimura, M.: Prediction of information diffusion probabilities for independent cascade model. In: Lovrek, I., Howlett, R.J., Jain, L.C. (eds.) KES 2008. LNCS, vol. 5179, pp. 67–75. Springer, Heidelberg (2008). doi:10.1007/978-3-540-85567-5_9

18. Tang, J., Sun, J., Wang, C., Yang, Z.: Social influence analysis in large-scale networks. In: Proceedings of the 15th ACM SIGKDD International Conference on Knowledge Discovery and Data Mining, pp. 807–816. ACM (2009)

19. Han, M., Yan, M., Li, J., Ji, S., Li, Y.: Neighborhood-based uncertainty generation in social networks. J. Comb. Optim. **28**(3), 561–576 (2014)

20. Wang, C., Tang, J., Sun, J., Han, J.: Dynamic social influence analysis through time-dependent factor graphs. In: 2011 International Conference on Advances in Social Networks Analysis and Mining (ASONAM), pp. 239–246. IEEE (2011)

21. Wang, F., Camacho, E., Xu, K.: Positive influence dominating set in online social networks. In: Du, D.-Z., Hu, X., Pardalos, P.M. (eds.) COCOA 2009. LNCS, vol. 5573, pp. 313–321. Springer, Heidelberg (2009). doi:10.1007/978-3-642-02026-1_29

22. Wang, F., Du, H., Camacho, E., Xu, K., Lee, W., Shi, Y., Shan, S.: On positive influence dominating sets in social networks. Theoret. Comput. Sci. **412**(3), 265–269 (2011)

23. Zhu, X., Yu, J., Lee, W., Kim, D., Shan, S., Du, D.-Z.: New dominating sets in social networks. J. Global Optim. **48**(4), 633–642 (2010)

24. He, J.S., Ji, S., Beyah, R., Cai, Z.: Minimum-sized influential node set selection for social networks under the independent cascade model. In: Proceedings of the 15th ACM International Symposium on Mobile ad hoc Networking and Computing, pp. 93–102. ACM (2014)

25. Kaur, H., He, J.S.: Blocking negative influential node set in social networks: from host perspective. Trans. Emerg. Telecommun. Technol. (ETT) **28**(4)

26. Kempe, D., Kleinberg, J., Tardos, É.: Influential nodes in a diffusion model for social networks. In: Caires, L., Italiano, G.F., Monteiro, L., Palamidessi, C., Yung, M. (eds.) ICALP 2005. LNCS, vol. 3580, pp. 1127–1138. Springer, Heidelberg (2005). doi:10.1007/11523468_91

27. Leskovec, J., Krause, A., Guestrin, C., Faloutsos, C., VanBriesen, J., Glance, N.: Cost-effective outbreak detection in networks. In: Proceedings of the 13th ACM SIGKDD International Conference on Knowledge Discovery and Data Mining, pp. 420–429. ACM (2007)

28. Chen, W., Yuan, Y., Zhang, L.: Scalable influence maximization in social networks under the linear threshold model. In: 2010 IEEE 10th International Conference on Data Mining (ICDM), pp. 88–97. IEEE (2010)

29. Han, M., Li, J., Cai, Z., Qilong, H.: Privacy reserved influence maximization in gps-enabled cyber-physical and online social networks. SocialCom **2016**, 284–292 (2016)

30. Han, M., Han, Q., Li, L., Li, J., Li, Y.: Maximizing influence in sensed heterogenous social network with privacy preservation. Int. J. Sens. Netw

31. Du, D.-Z., Ko, K.-I.: Theory of Computational Complexity, vol. 58. Wiley, Hoboken (2011)

32. Technical report. http://ksuweb.kennesaw.edu/~jhe4/Research/MPINS

33. Leskovec, J., Adamic, L.A., Huberman, B.A.: The dynamics of viral marketing. ACM Trans. Web (TWEB) **1**(1) (2007). 5

34. Leskovec, J., Huttenlocher, D., Kleinberg, J.: Predicting positive and negative links in online social networks. In: Proceedings of the 19th International Conference on World Wide Web, pp. 641–650. ACM (2010)

35. Tang, J., Zhang, J., Yao, L., Li, J., Zhang, L., Su, Z.: Arnetminer: extraction and mining of academic social networks. In: Proceedings of the 14th ACM SIGKDD International Conference on Knowledge Discovery and Data Mining, pp. 990–998. ACM (2008)

36. Hopcroft, J., Lou, T., Tang, J.: Who will follow you back?: reciprocal relationship prediction. In: Proceedings of the 20th ACM International Conference on Information and Knowledge Management, pp. 1137–1146. ACM (2011)
37. Lou, T., Tang, J., Hopcroft, J., Fang, Z., Ding, X.: Learning to predict reciprocity and triadic closure in social networks. ACM Trans. Know. Discov. from Data (TKDD) **7**(2) (2013). 5
38. Wang, C., Chen, W., Wang, Y.: Scalable influence maximization for independent cascade model in large-scale social networks. Data Min. Know. Disc. **25**(3)

Guided Genetic Algorithm for the Influence Maximization Problem

Pavel Krömer[(✉)] and Jana Nowaková

Department of Computer Science, VŠB Technical University of Ostrava,
Ostrava, Czech Republic
{pavel.kromer,jana.nowakova}@vsb.cz

Abstract. Influence maximization is a hard combinatorial optimization problem. It requires the identification of an optimum set of k network vertices that triggers the activation of a maximum total number of remaining network nodes with respect to a chosen propagation model. The problem is appealing because it is provably hard and has a number of practical applications in domains such as data mining and social network analysis. Although there are many exact and heuristic algorithms for influence maximization, it has been tackled by metaheuristic and evolutionary methods as well. This paper presents and evaluates a new evolutionary method for influence maximization that employs a recent genetic algorithm for fixed–length subset selection. The algorithm is extended by the concept of guiding that prevents selection of infeasible vertices, reduces the search space, and effectively improves the evolutionary procedure.

Keywords: Influence maximization · Information diffusion · Social networks · Genetic algorithms

1 Introduction

The propagation of various phenomena through networks and linked environments is nowadays subject of intense research. An increasing number of real–world domains can be perceived and modelled as networks and analyzed by network and graph methods. Contemporary technological and social networks are complex distributed environments where phenomena can be carried from one node to another and spread along different routes and pathways. The knowledge of information propagation patterns and network structures that are associated with information diffusion can be used for a number of purposes including opinion making, viral marketing, expert (authority) identification, and e.g. robustness and stability analysis.

Influence maximization is a common term for a family of network problems that consist in the search for a set of seed nodes from which information spreads through a network most efficiently. The seed nodes need to be selected so that the information they release to the network affects the maximum of the remaining network nodes under an assumed propagation model. Influence maximization has been formulated as a combinatorial optimization problem and shown

© Springer International Publishing AG 2017
Y. Cao and J. Chen (Eds.): COCOON 2017, LNCS 10392, pp. 630–641, 2017.
DOI: 10.1007/978-3-319-62389-4_52

NP–hard [9]. It has been addressed by a number of exact and heuristic algorithms and, due to its hardness and the limitations of traditional methods, also by various stochastic and metaheuristic approaches.

Evolutionary methods have been successfully employed to solve a wide range of combinatorial optimization problems. They are often used in place of traditional exact and heuristic methods because the search and optimization strategies they implement are flexible and adapt very well to a wide range of conditions associated with different instances of solved problems. Evolutionary metaheuristics facilitate an efficient search process in a search space composed of encoded candidate problem solutions. An appropriate encoding that would translate candidate problem solutions to a form required by the employed metaheuristic algorithm efficiently is a crucial part of every metaheuristic problem solving strategy. A good encoding is compact and does not allow construction of invalid problem solutions through the artificial evolution.

In this work, a recent genetic algorithm for fixed–length subset selection is employed to address the influence maximization problem. The algorithm used previously for sampling from scale–free networks [11] and feature subset selection [12,13], is in this work extended by the concept of guiding to increase its performance for the influence maximization problem. The rest of this paper is organized in the following way. Section 2 introduces the influence maximization problem and outlines several heuristic and metaheuristic algorithms that were proposed to solve it. Section 3 presents the proposed evolutionary approach to influence maximization. It briefly outlines the basic principles of genetic algorithms, summarizes genetic algorithm for fixed–length subset selection, and suggests its new variant suitable for network problems such as influence maximization. The proposed approach is experimentally evaluated in Sect. 4 and the work is concluded in Sect. 5.

2   Influence Maximization Problem

Influence maximization problem (IMP) is a high–level problem that involves analysis of influence spreading through a (social) network and identification of the most important network nodes taking part in this process [9]. An IMP instance is defined by three parameters: a graph, $\mathcal{G} = (V, E, W)$, with a set of vertices V, a set of edges E, a set of edge weights W, a discrete–time information propagation model, $\mathcal{M}$, and the number of seed nodes, k [1]. The goal of the IMP is to select a set of seed nodes from which information spreads through the graph (network) most efficiently and propagates to the maximum of remaining network nodes.

Linear threshold and independent cascade are two most used progressive information propagation (diffusion) models [2,9]. Under the linear threshold model, each node, i, chooses a random threshold, $\theta_i \in [0,1]$, that corresponds to the weighted fraction of its neighbors that must be active for i to activate. The independent cascade model, on the other hand, takes a different approach. When a node, i, becomes active, it becomes a single chance to activate its currently inactive neighbours. The activation of an inactive neighbor, j, happens with probability p_{ij} and is often associated with the weight of the edge between i and j.

The IMP has many applications in a number of areas and especially in social network analysis [1,2,4–7,9,14,18,19,21]. It is known that an optimum IMP solution under linear threshold and independent cascade propagation models is NP–hard and a number of heuristic and metaheuristic approximate algorithms has been proposed to tackle it.

2.1   Recent Algorithms for the Influence Maximization Problem

An IMP algorithm designed for modular social networks was proposed in [2]. The problem was cast as an optimal resource allocation problem and a quasi-optimal dynamic programming method called OASNET (Optimal Allocation in Social NETwork) was designed. A scalable algorithm named SIMPATH, based on a "lazy–forward" (CELF) optimization, was introduced in [7]. In this approach, seeds are chosen iteratively in a lazy–forward manner and the IMP under the linear threshold model is solved by an alternative algorithm that evaluates information spreading by exploring simple paths in node neighborhood.

Two algorithms for influence estimation in continuous–time diffusion networks were presented in [4]. The first one, based on continuous–time Markov chains, was designed to estimate influence spreading and to identify a set of nodes with a guaranteed influence. It was, however, not scalable to large networks and another randomized method, able to find in networks with a general transition function sets of nodes with influence reduced by a constant factor only, was proposed.

Another work introduced a distributed influence propagation model with multiagent architecture [14]. It was used to discover influence propagation patterns in simulated evolving networks with seed sets detected by an evolution-based backward algorithm. A game–theoretic approach to the IMP was proposed in [10]. The IMP is in this approach cast as a game in which agents maximize the diffusion of their information. The Nash equilibrium of the game is sought in order to find an optimum seed set. An analysis of this approach showed that it is most affected by the social skills (i.e. connectedness) of the agent. Because of the computational complexity, an approximate algorithm was used for scenarios with larger seed sets.

A recent study [17] showed that the popular independent cascade model does not correspond to information diffusion patterns observed in real–world networks. An alternative diffusion model called three steps cascade was proposed and shown more reliable and robust than the independent cascade. It was found especially suitable for modelling information diffusion in online social networks. An adaptive method for seed set selection was proposed in [20]. It is based on another dynamic modification of the independent cascade model suitable for analysis of diffusion in dynamic networks with uncertainty. Finally, a novel probabilistic multi–hop information diffusion strategy was used in context of IMP as well [16].

Several metaheuristic and hybrid methods have been proposed for the IMP in the past. A high–performance algorithm based on a combination of genetic algorithms and a greedy method has been introduced in [21]. The greedy approach was used to refine the results obtained by the genetic search. An experimental evaluation showed that the introduction of local search improved the results obtained by the evolutionary method by approximately 10%.

A straightforward genetic algorithm for the IMP was introduced in [1]. It used only simple genetic operators and showed that such approach is well–comparable with traditional heuristics for the IMP. The work also concluded that the this approach is widely applicable and not limited to networks with special properties only. Swarm [6] and memetic [5] approaches have been proposed for the IMP recently as well. Clearly, nature–inspired methods and evolutionary metaheuristics in particular are at the present time intensively investigated in context of the IMP.

3   Genetic Algorithm for the Influence Maximization Problem

Genetic algorithms (GA) form a family of stochastic, population-based, metaheuristic optimization methods [8, 15]. They can be used to tackle hard optimization problems by an iterative evolution of a population of encoded candidate solutions. The candidate solutions are in each iteration of the algorithm (generation) evaluated by a problem specific objective (fitness) function and compete for survival in the population. This high–level procedure, implemented by the application of so–called genetic operators, yields an iterative problem–solving strategy inspired by the survival–of–the–fittest model of genetic evolution.

Problem encoding is an important part of the genetic search. It translates candidate solutions from the problem domain (phenotype) to the encoded search space (genotype) of the algorithm. It defines the internal representation of the problem instances used during the optimization process. The representation specifies the chromosome data structure and the decoding function [3]. The data structure, in turn, defines the actual size and shape of the search space.

Crossover (recombination), mutation, and selection are the most important genetic operators used by GAs. Crossover leads to creation of new (offspring) problem solutions from selected existing (parent) solutions and mimicks sexual reproduction. It is the main operator that distinguishes GAs from other population–based stochastic search methods [8, 15]. Its role in GAs has been thoroughly investigated and it has been identified as the primary creative force in the evolutionary search process. It propagates so called building blocks (solution patterns with above average fitness) from one generation to another and creates new, better performing, building blocks through their recombination. It can introduce large changes in the population with small disruption of these building blocks [22]. In contrast, mutation is expected to insert new material into the population by random perturbation of the chromosome structure and is essential for the introduction of new genetic material into the pool of candidate problem solutions.

3.1   Genetic Algorithms for Fixed–length Subset Selection

The application of GA to a particular problem involves definition of encoding, fitness function, and operator implementation. The IMP is a special case of fixed–length subset selection problem. Fixed–length subset selection is a combinatorial optimization problem that involves search for a subset of k distinct elements from a (larger) finite set of n elements. The subset is formed with respect to a specific criterion (property) and the process can be cast as an optimization problem.

Although many GAs for subset selection exist, a compact and computationally inexpensive representation of a fixed–length subsets is not straightforward. Nevertheless, an efficient GA for the class of fixed–length subset selection problems has been introduced recently [11,12]. It involved a suitable chromosome encoding and customized genetic operators. The encoding and genetic operators were designed with the aim to use compact chromosomes, to exploit both crossover and mutation, and to avoid the creation of invalid individuals during the artificial evolution [11,12]. Later, it was extended by a new, more universal, crossover operator and successfully applied to image classification [13].

A fixed-length subset of k elements can be defined by the indices of the elements selected from the full set of n elements. No element can appear more than once in order to achieve the required subset size k, A chromosome representing k elements, $\boldsymbol{c}$, can be defined by:

$$\boldsymbol{c} = (c_1, c_2, \dots c_k), \quad \forall (i,j) \in \{0, \dots n\} : c_i \neq c_j \tag{1}$$

Apparently, such encoding can spawn invalid individuals in case the traditional genetic operators such as the 1–point crossover or uniform mutation are applied. Equation (2) shows an example of two valid chromosomes forming an invalid offspring ($\boldsymbol{o_1}$) through 1–point mutation.

$$\begin{array}{ccc}
\boldsymbol{c} = (\ 9\ |\ 2\ 7\ ) & & \boldsymbol{o_1} = (\ 9\ |\ 8\ \ 9\ ) \\
\times & \implies & \\
\boldsymbol{d} = (\ 5\ |\ 8\ 9\ ) & & \boldsymbol{o_2} = (\ 5\ |\ 2\ \ 7\ )
\end{array} \tag{2}$$

To avoid this undesired behaviour, the encoding can be modified by sorting the indices within each chromosome [11]

$$\boldsymbol{c} = (c_1, c_2, \dots c_k), \forall (i,j) \in \{0, \dots n\} : i < j \implies c_i < c_j. \tag{3}$$

Then, the generation of random fixed–length subset individuals requires two steps. First, k unique elements out of all n possible elements are selected. Then, the indices in each individual are sorted in increasing order. On top of that, special mutation and crossover operators that respect the ordering of the indices within the chromosome are used. The operators are based on the traditional mutation and crossover operators that were modified so that they do not disrupt the ordering of the chromosomes and do not create invalid offspring [12,13].

The order–preserving mutation [11] replaces ith gene c_i in chromosome c with a value from the interval defined by its left and right neighbour as defined in:

$$mut(c_i) = \begin{cases} urand^*(0, c_{i+1}), & \text{if } i = 0 \\ urand(c_{i-1}, c_{i+1}), & \text{if } i \in (0, n-1) \\ urand(c_{i-1}, N), & \text{if } i = N - 1 \end{cases} \qquad (4)$$

where $i \in \{0, \dots, n\}$ and $urand(a, b)$ selects a uniform pseudo–random integer from the interval (a, b) (whereas $urand^*(a, b)$ selects a uniform pseudo–random integer from the interval $[a, b)$). The operator guarantees that the ordering of the indices within the chromosome remains valid. However, it has no effect for the ith gene in chromosomes for which it holds that $(c_{i-1} + 1) = c_i = (c_{i+1} - 1)$.

The order–preserving crossover [11] is based on the traditional one–point crossover operator [15]. It selects a random position, i, in parent chromosomes c and d and checks if the position can be used for crossover, i.e. whether (5)

$$c_i < d_{i+1} \wedge d_i < c_{i+1} \qquad (5)$$

holds. If Eq. (5) does not hold for i, the remaining positions in the chromosomes are sequentially scanned in the search for a suitable crossover point for which (5) holds. The order–preserving crossover closely imitates the traditional one–point recombination strategies known from many variants of genetic algorithms. However, it cannot be applied to all possible pairs of parent chromosomes and an alternative crossover strategy, employing the Mergesort algorithm, was proposed in [13]. A Mergesort–based recombination of two ordered chromosomes representing fixed–length subsets of k elements is performed in two steps. First, the Mergesort algorithm is employed to create from the parent chromosomes of length k, c, d, one temporary sorted array, t, of length $2k$. Then, offspring chromosomes, o_1 and o_2, are formed from t according to

$$o_1 = (t_0, t_2, \dots, t_{2k-1}), \qquad (6)$$
$$o_2 = (t_1, t_3, \dots, t_{2k}). \qquad (7)$$

This approach guarantees that valid offspring chromosomes are created from every possible combination of two valid parent chromosomes. Offspring chromosomes remain sorted and cannot contain duplicate column indices. The Mergesort–based crossover is illustrated in Eq. (8).

$$\begin{array}{ll} c = (\ 1\ 2\ 4\ 5\ 8\ ) & o_1 = (\ 0\ 1\ 3\ 5\ 8\ ) \\ \quad\times \quad \overset{merge}{\Longrightarrow} (\ 0\ 1\ 1\ 2\ 3\ 4\ 5\ 8\ 8\ 9\ ) \overset{split}{\Longrightarrow} & \\ d = (\ 0\ 1\ 3\ 8\ 9\ ) & o_2 = (\ 1\ 2\ 4\ 8\ 9\ ) \end{array}$$
$$(8)$$

3.2 Guided Genetic Algorithm for the Influence Maximization Problem

As a fixed–length subset selection problem, the IMP can be solved directly by the GA described in Sect. 3.1. This approach uses artificial evolution to search for a set of k seed nodes that would activate the maximum of remaining nodes in a network with fixed structure under a fixed activation (diffusion) model. However, the IMP is a network problem. The network structure, associated with each IMP instance, provides an additional information that can be used to guide the GA and improve its ability to find good IMP solutions. Various metrices can be computed for each network node and used to guide the GA towards nodes with higher information diffusion potential. For example, it is easy to see that a sink (i.e. a node no outgoing edges) plays in the information diffusion process a different role than a hub (i.e. a node with a large number of outgoing edges) and the knowledge of node outdegree can be therefore useful for the algorithm.

The *guided GA for IMP* (G^2A) takes advantage of the auxiliary information hidden in network structure and uses it whenever feasible. In G^2A, node outdegree it is used to generate initial population and as part of the mutation operator. In particular, the probability of a node, $i \in V$, being included in a randomly generated initial solution and being inserted to a chromosome by a mutation operator is proportional to its outdegree. Because the network structure is constant, the probabilities associated with each node can be computed beforehand and the computational complexity of the algorithm increases only marginally.

The fitness function used by both GAs for IMP is based on the independent cascade diffusion model. Because the diffusion process is stochastic, it was executed as a series of independent Monte Carlo iterations (information propagations) and the average number of activated nodes was used as a fitness value. In the remainder of this work, the ability of GA and G^2A to find IMP solutions is evaluated experimentally.

4   Experiments

In order to assess the ability of GA and G^2A to find IMP solutions, a set of weighted *test networks* with 50, 100, 150, and 200 nodes was generated randomly. They are labeled G50, G100, G150, and G200, respectively. Examples of the test networks are shown in Fig. 1.

GA and G^2A were used to find in each test graph seed sets of 10, 20, 30, and 40 seed nodes, respectively. Both algorithms were executed with the same fixed parameters, based on past experience and initial trial–and–error runs. They implemented a steady–state genetic algorithm with population gap 2 [15] and evolved a population of 100 candidate solutions. The crossover probability was set to 0.8 and mutation probability to 0.02. The maximum number of generations was set to 10,000 and the information propagation model was independent cascade with 1,000 Monte Carlo iterations. The average number of nodes activated by selected seed sets was used as a fitness function. However, because of the stochastic nature of genetic algorithms, each experiment was executed 31 times independently and average results are reported in Table 1.

The table displays for each experiment (i.e. a combination of data set and seed set size) minimum, mean, and maximum average number of nodes activated by

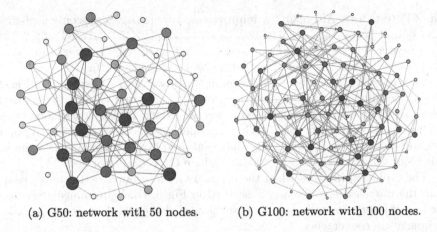

(a) G50: network with 50 nodes.   (b) G100: network with 100 nodes.

Fig. 1. Examples of test networks. Node size and color is proportional to its outdegree, edge thickness and color is proportional to its weight. The lighter the lower. Note that the sizes and colors are relative to other nodes within the same network (i.e. subfigure).

Table 1. Average number of activated nodes (final coverage).

dataset	k	Original GA			G^2A		
		min	mean (σ)	max	min	mean (σ)	max
G50	10	24.782	25.189 (0.407)	25.5	25.590	**25.683** (0.092)	**25.7**
	20	35.843	36.193 (0.350)	36.5	36.284	**36.445** (0.161)	**36.6**
	30	43.380	**43.435** (0.055)	**43.4**	42.462	42.521 (0.059)	42.5
	40	47.224	**47.421** (0.198)	**47.6**	45.336	45.342 (0.007)	45.3
G100	10	30.981	31.444 (0.463)	**31.9**	31.677	**31.783** (0.106)	31.8
	20	47.953	48.296 (0.343)	48.6	48.324	**49.155** (0.831)	**49.9**
	30	61.063	61.115 (0.052)	61.1	61.495	**61.640** (0.145)	**61.7**
	40	69.910	70.154 (0.243)	70.3	69.862	**70.330** (0.468)	**70.7**
G150	10	39.897	40.038 (0.141)	40.1	39.683	**40.169** (0.486)	**40.6**
	20	60.762	60.970 (0.208)	61.1	62.005	**62.054** (0.049)	**62.1**
	30	75.362	76.012 (0.650)	76.6	76.629	**76.940** (0.310)	**77.2**
	40	86.346	87.381 (1.035)	88.4	86.865	**87.989** (1.124)	**89.1**
G200	10	38.304	40.516 (2.212)	42.7	42.402	**42.988** (0.586)	**43.5**
	20	65.104	65.665 (0.561)	66.2	64.271	**66.162** (1.891)	**68.0**
	30	82.861	84.508 (1.647)	**86.1**	85.547	**85.809** (0.262)	86.0
	40	95.369	96.127 (0.757)	96.8	96.285	**97.432** (1.147)	**98.5**

seed sets obtained by GA and G^2A (the higher the better). The better mean and maximum results are typed in bold face. The results show that the proposed approach delivers average IMP solutions better than plain GA for fixed–length subset selection in 10 out of 12 test cases and better best (maximum) IMP solutions in 8 out of 12 test cases. Although the improvements are in some cases only negligible (for example, an average of 0.5 more activated nodes for G50 and seed set size 10), G^2A discovered in some other cases seed sets that activated on average more than 2 extra nodes (as in the experiment with G200 and seed set 10).

The course of the artificial evolution in these two cases is displayed in Fig. 2. The figures illustrate that G^2A is superior to the original GA for fixed–length subset selection during the whole evolution procedure. It also clearly shows that the initialization strategy used by G^2A provides the algorithm with initial population of significantly better quality than that used by the original GA. A similar behaviour was observed in all test cases where G^2A outperformed GA.

The evolution in cases when the original GA has found better mean result than the proposed algorithm is illustrated in Fig. 3. One can immediately note the lack of improvement in Fig. 3b. It suggests that the algorithm might be prone to premature convergence.

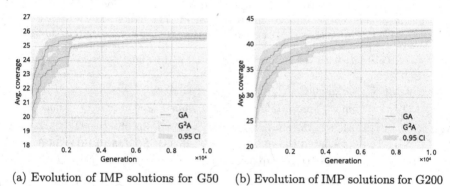

(a) Evolution of IMP solutions for G50 with seed set of 10 nodes   (b) Evolution of IMP solutions for G200 with seed set of 10 nodes

Fig. 2. Examples of IMP solution evolution (cases when G^2A outperforms GA). The displayed range corresponds to a 95% confidence interval for average fitness values.

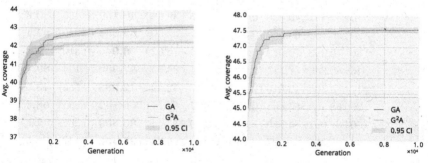

(a) Evolution of IMP solutions for G50 with seed set of 30 nodes   (b) Evolution of IMP solutions for G100 with seed set of 10 nodes

Fig. 3. Examples of IMP solution evolution (cases when GA outperforms G^2A). The displayed range corresponds to a 95% confidence interval for average fitness values.

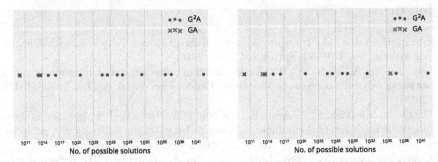

(a) Algorithm delivering better mean results as a function of problem space size.

(b) Algorithm delivering better best results as a function of problem space size.

Fig. 4. Results of experiments in relation to search space size. Note the logarithmic scale of the x–axes.

Finally, the relation between the search space size (i.e. problem complexity) and obtained results is illustrated in Fig. 4. The figures clearly show that the proposed G^2A algorithm outperforms the original GA especially for larger search spaces. It also illustrates that the fact that the original GA has found better best solution for G200 with sees set size 30 is an exception rather than a trend (note that the G^2A has fund better mean result also for this case).

5   Conclusions

A new genetic algorithm for the influence maximization problem was proposed and experimentally evaluated in this work. The proposed approach is an extension of a previous genetic algorithm for fixed–length subset selection that can use information from the underlying network structure, associated with an IMP problem instance, to guide the metaheuristic algorithm towards potentially better problem solutions.

The algorithm was experimentally evaluated on a series of randomly generated test networks and obtained results confirm that it outperforms the original genetic method in the majority of cases. The analysis of experimental results shows that it is especially useful when the search space is large. However, it also shows that it might be prone to premature convergence. Future work on this topic will include developement of measures to mitigate this problem, evaluation of the usefulness of different network metrices as auxiliary information, and an evaluation of the proposed approach on real–world IMP instances from the areas of social networks and others.

Acknowledgement. This work was supported by the Czech Science Foundation under the grant no. GA15-06700S, and by the projects SP2017/100 and SP2017/85 of the Student Grant System, VŠB-Technical University of Ostrava.

References

1. Bucur, D., Iacca, G.: Influence Maximization in Social Networks with Genetic Algorithms, pp. 379–392. Springer, Cham (2016)
2. Cao, T., Wu, X., Wang, S., Hu, X.: Maximizing influence spread in modular social networks by optimal resource allocation. Expert Syst. Appl. **38**(10), 13128–13135 (2011)
3. Czarn, A., MacNish, C., Vijayan, K., Turlach, B.A.: Statistical exploratory analysis of genetic algorithms: the influence of gray codes upon the difficulty of a problem. In: Australian Conference on Artificial Intelligence, pp. 1246–1252 (2004)
4. Gomez-Rodriguez, M., Song, L., Du, N., Zha, H., Schölkopf, B.: Influence estimation and maximization in continuous-time diffusion networks. ACM Trans. Inf. Syst. **34**(2), 1–33 (2016)
5. Gong, M., Song, C., Duan, C., Ma, L., Shen, B.: An efficient memetic algorithm for influence maximization in social networks. IEEE Comput. Intell. Mag. **11**(3), 22–33 (2016)
6. Gong, M., Yan, J., Shen, B., Ma, L., Cai, Q.: Influence maximization in social networks based on discrete particle swarm optimization. Inf. Sci. **367–368**, 600–614 (2016)
7. Goyal, A., Lu, W., Lakshmanan, L.V.S.: Simpath: an efficient algorithm for influence maximization under the linear threshold model. In: 2011 IEEE 11th International Conference on Data Mining, pp. 211–220, December 2011
8. Holland, J.H.: Adaptation in Natural and Artificial Systems. University of Michigan Press, Ann Arbor (1975)
9. Kempe, D., Kleinberg, J., Tardos, E.: Maximizing the spread of influence through a social network. In: Proceedings of the Ninth ACM SIGKDD International Conference on Knowledge Discovery and Data Mining, pp. 137–146, KDD 2003. ACM, New York (2003)
10. Kermani, M.A.M.A., Ardestani, S.F.F., Aliahmadi, A., Barzinpour, F.: A novel game theoretic approach for modeling competitive information diffusion in social networks with heterogeneous nodes. Physica A: statistical mechanics and its applications **466**, 570–582 (2017)
11. Krömer, P., Platos, J.: Genetic algorithm for sampling from scale-free data and networks. In: Genetic and Evolutionary Computation Conference, GECCO 2014, Vancouver, BC, Canada, 12–16 July 2014, pp. 793–800 (2014)
12. Krömer, P., Platos, J.: Evolutionary feature subset selection with compression-based entropy estimation. In: Proceedings of the 2016 on Genetic and Evolutionary Computation Conference, Denver, CO, USA, 20–24 July 2016, pp. 933–940 (2016)
13. Krömer, P., Platoš, J., Nowaková, J., Snášel, V.: Optimal column subset selection for image classification by genetic algorithms. Ann. Oper. Res. 1–18 (2016)
14. Li, W., Bai, Q., Zhang, M.: Agent-based influence propagation in social networks. In: 2016 IEEE International Conference on Agents (ICA), pp. 51–56, September 2016
15. Mitchell, M.: An Introduction to Genetic Algorithms. MIT Press, Cambridge (1996)
16. Nguyen, D.L., Nguyen, T.H., Do, T.H., Yoo, M.: Probability-based multi-hop diffusion method for influence maximization in social networks. Wirel. Pers. Commun. **93**(4), 1–14 (2017)

17. Qin, Y., Ma, J., Gao, S.: Efficient influence maximization under tscm: a suitable diffusion model in online social networks. Soft Comput. **21**(4), 827–838 (2017)
18. Samadi, M., Nikolaev, A., Nagi, R.: The temporal aspects of the evidence-based influence maximization on social networks. Optim. Methods Softw. **32**(2), 290–311 (2017)
19. Shi, Q., Wang, H., Li, D., Shi, X., Ye, C., Gao, H.: Maximal Influence Spread for Social Network Based on MapReduce. In: Wang, H., et al. (eds.) ICYCSEE 2015. Communications in Computer and Information Science, vol. 503, pp. 128–136. Springer, Berlin, Heidelberg (2015)
20. Tong, G., Wu, W., Tang, S., Du, D.Z.: Adaptive influence maximization in dynamic social networks. IEEE/ACM Trans. Netw. **25**(1), 112–125 (2017)
21. Tsai, C.W., Yang, Y.C., Chiang, M.C.: A genetic newgreedy algorithm for influence maximization in social network. In: 2015 IEEE International Conference on Systems, Man, and Cybernetics, pp. 2549–2554, October 2015
22. Wu, A.S., Lindsay, R.K., Riolo, R.: Empirical observations on the roles of crossover and mutation. In: Bäck, T. (ed.) Proceedings of the Seventh International Conference on Genetic Algorithms, pp. 362–369. Morgan Kaufmann, San Francisco (1997)

Optimal Local Routing Strategies for Community Structured Time Varying Communication Networks

Suchi Kumari[1], Anurag Singh[1(✉)], and Hocine Cherifi[2]

[1] Department of Computer Science and Engineering,
National Institute of Technology Delhi, New Delhi 110040, India
{suchisingh,anuragsg}@nitdelhi.ac.in
[2] University of Burgundy, LE21 UMR CNRS 6306, Dijon, France
hocine.cherifi@gmail.com

Abstract. In time varying data communication networks (TVCN), traffic congestion, system utility maximization and network performance enhancement are the prominent issues. All these issues can be resolved either by optimizing the network structure or by selecting efficient routing approaches. In this paper, we focus on the design of a time varying network model and propose an algorithm to find efficient user route in this network. Centrality plays a very important role in finding congestion free routes. Indeed, the more a node is central, the more it can be congested by the flow coming from or going to its neighborhood. For that reason, classically, routes are chosen such that the sum of centrality of the nodes coming in user's route is minimum. In this paper, we show that closeness centrality outperforms betweenness centrality in the case of community structured time varying networks. Furthermore, Kelly's optimization formulation for a rate allocation problem is used in order to compute optimal rates of distinct users at different time instants.

Keywords: Data communication networks model · System utility · Community structure · Closeness and betweenness centrality

1 Introduction

Well structured network topology and efficient routing can lead to a congestion free system together with an improved network capacity. The aim of this paper is to develop a network model by considering all the aspects of growth (network size) and alteration (rewiring & removal of link) of the network. In the designed network, nodes establish dedicated connections to the nodes they want to communicate with. These connections are formed such that they are able to manage traffic efficiently and minimize congestion in the network. In the communication networks, communities are formed and these structures can be exploited in order to send data more efficiently. As the network topology is evolving trough time, it is important to study the strategies used for routing data between user's source and

Y. Cao and J. Chen (Eds.): COCOON 2017, LNCS 10392, pp. 642–653, 2017.
DOI: 10.1007/978-3-319-62389-4_53

destination pair. In this work we consider approaches based on the importance of the nodes. The importance of the node is related to various centrality measures: (a) calculate the betweenness centrality of nodes using shortest path. (b) calculate the closeness centrality of the nodes. It is of prime interest to compare both routing strategies. Indeed, if the network is least congested then users can send a maximum amount of data to the destination node through the established route and can maximally utilize the system within their decided budget.

Real-world networks, such as social networks, communication networks, citation networks, traffic networks etc. [15] are growing with time, hence there is a need to include time with the nodes and links of the basic network structure. $G(N, E, t)$. Many models have been proposed for time-varying networks (TVNs) [3,18]. Wehmuth *et al.* [18] have proposed a unifying model for representing TVNs. TVN nodes are considered as independent entities at each time and it is analyzed that the proposed TVN is isomorphic to a static graph. A framework to represent mobile networks dynamically in a spatio-temporal fashion is designed and algebraic structural representation is also provided [10]. Most of the real world networks are scale free and follow a power law degree distribution with power law exponent α in the range of $(2, 3]$. In the Barabasi Albert (BA) model [1], new links are added when new nodes are added into the system and this links preferentially attach the new nodes to high degree nodes. Many models have been developed as a variation of the BA model. Tadic [17] has proposed a model considering both outflowing and inflowing links and has shown correlation between them. She has considered network expansion as well as update within the network. Apart from addition and update of links, removal of infrequently used links is considered in [9]. Addition of new links is based on preferential attachment, while preference for update is given to most active nodes at time t.

Some researchers have tackled the issue of improving the capacity of the network by restructuring the network according to traffic and congestion in the network. Zhao *et al.* [19] have redistributed the load of heavily loaded nodes to the other nodes in the network. Once a network is designed, then a connection is established between source and destination. In different networks connections are established for different purposes. In communication networks, data packets are sent between source and destination, goods are supplied from vendor to customer in transportation networks, signals are passing in electric network. In all these networks, objects are sent such that routing cost as well as congestion is minimized. In communication networks, two rates are associated with each node: packet generation rate λ and packet forwarding rate μ. There is a critical rate of packet generation (λ_c), below which network is free from congestion. Two models are proposed by Zhao *et al.* [19] for finding packet delivery rate of a node, (i) based on node's links and (ii) number of the shortest path passing through the node. Du *et al.* [4] have developed a model in which packets propagate to minimize the load and are processed by the Shortest Remaining Packet First (SRPF) strategy. A heuristic routing strategy is used by Jiang *et al.* [7], in which routing process is divided into N (size of network) steps and routes are based on the betweenness centrality and node degree information. All the proposed strategies are used to enhance the capacity of the system.

In all the previously discussed approaches, global information is needed for addition, update and removal of links. It is not easy to get this global information about the network as it is evolving and its size is usually very large. Therefore local information can play an important role. Various local routing strategies have been proposed by Lin *et al.* [13]. It includes node duplication avoidance, searching next nearest neighbors and restricted queue length. In Internet network, a router searches for the connection of shortest distance when establishing a new connection. Thus, the routers within the same region will have more connections than the others. Hence, to capture the localization property of these kind of networks, a local world model is proposed [20]. This local inhomogeneity leads to the formation of communities in the networks. Community structure is an important characteristic of networks structure; it is observed in many real world network such as social, biological, computer network, Internet etc. A group of nodes which have common properties or play similar role within the network form communities. Even if there is no clear definition of communities, it is generally admitted that a network is community structured, if group of nodes are densely connected within the community, and sparsely connected between the communities [5]. To understand the strength and weakness of all the available community detection methods, Fortunato *et al.* [6] have surveyed the most influential detection algorithms. In community structured network, dedicated routing approaches may be applied for sending user's data. As the size of the communication networks are increasing day by day, so one of the challenging task for service providers is to allocate the resource bandwidth among the users fairly and acquiring global information about the network topology. As Internet is a public domain, so pricing mechanism is also introduced to control the action of users and to provide quality of service among users.

Kelly [8] has considered rate allocation problem as an assignment of optimal rate to each user for maximizing individual user's utility as well as system utility. Users are assumed to generate elastic traffic and they control their rates based on the feedback provided by the network. For large network, it is quite difficult to solve the system utility, hence he divided the whole problem in two subproblem: network and user's optimization problem. The network tends to optimize its cost whereas the user wants to maximize its rate based on system utility. La *et al.* [11] have introduced a suitable pricing scheme for each user where users update their target queue size when certain constraints are fulfilled. Mo *et al.* have proposed an algorithm in which computation of the optimal target queue sizes are not needed, but the users update their window size implicitly. The concept of self updating of window size is taken by the paper based on fluid model of network and is developed by Mo *et al.* [14].

This paper addresses the problem to discover suitable route in time varying networks with or without congestion. For routing, centrality of nodes is considered as its plays an important role in network congestion. In community structured network, we rely on closeness and betweenness centrality of each node individually within each community. Closeness centrality of a node u is the reciprocal of sum of the shortest path distances from all other nodes $N_\Omega(t) - 1$ to u

within the community Ω and it is normalized by the sum of all possible short distance i.e.,

$$C(u) = \frac{N_\Omega(t) - 1}{\sum_{v \neq u \in \Omega} d(v, u)}$$

where $d(v, u)$ is the shortest path distance between v and u and $N_\Omega(t)$ is the total number of nodes within the community at time t.

Similarly, Betweenness centrality of a node v is equal to the number of shortest paths from all node pairs of the community pass through that node v and is given by

$$g(v) = \sum_{s \neq v \neq d \in \Omega} \frac{\sigma_{s \to d}(v)}{\sigma_{s \to d}}$$

where s, d and v lies within the community Ω. In the proposed approach, the route of the user is chosen such that the sum of closeness centrality of the nodes coming in user's shortest route should be minimum to avoid congestion. If a node is closer to more number of nodes then the probability of appearance of the node in the user's routes increases and hence chances of congestion at that node increases accordingly. Our aim is to check availability of alternate routes to minimize congestion in the system. User's optimal data rate of the chosen route is calculated by using Kelly's rate control algorithm [8].

Section 2 describes the mathematical model used in the analysis of rate control behavior of the user's route. Section 3 introduces model for growth dynamics of TVCN. Section 4 presents the proposed routing strategies for communication network. Section 5 reports simulation results, and in Sect. 6, conclusion and future directions of the work are discussed.

2 Mathematical Model for Rate Allocation Problem

Network structure is changing with time as nodes and links are added and rewired into the system. Let the network consists of N nodes and E links and life expectancy of the network is considered as T. A set of R users are willing to access and send data to the desired location through the network. A link e_{ab} establishes a connection between node a and node b and can send maximum $C_{e_{ab}}$ units of data through it, where, $C_{e_{ab}}$ is the capacity of the link e_{ab} and $e_{ab} \in E$. Each user r is assigned a route $r \in R$ in the duration of time $t_{(i-1)} \in T$ to $t_i \in T$. At the end of t_i^{th} time, a zero-one matrix A of the size $(N \times N)_{t_i}$ is defined where, $A_{n_i, n_j}(t_i) = 1$, if node n_i and n_j are connected in the duration of time $t_{(i-1)}$ to t_i, otherwise $A_{n_i, n_j}(t_i) = 0$. System utility for each user $r \in R$ is dependent on the data rate $x_r(t_i)$ of the user r and is denoted by $U_r(x_r(t_i))$. It is a increasing, strictly concave function of $x_r(t_i)$ over the range $x_r(t_i) \geq 0$. Each user calculate its utility by using a suitable utility function. Once utility of each user r is calculated then aggregate utility is calculated by summing all the utilities of users and is denoted as $\sum_{r \in R} U_r(x_r(t_i))$. The rate allocation problem can be formulated as the following optimization problem.

$$SYSTEM(U(t_i), A(t_i))$$
$$maximize \sum_{r \in R, t_i} U_r(t_i)(x_r(t_i)) \tag{1}$$
$$A^T(t_i)x(t_i) \leq C(t_i) \text{ and } x(t_i) \geq 0$$

$A(t_i)$ is a matrix at time interval t_{i-1} to t_i. The given constraint states that a link can not send data more than its capacity and the value of the flow, $x(t_i)$, should be non-negative [8]. As the system does not know the utility function of each user so it is quite difficult and unmanageable for the system to find optimal rate of distinct users in the network. Hence, Kelly [8] has divided this problem into two simpler problems named as user's optimal problem and network's optimal problem [8].

At each time instant if a user wants to access a link $e_{ab} \in E$ then, the cost will depend on the total data flow through the link at that time. It is given by $\psi_{e_{ab}}(t_i) = \varsigma_{e_{ab}}(\sum_{r:e_{ab} \in E} x_r(t_i))$ where, $\varsigma_{e_{ab}}(\bullet)$ is an increasing function of the data rate. The optimal data rate (x^*) can be obtained by considering the following system of differential equation

$$\frac{dx_r(t_i)}{dt_i} = \vartheta_r(\mathcal{P}_r(t_i) - x_r(t_i) \sum_{e_{ab} \in r} \psi_{e_{ab}}(t_i)) \tag{2}$$

Here, ϑ_r is a proportionality constant. Equation (2) consists of two components: a steady increase in the rate proportional to $\mathcal{P}_r(t_i)$ and a steady decrease in the rate proportional to the response $\psi_{e_{ab}}(t_i)$ provided by the network.

3   Proposed Time Varying Network Model

Time varying data communication networks are designed by maintaining specific set of rules where nodes are connected via directed links. Directed links may belong to two categories: links which come out from a node (out-flowing) and links that are incidental to a node (in-flowing). As the network is scale free hence, the degree distributions of both type of links follow a power law with distinct power law exponents. In the proposed model, a node prefer to attach with the node which is endorsed by the maximum number of existing nodes in the network and hence, the preferences are assigned to in-flowing links [1,17]. A communication network model is proposed based on the BA model [1]; the concept of rewiring of links is taken from the model proposed in [17]. Apart from the concept of rearrangement of links, we have considered the removal of inefficient links from the network. At each time instant $t_i \in T$, a new node $n(t_i)$ is added to the network (expansion) and a number $X(\leq n_0)$ is selected for network growth and link rearrangement and removal, where n_0 is the initial number of nodes present in the seed network. Links are divided into three categories: newly added, rewired and removed links. Distribution of links is done using the given set of rules [9].

Notations:

(a) β = fraction of links out-flowing from new incoming node $n(t_i)$ at time t_i, $0 < \beta < 1$.

(b) γ = fraction of links, those are rearranged in the existing network, $0.5 < \gamma \leq 1$.

Using the above notations, the following set of rules may be defined:

(i) Total number of newly appearing out-flowing links from the new appearing node, $n(t_i)$

$$f_{add}(t_i) = \beta X$$

(ii) Total rewired links within existing network,

$$f_{rewire}(t_i) = \gamma(X - f_{add}(t_i)) = \gamma(1 - \beta)X$$

(ii) Remaining part of X is used for deleting the most infrequently used links.

$$f_{delete}(t_i) = X - f_{add}(t_i) - f_{rewire}(t_i) = (1 - \gamma)(1 - \beta)X$$

While studying the behavior of communication networks, the concept of preferred linking to a node is driven by the demand of the node to outflow the data into the network. Only few influential nodes have the right to rewire their links. Nodes rewire their infrequently used links and connect it to preferred nodes for data communication. The preference is given to the infrequently used links for the removal of links. Hence, the links attached to a node having lower degree will be preferred for removal. Once the network is formed, we use the community detection algorithm proposed in [2] in order to uncover its community structure. This community detection method is based on modularity optimization and it performs well in terms of computation time and quality. The method is divided in two parts: in first part, a different community is assigned to each node and in second part, new network is built based on the results obtained during the first part and both parts are called recursively. In the following we refer to it as the community structured network. Further routing strategies for each user is formed such that network congestion is minimized.

4   Routing Strategies Using Closeness Centrality

Here, the capacity of a node is calculated by the in-degree of the node and over-all capacity (C) is calculated by summing up the capacity of all the individual nodes in the communication network. The load, l_n at any node n, is calculated by summing up the packet generation rate of the in-flowing link at that node. Aggregate values of the packet generation rate of all the nodes are termed as the load of the network. When, load (L) exceeds the capacity (C) of the network then, it becomes congested with the overloaded packets. In large networks, multiple path exists between the pair of nodes. We want to find efficient short path among them. Multiple users try to access the network and want to send data from source s to destination d. As the network is community structured, user's source and destination may lie in different communities. The betweenness centrality and closeness centrality of each node are calculated within each community. If a node serves as the only way to reach some other nodes then that node will have the highest betweenness centrality. It becomes an articulation point and removing it would disconnect the network. If most of the shortest paths pass through a node, then its betweenness centrality is high and sending data through it will lead to the congestion in the network. When a node has the shortest paths to all other nodes then closeness centrality of the node has the highest value and it becomes a key node to send data efficiently across the network; indeed, it will also be congested. Therefore, it is important to investigate a shortest path $\sigma(s \rightarrow d)$ between the user pairs such that the overall value of closeness centralities of the nodes appear in the path.

$$\theta_C[\sigma_z(s \rightarrow d)] = \sum_{v:v \in \sigma_z(s \rightarrow d)} C_v$$

here, $C_v =$ closeness of node v

should be minimum. It is defined by,

$$min \ \{\forall z : \theta_C(\sigma_z(s \rightarrow d))\}$$

A similar formulation can be given for finding minimum sum of betweenness centrality θ_g of user paths. Capacity of a link $C_{e_{ab}}$ is dependent on the degrees k_a and k_b of nodes a and b. It can be approximated by a power-law dependence [16].

$$C_{e_{ab}} = b(k_a k_b)^\alpha$$

$$\text{here, } b = \frac{1}{K} \text{ and } K = \frac{\sum_{n \in N} deg^-(n)}{|N|}, \text{ average degree of the network}$$

$$\alpha = 1$$

$$(3)$$

α is the degree influence exponent which depends on the type of networks and b is a positive quantity. Three possible relations between capacity $C_{e_{ab}}$ and exponent α: (i) if $\alpha > 0$ the data is transmitted through high degree nodes (ii) if $\alpha < 0$ the data will be sent through low degree nodes and (iii) $\alpha = 0$ shows degree independent transmission. In the proposed model, we have taken a positive value of α. Algorithmic steps for expansion, rewiring and removal of links in the TVCN is given in [9]. In the time varying network, various users want to communicate data from a source s to destination d. User gives information about the source and destination, and accordingly $s \rightarrow d$ pairs are generated. Increments in the number of users lead to the network being into a congested state. Therefore, our aim is to find efficient routing paths such that a maximum number of users are getting benefited with a unique stable value of the optimal data sending rate x_r^* and corresponding convergence vectors will be $x^* = x_r^*$, $r \in R$ using Eq. (2).

After finding paths of all the users, optimal data sending rate of each user is calculated by using rate allocation Eq. (2). Detailed description for selecting shortest path with lowest closeness centrality along with optimal rate is given in Algorithm 1.

Algorithm 1. Finding shortest path having lowest (highest) closeness centrality and optimal rate for each user

---

1: **Input:** All user's $s - d$ pairs $\mathcal{N}_{sd}$, a and b such that $a > 0$ & $0 < b < 1$.
2: **for** $i := 1$ to $length(\mathcal{N}_{sd})$ **do**
3:     Evaluate all shortest paths χ_i of user i.
4:     Select shortest paths $\chi_i(s \rightarrow d)$ having maximum and minimum value of $\theta_C[\sigma_m(\chi_i(s \rightarrow d))]$.
5: **end for**
6: **for** $i := 1$ to $length(\mathcal{N}_{sd})$ **do**
7:     **for** $k := 1$ to $length(\mathcal{N}_{sd}(i))$ **do**
8:         Update network feedback ψ_k for each element k.
9:     **end for**
10:     $x_r = min(x_{\mathcal{N}_{sd}(i)})$;
11:     $A(r) = rand(1, 10)$;
12:     $\mathcal{P}_r = x_r * (\frac{a}{x_r + b})$;
13:     $\psi_r = \psi_d\{\forall d : d \in \mathcal{N}_{sd}(i)\}$;
14: **end for**
15: Use the value of x_r, $A(r)$, $\mathcal{P}_r$ and ψ_r to find the rate of convergence of each user.

---

5   Simulation and Results

The simulation is started by establishing the structure of the network according to the TVCN model proposed in the Sect. 3. It is aimed for representing real-world networks, hence, the degree distribution is found power law. In the simulation, the parameters are set to be seed node $n_0 = 5$, number $X = 5$, fraction of newly added links β range in $(0, 1)$, fraction of rewired links γ is in the range of $(0.5, 1)$, with network size ranging from $N = 10^3$ to 10^4. Any pair of nodes in the network may be considered as user's $s - d$ sets or may participate in routing also. Data forwarding capacity of a node n, C_n is defined by in-flowing degree $deg^-(n)$ of the node n. Capacity of a link $C_{e_{mn}}$ is obtained by Eq. (3). At each step, the degree of the nodes will be different, hence capacity of the nodes as well as links change accordingly.

In the communication networks, multiple paths are available among desired source destination pairs for sending packets for each user. All routes are assumed as equally weighted hence, users can select any of the available routes for sending the packets. Each user can send data along one of the shortest paths to the destination with a maximum flow rate of individual links. The data sending rate is reduced as multiple users want to share the common resources. Furthermore, user may not send more data with maximum rate. User's rate depends on two parameters; it's own willingness to pay $\mathcal{P}_r(t_i)$ and network's feedback $\psi_r(t_i) = \sum_{e \in r} \psi_e(t_i)$. Using rate control given in Eq. (2), an optimal data sending rate of each user is obtained and finally the system utility is calculated by using Eq. (1). User rates depend on the demand of particular resources coming in the shortest route. If demand is high then, data sending rate will be low.

In Table 1, the average value of optimal user's rate for highest and lowest value of θ_g and θ_C are shown for both community and non-community structured network. Betweenness centrality and closeness centrality are considered. Community structured network, efficiently analyze the traffic condition in the network and helps user to get more value of optimal rates for both highly congested and least congested shortest paths. Average optimal rate obtained through the path by considering betweenness centrality is lower for both type of networks. The betweenness centrality approach in non community structured network gives the lowest value.

Table 1. Average value of user's highest and lowest rate in community and non community structured networks using betweenness and closeness centrality measure when $N = 10^3$.

	Avg. user rate lowest	Avg. user rate highest
Betweenness with community	2.4537	3.2563
Betweenness without community	2.3438	2.8541
Closeness with community	2.5043	3.3241
Closeness without community	2.3992	3.2861

Total $\frac{1}{10}$ fraction of total nodes are considered as total number of user pairs. In the community and non community structured networks, firstly we get user's source destination $(s - d)$ information and then all the shortest paths between $(s - d)$ pair is calculated. Among all the available shortest routes, highest and lowest value of $\theta_C[\sigma(s \rightarrow d)]$ and $\theta_g[\sigma_z(s \rightarrow d)]$ are calculated and then optimal data rate (highest and lowest) of each user is calculated. Simulation is performed in python 2 with canopy editor and configuration of the system is given as Intel® Core ^{TM} i5-5200U CPU @ 2.20 GHz * 4 with memory size 7.7 GiB. Overall computation time for getting optimal rate of all users with varying network size is shown in Table 2. Betweenness and closeness centrality measures in non-community structured network takes 3.81 and 0.4734 times more time than the network with community structure respectively. While comparing the computation time through betweenness centrality and closeness centrality measures in both kind of networks, betweenness centrality takes 0.0245 and 2.1760 times more time than closeness centrality in community and non-community structured networks respectively. In community structured network, we study betweenness and closeness centrality of nodes within community. Size of the community will always far less than the size of the network N and we know that time complexity of both the centralities depend on size of the number of nodes in the network. From the results of both Tables 1 and 2, it is shown that performance of betweenness centrality measure in non-community structured networks, is not efficient.

Table 2. Time taken by betweenness and closeness centrality measures in community and non community structured networks for network size ranging from 10^3 to 10^4.

Network size	Betweenness centrality		Closeness centrality	
	Within community	Without community	Within community	Without community
	Time (in Seconds)			
1000	3.240	11.412	3.092	4.462
2000	13.822	44.712	13.097	19.365
3000	33.845	101.519	32.275	49.035
4000	58.997	192.232	55.242	85.452
5000	95.512	411.987	95.216	137.906
6000	163.573	702.687	164.549	215.222
7000	178.294	992.269	170.104	316.451
8000	233.572	1055.539	235.651	364.511
9000	306.327	1490.951	290.613	443.249
10000	400.853	1917.571	392.571	543.462

Size of the network N is 10^3 and total $\frac{1}{10}$ of N i.e., 10^2 users want to access the networks. Here, 49 users' source destination pairs are taken and optimal rates of each user in community and non-community structured networks are shown in Fig. 1. Optimal data sending rates of the 49 users with highest and lowest closeness centrality are calculated and the result is shown in Fig. 1(a). User's optimal data rate is sorted according to the highest closeness value within the community and the value obtained through the path with lowest closeness within community and without community structured networks are plotted accordingly. Apart from closeness centrality, there is one more measure named as betweenness centrality, effects optimal data rate x^* of each user. Results of optimal data rates (x^*) of 49 users with aggregate sum of most and least betweenness central nodes in community and non-community structured networks are displayed in Fig. 1(b). User's optimal data rate is sorted according to the highest betweenness centrality value within the community and it is shown that the deviation of other optimal date rate values of that user is more. Data sending rate of the user path having lowest or highest value for the aggregate sum of betweenness and lowest or highest total sum of closeness, is always maximum or minimum.

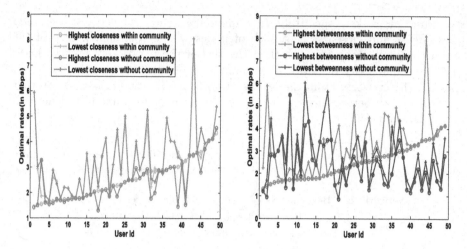

Fig. 1. User's optimal rates of user paths with (**a**) highest & lowest closeness and (**b**) betweenness (highest & lowest) in community and non-community structured networks

6   Conclusion and Future Directions

A time varying communication networks (TVCN) model is designed by applying variation on BA model [1]. We consider the case where nodes are added continuously and links may appear, disappear and rewired randomly with time by using some predefined probability condition. Using a community detection algorithm, we partition the proposed TVCN into communities and then find betweenness and closeness centrality of each node within community. For dynamic networks, closeness centrality measure performs better in terms of time complexity [12]. Time complexity for finding user optimal path on the basis of betweenness centrality measure for the case of non community structured networks is much higher than the closeness centrality approach in all kind of networks (community and non community structured). Computation time for both kind of measures are studied at different time instants with increased network size N. Interval of the two time instants (Δt) is taken as $\Delta t = 10^3$. For each user's path optimal data sending rate is obtained by using rate control equation proposed by Kelly [8]. Shortest path with lowest and highest sum of closeness centrality are chosen for routing and observed that the value of optimal rate will always be maximum for the path containing least closeness central nodes.

In this paper, system utility may be calculated with consideration of all kind of delays in the network. Utility function, itself can be replaced by using some other Lyapunov function. Some window size update method can also be studied.

References

1. Barabási, A.L., Albert, R., Jeong, H.: Mean-field theory for scale-free random networks. Phys. A **272**(1), 173–187 (1999)
2. Blondel, V.D., Guillaume, J.L., Lambiotte, R., Lefebvre, E.: Fast unfolding of communities in large networks. J. Stat. Mech. Theory Exp. **2008**(10), P10008 (2008)
3. Casteigts, A., Flocchini, P., Quattrociocchi, W., Santoro, N.: Time-varying graphs and dynamic networks. Int. J. Parallel Emergent Distrib. Syst. **27**(5), 387–408 (2012)
4. Du, W.B., Wu, Z.X., Cai, K.Q.: Effective usage of shortest paths promotes transportation efficiency on scale-free networks. Phys. A **392**(17), 3505–3512 (2013)
5. Fortunato, S.: Community detection in graphs. Phys. Rep. **486**(3), 75–174 (2010)
6. Fortunato, S., Hric, D.: Community detection in networks: a user guide. Phys. Rep. **659**, 1–44 (2016)
7. Jiang, Z.Y., Liang, M.G.: Incremental routing strategy on scale-free networks. Phys. A **392**(8), 1894–1901 (2013)
8. Kelly, F.P., et al.: Mathematical modelling of the internet. In: Engquist, B., Schmid, W. (eds.) Mathematics Unlimited-2001 and Beyond 1, pp. 685–702. Springer, Heidelberg (2001)
9. Kumari, S., Singh, A.: Modeling of data communication networks using dynamic complex networks and its performance studies. In: Cherifi, H., Gaito, S., Quattrociocchi, W., Sala, A. (eds.) Complex Networks & Their Applications V. SCI, vol. 693, pp. 29–40. Springer, Cham (2017). doi:10.1007/978-3-319-50901-3_3
10. Kumari, S., Singh, A., Ranjan, P.: Towards a framework for rate control on dynamic communication networks. In: Proceedings of the International Conference on Internet of Things and Cloud Computing, p. 12. ACM (2016)
11. La, R.J., Anantharam, V.: Utility-based rate control in the internet for elastic traffic. IEEE/ACM Trans. Netw. **10**(2), 272–286 (2002)
12. Lee, M.J., Lee, J., Park, J.Y., Choi, R.H., Chung, C.W.: Qube: a quick algorithm for updating betweenness centrality. In: Proceedings of the 21st International Conference on World Wide Web, pp. 351–360. ACM (2012)
13. Lin, B., Chen, B., Gao, Y., Chi, K.T., Dong, C., Miao, L., Wang, B.: Advanced algorithms for local routing strategy on complex networks. PLoS ONE **11**(7), e0156756 (2016)
14. Mo, J., Walrand, J.: Fair end-to-end window-based congestion control. IEEE/ACM Trans. Netw. (ToN) **8**(5), 556–567 (2000)
15. Newman, M.: Complex systems: a survey. arXiv preprint arXiv:1112.1440 (2011)
16. Onnela, J.P., Saramäki, J., Hyvönen, J., Szabó, G., Lazer, D., Kaski, K., Kertész, J., Barabási, A.L.: Structure and tie strengths in mobile communication networks. Proc. Natl. Acad. Sci. **104**(18), 7332–7336 (2007)
17. Tadić, B.: Dynamics of directed graphs: the world-wide web. Phys. A **293**(1), 273–284 (2001)
18. Wehmuth, K., Ziviani, A., Fleury, E.: A unifying model for representing time-varying graphs. In: IEEE International Conference on Data Science and Advanced Analytics (DSAA), 36678 2015, pp. 1–10. IEEE (2015)
19. Zhao, L., Lai, Y.C., Park, K., Ye, N.: Onset of traffic congestion in complex networks. Phys. Rev. E **71**(2), 026125 (2005)
20. Zhou, M., Yang, J.H., Liu, H.B., Wu, J.P.: Modeling the complex internet topology. J. Softw. **20**(1), 109–123 (2009)

Graph Construction Based on Local Representativeness

Eliska Ochodkova, Sarka Zehnalova, and Milos Kudelka[✉]

FEI, VSB, Technical University of Ostrava,
17. Listopadu 15, 708 33 Ostrava, Czech Republic
{eliska.ochodkova,sarka.zehnalova,milos.kudelka}@vsb.cz

Abstract. Graph construction is a known method of transferring the problem of classic vector data mining to network analysis. The advantage of networks is that the data are extended by links between certain (similar) pairs of data objects, so relationships in the data can then be visualized in a natural way. In this area, there are many algorithms, often with significantly different results. A common problem for all algorithms is to find relationships in data so as to preserve the characteristics related to the internal structure of the data. We present a method of graph construction based on a network reduction algorithm, which is found on analysis of the representativeness of the nodes of the network. It was verified experimentally that this algorithm preserves structural characteristics of the network during the reduction. This approach serves as the basis for our method which does not require any default parameters. In our experiments, we show the comparison of our graph construction method with one well-known method based on the most commonly used approach.

Keywords: Graph construction · Machine learning · Nearest neighbor · Local representativeness

1 Introduction

Simply put, the methods of data mining analyze data arranged in a table whose rows are data objects and columns are attributes (features). Typical tasks include, e.g., clustering and classification. These two areas have different objectives, but both are based on the assumption that there are internal data structures which can be utilized. These structures are then meant for solving clustering, classification, and other problems. While examining structures, the distance (similarity) of individual data objects, in particular, is examined. Assuming that we are working with numeric attributes, using e.g. the Euclidean distance or Gaussian function, the distance or similarity, respectively, can be precisely measured. Due to these measurable properties, we can apply many different algorithms that reveal information about the structures mentioned above.

However, it is not easy to find the way to visualize such data and structures because usually we work with more than three dimensions. Visualization is one

© Springer International Publishing AG 2017
Y. Cao and J. Chen (Eds.): COCOON 2017, LNCS 10392, pp. 654–665, 2017.
DOI: 10.1007/978-3-319-62389-4_54

of the essential tools allowing for an easier understanding of the data. Graphs (networks) have the property that, thanks to various layout algorithms, provide a clear visualization depicting the natural structure of data, particularly for small networks (for large-scale networks with higher computational costs). Besides visualization, networks also provide space for the application of analytical methods which can describe other characteristics (node ranking, local and global clustering coefficient, community structure, etc.).

As a graph (or network), we understand an ordered pair $G = (V, E)$ (undirected, unweighted graph) of a set V of nodes and a set E of edges which are unordered pairs of nodes from G.

The basic problem of graph construction is to find the parameters for conversion, which retain the essential properties of transferred data. The nodes in the constructed graph represent individual objects where interconnection between nodes can be done in various ways. The result of conversion should be a graph in which the clusters, the nearest neighbors (in the sense of distance or similarity, respectively), outliers, etc., are preserved. One of the known approaches based on the nearest neighbor analysis published Huttenhower et al. [9]. In this approach, in addition to graph construction, the main objective is to find strongly interconnected clusters in the data. However, the method assumes that the user must specify the number of nearest neighbors with which the algorithm works. Methods using the principle based on the use of k nearest neighbors are referred to as the k-NN networks and assume the k parameter to be a previously known value.

In our method, we use the nearest neighbors in another way. We assume that data objects (and consequently future graph nodes) have a different representativeness. The representativeness is a local property based on the number of objects that are the nearest neighbors of a selected node. Zehnalova et al. [17,18]) showed in their experiments that local representativeness could be used for data reduction in which the structure and properties of the original data are very well preserved.

In our approach we use representativeness in the construction of the graph so that (1) we create edges between all pairs of nearest neighbors, and (2) we create additional edges between the individual data objects in the number proportional to the representativeness of these objects. Representativeness of nodes in the constructed graph then corresponds, approximately, to the representativeness of objects in the data. This makes a natural graph representation of the original data, which preserves its local properties.

The paper is organized as follows: In Sect. 2, we discuss the related work. The proposed method is presented in Sect. 3. In Sect. 4, we focus on the experiment and on its results. Section 5 concludes the paper.

2   Related Work

Graphs are powerful tools for data analysis, and graph-based algorithms are currently widely used in various fields. Construction of a graph from vector data represents a precursor of many tasks in machine learning, computer vision, and signal processing. Examples include semi-supervised learning (SSL) or spectral clustering. For graph-based SSL algorithms, a graph is a central object on which the subsequent learning takes place. The nodes of the graph represent data objects, and weighted edges express the relationships between the nodes.

For some application domains, the relationships between data are natural (e.g. social networks, collaboration networks, citation networks, the web). In this case, graph-based methods, such as methods of social network analysis (SNA) which use a graph as the basic representation of data, can be applied straightforwardly. However, most data mining tasks consider data to be independent and identically distributed (i.i.d.) variables so we do not have access to any explicit graph structure. Therefore, a graph of independent objects is often generated first, and then graph-based machine learning methods are used. Even though graph-based machine learning methods are currently very popular, little attention is paid to the impact of graph construction methods and their results [7] and there is not a general way to establish high-quality graphs [11]. As an example study of the impact of graph construction see Maier et al. [12], where the authors discuss the influence of graph construction on graph-based clustering measures such as the normalized cut. However, little attention is paid to analyzing the properties of the resulting graph using SNA methods.

From a wider perspective, graph construction techniques can be divided into the following two groups [14]:

1. **Task-dependent graph construction**: algorithms in this group use both labeled and unlabeled data to construct the graph. Labeled data can be used for adapting the graph to the task at hand. This group of algorithms receives less attention. An example is the semi-supervised metric learning algorithm IDML [6] which shows how the available label information can be used for graph construction in graph-based SSL. In particular, authors focus on learning a distance metric using available label information, which can then be used to set the edge weights on the constructed graph.
2. **Task-independent graph construction methods**: algorithms in this group do not use labeled data, so these methods can be considered unsupervised. Their advantage is that they are applicable to any graph-based learning tasks. The disadvantage is that they do not necessarily help subsequent learning tasks.

Graph construction, in general, consists of two consecutive steps. Once a neighborhood graph is constructed, the edge weight assignment step follows. The basic edge weighting methods include Gaussian function (Heat kernel) [1], inverse Euclidean distance, local reconstructive relationship [15] or e.g. sparse representation (SR) [16].

Construction of a graph in which both the selection of neighbors and edge weighting take place in one step, minimizing the weighted sum of the squared distance from each node to the weighted average of its neighbors, was proposed by [4] as a concept of hard and $\alpha-$soft graphs.

A graph-based algorithm to generate clusters of genes with similar expression profiles was proposed in [9]. The k-NN approach is used to construct the graph together with the detection of overlapping cliques of a given size (similarly to [5]).

3   Theoretical Background

In this section, we define the basic ideas behind the proposed graph construction algorithm. Zehnalova et al. [18] present an algorithm for network reduction with linear (quadratic in the worst case) complexity, which is based on the so-called local representativeness. The algorithm is based on the idea that objects which are the nearest neighbors of other objects are the important ones in a dataset. This algorithm is used in [17] for the reduction of vector data. In both cases, the reduction retains the structural properties of the original data very well. The following is a description of that approach.

In the following text we assume that $D = (O, s)$ is a dataset, where O is a set of objects and $s : O \times O \to \mathbb{R}_{\geq 0}$ is a similarity function that quantifies the similarity between two objects. A similarity function is dependent on different kinds of data, e.g. weighted and unweighted network data and vector data. The similarity measure is often based on the Gaussian function for the vector data.

Objects O_j for which $s(O_i, O_j) > 0$ are *neighbors*. The *neighborhood* of the object O_i is a set of all its neighbors. The *nearest neighbor* of object O_i is an object O_j which is most similar to the object O_i. When more such object exists, object O_i has more nearest neighbors.

Definition 1 (*Local degree*). Let $d(O_i)$ be the number of objects for which object O_i is a neighbor. Then $d(O_i)$ is a *local degree* of object O_i. Objects for which $d(O_i) = 0$ are *outliers*.

Definition 2 (*Local significance*). Let $g(O_i)$ be the number of objects for which object O_i is the nearest neighbor. Then $g(O_i)$ is *local significance* of object O_i. Object (O_i) is *locally significant* if $g(O_i) > 0$, otherwise it is *locally insignificant*.

Definition 3 (*x-representativeness*). Let $x \in \mathbb{R}_{>1}$ and $d(O_i) > 0$. Then $r_x(O_i)$ is *x-representativeness* of object O_i defined as follows:

$$r_x(O_i) = \frac{g(O_i)}{\log_x(1 + d(O_i))} \tag{1}$$

An important parameter of the x-representativeness is the base of the logarithm x. Informally speaking, x affects the degree of representativeness of the object in relationship to its surroundings.

Definition 4 (*x-representativeness base*). Let $x \in \mathbb{R}_{x>1}$ and $r_x(O_i) = 1$. Then $x = b(O_i)$ is *x-representativeness base* of object O_i defined as follows:

$$b(O_i) = (1 + d(O_i))^{\frac{1}{g(O_i)}} \tag{2}$$

The x-representativeness base represents the threshold of x-representativeness of the object. The smaller the x-representativeness base of the object is, the higher its local importance is.

Locally insignificant objects and outliers do not have any x-representativeness base, i.e. there is no x, for which locally insignificant objects would be x-representatives of dataset D.

3.1   Local Representativeness

To construct a graph, we can use local information about individual objects in the dataset. Based on the definitions above, we further assume that local importance of the object is related to its x-representative base. For the graph construction algorithm let's define representativeness of the object and its neighbor, respectively, as follows:

Definition 5 (*local representativeness*). Let $b(O_i)$ be *x-representativeness base* of object O_i. Then $lr(O_i)$ is *local representativeness* of object O_i defined as follows:

$$lr(O_i) = \frac{1}{b(O_i)} for\ locally\ significant O_i, \tag{3}$$

$$lr(O_i) = 0 otherwise \tag{4}$$

Definition 6 (*representative neighbor*). Let

$$k = \text{ROUND}\ (lr(O_i) \cdot d(O_i)) \tag{5}$$

and $K(O_i)$ be a set of k neighbors O_j of object O_i with highest similarity $s(O_i, O_j)$. Then *representative neighbors* of object O_i are objects $O_j \in K(O_i)$.

3.2   Proposed *LRNet* algorithm

The LRNet algorithm for the construction of the weighted graph utilizing local representativeness is composed of four steps:

1. Create a similarity matrix S of dataset D.
2. Calculate the representativeness of all objects O_i.
3. Create the set V of nodes of graph G so that node v_i of graph G represents object O_i of dataset D.
4. Create the set of edges E of graph G so that E contains an edge e_{ij} between nodes v_i and v_j ($i \neq j$) if O_j is the nearest neighbor of O_i or O_j is the representative neighbor of O_i.

The time complexity of the algorithm is $O(n^2)$, where $n = |D|$. This is based on following facts:

– The calculation of the similarity matrix (the first step of the algorithm) has $O(n^2)$ complexity (if we neglect the complexity of calculating the similarity of two objects).

- The calculation of representativeness, as mentioned above, has in the worst case also $O(n^2)$ complexity (the second step of the algorithm). The worst case is a situation where the similarity matrix is dense; this can be assumed for vector data. For data represented by a sparse similarity matrix, the complexity is $O(n)$; for details see [17].
- The complexity of selecting representative neighbors of objects from dataset D (the fourth step of the algorithm) depends on the average number of representative neighbors $\overline{k}$ that we need to select from (in general) n neighbors. In the graph construction task, we can assume that the constructed graph is sparse; then it holds that $\overline{k} \ll n$ and complexity is $O(n^2)$.

In the proposed LRNet algorithm, each object O_i of dataset D contributes to the graph with the number of edges corresponding to its local representativeness. Regarding local properties and structures in the original data, we understand this principle as crucial to maintaining these properties.

4   Experiment

It is not possible to clearly state what properties the constructed graph should have. It depends on the task in which the graph would be utilized. In our approach, we assume that the constructed graph should be designed to best match the converted data. Moreover, because the data are an image of reality, the constructed graph should reflect this reality. Therefore, to assess graph properties, we use methods of social network analysis. The results should then lead to the properties observed in real-world networks.

For comparison, we chose the often used k-NN network method [9]. This method requires a pre-selected number of nearest neighbors k. For all of the analyzed datasets, we chose $k = Sqrt(N)$ where N is the number of objects in the dataset (SqrtNN). For evaluation, we selected three datasets meant for classification. The reason for the analysis of labeled data is to assess how the compared methods reflect known information about classes in the data. The chosen datasets are 'Cortex Nuclear' [8], 'Ecoli' [10] and 'Seeds' [3]. For all of the datasets, rescaling the range of features to scale the range in $[0, 1]$ and Gaussian function as a similarity measure were used.

The key properties of real-world networks are so-called small-world (small average shortest path length and diameter of the graph), scale-free structure (power-law degree distribution), high clustering coefficient and community structure. Another property is assortativity, which has positive values for social networks but negative values for e.g. biological or technological networks (see Newman [13]). Table 1 shows the properties of graphs constructed from analyzed datasets. The measured properties include the number of nodes n and edges m, average degree <k>, average shortest path length <l>, diameter L_{max}, average clustering coefficient CC, assortativity r, number of communities com_L detected by Louvain [2] algorithm, and the corresponding modularity Q_L.

It is evident from Table 1 that the pairs of graphs constructed by different methods have a similar number of edges, and therefore also an average degree

Table 1. Global properties of constructed graphs

Graph	n	m	$<k>$	$<l>$	l_{max}	CC	r	com_L	Q_L
Cortex: LRNet	1080	10648	19.719	3.212	7	0.519	−0.212	8	0.607
Cortex: SqrtNN	1080	11934	22.162	4.407	10	0.528	0.476	10	0.741
Ecoli: LRNet	336	2214	13.179	3.519	9	0.599	−0.352	5	0.624
Ecoli: SqrtNN	336	1986	11.857	4.834	12	0.497	0.521	9	0.703
Seeds: LRNet	210	997	9.495	3.472	7	0.662	−0.397	5	0.635
Seeds: SqrtNN	210	1040	9.905	5.207	13	0.509	0.484	7	0.707

and density. However, graphs constructed by the k-NN algorithm, unlike graphs constructed by LRNet, do not have some properties known for real-world networks. This is, in particular, longer average shortest path, large diameter, and unnaturally high assortativity. Another important property observed in real-world networks, that k-NN graphs do not have, is scale-free structure. The consequence of scale-freeness is so-called hubs, which are nodes with a very high degree, and also the fact that most of the nodes in real-world networks have a low degree. Degree distribution for each of the graphs on a log-log scale is for the LRNet method in Fig. 1 and for the SqrtNN method in Fig. 2. In graphs constructed by the SqrtNN method, most nodes have a higher degree and hubs are missing. Also, only a few nodes have a lower degree. This characteristic is also clearly documented by constructed graphs in Figs. 3 and 4. The size of the nodes in those graphs corresponds to their degree; colors represent classes (labels).

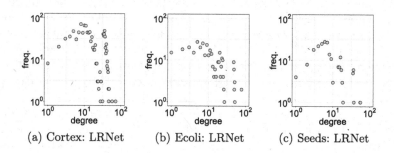

(a) Cortex: LRNet   (b) Ecoli: LRNet   (c) Seeds: LRNet

Fig. 1. Degree distribution, LRNet

4.1 Analysis Using Classes

All three pairs of analyzed graphs are constructed from labeled data. This allows us to measure how precisely the constructed graphs connect nodes belonging to the same class. In Table 2, three different values are calculated. The first is the average weighted precision which is the average value of

$$w = \frac{w_{pos}}{w_{all}} \tag{6}$$

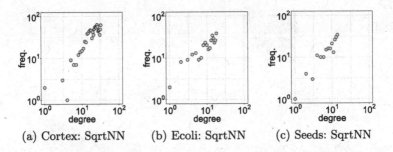

(a) Cortex: SqrtNN   (b) Ecoli: SqrtNN   (c) Seeds: SqrtNN

Fig. 2. Degree distribution, SqrtNN

Table 2. Classification accuracy of constructed graphs

Graph	Weighted precission	Precission	Modularity by classes
Cortex: LRNet	0.607	0.870	0.376
Cortex: SqrtNN	0.643	0.887	0.457
Ecoli: LRNet	0.764	0.842	0.370
Ecoli: SqrtNN	0.777	0.869	0.343
Seeds: LRNet	0.872	0.914	0.341
Seeds: SqrtNN	0.891	0.929	0.312

for all nodes of the graph, where w_{pos} is the sum of edge weights with neighbors of the same class and w_{all} is the sum of edge weights to all neighbors. The second value is the average precision, which is the average value of p. This value $p = 1$ for the selected node if the sum of the of edge weights with neighbors in the same class is higher than the sum of edge weights with neighbors in each of the other classes. Otherwise, $p = 0$. The third value is modularity Q (Blondel et al. [2]) by classes defined as follows:

$$Q = \frac{1}{2m} \sum_{ij} \left[w_{ij} - \frac{k_i k_j}{2m} \right] \delta(c_i, c_j), \tag{7}$$

where w_{ij} is the edge weight between nodes v_i and v_j, k_i and k_j are the sums of the weights of the edges between nodes v_i and v_j, m is the sum of all of the edge weights in the graph, c_i and c_j are classes of the nodes v_i and v_j, and δ is Kronecker delta.

Furthermore, Figs. 5 and 6 show ROC (receiver operating characteristic) curves with calculated values of AUC (area under the curve) describing the accuracy of assigning objects to each class. ROC curves are calculated from the values of weighted precision w of individual nodes of constructed graphs.

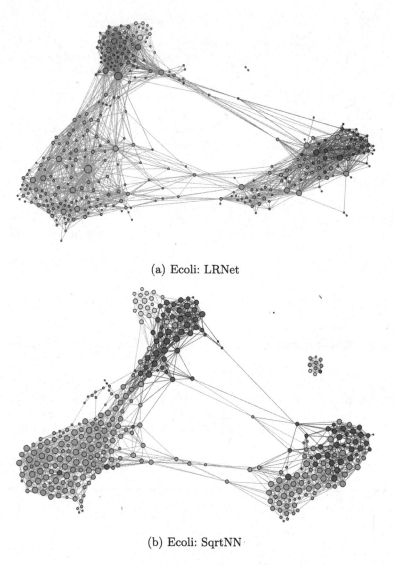

(a) Ecoli: LRNet

(b) Ecoli: SqrtNN

Fig. 3. Constructed graphs: Ecoli dataset

The results show that the accuracy of connecting the nodes to the nodes of the same class for the two methods differ a little. The advantage of the LRNet method is, however, that for the construction of the graph, no parameter has to be selected, which is a required step for the k-NN method.

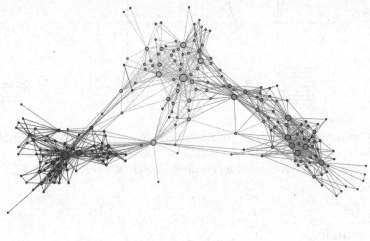

(a) Seeds: LRNet

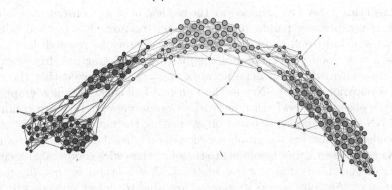

(b) Seeds: SqrtNN

Fig. 4. Constructed graphs: Seeds dataset

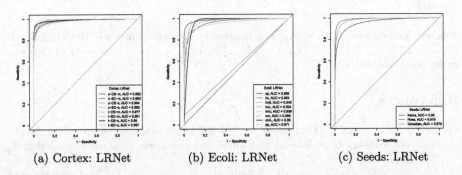

(a) Cortex: LRNet   (b) Ecoli: LRNet   (c) Seeds: LRNet

Fig. 5. Receiver operating characteristic (ROC) curves, LRNet

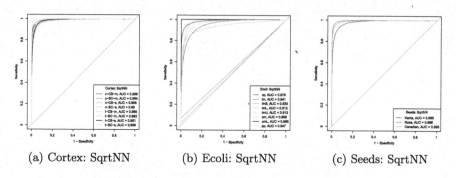

(a) Cortex: SqrtNN   (b) Ecoli: SqrtNN   (c) Seeds: SqrtNN

Fig. 6. Receiver operating characteristic (ROC) curves, SqrtNN

5   Conclusions

The aim of this paper was to introduce the method of graph construction, LRNet, that does not use any parameters for the construction. The method is based on the use of local representativeness, which, as previous research has shown, ensures a very good preservation of the internal data structure. In our experiments we used methods of social network analysis and showed that the result of the application of the LRNet method on real biological data is a graph with properties observed in real-world networks. Comparing with the most commonly used k-NN method, which is based on estimating the number of nearest neighbors needed to construct the graph, we showed on labeled data that the resulting graphs interconnect nodes belonging to the same class with comparable accuracy.

One of the interesting features of the LRNet method is that due to the properties of the constructed graph, we are able to detect objects with high representativeness. These objects are in the network represented by high-degree nodes and can be of two types. The first type is the hubs, the nodes connecting the clusters. The second type is the centers of these clusters. A more detailed analysis of the role of these objects in the data will be the subject of further research.

Acknowledgments. This work was supported by grant of Ministry of Health of Czech Republic (MZ CR VES16-31852A) and by SGS, VSB-Technical University of Ostrava, under the grant no. SP2017/85.

References

1. Belkin, M., Niyogi, P., Sindhwani, V.: Manifold regularization: A geometric framework for learning from labeled and unlabeled examples. J. Mach. Learn. Res. **7**, 2399–2434 (2006)
2. Blondel, V.D., Guillaume, J.L., Lambiotte, R., Lefebvre, E.: Fast unfolding of communities in large networks. J. Stat. Mech. Theory Exp. **2008**(10), P10008 (2008)

3. Charytanowicz, M., Niewczas, J., Kulczycki, P., Kowalski, P.A., Łukasik, S., Żak, S.: Complete gradient clustering algorithm for features analysis of X-ray images. In: Piętka, E., Kawa, J. (eds.) Information Technologies in Biomedicine. Advances in Intelligent and Soft Computing, vol. 69, pp. 15–24. Springer, Heidelberg (2010)
4. Daitch, S.I., Kelner, J.A., Spielman, D.A.: Fitting a graph to vector data. In: Proceedings of the 26th Annual International Conference on Machine Learning, pp. 201–208. ACM (2009)
5. Derényi, I., Palla, G., Vicsek, T.: Clique percolation in random networks. Phys. Rev. Lett. **94**(16), 160202 (2005)
6. Dhillon, P.S., Talukdar, P.P., Crammer, K.: Inference-driven metric learning for graph construction. In: 4th North East Student Colloquium on Artificial Intelligence (2010)
7. Dornaika, F., Bosaghzadeh, A.: Adaptive graph construction using data self-representativeness for pattern classification. Inf. Sci. **325**, 118–139 (2015)
8. Higuera, C., Gardiner, K.J., Cios, K.J.: Self-organizing feature maps identify proteins critical to learning in a mouse model of down syndrome. PLoS one **10**(6), e0129126 (2015)
9. Huttenhower, C., Flamholz, A.I., Landis, J.N., Sahi, S., Myers, C.L., Olszewski, K.L., Hibbs, M.A., Siemers, N.O., Troyanskaya, O.G., Coller, H.A.: Nearest neighbor networks: clustering expression data based on gene neighborhoods. BMC Bioinform. **8**(1), 250 (2007)
10. Lichman, M.: UCI machine learning repository (2013). http://archive.ics.uci.edu/ml
11. Liu, W., Chang, S.F.: Robust multi-class transductive learning with graphs. In: 2009 IEEE Conference on Computer Vision and Pattern Recognition, CVPR 2009, pp. 381–388. IEEE (2009)
12. Maier, M., Von Luxburg, U., Hein, M.: Influence of graph construction on graph-based clustering measures. In: NIPS, pp. 1025–1032 (2008)
13. Newman, M.E.: Assortative mixing in networks. Phys. Rev. Lett. **89**(20), 208701 (2002)
14. Subramanya, A., Talukdar, P.P.: Graph-based semi-supervised learning. In: Synthesis Lectures on Artificial Intelligence and Machine Learning, vol. 8(4), pp. 1–125 (2014)
15. Wang, F., Zhang, C.: Label propagation through linear neighborhoods. IEEE Trans. Knowl. Data Eng. **20**(1), 55–67 (2008)
16. Yan, S., Wang, H.: Semi-supervised learning by sparse representation. In: Proceedings of the 2009 SIAM International Conference on Data Mining, SIAM, pp. 792–801 (2009)
17. Zehnalova, S., Kudelka, M., Platos, J.: Local representativeness in vector data. In: 2014 IEEE International Conference on Systems, Man and Cybernetics (SMC), pp. 894–899. IEEE (2014)
18. Zehnalova, S., Kudelka, M., Platos, J., Horak, Z.: Local representatives in weighted networks. In: 2014 IEEE/ACM International Conference on Advances in Social Networks Analysis and Mining (ASONAM), pp. 870–875. IEEE (2014)

SHADE Algorithm Dynamic Analyzed Through Complex Network

Adam Viktorin$^{(\boxtimes)}$, Roman Senkerik, Michal Pluhacek, and Tomas Kadavy

Faculty of Applied Informatics, Tomas Bata University in Zlin,
T. G. Masaryka 5555, 760 01 Zlin, Czech Republic
{aviktorin,senkerik,pluhacek,kadavy}@fai.utb.cz

Abstract. In this preliminary study, the dynamic of continuous optimization algorithm Success-History based Adaptive Differential Evolution (SHADE) is translated into a Complex Network (CN) and the basic network feature, node degree centrality, is analyzed in order to provide helpful insight into the inner workings of this state-of-the-art Differential Evolution (DE) variant. The analysis is aimed at the correlation between objective function value of an individual and its participation in production of better offspring for the future generation. In order to test the robustness of this method, it is evaluated on the CEC2015 benchmark in 10 and 30 dimensions.

Keywords: Differential evolution · SHADE · Complex network · Centrality

1 Introduction

Heuristic algorithms based on the original Differential Evolution (DE) [1] are constantly ranked high in competitions on continuous optimization during the last years [2–7]. The common aspect of successful variants is the ability to adapt the optimization process to the given problem. The original DE from 1995 had three static control parameters, which were required to be set by the user. Those parameters are population size NP, scaling factor F and crossover rate CR. Adaptive algorithms overcome the issue of fine-tuning of those parameters by initializing them to predefined values and changing them during the optimization process in accordance with successful values of these parameters from previous generations. The adaptivity of these algorithms is trying to overcome the famous No Free Lunch (NFL) theorem [8].

A. Viktorin—This work was supported by Grant Agency of the Czech Republic – GACR P103/15/06700S, further by the Ministry of Education, Youth and Sports of the Czech Republic within the National Sustainability Programme Project no. LO1303 (MSMT-7778/2014). Also by the European Regional Development Fund under the Project CEBIA-Tech no. CZ.1.05/2.1.00/03.0089 and by Internal Grant Agency of Tomas Bata University under the Projects no. IGA/CebiaTech/2017/004.

Y. Cao and J. Chen (Eds.): COCOON 2017, LNCS 10392, pp. 666–677, 2017.
DOI: 10.1007/978-3-319-62389-4_55

An overview of DE based algorithms can be found in a recent survey [9], but some of the most interesting variants are SDE [10], JADE [11], SHADE [12] and L-SHADE [13]. The last two named were successful in recent competitions in continuous numerical optimization [14–17] and therefore, the SHADE algorithm was selected as a candidate for the analysis in this paper.

The main motivation is to understand the inner dynamic of the algorithm and use this information for the future changes in its design because there is still a lot of room for improvement, as suggested in [18].

In this work, the dynamic of the algorithm is transformed into the Complex Network (CN) and node degree values are used for the sake of analysis of the algorithm performance and in an effort to detect its possible weak spots in the design. Similar CN creation was done in recently published papers [19,20]. Since the analysis is based on the activity of an individual in evolution, such simple network design is sufficient.

In order to provide a robust overview of the method, CEC2015 benchmark set of 15 test functions [21] was used for the analysis in two dimensional settings – 10 and 30.

The remainder of the paper is structured as follows: Sects. 2 and 3 depict the DE and SHADE algorithms, Sect. 4 describes the network creation, Sect. 5 defines the experimental settings, Sect. 6 provides results and discussion. Finally, concluding remarks can be found in the last section.

2   Differential Evolution

The DE algorithm is initialized with a random population of individuals P, that represent solutions of the optimization problem. The population size NP is set by the user along with other control parameters – scaling factor F and crossover rate CR.

In continuous optimization, each individual is composed of a vector x of length D, which is a dimensionality (number of optimized attributes) of the problem, and each vector component represents a value of the corresponding attribute, and of objective function value $f(x)$.

For each individual in a population, three mutually different individuals are selected for mutation of vectors and resulting mutated vector v is combined with the original vector x in crossover step. The objective function value $f(u)$ of the resulting trial vector u is evaluated and compared to that of the original individual. When the quality (objective function value) of the trial individual is better, it is placed into the next generation, otherwise, the original individual is placed there. This step is called selection. The process is repeated until the stopping criterion is met (e.g. the maximum number of objective function evaluations, the maximum number of generations, the low bound for diversity between objective function values in population).

The following sections describe four steps of DE: Initialization, mutation, crossover and selection.

2.1   Initialization

As aforementioned, the initial population $\boldsymbol{P}$, of size NP, of individuals is randomly generated. For this purpose, the individual vector $\boldsymbol{x}_i$ components are generated by Random Number Generator (RNG) with uniform distribution from the range which is specified for the problem by *lower* and *upper* bound (1).

$$x_{j,i} = U\left[lower_j,\ upper_j\right] \text{ for } j = 1,\ \ldots,\ D \tag{1}$$

where i is the index of a current individual, j is the index of current attribute and D is the dimensionality of the problem.

In the initialization phase, a scaling factor value F and crossover value CR has to be assigned as well. The typical range for F value is $[0, 2]$ and for CR, it is $[0, 1]$.

2.2   Mutation

In the mutation step, three mutually different individuals $\boldsymbol{x}_{r1}$, $\boldsymbol{x}_{r2}$, $\boldsymbol{x}_{r3}$ from a population are randomly selected and combined in mutation according to the mutation strategy. The original mutation strategy of canonical DE is "rand/1" and is depicted in (2).

$$v_i = \boldsymbol{x}_{r1} + F\left(\boldsymbol{x}_{r2} - \boldsymbol{x}_{r3}\right) \tag{2}$$

where $r1 \neq r2 \neq r3 \neq i$, F is the scaling factor value and v_i is the resulting mutated vector.

2.3   Crossover

In the crossover step, mutated vector $\boldsymbol{v}_i$ is combined with the original vector $\boldsymbol{x}_i$ and produces trial vector $\boldsymbol{u}_i$. The binary crossover (3) is used in canonical DE.

$$u_{j,i} = \begin{cases} v_{j,i} \text{ if } U\left[0,1\right] \leq CR \text{ or } j = j_{rand} \\ x_{j,i} \qquad\qquad \text{otherwise} \end{cases} \tag{3}$$

where CR is the used crossover rate value and j_{rand} is an index of an attribute that has to be from the mutated vector $\boldsymbol{v}_i$ (ensures generation of a vector with at least one new component).

2.4   Selection

The selection step ensures, that the optimization progress will lead to better solutions because it allows only individuals of better or at least equal objective function value to proceed into next generation $G+1$ (4).

$$x_{i,G+1} = \begin{cases} \boldsymbol{u}_{i,G} \text{ if } f\left(\boldsymbol{u}_{i,G}\right) \leq f\left(\boldsymbol{x}_{i,G}\right) \\ \boldsymbol{x}_{i,G} \qquad\quad \text{otherwise} \end{cases} \tag{4}$$

where G is the index of current generation.

The whole DE algorithm is depicted in pseudo-code below.

Algorithm 1. DE

1: Set NP, CR, F and stopping criterion;
2: $G = 0$, $x_{best} = \{\}$;
3: Randomly initialize (1) population $\boldsymbol{P} = (\boldsymbol{x}_{1,G}, \ldots, \boldsymbol{x}_{NP,G})$;
4: $\boldsymbol{P}_{new} = \{\}$, $\boldsymbol{x}_{best}$ = best from population $\boldsymbol{P}$;
5: **while** stopping criterion not met **do**
6:     **for** $i = 1$ to NP **do**
7:         $\boldsymbol{x}_{i,G} = \boldsymbol{P}[i]$;
8:         $\boldsymbol{v}_{i,G}$ by mutation (2);
9:         $\boldsymbol{u}_{i,G}$ by crossover (3);
10:        **if** $f(\boldsymbol{u}_{i,G}) < f(\boldsymbol{x}_{i,G})$ **then**
11:            $\boldsymbol{x}_{i,G+1} = \boldsymbol{u}_{i,G}$;
12:        **else**
13:            $\boldsymbol{x}_{i,G+1} = \boldsymbol{x}_{i,G}$;
14:        **end if**
15:        $\boldsymbol{x}_{i,G+1} \rightarrow \boldsymbol{P}_{new}$;
16:     **end for**
17:     $\boldsymbol{P} = \boldsymbol{P}_{new}$, $\boldsymbol{P}_{new} = \{\}$, $\boldsymbol{x}_{best}$ = best from population $\boldsymbol{P}$;
18: **end while**
19: **return** $\boldsymbol{x}_{best}$ as the best found solution

---

3 Success-History Based Adaptive Differential Evolution and Enhanced Archive

In SHADE, the only control parameter that can be set by the user is population size NP, other two (F, CR) are adapted to the given optimization task, a new parameter H is introduced, which determines the size of F and CR value memories. The initialization step of the SHADE is, therefore, similar to DE. Mutation, however, is completely different because of the used strategy "current-to-$p$best/1" and the fact, that it uses different scaling factor value F_i for each individual. Mutation strategy also works with a new feature – external archive of inferior solutions. This archive holds individuals from previous generations, that were outperformed in the selection step. The size of the archive retains the same size as the size of the population by randomly discarding its contents whenever the size overflows NP.

Crossover is still binary, but similarly to the mutation and scaling factor values, crossover rate value CR_i is also different for each individual.

The selection step is the same and therefore following sections describe only different aspects of initialization, mutation and crossover.

3.1 Initialization

As aforementioned, initial population $\boldsymbol{P}$ is randomly generated as in DE, but additional memories for F and CR values are initialized as well. Both memories have the same size H and are equally initialized, the memory for CR values is titled $\boldsymbol{M}_{CR}$ and the memory for F is titled $\boldsymbol{M}_F$. Their initialization is depicted in (5).

$$M_{CR,i} = M_{F,i} = 0.5 \text{ for } i = 1,\ldots,H \qquad (5)$$

Also, the external archive of inferior solutions $\boldsymbol{A}$ is initialized. Since there are no solutions so far, it is initialized empty $\boldsymbol{A} = \emptyset$ and its maximum size is set to NP.

3.2   Mutation

Mutation strategy "current-to-$p$best/1" was introduced in [11] and unlike "rand/1", it combines four mutually different vectors, where $pbest \neq r1 \neq r2 \neq i$ (6).

$$\boldsymbol{v}_i = \boldsymbol{x}_i + F_i \left(\boldsymbol{x}_{pbest} - \boldsymbol{x}_i \right) + F_i \left(\boldsymbol{x}_{r1} - \boldsymbol{x}_{r2} \right) \qquad (6)$$

where $\boldsymbol{x}_{pbest}$ is randomly selected from the best $NP \times p$ best individuals in the current population. The p value is randomly generated for each mutation by RNG with uniform distribution from the range $[p_{min}, 0.2]$. where $p_{min} = 2/NP$. Vector $\boldsymbol{x}_{r1}$ is randomly selected from the current population and vector $\boldsymbol{x}_{r2}$ is randomly selected from the union of current population $\boldsymbol{P}$ and archive $\boldsymbol{A}$. The scaling factor value F_i is given by (7).

$$F_i = C\left[M_{F,r}, 0.1\right] \qquad (7)$$

where $M_{F,r}$ is a randomly selected value (by index r) from $\boldsymbol{M}_F$ memory and C stands for Cauchy distribution, therefore the F_i value is generated from the Cauchy distribution with location parameter value $M_{F,r}$ and scale parameter value 0.1. If the generated value $F_i > 1$, it is truncated to 1 and if it is $F_i \leq 0$, it is generated again by (7).

3.3   Crossover

Crossover is the same as in (3), but the CR value is changed to CR_i, which is generated separately for each individual (8). The value is generated from the Gaussian distribution with mean parameter value of $M_{CR,r}$, which is randomly selected (by the same index r as in mutation) from $\boldsymbol{M}_{CR}$ memory and standard deviation value of 0.1.

$$CR_i = N\left[M_{CR,r}, 0.1\right] \qquad (8)$$

3.4   Historical Memory Updates

Historical memories $\boldsymbol{M}_F$ and $\boldsymbol{M}_{CR}$ are initialized according to (5), but its components change during the evolution. These memories serve to hold successful values of F and CR used in mutation and crossover steps. Successful in terms of producing trial individual better than the original individual. During one generation, these successful values are stored in corresponding arrays $\boldsymbol{S}_F$ and $\boldsymbol{S}_{CR}$.

After each generation, one cell of M_F and M_{CR} memories is updated. This cell is given by the index k, which starts at 1 and increases by 1 after each generation. When it overflows the size limit of memories H, it is again set to 1. The new value of k-th cell for M_F is calculated by (9) and for M_{CR} by (10).

$$M_{F,k} = \begin{cases} \text{mean}_{WL}\left(\boldsymbol{S}_F\right) \text{ if } \boldsymbol{S}_F \neq \emptyset \\ M_{F,k} \quad\quad \text{otherwise} \end{cases} \tag{9}$$

$$M_{CR,k} = \begin{cases} \text{mean}_{WA}\left(\boldsymbol{S}_{CR}\right) \text{ if } \boldsymbol{S}_{CR} \neq \emptyset \\ M_{CR,k} \quad\quad \text{otherwise} \end{cases} \tag{10}$$

where $\text{mean}_{WL}()$ and $\text{mean}_{WA}()$ are weighted Lehmer (11) and weighted arithmetic (12) means correspondingly.

$$\text{mean}_{WL}\left(\boldsymbol{S}_F\right) = \frac{\sum_{k=1}^{|S_F|} w_k \bullet S_{F,k}^2}{\sum_{k=1}^{|S_F|} w_k \bullet S_{F,k}} \tag{11}$$

$$\text{mean}_{WA}\left(\boldsymbol{S}_{CR}\right) = \sum_{k=1}^{|S_{CR}|} w_k \bullet S_{CR,k} \tag{12}$$

where the weight vector $\boldsymbol{w}$ is given by (13) and is based on the improvement in objective function value between trial and original individuals.

$$w_k = \frac{\text{abs}\left(f\left(\boldsymbol{u}_{k,G}\right) - f\left(\boldsymbol{x}_{k,G}\right)\right)}{\sum_{m=1}^{|S_{CR}|} \text{abs}\left(f\left(\boldsymbol{u}_{m,G}\right) - f\left(\boldsymbol{x}_{m,G}\right)\right)} \tag{13}$$

And since both arrays $\boldsymbol{S}_F$ and $\boldsymbol{S}_{CR}$ have the same size, it is arbitrary which size will be used for the upper boundary for m in (13). Complete SHADE algorithm is depicted in pseudo-code below.

4   Network Design and Metrics

The network creation is designed to be as simple as possible, therefore edges in the network are undirected and unweighted.

The network is represented by a multigraph $G = (V, E)$, where V is a set of vertices and E is a set of edges. Vertices mirror individuals in the population, therefore the size of V is the same as population size NP. Undirected edges are created after each successful evolution (produced trial individual $\boldsymbol{u}_i$ has better objective function value than the active individual $\boldsymbol{x}_i$) between active individual $\boldsymbol{x}_i$ and three individuals form the mutation step – $\boldsymbol{x}_{pbest}$, $\boldsymbol{x}_{r1}$ and $\boldsymbol{x}_{r2}$. And the edges are denoted as $e_{i,pbest}$, $e_{i,r1}$ and $e_{i,r2}$. The scheme is depicted below in Fig. 1.

The maximum number of edges created in one generation is equal to $3 \times NP$ and this occurs only if all individuals in the population were improved in this generation.

The optimization process is captured in snapshots of each generation. Therefore, for each generation a new multigraph G_n is created, where n stands for generation number. There is no transfer of information between generations.

For the purpose of the analysis of CN two metrics were introduced.

Algorithm 2. SHADE

1: Set NP, H and stopping criterion;
2: $G = 0$, $\boldsymbol{x}_{best} = \{\}$, $k = 1$, $p_{min} = 2/NP$, $\boldsymbol{A} = \emptyset$;
3: Randomly initialize (1) population $\boldsymbol{P} = (\boldsymbol{x}_{1,G}, \ldots, \boldsymbol{x}_{NP,G})$;
4: Set $\boldsymbol{M}_F$ and $\boldsymbol{M}_{CR}$ according to (5);
5: $\boldsymbol{P}_{new} = \{\}$, $\boldsymbol{x}_{best} = $ best from population $\boldsymbol{P}$;
6: **while** stopping criterion not met **do**
7:   $\boldsymbol{S}_F = \emptyset$, $\boldsymbol{S}_{CR} = \emptyset$;
8:   **for** $i = 1$ to NP **do**
9:     $\boldsymbol{x}_{i,G} = \boldsymbol{P}[i]$;
10:    $r = U[1, H]$, $p_i = U[p_{min}, 0.2]$;
11:    Set F_i by (7) and CR_i by (8);
12:    $\boldsymbol{v}_{i,G}$ by mutation (6);
13:    $\boldsymbol{u}_{i,G}$ by crossover (3);
14:    **if** $f(\boldsymbol{u}_{i,G}) < f(\boldsymbol{x}_{i,G})$ **then**
15:      $\boldsymbol{x}_{i,G+1} = \boldsymbol{u}_{i,G}$;
16:      $\boldsymbol{x}_{i,G} \rightarrow \boldsymbol{A}$;
17:      $F_i \rightarrow \boldsymbol{S}_F$, $CR_i \rightarrow \boldsymbol{S}_{CR}$;
18:    **else**
19:      $\boldsymbol{x}_{i,G+1} = \boldsymbol{x}_{i,G}$;
20:    **end if**
21:    **if** $|\boldsymbol{A}| > NP$ **then**
22:      Randomly delete an individual from $\boldsymbol{A}$;
23:    **end if**
24:    $\boldsymbol{x}_{i,G+1} \rightarrow \boldsymbol{P}_{new}$;
25: **end for**
26: **if** $\boldsymbol{S}_F \neq \emptyset$ **and** $\boldsymbol{S}_{CR} \neq \emptyset$ **then**
27:    Update $\boldsymbol{M}_{F,k}$ (9) and $\boldsymbol{M}_{CR,k}$ (10), k++;
28:    **if** $k > H$ **then**
29:      $k = 1$;
30:    **end if**
31: **end if**
32: $\boldsymbol{P} = \boldsymbol{P}_{new}$, $\boldsymbol{P}_{new} = \{\}$, $\boldsymbol{x}_{best} = $ best from population $\boldsymbol{P}$;
33: **end while**
34: **return** $\boldsymbol{x}_{best}$ as the best found solution

---

4.1   Objective Function Value Rank

This metric is abbreviated to *ofvRank* and is evaluated for each individual in the population and corresponds to the number of individuals, that have worse objective function value 14.

$$ofvRank_{i,G} = |\mathbf{S}|, \ \mathbf{S} \in \forall \boldsymbol{x}_{j,G} : \ f(\boldsymbol{x}_{i,G}) < f(\boldsymbol{x}_{j,G}), \ j = 1, \ldots, NP \qquad (14)$$

where $\boldsymbol{S}$ is a set of individuals with worse objective function value than that of the evaluated individual.

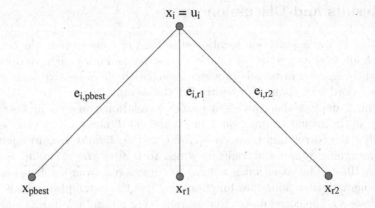

Fig. 1. Schema of the network creation for one successful evolution.

4.2 Centrality Rank

This metric is abbreviated to *cenRank* and mirrors the individual's position in the node degree centrality in the population 15.

$$cenRank_{i,G} = |\mathbf{S}|, \ \mathbf{S} \in \forall \boldsymbol{x}_{j,G} : \ deg(\boldsymbol{x}_{i,G}) < deg(\boldsymbol{x}_{j,G}), \ j = 1, \ldots, NP \quad (15)$$

where *deg* stands for node degree centrality and $\boldsymbol{S}$ is a set of individuals with lower node degree centrality than that of the evaluated individual.

5 Experimental Setting

The SHADE algorithm was run 51 times on each of the test functions of CEC2015 benchmark set and this was done for two different dimensional settings – $D = 10$ and $D = 30$. In each run, the stopping criterion was set according to the benchmark requirements to $10,000 \times D$. The population size NP was set to 100 and the size of historical memories H was set to 10.

For the purpose of the analysis of CN, node degree centrality values in each generation were recorded and two above mentioned metrics were evaluated.

The greedy approach is quite common for various optimization techniques and SHADE algorithm is no exception. There is the $\boldsymbol{x}_{pbest}$ individual in the mutation step, which is selected from the best subset of the population and therefore provides the greediness. This greediness is balanced by the use of the external archive of inferior solutions A. However, the basic assumption in this approach is that individuals with better objective function values are the ones which lead the optimization process and their search direction should be followed. In order to test that assumption, the Pearson correlation coefficient between the two metrics *ofvRank* and *cenRank* was calculated and the results are presented in the next section.

6   Results and Discussion

According to the assumption mentioned in the previous section, the correlation between *ofvRank* and *cenRank* should be positive and quite high. In this section, selected correlation history is provided along with the objective function value history, in order to confirm or contradict this assumption.

Figure 2 depicts the convergence and correlation comparison between the average optimization of function 1 in 10 and 30 dimensions. It can be clearly seen, that the correlation between *ofvRank* and *centrRank* is quite high during the convergence phase and logically drops to 0 after the optimum was found (only in 10*D*). The situation is similar for function 2 from the benchmark set, which suggests that unimodal functions confirm the assumption about correlation. However, the correlation value is stagnating around 0.5 during the convergence, which is on the edge between medium and high correlation.

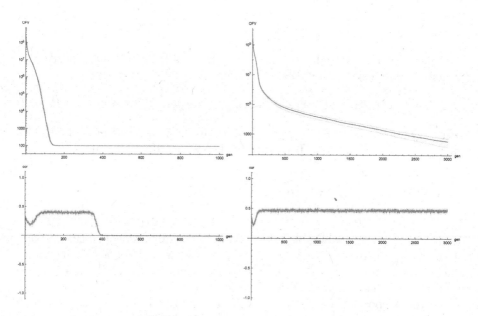

Fig. 2. Average correlation and convergence history (with 95% confidence interval - lighter color) for test function 1 in two different dimensional settings. Top left – convergence of f1 in 10*D*, top right – convergence of f1 in 30*D*, bottom left – correlation of f1 in 10*D*, bottom right – correlation of f1 in 30*D*.

In most cases of the simple multimodal functions (functions 3 to 9 in the benchmark set), the convergence and correlation graphs are quite similar in both dimensional settings (see Fig. 3, left part), except for the function 6. The correlation and convergence history of the optimization of this function are depicted in Fig. 3. In the case of 30 *D*, the correlation history is similar to the one from

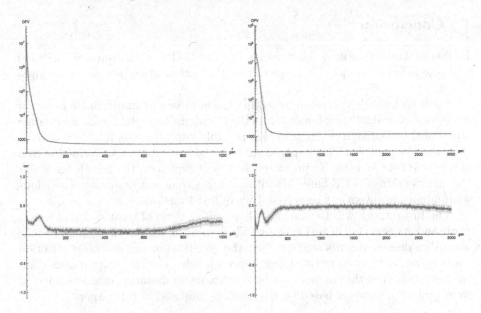

Fig. 3. Average correlation and convergence history (with 95% confidence interval - lighter color) for test function 6 in two different dimensional settings. Top left – convergence of f6 in 10D, top right – convergence of f6 in 30D, bottom left – correlation of f6 in 10D, bottom right – correlation of f6 in 30D.

simple unimodal functions, whereas in 10D it is similar to the correlation history on simple multimodal functions. This suggests, that the algorithm in 30 D converges fast to the local optima and the divergence in the population is low.

As for other simple multimodal functions, the correlation is around 0 almost from the beginning of the optimization and therefore it contradicts the assumption about positive correlation. This means, that the individuals that are most active in the population and lead the evolution might not be the one with the best objective function values and therefore the greedy approach might not be suitable.

Because of the limited length of this contribution, the rest of the results is omitted, but all correlation and convergence graphs along with the statistical results of the optimization can be found here [22].

The correlation history for hybrid (functions 10 to 12) and composition functions (functions 13 to 15) is similar to the correlation history of simple multimodal functions with the exception of functions 10, 11 and 15 in 30D, where the correlation history is of the same type as on simple unimodal functions.

Overall, the assumption about correlation stands only for simple unimodal functions and some exceptions from other types of functions in the benchmark set. Therefore, the greedy approach might be labeled as insufficient and a different type of approach to *pbest* set might be more beneficial.

7   Conclusion

In this preliminary study, the dynamic of the SHADE algorithm was analyzed with the help of a simple CN created from the optimization process of the algorithm.

Findings from this paper suggest, that the approach of translating a behavior of evolutionary algorithm dynamic into the CN might be useful for the succeeding study of the algorithm design and that possible improvements might be derived from the CN analysis. This paper should be understood as an introduction into this hybrid field and tries to promote the idea of exploiting the knowledge about the inner workings of DE based algorithms rather than just proposing algorithms with minor improvements on selected set of test functions.

The future work will be aimed at thorough analysis of created complex network and its features in order to provide beneficial search directions into the algorithm design. Results suggest, that the greedy approach for *pbest* selection could be replaced by more intelligent approach using the information from complex network. Also the combination of individuals for mutation could be selected from ranked population based on the ranking proposed in this paper.

References

1. Storn, R., Price, K.: Differential evolution-a simple and efficient adaptive scheme for global optimization over continuous spaces, vol. 3. ICSI, Berkeley (1995)
2. Brest, J., Greiner, S., Boskovic, B., Mernik, M., Zumer, V.: Self-adapting control parameters in differential evolution: a comparative study on numerical benchmark problems. IEEE Trans. Evol. Comput. **10**(6), 646–657 (2006)
3. Qin, A.K., Huang, V.L., Suganthan, P.N.: Differential evolution algorithm with strategy adaptation for global numerical optimization. IEEE Trans. Evol. Comput. **13**(2), 398–417 (2009)
4. Das, S., Abraham, A., Chakraborty, U.K., Konar, A.: Differential evolution using a neighborhood-based mutation operator. IEEE Trans. Evol. Comput. **13**(3), 526–553 (2009)
5. Mininno, E., Neri, F., Cupertino, F., Naso, D.: Compact differential evolution. IEEE Trans. Evol. Comput. **15**(1), 32–54 (2011)
6. Mallipeddi, R., Suganthan, P.N., Pan, Q.K., Tasgetiren, M.F.: Differential evolution algorithm with ensemble of parameters and mutation strategies. Appl. Soft Comput. **11**(2), 1679–1696 (2011)
7. Brest, J., Korošec, P., Šilc, J., Zamuda, A., Bošković, B., Maučec, M.S.: Differential evolution and differential ant-stigmergy on dynamic optimisation problems. Int. J. Syst. Sci. **44**(4), 663–679 (2013)
8. Wolpert, D.H., Macready, W.G.: No free lunch theorems for optimization. IEEE Trans. Evol. Comput. **1**(1), 67–82 (1997)
9. Das, S., Mullick, S.S., Suganthan, P.N.: Recent advances in differential evolution-an updated survey. Swarm Evol. Comput. **27**, 1–30 (2016)
10. Omran, M.G.H., Salman, A., Engelbrecht, A.P.: Self-adaptive differential evolution. In: Hao, Y., Liu, J., Wang, Y., Cheung, Y., Yin, H., Jiao, L., Ma, J., Jiao, Y.-C. (eds.) CIS 2005. LNCS, vol. 3801, pp. 192–199. Springer, Heidelberg (2005). doi:10.1007/11596448_28

11. Zhang, J., Sanderson, A.C.: JADE: adaptive differential evolution with optional external archive. IEEE Trans. Evol. Comput. **13**(5), 945–958 (2009)
12. Tanabe, R., Fukunaga, A.: Success-history based parameter adaptation for differential evolution. In: 2013 IEEE Congress on Evolutionary Computation (CEC), pp. 71–78. IEEE, June 2013
13. Tanabe, R., Fukunaga, A.S.: Improving the search performance of SHADE using linear population size reduction. In: 2014 IEEE Congress on Evolutionary Computation (CEC), pp. 1658–1665. IEEE, July 2014
14. Brest, J., Maučec, M.S., Bošković, B.: iL-SHADE: improved L-SHADE algorithm for single objective real-parameter optimization. In: 2016 IEEE Congress on Evolutionary Computation (CEC), pp. 1188–1195. IEEE, July 2016
15. Viktorin, A., Pluhacek, M., Senkerik, R.: Success-history based adaptive differential evolution algorithm with multi-chaotic framework for parent selection performance on CEC2014 benchmark set. In: 2016 IEEE Congress on Evolutionary Computation (CEC), pp. 4797–4803. IEEE, July 2016
16. Poláková, R., Tvrdík, J., Bujok, P.: L-SHADE with competing strategies applied to CEC 2015 Learning-based Test Suite. In: 2016 IEEE Congress on Evolutionary Computation (CEC), pp. 4790–4796. IEEE, July 2016
17. Awad, N.H., Ali, M.Z., Suganthan, P.N., Reynolds, R.G.: An ensemble sinusoidal parameter adaptation incorporated with L-SHADE for solving CEC 2014 benchmark problems. In: 2016 IEEE Congress on Evolutionary Computation (CEC), pp. 2958–2965. IEEE, July 2016
18. Tanabe, R., Fukunaga, A.: How far are we from an optimal, adaptive DE? In: Handl, J., Hart, E., Lewis, P.R., López-Ibáñez, M., Ochoa, G., Paechter, B. (eds.) PPSN 2016. LNCS, vol. 9921, pp. 145–155. Springer, Cham (2016). doi:10.1007/978-3-319-45823-6_14
19. Viktorin, A., Pluhacek, M., Senkerik, R.: Network based linear population size reduction in SHADE. In: 2016 International Conference on Intelligent Networking and Collaborative Systems (INCoS), pp. 86–93. IEEE, September 2016
20. Skanderova, L., Fabian, T.: Differential evolution dynamics analysis by complex networks. Soft Comput. 1–15 (2015)
21. Chen, Q., Liu, B., Zhang, Q., Liang, J.J., Suganthan, P.N., Qu, B.Y.: Problem definition and evaluation criteria for CEC 2015 special session and competition on bound constrained single-objective computationally expensive numerical optimization. Computational Intelligence Laboratory, Zhengzhou University, China and Nanyang Technological University, Singapore, Technical Report (2014)
22. Viktorin, A., Senkerik, R., Pluhacek, M., Kadavy, T.: CSoNet 2017 data (2017). https://owncloud.cesnet.cz/index.php/s/GLw9XggT0cBky1N

An Efficient Potential Member Promotion Algorithm in Social Networks via Skyline

Siman Zhang[1(✉)] and Jiping Zheng[1,2(✉)]

[1] College of Computer Science and Technology, Nanjing University
of Aeronautics and Astronautics, Nanjing, People's Republic of China
{zhangsiman,jzh}@nuaa.edu.cn
[2] Collaborative Innovation Center of Novel Software Technology
and Industrialization, Nanjing, People's Republic of China

Abstract. With the development of skyline queries and social networks, the idea of combining member promotion with skyline queries attracts people's attention. Some algorithms have been proposed to deal with this problem so far, such as skyline boundary algorithms in unequal-weighted social networks. In this paper, we propose an improved member promotion algorithm by presenting *reputation level* based on eigenvectors as well as in/out-degrees and providing skyline distance and sort-projection operations. The added reputation level helps a lot to describe the importance of a member, the skyline distance and sort-projection operations help us to obtain the necessary situations for not being dominated so that some meaningless plans can be pruned. Then we verify some promotion plans in minimum time cost based on dominance. Experiments on the DBLP dataset verify the effectiveness and efficiency of our proposed algorithm.

Keywords: Social networks · Member promotion · Reputation level · Skyline distance

1 Introduction

Different members play different roles in Social Networks (SN for short). Some members play an important role, and others who seems ordinary for the moment, but it may be outstanding in the future. Our goal is to find those members who have the potential to be a "star" in the future.

When skyline was first used to promoting in SNs [9], the author proposed the definition of promotion and the brute-force algorithm to realize it. However, this algorithm was inadvisable for a waste of time and space. After that, the author took the *in-degree* and *out-degree* as the measurements and in the form of adding directed edges to get the potential members via skyline queries. Nevertheless, the final result was unsatisfactory on account of the simple measurement of importance.

In this paper, we mainly study directed social graph with the knowledge of graph theory [10], taking the *in-degree*, *out-degree* and *reputation level* as the

© Springer International Publishing AG 2017
Y. Cao and J. Chen (Eds.): COCOON 2017, LNCS 10392, pp. 678–690, 2017.
DOI: 10.1007/978-3-319-62389-4_56

measurement of member's importance. What's more, we intuitively employ edge addition as the promotion manner to change the walk structure. Generally, it will take some costs to add new edges into the SNs. Therefore, the problem of member promotion in SNs is defined as to excavate the most appropriate non-skyline member(s) which can be promoted to be skyline member(s) by adding new edges with the minimum cost. However, the calculation of *reputation level* involves series of mathematical operations, it may need enormous computational cost.

To tackle the above challenges, we only consider the changes of *in-degree* and *out-degree* due to the great number of members in large SNs in the process of edge addition. When calculating a point's *reputation level*, we need to take the total number of the members as denominator. Apparently, for the great changes of the denominator (we assume the SN is dynamic), the weak changes of numerator can be ignored. If we combine the *in-degree*, *out-degree* and *reputation level* to get a *skyline set* before promotion, we can get the *non-skyline set* and take it as the *candidate set* [1], which helps to reduce the number of candidates greatly.

Therefore, the contributions of this paper are summarized as follows.

- We add the *reputation level* as a measure attribute, it helps to improve the accuracy of the prediction.
- The *skyline distance* and the sort-projection operation are utilized to obtain the necessary situations for not being dominated and prune the search space.
- Experiments on DBLP dataset are conducted to show the effectiveness and efficiency of our approach.

The rest of this paper is organized as follows. Section 2 reviews related work. In Sect. 3, we introduce several preliminary concepts. Then we bring forward the problem and propose the algorithm with analysis in Sect. 4. The results of the experiments are presented to show the effectiveness and efficiency of our algorithms in Sect. 5. Finally, we conclude our work in Sect. 6.

2 Related Work

2.1 Skyline Queries

Skyline query processing has been studied extensively in recent years. The skyline operator was first introduced by Borzsony et al. [1] in 2001. Then some representative algorithms for skyline computation including Block Nested Loops (BNL) and Divide-and-Conquer (D&C) [1], Bitmap and Index [12], Nearest Neighbor (NN) [4], and the Branch and Bound Skyline (BBS) algorithm [7] are proposed. Both BNL and D&C need to process the entire object set before reporting skyline data. The bitmap-based method transforms each object to bit vectors. In each dimension, the value is represented by the same number of '1'. However, it cannot guarantee a good initial response time and the bitmaps would be very large for large values. Therefore another method which transforms high dimension into a single dimension space where objects are clustered and indexed using a B^+ tree is raised. It helps a lot to save the time because skyline points can be determined without examining the rest of the object set not accessed yet. The NN algorithm proposed by Kossmann et al. [4] can process an online skyline query which can progressively report the skyline points in an order according to user's preferences. However, one object may be processed many times until being dominated. To remedy this problem, Papadias et al. [7] proposed an R-tree based algorithm, BBS, which retrieves skyline points by browsing the R-tree based on the best-first strategy. There are also numerous studies on skyline variations for different applications, such as subspace skylines [8], probabilistic skyline computation on uncertain data [5], weighted attributes skylines [6], skyline computation in partially ordered domains [11] and using skylines to mine user preferences, making recommendations [2].

2.2   Member Promotion

Promotion, as an important concept in marketing, has been introduced into data management applications recently. Wu et al. [13] proposed a method to find such appropriate subspaces that the target can be prominent in these subspaces. The authors bring forward an effective framework, namely "PromoRank", for promotion query and propose a "Promotion Cube" based on the concept of "Cube" in OLAP for further speeding up the query. Furthermore, they have extended the algorithm by developing a cell clustering approach that further achieves better tradeoff between offline materialization and online query processing, and thus greatly speed up the query processing.

Member promotion in SNs was first proposed and studied in [9]. The formal definition was provided and a brute-force algorithm was proposed at first. Based on the characteristics of the skyline and the promotion process, several optimization strategies were proposed to improve the efficiency and led to the IDP (Indexed Dynamic Program) algorithm [10]. Later, in the unequal-weighted SNs, the author proposed an effective promotion boundary-based pruning strategy to prune the search space and a cost-based pruning strategy based on the permutation and combination theories to verify the plans in ascending order of cost. So as to the equal-weighted SNs, the author optimized the cost model and put forward a new concept named "Infra-Skyline" to remarkably prune the candidate space. However, those algorithms were limited for only metrics like *in-degree* and *out-degree* which couldn't describe the member's importance entirely, so the prediction of promoting was not very satisfying.

A major distinction between our model and existing methods is that we add the *reputation level* as a measure attribute, which works a lot to describe a member's characteristic. With an upgrade of the measurement, our work shows more efficient.

3   Preliminaries

In this paper, an SN, is modeled as a weighted directed graph $G(V, E, W)$. The nodes in V represent the members in the SN. Those elements of E are the existing directed edges between the members. Each $w \in W : V \times V \rightarrow R^+$ denotes the cost for establishing the directed edge between any two different members. Assume $d_{in}(v)$, $d_{out}(v)$ and $P(v)$ represent the *in-degree*, *out-degree* and *reputation level* of node v in V respectively. We consider the larger the $d_{in}(v)$, $d_{out}(v)$ and $P(v)$ ($P(v)$ is defined in Definition 3), the better the v is.

Definition 1 (Social relationship matrix). Given an SN $G(V, E, W)$, the *social relationship matrix* is an adjacency matrix which expresses the links between the members in the SN, denoted as M.

Definition 2 (Normalization social matrix). If a *social relationship matrix* is M, then its *normalization social matrix* is a matrix which has sum of the elements for any column is 1, we denote the normalization matrix as M'.

Definition 3 (Reputation level). Given a node v in a SN $G(V, E, W)$, the *reputation level* of v, marked as $P(v)$, is the value of the corresponding component in the *eigenvector* of the *normalized social relationship matrix* whose *eigenvalue* is 1.

Example 1. Suppose that there are three nodes in the SN, let the nodes be v_1, v_2, v_3, the *eigenvector* of the *normalized social relationship matrix* whose *eigenvalue* is 1 is $p = (p_1, p_2, p_3)$, then the v_1, v_2, v_3's *reputation level* is p_1, p_2 and p_3 respectively.

Definition 4 (Dominator set). Given an SN $G(V, E, W)$, if v_1 dominates v_2, we call v_1 a dominator of v_2. Correspondingly, all dominators of a member v, marked as $\delta(v)$, is denoted as the *dominator set* of v.

Definition 5 (Infra-Skyline). Given an SN $G(V, E, W)$, the *Infra-Skyline* of G is the skyline of the set of all non-skyline members of G, namely, where is a new SN generated from G by eliminating its skyline members.

Example 2. Given an SN consists of seven members, namely $\{A, B, C, D, E, F, G\}$, suppose that the *skyline set* is $\{A, B, D\}$, then the *Infra-Skyline* in the SN is $\{C, E, F, G\}$.

3.1 Reputation Level

In SNs, if a member's moments are composed of prestigious people, then we think she/he is reputable. Otherwise, her/his reputation should be low.

From the point of mathematics, members' *reputation level* depends on the reputation of those members who follow them. The *reputation level* of those members who are following depends on those who follow them, and the subsequent process can be implemented in the same manner. So, for solving this kind of "infinite regress", we define $P(v_i)$ as the *reputation level* of member i, and we noticed that the ith column of the *social relationship matrix* shows those members who follow it. Therefore, we can get v_i's *reputation level* by adding these products between the relation state and the *reputation level* of all other members, namely

$$P(v_i) = x_{1i}P(v_1) + x_{2i}P(v_2) + \dots + x_{gi}P(v_g) \tag{1}$$

where the coefficient x_{ij} denotes the relation state between the member i and j and g denotes the number of the members.

Example 3. If there are 7 members in an SN, the member v_2 is followed by v_5 and v_7, then the rest entries of the second column in the *social relationship matrix* are all be 0s. Thus, we consider v_2's *reputation level* is $p(v_5) + p(v_7)$.

From Example 3, we know that if the member v_5 and v_7 have a high *reputation level*, so does v_2. It means that if a member is followed by some prestigious, the member may be reputable as well.

Therefore, we have g formulas like Eq. (1), and we have a system of g linear equations. If we compute the *social relationship matrix* M and put the value of the *reputation level* into the vector, the whole formula system is expressed as

$$P = M^T P \tag{2}$$

where P represents the vector consisting of the corresponding *reputation level* of each member in the limited state.

By organizing these formulas, we obtain the formula $(I - M^T)P = 0$, where I represents a g dimensional unit matrix, P and 0 both represent vectors with the length of g. In order to solve this formula, Katz [3] supposed that the *social relationship matrix* should be normalized. If the matrix has been normalized [5], the eigenvector P' whose *eigenvalue* is 1 represents the *reputation level* of the members [6].

3.2   The Property of Reputation Level

It should be noticed that a point's *reputation level* is partially consistent with its *in-degree*. However, by this property alone cannot show the difference between the top and the next. When some members have the same *in-degrees*, the *out-degree* also affects the *reputation level*.

Example 4. Given 7 members in the SN, as shown in Fig. 1, its corresponding *social relationship matrix* M and its form of normalization M' are as follows.

$$M = \begin{bmatrix} 0 & 1 & 1 & 1 & 1 & 0 & 1 \\ 1 & 0 & 0 & 0 & 0 & 0 & 0 \\ 1 & 1 & 0 & 0 & 0 & 0 & 0 \\ 1 & 1 & 1 & 0 & 1 & 0 & 0 \\ 1 & 0 & 1 & 1 & 0 & 1 & 0 \\ 1 & 0 & 0 & 0 & 1 & 0 & 0 \\ 0 & 0 & 0 & 0 & 1 & 0 & 0 \end{bmatrix} \quad M' = \begin{bmatrix} 0 & \frac{1}{3} & \frac{1}{3} & \frac{1}{2} & \frac{1}{4} & 0 & 1 \\ \frac{1}{5} & 0 & 0 & 0 & 0 & 0 & 0 \\ \frac{1}{5} & \frac{1}{3} & 0 & 0 & 0 & 0 & 0 \\ \frac{1}{5} & \frac{1}{3} & \frac{1}{3} & 0 & \frac{1}{4} & 0 & 0 \\ \frac{1}{5} & 0 & \frac{1}{3} & \frac{1}{2} & 0 & 1 & 0 \\ \frac{1}{5} & 0 & 0 & 0 & \frac{1}{4} & 0 & 0 \\ 0 & 0 & 0 & 0 & \frac{1}{4} & 0 & 0 \end{bmatrix} \tag{3}$$

Then we obtain the eigenvector $\alpha = (0.304, 0.166, 0141, 0.105, 0.179, 0.045,$
$0.061)^T$ of M' when the *eigenvalue* is 1, we think the top one whose ID is 1 has the
highest *reputation level* because she/he gains all the reputation from ID 2. What's
more, ID 1 has the highest *in-degree* and *out-degree*, thus we consider she/he is the
most popular one in the SN. On the other hand, ID 2 and ID 3 have the same *in-degree* but ID 2's *out-degree* is smaller than ID 3, we can see that ID 2's *reputation level* is larger than ID 3.

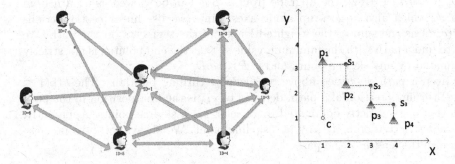

Fig. 1. A social network example   **Fig. 2.** A skyline distance example

4   Prediction of Promoting Members in the SNs

4.1   Problem Statement

The problem we study in this paper is to locate the most "appropriate" member(s) for promotion by means of elevating it (them) into skyline. Suppose we
have two datasets D_1 and D_2. The D_1 represents some data of previous years
and the D_2 represents that of the following several years. If $S_1 = SKY(D_1)$,
$S'_1 = SKY(D_1 - S_1)$, $S_2 = SKY(D_2)$, where the $SKY()$ represents the *skyline
set* of the dataset, then S'_1 is the *candidate set* in our algorithm. After promoting
towards each point in S'_1, if there exist some of points in S'_1 appearing in S_2, the
prediction is successful, otherwise, it fails.

4.2   The Sort-Projection Operation

We consider projecting all the members into a 2-dimensional Cartesian coordinate
system, where the axe x represents the *in-degree* and the axe y represents the
out-degree. Taking the candidate c as an example, suppose that c is dominated
by t skyline points, and we simply sort the skyline points in ascending order on
axe x.

Definition 6 (Strictly dominate). If the point p_1 is not smaller than p_2 on
each dimension and larger than p_2 on at least one dimension, we say p_1 dominate
p_2, denoted by $p_1 \prec p_2$. Furthermore, if p_1 is larger than p_2 on each dimension,
we say p_1 strictly dominate p_2, denoted by $p_1 \prec\prec p_2$.

Definition 7 (Skyline distance). Given a set DS of points in a 2-dimenional space D, a candidate c, and a path function $Path(.,.)$, the *skyline distance* of c is the minimum $Path(c, c')$, where c' is a position in D such that $c'.D_i \geq c.D_i$, $i \in [1, 2]$, and c' is not strictly dominated by any point in DS. We denote the *skyline distance* as $SkyDist()$.

Suppose that c is strictly dominated by t skyline points in $SKY(DS)$. For any position c' which is not strictly dominated by any point in DS satisfies $c'.D_i \geq c.D_i$, $i \in [1, 2]$, the upgrade from c to c' can be viewed as a path from c to c', which always goes up along axes. Since we use linear cost functions $cost(c, c')$ as the sum of the weighted length of the segments on the path. We aim to find a path with the minimum value so that the end point c' is not strictly dominated by any skyline point, and $c'.D_i \geq c.D_i$, $i \in [1, 2]$.

Given a path described above, we define t turning positions. The kth$(1 \leq k \leq t)$ turning position of a path, denoted by c_k, is the first position in the path such that c_k, is strictly dominated by at most $t - k$ skyline points. Apparently, c_t is not strictly dominated by any skyline point. We always set c' to be c_t to minimize the length of the path.

Definition 8 (Skyline boundary). Given a set SKY of skyline points in DS, we say a point p is on the skyline boundary if there exists a point u such that $u \prec p$ and there does not exist a point $u' \in SKY$, such that $u' \prec\prec p$.

From the definition of skyline boundary, we know that each point on the skyline boundary has a *skyline distance* 0. In the 2-dimensional space D, if there is an intersection point in the skyline boundary, then we call this intersection "boundary intersection point".

Property 1 (Dominance in boundary intersection points). Let p is a point on the skyline boundary $w.r.t.$ a set DS of points and p is not a boundary intersection point. Then, there exists a boundary intersection point p' such that $p \prec p'$.

We call a boundary intersection point a local optimal point if it does not dominate any other boundary intersection point.

Consider a candidate c dominated by t skyline points $s_1, s_2, ..., s_t$. Let $p_1, ..., p_r$ be the r local optimal points determined by c and $s_1, s_2, ..., s_t$, then the *skyline distance* of c is the minimum path from c to p_i.

In the 2-dimensional space D, for the candidate c and the t skyline points $s_1, s_2, ..., s_t$, we have $s_1.D_1 < s_2.D_1 <, ..., < s_t.D_1$. Without loss of generality, we know $s_1.D_2 > s_2.D_2 >, ..., > s_t.D_2$. Clearly, there are $t + 1$ local optimal points and the ith one p_i is given by the following formula,

$$P_i = \begin{cases} (c.D_1, s_1.D_2), & i = 1; \\ (s_{i-1}.D_1, s_i.D2), & 2 \leq i \leq t; \\ (s_t.D_1, c.D2), & i = t + 1. \end{cases} \qquad (4)$$

Example 5. There is a candidate c and s_1, s_2, s_3 are skyline points which dominate c, as shown in Fig. 2, we can obtained the 4 local optimal points p_1, p_2, p_3 and p_4 by Eq. (4), by comparing the path between c and p_i, we can get the *skyline distance* of c.

Algorithm 1 gives the pseudo-codes of the sort-projection operation.

Algorithm 1. The sort-projection algorithm $SP(c, SKY)$

Input: $SKY()$
Output: $SkyDist(c)$

1 sort points in $SKY()$ in ascending order on dimension D_1;
2 $P = \{p_i \mid l\}$ where p_i is given by Eq. (4);
3 return $\min\{Path(c, P) \mid P\}$;

4.3   Pruning Based on Cost and Dominance

After obtaining the *skyline distance* of a candidate, we know the necessary situations for not being dominated by skyline points. Taking the candidate c as an example, if c' is the end point in the *skyline distance* of c, we both plus 1 on the axe x and axe y of the c' (we denote the increased c' as c''). It is obvious that c'' could not be dominated by any point at all. If we call the position where a candidate will not be dominated as $GoodPosition()$, we say $c'' \in GoodPosition()$. Besides c'', all points in skyline will not be dominated either. Thus the *dominator set* of c belongs to $GoodPosition(c)$.

In view of unequal costs for establishing different edges, it probably takes different costs to promote c by different plans. If we respectively organize all the edges which can be added in the plans against each candidate c, denoted as E_c, sorted in ascending order of weights. Then we can easily locate the plans from E_c with minimum costs. Once the plan is verified to be successful to promote the candidates, the process of promotion could be ended.

Observation 1. The successive plans are introduced by the following rules:

- If the edge e_0 with minimum cost in E_c is not included in the current plan, add e_0.
- If any successive edge of e_i, namely e_{i+1}, does not belong to the current plan, replace e_i with e_{i+1}.

Observation 2. The prunable plans are introduced by the following rules:

- If adding an edge e connecting node v_i and the candidate node c still cannot promote c into the *skyline set*, all the attempts of adding an edge e' connecting the node v_j and c with the same direction as e is not able to successfully promote c, where $v_j \in \delta(v_i)$.
- If a plan $p(e_1, ..., e_w)$ cannot get its target candidate c promoted, all the plans with w edges which belong to can be skipped in the subsequent verification process against c, where for each e_i connecting v_i and c, l is a list containing all the non-existing edges each of which links one member of $\delta(v_i)$ and c with the same direction as e_i ($i = 1, 2, ..., w$), is the Cartesian product of l.

4.4   Verification of the Result

We notice that those points which don't dominate the candidate point before promoting wouldn't dominate it after promotion. Thus we can ignore it in the verification process. Therefore, after pruning, we should just consider the following situations when verifying,

1. The points which dominate the candidate before promoting.
2. The points which related to the promotion plans.

4.5   The PromSky Algorithm

The whole process of member promotion in an SN is presented in Algorithm 2. Line 4–5 represents a preprocessing phase by generating the sorted available edge. Then $GoodPosition()$ is generated in line 5–9. Line 10–11 shows that the corresponding promotion plans are generated and put into the priority queue Q. Once the queue is not empty, we fetch the plan with minimum cost for further verification. Line 12 shows that before verifying the plan, we first generate its *children plans* by Observation 1 so that we can verify all the possible plans in ascending order of cost. Line 14–15 represents that after checking based on the result verification strategy the result will be output if the promotion succeeds. If not, some prunable plans will be generated. The generation of prunable plans are showed in Line 17. Line 16 represents that if the plan is in the prunable list, there is no need of further verification. Line 13 shows that after a successful promotion, the process will halt once we encounter a plan with the higher cost.

Algorithm 2. The promotion algorithm $PromSky(G)$

Input: social network $G(V, E, W)$.
Output: optimal members for promotion and corresponding plans.

```
 1  initialize a priority queue Q;
 2  C = V − SKY(V);
 3  for each c ∈ C do
 4      E=genCandidateEdgeSet(c);
 5      sortByWeight(E);
 6      SkyDist(c) = Path(c, SKY);
 7      good.x = DistSky(c.x + 1);
 8      good.y = DistSky(c.y + 1);
 9      GetDominator(c);
10      P=getMinCostPlan(good, GetDominator(c), E) and add P into Q;
11      while p=ExtractMin(Q) do
12          p_child = GenerateChildren(p, E) and add it into Q;
13          if p_succ is NULL and cost(c) > cost(p_succ) then
14              break;

15          if c is promoted by p then
16              return c and p;
17              if p_succ =∅ then
18                  p_succ = p;

19          if p ∈ prunableList then
20              continue;

21          else
22              genPrunablePlans(p, c);
```

5   Experimental Analysis

5.1   Experimental Setup

We take the evaluation of the experiments on the dataset of DBLP[1] each record of which consists of authors' name, paper title and published year. We collect all the records from 1992 to 2016, in order to properly show the value of the authors in their own eras, for each year's data, we consequently combine the current yearly data with its previous 4 years' data to generate a 5-year sub-network. Then we run our promotion algorithm on the 5-year sub-networks (from 1996 to 2016 with 4 sub-networks), so as to make comparisons between the corresponding yearly potential stars and those skyline authors in the following couple of years.

The experiments are implemented using C++ with Visual Studio 2010 and conducted on an Intel Core CPU i75500U@2.4GHZ machine with 8G RAM and 1Tbytes Hard disk running on Windows 7.

5.2   Results

Successful rate comparison. In this section, we record the predicted potential stars and the skyline authors detected by our algorithm from 1996 to 2016. We can get the *successful rate* by using the number of potential stars promoted into *skyline* in the next few years divided by the size of the whole potential star set, namely

$$r = PN/CS \tag{5}$$

where "r" denotes the *successful rate*, "PN" and "CS" are the number of successfully promoted members and the number of all the candidates respectively.

From Table 1, we know that the number of the potential candidates is 20 (by merging the duplicated potential stars and removing the potential stars of the year 2016 because it is unable to be verified), and the number of the potential candidates who appear in the next skyline authors is 13 (some successfully promoted candidates are marked), which means the *successful rate* is 65%. However, in the previous research [10], when conducting the experiments on the dataset from 1971 to 2012, we just find the *successful rate* is only 48%. It shows that our algorithm is more accurate than the previous. The skyline authors and potential stars for each year are illustrated in Table 1.

Successful promotion rate comparison based on network scales. At the same time, we conduct our PromSky and SkyBoundary algorithms in various network scales. From Fig. 3, we know that the *successful rate* of our promotion algorithm is apparently higher than the previous algorithm in various network scales. This is because we add more attributes in our PromSky algorithm for a skyline member that it should increase the number of skyline set. Thus our successful promotion rate is higher in various network scales.

Effect of the PromSky algorithm. Figure 4 shows the average processing time under different network scales (varying from 1 K to 1000K). We can see that

---

[1] http://dblp.org/.

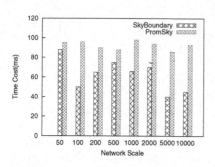 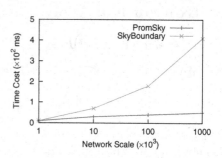

Fig. 3. Successful rate comparison on various network scales

Fig. 4. Effect comparison with the SkyBoundary algorithm

Table 1. Skyline authors and potential stars from 1996 to 2016

Year	Skyline	Potential skyline
1996	Robert L. Glass, David Wilczynski	Robert W. Floyd
1997	Noga Alon, Jean P, Caxton Foster	Peter Kron
1998	Noga Alon, Robert L. Glass, V. Kevin M	Carl Hewitt, Bill Hancock
1999	**Robert W. Floyd**, Noga Alon, Honien Liu	Paul A.D., Alan G. Merten
2000	**Bill Hancock, Peter Kron**	Paul A.D.
2001	**Bill Hancock**, Nan C. Shu	Pankaj K. Agarwal
2002	**Bill Hancock**, Charles W. Bachman, Daniel L. Weller	Pankaj K. Agarwal
2003	**Bill Hancock**, Daniel L. Weller	Elisa Bertino, Alan G. Merten
2004	**Pankaj K. Agarwal**, Morton M. Astrahan, David R. Warn	Elisa Bertino, Mary Zosel
2005	Gary A. Kildall, Diane Crawford, Hans-Peter Seidel, **Erik D.Demaine**	Carl Hewitt
2006	Noga Alon, Diane Crawford, **Pankaj K. Agarwal**	Ingo H. Karlowsky, Louis Nolin
2007	**Elisa Bertino**, G. RuggiuW, J. Waghorn, M.H. Kay, Erik D. Demaine	T. William Olle
2008	Diane Crawford, **Paul A.D.**	B.M. Fossum
2009	Wen Gao, Xin Li, Jun Wang, P.A. Dearnley, Giampio Bracchi, Paolo Paolini, Ajith Abraham	H. Schenk, Gordon E. Sayre
2010	Xin Li, **B.M. Fossum**, J.K. Iliffe, Wen Gao, **Mary Zosel**, Wei Wang	Paul Mies, Ingo H. Karlowsky
2011	Xin Li, Gordon E. Sayre, T. William Olle	Peter Sandner
2012	H. Vincent Poor, **Peter Sandner**, Ulrich Licht	Yan Zhang
2013	**Ingo H. Karlowsky**, Heidi Anlauff, Günther Zeisel	Guy G. Boulaye
2014	**Yan Zhang**, Yu Zhang, Gordon E. Sayre, Witold Pedrycz	Carl Hewitt
2015	Harold Joseph Highland, Bernard Chazelle	Won Kim
2016	**Won Kim**, Dale E. Jordan, **B.M. Fossum**	Nan C. Shu

as the network scale increases, the processing time of our PromSky algorithm increases slightly, which performs much better than the SkyBoundary algorithm. This is because the candidates in SkyBoundary algorithm are all the *non-skyline set* but we conduct the skyline query over the *non-skyline set* and take the *infra-skyline* as the candidates thus remarkably reducing the size of the candidates and controlling the result in a reliable range to a great extent. Besides, by bringing forward the *skyline distance*, we can get the minimum cost from the candidates to the position where not being dominated.

6   Conclusions

In this paper, we propose an improved member promotion algorithm in SNs, which aims at discovering the most potential stars which can be promoted into the skyline with minimum cost. By adding the attribute of *reputation level*, we describe members' importance more precisely. Then we introduce the skyline distance and the sort-projection operations to obtain the necessary situations for not being dominated. At the same time it also helps a lot to reduce the number of promotion plans. What's more, we prune some extra plans based on the dominance while verifying. Finally, experiments on the DBLP dataset illustrate the effectiveness and efficiency of our approach.

Acknowledgment. This work is partially supported by Natural Science Foundation of Jiangsu Province of China under grant No. BK20140826, the Fundamental Research Funds for the Central Universities under grant No. NS2015095, Funding of Graduate Innovation Center in NUAA under grant No. KFJJ 20161606.

References

1. Börzsönyi, S., Kossmann, D., Stocker, K.: The skyline operator. In: ICDE, pp. 421–430 (2001)
2. Jiang, B., Pei, J., Lin, X., Cheung, D.W., Han, J.: Mining preferences from superior and inferior examples. In: SIGKDD, pp. 390–398 (2008)
3. Katz, L.: A new status index derived from sociometric analysis. Psychometrika 18(1), 39–43 (1953)
4. Kossmann, D., Ramsak, F., Rost, S.: Shooting stars in the sky: an online algorithm for skyline queries. In: VLDB, pp. 275–286 (2002)
5. Lian, X., Chen, L.: Monochromatic and bichromatic reverse skyline search over uncertain databases. In: SIGMOD, pp. 213–226 (2008)
6. Mindolin, D., Chomicki, J.: Discovering relative importance of skyline attributes. PVLDB 2(1), 610–621 (2009)
7. Papadias, D., Tao, Y., Fu, G., Seeger, B.: Progressive skyline computation in database systems. ACM Trans. Database Syst. 30(1), 41–82 (2005)
8. Pei, J., Jiang, B., Lin, X., Yuan, Y.: Probabilistic skylines on uncertain data. In: VLDB, pp. 15–26 (2007)
9. Peng, Z., Wang, C.: Discovering the most potential stars in social networks with skyline queries. In: Proceedings of the 3rd International Conference on Emerging Databases (2011)

10. Peng, Z., Wang, C.: Member promotion in social networks via skyline. World Wide Web **17**(4), 457–492 (2014)
11. Sacharidis, D., Papadopoulos, S., Papadias, D.: Topologically sorted skylines for partially ordered domains. In: ICDE, pp. 1072–1083 (2009)
12. Tan, K., Eng, P., Ooi, B.C.: Efficient progressive skyline computation. In: VLDB, pp. 301–310 (2001)
13. Wu, T., Xin, D., Mei, Q., Han, J.: Promotion analysis in multi-dimensional space. PVLDB **2**(1), 109–120 (2009)

Author Index

Printed in the United States
by Book...

Printed in the United States
By Bookmasters